Handbook of Research on Tribology in Coatings and Surface Treatment

Amirhossein Pakseresht
FunGlass - Centre for Functional and Surface Functionalized Glass, Alexander Dubček University of Trenčín, Slovakia

Omid Sharifahmadian
FunGlass - Centre for Functional and Surface Functionalized Glass, Alexander Dubček University of Trenčín, Slovakia

IGI Global
PUBLISHER of TIMELY KNOWLEDGE

A volume in the Advances in Chemical and Materials Engineering (ACME) Book Series

Published in the United States of America by
IGI Global
Engineering Science Reference (an imprint of IGI Global)
701 E. Chocolate Avenue
Hershey PA, USA 17033
Tel: 717-533-8845
Fax: 717-533-8661
E-mail: cust@igi-global.com
Web site: http://www.igi-global.com

Library of Congress Cataloging-in-Publication Data

Names: Pakseresht, Amir Hossein, 1981- editor. | Sharifahmadian, Omid, 1987- editor.
Title: Handbook of research on tribology in coatings and surface treatment / Amirhosssein Pakseresht, and Omid Sharifahmadian, editors.
Description: Hershey, PA : Engineering Science Reference, [2022] | Includes bibliographical references and index. | Summary: "This book evaluates the latest advances in applying the coatings and surface treatments by different techniques to reduce the damages from tribology, giving readers more detailed information to compare the capability of each coating process in wear resistant and lubrication applications"-- Provided by publisher.
Identifiers: LCCN 2021058147 (print) | LCCN 2021058148 (ebook) | ISBN 9781799896838 (h/c) | ISBN 9781799896852 (eISBN)
Subjects: LCSH: Surfaces (Technology) | Contact mechanics. | Coatings. | Tribology.
Classification: LCC TA418.72 .T75 2022 (print) | LCC TA418.72 (ebook) | DDC 620/.44--dc23/eng/20220110
LC record available at https://lccn.loc.gov/2021058147
LC ebook record available at https://lccn.loc.gov/2021058148

This book is published in the IGI Global book series Advances in Chemical and Materials Engineering (ACME) (ISSN: 2327-5448; eISSN: 2327-5456)

British Cataloguing in Publication Data
A Cataloguing in Publication record for this book is available from the British Library.

For electronic access to this publication, please contact: eresources@igi-global.com.

Advances in Chemical and Materials Engineering (ACME) Book Series

J. Paulo Davim
University of Aveiro, Portugal

ISSN:2327-5448
EISSN:2327-5456

Mission

The cross disciplinary approach of chemical and materials engineering is rapidly growing as it applies to the study of educational, scientific and industrial research activities by solving complex chemical problems using computational techniques and statistical methods.

The **Advances in Chemical and Materials Engineering (ACME) Book Series** provides research on the recent advances throughout computational and statistical methods of analysis and modeling. This series brings together collaboration between chemists, engineers, statisticians, and computer scientists and offers a wealth of knowledge and useful tools to academics, practitioners, and professionals through high quality publications.

Coverage

- Multifuncional and Smart Materials
- Computational methods
- Ductility and Crack-Resistance
- Metallic Alloys
- Thermo-Chemical Treatments
- Heat Treatments
- Artificial Intelligence Methods
- Wear of Materials
- Industrial Chemistry
- Coatings and surface treatments

IGI Global is currently accepting manuscripts for publication within this series. To submit a proposal for a volume in this series, please contact our Acquisition Editors at Acquisitions@igi-global.com or visit: http://www.igi-global.com/publish/.

The Advances in Chemical and Materials Engineering (ACME) Book Series (ISSN 2327-5448) is published by IGI Global, 701 E. Chocolate Avenue, Hershey, PA 17033-1240, USA, www.igi-global.com. This series is composed of titles available for purchase individually; each title is edited to be contextually exclusive from any other title within the series. For pricing and ordering information please visit http://www.igi-global.com/book-series/advances-chemical-materials-engineering/73687. Postmaster: Send all address changes to above address. ©

Titles in this Series

For a list of additional titles in this series, please visit: http://www.igi-global.com/book-series/advances-chemical-materials-engineering/73687

Advanced Manufacturing Techniques for Engineering and Engineered Materials
R. Thanigaivelan (Muthayammal Engineering College, India) N. Rajan (Vinayaga Mission Kripananda Variyar Engineering College, India) and T.G. Argul (St. Mary's Engineering College, India)
Engineering Science Reference • © 2022 • 351pp • H/C (ISBN: 9781799895749) • US $245.00

Handbook of Research on Green Synthesis and Applications of Nanomaterials
Rajni Garg (Rayat Bahra University, India) Rishav Garg (Galgotias College of Engineering and Technology, India) and Nnabuk Okon Eddy (Department of Pure and Industrial Chemistry, University of Nigeria, Nsukka, Nigeria)
Engineering Science Reference • © 2022 • 569pp • H/C (ISBN: 9781799889366) • US $295.00

Handbook of Research on Nanoemulsion Applications in Agriculture, Food, Health, and Biomedical Sciences
Karthikeyan Ramalingam (B.S. Abdur Rahman Crescent Institute of Science and Technology, India)
Engineering Science Reference • © 2022 • 671pp • H/C (ISBN: 9781799883784) • US $295.00

Handbook of Research on Advancements in the Processing, Characterization, and Application of Lightweight Materials
Kaushik Kumar (Birla Institute of Technology, India) B. Sridhar Babu (CMR Institute of Technology, India) and J. Paulo Davim (University of Aveiro, Portugal)
Engineering Science Reference • © 2022 • 458pp • H/C (ISBN: 9781799878643) • US $285.00

Emerging Applications and Implementations of Metal-Organic Frameworks
Shimaa Mohamed Elsaeed (Egyptian Petroleum Research Institute, Egypt) Elsayed Zaki (Egyptian Petroleum Research Institute, Egypt) and Abdel-Azim Abdel-Azim (Egyptian Petroleum Research Institute, Egypt)
Engineering Science Reference • © 2021 • 254pp • H/C (ISBN: 9781799847601) • US $225.00

Normal Partitions and Hierarchical Fillings of N-Dimensional Spaces
Gennadiy Vladimirovich Zhizhin (Russian Academy of Natural Sciences, Russia)
Engineering Science Reference • © 2021 • 280pp • H/C (ISBN: 9781799867685) • US $195.00

Applications of Nanomaterials in Agriculture, Food Science, and Medicine
Mohd Amin Bhat (Higher Education Department, Government of Jammu and Kashmir, India) Irshad Ahmad Wani (Government Degree College for Boys, Anantnag, India) and Shah Ashraf (National Institute of Technology Srinagar (NIT Srinagar), India)
Engineering Science Reference • © 2021 • 442pp • H/C (ISBN: 9781799855637) • US $215.00

IGI Global
PUBLISHER of TIMELY KNOWLEDGE

701 East Chocolate Avenue, Hershey, PA 17033, USA
Tel: 717-533-8845 x100 • Fax: 717-533-8661
E-Mail: cust@igi-global.com • www.igi-global.com

Editorial Advisory Board

List of Contributors

Table of Contents

Detailed Table of Contents

Mahdi Alebrahim, FunGlass – Centre for Functional and Surface Functionalized Glass, Alexander Dubček University of Trenčín, Slovakia
Amirhossein Pakseresht, FunGlass – Centre for Functional and Surface Functionalized Glass, Alexander Dubček University of Trenčín, Slovakia
Reza Jafari, Tampere University, Finland
Dusan Galusek, FunGlass – Centre for Functional and Surface Functionalized Glass, Alexander Dubček University of Trenčín, Slovakia
Fariborz Sharifian Jazi, The University of Georgia, Georgia

Thermal barrier coatings (TBCs) are intended to defend the engines against excessive heat. The surface of a component is subjected to mechanical degradation, such as wear, in a variety of engineering applications. The majority of equipment used in industrial processes is subjected to abrasive wear at elevated temperatures. In the engineering industry, surface modification and deposition of various coatings are used to improve the tribological characteristics of materials. Reinforced materials offer superior mechanical and thermal qualities as compared to conventional materials. The ceramic matrix and TBCs have low fracture toughness and weak cohesive strength of the ceramic layer, limiting their applicability range. A technique for improving the mechanical properties of ceramic coatings is to incorporate reinforcing materials, such as ceramic whiskers or fibers, into the coating structure, which can improve the coating's wear behavior and mechanical properties through a variety of mechanisms, including fiber pullout and breakage.

Magnificently developed Al2O3, 3YSZ, 8YSZ, and CNT-based thermal barrier coatings (TBCs) were subjected to study multi-length scale tribological behaviour (fretting wear and micro-scratch) of the composite coatings. Subsequently, the role of constituents of the composite on the tribological behaviour of the coatings has been recognized. Fretting wear rate and the dominative wear mechanism are identified. The fretting wear behaviour is evaluated with a distinct representation (frictional force mapping) to understand the transition of regimes. Further, micro-scratching is used to assess the composite coatings against the sharp edges. The critical load of failure and scratch characteristics (scratch hardness, toughness, and scratch resistance) are appraised to find the suitability of the composite in TBC. Notably, the role of CNT in a multi-length scale is reported quantitatively. Thus, the chapter provides a comprehensive overview of the Al2O3-based composites that deal with the understanding of the multi-length scale tribological properties at room temperature.

In the present work, the alumina-induced thermal barrier coatings with LaTi2Al9O19 (LTA) and yttria-stabilized zirconia (YSZ), LTA/YSZ double ceramic layer (DCL) are studied for oxidation and wear tests. Different coatings combinations with varying thickness of LaTi2Al9O19 (LTA) top coat layer are developed using plasma spray method and are tested for isothermal oxidation and wear test. An Alumina layer is induced after the bond coat layer to provide a readily available oxide layer. The activation energy is calculated using the Arrhenius equation. Arrhenius plots are developed using oxidation kinetics. Coatings are tested for wear performance also. The coating combination with a higher thickness of LTA proved best for both oxidation and wear performances. Surface characterization is done using EDS and XRD analysis.

Severe wear is the root cause for the failures of components, which resulted in the loss of functionality and dimensional accuracy of systems. An estimation of durability and reliability of components is governed by failure analysis under rigorous operational environments with a range of wear mechanisms. The researchers have two ways to minimise the wear: either develop new anti-wear materials or improve wear resistance of existing materials by surface engineering techniques. Thermally sprayed tungsten carbide-based coatings are often feasible approaches to improve the wear resistance of a metallic surface. Therefore, these coatings on the components enhance the useful lifespan. The investigation focuses on the wear characteristics in terms of coefficient of friction and wear rate for tungsten carbide coatings. The focus of the chapter is on abrasive wear and erosion since these two types of wear have the maximum contribution to the failure rate in real-life problems. The applications, benefits, and detriments of the coating deposition techniques are summarised.

In recent years, the demand of wear resistance material and coating is increasing very rapidly as it reduces substantial energy losses resulting from wear and friction. To overcome these energy losses, surface engineering is employed. Surface engineering is the process of coating or modifying the surface of part to minimize wear, friction, corrosion, as well as to enhance the lifespan of machine components and reduce the manufacturing cost. Recently, numerous coating methods are available for distinct material (pure metals to alloys, carbides, composites, and ceramics) applications. Hence, this chapter provides an overview on the prevention of tribo-surfaces through distinct methods of surface modification such as thermal, physical, and chemical methods of coating. Further, distinct coating properties, applications, future scope, and challenges are described.

This chapter is focused on cold spray deposition of particles for surface modification. The method, which has been recently proven to have wide applicability in the domain of tribology and wear-resistant coatings, relies on supersonic gaseous jets to deposit the particle without phase change. The chapter aims at examining the influence of the unique gas dynamic characteristics of the high-speed jets on the deposition process. The general structure of the supersonic jets, including the velocity field, pressure gradients, and the impingement behaviour, is discussed with specific attention to the requirements of the sprays for tribological coatings. Results of detailed numerical simulation of the impingement process are made use of to demonstrate the parametric influence of the supersonic jet structure on critical spray characteristics, like the particle velocity. The study also examines various aspects of the energy conversion as applied to the basic nature of the supersonic jet as well as its interaction with the microparticles.

Section 2

Wear, plastic deformation, and mechanical fatigue of dies are the most common failures found during hot forming operations at elevated temperatures. The change in frictional behaviour also happens. The performance of the forming operation is affected. To ensure the quality of the end products and productivity, it has become very important to control the wear and friction of die materials. Surface treatment techniques with superior wear properties and good performance can enhance the life and functionality of dies. Plasma nitriding is the most rapidly developing technique for hot forming dies. It is a cost-effective technique and improves the mechanical properties of the die surfaces. This chapter explains the tribology of hot forming dies, the plasma nitriding technique, and the procedures to develop plasma nitriding on the die steels. Thereafter, the tribological behaviour of AISI H11 and AISI H13 plasma nitrided die steels has been reported. Plasma nitriding was found to be most promising and effective in reducing wear and friction at elevated temperatures.

Physical vapor deposition (PVD) technologies are widely used to produce wear and corrosion resistant coatings for a variety of industrial applications. In recent years, there has been remarkable interest in the development of novel wear resistant coatings prepared through PVD methods, which helps to reduce friction and wear, as a result of recovering energy losses up to 30% due to friction and economy loss due to wear. This chapter provides comprehensive data of recent progress in wear resistant coatings prepared using PVD methods, starting with the introduction of it needs, significance, physiochemical properties, and the selection criteria of wear resistant coatings. The applications, physical, and chemical properties of superhard materials such as diamond like carbon (DLC), titanium nitride (TiN), chromium nitride (CrN), and tantalum nitride (TaN) are also presented.

The chapter presented a short review related to the factors that dictate the wear and friction behavior of various coatings obtained by PVD techniques used for various applications. An important parameter with high impact on the final properties of the coatings prepared by cathodic arc evaporation is the reactive atmosphere. The friction and wear performance of the carbide coatings were strongly dependent on the carbon content which can be controlled by varying the C2H2 mass flow rate and the arc current on the cathode. Regarding the carbonitrides coatings, the ratio of C/N plays an important role; the coating with a low C/N ratio showed reduced friction coefficients, while for the coatings with a high C/N ratio the wear was improved. For biomedical applications, the magnetron sputtering deposition technique was employed to enhance the tribological performance of Ti6Al4V alloy in Ringer solution using various types of coatings such as carbonitrides, carbide, or hydroxyapatite.

To achieve a more important service life of hammers, used in crushing process, a Ni-B-TiO2 composite coating was electrodeposited on heat treated AISI P20 using conventional and novel methods. The prepared coatings underwent different tribological tests to quantify the coating that offers the best resistance against wear. For this reason, abrasive wear tests such as pin-on-disk test and multi-pass scratch test were performed to evaluate the abrasive wear resistance of the coatings under a round counterbody (alumina ball) and a sharper contrerbody (sphero-conical indenter), respectively. In addition, the impact-sliding test was also performed to assess the impact resistance of the composite coatings. The obtained results showed that the novel method promotes the best mechanical and tribological properties of the elaborated Ni-B-TiO2 composite coating. This is attributed to the fact of adding TiO2 sol into Ni-B electrolyte which enhances the dispersive strength of the formed TiO2 nanoparticles, contrary to adding solid TiO2 nanoparticles into the electroplating bath.

Commercially available metallic orthopaedic implant materials cause major problems like stress shielding and the release of harmful ions due to corrosion and wear. Also, the secondary operation is a must for the implant removal. Therefore, the biodegradable and biocompatible magnesium (Mg) implant materials have been investigated. Mg shows favorable biological properties and matching mechanical properties with the natural bone. Mg alloys rapidly corrode in the physiological environment, which cause failure of the implant before completing the expected function. Surface coating is the most effective method for improving the corrosion performance of Mg and its alloys. Hydroxyapatite (HA), being the most stable phase of calcium phosphates in physiological conditions, is preferred as a coating material. The chapter focuses on the tribo-corrosion and characterization of HA coatings prepared by electrodeposition process on Mg alloys. The results are useful for the designer community in the selection of biocompatible coatings and process parameters to maximize the life of bio-implants.

This chapter summarizes key issues in the fields of multifunctional superhydrophobic coatings with an analysis of their tribological properties. In this respect, the authors explore a simple sol-gel process strategy and tribological properties controlled through a reaction of ORMOSIL-based polymers that generate multifunctional minimum free energy structures of micro- to nano-scale siloxane chains. Different compositions and dimensions of solid materials (contact angle = 150° and sliding angle = 10°) can be superhydrophobic fabricated through various deposition methods. The complete waterproof layering has been demonstrated to have excellent cost, scalability, and especially the ability to encapsulate other functional groups. The perspectives have established many significant functionalities with better tribological properties for the next generation of smart multifunctional superhydrophobic coatings.

Section 3

Moving mechanical systems create a lot of friction; therefore, a lot of the energy produced is used to overcome it. It is vital to find unique ways to develop lubricants that allow for more effective control or decrease friction to reach a sustainable future. High friction, if not efficiently reduced or controlled, can result in higher wear losses and, as a result, shorter system life and lower reliability. Two-dimensional (2D) materials have distinct friction and wear properties from their three-dimensional (3D) counterparts. They can be used as additives in oils and composites to reduce stiction, friction, and wear, even though they are ultra-thin even with numerous layers. The role of these materials in superlubricity on surfaces prone to friction and wear is discussed in this chapter. These materials are solid two-dimensional lubricants that can address developing needs in mechanical systems in motion in current and emergent real-world applications.

The materials research community has been very interested in bulk metallic glasses (BMGs) over the past two decades because of their demonstrated friction and wear properties, as well as their potential for use in a variety of important tribological applications. Because of their superior chemical, physical, and mechanical properties, BMGs are a promising candidate material for advanced engineering applications. Compared to conventional crystalline metals and alloys, BMGs have higher strength, higher elastic strain, and higher hardness, making them a promising material class for tribological applications. A unique deformation process is realized in these materials due to the lack of a crystalline structure and faults such as misalignments, which display high strength, hardness, strong wear resistance, massive plastic deformation, corrosion-resistant, and hardness. In this chapter, the authors describe the research achievements in the field of BMGs, the tribological properties, structure, and applications.

Electrochemical applications had their first major impact in the late 20th century with the development of improved energy storage and conversion systems such as lithium-ion batteries, organic-inorganic dye-sensitized solar cells, and even e-ink displays. Depending on the requirements, the electrodes can be made of different materials, such as metal or alloy sheets, foils, bars or conductive ceramics, conductive polymers, etc. In this chapter, methods for surface functionalization and characterization of metallic and non-metallic surfaces used as electrode substrates are presented. The focus is on the use of metal foils in lithium-ion batteries and especially in the novel architecture of optoelectronic devices – from electrochromic and photovoltaic devices to biosensors.

Preface

INTRODUCTION

Tribology is a new field of science and engineering, involving wear, lubrication, and friction between surfaces that are in contact and relative motion with each other. Tribology originates from the Greek word 'tribos' which means sliding. Nowadays, it is unanimously agreed that the influence of tribological activities on our lives is inevitable and it has a greater effect than previously imagined. Wear is the main cause of deterioration of mechanical properties and durability of engineering parts. The main reason for wear is friction, which causes energy dissipation. Therefore, controlling friction can play a key role in improving durability and saving the engineering products considerably from failure. Lubrication is one of the best-known means of controlling friction and wear. Lubrication, defined as a thin layer of material in different states (liquid, gas or solid) with a low friction coefficient, has been applied between surfaces that are in contact with each other. However, the practical goal of tribology is not just to reduce friction and wear, which is referred to as detrimental solid-solid contact. In some situations, designers alter the surfaces; minimal friction and maximum wear (for example with pencils), but for brakes, the opposite is true (maximum friction, but minimal wear). On the other hand, maximizing both wear and friction is favorable in some applications, such as erasers. So, understanding the tribological response behind the interactions of any engineered surface that slides over or rubs against another is very important and is crucial to improve the efficiency of the engineered surfaces.

It is also necessary to identify how tribology affects the economy of countries. Every year, it has been shown that the cost of wear and friction imposes a heavy burden on the world economy. The cost of energy and material losses to manufacturing due to friction and wear are not very high for an individual machine. However, the total cost to a whole country's economy can be significant. Therefore, scientific research in tribology has evolved dramatically from quality to quantity, and from static to dynamic. In addition, research based on control of tribological behavior has grown extensively compared to research related to the analysis of tribological phenomena. Although most of the tribology research in the past focused more on equipment maintenance, now it is mainly focused on the design of mechanical products. In recent years, studies have been carried out on the development of new coatings and surface treatment technologies to introduce a new approach for the improvement of the tribological performance of industrial products. In terms of the new generation of surface engineering using chemical, mechanical, and physical methods, the field of tribology study has evolved from metallic to non-metallic materials, such as ceramics, polymers, and composites.

Tribology is a multi-disciplinary science linked to evaluating the wear mechanism, friction, lubrication, and other tribological phenomena. Bio-tribology discusses contact mechanisms in biological systems to improve human life expectancy. Additionally, tribo-corrosion is another field of study related to the mechanical interactions and electrochemical/chemical reactions that occur between different elements of the tribo-system in a corrosive environment. Nano-tribology is a new discipline which provides a new approach from macro- to nanoscales to investigate in depth the tribological mechanisms in atomic and molecular level interactions. In the past few decades, there has been considerable development in tribology at the interdisciplinary level. The advancement in tribology has been extended to the automotive industry, aerospace, geology, and even ecology, which will become appealing fields of study in the near future.

This book aims to present the scientific fundamentals of tribology for controlling friction and wear in various applications. The reader will find some practical examples of new coatings and surface treatments applied to the surface of engineering components to reduce the damage that originates due to tribological contact. It covers different types of tribology studies that include tribo-corrosion, bio-tribology, tribo-oxidation, high-temperature tribology, lubrication, and so on. In addition, the book is focused on a wide variety of deposition technologies that produce an engineered surface with inherent characteristics that are evaluated from the tribological perspective. The contributing authors of the book are experienced professors, young scientists and researchers from premier research institutes across the world. This book will appeal to a wide range of readers from an academic to an industrial background. The subject matter of the book is multi-disciplinary and is aimed at researchers and engineers interested in tribology. Additionally, it is designed to inspire graduate students with the fundamentals and solutions for real problems in tribology.

ORGANIZATION OF THE BOOK

This book is principally concerned with the evaluation of the latest methods to improve the tribological behavior of materials for various applications. The book is categorized into three sections. The first part of the book, Chapters 1–6 is about the utilization of different thermal spray techniques to produce a wear resistant coating for various working conditions, such as high temperature and corrosive environments. The first three chapters (Chapters 1–3) focus on thermal barrier coatings (TBCs) which have a great influence on the efficiency of turbine engines and combustion chambers at high-temperature working conditions. The second section (Chapters 7–12) of the book is related to thin films and surface treatments which can reduce friction and improve the wear resistance of the materials. In this section, plasma-assisted chemical vapor deposition (PACVD), physical vapor deposition (PVD), electrodeposition and sol-gel methods that influence the tribological behavior of the materials have been described in detail. The last part (Chapters 13–15) of the book is focused mostly on the tribological properties of some advanced materials, such as two-dimensional lubricants, metallic glasses and surface-functionalized metal foils.

Gas turbine components mainly fail because of oxidation, hot corrosion and erosion. In the case of erosion, sand particles and fly ash damage them and lead to failure in gas turbine which is described in the first chapter. Chapter 1 discusses the incorporation of ceramic whiskers or fibers into the thermal barrier coatings. The improvement in wear resistance of the reinforced TBCs is related to the decrease of interlamellar cracks in the top-coat and interlocking interfaces between splats. This chapter is physically separated into three sections, starting with an introduction of TBCs and their failure mechanism, erosion in TBCs, and, finally, fiber reinforced TBCs. Another tribological aspect of thermal barrier coating, including fretting wear and micro-scratch, is evaluated in Chapter 2. The role of carbon nanotube on multi-length scale tribological behavior of Al_2O_3-based thermal barrier coatings is reported quantitatively, and the dominant wear mechanism is identified and described. In this chapter, frictional force mapping is shown to investigate the fretting wear behavior and understand the transition regimes. In Chapter 3, the wear resistance and oxidation behavior of $LaTi_2Al_9O_{19}$/YSZ double-layered ceramic as a new and modern TBC are investigated. Here, the plasma spray technique is employed to fabricate the coatings with varying thickness of $LaTi_2Al_9O_{19}$ as the top layer of the TBC system. Chapter 4 is about the WC-based coatings produced by thermal spray techniques. These coatings are processed and utilized for wear resistance purposes by decreasing the coefficient of friction and wear rate of the components. Also, tribological properties of thermally sprayed WC-based coatings are comprehensively assessed in terms of erosion and abrasive wear. Moreover, the tribological behavior and wear mechanisms of WC-based coatings at elevated temperatures are summarized in Chapter 4. In Chapter 5, the aim is to provide an overview of the prevention of tribo-surfaces through distinct deposition techniques. It is depicted with the general principle of thermal spray coating, PVD, CVD and sol-gel coatings. Furthermore, the coating properties, the application of deposition techniques, and associated challenges are described. Another deposition method to reduce the wear rate and friction is cold spray described in Chapter 6. This chapter aims to examine the effect of the unique gas-dynamic characteristics of supersonic jets on the deposition process. Additionally, the influence of some properties of supersonic jets, such as pressure gradients, velocity fields, and impingement behavior on the tribological behavior of the coatings is discussed.

As cited earlier, the second part of the book discusses surface treatments and thin films to improve the wear resistance of the materials. Over the last decades, surface treatment has been one of the most appealing approaches to improving the wear-resistance of metallic components. Among the surface treatments, plasma nitriding, which is a widely used technique in various industries, is discussed in the first chapter (Chapter 7) of this section. The die steels, such as AISI H11 and AISI H13 are plasma nitrided, and their tribological properties are investigated. As can be inferred in this chapter, plasma nitriding is a promising method of reducing the wear and friction of the hot forming die at elevated temperatures. Additionally, physical vapor deposition (PVD) techniques have been widely used in the deposition of corrosion and wear-resistant coatings for different industrial applications. Chapter 8 provides a critical review of recent advancements in the wear-resistant coatings prepared by PVD methods, such as diamond-like carbon (DLC), titanium nitride (TiN), chromium nitride (CrN) and tantalum nitride (TaN). Also, the influence of the physical and chemical properties of the deposited coatings on their tribological properties is examined. The objective of Chapter 9 is to provide an overview of the tribological properties of carbide and carbo-nitride based coatings which are fabricated by cathodic arc evaporation and magnetron sputtering techniques. The C/N ratio in the carbo-nitride coatings, which plays a key role in the reduction of friction coefficient, and wear rate have been explained in this chapter. Electrodeposition has different applications and is one of the most attractive methods of fabricating wear-resistant coatings, which is discussed in Chapters 10 and 11. Chapter 10 introduces a novel electrodeposition method for deposition of Ni-B-TiO_2 composite coating on heat-treated AISI P20. The main purpose of this electroplated coating is to increase the service life of hammers used in the crushing process. After this, the pin-on-disk test and multi-pass scratch test are performed to investigate the abrasive wear resistance of the deposited coatings. The result indicates that this novel method increased the mechanical and tribological properties of the composite coatings. Chapter 11 contains useful review of tribo-corrosion of hydroxyapatite coating prepared by the electrodeposition process on Mg alloys. It describes the influence of electrodeposition parameters on the mechanical and structural properties of the hydroxyapatite coating. Furthermore, the relationship between process parameters and biocompatibility are comprehensively discussed. Chapter 12 provides an overview of the relationship between hydrophobicity and tribological performance of multi-functional superhydrophobic coatings. Additionally, other types of functionalities of coatings like self-cleaning, transparency, anti-icing and anti-corrosion are discussed and their tribological properties compared.

As previously mentioned, the last part of the book discusses the role of advanced materials on tribological properties of engineered surfaces. If the high friction in tribological systems is not controlled efficiently, it can reduce the system life and reliability of the industrial parts. For this reason, the first part of the last section (Chapter 13) is related to the introduction of new generation lubricants, called two-dimensional (2D) materials, which can also be used as additives in oils and composites; different aspects of these solid 2D lubricant materials on tribological performance are evaluated. Chapter 14 is also devoted to the latest achievements in the tribological performance of metallic glasses. This chapter describes the importance of the correlation between structural evolution and mechanical properties of metallic glasses to design high yield tribological systems. Finally, Chapter 15 discusses the different methods for surface functionalization and characterization of metallic and non-metallic surfaces used as electrode substrates. In addition, the major techniques for thin-film fabrication and surface characterization of metal foils are studied particularly for electrochromic and photovoltaic applications.

CONCLUSION

Professors in academia, lecturers, researchers, company managers, engineers, students and libraries from the four corners of the world involved in material science, mechanical engineering, physics, and applied chemistry, are considered the target audience of the book. The aim is to shed light on the principles and application concepts of tribology in coatings and surface treatments. Furthermore, it enlightens the readers about the latest advances in this subject area. Lastly, the book covers new information around the topic, with the aim of explaining all the research areas, to pass the knowledge on and to encourage other researchers regarding this developing subject.

Amirhossein Pakseresht
FunGlass – Centre for Functional and Surface Functionalized Glass, Alexander Dubček University of Trenčín, Slovakia

Omid Sharifahmadian
FunGlass – Centre for Functional and Surface Functionalized Glass, Alexander Dubček University of Trenčín, Slovakia

Acknowledgment

The editors would like to acknowledge the help of all the people involved in this project and, more specifically, to the authors, reviewers and editorial advisory board members that took part in the review process. Without their help and support, this book would not have become a reality.

First, the editors would like to express our deep appreciation to each one of the authors for their contributions. Our sincere gratitude goes to the authors of each chapter who contributed their time and expertise to this book. It is an honor to have each of you included in this project. Second, the editors wish to acknowledge the valuable contributions of the reviewers regarding the improvement of the quality, coherence, and content presentation of each chapter. Your attention and expertise are appreciated. Most of the authors also served as reviewer; we greatly appreciate their devotion to this double task. Third, the editors would like to thank you, the reader for providing the inspiration for this work. We hope you find it useful and interesting for your studies and research. Special thanks are also due to all IGI Global staff for their kind help and cooperation from the very beginning of this project. This book would not have been possible without their support.

Finally, the editors are deeply grateful to the Funglass center which provided us with a stimulating atmosphere in order to finish this book. This work is a part of dissemination activities of project FunGlass. This project has received funding from the European Union´s Horizon 2020 research and innovation program under grant agreement No 739566. Also, this work was supported by the VEGA grant no. 1/0171/21. This work was also created in the frame of the project Centre for Functional and Surface Functionalized Glass (CEGLASS), ITMS code is 313011R453, operational program Research and innovation, co-funded from European Regional Development Fund.

Amirhossein Pakseresht
FunGlass – Centre for Functional and Surface Functionalized Glass, Alexander Dubček University of Trenčín, Slovakia

Omid Sharifahmadian
FunGlass – Centre for Functional and Surface Functionalized Glass, Alexander Dubček University of Trenčín, Slovakia

Section 1

Chapter 1
The Effect of Additional Ceramic Fiber on Wear and Mechanical Properties of Thermal Barrier Coatings

Mahdi Alebrahim
FunGlass – Centre for Functional and Surface Functionalized Glass, Alexander Dubček University of Trenčín, Slovakia

Amirhossein Pakseresht
FunGlass – Centre for Functional and Surface Functionalized Glass, Alexander Dubček University of Trenčín, Slovakia

Reza Jafari
Tampere University, Finland

Dusan Galusek
FunGlass – Centre for Functional and Surface Functionalized Glass, Alexander Dubček University of Trenčín, Slovakia

Fariborz Sharifian Jazi
The University of Georgia, Georgia

ABSTRACT

Thermal barrier coatings (TBCs) are intended to defend the engines against excessive heat. The surface of a component is subjected to mechanical degradation, such as wear, in a variety of engineering applications. The majority of equipment used in industrial processes is subjected to abrasive wear at elevated temperatures. In the engineering industry, surface modification and deposition of various coatings are used to improve the tribological characteristics of materials. Reinforced materials offer superior mechanical and thermal qualities as compared to conventional materials. The ceramic matrix and TBCs have low fracture toughness and weak cohesive strength of the ceramic layer, limiting their applicability range. A technique for improving the mechanical properties of ceramic coatings is to incorporate reinforcing materials, such as ceramic whiskers or fibers, into the coating structure, which can improve the coating's wear behavior and mechanical properties through a variety of mechanisms, including fiber pullout and breakage.

DOI: 10.4018/978-1-7998-9683-8.ch001

INTRODUCTION AND BACKGROUND

Gas turbines should be operated at as high a temperature as possible to maximize efficiency. Over the last few decades, improved alloy design, the production of blades comprised of textured microstructures, and later internal cooling and single crystals using air circulation via internal channels built within the blade have all led to consistent increases in operating temperatures. Recently, TBCs deposited on engine components have permitted recent temperature increases (Stern, 1996). The technology for thermal barrier coating comes mostly from the aircraft and land-based turbine engine sectors. TBCs are flexible, thick refractory coatings that shield metal from the gas's severe temperatures (Clarke & Phillpot, 2005).

Raising the turbine inlet temperature (TIT) or minimizing the need for cooling in gas turbines is the main reason for employing thermal barrier coatings, resulting in increased engine efficiency, reduced emissions, and better performance. Commercially available TBCs are generally two-layered systems with a metallic bond coat and a ceramic top coat placed on a substrate of superalloy. The ceramic top layer is the true thermal barrier, and its primary role is to reduce heat transmission to the metallic substrate (Bose, 2007).

Coatings allow metallic materials to be employed at temperatures well above their melting points in many gas turbine engines. The coating's thermal conductivity determines the temperature decrease across the TBC under these conditions (Clarke & Phillpot, 2005).

A ceramic layer with low-thermal conductivity is placed over an MCrAlY bond coat to create thermal barrier coatings (TBCs). Since the early 1960s, such coating systems have been extensively studied, and TBCs have evolved steadily ever since. TBCs are mainly utilized as thermal barriers, but the extreme thermomechanical conditions wherein they are often used require that they also meet additional severe performance criteria. To withstand the thermal expansion pressures associated with heating and cooling, whether during normal operation or during a "flame-out," the coatings must be able to withstand enormous strains without failing (Clarke & Phillpot, 2005).

Another less rigorous but still rather practicable condition is that there would be no phase transitions for the material while it cycles between room and elevated temperatures. These phase transitions are frequently followed by expansion and contraction, which reduce the coating's strain tolerance and reversibility, as well as its ability to withstand repeated temperature cycling. Functional TBC materials should be resistant to erosion too, demanding a high degree of fracturing and deformation resistance. For air-breathing engines, coatings would have to be able to endure continuous high temperatures in an oxidizing environment. To meet this need, the quest for innovative TBC materials is focused on refractory oxides (Pakseresht et al., 2016).

Although zirconia (ZrO_2) is the preferred ceramic for TBCs, the phase transition in pure zirconia leads to a significant volume change that may cause internal stresses and result in early coating failure (Pakseresht et al., 2020). To avoid this kind of failure, oxide stabilizers such as yttria, calcia, magnesia, or ceria are added to the zirconia. At the moment, yttria (Y_2O_3) is the most commonly used stabilizer for TBCs, and the material is generally referred to as yttria-stabilized zirconia (YSZ) (Bose, 2007).

In recent years, several studies have focused on three main areas to improve conventional TBCs. First, doping several metallic stabilizers including ceria and yttria into traditional powder coating (ZrO2); second, seeking novel materials with the structures of pyrochlore, fluorite, and perovskite that can be used as TBCs at higher temperatures; and third, expanding TBC architectural design according to double ceramic layered TBCs and functionally graded TBCs (FG-TBCs) (Chen et al., 2011).

This chapter will discuss the mechanical and wear properties of thermal barrier coatings, the production processes for wear-resistant TBCs, and the degradation mechanisms associated with wear in these materials. Additionally, the impact of adding ceramic fibers to thermal barrier coatings will be investigated, as well as the tribological characteristics of the resulting depositions. Finally, the chapter will examine state-of-the-art research on enhancing the YSZ plasma sprayed coatings' wear behavior by including Al_2O_3 whiskers.

CONVENTIONAL THERMAL BARRIER COATINGS

Zirconia-based coatings have been popular because of their unique characteristics, including strong stability of the microstructure, comparatively high coefficient of thermal expansion, low thermal conductivity, and superior resistance against thermal shock and oxidation (Pakseresht, et al., 2016). At around 1443 K, a monoclinic to tetragonal phase change takes place with a 4% volume reduction, while the transformation from tetragonal to cubic phase happens at about 2643 K. This volume change (due to temperature cycling) might cause the coating to spall, resulting in component failure. As a result, ZrO_2 can be blended with phase-stabilizing agents such as Al_2O_3, CaO, CeO_2, MgO, and Y_2O_3 to reduce the volume change caused by phase changes at higher temperatures (Chen, 2006).

Because of its high fracture toughness, comparatively high thermal expansion coefficient, chemical stability, and low thermal conductivity, manufacturers have extensively employed yttria-stabilized zirconia as the top coat. A standard TBC has two layers: first, an exterior insulating ceramic top coat (TC) that reduces heat transmission to the substrate. The second layer is a metallic bond coat (BC) that primarily impacts resistance against oxidation and serves as the real thermal barrier. Straight onto the metallic material surface, BC is deposited first, followed by TC. When the BC is oxidized at high temperatures, thermally grown oxide (TGO) is formed between the BC and the TC during operation (Pakseresht et al., 2021). The working temperature of the engine can rise because the top ceramic layer can withstand a higher temperature than the metallic substrate, extending the life of metal components exposed to high temperatures and pressures and reducing cooling requirements, resulting in improved engine performance, efficiency, and durability (Guo et al., 2002).

The next section discusses the three most often used manufacturing procedures for depositing TBCs onto substrates.

Introduction of Thermal Spray Coating

Thermal spray refers to specific industrial methods, known as coating processes, for the deposition of metallic or nonmetallic materials onto the surface to enhance their performance. A vast range of materials (metal, cermet, ceramic, and polymer) in different shapes (in the form of a wire, rod, or powder) can be deposited on various types of substrates (Cotell et al. 1994). The main purpose of thermal spray techniques is to increase the materials' endurance in comparison with their structural failure mechanisms or to establish a particular characteristic. The widespread application of thermal sprayed coatings in corrosive, erosive, and harsh environments has made this technological field a growing global market (Vardelle et al. 2016). Additionally, these approaches provide cost-effective solutions for engineering product wear and corrosion issues and are used extensively in gas turbine engines (Khan et al., 2015).

Air Plasma Spray

The plasma-spray procedure is the primary approach for generating thermal barrier coatings. This method is popular due to its ability to provide long-lasting and repeatable coatings at a reasonable cost (Miler, 1987). One of the most common procedures for producing thermal barrier coatings is air plasma spray (APS). The APS method is used to coat big stationary engine elements like stator vanes and the combustion chamber since it is less expensive and allows for a wider choice of materials to be sprayed than Electron Beam-Physical Vapor Deposition Method (EB-PVD) . Furthermore, the thermal conductivity of APS TBCs is lower than that of EB-PVD (Hu et al., 2018).

The air plasma spray procedure (also known as conventional plasma spray) may heat the powder to temperatures ranging from 6000 to 15,000 °C, which is hot enough to melt any solid. Plasma is formed by superheating inert gases such as argon (low energy) or hydrogen-argon mixtures with a direct current arc. By using hot gas as a carrier, the plasma jet warms and pushes the powder onto the substrate. When molten particles come into contact with the substrate, they quickly cool and form primary splats. This technique is repeated alternately until the desired thickness for the coating is obtained (Malmberg & Heberlein, 1993; Dykhuizen et al., 1999; Pakseresht et al., 2016). To protect the coating from residual stresses, the substrate temperature should be kept between 100 and 200 °C, especially when spraying ceramic powders. Spray device design, feedstock powder composition, and plasma gases all have an effect on the spray rate, which in turn has an effect on the substrate temperature. The primary drawback of this technique is that it results in oxidation due to the interaction of melted particles with the surrounding environment (Malmberg & Heberlein, 1993; Dykhuizen et al., 1999). Splats, as well as other microstructure characteristics including intersplat gaps, intrasplat fractures, and interlamellar and globular holes, describe the sprayed coating microstructure. The ceramic top coating microstructure is intimately connected to the lifespan, mechanical, and thermal characteristics of TBCs (Stern, 1996).

This method has a number of significant benefits, including high deposition rates and a variety of oxide and metallic coatings. However, the spray technique's inability to provide a homogenous and dense microstructure is its drawback. Additionally, spray procedures provide a somewhat rough coating surface. Furthermore, coatings contain flaws such as unbonded interfaces, microcracks, and microporosity, resulting in a reduction in mechanical qualities including corrosion, oxidation resistance, impact, ductility, and wear (Dorfman, 2018). The prevalent flaws in sprayed coatings decline not only the coating's physical characteristics, but also the sprayed film's mechanical properties (Pawlowski, 2009).

High-Velocity Oxy-Fuel Coating

In the 1950s decade, Union Carbide, the detonation gun technique developer, introduced High-Velocity Oxy-fuel (HVOF), and later in the 1970s decade, this method attracted interest and became widely used. Despite the similarity with detonation gun spraying in having a blast chamber inside, HVOF is different in operating continuously, which means it works more efficiently. One of the widespread utilizations of HVOF is for cermet (ceramic–metallic) material deposition. The gases employed in this process as fuel are generally acetylene, hydrogen, propylene, kerosene, and propane, combined with oxygen in the combustion chamber. The blast product expands into a nozzle and reaches supersonic speed; the powder then axially or radially meets the heated gas and propels with nitrogen or hydrogen gases. The superheated gas causes melting or semi-melting of the powder where oxidation might happen depending on the powder material. Feedstock powder can be chosen from all kinds of ceramics or metals, but the fuel needs to be acetylene to produce an excellent and successful coating when it comes to ceramics (Cotell et al. 1994; Davis, 2004).

HVOF can deposit high-bond-strength coatings with excellent microstructure density, making them excellent for anti-wear ceramic and composite coatings, as well as metal coatings for protecting against aggressive media at high temperatures (Giovanni & Serra, 2015).

Introduction of the Electron Beam-Physical Vapor Deposition Method

Another popular method for creating thermal barrier coatings is electron beam physical vapor deposition (EB-PVD). It is common for aviation engines to utilize TBCs with a high degree of strain tolerance on the moving blades of their engines (Hu et al., 2018).

In this process, a high-energy electron beam capable of fast melting and evaporation is concentrated on the base material. Then, the vapor is applied to specific substrates, where it might evaporate both metals and ceramics. Its fast deposition rate creates an advantage in a variety of sectors, including semiconductors, filters, optical prisms, and high-value thermal barrier coating devices, among others (Xu & Guo, 2011; Alehojat et al., 2020).

This technique has been used to prolong the lifespan of turbine components and address a number of the drawbacks related to thermal spraying because of the weak bonding among splats, voids, inhomogeneous microstructure, unmelted particles, and non-bonded interfaces formed during the APS process (Wolfe & Singh, 1998). EB-PVD TBCs outperform PS TBCs in the categories listed (Xu et al., 2008):

(1) the probability of forming a ceramic coating with a columnar structure and higher strain acceptance compared to sprayed coatings, prolonging the thermal cycling life span; (2) the opportunity of establishing a clear surface to prevent interferention with the aerodynamic flow around the turbine blades; (3) markedly enhanced interfacial interaction between the metallic bond coat and ceramic topcoat; and (4) the capability to change the deposition microstructures. EB-PVD TBCs, on the other hand, have a higher thermal conductivity than PS coatings because the intervals between the columns are parallel to the direction of the heat flow, resulting in less effective thermal insulation. Due to their excellent thermal shock resistance (Stern, 1996), EB-PVD TBCs are often employed in the most demanding applications, including aircraft engine vanes and blades.

Modern TBCs were implemented to avoid failure owing to common defects and high operating temperatures, to extend the lifespan of TBCs, and to enhance the wear and mechanical properties of the YSZ coatings. TBCs with complicated crystal structures, including perovskites and pyrochlores, and novel architectures, such as multilayer and functionally graded TBCs, are referred to as new and modern TBCs. A recent study has shown these advanced TBCs to offer superior physical and mechanical qualities over traditional ones (Pakseresht et al., 2018).

Other Techniques

In addition to the developed coating processes, there are several other emerging technologies, including high velocity air fuel (HVAF) and cold spray coating depositions. These methods generally operate at a lower temperature, virtually below the melting point of conventional powder feedstock materials. In other words, kinetic energy derived from particles' high speeds also plays a part in the deposition process by deforming the impacted particles onto the substrate surface (Ichikawa et al., 2017; Richer et al., 2010; Szymanski et al., 2015). Therefore, they are considered potential techniques to produce the metallic bond coat for TBC applications while conserving the initial composition and phase structure of precursors. Despite these promising features and recent advances, there are still challenges in dense coating fabrication from hard, brittle, high melting point feedstock that impose limitations on wide application for TBC and erosion resistance coating fabrication (Guo et al., 2021; Jafari & Sadeghi, 2019).

FAILURE MECHANISM OF THERMAL BARRIER COATINGS

Thermal barrier coating failure is a highly multifactorial process, including both inherent and extrinsic factors. Intrinsic failure mechanisms arise as a result of the formation of a TGO layer at the BC/TC boundary and the modifications caused by thermal stresses related to the kinetically driven processes. Additionally, sintering of the porous ceramic with prolonged temperature exposure results in a decrease in compliance, especially with plasma sprayed TBCs (Viswanathan et al., 2014). Moreover, the thermo-mechanical characteristics of the TC, BC, and substrate vary.

Thermal stresses are generally generated on the surface layer and at interfaces as a result of thermal expansion mismatches and imposed temperature fluctuations. Surface and interface cracking may occur as a result of these factors (Kokini et al., 2002). Thermal stresses applied repeatedly cause the coating to develop microcracks. As the surface fractures propagate farther, an increased driving force is created, culminating in the formation of the interface crack (Rangaraj & Kokini, 2003).

Additionally, TBC systems have a number of intrinsic drawbacks, including low toughness at the interface of TC and BC, a considerable misalignment of the TC and BC characteristics, and weak top coat material fracture toughness. The APS and EB-PVD approaches have different underlying failure mechanisms, in part owing to the microstructures of their end products. In the latter situation, delamination happens at the BC–substrate contact as a result of the high adhesion of TC and BC. Despite this, the YSZ layer fails in the sprayed coating due to TC spallation at the BC-ceramic layer contact (Lee et al. 1996). Additionally, the stiffness increase due to the interface–crack sintering might lead to the multilayer coating losing its compliance (Viswanathan et al., 2014).

Extrinsic mechanisms of failure are more unpredictable and are caused by erosion and foreign particle damage. These variables are the same for both techniques (Viswanathan et al., 2014). In general, thermal stresses associated with high heat flux increase the development of TBC–BC interface cracks, resulting in spalling and delamination along the TC–BC interface as well as a reduction in structural integrity for such coatings (Xiong et al., 2005). Thus, the creation of multilayer and functionally graded TBCs may be a potential technique for reducing the motivating factor in TC–BC interface fractures since the thermal stresses produced at the TC–BC interfaces are reduced by the FGM structure. As a result, it is thought that functionally graded and multilayer TBCs are more resistant to delamination and thus longer-lasting (Rangaraj & Kokini, 2003).

WEAR RESISTANT COATINGS

The undesirable disappearance of material from a surface as a consequence of mechanical activity is known as wear. Managing the wear process requires monitoring a wide range of products and processes. This begins with a thorough knowledge of the product: the wear kind that it is going to experience; the conditions it will be exposed to (temperature and gas), the history of the material (history of the manufacturing, chemistry, hardness of the surface, alloy, and grain size), and the environment. Erosion, abrasion, sliding, and fretting are some of the types of wear listed here. Each mechanism has a distinct feature for each kind of wear. More often than not, wear is related to many mechanisms. Validating the product and testing the field are necessary prior to placing new components into service. Coating toughness becomes increasingly critical for solid impingement angles of around 90 degrees. In general, a coating with strong surface fatigue resistance is required for cavitation and liquid impingement (Dorfman, 2018).

Erosive Wear

The impact of solid particles on a surface, causing material to be removed over time, is the process that occurs during erosion. Coatings with high hardness would be needed when the impact angle is low (Dorfman, 2018). The angle of impact, particle size, and physical characteristics of the debris define the wear process when engineering components are eroded (Dorfman, 2018). Wear may also be influenced by temperature and the surrounding environment (Dorfman, 2018). On steam turbine blades, solid particle erosion is a significant factor. Stainless steel blades of the 400 series are commonly subjected to temperatures of up to 540 °C (1000 °F), resulting in the formation of tiny iron chromite particles. Because the expense of replacing these blades has historically been high, researchers have investigated coating them using unique material systems and techniques. To overcome these application challenges, high-energy plasma, DGun, or HVOF technologies are being employed to coat NickelChromium/Chromium Carbide and FeCrAlY-Chrome Carbide overlay coatings (Dorfman, 2018).

Fretting Wear

Fretting wear is seen as a result of compressor and fan blade vibration in aviation jet engines. A damper at the mid-span functioning as a point of contact between the blades, restricting motion that would otherwise cause blade resonance, are used to control, and partly eliminate the fretting wear.

Adhesive/Sliding Wear

The movement of two objects wherein their surface velocities are varied in size and/or direction at the contact point is described as sliding wear. Mechanical loading, temperature, load type, chemical condition, and lubrication are all important factors in sliding wear, which is noticeable in reciprocating and spinning equipment. Modified carbide-containing coatings were studied to decrease sliding wear in large cylinder applications (Dorfman, 2018).

Abrasive Wear

Abrasive wear refers to the phenomenon when particles travel over a surface, removing material. The particles may be loose or may have been a component of another surface that came into contact with the worn surface. High-or low-velocity oxy-fuel methods are employed to produce blades used for cutting made of cobalt-tungsten carbide or self-fluxing alloys. Alumina-titania, tungsten carbide, and steel coatings have been utilized to reduce or eliminate abrasive wear in construction equipment. The matrix hardness and chemical composition of the coating are critical factors in abrasive wear. The addition of chromium and molybdenum improves the corrosion resistance of nickel, cobalt, and iron (Dorfman, 2018).

EROSION IN THERMAL BARRIER COATINGS

A common kind of TBC failure is solid particle erosion, which may occur in land-based gas turbines as a consequence of particles escaping through filters or as a result of particle production inside the combustion chamber or compressors, depending on the engine's components and operational circumstances. In particular, solid particle erosion can occur in aero gas turbines that operate in sandy (or ashy) conditions. Solid particles do not travel following flow streamlines owing to their inertia, and as a consequence, they come into contact with components, eroding protective coatings and causing corrosion. The principal impacts of erosion in gas turbines are alteration in blade geometry, pressure drop, and heat generation from the base metals (Cernuschi et al., 2011). Because of their differing microstructures, the erosion processes in EB-PVD and APS coatings vary dramatically, as discussed in the following section.

The Mechanisms of Erosion in TBCs

Coatings Made Using the APS Method

According to Eaton and Novak's research (Eaton & Novak, 1987), there are three forms of solid particle erosion in APS TBCs: The most noticeable characteristics on the erosion surface are early scars (low erosion rate), followed by the emergence of cracks on the coating surface around the impact zone (moderate erosion rate), and the formation of surface tunnels (high erosion rate).

Impacting particles leave indentations on the surface in the first scenario, and repetitive collisions on the deformed material result in erosion. Impact causes fracture development across splat boundaries in the second kind of mechanism. In the third scenario, the particles' kinetic energy is sufficient to link pre-existing pores within the TBC, resulting in the degradation of clusters of multiple splats each time. The authors also discovered a link between TBC strength and the rate of degradation, as determined by a four-point bending test (the higher the former, the lower the latter). While the total porosity stays constant, the rate of erosion increases as a function of the porosity's specific surface area (Eaton & Novak, 1987). Regardless of the porosity fraction, Janos et al. (1999) discovered that the coating to particle micro-hardness ratio has a significant impact on the erosion rate of ceramics. The coating microhardness is all that counts after the particle microhardness is set. Wellman et al. (1999) and Li et al. (2006) explain APS TBC erosion as a result of erodent particles spalling from the surface lamella. As a consequence, fracture propagation at the interface between neighboring lamellae controls coating deterioration. To put it another way, APS fails because fractures spread across the microcrack network and over splat boundaries. This indicates that the lower the erosion rate, the greater the proportion and/or strength of the bonded contact between lamellae. Because the sintering process causes lamellae to connect, the erosion resistance of APS TBC has been demonstrated in the literature for older samples (Li et al., 2006).

Coatings Made Using the EB-PVD Method

Damage processes in EB-PVD coatings differ from those observed in APS coatings and bulk ceramic materials due to their columnar structure. Wellman and Nicholls (2007), as well as Chen et al. (2003), suggest three possible paths.

The most prevalent kind of cracking is Mode I (near-surface cracking or lateral cracking). The TBC's reaction is simply elastic in this experimental context, and cracking parallel to the surface is produced by tensile stresses generated by elastic waves propagating forward and backward along either single column all around the impingement point. Slower erosion rates define this damage phase. Even though the erosion rates differ at high temperatures due to the fact that TBC's fracture toughness and elastic modulus hardness are temperature dependent, this degradation process has been seen at both room temperature and high temperatures (Chen et al., 2003).

Mode II is the most advanced mode (compaction damage). A densification of a shallow top layer of the TBC generated by plastic deformation may be seen due to the impingement of particles with somewhat greater momentum (speed and perhaps mass) than in the prior example. This is a frequent situation when several particles are involved. In reality, despite the fact that each impact does not cause cracking, a massive, densified top layer forms as a result of repeated blows. As a consequence of the stress concentration produced by the following impacts, cracks may develop and propagate at the contact between each underlying column and the densified layer. There are generally some overlapping areas when various modes occur in the same sample (Chen et al., 2004; Wellman & Nicholls, 2007).

Foreign object damage (FOD) is the most advanced mode. When high-velocity large particles smash into the EB-PVD TBC, plastic deformation and densification bending absorb the bulk of the kinetic energy. Kink bands emerge around the periphery of the plastic zone as a result of deformation. These crack bands progress conically toward the coating and bend at the TBC/bond coat contact, resulting in delamination.

At elevated temperatures, a novel kind of FOD has been reported. Column buckling has been observed, as well as concurrent cracking and densification. Further strikes in this condition will increase the degree of buckling to the point where it stimulates cracking and, as a result, material loss. The presence of both types of FOD in the same population suggests that they do not have to be totally exclusive (Wellman & Nicholls 2007; Chen et al. 2003).

Cross-sections of EB-PVD materials eroded at high temperatures (usually 1150 °C) have shown a variety of material removal mechanisms that are controlled by temperature, velocity, particle size, and the kind of material being removed (Chen et al., 2004). Unlike APS coatings, the erosion rate of EB-PVD coatings increases with ageing at high temperatures. In fact, in all three damage phases, sintering permits fractures to spread quickly along additional proximal sintered columns (Cernuschi et al., 2011). Wellman et al. (1999) found a linear relationship between impingement speed and erosion rate for both APS and EB-PVD coatings, with the former having a seven to ten times greater erosion rate. EB-PVD and thick vertically broken APS TBC, on the other hand, have comparable degradation rates.

CERAMIC REINFORCEMENT BY FIBER

Ceramic materials offer several advantages, such as a high working temperature and a low density. On the other hand, these materials, have a low hardness and are vulnerable to major failure, restricting their usage. The development of advanced continuous ceramic fiber toughening phases has altered the intrinsic brittleness of ceramic matrix materials (Schneider et al., 2008).

Ceramic fibers are used as reinforcing elements in composites because of their distinctive characteristics of high temperature endurance and high elastic modulus. Their qualities make them ideal for use in automotive and aviation applications requiring strong heat resistance. There are two different kinds of ceramic fibers called continuous (long length) and discontinuous (short length). Ceramic matrix composites (CMCs) are essentially classed as oxide-free or oxide-based composites. The sol-gel method produces continuous oxide fibers formed of alumina and silicate, whereas the melt-spinning method produces short oxide fibers. Nonoxide ceramic fibers based on silicon and boron, on the other hand, are presently being investigated and made by the thermal conversion of polymer precursors (Yalamac et al., 2017).

The fibers are organized as satin weave textiles or 2D stacked plains in numerous ceramic matrix nanocomposites. As a consequence, the material is anisotropic or, to put it another way, orthotropic. A fracture between the layers cannot be bridged by fibers. As a consequence, interlaminar shear strength and strength perpendicular to the 2D fiber orientation are low in these materials. Delamination is a frequent occurrence under certain mechanical loads. When fiber is utilized in ceramic matrix nanocomposites, the wear resistance increases dramatically (Carruthers et al., 2002).

Oxide Fibers

Since the 1970s, ceramic oxide fibers in both long and short lengths have been commercially accessible. Due to the high melting points of alumina (Al_2O_3) and alumina-silica (Al_2O_3-SiO_2) combinations, these fibers are often utilized in high-temperature applications (Krenkel, 2008). These oxide fibers have gained significant attention in applications of power engineering and aerospace over the past decade due to their great corrosion resistance and inexpensive cost.

Furthermore, because of their microstructural consistency, superior mechanical strength, and long-term creep behavior at extremely high temperature, they are chosen in engineering applications. The inclusion of crystalline phases such as alumina, zirconia, yttria-doped alumina, and mullite results in these improved characteristics (Krenkel, 2008).

The most prevalent kind of oxide fiber is alumina (Al_2O_3) fiber. These fibers have excellent creep and thermal shock resistance, high temperature strength, low coefficient of thermal expansion, outstanding dimensional stability, and superior dielectric properties. Furthermore, Al_2O_3-based fibers are preferred due to their high chemical stability in oxidation and reduction environments with temperatures greater than or equal to 1000 °C. These fibers are employed in a wide range of applications, including rocket boosters and industrial furnaces. Insulation, reinforcement of composites, and flame, heat, and impact shields for aviation are just a few of the uses for these fibers (Cooke, 1991).

Alumina-based fibers are often employed to improve the structural integrity of a range of different material composites, thus increasing their stiffness and strength (Tan, 2011).

Nonoxide Fibers

Due to their high melting temperatures and susceptibility to densification, manufacturing nonoxide fibers is challenging. Their primary deficit is often low resistance to oxidation. An extensive study has been undertaken and published on the production, mechanical, microstructure, and thermal durability of these fibers. These ceramic fibers are available in two diameters: fine and thick (Yalamac et al. 2017).

WEAR RESISTANT THERMAL BARRIER COATINGS

One of the most serious concerns in recent decades has been extending the lifetime of engineering products. The contact of surface to surface between components is among the most important reasons for limiting the life of equipment in the industry (Sharifahmadian & Mahboubi, 2019).

Various sorts of surface treatments and coatings have been employed in the industry to enhance the tribological characteristics of engineering systems (Khan et al., 2017). Anti-wear coatings have been one of the industry's top focuses in recent years (Li et al., 2004). Thermal spraying and laser/electron beam techniques are two main methods used to create wear-resistant coatings (Khan et al., 2017).

TBCs are subjected to hot corrosion, high-temperature oxidation, and severe wear, including erosion, while working at high temperatures. Wear processes are accelerated by the synergistic impact of hot corrosion and oxidation. As a result, anti-wear coatings are critical for increasing the performance and efficiency of metal substrates under severe conditions (Bolleddu et al., 2017). New TBCs have been designed to eliminate failures caused by frequent defects, improve the wear and mechanical properties of the YSZ coating, and eventually increase the life of TBCs (Pakseresht et al., 2016).

Modern TBCs have complicated crystal structures like pyrochlore and perovskites and innovative designs like multilayer and functionally graded TBCs (Pakseresht et al., 2016). These TBCs outperform traditional ones in terms of mechanical and physical properties (Pakseresht et al., 2015).

Previous research has mostly concentrated on the wear resistance and microstructure of TBC deposited through the APS technique. They demonstrated that the friction and wear properties of TBCs are highly dependent on three critical factors: 1) the parameters of plasma spraying, including spray distance and power; 2) the conditions of wear, like temperature and sliding speed; and 3) coating microstructures, including nano and multilayer TBCs (Li et al., 2004; Sahith et al. 2018).

Scientists have investigated nanostructured YSZ coatings, concluding that their tribological properties have been enhanced in terms of wear rate and friction coefficient compared to traditional films (Kossman et al., 2014). Ghasemi et al. (2013) compared the wear and mechanical properties of regular and nano-structured TBCs. They discovered that nanocoatings had superior properties in comparison to conventional coatings. In addition, the friction coefficient was found to be decreased in nanostructured YSZ coating deposited on a TiAl intermetallic alloy in the study conducted by Wang et al. (2012).

Li et al. (2004) compared the tribological properties of plasma sprayed YSZ thermal barrier coating prepared with conventional and nanopowder feedstock by sliding test against 100C6 steel. The results demonstrated that the wear resistance of the deposited coatings was improved by using nanopowder feedstock, which may be attributed mostly to the reduction of micrometer-sized flaws in the coatings. All of the coatings' wear mechanisms were described in terms of lamellar fracture and adhesion-induced spallation.

According to Cui et al. (2007), nanostructured TBC has superior wear qualities over traditional TBCs. They also discovered that nanostructured coatings outperformed typical YSZ top coats in terms of micro-cracking resistance, adhesive strength, and thermal shock resistance. According to Chen et al. (2002), zirconia nanostructured top coats demonstrated higher wear resistance than their traditional counterparts, and the wear rate of nanostructured zirconia TBCs was around two-fifths of that of the conventional zirconia coating. The enhanced wear resistance can be due to the microstructure optimization and mechanical characteristics improvement, which increase the plastic deformation ability. Chen et al. (2003) also examined a zirconia top coat's wear resistance against an AISI316 stainless steel ball by applying four loads (20, 30, 40, and 50 N) at an oscillating frequency of 25 Hz, an amplitude of 1 mm, and a duration of 15 minutes under lubricated settings. They discovered that nano YSZ outperformed conventional YSZ in terms of wear resistance by four to five times. Ghasemi et al. (2013) compared the wear and mechanical parameters of conventional and nanostructured-TBC and found that the latter had superior mechanical and wear resistance. Wang et al. (2012) sprayed conventional and nanostructured YSZ over TiAl intermetallic alloy, demonstrating that the nanostructured coating had a weight loss of half that of the conventional coating. According to Cui et al. (2012), nanostructured zirconia TBCs have been shown to have improved wear qualities than traditional TBCs. The nanostructured zirconia TBCs were demonstrated to be more micro-crack and thermal shock resistant and showed superior adhesion to the substrate.

WEAR RESISTANT FIBER REINFORCED THERMAL BARRIER COATINGS

Pakseresht et al. (2019) studied the influence of Al_2O_3 ceramic whiskers on the microstructure and wear characteristics of yttria-stabilized zirconia coated by plasma spray technique on an Inconel substrate using optimum process settings. They compared the top surface, cross-section, and fracture cross-section of whisker reinforced coatings with conventional ones. The results showed that the addition of the whiskers improved the adhesion strength of the coated samples from 29.6 Mpa for conventional coatings to 40.4 Mpa for reinforced ones. They concluded that adding 10% alumina can reduce the wear rate of coating surface by up to 25%. The SEM image of the alumina whisker morphology is shown in Fig. 1 indicating that during the spraying procedure, the whiskers have maintained their integrity. They were also evenly dispersed throughout the coating, especially in un-melted or partly melted regions.

Figure 1. Top surface morphology of as-sprayed YSZ reinforced with Al2O3 whisker

Whiskers are designed to connect lamellae in the top coat, which is why introducing them to the top coat results in the composite structure's maximal strengthening. Ceramic fibers may operate as an interfacial locker at the lamellar interface between the YSZ top layer and the interface between the metallic layer and YSZ in this kind of coating, improving interfacial bonding (Krishnamurthy et al., 2012). The whisker reinforced thermal barrier coating's fractured cross-sectional morphology is shown in Fig. 2, which displays the whiskers bridging lamellae in the top coat.

Figure 2. The fractured cross-sectional morphology of fiber reinforced TBC, showing the bridging mechanism which results in interfacial bonding improvement by fibers.

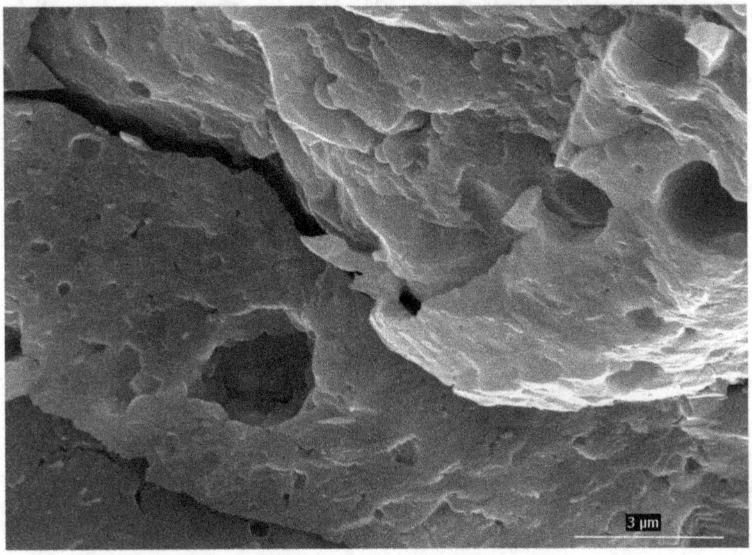

Fibers are also used to strengthen the network across the ceramic coating structure by incorporating them into the lamellar structure. The major goal is to increase the YSZ top coat's cohesive strength (Krishnamurthy et al., 2012). To put it another way, when the lamellar interface bonding ratio increases, the physical and mechanical characteristics of the sprayed films improve (Pakseresht et al., 2019).

The mechanism of the wear was described as exfoliation and degradation, which may readily be triggered by adding additional micrometer-sized flaws to a typical coating. Moreover, the mechanisms of crack deflection, breakage of whiskers, and interfacial locking, which contribute to higher delamination resistance of splats, were interpreted to be the reasons for the improved wear behavior of whisker-reinforced thermal barrier coatings.

CONCLUSION

Wear and other tribological issues may cause most engineering products, as well as thermal barrier coatings, to fail, resulting in system deterioration and ultimate collapse. Several studies over the past several decades have concentrated on improving the mechanical and wear features of traditional TBCs. In general, three important elements influence the friction and wear qualities of TBCs: plasma spraying parameters, such as spray distance and power; wear testing standards, such as temperature and sliding speed; and coating microstructures, such as nano and multilayer TBCs. Nano-structured TBCs have been shown to have better wear and mechanical resistance than traditional ones due to mechanisms related to grain size and micrometer-sized flaws. Scientists are also interested in fiber-reinforced thermal barrier coatings. According to current research, the inclusion of alumina whiskers may significantly improve the wear properties of TBCs due to suppression of exfoliation and degradation.

ACKNOWLEDGMENT

This work was supported by the VEGA grant no. 1/0171/21. This work is a part of dissemination activities of project FunGlass. This project has received funding from the European Union´s Horizon 2020 research and innovation programme under grant agreement No 739566.

REFERENCES

Alehojat, M., Jafari, R., Karimi, P., & Sadeghi, E. (2020). Electron beam-powder bed fusion of Alloy 718: Effect of hot isostatic pressing and thermal spraying on microstructural characteristics and oxidation performance. *Surface and Coatings Technology*, *404*, 126626. doi:10.1016/j.surfcoat.2020.126626

Aliofkhazraei, M. (2014). *Anti-abrasive nanocoatings*. Elsevier.

Bolleddu, V., Racherla, V., & Bandyopadhyay, P. P. (2017). Characterization of air plasma-sprayed yttria-stabilized zirconia coatings deposited with nitrogen. *International Journal of Advanced Manufacturing Technology*, *90*(9–12), 3437–3449. doi:10.100700170-016-9613-1

Bose, S. (2007). *High temperature coatings*. Elsevier Butterworth-Heinemann.

Carruthers, W. D., van Roode, M., Becher, P. F., Ferber, M. K., & Pollinger, J. (2002). Advances in the Development of Silicon Nitride and Other Ceramics. *Turbo Expo 2002,* 67–76. doi:10.1115/GT2002-30504

Cernuschi, F., Lorenzoni, L., Capelli, S., Guardamagna, C., Karger, M., Vaßen, R., von Niessen, K., Markocsan, N., Menuey, J., & Giolli, C. (2011). Solid particle erosion of thermal spray and physical vapour deposition thermal barrier coatings. *Wear*, *271*(11–12), 2909–2918. doi:10.1016/j.wear.2011.06.013

Chen, H., Ding, C., Zhang, P., La, P., & Lee, S. W. (2003). Wear of plasma-sprayed nanostructured zirconia coatings against stainless steel under distilled-water conditions. *Surface and Coatings Technology*, *173*(2–3), 144–149. doi:10.1016/S0257-8972(03)00311-6

Chen, H., Zhang, Y., & Ding, C. (2002). Tribological properties of nanostructured zirconia coatings deposited by plasma spraying. *Wear*, *253*(7–8), 885–893. doi:10.1016/S0043-1648(02)00221-1

Chen, L. B. (2006). Yttria-stabilized zirconia thermal barrier coatings— A review. *Surface Review and Letters, 13*(05), 535–544. doi:10.1142/S0218625X06008670

Chen, X., Gu, L., Zou, B., Wang, Y., & Cao, X. (2012). New functionally graded thermal barrier coating system based on LaMgAl11O19/YSZ prepared by air plasma spraying. *Surface and Coatings Technology, 206*(8–9), 2265–2274. doi:10.1016/j.surfcoat.2011.09.076

Chen, X., He, M. Y., Spitsberg, I., Fleck, N. A., Hutchinson, J. W., & Evans, A. G. (2004). Mechanisms governing the high temperature erosion of thermal barrier coatings. *Wear, 256*(7–8), 735–746. doi:10.1016/S0043-1648(03)00446-0

Chen, X., Wang, R., Yao, N., Evans, A. G., Hutchinson, J. W., & Bruce, R. W. (2003). Foreign object damage in a thermal barrier system: Mechanisms and simulations. *Materials Science and Engineering A, 352*(1–2), 221–231. doi:10.1016/S0921-5093(02)00905-X

Clarke, D. R., & Phillpot, S. R. (2005). Thermal barrier coating materials. *Materials Today, 8*(6), 22–29. doi:10.1016/S1369-7021(05)70934-2

Cooke, T. F. (1991). Inorganic Fibers-A Literature Review. *Journal of the American Ceramic Society, 74*(12), 2959–2978. doi:10.1111/j.1151-2916.1991.tb04289.x

Cui, T., Wang, J., Guan, R., Chen, L., & Qiu, G. (2007). Microstructures and Properties of Thermal Barrier Coatings Plasma-Sprayed by Nanostructured Zirconia. *Journal of Iron and Steel Research International, 14*(5), 116–120. doi:10.1016/S1006-706X(08)60063-1

Di Girolamo, G., & Serra, E. (2015). Thermally sprayed nanostructured coatings for anti-wear and TBC applications. In *Anti-Abrasive Nanocoatings* (pp. 513–541). Elsevier. doi:10.1016/B978-0-85709-211-3.00020-0

Dorfman, M. R. (2018). Thermal Spray Coatings. In *Handbook of Environmental Degradation of Materials* (pp. 469–488). Elsevier. doi:10.1016/B978-0-323-52472-8.00023-X

Dykhuizen, R. C., Smith, M. F., Gilmore, D. L., Neiser, R. A., Jiang, X., & Sampath, S. (1999). Impact of High Velocity Cold Spray Particles. *Journal of Thermal Spray Technology, 8*(4), 559–564. doi:10.1361/105996399770350250

Eaton, H. E., & Novak, R. C. (1987). Particulate erosion of plasma-sprayed porous ceramic. *Surface and Coatings Technology, 30*(1), 41–50. doi:10.1016/0257-8972(87)90006-5

Ghasemi, R., Shoja-Razavi, R., Mozafarinia, R., & Jamali, H. (2013). Comparison of microstructure and mechanical properties of plasma-sprayed nanostructured and conventional yttria stabilized zirconia thermal barrier coatings. *Ceramics International, 39*(8), 8805–8813. doi:10.1016/j.ceramint.2013.04.068

Guo, D., Wang, Y., Fernandez, R., Zhao, L., & Jodoin, B. (2021). Cold spray for production of in-situ nanocrystalline MCrAlY coatings – Part I: Process analysis and microstructure characterization. *Surface and Coatings Technology, 409*, 126854. doi:10.1016/j.surfcoat.2021.126854

Guo, H., Xu, H., Bi, X., & Gong, S. (2002). Preparation of Al2O3–YSZ composite coating by EB-PVD. *Materials Science and Engineering A, 325*(1–2), 389–393. doi:10.1016/S0921-5093(01)01476-9

Hu, Y., Cai, C., Wang, Y., Yu, H., Zhou, Y., & Zhou, G. (2018). YSZ/NiCrAlY interface oxidation of APS thermal barrier coatings. *Corrosion Science*, *142*, 22–30. doi:10.1016/j.corsci.2018.06.035

Ichikawa, Y., Horiuchi, S., Ogawa, K., Oikawa, M., Tatsuki, T., & Yamazaki, H. (2017). Microstructural Change and Delamination Resistance of the Thermal Barrier Coatings with Cold Sprayed Ce-Content Bond Coats. *Journal of the Society of Materials Science, Japan*, *66*(2), 142–149. doi:10.2472/jsms.66.142

Jafari, R., & Sadeghi, E. (2019). High-temperature corrosion performance of HVAF-sprayed NiCr, NiAl, and NiCrAlY coatings with alkali sulfate/chloride exposed to ambient air. *Corrosion Science*, *160*, 108066. doi:10.1016/j.corsci.2019.06.021

Janos, B. Z., Lugscheider, E., & Remer, P. (1999). Effect of thermal aging on the erosion resistance of air plasma sprayed zirconia thermal barrier coating. *Surface and Coatings Technology*, *113*(3), 278–285. doi:10.1016/S0257-8972(99)00002-X

J.R. Davis & Associates & ASM International. (2004). *Handbook of thermal spray technology*. ASM International. https://app.knovel.com/hotlink/toc/id:kpHTST0006/handbook-of-thermal

Khan, M. A., Sundarrajan, S., Duraiselvam, M., Natarajan, S., & Kumar, A. S. (2017). Sliding wear behaviour of plasma sprayed coatings on nickel based superalloy. *Surface Engineering*, *33*(1), 35–41. doi:10.1179/1743294415Y.0000000087

Kokini, K., DeJonge, J., Rangaraj, S., & Beardsley, B. (2002). Thermal shock of functionally graded thermal barrier coatings with similar thermal resistance. *Surface and Coatings Technology*, *154*(2–3), 223–231. doi:10.1016/S0257-8972(02)00031-2

Kossman, S., Chicot, D., Decoopman, X., Iost, A., van Gorp, A., Meillot, E., Puchi-Cabrera, E. S., Santana, Y. Y., & Staia, M. H. (2014). Sliding Wear Response of Nanostructured YSZ Suspension Plasma-Sprayed Coating. *Journal of Thermal Spray Technology*, *23*(8), 1350–1361. doi:10.100711666-014-0146-6

Krenkel, W. (Ed.). (2008). *Ceramic Matrix Composites: Fiber Reinforced Ceramics and their Applications* (1st ed.). Wiley. doi:10.1002/9783527622412

Krishnamurthy, N., Prashanthareddy, M. S., Raju, H. P., & Manohar, H. S. (2012). A Study of Parameters Affecting Wear Resistance of Alumina and Yttria Stabilized Zirconia Composite Coatings on Al-6061 Substrate. *ISRN Ceramics*, *2012*, 1–13. doi:10.5402/2012/585892

Kutz, M. (Ed.). (2018). Handbook of environmental degradation of materials (3rd ed.). William Andrew/ Elsevier.

Lee, W. Y., Stinton, D. P., Berndt, C. C., Erdogan, F., Lee, Y.-D., & Mutasim, Z. (1996). Concept of Functionally Graded Materials for Advanced Thermal Barrier Coating Applications. *Journal of the American Ceramic Society*, *79*(12), 3003–3012. doi:10.1111/j.1151-2916.1996.tb08070.x

Li, C.-J., Yang, G.-J., & Ohmori, A. (2006). Relationship between particle erosion and lamellar microstructure for plasma-sprayed alumina coatings. *Wear*, *260*(11–12), 1166–1172. doi:10.1016/j.wear.2005.07.006

Li, J. F., Liao, H., Wang, X. Y., Normand, B., Ji, V., Ding, C. X., & Coddet, C. (2004). Improvement in wear resistance of plasma sprayed yttria stabilized zirconia coating using nanostructured powder. *Tribology International*, *37*(1), 77–84. doi:10.1016/S0301-679X(03)00138-5

Malmberg, S., & Heberlein, J. (1993). Effect of plasma spray operating conditions on plasma jet characteristics and coating properties. *Journal of Thermal Spray Technology*, *2*(4), 339–344. doi:10.1007/BF02645862

Miller, R. A. (1987). Current status of thermal barrier coatings—An overview. *Surface and Coatings Technology*, *30*(1), 1–11. doi:10.1016/0257-8972(87)90003-X

Pakseresht, Rahimipour, Alizadeh, Hadavi, & Shahbazkhan. (2016). Concept of Advanced Thermal Barrier Functional Coatings in High Temperature Engineering Components. In Research Perspectives on Functional Micro- and Nanoscale Coatings (pp. 396–419). IGI Global. doi:10.4018/978-1-5225-0066-7.ch015

Pakseresht, A. H. (Ed.). (2018). *Production, Properties, and Applications of High Temperature Coatings*. IGI Global. doi:10.4018/978-1-5225-4194-3

Pakseresht, A. H., Kimiayi, A., Alizadeh, M., Nuranian, H., & Faeghinia, A. (2020). Microstructural study and hot corrosion behavior of bimodal thermal barrier coatings under laser heat treatment. *Ceramics International*, *46*(11), 19217–19227. doi:10.1016/j.ceramint.2020.04.259

Pakseresht, A. H., Mousavi, S. M., Saremi, M., Ghasali, E., & Rajaei, H. (2021). Microstructure and mechanical properties of YSZ-alumina composites designed for thermal barrier coatings. *Materials at High Temperatures*, *38*(1), 23–30. doi:10.1080/09603409.2020.1837414

Pakseresht, A. H., Rahimipour, M. R., Vaezi, M. R., & Salehi, M. (2015). Effect of splat morphology on the microstructure and dielectric properties of plasma sprayed barium titanate films. *Applied Surface Science*, *324*, 797–806. doi:10.1016/j.apsusc.2014.11.041

Pakseresht, A. H., Rahimipour, M. R., Vaezi, M. R., & Salehi, M. (2016). Thermal plasma spheroidization and spray deposition of barium titanate powder and characterization of the plasma sprayable powder. *Materials Chemistry and Physics*, *173*, 395–403. doi:10.1016/j.matchemphys.2016.02.028

Pakseresht, A. H., Saremi, M., Omidvar, H., & Alizadeh, M. (2019). Micro-structural study and wear resistance of thermal barrier coating reinforced by alumina whisker. *Surface and Coatings Technology*, *366*, 338–348. doi:10.1016/j.surfcoat.2019.03.059

Pawlowski, L. (2009). Suspension and solution thermal spray coatings. *Surface and Coatings Technology*, *203*(19), 2807–2829. doi:10.1016/j.surfcoat.2009.03.005

Rangaraj, S., & Kokini, K. (2003). Interface thermal fracture in functionally graded zirconia–mullite–bond coat alloy thermal barrier coatings. *Acta Materialia*, *51*(1), 251–267. doi:10.1016/S1359-6454(02)00396-8

Richer, P., Yandouzi, M., Beauvais, L., & Jodoin, B. (2010). Oxidation behaviour of CoNiCrAlY bond coats produced by plasma, HVOF and cold gas dynamic spraying. *Surface and Coatings Technology*, *204*(24), 3962–3974. doi:10.1016/j.surfcoat.2010.03.043

Sahith, M. S., Giridhara, G., & Kumar, R. S. (2018). Development and analysis of thermal barrier coatings on gas turbine blades – A Review. *Materials Today: Proceedings*, *5*(1), 2746–2751. doi:10.1016/j.matpr.2018.01.060

Schneider, H., Schreuer, J., & Hildmann, B. (2008). Structure and properties of mullite—A review. *Journal of the European Ceramic Society*, *28*(2), 329–344. doi:10.1016/j.jeurceramsoc.2007.03.017

Sharifahmadian, O., & Mahboubi, F. (2019). A comparative study of microstructural and tribological properties of N-DLC/DLC double layer and single layer coatings deposited by DC-pulsed PACVD process. *Ceramics International*, *45*(6), 7736–7742. doi:10.1016/j.ceramint.2019.01.076

Szymański, K., Góral, M., Kubaszek, T., & Monteiro, P. C. (2015). Microstructure of TBC Coatings Deposited by HVAF and PS-PVD Methods. *Diffusion and Defect Data, Solid State Data. Part B, Solid State Phenomena*, *227*, 373–376. doi:10.4028/www.scientific.net/SSP.227.373

Tan, H. (2011). Preparation of long alumina fibers by sol-gel method using tartaric acid. *International Journal of Minerals Metallurgy and Materials*, *18*(6), 691–694. doi:10.100712613-011-0498-6

(1994). Thermal Spray Coatings. InCotell, C. M., Sprague, J. A., & Smidt, F. A. (Eds.), *Surface Engineering* (pp. 497–509). ASM International. doi:10.31399/asm.hb.v05.a0001282

Vardelle, A., Moreau, C., Akedo, J., Ashrafizadeh, H., Berndt, C. C., Berghaus, J. O., Boulos, M., Brogan, J., Bourtsalas, A. C., Dolatabadi, A., Dorfman, M., Eden, T. J., Fauchais, P., Fisher, G., Gaertner, F., Gindrat, M., Henne, R., Hyland, M., Irissou, E., ... Vuoristo, P. (2016). The 2016 Thermal Spray Roadmap. *Journal of Thermal Spray Technology*, *25*(8), 1376–1440. doi:10.100711666-016-0473-x

Viswanathan, V., Dwivedi, G., & Sampath, S. (2014). Engineered Multilayer Thermal Barrier Coatings for Enhanced Durability and Functional Performance. *Journal of the American Ceramic Society*, *97*(9), 2770–2778. doi:10.1111/jace.13033

Wang, D. S., Tian, Z. J., Wang, S. L., & Shen, L. D. (2012). The Friction and Wear Properties of Conventional and Nanostructured ZrO_2-7wt.%Y_2O_3 Thermal Barrier Coatings Deposited on TiAl Intermetallic Alloy by Plasma Spraying. *Advanced Materials Research*, *538–541*, 336–339. doi:10.4028/www.scientific.net/AMR.538-541.336

Wellman, R. G., Deakin, M. J., & Nicholls, J. R. (2005). The effect of TBC morphology on the erosion rate of EB PVD TBCs. *Wear*, *258*(1–4), 349–356. doi:10.1016/j.wear.2004.04.011

Wellman, R. G., & Nicholls, J. R. (2007). A review of the erosion of thermal barrier coatings. *Journal of Physics. D, Applied Physics*, *40*(16), R293–R305. doi:10.1088/0022-3727/40/16/R01

Wolfe, D., & Singh, J. (1998).. . *Journal of Materials Science*, *33*(14), 3677–3692. doi:10.1023/A:1004675900887

Xiong, H.-P., Kawasaki, A., Kang, Y.-S., & Watanabe, R. (2005). Experimental study on heat insulation performance of functionally graded metal/ceramic coatings and their fracture behavior at high surface temperatures. *Surface and Coatings Technology*, *194*(2–3), 203–214. doi:10.1016/j.surfcoat.2004.07.069

Xu, H., & Guo, H. (Eds.). (2011). *Thermal barrier coatings*. Woodhead Pub Ltd. doi:10.1533/9780857090829

Xu, H., Guo, H., & Gong, S. (2008). Thermal barrier coatings. In *Developments in High Temperature Corrosion and Protection of Materials* (pp. 476–491). Elsevier. doi:10.1533/9781845694258.2.476

Yalamaç, E., Sutcu, M., & Basturk, S. B. (2017). Ceramic fibers. In *Fiber Technology for Fiber-Reinforced Composites* (pp. 187–207). Elsevier. doi:10.1016/B978-0-08-101871-2.00009-6

KEY TERMS AND DEFINITIONS

APS: Air Plasma Spraying (APS) is a thermal spraying procedure that uses a high-energy heat source to melt and accelerate tiny particles onto a prepared surface in the ambient environment.

Bond Coat: A metallic coating that is applied between the substrate and the thermal barrier coating's thermally insulating top layer to enhance bonding strength and oxidation resistance.

EB-PVD: Electron Beam-Physical Vapor Deposition (EB-PVD) is a method that involves heating feedstock material with a high-energy electron beam and depositing it in molecular form on the surface of a solid under high vacuum conditions, which is primarily used for the preparation of gas turbine engine blades.

Failure Mechanisms: The processes, such as rapid growth of TGO and residual stress creation, that cause coatings to fracture are known as failure mechanisms.

TBC: TBC is a multilayer coating structure that includes a ceramic top coat (TC), a metallic bond coat (BC), and a thin layer of thermally generated oxide (TGO) that occurs at the interface between the TC and the BC during operation. These coatings are often applied to metallic surfaces, such as gas turbine components, to raise their operating temperature and preserve them from the aggressive environment.

Wear: Wear is the unwanted removal of material from a surface as a result of mechanical action.

YSZ: Yttria-stabilized zirconia (YSZ) is one of the most commonly used and investigated materials because of its strong ionic conductivity at high temperatures, great thermal stability, and high hardness. For decades, it has been utilized as an excellent ceramic coating material.

Chapter 2
Role of Carbon Nanotube on Multi-Length Scale Tribological Properties of Al^2O^3-Based Thermal Barrier Coating

Ariharan S.
FunGlass – Centre for Functional and Surface Functionalized Glass, Alexander Dubček University of Trenčín, Slovakia

Rita Maurya
Department of Materials Science and Engineering, National Institute of Technology-Hamirpur, India

ABSTRACT

Magnificently developed Al^2O^3, 3YSZ, 8YSZ, and CNT-based thermal barrier coatings (TBCs) were subjected to study multi-length scale tribological behaviour (fretting wear and micro-scratch) of the composite coatings. Subsequently, the role of constituents of the composite on the tribological behaviour of the coatings has been recognized. Fretting wear rate and the dominative wear mechanism are identified. The fretting wear behaviour is evaluated with a distinct representation (frictional force mapping) to understand the transition of regimes. Further, micro-scratching is used to assess the composite coatings against the sharp edges. The critical load of failure and scratch characteristics (scratch hardness, toughness, and scratch resistance) are appraised to find the suitability of the composite in TBC. Notably, the role of CNT in a multi-length scale is reported quantitatively. Thus, the chapter provides a comprehensive overview of the Al2O3-based composites that deal with the understanding of the multi-length scale tribological properties at room temperature.

DOI: 10.4018/978-1-7998-9683-8.ch002

INTRODUCTION

The application of thermal barrier coatings (TBCs) on turbine engine components, such as combustors, high-temperature turbine (HPT) blades and nozzles is increasing in commercial and military jet engines. The insulating capability of TBC enables high operating temperatures, thereby improving efficiency, reducing emission and increasing the thrust/weight ratio of the turbine and combustors (Bikramjit & Kantesh, 2013; Hassanzadeh et al., 2018; Musalek et al., 2020; Amir Hossein Pakseresht et al., 2016). The structure of TBC is a complex system of functionally graded materials, that contains a metallic bond coat and ceramic coat over a metallic substrate (engine components) (Amir Hossein Pakseresht, 2018). The TBC aims to improve the structural stability of the metallic components which are exposed to a high-temperature or extreme condition. It mandates the utilization of high thermal insulation to the metallic components and the structural stability of these components are eventually conserved. So, thermally insulating TBC materials must possess sufficient thickness with durability to withstand at an exposed temperature of the load-bearing materials. Ultimately, it should possess good mechanical, thermal and tribological properties at high-temperature or working temperature. Among the aforementioned properties, thermal properties are the prerequisite for the selection of TBC materials. Low thermal conductivity and thermal expansion coefficient (CTE) mismatch with the substrate are the prime assets for TBCs. Low thermal conductivity will increase the temperature gradient across the coating and ultimately decreases the exposure temperature. Thus it improves the efficiency or working temperature of the engine by giving structural stability to the engine components. After several working hours of engine, the TBCs lose their properties and start to fail.

The failure is caused in two ways: (i) internally and (ii) externally induced failures. Among the internally induced failure, the formation of thermally grown oxide (TGO) and stress accumulation at the interface due to relatively high CTE mismatch with the underneath layer is the major cause for failure. Especially, high CTE mismatch between the ceramic top coat (ZrO_2 based ceramic) and TGO (continuous & thin layer of Al_2O_3) lead to premature failure of coating at the interface. But, the existence of TGO is unavoidable and it is intensely grown layer to avoid the oxidation of underneath bond coat and substrate. Thin and continuous layer of TGO will act as a barrier to the atmospheric oxygen. So, the main objective of the TBC system i.e. protection at high-temperature will be attained. On the other hand, there will be accumulation of stresses at TGO-ceramic top coat interface due to the CTE mismatch that leads to failure of the coating. So, the scientist have put efforts to increase the life span of the TBCs by tailoring the microstructure, chemistry and other properties of the interface. One of the ways to improve the life span of the TBCs is tailoring the chemistry of the top ceramic layer without compromising its prime characteristics, such as thermal insulation and thermal stability. The scientific community has studied several materials for the top layer of TBCs. After controlling several process parameters, several materials have been used in place of conventional TBCs, such as $BaTiO_3$, $La_2Ce_2O_7$, SiO_2, and Al_2O_3 reinforced composites (Ariharan et al., 2013; Hassanzadeh et al., 2018; Musalek et al., 2020; A H Pakseresht et al., 2015, 2016; Y. Wang et al., 2009). Based on the primary prerequisite properties for TBC, conventional Y_2O_3 stabilized ZrO_2 (YSZ) was replaced with Al_2O_3-YSZ based ceramic matrix composites (CMCs). It showed excellent stability against the impact of foreign particles and corrosive salt (hot corrosion) (A H Pakseresht et al., 2020; X. Wang et al., 2015).

Ceramic-matrix composites (CMCs) with customized properties are also can replace the ceramic materials. Al_2O_3 based composites are chosen as the top insulating layer to achieve chemical integrity over the TGO layer (A H Pakseresht et al., 2020, 2021). So, the TGO and top layer will be assumed as the functionally graded structure. It will enhance the allowable limit of stress accumulation at the interface and the life span of the TBC system will be improved eventually. An earlier report on Al_2O_3-based coating shows outstanding thermal and tribological properties (Balani et al., 2008; A H Pakseresht et al., 2019) which will be supportive of an alternative for conventional top ceramic layer. Researchers are developed Al_2O_3-based composite coatings reinforced with YSZ, carbon nanotube (CNT), graphene platelets and graphite that shows determined properties for TBCs. So, an effort is made to replace the convention YSZ with Al_2O_3-based CMCs with improved mechanical and tribological properties, which assure the structural stability of TBCs.

BACKGROUND

YSZ is the ideal structural material for TBC application because of its high thermal stability, low thermal conductivity (~2 $Wm^{-1}K^{-1}$), low CTE mismatch with bond coat and high erosion resistance at elevated temperatures (Balani et al., 2007; Mahato et al., 2013). In ZrO_2, tetragonal (t-ZrO_2) to monoclinic (m-ZrO_2) phase transformation is martensitic in nature. The associated volume expansion (~4.9%) due to the phase transformation can be used to improve fracture toughness and fracture strength of the coatings, moderately (Narayan & Colombo, 2009). 3 mol.% and 8 mol.% of Y_2O_3 doping in ZrO_2, (referred to as 3YSZ and 8YSZ, respectively), reduces the grain growth and stabilizes the t-ZrO_2 and the failure induced by the phase transformation can be avoided (Leib et al., 2015). Moreover, both 3YSZ and 8YSZ possess low thermal conductivity (< 2 $Wm^{-1}K^{-1}$) and may serve as an ideal material due to their high durability in turbines at elevated temperatures (~1400 °C). But, the mechanical (hardness and elastic modulus) and tribological properties of YSZ is poor. So the externally induced failure will be dominative. Thus, Al_2O_3 based composite coating is chosen as the choice for the top layer of the TBCs.

Novel ceramic-based composite of Al_2O_3 fibres, glass beads and hollow silica powders with good thermal insulation with the thermal conductivity of 0.12 $Wm^{-1}K^{-1}$ has been fabricated (A H Pakseresht et al., 2019; Xu et al., 2017a). But their usage at ~1400 °C may be constrained due to thermal stability (Xu et al., 2017b). Al_2O_3 is known for its high hardness and elastic modulus that will shield the TBCs from the externally induced failure, such as foreign body impact, erosion etc. So, hard Al_2O_3 is used as a matrix material; YSZ is reinforced into the Al_2O_3 matrix for thermal insulation. However, the poor fracture toughness of Al_2O_3 (~3.2 $MPa.m^{0.5}$) and YSZ (K_{IC} ~5-7 $MPa.m^{0.5}$) limits its usage in the structural application (Chun-Hong & Hideo, 2007). So, efforts have been carried out to improve the toughness of Al2O3-YSZ composites for their continuous and prolonged use in structural applications (Ariharan et al., 2013; Bakshi et al., 2008a; Balani et al., 2007; Y. Chen et al., 2007). For instance, carbon nanotubes (CNTs) exhibit an extraordinary performance among other well-known materials (graphite, Kevlar fibers and ductile materials) in enhancing the toughness of the composites (Carvalho et al., 2016; F. Yang et al., 2010). Therefore, an optimum amount of CNT is reinforced in the Al_2O_3-YSZ based composites to improve the fracture toughness. Further, researchers are (Y. Chen et al., 2007) reported that the addition of 5-8 wt.% CNT enhances (~42%) the fracture toughness of Al_2O_3 composites via energy dissipation mechanisms (CNT pull-out, branching, and bridging) with a marginal increase in the thermal conductivity (Bakshi et al., 2008b; Y. Chen et al., 2009). Also, the CNT was stable in a non-oxidizing environment at 1000 °C (L. W. Yang et al., 2017). Besides, composite of carbon fibre reinforced SiC joints are used for thermal protection with Al_2O_3-graphite-ZrO_2-$ZrSiO_4$ based commercial adhesive was able to resist the thermal shock at 1100 °C (Triantou et al., 2017). Afterwards, CMCs reinforced with CNT have become an intense research area to enhance the mechanical and thermal properties (Carvalho et al., 2016; Triantou et al., 2017).

The CNTs for thermal barrier coatings is not been a logical choice due to their high thermal conductivity (~3000 $Wm^{-1}K^{-1}$ along the CNT axis). Only a few researchers have testified the thermal conductivity of CNT reinforced composites (T Tojo et al., 1999; Triantou et al., 2017). Multi-walled CNT (MWNT) reinforced SiO_2-10 vol.% MWNT composite showed an increase (~ 65%) in thermal conductivity compared to that of pure SiO_2 (2.47 $Wm^{-1}K^{-1}$) (Takeo Tojo et al., 1999). It is attributed to the cleaning effect that results in the removal of impurities and amorphous phases present at the SiO_2 and CNT interface. It leads to a high thermal transport at the interface with insufficient phonon scattering to conduct the thermal energy. On the other hand, a significant decrease (up to 73%) in thermal conductivity of Al_2O_3-10vol.% CNT reinforced dense (<87.0%) nanocomposite (6.0-11.4 $Wm^{-1}K^{-1}$) due to strong coupling, high crystallite boundary, inter-splat thermal resistance, bending and twisting of CNT (Balani et al., 2007; Leclercq et al., 2003) are the leading obstacles to the thermal transport with high phonon scattering. Moreover, the overestimated (3-4 times) thermal conductivity of Al_2O_3-CNT nanocomposites is due to the high thermal resistance of the Al_2O_3 and CNT interface (Balani et al., 2007).

The main reason for the decrease in thermal conductivity of CNT reinforced CMCs is not well understood. But, the decrease in the thermal conductivity of highly flow-able polymer-CNT composites is due to the anisotropic thermal conductivity of aligned CNT during processing (Bastwros et al., 2014; Fauchais, 2004). Electric, magnetic fields (C-W Nan et al., 2003; Smith et al., 2003), shear flow, mechanical ordering (Hepplestone et al., 2006) and aligned growth (Ce-Wen Nan et al., 2004; H.-S. Yang et al., 2002) are the several processing conditions that lead to align of CNT is a preferred direction in the polymer-based matrix. Similarly, researchers have been developed process-induced alignment of CNT in Al_2O_3-YSZ based CMC coatings (Shenogin et al., 2004). It has been shown the thermal conductivity with the CNT orientation factor is estimated as 1.3-6.6 $Wm^{-1}K^{-1}$, which is similar to that of the experimental value (< 3.5 $Wm^{-1}K^{-1}$) (Shenogin et al., 2004). So, the tailored materials of Al_2O_3-YSZ based composite with aligned CNT is qualified for TBCs by satisfying the prime objective, high thermal insulation. Although the primary objective is achieved, the tribological properties in a multi-length scale are also an equivalent important objective for the utilization of the Al_2O_3-YSZ-CNT based composite as the top layer in the TBC. Multi-length scale tribological properties, such as fretting wear and micro-scratch are important tests to conduct on the Al_2O_3-YSZ-CNT based composite and studies its response for the successful alternative of conventional TBC materials.

FRETTING WEAR

Fretting Wear Rate of Al_2O_3-YSZ Based Composite Coatings

The top ceramic coatings of TBCs encountered with other structural parts of gas-turbine engine components in a reciprocating motion and the contact failure mechanism that is associated with such kind of oscillating motion is called fretting wear. A constant normal of 9.81 N is applied through the counter body on the surface of the coatings. The frictional force during reciprocating motion of the counter body (WC ball of 6 mm diameter, frequency: 2.5 Hz, and amplitude: 400 μm) on the polished surface of the atmospheric plasma sprayed Al_2O_3-based composite coatings is monitored. The frictional force during fretting wear of Al_2O_3-based composite coatings is continually recorded and the coefficient of friction

(CoF) is presented for 15700 cycles with a total distance of 12.56 m (**Figure 1**). It is noted that there are two distinct regimes, such as running-in and steady-state regimes noticed from the CoF vs cycle/distance plot. At the beginning of the fretting wear test, the CoF is rapidly changing to a constant value (CoF, **Table 1**).

The rapid change of CoF is due to initial hindrance by the surface roughness (R_a, ~3.0 µm) of the coating. The initial roughness has been changed due to the continuous reciprocating movement of the WC ball on the surface of the coating. Soon after the reduction of roughness to a threshold value, the resistance exerted by the coating against the counter body will be reduced and the frictional force reaches a constant maximum value of the running-in regime. Later, the instantaneous properties of the surface which is in contact with the WC ball are responsible for the wear resistance. So, the CoF after the running-in period shows a constant value, irrespective of the initial hindrance by the sample surface and it reaches 0.24-0.54 for Al_2O_3-based composite coatings, called steady-state regime. The saturated maximum CoF has been reached within 1000 cycles (~0.8 m of sliding distance) for composites without CNT and more than 1000 cycles for the composites with CNT reinforcement.

Figure 1. Coefficient of friction during fretting wear of Al_2O_3-YSZ-CNT composites against WC ball (Ariharan & Balani, 2021)

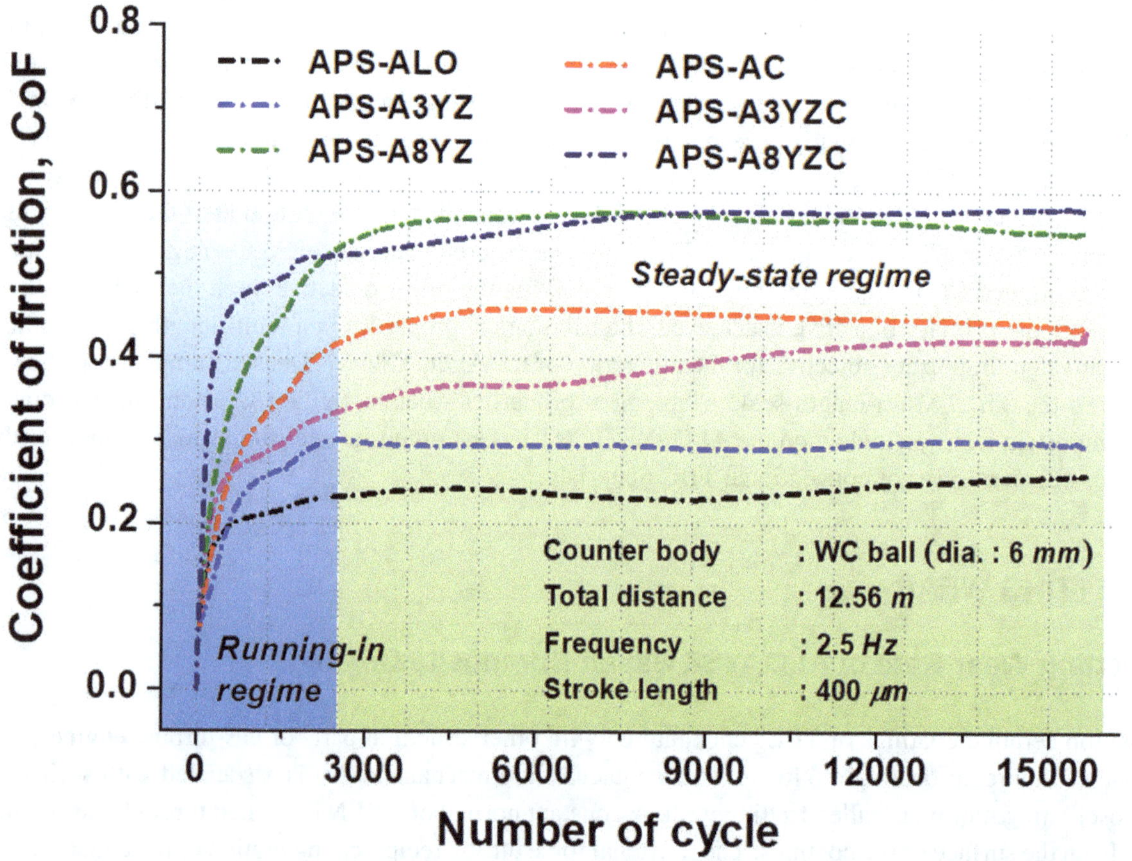

The APS-ALO coating (monolithic Al_2O_3) gives less resistance to reciprocating motion of WC ball that is signified with a minimum CoF (0.24) among the composite coatings due to an average mechanical property (hardness: 12.97 GPa and elastic modulus: 136.9 GPa). The reinforcement of 20 wt.% 3YSZ and 8YSZ increases the steady-state CoF to the maximum, up to 0.50, which is attributed to resistance by the coatings with enhanced mechanical properties (hardness: 13.66-14.25 GPa and elastic modulus: 162.7-171.5 GPa). Further, CNT addition (4 wt.%) along with 20 wt.% of 3YSZ and 8YSZ showed an increase in the average CoF value from 0.24 to 0.39 due to improved mechanical properties of the composite coatings.

The resistance exerted by the composite coatings against the movement of the WC ball affects the wear volume created due to fretting wear. The wear volume loss, V_W during fretting wear against WC ball for the total travelling distance of 12.56 m with the constant normal load of 1000 g (9.81 N) is measured using an optical profiler. The wear volume loss for Al_2O_3-YSZ based composite coatings is observed as $1.2\text{-}2.3\times10^{-6}$ mm^3 and the corresponding wear rate ($W_{rate\ (exp.)}$, **equation (1)**) is calculated as $0.97\text{-}1.87\times10^{-6}$ $mm^3N^{-1}m^{-1}$ (**Table 1**).

$$W_{rate(exp.)} = \frac{V_W}{F.L} \tag{1}$$

where, V_W is the wear volume loss (in mm^3) due to fretting wear against the WC ball with the normal force of F (in N) for the total length of L (in m). Among the composite coatings, APS-AC shows high $W_{rate\ (exp.)}$ (**Table 1**) and thus low wear resistance. It is attributed to the reciprocating motion of the WC ball over APS-AC debilitate the bonding between the Al_2O_3 matrix and CNT by CNT pull-out due to repeated oscillation of the WC ball on the interface between Al_2O_3 and CNT (Ahmad et al., 2009; Zheng et al., 1999). Thus, the fretting wear rate of APS-AC is ~15.4% higher than APS-ALO.

Among the composite coatings, APS-A3YZ shows improved wear resistance (low wear rate) due to improved mechanical properties with 20 wt.% of 3YSZ (**Table 1**) by improving the relatively strong interface between the reinforcement and the matrix that resists the movement of the WC ball. Further, the reinforcement of 4 wt.% CNT degraded the wear properties of APS-A3YZC with a 34.0% high wear rate than of its counterpart (APS-A3YZ). The detriment of relatively poor wear resistance with CNT reinforcement in the Al_2O_3 based composite system is endorsed by the nature of the complaint and the bonding between the CNT with matrix (Al_2O_3) or other reinforcement (3YSZ/8YSZ). The covalent bonding between the carbon atoms in CNT makes it more saturated and less attentive in chemical bonding with highly stable ceramics, such as Al_2O_3, 3YSZ and 8YSZ (Ahmad et al., 2009; Lee et al., 2011; Zheng et al., 1999).

Consequently, the bonding of CNT with other constituents of Al_2O_3 based composite coatings is more of mechanical bonding, particularly in the non-equilibrium processing technique, atmospheric plasma spraying. So, the interface can be easily altered due to mechanical disturbance. Thus, the CNT reinforced composite coating shows a higher wear rate than that of its counterpart without CNT. But, the experimentally calculated wear rate of APS-A8YZC shows a lower value than that of APS-A8YZ. It is endorsed to the high hardness of the sample APS-A8YZC (13.79 GPa) was controlling the CNT's nature on bonding with other constituents. But, the estimated fretting wear rate, $W_{rate(th.)}$ (**Equation (2)**) were higher (410-660%) than that of the experimentally calculated wear rate. It indicates that the surface of the wear scar dynamically changes over the fretting wear test and it is prone to wear resistance.

$$W_{rate(th.)} = CoF\frac{\frac{E}{H_V}F^{1/8}}{K_{IC}^{1/2}H_V^{5/8}} \quad (2)$$

where, H_V: Bulk hardness, K_{IC}: Fracture toughness, CoF: Coefficient of friction, F: Applied normal load.

Table 1. Al_2O_3-YSZ based coatings during fretting wear against WC ball

Sample	Composition, wt.%		Porosity (%)	H (GPa) (Ariharan et al., 2017)	H^3/E^2 (MPa)	CoF	V_W ($\times10^{-6}$ mm^3)	$W_{rate (exp.)}$ ($\times10^{-6}$ mm^{3}N^{-1}m^{-1})
	3YSZ/8YSZ	CNT						
APS-ALO	--	--	8.1 ± 0.3	12.97 ± 0.06	116.4	0.24 ± 0.07	2.0 ± 0.1	1.62 ± 0.07
APS-A3YZ	20	--	5.5 ± 0.8	14.25 ± 0.23	109.3	0.28 ± 0.05	1.2 ± 0.2	0.97 ± 0.14
APS-A8YZ	20	--	7.4 ± 0.6	13.66 ± 0.16	86.7	0.50 ± 0.04	1.5 ± 0.4	1.22 ± 0.28
APS-AC	--	4	4.0 ± 1.1	12.39 ± 0.23	62.0	0.39 ± 0.05	2.3 ± 0.4	1.87 ± 0.28
APS-A3YZC	20	4	6.3 ± 0.8	13.65 ± 0.25	75.9	0.36 ± 0.04	1.6 ± 0.3	1.30 ± 0.21
APS-A8YZC	20	4	5.3 ± 1.4	13.79 ± 0.29	70.6	0.54 ± 0.02	1.3 ± 0.3	1.06 ± 0.21

H: Hardness, H^3/E^2: Plasticity index, CoF: Coefficient of friction at steady-state regime, V_W: Wear volume loss, $W_{rate (exp.)}$: Experimental wear rate.

The enhancement in the actual or experimental wear resistance is due to two major reasons, (a) altered surface of the fretting wear scar with high hardness, and (b) resistance offered by the wear debris formed during the fretting wear process (Pandey & Biswas, 2016). The reciprocating motion of the counter body keeps the wear debris stay back at the wear scar region as a third body.

Fretting Wear Mechanism in Al_2O_3-YSZ Based Composite Coatings

The microstructure of the fretting wear scar is shown in **Figure 2** to understand the active wear mechanism during the reciprocating motion of the Al_2O_3-YSZ based composite coatings. Micro-cracks are present at the wear scar of YSZ and CNT reinforced composite coatings. The cracks on the wear scar are called fatigue cracks due to their formation during the reciprocating motion of the hard WC ball on the surface of the composite coatings.

Figure 2. Scanning electron micrograph (back-scattered electron mode image) of wear scar during fretting wear of Al_2O_3-YSZ-CNT composites against WC ball

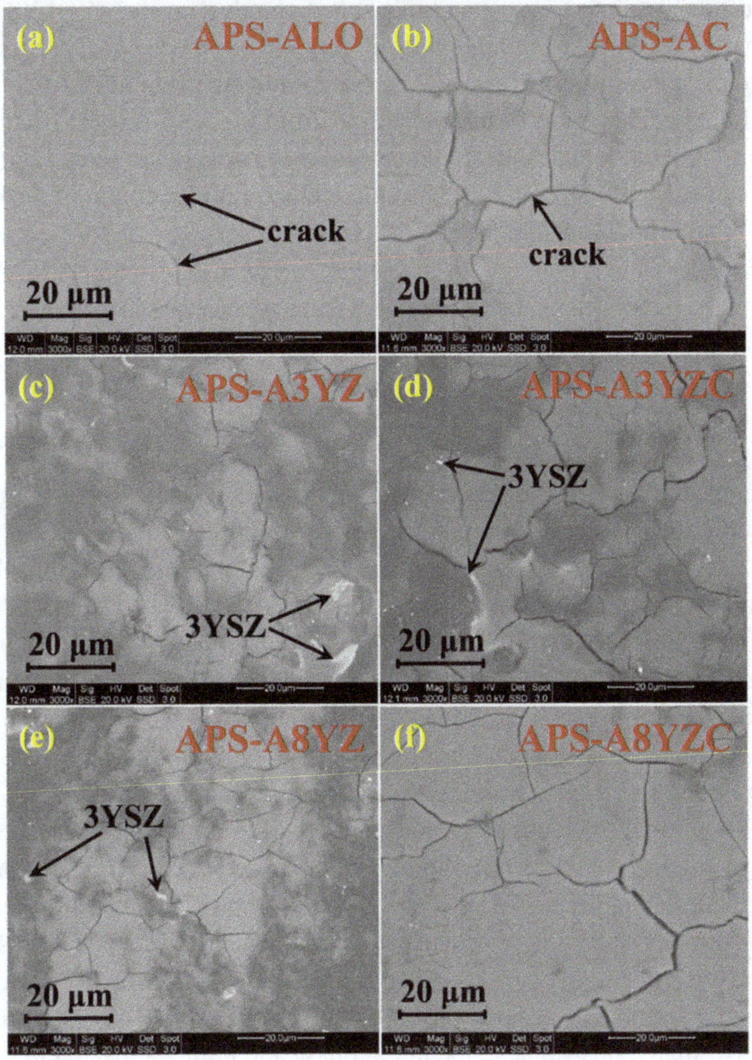

Unlike other composite coatings, the APS-ALO sample has no cracks on the wear scar and it is attributed to the high (116.4 MPa) plasticity index, H^3/E^2. Other composites with 3YSZ/8YSZ and CNT reinforcement are showing highly altered surface (**Figure 2**) during the wear test. Further, these altered wear scar surface acts as a protective layer against the fretting wear. The altered surface with fatigue cracks is called protective tribo-layer [38] and it significantly reduces the wear rate of the composite coatings from further wear.

The improvement in wear resistance is revealed from the overestimation of wear rate, $W_{rate(th.)}$ up to 660% compared to that of experimental wear rate. The low plasticity index value of the composite coatings tends to alter the surface more easily. Also, it initiates the cracks in the recovered strain region. The accumulated stresses start the micro-cracks to propagate along with the weak interface of the composite coatings. As a result of the accumulated tensile stresses due to repetitive motion of the counter-body, the micro-cracks appear on the surface. The relatively high strained surface will be detached from the worn surface after an extended number of cycle.

The initial contact diameter and corresponding contact pressure for the Al_2O_3-YSZ based composite coatings are marginally varied due to the negligible difference in the mechanical properties. Nevertheless, the stress distribution is almost uniform at the wear surface during the contact of the WC ball. So there is no substantial failure or removal of coating materials from the wear scar and eventually, there is no spallation of coatings. Although optical profile images show that there are some regions with spallation of coatings (of tens of μm) in 3YSZ and 8YSZ reinforced composite coatings. The spallation can be also categorized as a third body formation. The third body or wear debris that are present in the wear scar will experience excess stress than that of the parent surface. So, the impact of the WC ball on the composite coatings will be reduced by the formation of these wear debris. Thus, there will be a considerable improvement in the wear resistance by the formation of wear debris and its associated mechanism

Frictional Force Mapping During Fretting Wear

The variation in the different regimes and wear mechanisms can be addressed using the frictional force mapping during the fretting wear test. **Figure 3** shows the frictional force mapping for the amplitude (stroke length: 400 μm) that includes both directions. The frictional force mapping shows a lower with less fluctuation in the frictional force for APS-ALO and APS-A3YZ (~13.5 and ~13.7 N, respectively) than other composites (**Figure 3**).

The starting of the stroke length is noted as '0 μm' and the end is noted as '400 μm'. It is noted that the frictional force of CNT reinforced composites (APS-AC and APS-A3YZC) shows marginally higher (**Figure 3**) than the composite coatings with no CNT, such as APS-ALO and APS-A3YZ. The high resistance exerted through the frictional force leads to intense damage on APS-AC and APS-A3YZC coatings with a high wear rate. It is also in support of the earlier argument of composite coatings with moderate mechanical properties and correspondingly poor wear resistance (high wear rate) (X. Chen et al., 2004; Zhu & Miller, 2000). Similarly, APS-A8YZ shows a non-fluctuation in frictional force with a maximum of ~14.3 N to that of coatings, APS-ALO and APS-A3YZ. But, CNT reinforced composites showed an evidence of fluctuations with a gradual increase in the frictional force during the fretting wear. Further, the frictional force map shows the high fluctuations at the edges (starting and end) of each fretting wear cycle. Normally, it is vague in the initial phase and highly appeared in the advanced stage. It is endorsed to the accumulation of third body (wear debris) at the end of each fretting wear cycle due to continuous motion of WC ball on coatings surface.

Figure 3. Frictional force mapping (colour code) during fretting wear of Al_2O_3-YSZ-CNT composites against WC ball in both (forward and reverse) direction
(Ariharan & Balani, 2021)

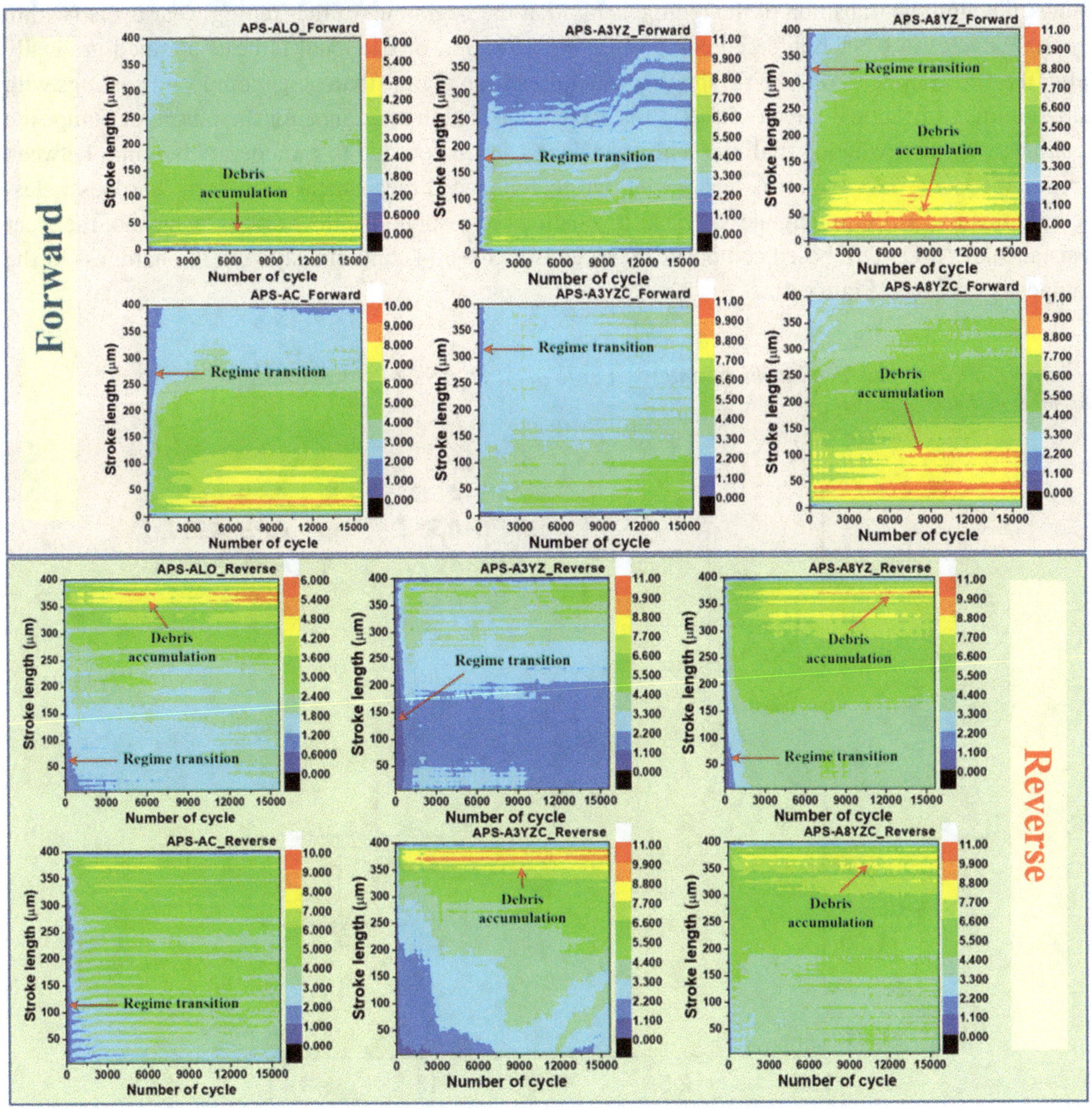

The continuous reciprocating motion of the counter body collects the debris in the vicinity of the wear scar. The fluctuations in the frictional force are separated and represented in **Figure 3** for a better understanding. It is noticed that frictional force reached saturation (~5.0 N) rapidly in the case of monolithic Al_2O_3 coating, APS-ALO. The transition of the regime from running-in to steady-state has occurred around the 100th cycle. The APS-A3YZ and APS-A8YZ shows a maximum frictional force of ~5.5 N and ~10.7 N, respectively (**Figure 3**). Also, the APS-A3YZ and APS-A8YZ shows the transition of regimes after the 100th cycle and attained a constant frictional force. The rapid change and saturation of frictional force in the coatings signify the poor resistance employed against the WC ball. So, the wear rate is high in the case of APS-ALO, APS-A3YZ and APS- A8YZ (**Table 1**). On the other hand, CNT reinforcement in the Al_2O_3-YSZ (**Figure 3**) shows a continuous and slow change in the frictional force values. Though the transition of the regime is rapid at the beginning of the fretting wear cycle (within the 100th cycle) in CNT reinforced composite, the saturation of frictional force is attained gradually (after the 1000th cycle). APS-A3YZ shows a wider transition regime than other composite coatings with a low saturated frictional force (~5.5 N). It is ascribed as a high resistance for the failure of composite coatings, such as spallation, pull-out and wear debris formation due to a moderate bonding between CNT and Al_2O_3/YSZ. Also, it is recognized that the enhanced mechanical properties (hardness, elastic modulus and fracture toughness) resist the coatings to respond against fretting wear. So, the wear resistance of Al_2O_3-YSZ based composite coatings is increasing continuously with the hardness of the composite coatings (**Figure 4**).

Figure 4. Relation between hardness and wear rate of Al_2O_3-YSZ-CNT composites
(Ariharan & Balani, 2021)

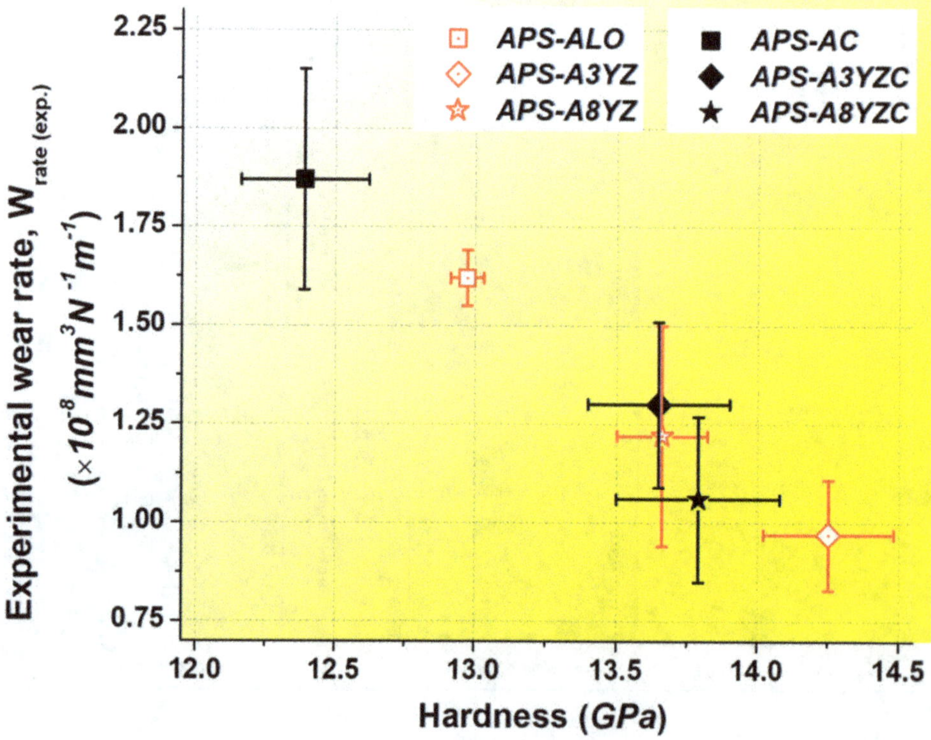

Based on the fretting wear behaviour (wear resistance, the transition of regime, saturation and mapping of frictional force), Al_2O_3 with 3YSZ (20 wt.%) and CNT (4 wt.%) is the optimal composition that resists against WC ball (diameter: 6 mm radius) during the fretting wear.

MICRO-SCRATCHING

Frictional Force During Progressive Load Scratching on the Coatings

The Al_2O_3-YSZ based composites are assessed against the sharp edge (Rockwell indenter tip radius: 100 µm) with the normal progressive load of 0.03-30 N (loading rate: ~9.9 N/min). The frictional force during progressive load scratching on the top polished surface has been continuously monitored and the scratch coefficient is measured (**Figure 5**). Above the normal load of 7.5 N, it reveals that there are sharp points in the frictional force plot. In all frictional force plots of Al_2O_3-YSZ based composite, there is a minimal variation in the frictional force up to the normal load of 7.5 N. It is attributed to the minimal damage or wear of the composite coating against the scratch tip at starting of the progressive load scratching. No quick variation in the frictional force confirms the gradual wear or failure of coatings.

Figure 5. Scratch coefficient of Al_2O_3-YSZ-CNT composites during progressive load scratching with Rockwell tip

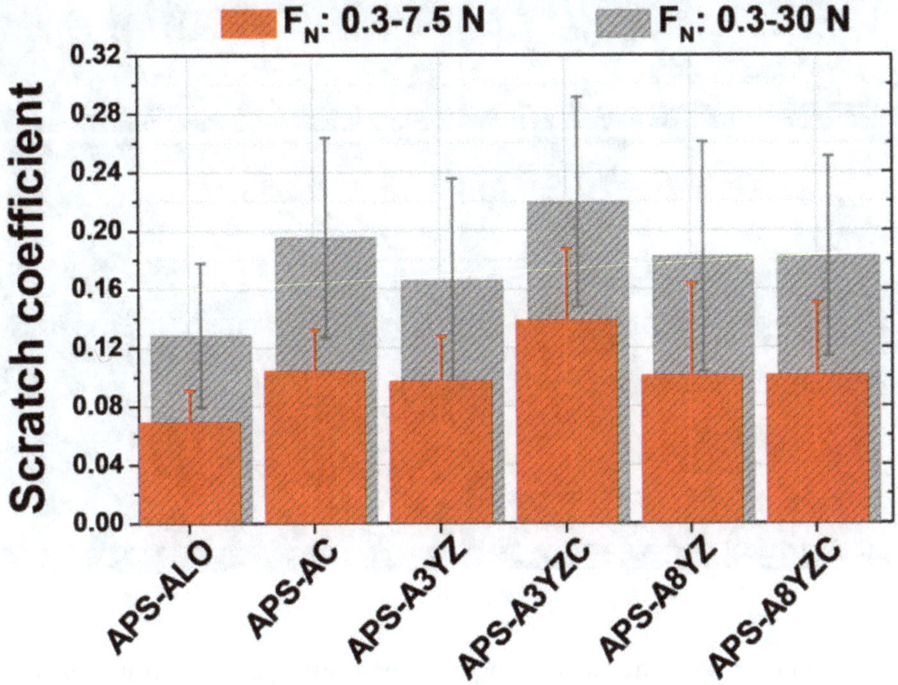

The continuous and slow loss in coating removal and accommodation of detached materials (debris) at the pores (3.4-8.1%) due to compaction during scratching is attributed to the initial stage of progressive load scratching. Later stage (normal load: >7.5 N), it is noticed in all composite coatings at a high normal load that enormous fluctuation (120-240%) in the frictional force that of its average value. The considerable fluctuations in the frictional force at the later stage are due to the various failures of composite coatings due to the high normal load. The scanning electron micrograph of scratch groves has depicted the various failure mechanisms that are activated at different stages of the progressive load scratching. The damage that occurred at normal load <7.5 N is comparatively more secure than that of high normal load (> 7.5 N). The abrupt drop in frictional force observed at the normal loads >7.5 N, indicates the failures which are identified as ploughing, flaking and chipping (**Figure 6**).

Figure 6. Scanning electron micrograph of scratch groves on Al_2O_3-YSZ-CNT composites due to progressive load scratching
(Ariharan & Maurya, 2021)

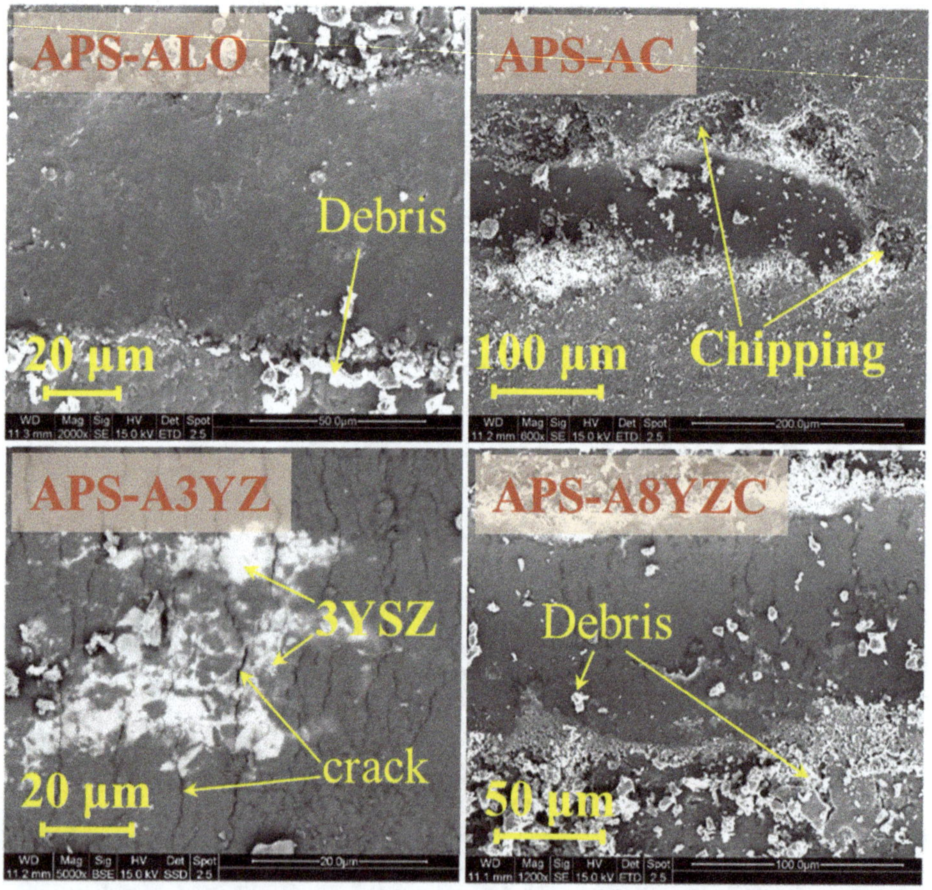

The reinforcement of YSZ in the Al_2O_3 reduces the porosity in the composite coatings due to the high sinterability of the YSZ during processing using atmospheric plasma spraying. So, it leads to a high conflict against scratching. The improved mechanical properties (hardness, elastic modulus and fracture toughness) of 3YSZ/8YSZ and CNT reinforced Al_2O_3 are known for the improvement in the bonding between the constituents in the composite coating. The collective properties of the composite coatings will be in support of scratching behaviour of the Al_2O_3-YSZ based composites.

Scratch Resistance of Al_2O_3-YSZ Based Composite Coatings

The resistance against progressive load scratching, scratch resistance is evaluated by measuring the scratch grove volume due to the scratching and normalized with scratching length and load. Generally, the scratch grove is directly related to the penetration depth of the scratch tip that changes accordingly to the hardness of the composite. The maximum penetration of the scratch tip on the Al_2O_3-YSZ based coatings at a normal load of 30 N is observed as 43.8 μm (**Table 2**). Though the composite coatings possess high hardness (13.7-14.3 GPa, they showed a high penetration depth (~44.0 μm) compared to APS-ALO (~23.0 μm, **Table 2**). It is attributed to the low plasticity index, H^3/E^2 (62.0-109.3 MPa) of the composites compared to that of APS-ALO, which leads to accommodating the deformed composite material during the scratching. Correspondingly, the grooves width (94.9 μm) and the average frictional force ($F_{T(max)}$, 7.90 N) are low for APS-ALO compared to that of composites (>112.3 μm and >8.36 N, respectively). Accordingly, the groves volume (W_{vol}) due to scratching changes linearly with penetration of scratch tip and it is measured using an optical profiler.

The average groves volume for the composite coating is in the range of 1.04-2.07×10^{-2} mm^3 and the corresponding scratch rate ($W_{S(rate)}$) of 0.35-0.69 mm^{3}N^{-1}m^{-1}, which is lower than that of monolithic Al_2O_3 coatings. The reinforcement of YSZ in Al_2O_3 increases the groves volume up to1.89×10^{-2} mm^3. It corresponds to a maximum scratch rate of 0.63 mm^{3}N^{-1}m^{-1}. It is attributed to the following of materials due to scratching on the materials with relatively low plasticity index value (83.8-109.3 MPa). The low plasticity index has its tendency to distort the material easily to its ductile nature compared to that of high plasticity indexed APS-ALO coating. Though the hardness of composite coatings possesses improved hardness, it demonstrates a poor resistance against progressive scratching with a high scratch rate (0.56-0.69 mm^{3}N^{-1}m^{-1}). So, the decrease in the scratch resistance (60-97%) in 3YSZ/8YSZ and CNT reinforced composite is attributed to the combined effect of a low plasticity index with an optimum hardness that accommodates porosity.

Scanning electron micrograph of groves reveals that there is the debris of size, roughly <10 μm were collected on either side of the scratch groves without any pits on the grove's surface (**Figure 6**). But, the surface before scratching comprises pits and porosity. These pores are occupied with the distorted or displaced volume of materials. Further, the CNT reinforced composites behaved uniquely during the scratching. It bears up the progressive load for a specific load against failure due to scratching up to a threshold load, referred to as the critical load of failure (L_C). The critical load of failure depends on the friction between the phases, such as CNT, 3YSZ/8YSZ and Al_2O_3 matrix. After CNT experiences the critical load of failure, the debonding of CNT from other constituents of the composites (Al_2O_3, 3YSZ and 8YSZ) will occur in the Al_2O_3-YSZ based composites. The debonding of CNT leads to failures, such as flaking and/or chipping of the lamella in CNT reinforced composites (**Figure 6**). The failure is attributed to spallation due to the formation of cracks and the excess stress exerted by the scratch tip on the scratch groves.

The occurrence of flaking and the development of cracks that are in the direction normal to the scratching are due to the dissipation of accumulated deformation energy due to the progressive movement of the scratch tip (**Figures 6**). These cracks are named tensile cracks that are formed at the rear of the moving indenter during progressive load scratching (**Figure 6**) (Burnett & Rickerby, 1987). The cracks in the grove increase significantly with the progressive normal load. Also, there are almost no cracks present in the groves at the lower normal load. It started nucleate at high loads due to the accumulation of stresses and increase of shear tractions at the rear side of the moving tip in the groves (Burnett & Rickerby, 1987).

Scratch Hardness and Toughness of Al_2O_3-YSZ Based Composite Coatings

The scratch hardness (**equation (3)**) and toughness (**equation (4)**) of the Al_2O_3-YSZ based composites is measured using the projected area (A_p) of the scratch groves.

$$H_S = \frac{8F_n}{\pi w^2} \quad (3)$$

$$K_{C-S} = \frac{F_T}{\left(2pA_P\right)^{1/2}} \quad (4)$$

Where, Projected area,

$$A_P = R^2 \cos^{-1}\left\{1 - \frac{P_D}{R}\right\} - \left(R - P_D\right)\sqrt{2RP_D - P_D^2},$$

where, R: Radius of the Rockwell indenter (100 μm), P_D: Penetration depth, w: Maximum scratch width, F_N: Normal load, F_T: Frictional force and p: Perimeter of the indenter. The scratch hardness (H_S) of the composites was calculated as 4.29-8.55 GPa with a maximum value for APS-ALO (8.55 GPa) (**Table 2**). The high scratch hardness is ascribed to the strong chemical bonding and mechanical interlocking between the lamella. Further, CNT and 3YSZ/8YSZ reinforcement individually in the Al_2O_3 matrix declines (29-45%) the scratch hardness.

The collective reinforcement of CNT with 3YSZ/8YSZ, synergistically decreases the scratch hardness up to 50%. But, the influence of CNT alone improved (up to 24-28%) scratch hardness of APS-A3YZC and APS-A8YZC compared to that of APS-A3YZ and APS-A8YZ, respectively. The improvement with CNT is attributed to the strong interface between CNT with 3YSZ/8YSZ. It is due to the comparatively elastic nature of YSZ that accommodate the CNTs well during the processing using atmospheric plasma spraying. Also, the accumulation of stress has been reduced in the CNT and 3YSZ/8YSZ reinforced composites due to transformation toughening (Ariharan et al., 2017). So, the contribution of residual stress to improve scratch hardness is reduced in the CNT and 3YSZ/8YSZ reinforced composites. Further, 3YSZ, 8YSZ and CNT phases enhance the ductility (low plasticity index) of the composite. It leads to high penetration of scratch tip during progressive load scratching that decreases the scratch hardness of the composite.

The scratch toughness, K_{C-S} (**equation (4)**) of the composite coatings is calculated using the projected area of the cross-section of scratch groves and it is measured as 1.05-3.97 $MPam^{0.5}$. Among all the coatings, the monolithic Al_2O_3 sample showed an average scratch toughness, ~2.4 $MPam^{0.5}$. Scratch toughness of 3YSZ and 8YSZ reinforced composite shows 68% and 11% increase compared to that of APS-ALO. But, the addition of CNT declines the scratch toughness of the composite by 44-51%. It validates the contribution of CNT is almost nil in scratch toughness, But, the CNT addition improves the bulk fracture toughness (K_{IC}) of the composite coatings up to 54%. Since the scratching is done in the lateral direction to the lamella in the coatings (Ariharan et al., 2019), the interaction of aligned CNT is influenced less to resist the movement of the scratch tip. Furthermore, the scratch resistance is directly related to the critical load of failure that corresponds to the scratch hardness and toughness of the composite coatings. The critical load of failure during scratching is calculated to assess the capabilities and adhesion between the constituents of the composites for structural application.

Adhesion Between the Lamella in the Al_2O_3-YSZ Based Composite Coatings

The adhesion of thermal insulate ceramic coatings in the TBC system and thus with the structural part is the most characteristic to decides the service life or life span of TBCs. Similarly to the adhesion of TBCs with the substrate, adhesions between the splats are the primary concern. So, the strong integration of splats or lamella is more noteworthy to decide the integrity of the desirable composite for TBCs. The strength of adhesion can be evaluated using various methods, such as pull-off, indentation, scratch and bending tests. These methods have their own merits and demerits. Among these, the scratch test is one of the methods to provide a simple, dynamic and quick way to understand the adhesion of coatings and associated mechanisms.

Table 2. Al_2O_3-YSZ based coatings during progressive load micro-scratching

Sample	P_D (µm)	$F_{T(max)}$ (N)	W_R ($mm^3N^{-1}m^{-1}$)	L_C (N)	H_S (GPa)	K_{C-S} ($MPa.m^{0.5}$)	C_A ($\times10^3$ J)
APS-ALO	17.9 ± 1.7	5.3 ± 0.5	0.35 ± 0.03	16.1 ± 1.2	8.65 ± 1.30	2.37 ± 0.07	2.91 ± 0.73
APS-A3YZ	31.4 ± 3.9	6.6 ± 0.7	0.49 ± 0.04	16.1 ± 1.0	6.11 ± 0.90	3.97 ± 0.09	2.75 ± 0.72
APS-A8YZ	43.0 ± 2.1	7.0 ± 0.8	0.63 ± 0.05	10.9 ± 0.8	4.69 ± 0.60	2.62 ± 0.05	2.31 ± 0.56
APS-AC	34.2 ± 2.3	7.5 ± 0.7	0.52 ± 0.05	11.8 ± 1.3	5.73 ± 0.99	1.05 ± 0.07	1.16 ± 0.25
APS-A3YZC	39.7 ± 3.7	8.1 ± 0.7	0.56 ± 0.02	9.4 ± 0.5	5.28 ± 0.36	2.04 ± 0.06	2.00 ± 0.33
APS-A8YZC	45.7 ± 3.2	7.3 ± 0.7	0.69 ± 0.06	10.9 ± 0.8	4.29 ± 0.63	1.14 ± 0.03	1.73 ± 0.35

P_D: Penetration depth, $F_{T(max)}$: Maximum frictional force, W_R: Scratch rate, L_C: Critical load of failure, H_S: Scratch hardness, K_{C-S}: Scratch toughness, C_A: Work of adhesion of lamella

The critical load of failure can be measured only from the progressive load scratching. The progressive load scratching comprises multiple scratching with constant load. So, the transition (safe to failure regime) can be easily identified in a single progressive load scratching with a wide range of normal loads. Thus, the critical load of failure can be directly related to the adhesive strength between the splats or lamella of the composite coatings. The critical load is recognized from the frictional force and acoustic emission plot (**Figure 7**). The detachment of splat or lamella in the coating scratch leads to a sharp change in the acoustic signal that appeared as a sharp peak in **Figure 7**.

Figure 7. Assessment of Al_2O_3-YSZ-CNT composites using acoustic signal collected during scratching and corresponding penetration depth

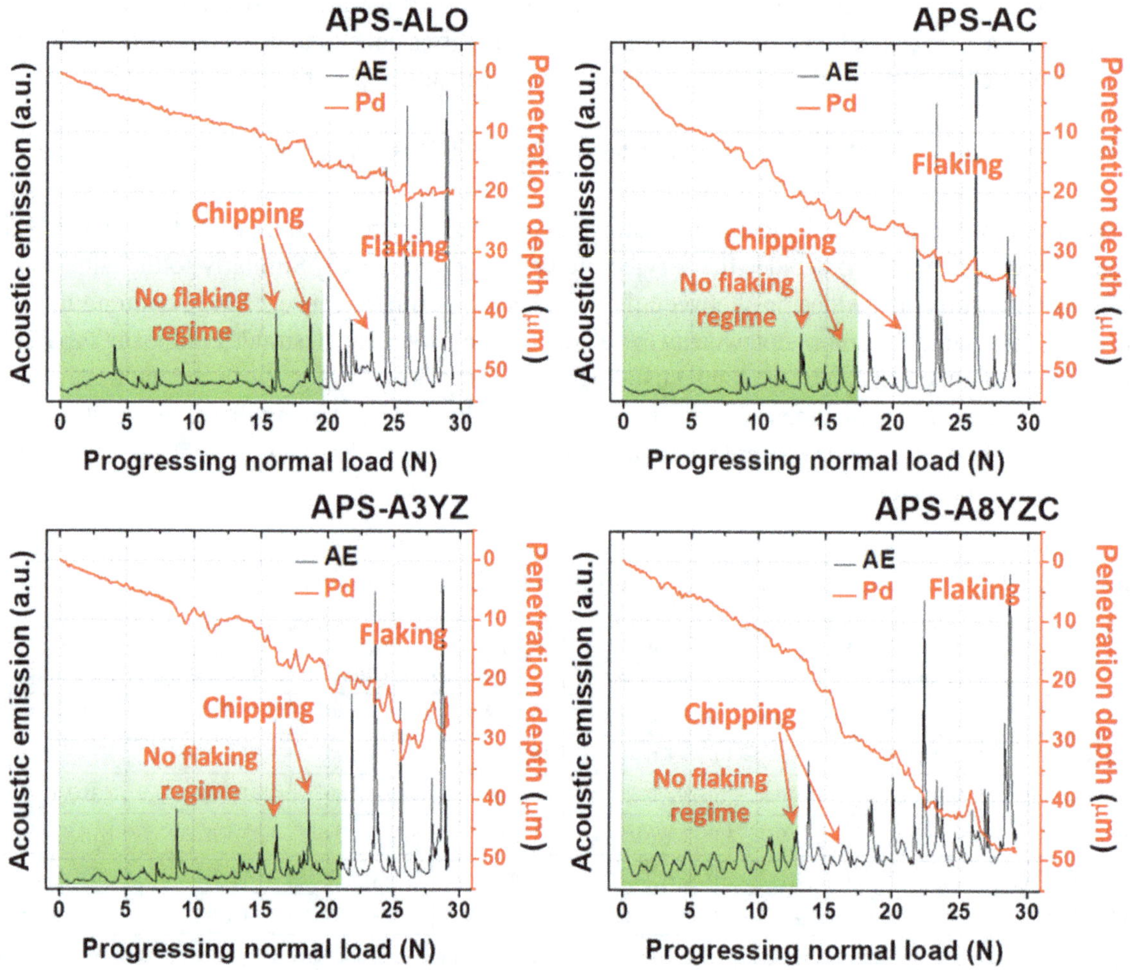

The critical load of failure links with an abrupt fall of frictional force along with a sharp rise in the acoustic signal. It is corresponding to an earlier drop in frictional force with a minimum of 0.5 N during progressive load scratching. The CNT reinforced composites showed lower L_C than the coatings without CNT. It designates that the CNT in the composite is not supportive to improve the essential suitability of the coating for the development of an engineered surface. The sudden failure of CNT reinforced composites coatings are attributed to the poor interface of CNT with other constituents of the coatings, such as Al_2O_3, 3YSZ and 8YSZ. The critical load leads to various failures, such as flaking, chipping and spallation in the coatings. The minimum load corresponds to the critical load for adhesion and the adhesive strength of the coating (**Equation. (5)**, (Bull et al., 1988)):

$$C_A = \frac{\mu^2 L_c^2 \nu^2}{2A_p^2 E} \tag{5}$$

where, A_p: Projected area of the scratch grove, ν: Estimated range of Poisson's ratio, μ: Scratch coefficient, and E: Elastic modulus of the coatings. The adhesion strength for the composite coating is calculated as 1.16-2.91×10^3 J. The CNT reinforced composite showed a lower adhesion strength (1.16-2.00×10^3 J) than that of the composite with no CNT (2.31-2.91×10^3 J). It designates that the energy required for detaching the lamella or splat in CNT reinforced composite is low.

Role of CNT in Failure of Al_2O_3-YSZ Based Composite Coatings

The discontinuity with a saw-like variation of frictional force is observed in the frictional force plot (**Figure 7**) is attributed to the detachment of coatings due to the presence of pores, different phases, and other failures, such as flaking, chipping, spallation of the lamella or splat. It is noticed commonly in the thermal sprayed coatings due to the collapse of pores underneath the scratch tip. Generally, it is recognized in coatings with a low plasticity index (relatively ductile). But, the CNT reinforcement in the Al_2O_3 inclines to resist the nucleation and propagation of the crack, chipping, flaking or spallation through the bridging and pull-out mechanism. So, it is predicted that the CNT improves the critical load of failure to the high normal load. Instead of pushing the critical load and adhesion strength to high, there is a drop in values in CNT reinforced composites (**Table 2**). It is ascribed to the structural damage that occurred marginally in the CNTs due to high-temperature exposure during plasma spraying. It changes the characteristics of the CNTs and effective utilization is minimized in the composite coatings (Ariharan et al., 2017). But, CNT resists the failure of coatings that is noticeable in **Figure 8**. The abrupt fall in the frictional force signifies the chipping, fragments, flaking and spallation of composite coatings that are appeared in the monolithic Al_2O_3 (**Figure 8**). Deviation in the frictional force from the average F_T with <1 N is referred to as chipping and >1 N is referred to as flaking. The CNT reinforcement in the composites has steadily delayed the drops in frictional force (**Figure 8**) due to its usual toughening mechanisms, such as bridging and pull-out.

The CNT, 3YSZ and 8YSZ are acted as the toughening phases and it gradually enhances the suitability of the coatings by avoiding various failure modes through branching, bridging and pull-out mechanism. The presence of toughening phases tolerates the progressive loading during the scratching and delay the failure due to the detachment of materials from the composite coatings (**Figure 8**). The sharp peaks in **Figure 8** imply the failure as chipping and flaking with a drop in frictional force of <1 N and >1 N, respectively. But, CNT reinforced composites shows a dull or blunted peak due to resistance employed by CNT against failure (**Figure 8**). It is ascribed to the bridging and pull-out of CNTs. The trend of the blunted peak is not in the peaks due to not uniform distribution of CNT in the composite. Thus, the CNT reinforced composites contains a sharp drop in F_T as well.

Figure 8. Role of CNT in scratch resistance of the Al_2O_3-3YSZ-CNT composite
(Ariharan & Maurya, 2021)

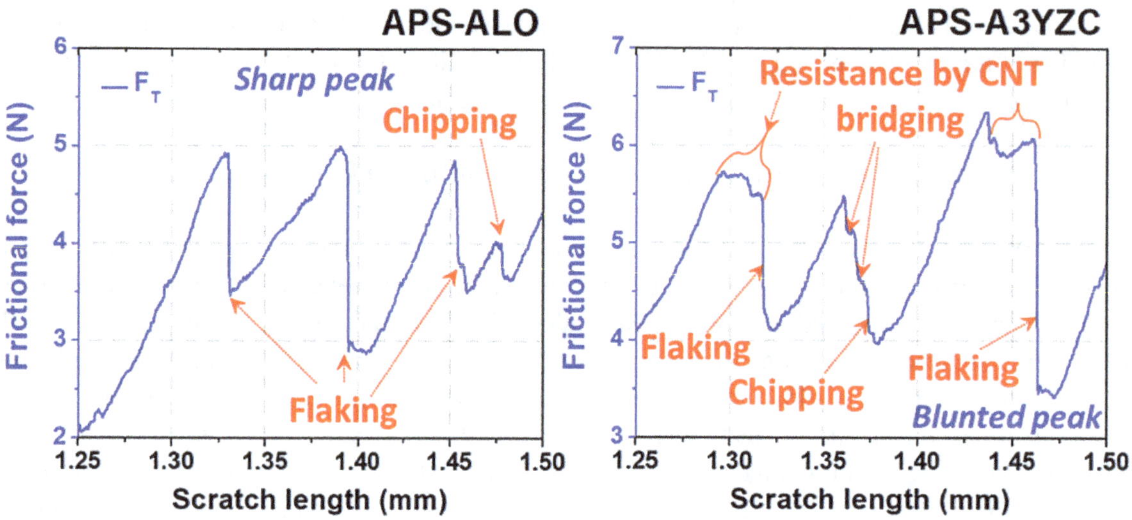

The high (~16.0 N) critical load of failure in the APS-ALO is ascribed to enhanced scratch hardness (**Table 2**) with a relatively good plasticity index (116.4 MPa) than other compositions. Except for APS-A3YZ, reinforcement of 8YSZ and CNT in the Al_2O_3 push the critical load of failure to the lowest value (9.4-11.8 N). Successively, 3YSZ, 8YSZ and CNT reinforcement decrease the critical load of failure and thus the scratch hardness, toughness due to a decrease in the plasticity index and bulk hardness of the composites (Ariharan et al., 2017). Though the CNT improves some of the coating characteristics and extends the life of the coatings from failure (such as flaking, chipping and delamination), it shows poor scratch properties (such as scratch hardness and toughness) with marginally poor adhesion strength. In contrast, it is simplified that the composites with CNT shows an improvement in the coating characteristics and improves it from failure (below the regime of critical load) by the CNT's usual mechanism, such as bridging and pull-out. Also, the scratch hardness and toughness are relatively poor for CNT reinforced composites compared to that of its counterpart. It is attributed to the poor distribution of structurally damaged and aligned CNT in the composite that offers low resistance to the progressive load scratching.

Generally, the CNT reinforced Al_2O_3-YSZ based composites are not suitable for the application of TBCs to withstand the failure induced by the sharp tip or edges in contact with the coatings. Among the Al_2O_3-3YSZ/8YSZ-CNT based composites, 20wt.% 3YSZ reinforced in Al_2O_3 matrix is suitable for better resistance against the sharp tip during progressive load scratching with optimum scratch rate, scratch hardness, scratch toughness and adhesion strength.

SCOPE FOR THE FUTURE WORK

Tailoring the properties by adjusting the processing parameters and improving the distribution of the reinforcements are the major scope for future research in improving the tribological properties of the Al^2O^3-YSZ-CNT based composite coatings. Further, multiple reinforcements of different morphology can also be effectively utilized to achieve better properties.

CONCLUSION

Atmospheric plasma sprayed Al_2O_3-YSZ-CNT composite coatings are compatible with multi-length scale externally induced failures, such as the reciprocating motion of foreign hard particles and sharp edges. The high plasticity index and moderate mechanical properties of monolithic Al_2O_3 showed poor resistance against fretting wear and progressive load scratching. CNT addition in Al_2O_3 further declines (15.4%) the fretting wear resistance due to poor hardness and interfaces. But, the addition of both 3YSZ/8YSZ and CNT in Al_2O_3 synergistically improves (40.1%) the fretting wear resistance of the composites. Further, it reveals the presence of fatigue crack in the composites due to the reciprocating motion of the counter body on the surface that possesses a relatively high ductile nature (low plasticity index). Further, the overestimation (4.1-6.6 times) of wear resistance is attributed to the transformed surface at the wear scar due to repetitive motion of the counter-body and the altered surface acts as a protective layer with high hardness to resist wear. The indication of drastic change in the frictional force at the end of each fretting wear cycle confirms the presence of wear debris. The wear debris collected in the regime of wear scar supports the wear resistance enhancement through the third body wear mechanism during the fretting wear. The progressive load micro-scratching investigation reveals that the monolithic Al_2O_3 coatings showed improved scratch resistance (up to 97%) due to the synergistic effect of mechanical properties and the bonding between the phases. Al_2O_3-3YSZ exhibited a good scratch toughness (~68%) compared to that of the monolithic coating due to the relatively low plasticity index that leads to materials to deforming and accommodating in the pores and pits in the coatings. But, CNT addition declines the scratch toughness (up to 56%) of the composite coatings. It validates that there is no effect of CNT on scratch toughness. Moreover, the assessment of structural integrity is characterized using the critical load of failure and adhesion strength of the splats in the composites. The composites with CNT showed a lower critical load of failure (up to 26.8%) than the coatings without CNT. It indicates that the contribution of CNT to improve the coating stability is not well transferred due to weak interfaces between the phases, non-uniform distribution and structural damage of CNT. The failure of coating, such as flaking, chipping and spallation of the coatings corresponds to the adhesion strength of splats or lamella in the coating. The adhesion strength of the splats is 1.5-2.5 times weaker than the monolithic Al_2O_3 due to the weak bonding of oriented CNTs with Al_2O_3, 3YSZ and 8YSZ. On the other hand, the failure (flaking, chipping and spallation) were counterattacked below the critical load regime by the usual toughening mechanism, such as branching, bridging and pull-out mechanism. Among the atmospheric plasma-sprayed Al_2O_3-3YSZ-8YSZ-CNT based composite coatings, Al_2O_3-20wt.% 3YSZ with a good wear rate (up to 40.1%) scratch resistance (29%), scratch toughness (68%) and adhesion strength (~2.4 times) can be a good choice to resist the failure that induced externally in TBCs.

REFERENCES

Ahmad, I., Fay, M., Kennedy, A., & Zhu, Y. (2009). Interfacial investigations and mechanical properties of carbon nanotube reinforcing Al2O3 nanocomposites. *ICCM International Conferences on Composite Materials*.

Ariharan, S., & Balani, K. (2021). Fretting wear behaviour and frictional force mapping of Al2O3 based thermal barrier coatings. *International Journal of Refractory Metals & Hard Materials*, *98*, 105525. doi:10.1016/j.ijrmhm.2021.105525

Ariharan, S., Gupta, A., Keshri, A., Agarwal, A., & Balani, K. (2013). Size effect of yttria stabilized zirconia addition on fracture toughness and thermal conductivity of plasma sprayed aluminum oxide composite coatings. *Nanoscience and Nanotechnology Letters*, *4*(3), 323–332. doi:10.1166/nnl.2012.1317

Ariharan, S., & Maurya, R. (2021). Assessment of plasma sprayed carbon nanotube reinforced Al2O3 -based nanocomposite with micro-scratching. *Surface and Coatings Technology*, *418*, 127216. doi:10.1016/j.surfcoat.2021.127216

Ariharan, S., Nisar, A., Balaji, N., Aruna, S. T., & Balani, K. (2017). Carbon nanotubes stabilize high temperature phase and toughen Al2O3 -based thermal barrier coatings. *Composites. Part B, Engineering*, *124*, 76–87. doi:10.1016/j.compositesb.2017.05.032

Ariharan, S., Wangaskar, B., Xavier, V., Venkateswaran, T., & Balani, K. (2019). Process induced alignment of carbon nanotube decreases longitudinal thermal conductivity of Al2O3 based porous composites. *Ceramics International*, *45*(15), 18951–18964. doi:10.1016/j.ceramint.2019.06.133

Bakshi, S. R., Balani, K., & Agarwal, A. (2008a). Thermal Conductivity of Plasma-Sprayed Aluminum Oxide-Multiwalled Carbon Nanotube Composites. *Journal of the American Ceramic Society*, *91*(3), 942–947. doi:10.1111/j.1551-2916.2007.02081.x

Bakshi, S. R., Balani, K., & Agarwal, A. (2008b). Thermal Conductivity of Plasma-Sprayed Aluminum Oxide—Multiwalled Carbon Nanotube Composites. *Journal of the American Ceramic Society*, *91*(3), 942–947. doi:10.1111/j.1551-2916.2007.02081.x

Balani, K., Bakshi, S. R., Chen, Y., Laha, T., & Agarwal, A. (2007). Role of powder treatment and carbon nanotube dispersion in the fracture toughening of plasma-sprayed aluminum oxide-carbon nanotube nanocomposite. *Journal of Nanoscience and Nanotechnology*, *7*(10), 3553–3562. doi:10.1166/jnn.2007.851 PMID:18330173

Balani, K., Harimkar, S. P., Keshri, A., Chen, Y., Dahotre, N. B., & Agarwal, A. (2008). Multiscale wear of plasma-sprayed carbon-nanotube-reinforced aluminum oxide nanocomposite coating. *Acta Materialia*, *56*(20), 5984–5994. doi:10.1016/j.actamat.2008.08.020

Bastwros, M., Kim, G.-Y., Zhu, C., Zhang, K., Wang, S., Tang, X., & Wang, X. (2014). Effect of ball milling on graphene reinforced Al6061 composite fabricated by semi-solid sintering. *Composites. Part B, Engineering*, *60*, 111–118. doi:10.1016/j.compositesb.2013.12.043

Bikramjit, B., & Kantesh, B. (2013). Advanced Structural Ceramics. *Journal of Chemical Information and Modeling*, *53*(9).

Bull, S. J., Rickerby, D. S., Matthews, A., Leyland, A., Pace, R., & Valli, J. (1988). The use of scratch adhesion testing for the determination of interfacial adhesion: The importance of frictional drag. *Surface and Coatings Technology*, *36*(1–2), 503–517. doi:10.1016/0257-8972(88)90178-8

Burnett, P. J., & Rickerby, D. S. (1987). The relationship between hardness and scratch adhesion. *Thin Solid Films*, *154*(1–2), 403–416. doi:10.1016/0040-6090(87)90382-8

Carvalho, O., Buciumeanu, M., Madeira, S., Soares, D., Silva, F. S., & Miranda, G. (2016). Mechanisms governing the mechanical behavior of an AlSi–CNTs–SiCp hybrid composite. *Composites. Part B, Engineering*, *90*, 443–449. doi:10.1016/j.compositesb.2016.01.032

Chen, X., He, M. Y., Spitsberg, I., Fleck, N. A., Hutchinson, J. W., & Evans, A. G. (2004). Mechanisms governing the high temperature erosion of thermal barrier coatings. *Wear*, *256*(7), 735–746. doi:10.1016/S0043-1648(03)00446-0

Chen, Y., Balani, K., & Agarwal, A. (2007). Modified Eshelby tensor modeling for elastic property prediction of carbon nanotube reinforced ceramic nanocomposites. *Applied Physics Letters*, *91*(3), 31903. doi:10.1063/1.2756360

Chen, Y., Samant, A., Balani, K., Dahotre, N. B., & Agarwal, A. (2009). Effect of laser melting on plasma-sprayed aluminum oxide coatings reinforced with carbon nanotubes. *Applied Physics. A, Materials Science & Processing*, *94*(4), 861–870. doi:10.100700339-008-4990-4

Chun-Hong, C., & Hideo, A. (2007). Temperature dependence of mechanical properties of aluminum titanate ceramics. *Journal of the European Ceramic Society*, *27*(1), 13–18. doi:10.1016/j.jeurceramsoc.2006.04.182

Fauchais, P. (2004). Understanding plasma spraying. *Journal of Physics. D, Applied Physics*, *37*(9), R86–R108. doi:10.1088/0022-3727/37/9/R02

Hassanzadeh, M., Saremi, M., Valefi, Z., & Pakseresht, A. H. (2018). Investigation on Improved-Durability Thermal Barrier Coatings: An Overview of Nanostructured, Multilayered, and Self-Healing TBCs. *Production, Properties, and Applications of High Temperature Coatings*, 60–78.

Hepplestone, S. P., Ciavarella, A. M., Janke, C., & Srivastava, G. P. (2006). Size and temperature dependence of the specific heat capacity of carbon nanotubes. *Surface Science*, *600*(18), 3633–3636. doi:10.1016/j.susc.2005.12.070

Leclercq, B., Mevrel, R., Liedtke, V., & Hohenauer, W. (2003). Thermal conductivity of zirconia-based ceramics for thermal barrier coating. *Materialwissenschaft und Werkstofftechnik*, *34*(4), 406–409. doi:10.1002/mawe.200390084

Lee, K., Mo, C., Park, S. B., & Hong, S. H. (2011). Mechanical and electrical properties of multiwalled CNT-alumina nanocomposites prepared by a sequential two-step processing of ultrasonic spray pyrolysis and spark plasma sintering. *Journal of the American Ceramic Society*, *94*(11), 3774–3779. doi:10.1111/j.1551-2916.2011.04689.x

Leib, E. W., Vainio, U., Pasquarelli, R. M., Kus, J., Czaschke, C., Walter, N., Janssen, R., Müller, M., Schreyer, A., Weller, H., & Vossmeyer, T. (2015). Synthesis and thermal stability of zirconia and yttria-stabilized zirconia microspheres. *Journal of Colloid and Interface Science*, *448*, 582–592. doi:10.1016/j.jcis.2015.02.049 PMID:25796070

Mahato, N., Sharma, S., Keshri, A. K., Simpson, A., Agarwal, A., & Balani, K. (2013). Nanomechanical properties and thermal conductivity estimation of plasma-sprayed, solid-oxide fuel cell components: Ceria-doped, yttria-stabilized zirconia electrolyte. *JOM*, *65*(6), 749–762. doi:10.100711837-013-0601-8

Musalek, R., Tesar, T., Medricky, J., Lukac, F., Chraska, T., & Gupta, M. (2020). Microstructures and Thermal Cycling Properties of Thermal Barrier Coatings Deposited by Hybrid Water-Stabilized Plasma Torch. *Journal of Thermal Spray Technology*, *29*(3), 444–461. doi:10.100711666-020-00990-2

Nan, C.-W., Liu, G., Lin, Y., & Li, M. (2004). Interface effect on thermal conductivity of carbon nanotube composites. *Applied Physics Letters*, *85*(16), 3549–3551. doi:10.1063/1.1808874

Nan, C.-W., Shi, Z., & Lin, Y. (2003). A simple model for thermal conductivity of carbon nanotube-based composites. *Chemical Physics Letters*, *375*(5–6), 666–669. doi:10.1016/S0009-2614(03)00956-4

Narayan, R., & Colombo, P. (2009). Advances in Bioceramics and Porous Ceramics. *Ceramic Engineering and Science Proceedings*, *52*. doi:10.1002/9780470584354

Pakseresht, A. H., Rahimipour, M. R., Alizadeh, M., Hadavi, S. M. M., & Shahbazkhan, A. (2016). Concept of advanced thermal barrier functional coatings in high temperature engineering components. In Research perspectives on functional micro-and nanoscale coatings (pp. 396–419). IGI Global.

Pakseresht, A. H. (2018). *Production, Properties, and Applications of High Temperature Coatings*. IGI Global. doi:10.4018/978-1-5225-4194-3

Pakseresht, A. H., Kimiayi, A., Alizadeh, M., Nuranian, H., & Faeghinia, A. (2020). Microstructural study and hot corrosion behavior of bimodal thermal barrier coatings under laser heat treatment. *Ceramics International*, *46*(11), 19217–19227. doi:10.1016/j.ceramint.2020.04.259

Pakseresht, A. H., Mousavi, S. M., Saremi, M., Ghasali, E., & Rajaei, H. (2021). Microstructure and mechanical properties of YSZ-alumina composites designed for thermal barrier coatings. *Materials at High Temperatures*, *38*(1), 23–30. doi:10.1080/09603409.2020.1837414

Pakseresht, A. H., Rahimipour, M. R., Vaezi, M. R., & Salehi, M. (2015). Effect of splat morphology on the microstructure and dielectric properties of plasma sprayed barium titanate films. *Applied Surface Science*, *324*, 797–806. doi:10.1016/j.apsusc.2014.11.041

Pakseresht, A. H., Rahimipour, M. R., Vaezi, M. R., & Salehi, M. (2016). Thermal plasma spheroidization and spray deposition of barium titanate powder and characterization of the plasma sprayable powder. *Materials Chemistry and Physics*, *173*, 395–403. doi:10.1016/j.matchemphys.2016.02.028

Pakseresht, A. H., Saremi, M., Omidvar, H., & Alizadeh, M. (2019). Micro-structural study and wear resistance of thermal barrier coating reinforced by alumina whisker. *Surface and Coatings Technology*, *366*, 338–348. doi:10.1016/j.surfcoat.2019.03.059

Pandey, A. K., & Biswas, K. (2016). Effect of hydrothermal treatment on tribological properties of alumina and zirconia based bioceramics. *Ceramics International*, *42*(2), 2306–2316. doi:10.1016/j.ceramint.2015.10.026

Shenogin, S., Xue, L., Ozisik, R., Keblinski, P., & Cahill, D. G. (2004). Role of thermal boundary resistance on the heat flow in carbon-nanotube composites. *Journal of Applied Physics*, *95*(12), 8136–8144. doi:10.1063/1.1736328

Smith, D. S., Fayette, S., Grandjean, S., Martin, C., Telle, R., & Tonnessen, T. (2003). Thermal resistance of grain boundaries in alumina ceramics and refractories. *Journal of the American Ceramic Society*, *86*(1), 105–111. doi:10.1111/j.1151-2916.2003.tb03285.x

Tojo, T., Atake, T., Mori, T., & Yamamura, H. (1999). Excess heat capacity in yttria stabilized zirconia. *Journal of Thermal Analysis and Calorimetry*, *57*(2), 447–458. doi:10.1023/A:1010159807127

Tojo, T., Atake, T., Mori, T., & Yamamura, H. (1999). Heat capacity and thermodynamic functions of zirconia and yttria-stabilized zirconia. *The Journal of Chemical Thermodynamics*, *31*(7), 831–845. doi:10.1006/jcht.1998.0481

Triantou, K. I., Mergia, K., Perez, B., Florez, S., Stefan, A., Ban, C., Pelin, G., Ionescu, G., Zuber, C., Fischer, W. P. P., & Barcena, J. (2017). Thermal shock performance of carbon-bonded carbon fiber composite and ceramic matrix composite joints for thermal protection re-entry applications. *Composites. Part B, Engineering*, *111*, 270–278. doi:10.1016/j.compositesb.2016.12.020

Wang, X., Guo, L., Peng, H., Zheng, L., Guo, H., & Gong, S. (2015). Hot-corrosion behavior of a La2Ce2O7/YSZ thermal salt at 900 C barrier coating exposed to Na2SO4+V2O5 or V2O5. *Ceramics International*. Advance online publication. doi:10.1016/j.ceramint.2015.01.107

Wang, Y., Guo, H., & Gong, S. (2009). Thermal shock resistance and mechanical properties of La2Ce2O7 thermal barrier coatings with segmented structure. *Ceramics International*, *35*(7), 2639–2644. doi:10.1016/j.ceramint.2009.02.025

Xu, S., Chen, L., Gong, M., Hu, X., Zhang, X., & Zhou, Z. (2017). Characterization and engineering application of a novel ceramic composite insulation material. *Composites. Part B, Engineering*, *111*, 143–147. doi:10.1016/j.compositesb.2016.12.010

Yang, F., Zhao, X., & Xiao, P. (2010). Thermal conductivities of YSZ/Al2O3 composites. *Journal of the European Ceramic Society*, *30*(15), 3111–3116. doi:10.1016/j.jeurceramsoc.2010.07.007

Yang, H.-S., Bai, G.-R., Thompson, L. J., & Eastman, J. A. (2002). Interfacial thermal resistance in nanocrystalline yttria-stabilized zirconia. *Acta Materialia*, *50*(9), 2309–2317. doi:10.1016/S1359-6454(02)00057-5

Yang, L. W., Zhang, X. S., Liu, H. T., & Zu, M. (2017). Thermal resistant, mechanical and electrical properties of a novel ultrahigh-content randomly-oriented CNTs reinforced SiC matrix composite-sheet. *Composites. Part B, Engineering*, *119*, 10–17. doi:10.1016/j.compositesb.2017.03.039

Zheng, G., Sano, H., Suzuki, K., Kobayashi, K., Uchiyama, Y., & Cheng, H.-M. (1999). A TEM study of microstructure of carbon fiber/polycarbosilane-derived SiC composites. *Carbon*, *37*(12), 2057–2062. doi:10.1016/S0008-6223(99)00098-6

Zhu, D., & Miller, R. A. (2000). Thermal conductivity and elastic modulus evolution of thermal barrier coatings under high heat flux conditions. *Journal of Thermal Spray Technology*, *9*(2), 175–180. doi:10.1361/105996300770349890

KEY TERMS AND DEFINITIONS

Critical Load of Failure: It is the minimum load required to initiate the composite coatings to the failure, such as chipping, flaking and spallation during the progressive load scratching. The drop in the frictional force with > 0.5N during micro-scratching is considered for the presence of the listed failure.

Fretting Wear: It refers to wear damage of loaded surfaces that are in contact and encountering an oscillatory or reciprocating movement tangential to the surface of the composite coating. Also, it is caused by the adhesion of asperities and wear debris with the samples due to reciprocating motion and other testing conditions.

Frictional Force Map: The plot consists of variations in the frictional force to the fretting amplitude and the total fretting cycle. It is one of the data representations to compare the fretting wear nature of the composite coatings. The change of different regimes during the fretting wear can be easily identified using the frictional force mapping.

Progressive Load Scratching: Mainly, micro-scratching has two types of scratching conditions, such as (1) constant and (2) progressive load scratching. Progressive load scratching has the provision to change the load dynamically while conducting a scratching test on the composite coatings. In other words, it is the combination of multiple micro-scratching test with a constant normal load.

Scratch Hardness: It is the ability of the composite coatings that resistance to plastic deformation induced by the dynamic load applied during the scratching.

Scratch Toughness: It is the capability of the composite coatings to the amount of energy that can absorb per unit volume and deform plastically without failure, such as cracking, chipping, flaking and spallation etc. during the progressive load scratching.

Wear Debris: It is the detached portion of the composite coatings (of different dimensions) due to a high critical load for a particular testing condition during the fretting wear and micro-scratching test. Also, the presence of wear debris in the testing regime has the capability to change the mechanism during the tests.

Chapter 3
Oxidation and Tribology of Al2O3-Induced LaTi2Al9O19/YSZ Double Ceramic Layer Coatings:
Tribo-Oxidation

Pritee Purohit
https://orcid.org/0000-0002-5741-8497
Army Institute of Technology, India

Shashikant Tukaram Vagge
Government College of Engineering, Pune, India

ABSTRACT

In the present work, the alumina-induced thermal barrier coatings with LaTi2Al9O19 (LTA) and yttria-stabilized zirconia (YSZ), LTA/YSZ double ceramic layer (DCL) are studied for oxidation and wear tests. Different coatings combinations with varying thickness of LaTi2Al9O19 (LTA) top coat layer are developed using plasma spray method and are tested for isothermal oxidation and wear test. An Alumina layer is induced after the bond coat layer to provide a readily available oxide layer. The activation energy is calculated using the Arrhenius equation. Arrhenius plots are developed using oxidation kinetics. Coatings are tested for wear performance also. The coating combination with a higher thickness of LTA proved best for both oxidation and wear performances. Surface characterization is done using EDS and XRD analysis.

DOI: 10.4018/978-1-7998-9683-8.ch003

INTRODUCTION

At higher temperature the oxidation, wear and friction behavior of metals and different alloys is crucial in components of metal working processes, gas engines, internal combustion engines, aero engines and propulsion systems and also cutting tools. At elevated temperatures tremendous changes occur in surface reactivity, overall mechanical properties and thermo-physical properties (Blau P, 2010). The various components of a gas turbine and aero engines operate under a harsh high temperature environment. In these systems the operating temperature of usually varies from an ambient temperature nearly the melting point of metals and alloys. It affects the life span of components due to its degradation and failure. Many components need repair, refurbishment and even replacement due to change in their dimensional tolerance while operating in this aggressive environment. In needs very high cost (Rajendran R, 2012). In the development of high efficiency power generators the gas turbines are of prime importance. Many attempts are made by researchers to develop the most durable thermal barrier coating (TBC) with lower thermal conductivity and higher phase stability. The TBC are modified based on few criteria like use of different coating techniques, modifying the superalloys, by changing the chemical combination of coating architecture or by incorporating some protective elements. Self-healing is the most important desirable phenomenon to enhance the TBC life (Pakseresht A. H. et al., 2018).

Many gas turbine components fail due to erosion, friction, wear and oxidation. Sand particles and fly ash erosion against compressor blades damage it and causes premature engine failure. Components having surface contacts due to rotating and reciprocating motion, subject to high amount of wear and therefore needs some protection. Hot gases produced through burning of fuel causes oxidation and high temperature corrosion in the passages of combustion chamber.

Nuclear reactors are in high demand for the generation of electricity and hydrogen. Higher thermal efficiency is desirable and it increases with the increase in operating temperature (Rahman M. S. et al., 2018). Therefore all the hot sections, other surfaces of rotating as well as stationary components needs thermal and wear protection. In all these applications Nickel based super alloys with thermal barrier coatings (TBCs) are the ultimate solution. The TBC life is mainly affected by the factors like coating microstructure, top coat porosity and the residual stress developed in it. Residual stress contributes to different phenomenon like wear resistance, hardness, creep and fatigue behavior, development of micro-cracks and adhesion strength of coating. Thus the residual stress affects the overall performance of the TBC. The formation of residual stress is due to the mismatch stress caused by the varying coefficient of thermal expansion in the different coating layers and the superalloy substrates (Pakseresht A. H. et al., 2016).

As, lanthanum titanium aluminum oxide ($LaTi_2Al_9O_{19}$, LTA) is having best phase stability from room temperature to 1600°C. It is having low fracture toughness which results in poor thermal cycling behavior. LTA and yttria stabilized zirconia (LTA/YSZ) double ceramic layer (DCL) proved better. Its failure is observed due to decomposition of LTA coating exposed to high temperature (Xie et al., 2011). The decomposition of LTA coating is current challenge. It can be overcome by introducing an alumina layer as a protective oxide layer. It provides a protective oxide layer and contributes towards slower oxidation of bond coat. The intact alumina protects the bond coat and underlying substrate from accelerated oxidation. Also LTA also possesses good thermo-physical properties. Thus, combination of LTA/YSZ and alumina layer makes the thermal barrier coating wear and oxidation resistant coating. In present work the $LaTi_2Al_9O_{19}$ (LTA) powder is used to develop the top coat of varying thickness along with Zirconia stabilized Yttria (YSZ) using thermal plasma spray method. Also Al_2O_3 layer is incorporated on the bond coat to ensure the protection of underlying substrate from the ingress of oxygen. The LTA/YSZ double ceramic layer (DCL) coatings are tested for hardness, wear and oxidation resistance.

BACKGROUND

Thermal barrier coatings and wear resistant coatings are applied over Gas turbine blades and vanes which protects the parts from corrosion and oxidation. In case of gas turbine many types of coatings are used. Compressor blades and vanes are provided with erosion resistant coatings. Dovetail root are provided with an Anti-fretting coatings. The wear and oxidation resistance, hardness and coefficient of friction of coatings depends on various factors like coating techniques, coating combinations, doping or reinforcement of various elements and also on the composition of substrate alloy. With the rise in $Ti3AlC_2$ mass fraction the average coefficients of friction decreased in a Co-based alloy coating. It results in outstanding friction reduction. At high temperature the protective oxides formed by Co-based alloy coating and the intact friction layer produced by the Ti_3AlC_2 contributes to the friction reduction (Zhou J. et al., 2021).

Nickel based superalloy are most popular for their high temperature sustainability. Many researchers investigated the Nickel based superalloy for their wear resistance, erosion rate and oxidation and corrosion behavior. NiCr (Nickel chromium) superalloys are mostly used as a substrate in producing corrosion resistant coatings. CrC-NiCr (Chromium carbide and nickel chromium) superalloys are famous in wear resistant coatings at elevated temperature. Among various coatings method high velocity oxy fuel (HVOF) technique is used by many researchers. In this thermal spray the higher micro-hardness and lower porosity is obtained by maintaining higher oxygen flow rate (Derelizade et al., 2022).

Plasma thermal spray methods proved better on carbon steel for enhanced corrosion and wear resistance in case of oil and gas application (Ratia et al., 2019). Thermal barrier coatings also subjected to hot corrosion. The improved hot corrosion can be obtained by reduction in the specific coating surface area, by eliminating the molten salt penetration and by reducing the surface roughness. The nanostructured TBC showed higher hot corrosion resistance in the studies (Pakseresht A. H. et al., 2020).

Various coatings techniques are developed to protect the substrate alloy from wear, erosion and oxidation and from other failure mechanisms. Plasma electrolytic oxidation (PEO) is newly introduced coating technique. An intact ceramic coating layer can be produced at higher applied potentials which increase good abrasion, hardness, wear resistance and corrosion resistance with strong bond strength with the substrate material. In this method the porosity of the coating can be varied by changing the electrolyte chemical composition (Fattah-alhosseini et al., 2021). In plasma electrolytic oxidation (PEO) technique, additives like particle, powder or even sheet can be added into the solutions to produce composite coatings. These additives enhance the coating protection against wear by getting stuck into the cracks and pores present in the coating. With additive addition porosity size diminishes and the coating thickness and hardness increases drastically (Molaei et al., 2021). Thermally Sprayed Aluminum (TSA) coatings are also used to protect steel substrate from corrosion in marine application. The wear resistance of TSA can be increased by Plasma Electrolytic Oxidation (PEO) technique (Lopez-Ortega et al., 2018).

The coating film deposited by the sprayed powders shows higher micro-hardness compared to the coating film produced using the irregular granules (Pakseresht A. H. et al., 2016). The porosity of the coating plays a crucial role in determining the thermal conductivity of coating. It has found that the porosity of the coating developed at room temperature is better than the coating deposited on the pre heated substrates (Pakseresht A. H. et al., 2015). Whisker reinforced coating shows the better wear resistance in comparison to the conventional coating without modifying the friction coefficient. It is possible because of the interlocking splats and reduction in the intra-lamellar cracks developed in the top coat (Pakseresht A. H. et al., 2019).

In Al-doped Ti–12Si_3N_4 coatings the wear resistance, solid strengthening and hardness increased with an increase of Al content. At 700 °C the coating showed higher oxidation resistance due to the formation of external layer of protective Alumina and internal TiO_2 oxide layer (Wang L. et al., 2022). Is it studied and concluded in a research that the composite coatings have higher adhesion strength micro-hardness and toughness and lower porosity compared to the basic coatings (Zhou K. et al., 2021).

Many coating combinations were developed tested by researchers to know the effect of coating combination on wear and oxidation behavior. TiSiCN/ZrN coating combnation is compared with TiSiCN/ZrCN coating combination. Both the coatings were prepared by multi-arc ion plating method. It is found that addition of C contributes towards reduction of friction as it plays the role of lubricant but excess carbon percentage results in enhanced brittleness, cracks generation and propagation and also results in excess coating wear. The increased brittleness of TiSiCN/ZrCN coating is confirmed via mechanical properties (Dong et al., 2021).

High Entropy Alloys (HEAs) are recently used in many high temperature applications. These class materials retain high strength at high temperatures. It shows excellent fracture resistance, phase stability at high temperature and outstanding softening resistance. Oxidation studies were done on Al0.5CrCoFeNi HEA at the temperatures of 800, 900, and 1000°C and in the air. The results proved that the oxidation kinetics of these alloy follow the parabolic rate law (Abbaszadeh et al., 2020). In another study by Chen et al., 2022 the composite coatings having Al_2O_3 showed better wear resistance. Thermally Sprayed Aluminum (TSA) coatings are also used to protect steel substrate from corrosion in marine application. The wear resistance of TSA can be increased by Plasma Electrolytic Oxidation (PEO) technique (Lopez-Ortega et al., 2018). YSZ-alumina coatings are studied in the form of whiskers, particles or reinforced layer. It has shown better properties compared to conventional coatings (Pakseresht A. H. et al., 2021). Use of Alumina in the coating composition proved the increased resistance to wear and oxidation. When the plasma-sprayed Al_2O_3-40 wt% TiO_2 coating were investigated for tribological behavior resulted in remarkable wear resistance (Zavareh et al., 2014)

Coatings play a crucial role of providing the substrate a protection against oxidation, wear, friction and erosion. In present work thermal barrier coatings with varying top coat thickness of $LaTi_2Al_9O_{19}$ (LTA), are produced by Plasma spray method by inducing Al_2O_3 layer sandwiched in the LTA top coat and metallic bond coat. In present work thermal barrier coatings are tested to evaluate their oxidation and wear resistance. Also conventional thermal barrier coatings are developed and tested for oxidation and tribological performances.

MATERIALS AND METHODS

$LaTi_2Al_9O_{19}$ (LTA) spraying powder is prepared through a combination of La_2O_3-99%, TiO_2-99%, and Al_2O_3-99% with the size of 325 μm. The moisture from these powders is removed by heating these powders at for the duration of 10 h at the temperature of 200°C. After drying the powders are mixed properly and then ball milled for the duration of 5 h. Again the Mixture of these powders is dried for the duration of 5 h at the temperature of 200°C and then calcined for the duration of 24 h at the temperature of 1500°C. Later the powder is compacted at a pressure of 400 MPa followed by sintering at the temperature of

1600°C for the duration of 72 h (Xie X. et al., 2011). At the end the sintered mass of the LTA powder is crushed and then sieved using 90 μm sieves. The fluidity of LTA powder is maintain from 38 to 50 gm/min to make it ready for thermal spraying.

In the present study the LTA powder is prepared and used as a top ceramic layer with varying thickness. To determine the various oxides percentage chemical analysis of LTA powder is done. Oxides like Al_2O_3, TiO_2 and La_2O_3 are found in the chemical analysis. SEM analysis of LTA powder is done to study the powder morphology. XRD analysis is done to recognize the phases present in the LTA Powder. XRF technique is used for chemical analysis of LTA powder at Elca laboratory Pune. Shimadzu model EDX 720 is used for XRF study. The results obtained for chemical composition by XRF method are Al_2O_3-56.28%, La_2O_3-6.11% and TiO_2-33.58%.

In present work high temperature alloy Inconel 718 is used as a substrate super alloy. Samples of 10 mm x 10 mm x 4 mm are developed using wire cut EDM. After cutting the samples, their edges are ground to avoid the edge effect and also to get smoother samples. All Samples are then grit blasted in presence of alumina powder before actual coating process or spraying. The fluidity of all powders including $LaTi_2Al_9O_{19}$ powder is maintained 36 to 50 g/min to have smoother spraying. Grain sizes off all powders are kept at 90 μm. All coating combinations are developed using F4 Plasma spray gun.

The TBC thickness of all layers that is from bond coat to top coat is maintained between 320-370 μm. To remove the moisture contained in the $LaTi_2Al_9O_{19}$ powder it is preheated in microwave oven at 150°C before actual spraying. Two sets of coatings are finalized, 1. LTA 100 having thickness 100 μm of $LaTi_2Al_9O_{19}$ and 2. LTA 150 having 150 μm LTA thicknesses. These coatings are annealed for 5 h, at 1050°C in muffle furnace to have the recrystallization of TBC. For oxidation rate comparison YSZ coatings are also prepared. Testing are done on YSZ, LTA 100, LTA 150, LTA 150 annealed i.e. LTA 150A and LTA 100 annealed i.e. LTA 100A samples. Annealing of LTA 100 and LTA 150 samples were done and they were tested separately. It is done to know the difference between oxidation and wear resistance of as sprayed and annealed samples. Annealed samples are mentioned LTA 150A and LTA 100A in this article. In all sets of coatings NiCrAlY bond coat of 90 μm thickness is plasma sprayed on the substrate Inconel 718 super alloy. Al_2O_3 layer of 30 μm is sprayed immediately after bond coat followed by 150 μm layer of YSZ. Finally, a 100 μm and 150 μm top layer of LTA is sprayed on both samples. In case of YSZ coating the 200 μm YSZ layer is sprayed immediately after NiCrAlY bond coat.

Wear test is done using pin on disc apparatus to determine tribological performance. Friction data is also collected using the same apparatus. In this apparatus AISI 52100 steel disc kept moving against the test samples. All the Wear tests are conducted at dry condition for the speed of 0.1 meter for ones for the friction distance of about 1000 m at standard room temperature. The wear rate was estimated by measuring the sample weight after every round of test using an Afcoset electronic balance with the least count 1 μg. Every test is repeated for three times and the average is plotted. Hardness of the YSZ and annealed samples is measured using Vickers micro-hardness apparatus by keeping the weight of 25 gf. Graph plotted are determined by taking average mean of three sets of readings.

At the temperatures variation of 1000°C to 1200°C the isothermal oxidation tests are carried out for the duration of 10 days that is 240 h in a Therelek TE-4673 muffle furnace. Three sets of readings are taken under same condition and average is considered for reproducibility. After every 24 h the samples were taken out of the furnace for recording the mass of the coatings at the mentioned temperatures, using Afcoset electronic balance with the least count 1 μg. Ten sets are kept inside the furnace initially and thus one by one each set removed from furnace after every day. After first day first set is removed after 24 h. Similarly, after 48 h to 240 h the second to tenth set of coatings are taken out respectively.

Mass gain kinetics is determined using the data obtained from all these readings taken. Activation energy calculation is done using kp, rate constant and the gradient value obtained from Arrhenius plots. The coatings are analyzed for different phases generated using X-Ray diffraction instrument. The XRD results are determined using Cu Kα radiation of 2θ range varying from 10 to 70°. Different phases obtained are traced using X'Pert high Score plus software. Carl Zeiss Sigma FESEM is used to study the surface morphology of coatings. The SEM images are captured from 3 kV to 15 kV. The elemental percentage is determined from Energy Dispersive X-Ray analysis (EDS).

RESULTS AND DISCUSSION

Micro-hardness test are conducted on the LTA annealed and YSZ sample surfaces as well as on the cross sections of samples towards its depth. The hardness of YSZ sample was found 424 HV0.025. The LTA 150A had maximum hardness (1340 HV0.025). The results shown in Fig. 1 proved that the micro-hardness reduces from the top layer towards bond coat and substrate.

Figure 1. Vickers micro-hardness profiles of YSZ and annealed LTA 150 and LTA 100 samples

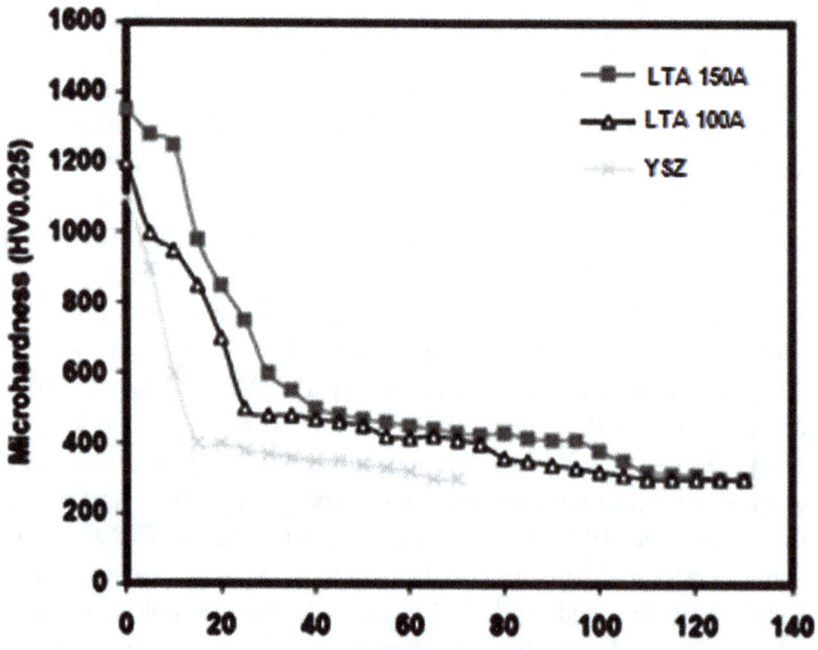

The value of friction coefficient for all LTA samples rubbing against the steel slider is depicted in Fig. 2. It's clear from the Figure that LTA 150A and LTA 100A that is annealed samples exhibits the higher value of friction coefficient compared to the as-sprayed LTA 100 and LTA 150 samples. Almost all annealed samples showed the similar pattern. The annealed samples have shown the higher coefficient of friction because they have higher surface roughness as a result of heat treatment and top most LTA layer.

The value of friction coefficient increases gradually with the friction distance because of the adhesive force between the surfaces in frictional contact. Due to this the surface roughness and value of friction coefficient increases and enhances overall wear rate. In the graph the discontinuity observed is due to intermediate weighing of samples. At this stage the samples are dissociated from the test set up of wear testing machine and then mounted on machine again.

Figure 2. The coefficient of friction graph for all LTA samples: A. annealed LTA 150A, B. annealed LTA 100A. C. LTA 150 and D. LTA 100

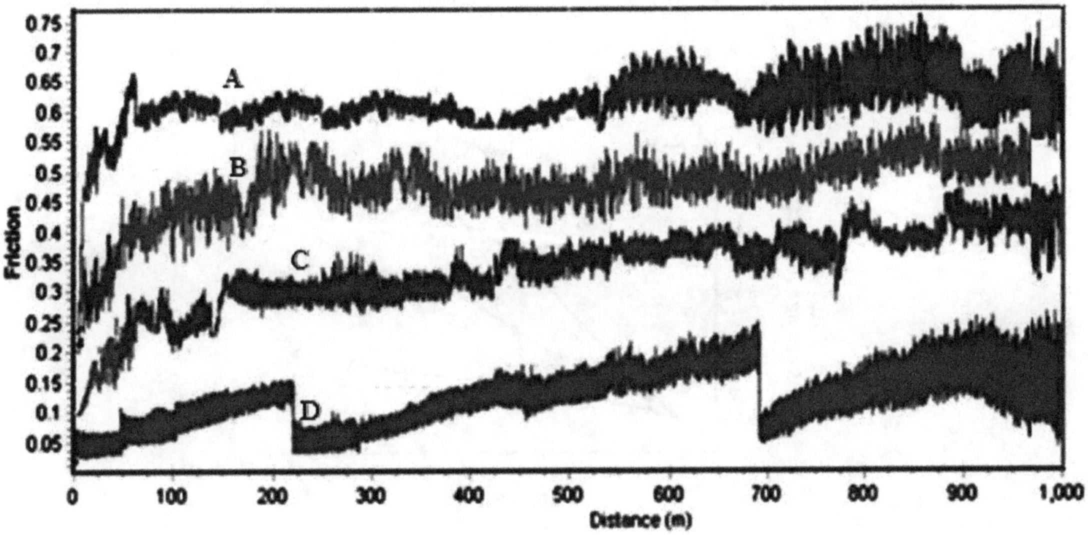

In oxidation tests, only annealed samples are tested therefore here onward only the annealed samples are mentioned as annealed LTA 150A and annealed LTA 100A. That means annealed LTA 150A and annealed LTA 100A samples are annealed samples. As-sprayed samples are not tested for oxidation test. At 1000°C annealed LTA 150A and annealed LTA 100A coating has shown negligible mass gain of 0.0096 mg/cm^2, 0.0250 mg/cm^2. YSZ coatings showed maximum amount of mass gain of 0.0398 mg/cm^2. At the temperature of 1100°C normal YSZ coating is survived upto 240 h that is 10 days with the maximum mass gain of 0.050 mg/cm^2. Both annealed LTA 100A as well as annealed LTA 150A sample showed the mass gain of 0.0320 mg/cm^2, 0.0148 mg/cm^2. YSZ coating is survived upto 4th day (96 h) at 1200°C. Complete spallation of YSZ sample is observed after the duration of 120 h. Hence, YSZ samples are removed and test is discontinued for those samples. Cracks are developed on LTA 100 sample after the duration of 240 h. annealed LTA 150A coating showed the excellent resistance to oxidation. After 10 days and at the temperature of 1200°C coating with 100 µm layer of $LaTi_2Al_9O_{19}$ that is annealed LTA 100A coating showed highest mass gain of 0.032 mg/cm^2 and annealed LTA 150A showed mass gain of 0.019 mg/cm^2. The mass gain plots at all temperature are shown in Fig. 3.

With increase in thermal exposure timing the atmospheric oxygen diffusion is observed through the top layer due to oxygen ionic movement, ingress of oxygen through internally connected pores and cracks. It results in the enhanced rate of oxidation of the metallic bond coat. The formulation and enhanced rate of TGO growth is result of bond coat oxidation and hence coating failure too. Also, parabolic oxidation of coatings enhances TGO thickening and results in the stresses generation, crack propagation and degradation of TBC. The TGO growth kinetics is expressed here in equation (1),

$$X^n = k_p . t \qquad (1)$$

Where, X is the TGO layer thickness μm, t is the oxidation time in h, k_p is the rate constant. TGO thickness is calculated using equation (2). The index n is taken as 2 in the X^n term.

Figure 3. Mass gain plots after isothermal oxidation for 10 days i.e. 240 h at: a) 1000°C, b) 1100°C, c) 1200°C.

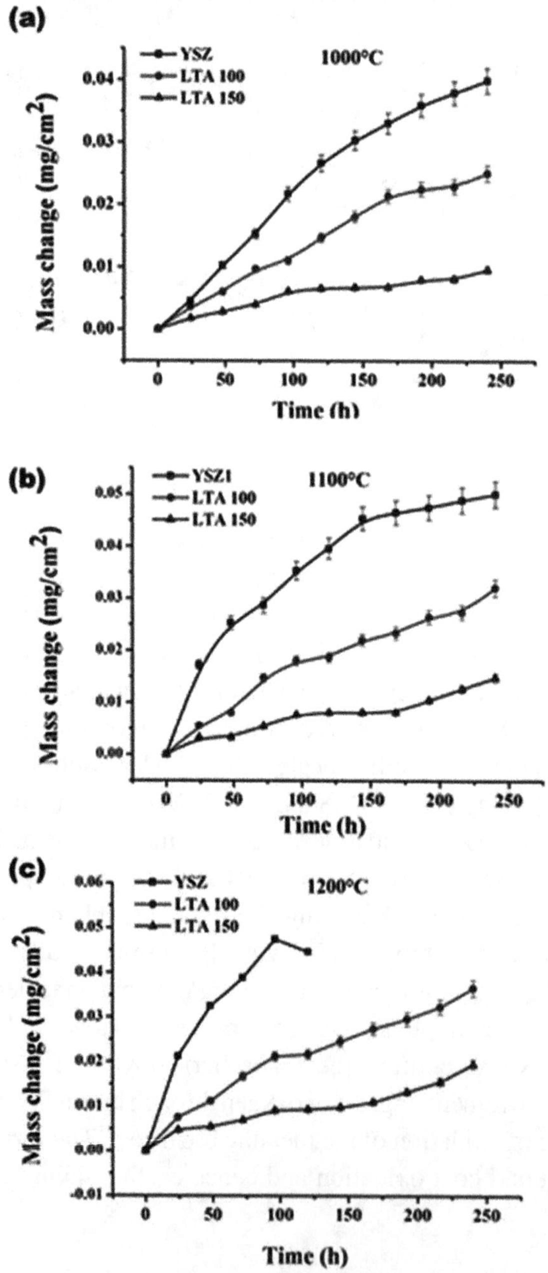

Fig. 4a, 4b and 4c are the plot of TGO layer thickness vs. time for isothermal oxidation at 1000°C, 1100°C and 1200°C. At all mentioned temperatures, the TGO layer growth kinetics of annealed $LaTi_2Al_9O_{19}$ coating is very slow. The induced Al_2O_3 layer acts as a barrier and resists the oxygen diffusion rate towards the bond coat. The TGO growth is steady and much slower compared to YSZ coating. The thicker top coat of annealed LTA 150A coating acts as an oxygen diffusion barrier. In case of all coatings the oxidation of metallic bond coat occurs as result of atmospheric oxygen the porous top coat of Zirconia. The diffusion follows many mechanisms when it happens through the top coat. Oxygen ionic movement is one of cause and the interconnected voids, cracks and pores allow the oxygen ingress. The protective oxide layers are forming throughout the coating interface intermittently and protect the underlying surface as an oxygen diffusion barrier. But the oxide scale simultaneously closes the pores and cracks. This causes a partial pressure drop around the bond coat and top coat interface of LTA coating compared to conventional YSZ coating. The reduced oxygen supply at the bond coat and top coat interface results in the lower oxidation in annealed LTA 150A and annealed LTA 100A coating compared to conventional YSZ coating. These parameters are responsible for the slower TGO growth rate in annealed LTA 150A and annealed LTA 100A coating (Nath S. et al., 2014). The TGO layer thickness variation vs time is obeying parabolic law.

From Fig. 4a to 4c it is clear that, at all mentioned temperatures YSZ coatings showed the maximum TGO layer thickening compared to annealed LTA 100A and annealed LTA 150A coatings. Fig. 4d, 4e and 4f shows the TGO thickness square plotted on Y-axis and oxidation time in hour for annealed $LaTi_2Al_9O_{19}$ 150A, annealed $LaTi_2Al_9O_{19}$ 100A and YSZ samples at all set of temperatures plotted on X-axis. Due to oxidation TGO is developed over the bond coat. The graph obtained is studied to determine how the thermally grown oxide layer grows with the increase in time duration of thermal exposure. Its growth rate-constants for all samples are calculated from the curve slope obtained from kinetics plots. It is showed in Fig. 4d, 4e and 4f.

Figure 4. Change in TGO thickness Vs time and change in square of TGO thickness Vs time

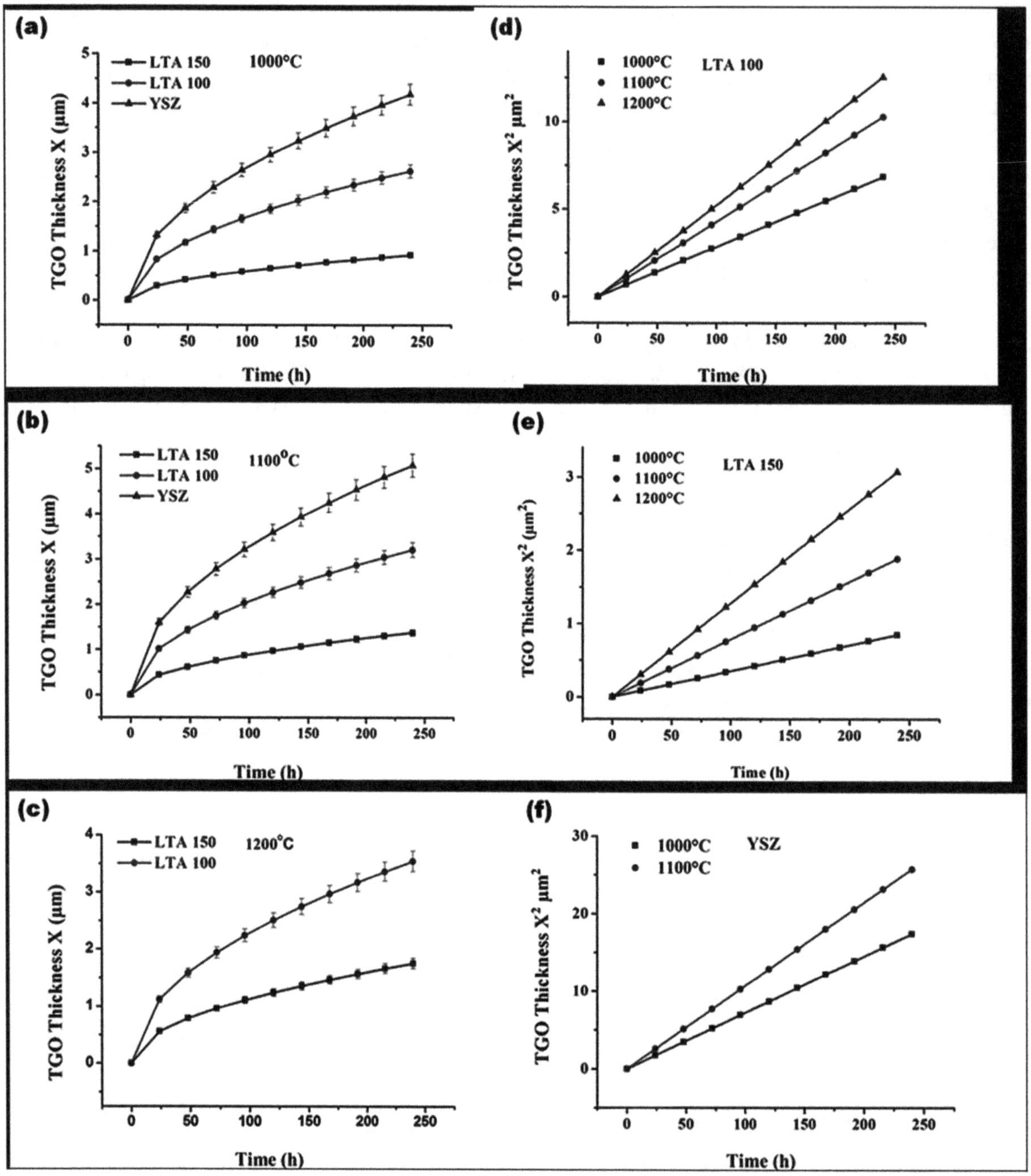

Fig. 5a and 5b is the Arrhenius plot. It is obtained by plotting the change in parabolic rate constant that is ln k_p on Y-axis and the inverse of temperature that is 1/T in Kelvin on X-axis. These values are mentioned in Table 1 and 2.

Figure 5. Arrhenius plots

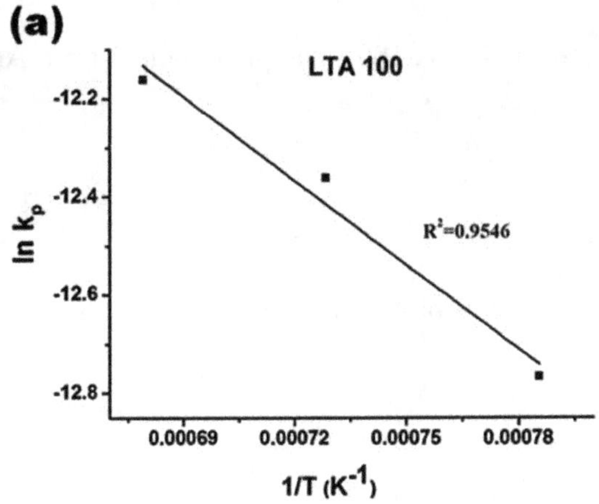

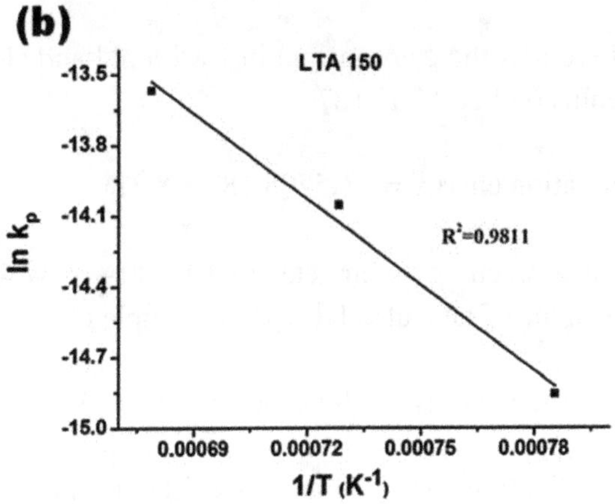

Table 1. Arrhenius plot data to calculate activation energy of coatings

Samples	T (K)	k_p (mg² cm⁻⁴ S⁻¹)	1/T (K⁻¹)	$\ln k_p$
LTA 100A	1273	3.05 x10^{-06}	0.00078	-12.7665
	1373	4.28 x10^{-06}	0.00072	-12.3616
	1473	5.23 x10^{-06}	0.00067	-12.1616
LTA 150A	1273	3.52 x10^{-06}	0.00078	-14.8585
	1373	8.45 x10^{-06}	0.00072	-14.0573
	1473	1.37 x10^{-06}	0.00067	-13.5693

The rate constants values are tabulated in Table 2. In this table the activation energy investigated by other researchers is also mentioned for comparison point of view. The value of activation energy obtained for annealed LTA 150A samples is much closer to the value obtained in other research. It is clear that value of rate constant k_p for annealed LTA 150A coating is much lower compared to annealed LTA 100A and YSZ samples. The annealed LTA 150A coatings have showed the lowest rate constant due to negligible rate of oxidation.

Table 2. Activation energy and Rate constant values calculated for all samples and their comparison with other research

Samples	Rate Constant, k_p (m² s⁻¹)			R^2	Activation Energy (kJ mol⁻¹)	Activation Energy Obtained to Different Researchers (kJ mol⁻¹)
	1000°C	1100°C	1200°C			
YSZ	7.28 x 10^{-14}	1.074 x 10^{-13}	-	-	-	129 (Nath et al., 2014) 139 (Sridharan et al., 2005) 102 (Seo et al., 2009) 130 (Yuan et al., 2008)
LTA 100A	2.87 x 10^{-14}	4.29 x 10^{-14}	5.24 x 10^{-14}	0.95	46.29	
LTA 150A	3.54 x 10^{-15}	7.87 x 10^{-15}	1.29 x 10^{-14}	0.98	102.47	

Activation energy = -slope of Arrhenius plot x R

Where R is the gas constant in J K^{-1} mol^{-1} and slope of annealed LTA 100A sample obtained from Arrhenius plot is -5708.187.

Activation energy = - (-5708.187 x 8.31)

Activation energy of annealed LTA 100A = 47.435 kJ/mol^{-1}
The slope of annealed LTA 150A sample obtained from Arrhenius plot is -12136.83

Activation energy = - (-12136.83 x 8.31)

Activation energy of annealed LTA 150A = 100.85 kJ/mol^{-1}

The values of activation energy obtain for annealed LTA 150A as well as annealed LTA 100A samples after oxidation for the duration of 240 h that is 10 days for the temperature of 1000°C as well as 1100°C and at 1200°C of are 100.85 kJ/mol as well as 47.44 kJ/mol. It shows that, annealed LTA 150A samples showed higher resistance to oxidation compared to annealed LTA 100A and YSZ coatings.

The XRD phase patterns of annealed $LaTi_2Al_9O_{19}$ 100A and annealed $LaTi_2Al_9O_{19}$ 150A samples pre and post oxidation are shown in Fig. 6. The phases obtained at the surface both annealed $LaTi_2Al_9O_{19}$ 100A and $LaTi_2Al_9O_{19}$ 150A samples before the oxidation are of LTA only. LTA 100 sample shows phases of LTA along with Al_2O_3, TiO_2 and also of Al_2TiO_5, $LaAlO_3$ and the spinel oxides of mainly Ni and Cr due to oxidation. The diffraction intensities of the peaks shows new phases of $LaAlO_3$ and Al_2TiO_5 are much stronger at 1200°C.

Figure 6. X-ray diffraction patterns of annealed LTA 100A on the left and annealed LTA 150A on the right

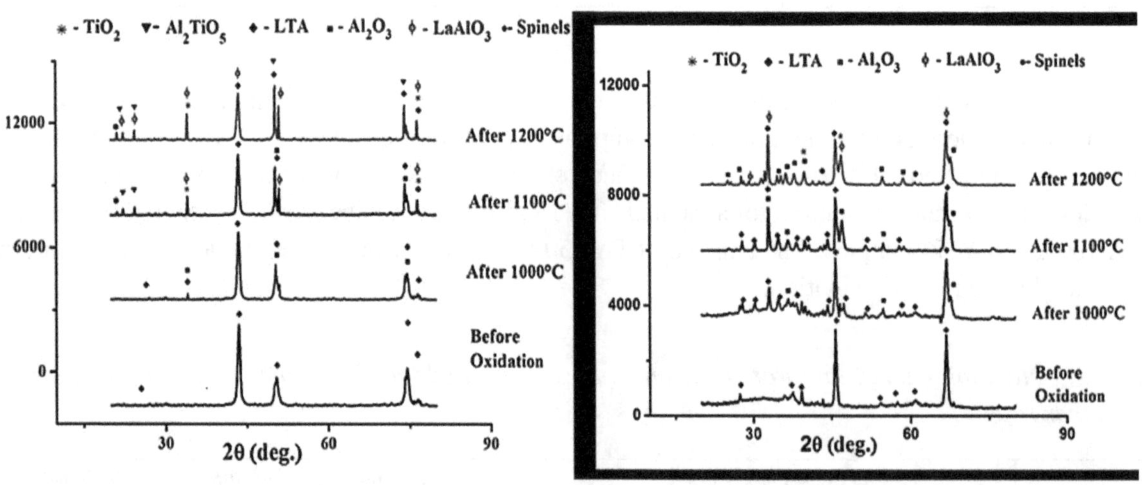

Annealed LTA 150A coatings, mainly the phases observed are of LTA as well as Al_2O_3 also slightly phases of TiO_2, $LaAlO_3$, phases of spinel oxides of Ni. LTA phases are intact even the duration of 10

days that is continuous 240 h of exposure to isothermal oxidation in both annealed LTA 100A as well as annealed LTA 150A coatings at 1200°C.

The microstructure evaluated through SEM of annealed $LaTi_2Al_9O_{19}$ 100 and annealed $LaTi_2Al_9O_{19}$ 150 coatings post oxidation for the duration of 10 days that is 240 h for temperature of 1000°C as well as 1100°C and at the maximum temperature of 1200°C is illustrated in Fig 7. EDS analysis of annealed $LaTi_2Al_9O_{19}$ 100 and annealed $LaTi_2Al_9O_{19}$ 150 coatings post oxidation for the duration of 10 days at all temperatures varying from 1000 °C to 1200 °C is tabulated in Table 3.

Table 3. EDX analysis (at. %) after thermal oxidation

Samples	Area	La	Ti	Al	O	Cr	Zr	Ni
LTA 150A at 1000°C	1	1.11	7.23	24.14	67.26	0.01	0.21	0.05
LTA 150A at 1100°C	2	0.65	5.38	26.94	65.91	0.06	0.68	0.25
LTA 150A at 1200°C	3	0.44	2.32	25.86	70.41	-	0.64	0.33
LTA 100A at 1000°C	4	2.17	12.58	23.40	60.98	0.56	0.17	0.15
LTA 100A at 1100°C	5	0.97	4.77	25.04	68.36	0.19	0.44	0.23
LTA 100A at 1200°C	6	1.32	6.31	27.10	64.83	0.03	0.21	0.19

The SEM microstructure of annealed LTA 150A and LTA 100A coating surface is illustrated in Fig. 7a to 7c and Fig. 7d to 7f respectively. It shows that, the porosity is not uniformly distributed throughout the surface of coating post oxidation for the duration of 10 days and maintained at a temperature of 1000°C. The coating surface porosity is the crucial factor that converts the coating surface into strain tolerant layer and enhances the coating surface life. The EDS study of elemental area 1 and 4 illustrated in Fig. 7a and 7d reports the occurrence of Lanthanum, Titanium, Aluminium, and Oxygen. It is due to the presence of LTA and Al_2O_3 post oxidation for the duration of 10 days that is 240 h at a temperature of 1000°C. XRD results show good agreement with the formation of LTA and Al_2O_3 phases.

The surface microstructure of annealed LTA 150A and annealed LTA 100A coating surfaces post oxidation studied through SEM for the duration 10 days that is 240 h at a temperature of 1100°C is shown in Fig. 7b and 7e. It shows the crack formation, oxidation prone areas and the micro crack propagation. Annealed LTA 150A sample shows irregular surface with intact and rough morphology after oxidation. The EDS study of annealed LTA 150A coating and annealed LTA 100A coatings post oxidation for the duration of 10 days and at a temperature 1100°C in pointed area 2 and 5, shown in Fig. 7b and 7e respectively. It indicates the occurrence of Lanthanum, Titanium, Aluminium, and Oxygen. Maximum amount of Oxygen and Aluminium proves the formation of Al_2O_3. In annealed LTA 100A coating the maximum amount of Oxygen, Lanthanum, Aluminium and Titanium proves the oxide formation of Al_2TiO_5 and $LaAlO_3$. The obtained EDS results are much similar to XRD analysis results.

Fig. 7c shows the negligible rate of micro cracks and grains of oxide on the coating surface of annealed LTA 150A post oxidation at the temperature of 1200°C. Annealed LTA 100A coating showed oxygen diffusion with the increase in oxidation timing. SEM illustrates the vertical cracks which are the oxygen diffusion path. Contributes to the development of $LaAlO_3$ phases and Al_2TiO_5 phases and hence the spallation failure of the coating surface at the temperature of 1200°C. EDS studies of elemental pointed area at 3 and 6 in Fig. 7c and 7f of both coating proves the presence of Oxygen, Aluminium, Titanium and Lanthanum. It proves the presence of $LaAlO_3$ phases and Al_2TiO_5 phases in annealed LTA 100A coatings and formation of $LaAlO_3$ phases in annealed LTA 150A coatings (Nagano M. et al., 1999). EDS results are in good agreement with the XRD analysis results.

Figure 7. Surface morphology of annealed LTA 150A and LTA 100A

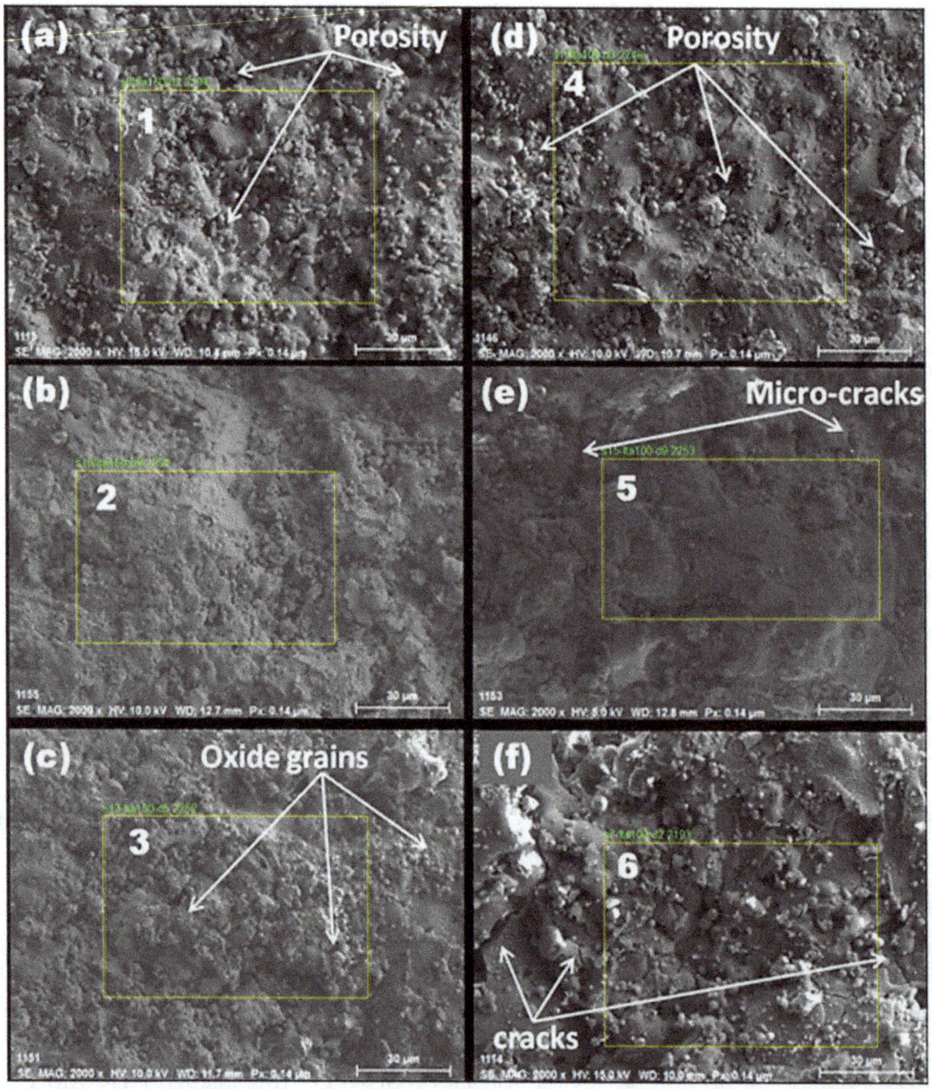

During thermal cycling test according to the phase diagram of ternary combination Al_2O_3-La_2O_3-TiO_2 and basic chemical composition, the Lanthanum and Titanium rich oxides on the surface of coating can form the solid solution phase of $La_{2/3}TiO_3$ and also the phases of $LaAlO_3$ (Skapin S. et al., 1993). As per the available literature $LaTi_2Al_9O_{19}$ is very stable compound at considerably higher temperatures. Formation of all the phases of $LaAlO_3$ is easy but the formation of all the higher aluminates is very

slow below 1500°C for more than 130 days. Therefore only $LaAlO_3$ phases are observed as an oxidation product below 1500°C.
Equation (2) is the possible chemical reaction of LTA after oxidation below 1500°C.

$$LaTi_2Al_9O_{19} \rightarrow LaAlO_3 + 2\ TiO_2 + 4\ Al_2O_3 \quad (2)$$

The phases of α-Al_2O_3 along with TiO_2 may be anatase or rutile forms phases of Al_2TiO_5. The chemical reaction is depicted in equation (3),

$$Al_2O_3 + TiO_2 \rightarrow Al_2TiO_5 \quad (3)$$

The activation energy of Al_2O_3 induced annealed LTA 150A coating is calculated from the Arrhenius plot as illustrated in Fig. 2. Its value is 100.85 kJ/mol. Thus, alumina induced annealed LTA 150A coating is the best TBC combination at 1200°C.

CONCLUSION

The tribological performance tests and isothermal oxidation test is carried out for YSZ and Al_2O_3 induced LTA/YSZ double ceramic layer coatings. Wear and hardness tests were performed. The top coat of LTA maintained at 100 μm and 150 μm is evaluated for oxidation at the temperatures of 1000°C, 1100°C and also at 1200°C for the duration of 10 days that is 240 h and few conclusions are made from this study:

The annealed LTA 150A had maximum hardness and have shown the higher coefficient of friction. Alumina induced LTA 150A annealed samples showed the best oxidation resistance with negligible mass gain. TGO growth observed is parabolic with very slow rate of TGO growth in both annealed LTA coatings. The calculated activation energy of samples of annealed LTA 100A and LTA 150A obtained is 47.435 kJ/mol and 100.85 kJ/mol. XRD results have revealed the phase formation of spinel oxides of Ni and Cr, also phases of LTA, Al_2O_3, TiO_2, Al_2TiO_5 and $LaAlO_3$. In Isothermal oxidation test at 1200°C annealed LTA 150A coatings are survived till 240 h and LTA 100A coating are failed after the time interval of 10 days that is 240 h.

REFERENCES

Abbaszadeh, S., Pakseresht, A. H., Omidvar, H., & Shafiei, A. (2020). Investigation of the High-Temperature Oxidation Behavior of the Al0.5CoCrFeNi High Entropy Alloy. *Surfaces and Interfaces*, *21*, 100724. doi:10.1016/j.surfin.2020.100724

Backman, D. G., & Williams, J. C. (1992). Advanced Materials for Aircraft Engine Applications. *Science*, *255*(5048), 1082–1087. doi:10.1126cience.255.5048.1082 PMID:17817782

Blau, P. J. (2010). Elevated temperature tribology of metallic materials. *Tribology International*, *43*(7), 1203–1208. doi:10.1016/j.triboint.2010.01.003

Cao, X. Q., Vassen, R., & Stoever, D. (2004). Ceramic Materials for Thermal Barrier Coatings. *Journal of the European Ceramic Society*, *24*(1), 1–10. doi:10.1016/S0955-2219(03)00129-8

Chen, Q., Yu, M., Cao, K., & Chen, H. (2022). Thermal conductivity and wear resistance of cold sprayed Cu-ceramic phase composite coating. *Surface and Coatings Technology*.

Derelizade, K., Rincon, A., Venturi, F., Wellman, R. G., Kholobystov, A., & Hussain, T. (2022). High temperature (900 °C) sliding wear of CrNiAlCY coatings deposited by high velocity oxy fuel thermal spray. *Surface and Coatings Technology*, *432*, 128063. doi:10.1016/j.surfcoat.2021.128063

Dong, M., Zhu, Y., Duan, J., Wang, C., Guo, W., Li, J., & Wang, L. (2021). Understanding wear mechanisms of TiSiCN/Zr(C)N coatings at elevated temperatures. *Materials Characterization*, *180*, 111411. doi:10.1016/j.matchar.2021.111411

Fattah-alhosseini, A., & Chaharmahali, R. (2021). Enhancing corrosion and wear performance of PEO coatings on Mg alloys using graphene and graphene oxide additions: A review. *FlatChem*, *27*, 100241. doi:10.1016/j.flatc.2021.100241

Hassanzadeh, M., Saremi, M., Valefi, Z., & Pakseresht, A. H. (2018). Investigation on improved-durability thermal barrier coatings- an overview of nanostructured, multilayered and self-healing TBCs. In A. H. Pakseresht (Ed.), *Production, Properties, and Applications of High Temperature Coatings* (pp. 60–78). IGI Global. doi:10.4018/978-1-5225-4194-3.ch003

Lopez-Ortega, A., Arana, J. L., Rodriguez, E., & Bayon, R. (2018). Corrosion, wear and tribocorrosion performance of a thermally sprayed aluminum coating modified by plasma electrolytic oxidation technique for offshore submerged components protection. *Corrosion Science*, *143*, 258–280. doi:10.1016/j.corsci.2018.08.001

Molaei, M., Babaei, K., & Fattah-alhosseini, A. (2021). Improving the wear resistance of plasma electrolytic oxidation (PEO) coatings applied on Mg and its alloys under the addition of nano- and micro-sized additives into the electrolytes: A review. *Journal of Magnesium and Alloys*, *9*(4), 1164–1186. doi:10.1016/j.jma.2020.11.016

Nagano, M., Nagashima, S., Maeda, H., & Kato, A. (1999). Sintering behaviour of Al_2TiO_5 base ceramics and their thermal properties. *Ceramics International*, *25*(8), 681–687. doi:10.1016/S0272-8842(98)00083-2

Nath, S., Manna, I., & Majumdar, J. D. (2014). Kinetics and mechanism of isothermal oxidation of compositionally graded yttria stabilized zirconia (YSZ) based thermal barrier coating. *Corrosion Science*, *88*, 10–22. doi:10.1016/j.corsci.2014.06.050

Pakseresht, A. H., Kimiayi, A., Alizadeh, M., Nuranian, H., & Faeghinia, A. (2020). Microstructural study and hot corrosion behavior of bimodal thermal barrier coatings under laser heat treatment. *Ceramics International, 46*(11), 19217-19227.

Pakseresht, A. H., Mousavi, S. M., Saremi, M., Ghasali, E., & Rajaei, H. (2021). Microstructure and mechanical properties of YSZ-alumina composites designed for thermal barrier coatings. *Materials at High Temperatures*, *38*(1), 23–30. doi:10.1080/09603409.2020.1837414

Pakseresht, A. H., Rahimipour, M. R., Alizadeh, M., Hadavi, S. M. M., & Shahbazkhan, A. (2016). Concept of advanced thermal barrier functional coatings in high temperature engineering components. In A. Zuzuarregui & Maria Carmen Morant-Miñana (Eds.), Research Perspectives on Functional Micro- and Nanoscale Coatings (pp. 396-419). IGI Global. doi:10.4018/978-1-5225-0066-7.ch015

Pakseresht, A. H., Rahimipour, M. R., Vaezi, M. R., & Salehi, M. (2015). Effect of splat morphology on the microstructure and dielectric properties of plasma sprayed barium titanate films. *Applied Surface Science*, *324*, 797–806. doi:10.1016/j.apsusc.2014.11.041

Pakseresht, A. H., Rahimipour, M. R., Vaezi, M. R., & Salehi, M. (2016). Thermal plasma spheroidization and spray deposition of barium titanate powder and characterization of the plasma sprayable powder. *Materials Chemistry and Physics*, *173*, 395–403. doi:10.1016/j.matchemphys.2016.02.028

Pakseresht, A. H., Saremi, M., Omidvar, H., & Alizadeh, M. (2019). Micro-structural study and wear resistance of thermal barrier coating reinforced by alumina whisker. *Surface and Coatings Technology*, *366*, 338–348. doi:10.1016/j.surfcoat.2019.03.059

Purohit, P., & Vagge, S. T. (2016). Evaluation of alumina incorporated combined ceramic layer thermal barrier coating. *Surface and Coatings Technology*, *307*, 871–878. doi:10.1016/j.surfcoat.2016.10.022

Rahmana, M. S., Ding, J., Beheshti, A., Zhang, X., & Polycarpou, A. (2018). Elevated temperature tribology of Ni alloys under helium environment for nuclear reactor applications. *Tribology International*, *123*, 372–384. doi:10.1016/j.triboint.2018.03.021

Rajendran, R. (2012). Gas turbine coatings-An overview. *Engineering Failure Analysis*, *26*, 355–369. doi:10.1016/j.engfailanal.2012.07.007

Ratia, V. L., Zhang, D., Daure, J. L., Shipway, P. H., McCartney, D. G., & Stewart, D. A. (2019). Sliding wear of a self-mated thermally sprayed chromium oxide coating in a simulated PWR water environment. *Wear, 426-427,* 1466-1473.

Ropp, R. C., & Libowitz, G. G. (1978). The nature of the alumina-rich phase in the System La_2O_3-$A1_2O_3$. *Journal of the American Ceramic Society*, *61*(11-12), 473–475. doi:10.1111/j.1151-2916.1978.tb16119.x

Seo, D., Ogawa, K., Nakao, Y., Miura, H., & Shoji, T. (2009). Influence of high-temperature creep stress on growth of thermally grown oxide in thermal barrier coatings. *Surface and Coatings Technology*, *203*(14), 1979–1983. doi:10.1016/j.surfcoat.2009.01.029

Skapin, S., Kolar, D., & Suvorov, D. (1993). X-ray diffraction and microstructural investigation of the Al_2O_3-La_2O_3-TiO_2 system. *Journal of the American Ceramic Society*, *76*(9), 2359–2362. doi:10.1111/j.1151-2916.1993.tb07777.x

Sridharan, S., Xie, L., Jordan, E. H., Gell, M., & Murphy, K. S. (2005). Damage evolution in an electron beam physical vapor deposited thermal barrier coating as a function of cycle temperature and time. *Materials Science and Engineering A*, *393*(1-2), 51–62. doi:10.1016/j.msea.2004.09.037

Tarasi, F., Medraj, M., Dolatabadi, A., Oberste-Berghaus, J., & Moreau, C. (2011). High-Temperature Performance of Alumina-Zirconia Composite Coatings containing Amorphous Phases. *Advanced Functional Materials*, *21*(21), 4143–4151. doi:10.1002/adfm.201100471

Vassen, R., Cao, X., Tietz, F., Basu, D., & Stöver, D. (2000). Zirconates as New Materials for Thermal Barrier Coatings. *Journal of the American Ceramic Society, 83*(8), 2023–2028. doi:10.1111/j.1151-2916.2000.tb01506.x

Wang, L., Huang, Y., Yuan, Y., Jia, C., & Yang, L. (2022). Microstructure, wear and oxidation resistance of Al-doped Ti–Si3N4 coatings by laser cladding. *Surface and Coatings Technology, 429*, 127942. doi:10.1016/j.surfcoat.2021.127942

Xie, X., Guo, H., & Gong, S. (2010). Mechanical properties of $LaTi_2Al_9O_{19}$ and thermal cycling behaviours of plasma-sprayed $LaTi_2Al_9O_{19}$/YSZ thermal barrier coatings. *Journal of Thermal Spray Technology, 19*(6), 1179–1185. doi:10.100711666-010-9529-5

Xie, X., Guo, H., Gong, S., & Xu, H. (2011). Thermal cycling behavior and failure mechanism of $LaTi_2Al_9O_{19}$/YSZ thermal barrier coatings exposed to gas flame. *Surface and Coatings Technology, 205*(17-18), 4291–4298. doi:10.1016/j.surfcoat.2011.03.047

Yuan, F. H., Chen, Z. X., Huang, Z. W., Wang, Z. G., & Zhu, S. J. (2008). Oxidation behavior of thermal barrier coatings with HVOF and detonation-sprayed NiCrAlY bond coats. *Corrosion Science, 50*(6), 1608–1617. doi:10.1016/j.corsci.2008.02.002

Zavareh, M. A., Diaa, A. A. Sarhan, M., Abd Razak, B. B., & Basirun, W. J. (2014). Plasma thermal spray of ceramic oxide coating on carbon steel with enhanced wear and corrosion resistance for oil and gas applications. *Ceramics International, 40*(9), 14267-14277.

Zhou, J., & Kong, D. (2021). Friction–wear performances and oxidation behaviors of Ti_3AlC_2 reinforced Co–based alloy coatings by laser cladding. *Surface and Coatings Technology, 408*, 126816. doi:10.1016/j.surfcoat.2020.126816

Zhou, K., Xie, F., Wu, X., & Wang, S. (2021). Fretting wear behavior of nano ZrO_2 doped plasma electrolytic oxidation composite coatings on TC21 titanium alloy. *Surface and Coatings Technology, 429*, 127429. doi:10.1016/j.surfcoat.2021.127429

KEY TERMS AND DEFINITIONS

DCL: It is a double ceramic layer coating. In present work the top coat is a DCL, that is combination of YSZ and LTA.

LTA: It is Lanthanum titanium aluminum oxide having chemical formula $LaTi_2Al_9O_{19}$. LTA is a ceramic coating powder having an outstanding.

LTA100: These coatings are developed on Inconel 718 LTA 100 having thickness 100 μm of $LaTi_2Al_9O_{19}$ (LTA). The NiCrAlY bond coat of 90 μm thickness is plasma sprayed on the substrate Inconel 718 super alloy. Al_2O_3 layer of 30 μm is sprayed immediately after bond coat followed by 150 μm layer of YSZ. Finally, a 100 μm layer of LTA is applied.

LTA100A: These are LTA 100 coatings annealed for 5 h, at 1050°C in muffle furnace to have the recrystallization of TBC.

LTA150: In this coating the LTA layer thickness is of 150 µm. The NiCrAlY bond coat of 90 µm thickness is plasma sprayed on the substrate Inconel 718 super alloy. Al_2O_3 layer of 30 µm is sprayed immediately after bond coat followed by 150 µm layer of YSZ. Finally, a 150 µm layer of LTA is applied.

LTA150A: These are LTA 150 coatings annealed for 5 h, at 1050°C in muffle furnace to have the recrystallization of TBC.

TBC: Thermal barrier coating.

YSZ: In case of YSZ coating the 200 µm YSZ layer is sprayed immediately after 90 µm NiCrAlY bond coat.

Chapter 4
Thermal Sprayed Coatings and Their Wear Characteristics

Digvijay G. Bhosale
Dr. D. Y. Patil Institute of Technology, India

T. Ram Prabhu
Defence R&D Organization, India

Walmik S. Rathod
Veermata Jijabai Technological Institute, India

ABSTRACT

Severe wear is the root cause for the failures of components, which resulted in the loss of functionality and dimensional accuracy of systems. An estimation of durability and reliability of components is governed by failure analysis under rigorous operational environments with a range of wear mechanisms. The researchers have two ways to minimise the wear: either develop new anti-wear materials or improve wear resistance of existing materials by surface engineering techniques. Thermally sprayed tungsten carbide-based coatings are often feasible approaches to improve the wear resistance of a metallic surface. Therefore, these coatings on the components enhance the useful lifespan. The investigation focuses on the wear characteristics in terms of coefficient of friction and wear rate for tungsten carbide coatings. The focus of the chapter is on abrasive wear and erosion since these two types of wear have the maximum contribution to the failure rate in real-life problems. The applications, benefits, and detriments of the coating deposition techniques are summarised.

INTRODUCTION

Over the years, improvement in the efficiency and durability of machinery leads to an increase in the operating temperature of the system. The systems were exposed to hostile operating conditions. This necessitates the development and extensive research on new materials having compatibility with high-temperature applications and can sustain under aggressive working environments (Tyagi et al., 2019). Surface coatings are the best feasible solution to enhance the useful life at a variety of extreme operating

DOI: 10.4018/978-1-7998-9683-8.ch004

conditions. They have added advantages of upgrading the surface properties such as hardness, wear, oxidation, and corrosion resistance (Huang et al., 2019). Over the past few decades, extensive research on coatings has been conducted by many researchers and scientists. Consequently, today number of technologies has been developed to deposit a coating on a substrate with cost-effectiveness (Vardelle et al., 2016). The coating acts as a barrier between the substrate material and the working environment of the component. Coating design includes a functional requirement of the surface to ensure a certain level of performance based on failure analysis and mechanism (Ozgurluk et al., 2018; X. Wang et al., 2015; Yuan et al., 2015). The selection of appropriate material is essential to enhance the performance of components and protect them against severe operating conditions.

Tungsten carbide-based are commonly used in numerous applications since they offer excellent wear, erosion, and corrosion (Meekhanthong & Wirojanupatump, 2014; Usmani et al., 1997). The transition metal carbides provide improved mechanical and physical properties such as high hardness and fracture toughness, excellent wear resistance, high melting point, good thermal stability and chemical inertness (Djafer et al., 2014; Jhi et al., 1999). Therefore, transition metal carbides coatings are used as wear-resistant and corrosion-resistant coatings (D. D. Kumar et al., 2017). The plasma spray (PS) and high-velocity oxy-fuel (HVOF) are favourable techniques among the thermal spray variants for the deposition of tungsten carbide-based coatings (Dorfman & Sharma, 2013; Zhe Geng et al., 2016; Hardwicke & Lau, 2013; V. Kumar & Balasubramanian, 2016; Vardelle et al., 2016) . The WC-based coatings are possible to deposit successfully on a wide range of substrate material using APS and HVOF techniques which are essential to extend application areas (Bolelli, Hulka, et al., 2014). The basic difference in the working principle of APS and HVOF techniques is related to the utilisation of kinetic and thermal energy. The plasma spray uses higher thermal energy and lower kinetic energy during the deposition of the coating. Moreover, HVOF techniques are based on higher kinetic energy with comparatively lower thermal energy. The WC-based coating compositions such as WC-Co (Hazra et al., 2012), WC-NiCr (Liu et al., 2008), WC (Zheng, 2013), WC-CoCr (Murthy & Venkataraman, 2006), WC-Cr_3C_2-Ni (Bhosale et al., 2019; Bolelli, Berger, et al., 2014; Ishikawa et al., 2007) has been deposited using thermal spray variants to improve the wear resistance of the substrate in numerous applications.

The study of the effect of degree of oxidation on the tribo-behaviour of coatings is one of the challenging tasks since argumentative results were reported by researchers. The change in the hardness and fracture toughness of phases in the coating with an increase in the temperature is noticeable, which added with hardness and fracture toughness of newly developed oxide phases and creates further complexities to predict the wear (Balamurugan et al., 2012; Federici et al., 2017; Mi et al., 2018; Myalska et al., 2019; Xie et al., 2013). The dry sliding behaviour of WC-based coatings in the inert gas environment at elevated temperature provides details about the contribution of the tribo-oxidation process in the wear mechanisms of the coating. The self-lubricated WC-based composite coatings developed by the researchers is helpful to lower the coefficient of friction (Bhosale & Rathod, 2020a).

The failure of components such as gas and steam turbine blades (Ilieva, 2016; Kazempour-Liacy et al., 2011; Laguna-Camacho et al., 2016), superheaters and boiler tubes (Firouzeh et al., 2018; S. Kumar et al., 2018) is due to the impingement of solid particles suspended in the fluid. Consequently, the shut-down of plants resulted in massive economic losses which are unavoidable during maintenance and replacement of the components. The stainless steel and alloys of nickel are most commonly used for these components. Due to inferior high-temperature erosion (Firouzeh et al., 2018; Ilieva, 2016; Kazempour-Liacy et al., 2011), the useful service life of these components is less. The material removal by high-temperature erosion is dependent on the hardness of the erodent and the hardness of the material under consideration (Babu et al., 2011). Consequently, the superior hardness of tungsten carbide-based coatings makes them favourable to enhance erosion resistance. The erosion behaviour of thermal sprayed carbide-based coatings is dominantly affected by the environmental conditions (Bergmann & Vicenzi, 2011; Sapate & Manish, 2015), processing technique (Ahmad Alidokht et al., 2017), properties of erodent such as size, shape, hardness (Babu et al., 2011; Vicenzi et al., 2006), impingement velocity (González et al., 2017; Vashishtha et al., 2017), temperature (Mathapati et al., 2017), and angle of attack (Ahmad Alidokht et al., 2017; González et al., 2017; Vashishtha et al., 2017).

The chapter focuses on a comprehensive overview of the wear resistance of WC-based coatings deposited using different spraying methods which include plasma spray (PS), high-velocity oxy-fuel (HVOF), detonation spray (DS) and other spraying techniques.

COATING DEPOSITION TECHNIQUES

Plasma Spray

The group of techniques involved in the plasma spray are generally used to deposit the thick coatings as per the requirement. **Figure 1** presents the plasma gun that generates the jet of molten material, which is usually in the form of the feedstock powder. The stream of molten powder accelerated towards the substrate with help of carrier gasses. The droplets of molten material solidify immediately after the impact on the substrate and develop the coating. The characteristics of deposited coatings are largely influenced by the process parameters such as stand-off distance, feed rate, type of feedstock powder, configuration and flow rate of carrier gas (Zhao et al., 2006). The deposited coatings consist of splats which are produced by the flattening effect of molten material (Pakseresht et al., 2015). The stacking of splats includes some inherent defects as voids, micro-cracks, and pores. The variants of plasma spray are air plasma spray (APS) (Bhosale, Prabhu, et al., 2020; Liao et al., 2000; Zhao et al., 2006), supersonic plasma spraying (SPS) (Ganvir et al., 2019), low-pressure plasma spraying (LPPS) (Vinayo et al., 1985; Young et al., 2000), vacuum plasma spray (VPS) and high-pressure plasma spraying (HPPS) (Quint & Kopech, 1999). Generally, the feedstock powder particle size from 10 μm to 100 μm are chosen for all the variants of plasma spray (Gadow et al., 2010; Pakseresht et al., 2016; Vinayo et al., 1985).

Figure 1. Schematic presentation of a plasma spray gun for plasma forming

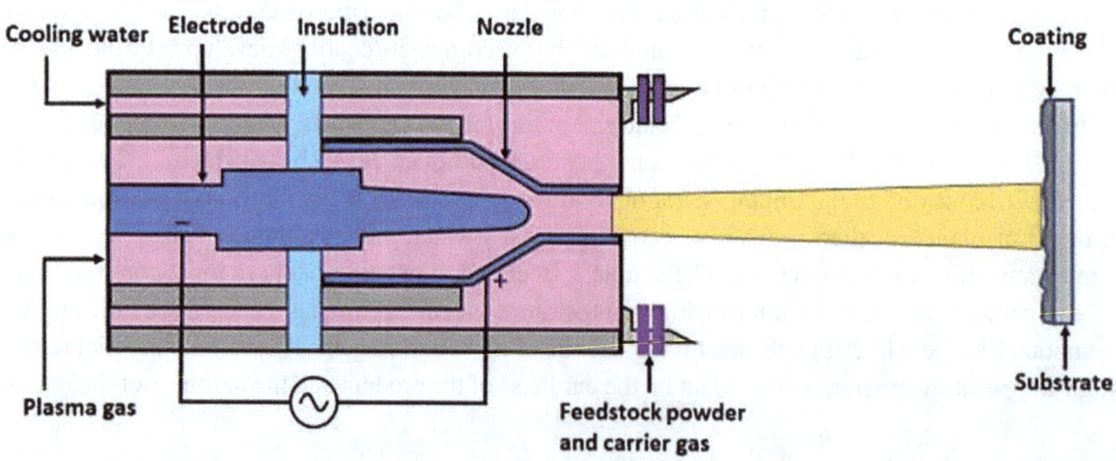

The modified version of plasma spray is the supersonic plasma spraying (SPS) technique, which has been developed to overcome a few drawbacks associated with the atmospheric plasma spray technique (Ganvir et al., 2019). The drawbacks of the APS include more porosity, lesser dense coating due to lower particle velocity, higher oxidation, and a higher degree of decarburisation during deposition of WC-based coatings (Bhosale & Rathod, 2021). In this thermal spray process, the acceleration of molten materials can increase significantly, and resulting coatings exhibited improved properties with a lesser degree of decarburisation.

A low-pressure plasma spraying (LPPS) and vacuum plasma spray (VPS) are two similar types of plasma spray variants that are extensively used for the deposition of oxidation-sensitive feedstock powders. The deposition is usually carried out in the argon environment chamber at lower pressure in order to minimise the degree of oxidation. The extent of oxidation degrades the mechanical properties of coatings. The oxide content in the coatings is negligible due to the elimination of interaction with the oxygen during the deposition. During the deposition, a limited amount of gas is consumed and therefore, the operating cost of LPPS and VPS is lower. The interface of substrate and coating is free from oxide inclusions and the porosity level in the coatings drops below 1% (Zhe Geng et al., 2016). Moreover, a denser coating with lower residual stresses can be developed with the LPPS techniques as compared to APS sprayed coatings (Young et al., 2000). In high-pressure plasma spraying (HPPS), the gas flow rate is very high as compared to the conventional plasma spray technique (Quint & Kopech, 1999). The significant improvement in the deposition efficiency could be achieved by HPPS which makes them a favourable process in industrial applications.

High-Velocity Oxy-Fuel Spray

Among the thermal spray variants, the high-velocity oxygen-fuel (HVOF) (Mi et al., 2018; Vashishtha et al., 2017) is famous technique for deposition of WC-based coatings. The HVOF method has a lesser percentage of decarburisation of the WC phase as compared to the plasma spray technique (Bhosale et al., 2019). Therefore, coatings are composed of a more retained hard WC phase with lesser porosity content which resulted in the improved wear-resistant coating (Bhosale, Prabhu, et al., 2020; T. Wang & Ye, 2018). The lesser in-flight time and lower temperature of the particle are the responsible factors for the reduction in the percentage of brittle W_2C phase in the coating as compared to plasma-sprayed coatings (Bhosale & Rathod, 2021). The development of a nanostructured coating with excellent mechanical properties is possible by the HVOF technique. The higher impact velocity and lower temperature are the parameters for the betterment of the HVOF process over the plasma spray process, which resulted in the minimal pores, micro-cracks and lower oxide content in the coating. However, based on gun design, type of oxidiser and type of fuel used, the HVOF technique is further classified in the variants as, high-velocity liquid-fuelled (HVOLF), high-velocity gas-fuelled (HVOGF) and high-velocity air-fuel (HVAF). The capability of the HVOF process to produce the WC-based coatings with significantly lower porosity and decomposition of the WC phase triggers further developments in the system.

The HVOF technique is also known as a high-velocity flame spray (HVFS) uses liquid fuel or gaseous fuel for melting purposes in the chamber of the spray gun with the help of two oxidisers as air or oxygen. The high kinetic energy at the expense of thermal energy is the key in production of dense, good bonded and less oxidation-content coatings. **Figure 2** shows a basic component included in the HVOF spray gun are convergent-divergent nozzle to achieve higher velocities 800-1200 m/s, combustion chamber to melt the feedstock powder, and mixture to blend the fuel and oxidiser in an appropriate proportion. The gun is usually provided with water-cooling or air-cooling jackets. In the HVOF spray, the continuous flow of a mixture of fuel (liquid or gaseous) and oxidiser (air and oxygen) is supplied to the combustion chamber. The combustion products coming out of the nozzle produces a jet of higher impact velocity. The stream of gas is then entered in the straight section of a barrel where feedstock powder is injected. The stream consists of gas and molten material directed with higher kinetic energy towards the substrate. The HVOF process is usually carried out in the ambient air atmosphere (Vashishtha et al., 2017). The gases used as fuel are propane, propylene, acetylene, hydrogen and ethylene.

Figure 2. Schematic presentation of a HVOF spray gun

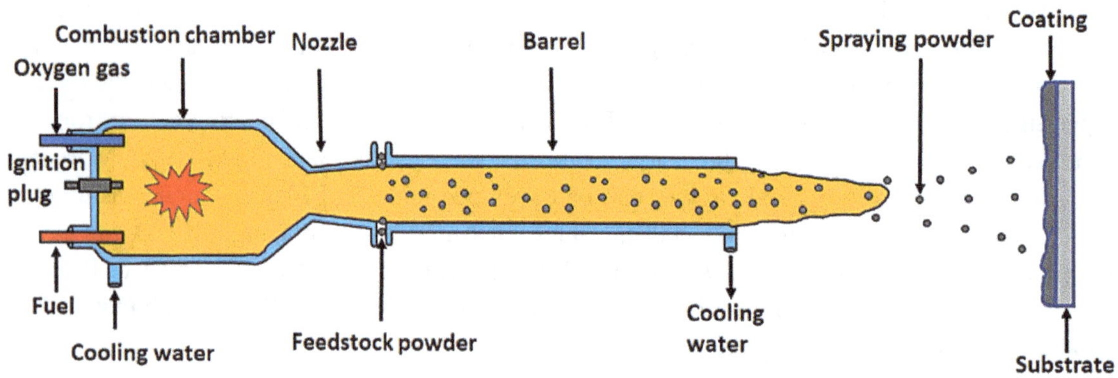

The high-velocity air-fuel (HVAF) technique (Verstak et al., 2004; Verstak & Baranovski, 2003) is a modified form of the conventional HVOF technique. The HVAF has a lower operational cost and manufacturing cost from a safety point of view since the mixture of gaseous fuel and air is used in the system. Moreover, the powder feed rate is comparatively higher than the HVOF process. The principal advantage of the process is the improved quality of coatings due to less oxide content. The third-generation gun was designed as shown in **Figure 3** for supersonic air-fuel technology which increases the particle velocity further (Vuoristo, 2014). The utilisation of nitrogen as carrier gas and lower temperature in the combustion chamber resulted in no oxidation environment during the deposition. The microstructural characteristics and phase composition of the feedstock powder are unaffected during the spraying because of comparatively lower temperatures. Based on the type of fuel used in the mixing chamber, the HVOF technique is further divided into the high-velocity liquid-fuelled (HVOLF) (if the kerosene is fuel) and high-velocity gas-fuelled (HVOGF) (if the gas is fuel). Consequently, both HVOF and HVAF deposition techniques provide coatings with higher surface hardness, minimal porosity and micro-cracks, good bonding strength and lower content of oxide.

Figure 3. Schematic presentation of a third-generation HVAF spray gun

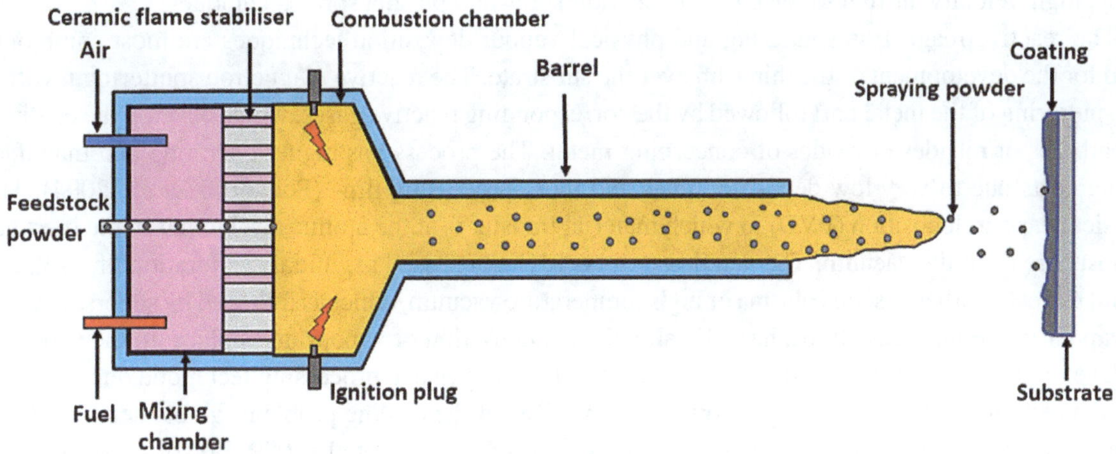

Other Deposition Techniques

Other techniques used to deposit the WC-based coatings includes detonation gun spray (DS) (Babu et al., 2011; Suresh Babu et al., 2017, 2018; Ulianitsky et al., 2017), cold spray (Grujicic, Zhao, et al., 2004; Lima et al., 2002; Maev & Leshchynsky, 2006; Papyrin, 2001; Van Steenkiste et al., 1999), reactive magnetron sputtering (Podgornik et al., 2004), physical vapour deposition (PVD) (Baragetti et al., 2009; Bobzin et al., 2009), laser cladding (Erfanmanesh et al., 2019; Fan & Zhang, 2020; Hulka et al., 2020; Mostajeran et al., 2020), supersonic laser deposition (SLD) (Jones et al., 2014; Luo et al., 2015). The detonation gun process is able to deposit very dense and strongly bonded coatings due to the high kinetic energy associated with the process (Babu et al., 2011; Suresh Babu et al., 2017, 2018). The ease in the control over the flame temperature is possible in the process by adjustment in the fuel gas and oxygen mixture (Ulianitsky et al., 2017). The development of the HVOF process limits the usage of detonation guns in industrial applications since the major drawback is a lower feed rate in the process (Vuoristo, 2014). The cold spray is one of the advanced thermal spray techniques that enable to development of very dense and minimal oxidised coatings as compared to any other thermal spray process (Lima et al., 2002). The process is entirely different from other thermal spray processes and depends on the plastic deformation of high kinetic energy particles. The energy associated with high pressure compressed gases is used to spray powder with very high velocities up to 1500 m/s. The specially designed divergent-convergent nozzle serves the purpose of extremely higher kinetic energy. The operating temperature of the process is always lower than the melting temperature of the powder to be sprayed. The lower temperature of spray powder is responsible for the unmelted state of powder during deposition and the process is also called as solid-state process (Grujicic, Cao, et al., 2004; Lima et al., 2002; Maev & Leshchynsky, 2006). The cold spray process is further classified into two types, a low-pressure cold spray which operates in the pressure range of 6 to 10 bar and the pressure range 10 to 50 bar is related to the high-pressure cold spray (Papyrin, 2001). The high degree of plastic deformation has a significant influence on the dense microstructure observed in cold sprayed coatings (S. W. Rathod et al., 2014; W. S. Rathod et al., 2014). The outcome from different thermal spray processes is the dense and oxide-free coatings are achievable by utilisation of higher kinetic energy and lower thermal energy. Consequently, the preference for WC-based coating deposition by thermal spray technique could be in the sequence as, gas dynamics cold spray, high-velocity air fuel followed by detonation spray and plasma spray techniques.

The reactive magnetron sputtering and physical vapour deposition techniques are most commonly used for the development of the thin film over the substrate. The reactive magnetron sputtering involves the sputtering of the metal part followed by the corresponding reactive gases, which develop the coatings of carbides or nitrides or oxides of concerning metal. The process has higher operating and manufacturing costs due to very low deposition rates and more processing time (Podgornik et al., 2004). The physical vapour deposition (PVD) in which material transfer is at the atomic level either in the inert gas atmosphere or in the vacuum. The detailed process is characterised as, initial evaporation of a solid or liquid target by either gaseous plasma or high-temperature vacuum, which is followed by supersaturation of vapour in the inert gas atmosphere. Finally, the condensation of vapour takes place on the substrate to develop a coating of desired thickness. The PVD is an expensive processing technique due to a very low deposition rate like in the magnetron sputtering. One of the leading problems associated with PVD is difficulties in coating the complex shaped components (Baragetti et al., 2009; Bobzin et al., 2009).

The laser cladding (Erfanmanesh et al., 2019; Fan & Zhang, 2020; Hulka et al., 2020; Mostajeran et al., 2020) is well-known for uniform coating with the lowest porosity. The coating material is in the form of wire or powder and a laser is used to melt the powder to form a thin layer of deposit. The process eliminates the distortion of the substrate, which is a common problem in the case of thermal spray coatings. During the cooling of the cladded material on the surface, natural quenching is activated and therefore, fine grain microstructure could be feasible in the laser cladding. The distinguished advantages of laser cladding are the capability to deposit a coating on complex areas and to avoid uninvited phase transitions such as oxidation. The supersonic laser deposition (SLD) (Jones et al., 2014; Luo et al., 2015) is a modified version of laser and cold spray processes. The process uses a convergent-divergent nozzle to increase the kinetic energy of feedstock powder and the supersonic velocity jet of solid powder carrying through nitrogen gas makes an impact on the substrate. The usage of laser heat reduces the impact velocities by almost half of the cold spray technique, and therefore, the operating cost of SLD is lower. The development of SLD techniques could deposit hard powders like WC with lower impact velocity and lower processing temperature. Consequently, the process eliminates undesired effects such as lower thermal stresses, minimal porosity, retention of phases during deposition.

Wear Resistance of Coatings

Wear mechanism and oxidation are dominant factors affecting useful service life machine components (Mi et al., 2018; Suresh Babu et al., 2018; T. Wang & Ye, 2018), which becomes even more severe for high temperature working environments. Most of the failure occurs due to excessive wear and oxidation mechanisms causing loss of dimensionality and functionality of the metal parts in the machine or systems. To trim down failure due to wear mechanism, researchers or designers pursue three feasible solutions as (Santecchia et al., 2015), 1) Use or develop new wear-resistant materials, 2) Use appropriate alloying elements to enhance anti-wear characteristics, 3) Apply surface engineering techniques to improve surface properties. The surface coating is the best solution to enhance the performance, durability and reliability of the machine components, especially in aggressive working conditions (Vardelle et al., 2016). Coatings on the system components enhance the useful life (Bolelli, Berger, et al., 2014) at elevated temperatures with added advantages of upgrading in the surface properties such as hardness, wear, oxidation, and corrosion resistance etc. (W. S. Rathod et al., 2014). Over the past few decades, extensive research on coatings has been conducted by many researchers and scientists. Consequently, today several technologies have been developed to deposit a coating on a variety of substrate materials with cost-effectiveness (Vardelle et al., 2016). Thermal spray is one of the technologies which is environmentally friendly and widely employed in real-world applications. Thermally sprayed WC coatings are commonly used in many industrial sectors including aerospace (Mayrhofer et al., 2015), oil and gas (Bhosale & Rathod, 2020a; Vardelle et al., 2016)), power generation (Firouzeh et al., 2018; Laguna-Camacho et al., 2016), marine (Bolelli et al., 2016; Wu et al., 2016), electrical (Zhang et al., 2015), automobiles (Bhosale & Rathod, 2021) etc. The thermally sprayed WC coatings are one of the most significant and extensively used coatings for room-temperature and high-temperature applications.

Wear Resistance of WC-based Coatings at Room Temperature

Table 1 is a summary of the deposition technique and wear characteristics with respect to abrasive wear at room temperature for WC based coatings that are reported in the literature.

Table 1. Summary of tribological characteristics reported for WC-based coatings at room temperature

Sr No	Deposition Technique	Coating Powder	Substrate	Counter Body	COF	Wear Rate	Type of Tribometer Used	Ref
1	high-velocity oxygen-fuel (HVOF)	WC–10Co–4Cr (Fine)	ASTM 1045 steel	flowing 200–300 µm particle size of silicon dioxide (arounded quartz grain sand, HV1058)	NA	2.08 mg/min	dry sand rubber wheel tribometer	[23]
		WC–10Co–4Cr (Biomodal - I)			NA	1.46 mg/min		
		WC–10Co–4Cr (Medium)			NA	1.25 mg/min		
		WC–10Co–4Cr(Biomodal - II)			NA	1.71 mg/min		
2	high-velocity oxygen-fuel (HVOF)	nanostructured WC–10Co– 4Cr	AISI 1045 steel	Al_2O_3 pin (Ø 6 mm)	0.45 (avg)	102.4 mm^3	pin-on-disc tribometer	[65]
3	supersonic plasma spraying (SPS)	WC-Co	gray casting iron	silicon nitride ball (Ø 6.35 mm)	0.1	9.5 x 10^{-4} mm^3	ball-on-disc tribometer	[66]
4	high-velocity liquid-fuelled (HVOLF)	WC-12Co	mild steel	Al_2O_3 ball (Ø 9.6 mm)	NA	53 x 10^{-15} mm^3/m	ball-on-disc tribometer	[67]
	high-velocity gas-fuelled (HVOGF)				NA	11 x 10^{-15} mm^3/m		
7	high-velocity oxygen-fuel (HVOF)	WC-Co	mild steel	Al_2O_3 ball (Ø 11mm)	0.36	4.7 x 10^{-4} mm^3	ball-on-disk tribometer	[69]
				Hard metal ball (Ø 11mm)	0.32	5.6 x 10^{-4} mm^3		
				Silicon nitride ball (Ø 11mm)	0.68	419 x 10^{-4} mm^3		
				marten sitic steel ball (Ø 11mm)	0.68	1723 x 10^{-4} mm^3		

Continued on following page

Table 1. Continued

Sr No	Deposition Technique	Coating Powder	Substrate	Counter Body	COF	Wear Rate	Type of Tribometer Used	Ref
8	reactive DC magnetron sputtering	WC	316L Stainless Steel	100Cr6 steel balls (Ø 6 mm)	0.98 (avg)	8.15 x 10^{-4} mm^3/Nm	linear reciprocating mode ball-on-disc micro tribo meter	[70]
		CrC			0.5 (avg)	1.3 x 10^{-3} mm^3/Nm		
		TiC			0.18 (avg)	4.89 x 10^{-5} mm^3/Nm		
9	high-velocity oxygen-fuel (HVOF)	WC–10Co–4Cr	mild steel	Al_2O_3 ball (Ø 10 mm)	NA	~15 cm^3	dry sand rubber wheel tribo meter	[19]
		Cr_3C_2–20NiCr			NA	~25 cm^3		
	detonation gun spray (DS)	WC–10Co–4Cr			NA	~10 cm^3		
		Cr_3C_2–20NiCr			NA	~20 cm^3		
10	vacuum plasma spray (VPS)	Nano WC-Co	steel	Al_2O_3 ball (Ø 10 mm)	0.32 (avg)	8 x 10^{-7} mm^3/Nm	reciprocating and vibrating test machine (SRV)	[71]
		Conventional WC-Co			0.37 (avg)	23 x 10^{-7} mm^3/Nm		
11	reactive DC magnetron sputtering	WC/C diamond-like carbon (DLC)	AISI M35 high-speed steel	steel ball	~0.04 in PAO oil	NA	ball-on-disk tribo meter	[72]
12	air plasma spray (APS)	WC-Co	25CD4 steel	Alumina grits with different grain sizes 100, 60 and 3 microns mixed with water	NA	~270 mg	abrasive tester ASTM B 611-85	[73]
	high-velocity oxygen-fuel (HVOF)				NA	~80 mg		
	vacuum plasma spray (VPS)				NA	~60 mg		
13	Supersonic laser deposition (SLD)	WC/ SS316L	stainless steel	Silicon nitride pin (Ø 4 mm)	0.39 (avg)	NA	pin-on-disc tribo meter	[74]
	Cold Spray (CS)				0.54 (avg)	NA		
14	laser cladding	WC-NiCrMo	SS 316L	Silicon nitride ball (Ø 6 mm)	NA	0.028 mm^3	ball-on-disk tribo meter	[75]
15	gas tungsten arc processing (GTA)	WC–FeCrC	AISI 1018 low-alloy steel	Tungsten carbide ball (Ø 6 mm)	NA	~0.005 g	pin-on-disc tribo meter	[76]
		FeCrC			NA	~0.007 g		
		FeCr			NA	~0.009 g		
16	physical vapour deposition (PVD)	WC/C diamond-like carbon (DLC)	high speed-steel (HSS)	Steel ball (Ø 6 mm)	~0.3 max	313 x 10^{-9} mm^3/Nm	ball-on-disk tribo-meter	[77]

Wear Resistance of WC-Based Coatings at Elevated Temperature

Table 2 is a summary of reported literature with the deposition technique and wear characteristics of WC-based coatings at elevated temperatures. The change in temperature affect the phase constitution of the coatings due to the accelerated oxidation process and therefore, oxidised phases detected in the phase analysis is also included in the summary.

The increase in the temperature is favourable for improvement in the wear resistance of coatings due to the developed tribo-oxide layer during the sliding process (Federici et al., 2017; Mi et al., 2018; Nieto et al., 2017; Xie et al., 2013). However, with an increase in temperature, the fracture toughness and hardness of existing and newly developed phases in coatings decreases. Furthermore, the ionic potential of every phase is a key aspect in the lubricious effect of oxides developed during the tribo-oxidation process. Therefore, the exact prediction of wear behaviour under various operating temperatures is a complex task. The ionic potential of oxides is another differentiating parameter, which either increases or decreases the lubricious effect at the interface between counter body and coating (Erdemir, 2000, 2005). Due to lower shear strength at high-temperature, the oxides which have higher ionic potential show better wear resistance. The composite tribo-film consist of mixed or binary oxides, more difference in their ionic potentials is essential for a lower friction coefficient. Overall tribo-film development is a very complex phenomenon that changes the tribo-behaviour of WC based coatings at elevated temperatures.

Table 2. Summary of tribological characteristics reported for WC-based coatings at different temperatures

Sr No	Deposition Technique	Coating Powder	Substrate	Counter Body	Temperature (°C)	COF	Wear Rate	Type of Tribo-meter Used	Ref
1	high-velocity oxygen-fuel (HVOF)	WC-(nano-WC-Co)	Mild Steel	Al_2O_3 pin (R 9.38 mm)	25	0.67	4.7×10^{-6} mm^3/N.m	pin-on-disk tribometer	[54]
					350	0.65	4.5×10^{-6} mm^3/N.m		
					450	0.6	2.9×10^{-6} mm^3/N.m		
					550	0.45	5.9×10^{-6} mm^3/N.m		
2	high-velocity oxygen-fuel (HVOF)	WC-Co	Mild Steel	Al_2O_3 pin (R, 9.38 mm)	25	0.53 (avg)	~4.4×10^{-6} mm^3/N.m	pin-on-disk tribometer	[55]
					350	0.62 (avg)	~5.9×10^{-6} mm^3/N.m		
					450	0.54 (avg)	~3.65×10^{-6} mm^3/N.m		
					550	0.43 (avg)	~6.1×10^{-6} mm^3/N.m		
3	high-velocity oxygen-fuel (HVOF)	WC-CoCr	ASTM A182 F51 stainless steel	WC pin (Ø 6 mm)	300	0.33	~1.9×10^{-7} mm^3/ N.m	pin-on-disk tribometer	[58]
					350	0.4	~4.2×10^{-7} mm^3/N.m		
4	high-velocity oxygen-fuel (HVOF)	Nanostructured WC-12Co	St 37 steel	Al_2O_3 pin (Ø 7 mm)	700	~0.45 to 0.65	91.7×10^{-4} mg/m	ball-on-disk test machine	[78]
		Electroless Ni-plated micro-structured WC-12Co			700	~0.4 to 0.47	28.8×10^{-4} mg/m		

Continued on following page

Table 2. Continued

Sr No	Deposition Technique	Coating Powder	Substrate	Counter Body	Temperature (°C)	COF	Wear Rate	Type of Tribo-meter Used	Ref
5	high-velocity oxygen-fuel (HVOF)	bimodal WC-Co	Mild Steel	Al_2O_3 pin (R, 9.38 mm)	450	0.61	2.96 x 10^{-6} mm^3/N.m	pin-on-disk high-temperature tribometer	[32]
		conventional WC-Co			450	0.63	3.6 x 10^{-6} mm^3/N.m		
6	high-velocity oxygen-fuel (HVOF)	WC-Co/ nanodiamond composite	1018 stainless steel	Silicon nitride ball (Ø 5mm)	300	~0.4	~ 0.028 mm^3	reciprocating dry sli ding tribometer	[33]
	air plasma spray (APS)				300	~0.5	~ 0.21 mm^3		
7	high-velocity oxygen-fuel (HVOF)	nanostructured WC-CoCr	AISI 1045 steel	Al_2O_3 pin (Ø 6 mm)	200	0.64 avg	4.53 x 10^{-5} mm^3/N.m	pin-on-disk high-temperature tribometer	[79]
					500	0.41 avg	8.71 x 10^{-5} mm^3/N.m		
8	high-velocity oxygen-fuel (HVOF)	WC-CoCr	S355 carbon steel	WC-CoCr coated pin	200 air	~1.0 max	1.96 x 10^{-5} mm^3/N.m	reciprocating type slid-ing tribometer	[34]
					200 N2	~0.8 max	0.24 x 10^{-5} mm^3/N.m		
9	air plasma spray (APS)	WC-12Co	annealed ASTM A283M Gr. D steel	Silicon nitride ball (Ø 6.35 mm)	350	0.59 max	~5.9 x 10^{-5} mm^3/N.m	GW/ML-MS high-temperature ball-on-disc tribometer	[16]
					450	0.48 max	~2.0 x 10^{-5} mm^3/N.m		
					550	0.27 max	~0.8 x 10^{-5} mm^3/N.m		
					650	0.19 max	~4.4 x 10^{-5} mm^3/N.m		
	low-pressure plasma spraying (LPPS)				350	0.37max	~0.5 x 10^{-5} mm^3/N.m		
					450	0.31 max	~0.4 x 10^{-5} mm^3/N.m		
					550	0.2 max	~0.8x 10^{-5} mm^3/N.m		
					650	0.18 max	~2.7 x 10^{-5} mm^3/N.m		
	high-velocity air-fuel (HVAF)				350	0.41 max	~0.3 x 10^{-5} mm^3/ N.m		
					450	0.42 max	~0.3 x 10^{-5} mm^3/N.m		
					550	0.29 max	~0.4 x 10^{-5} mm^3/N.m		
10	high-velocity oxygen-fuel (HVOF)	WC–10Co-4Cr	Low car bon Domex 355 steel	Al_2O_3 pin (Ø 6 mm)	400	0.9 max	0.56 x 10^{-7} mm^3/N.m	high-temperature ball-on-disc tribometer	[80]
	high-velocity air-fuel (HVAF)				400	1.25 max	0.7 x 10^{-7} mm^3/N.m		

Continued on following page

Table 2. Continued

Sr No	Deposition Technique	Coating Powder	Substrate	Counter Body	Temperature (°C)	COF	Wear Rate	Type of Tribo-meter Used	Ref
11	high-velocity oxygen-fuel (HVOF)	WC–10Co–4Cr	H13 steel	Silicon nitride ball (Ø 6.35 mm)	500	0.59 avg	2.52 x 10^{-6} $mm^3/N.m$	ball-on-disc tribometer	[63]
					600	0.44 avg	1.12 x 10^{-6} $mm^3/N.m$		
		Cr_3C_2–25NiCr			500	0.87 avg	1.7 x 10^{-6} $mm^3/N.m$		
					600	0.7 avg	1.31 x 10^{-6} $mm^3/N.m$		
12	high-velocity oxygen-fuel (HVOF)	WC– 12Co	annealed ASTM A283M Gr. D steel.	Silicon nitride ball (Ø 6.35 mm)	650(air)	NA	~1.1 mm^3	high-temperature ball-on-disc tribometer	[35]
					650 (Ar)		~1.0 mm^3		
		WC– 17Co			650 (air)	NA	~0.4 mm^3		
					650 (Ar)		~0.3 mm^3		
13	high-velocity oxygen-fuel (HVOF)	WC–FeCrAl	plain carbon steel	Al_2O_3 pin (Ø 6 mm)	400	~0.8 max	~0.47 x 10^{-6} $mm^3/N.m$	high-temperature ball-on-disc tribometer	[17]
	high-velocity air-fuel (HVAF)				400	~ 0.89 max	~1.00 x 10^{-6} $mm^3/N.m$		
14	high-velocity oxygen-fuel (HVOF)	WC–(W,Cr)$_2$C–Ni	X2CrNiMoN22-5-3 stainless steel	Al_2O_3 pin (Ø 6 mm)	400	0.81 avg	~1.0 x 10^{-6} $mm^3/N.m$	high-temperature ball-on-disc tribometer	[26]
					600	0.76 avg	~1.1 x 10^{-6} $mm^3/N.m$		
					750	0.62 avg	~0.9 x 10^{-6} $mm^3/N.m$		
		WC–CoCr			400	0.78 avg	~0.8 x 10^{-6} $mm^3/N.m$		
					600	0.52 avg	~0.5 x 10^{-6} $mm^3/N.m$		
					750	0.64 avg	~1.3 x 10^{-6} $mm^3/N.m$		
		WC–(W,Cr)$_2$C–Ni	Invar		750	0.62 avg	~0.8 x 10^{-6} $mm^3/N.m$		
15	air plasma spray (APS)	WC-Co	AISI 304 austenitic stainless steel	EN-24 medium carbon steel pin	100	~0.1	~0.06 gm	pin-on-disc tribometer	[28]
					200	~0.6	~0.04 gm		
					300	~0.5	0.024 gm		
16	air plasma spray (APS)	WC-NiCrBSi	carbon steel	Al_2O_3 pin (Ø 6 mm)	250	NA	~12 x10^{-9} Kg/s max	Reci procating type sliding tribometer	[81]
					500	NA	~15 x10^{-9} Kg/s max		
17	high-velocity oxygen-fuel (HVOF)	WC-17Co	stainless steel	Al_2O_3 pin (Ø 6 mm)	500	0.49 max	2.4 x10^{-15} $m^3/N.m$	high-temperature ball-on-disc tribometer	[31]
		WC-10Co-4Cr			500	0.42 max	16 x10^{-15} $m^3/N.m$		
		WC-12Co			500	0.41 max	11 x10^{-15} $m^3/N.m$		
		Cr_3C_2-25NiCr			500	0.51 max	8.2 x10^{-15} $m^3/N.m$		

Continued on following page

Table 2. Continued

Sr No	Deposition Technique	Coating Powder	Substrate	Counter Body	Temperature (°C)	COF	Wear Rate	Type of Tribo-meter Used	Ref
18	high-velocity oxygen-fuel (HVOF)	WC–12Co Fine carbide size distributions	steel (JIS SS 400)	Al_2O_3 pin (Ø 6 mm)	200	0.81 avg	~8 x10^{-7} mm^3/N.m	ring on-plate tribometer	[82]
					300	0.73 avg	~2 x 10^{-7} mm^3/N.m		
					400	0.82 avg	~2 x 10^{-7} mm^3/N.m		
		WC–12Co Medium carbide size distributions			200	0.75 avg	~10 x10^{-7} mm^3/N.m		
					300	0.70 avg	~4 x 10^{-7} mm^3/N.m		
					400	0.84 avg	~1.8x10^{-7} mm^3/N.m		
19	high-velocity oxygen-fuel (HVOF)	WC-(W,Cr)$_2$C-Ni	1Cr18Ni9Ti stain less steel	Silicon nitride ball (Ø 10 mm)	700	~0.8	~21 x 10^{-7} mm^3/N.m	reciprocating type sliding tribometer	[83]
					800	~0.7	~18 x 10^{-7} mm^3/N.m		
		WC-17Co			700	~0.9	~8.5x 10^{-7} mm^3/N.m		
					800	~0.9	~51 x 10^{-7} mm^3/N.m		
20	air plasma spray (APS)	WC–(W,Cr)$_2$C–Ni	1Cr18Ni9Ti stainless steel	Silicon nitride ball (Ø 3 mm)	200	0.83	~18 x 10^{-5} mm^3/N.m	high-temperature ball-on-disc tribometer	[36]
					400	0.91	~15.5 x 10^{-5} mm^3/N.m		
					600	0.72	~2 x 10^{-5} mm^3/N.m		
		WC–(W,Cr)$_2$C–Ni/ 5%Ag/5%BaF$_2$–CaF$_2$			200	0.66	~1.95 x 10^{-5} mm^3/N.m		
					400	0.78	~7.4 x 10^{-5} mm^3/N.m		
					600	0.69	~2.65 x 10^{-5} mm^3/N.m		
21	high-velocity air-fuel (HVAF)	WC-Co +1%TiC (Nano sized)		Al_2O_3 ball (Ø 6 mm)	400	0.39	4.82 x 10^{-8} mm^3/N.m	high-temperature ball-on-disc tribometer	[29]
		WC-Co +3%TiC (Nano sized)			400	0.5	2.51x 10^{-8} mm^3/N.m		
		WC-Co +5%TiC (Nano sized)			400	0.56	6.73 x 10^{-8} mm^3/N.m		
		WC-Co +7%TiC (Nano sized)			400	0.74	6.46 x 10^{-8} mm^3/N.m		
22	high-velocity oxygen-fuel (HVOF)	WC-Cr_3C_2-Ni	In conel 718	SUS 304 / SS 304 ball (Ø 9.5mm)	450	0.23 (avg)		reciprocating type tribometer	[84]

Wear Mechanisms of WC Based Coatings at Various Temperatures

A range of research has been carried out by researchers to study detailed wear behaviour of WC based coatings even at high-temperatures. The increase in the temperature is favourable for improvement in the wear resistance of coatings due to the developed tribo-oxide layer during the sliding process (Mi et al., 2018). The formation of a tribo-oxide layer at a higher temperature protected the underneath surface from being worn out and subsequently reduces the wear rate. The tribo-oxidation contributed to the development of WO_3 (Bolelli et al., 2015; Hong et al., 2017; Hou et al., 2011; Jafari et al., 2017; Papyrin, 2001), $CoWO_4$ (Bolelli et al., 2015; Papyrin, 2001; Yang et al., 2006; Zhang et al., 2015; Zhao et al., 2006), CoO_2 (Balamurugan et al., 2012), $NiWO_4$ (Bhosale & Rathod, 2020b; Fang et al., 2009; Hou et al., 2011), $CrWO_4$ (Fang et al., 2009; Hou et al., 2011) and Cr_2WO_6 (Bhosale & Rathod, 2020b; Fang et al., 2009; Hou et al., 2011) which act as self-lubricant. Generally higher temperature environment enhances anti-wear properties, and the reason could be the formation of oxide phases improves the hardness and fracture toughness of the coatings.

The tungsten carbide-based cermet coatings exhibited various abrasive wear mechanisms, which are dependent on the ample retention of the hard phases in the structure of the coating, binder mean free path, the size of carbide grains and proportion of binder phases in the coating. The detailed summary of wear mechanisms reported in the literature at various temperatures is in **Table 3**.

Table 3. Summary of wear mechanisms reported for WC-based coatings at different temperatures

Sr. No.	Temperature	Wear Mechanisms	Ref
1	Room Temperature	Mild abrasive grooving, plastic deformation, micro cracking, carbide pull outs, adhesion to counter body, grain fragmentation	[23] [47] [65-77]
2.	100 -300 °C	Carbide pull outs, small amount of oxidised region, relatively smooth tracks, cracks, adhesion to counter body, mild abrasive grooving, exfoliation, subsurface layer, fine scratches, plastic deformation, micro cracking	[16] [18] [28] [29] [32-34] [54] [55] [58] [79] [80]
3.	301-500°C	Brittle Cracking, fatigue delamination, clusters of oxidised material areas, pits (numerous carbides pull out), coating delamination, severe abrasive grooving, formation of plough, prow formation, mild adhesion, severe fracture, peeling off oxide film, shallow abrasive grooving, delamination pits	[17] [18] [26] [29] [33-36] [54] [55] [58] [78] [79] [80] [83] [87]
4.	501-700°C	severe brittle fracture, large spalling pits, delamination, oxide film delamination, dark region is mixture of Al_2O_3 and WO_3, excessive material transferred from Al_2O_3 ball due to adhesion, adhesion wear, deeper grooves.	[17] [26] [35] [36] [83] [87]

SUMMARY

1. The outcome from different thermal spray processes is the dense and oxide-free coatings are achievable by utilisation of higher kinetic energy and lower thermal energy. Consequently, the preference for WC-based coating deposition by thermal spray technique could be in the sequence as, gas dynamics cold spray, high-velocity air fuel followed by detonation spray and plasma spray techniques.

2. The wear performance of coatings has been examined with several testing configurations. The most commonly used tribometer to serve the purpose was pin or ball on disc since added experimental parameters variation could be studied with this configuration.
3. The results are helpful for the researcher and coating manufacturer for the selection of the appropriate tungsten carbide-based coating material, coating process and coating parameters for specific applications to enhance the life of components with minimised wear.
4. Overall common pool of results of different researchers has been reviewed, which will be useful for surface engineering related industries.

REFERENCES

Ahmad Alidokht, S., Vo, P., Yue, S., & Chromik, R. R. (2017). Erosive wear behavior of Cold-Sprayed Ni-WC composite coating. *Wear*, *376–377*, 566–577. doi:10.1016/j.wear.2017.01.052

Babu, P. S., Basu, B., & Sundararajan, G. (2011). The influence of erodent hardness on the erosion behavior of detonation sprayed WC-12Co coatings. *Wear*, *270*(11), 903–913. doi:10.1016/j.wear.2011.02.019

Balamurugan, G. M., Duraiselvam, M., & Anandakrishnan, V. (2012). Comparison of high temperature wear behaviour of plasma sprayed WC–Co coated and hard chromium plated AISI 304 austenitic stainless steel. *Materials & Design*, *35*, 640–646. doi:10.1016/j.matdes.2011.10.012

Baragetti, S., Lusvarghi, L., Bolelli, G., & Tordini, F. (2009). Fatigue behaviour of 2011-T6 aluminium alloy coated with PVD WC/C, PA-CVD DLC and PE-CVD SiOx coatings. *Surface and Coatings Technology*, *203*(20), 3078–3087. doi:10.1016/j.surfcoat.2009.03.040

Bergmann, C. P., & Vicenzi, J. (2011). *Protection against Erosive Wear Using Thermal Sprayed Cermet: A Review* (C. P. Bergmann & J. Vicenzi, Eds.). Springer Berlin Heidelberg. doi:10.1007/978-3-642-21987-0

Bhosale, D. G., Prabhu, T. R., Rathod, W. S., Patil, M. A., & Rukhande, S. W. (2020). High temperature solid particle erosion behaviour of SS 316L and thermal sprayed WC-Cr_3C_2–Ni coatings. *Wear*, *462–463*, 203520. doi:10.1016/j.wear.2020.203520

Bhosale, D. G., & Rathod, W. S. (2020a). Investigation on wear behaviour of SS 316L, atmospheric plasma and high velocity oxy-fuel sprayed WC-Cr_3C_2-Ni coatings for fracturing tools. *Surface and Coatings Technology*, *390*, 125679. doi:10.1016/j.surfcoat.2020.125679

Bhosale, D. G., & Rathod, W. S. (2020b). Tribological behaviour of atmospheric plasma and high velocity oxy-fuel sprayed WC-Cr_3C_2-Ni coatings at elevated temperatures. *Ceramics International*, *46*(8), 12373–12385. doi:10.1016/j.ceramint.2020.01.288

Bhosale, D. G., & Rathod, W. S. (2021). Tribo-behaviour of APS and HVOF sprayed WC–Cr_3C_2–Ni coatings for gears. *Surface Engineering*, *37*(1), 80–90. doi:10.1080/02670844.2020.1742988

Bhosale, D. G., Rathod, W. S., Ghorpade, U. S., & Rukhande, S. W. (2020). Nickel alloy C-263 protection by WCCr3C2Ni coatings against high-temperature wear in nuclear applications. *Surfaces and Interfaces*, *21*, 100689. doi:10.1016/j.surfin.2020.100689

Bhosale, D. G., Rathod, W. S., & Rukhande, S. W. (2019). Sliding wear behavior of high velocity oxy-fuel sprayed WC-Cr_3C_2-Ni coating for automotive applications. *Materials Today: Proceedings*, *19*, 339–343. doi:10.1016/j.matpr.2019.07.609

Bobzin, K., Bagcivan, N., Goebbels, N., Yilmaz, K., Hoehn, B.-R., Michaelis, K., & Hochmann, M. (2009). Lubricated PVD CrAlN and WC/C coatings for automotive applications. *Surface and Coatings Technology*, *204*(6), 1097–1101. doi:10.1016/j.surfcoat.2009.07.045

Bolelli, G., Berger, L.-M., Bonetti, M., & Lusvarghi, L. (2014). Comparative study of the dry sliding wear behaviour of HVOF-sprayed WC–(W, Cr) 2C–Ni and WC–CoCr hardmetal coatings. *Wear*, *309*(1–2), 96–111. doi:10.1016/j.wear.2013.11.001

Bolelli, G., Berger, L.-M., Börner, T., Koivuluoto, H., Lusvarghi, L., Lyphout, C., Markocsan, N., Matikainen, V., Nylén, P., Sassatelli, P., Trache, R., & Vuoristo, P. (2015). Tribology of HVOF-and HVAF-sprayed WC–10Co4Cr hardmetal coatings: A comparative assessment. *Surface and Coatings Technology*, *265*, 125–144. doi:10.1016/j.surfcoat.2015.01.048

Bolelli, G., Berger, L.-M., Börner, T., Koivuluoto, H., Matikainen, V., Lusvarghi, L., Lyphout, C., Markocsan, N., Nylén, P., Sassatelli, P., Trache, R., & Vuoristo, P. (2016). Sliding and abrasive wear behaviour of HVOF-and HVAF-sprayed Cr_3C_2–NiCr hardmetal coatings. *Wear*, *358*, 32–50. doi:10.1016/j.wear.2016.03.034

Bolelli, G., Hulka, I., Koivuluoto, H., Lusvarghi, L., Milanti, A., Niemi, K., & Vuoristo, P. (2014). Properties of WC–FeCrAl coatings manufactured by different high velocity thermal spray processes. *Surface and Coatings Technology*, *247*, 74–89. doi:10.1016/j.surfcoat.2014.03.021

Djafer, A. Z. A., Saoula, N., Madaoui, N., & Zerizer, A. (2014). Deposition and characterization of titanium carbide thin films by magnetron sputtering using Ti and TiC targets. *Applied Surface Science*, *312*, 57–62. doi:10.1016/j.apsusc.2014.05.084

Dorfman, M. R., & Sharma, A. (2013). Challenges and strategies for growth of thermal spray markets: The six-pillar plan. *Journal of Thermal Spray Technology*, *22*(5), 559–563. doi:10.100711666-013-9906-y

Du, L., Xu, B., Dong, S., Zhang, W., Zhang, J., Yang, H., & Wang, H. (2008). Sliding wear behavior of the supersonic plasma sprayed WC–Co coating in oil containing sand. *Surface and Coatings Technology*, *202*(15), 3709–3714. doi:10.1016/j.surfcoat.2008.01.009

Erdemir, A. (2000). A crystal-chemical approach to lubrication by solid oxides. *Tribology Letters*, *8*(2), 97–102. doi:10.1023/A:1019183101329

Erdemir, A. (2005). A crystal chemical approach to the formulation of self-lubricating nanocomposite coatings. *Surface and Coatings Technology*, *200*(5–6), 1792–1796. doi:10.1016/j.surfcoat.2005.08.054

Erfanmanesh, M., Shoja-Razavi, R., Abdollah-Pour, H., Mohammadian-Semnani, H., Barekat, M., & Hashemi, S. H. (2019). Friction and wear behavior of laser cladded WC-Co and Ni/WC-Co deposits at high temperature. *International Journal of Refractory Metals & Hard Materials*, *81*, 137–148. doi:10.1016/j.ijrmhm.2019.02.025

Fan, P., & Zhang, G. (2020). Study on process optimization of WC-Co50 cermet composite coating by laser cladding. *International Journal of Refractory Metals & Hard Materials*, *87*, 105133. doi:10.1016/j.ijrmhm.2019.105133

Fang, W., Cho, T. Y., Yoon, J. H., Song, K. O., Hur, S. K., Youn, S. J., & Chun, H. G. (2009). Processing optimization, surface properties and wear behavior of HVOF spraying WC–CrC–Ni coating. *Journal of Materials Processing Technology*, *209*(7), 3561–3567. doi:10.1016/j.jmatprotec.2008.08.024

Federici, M., Menapace, C., Moscatelli, A., Gialanella, S., & Straffelini, G. (2017). Pin-on-disc study of a friction material dry sliding against HVOF coated discs at room temperature and 300 C. *Tribology International*, *115*, 89–99. doi:10.1016/j.triboint.2017.05.030

Firouzeh, A., Ranjbar, K., Lari Baghal, S. M., Heidari Kaidan, A., & Mohemi, E. (2018). Failure assessment of ASTM A213-T12 superheater boiler tubes in a natural gas liquid plant. *Engineering Failure Analysis*, *89*, 15–27. doi:10.1016/j.engfailanal.2018.03.005

Gadow, R., Friedrich, C., Killinger, A., & Floristán, M. (2010). Development of Atmospheric Plasma Sprayed Dielectric Ceramic Coatings for High Efficiency Tubular Ozone Generators. *Journal of Water Resource and Protection*, *2*(9), 799–808. doi:10.4236/jwarp.2010.29094

Ganvir, A., Calinas, R. F., Markocsan, N., Curry, N., & Joshi, S. (2019). Experimental visualization of microstructure evolution during suspension plasma spraying of thermal barrier coatings. *Journal of the European Ceramic Society*, *39*(2), 470–481. doi:10.1016/j.jeurceramsoc.2018.09.023

Geng, Z., Hou, S., Shi, G., Duan, D., & Li, S. (2016). Tribological behaviour at various temperatures of WC-Co coatings prepared using different thermal spraying techniques. *Tribology International*, *104*, 36–44. doi:10.1016/j.triboint.2016.08.025

Geng, Z., Li, S., Duan, D. L., & Liu, Y. (2015). Wear behaviour of WC–Co HVOF coatings at different temperatures in air and argon. *Wear*, *330*, 348–353. doi:10.1016/j.wear.2015.01.035

Gong, T., Yao, P., Zuo, X., Zhang, Z., Xiao, Y., Zhao, L., Zhou, H., Deng, M., Wang, Q., & Zhong, A. (2016). Influence of WC carbide particle size on the microstructure and abrasive wear behavior of WC–10Co–4Cr coatings for aircraft landing gear. *Wear*, *362*, 135–145. doi:10.1016/j.wear.2016.05.022

González, M. A., Rodriguez, E. T., Mojardín, E., Jimenez, O., Guillen, H., & Ibarra, J. (2017). Study of the erosive wear behaviour of cryogenically and tempered WC-CoCr coating deposited by HVOF. *Wear*, *376-377*, 595–607. doi:10.1016/j.wear.2016.12.061

Grujicic, M., Cao, G., & Gersten, B. (2004). Atomic-scale computations of the lattice contribution to thermal conductivity of single-walled carbon nanotubes. *Materials Science and Engineering B*, *107*(2), 204–216. doi:10.1016/j.mseb.2003.11.012

Grujicic, M., Zhao, C. L., Tong, C., DeRosset, W. S., & Helfritch, D. (2004). Analysis of the impact velocity of powder particles in the cold-gas dynamic-spray process. *Materials Science and Engineering A*, *368*(1), 222–230. doi:10.1016/j.msea.2003.10.312

Hardwicke, C. U., & Lau, Y.-C. (2013). Advances in thermal spray coatings for gas turbines and energy generation: A review. *Journal of Thermal Spray Technology*, *22*(5), 564–576. doi:10.100711666-013-9904-0

Hazra, S., Sabiruddin, K., & Bandyopadhyay, P. P. (2012). Plasma and HVOF sprayed WC–Co coatings as hard chrome replacement solution. *Surface Engineering*, *28*(1), 37–43. doi:10.1179/17432944 10Y.0000000009

Hong, S., Wu, Y., Wang, B., Zhang, J., Zheng, Y., & Qiao, L. (2017). The effect of temperature on the dry sliding wear behavior of HVOF sprayed nanostructured WC-CoCr coatings. *Ceramics International*, *43*(1), 458–462. doi:10.1016/j.ceramint.2016.09.180

Hong, S., Wu, Y., Wang, B., Zheng, Y., Gao, W., & Li, G. (2014). High-velocity oxygen-fuel spray parameter optimization of nanostructured WC–10Co–4Cr coatings and sliding wear behavior of the optimized coating. *Materials & Design*, *55*, 286–291. doi:10.1016/j.matdes.2013.10.002

Hou, G.-L., Zhou, H.-D., An, Y.-L., Liu, G., Chen, J.-M., & Chen, J. (2011). Microstructure and high-temperature friction and wear behavior of WC-(W, Cr) 2C-Ni coating prepared by high velocity oxy-fuel spraying. *Surface and Coatings Technology*, *206*(1), 82–94. doi:10.1016/j.surfcoat.2011.06.047

Huang, B., Zhang, C., Zhang, G., & Liao, H. (2019). Wear and corrosion resistant performance of thermal-sprayed Fe-based amorphous coatings: A review. *Surface and Coatings Technology*, *377*, 124896. doi:10.1016/j.surfcoat.2019.124896

Hulka, I., Utu, D., Serban, V. A., Negrea, P., Lukáč, F., & Chráska, T. (2020). Effect of Ti addition on microstructure and corrosion properties of laser cladded WC-Co/NiCrBSi (Ti) coatings. *Applied Surface Science*, *504*, 144349. doi:10.1016/j.apsusc.2019.144349

Ilieva, G. I. (2016). Erosion failure mechanisms in turbine stage with twisted rotor blade. *Engineering Failure Analysis*, *70*, 90–104. doi:10.1016/j.engfailanal.2016.07.008

Ishikawa, Y., Kuroda, S., Kawakita, J., Sakamoto, Y., & Takaya, M. (2007). Sliding wear properties of HVOF sprayed WC–20% Cr_3C_2–7% Ni cermet coatings. *Surface and Coatings Technology*, *201*(8), 4718–4727. doi:10.1016/j.surfcoat.2006.10.007

Jafari, M., Han, J.-C., Seol, J.-B., & Park, C.-G. (2017). Tribological properties of HVOF-sprayed WC-Co coatings deposited from Ni-plated powders at elevated temperature. *Surface and Coatings Technology*, *327*, 48–58. doi:10.1016/j.surfcoat.2017.08.026

Jhi, S.-H., Ihm, J., Louie, S. G., & Cohen, M. L. (1999). Electronic mechanism of hardness enhancement in transition-metal carbonitrides. *Nature*, *399*(6732), 132–134. doi:10.1038/20148

Jones, M., Cockburn, A., Sparkes, M., O'Niell, W., & Lupoi, R. (2014). Supersonic Laser Deposition of Tungsten. *International Manufacturing Science and Engineering Conference, 45806*, V001T03A022.

Kazempour-Liacy, H., Abouali, S., & Akbari-Garakani, M. (2011). Failure analysis of a first stage gas turbine blade. *Engineering Failure Analysis*, *18*(1), 517–522. doi:10.1016/j.engfailanal.2010.09.040

Kumar, D. D., Kumar, N., Kalaiselvam, S., Radhika, R., Rabel, A. M., & Jayavel, R. (2017). Tribo-mechanical properties of reactive magnetron sputtered transition metal carbide coatings. *Tribology International*, *114*, 234–244. doi:10.1016/j.triboint.2017.04.031

Kumar, S., Kumar, M., & Handa, A. (2018). Combating hot corrosion of boiler tubes–A study. *Engineering Failure Analysis*, *94*, 379–395. doi:10.1016/j.engfailanal.2018.08.004

Kumar, V., & Balasubramanian, K. (2016). Progress update on failure mechanisms of advanced thermal barrier coatings: A review. *Progress in Organic Coatings*, *90*, 54–82. doi:10.1016/j.porgcoat.2015.09.019

Laguna-Camacho, J. R., Villagrán-Villegas, L. Y., Martínez-García, H., Juárez-Morales, G., Cruz-Orduña, M. I., Vite-Torres, M., Ríos-Velasco, L., & Hernández-Romero, I. (2016). A study of the wear damage on gas turbine blades. *Engineering Failure Analysis*, *61*, 88–99. doi:10.1016/j.engfailanal.2015.10.002

Liao, H., Normand, B., & Coddet, C. (2000). Influence of coating microstructure on the abrasive wear resistance of WC/Co cermet coatings. *Surface and Coatings Technology*, *124*(2), 235–242. doi:10.1016/S0257-8972(99)00653-2

Lima, R. S., Karthikeyan, J., Kay, C. M., Lindemann, J., & Berndt, C. C. (2002). Microstructural characteristics of cold-sprayed nanostructured WC–Co coatings. *Thin Solid Films*, *416*(1–2), 129–135. doi:10.1016/S0040-6090(02)00631-4

Liu, S., Sun, D., Fan, Z., Yu, H., & Meng, H. (2008). The influence of HVAF powder feedstock characteristics on the sliding wear behaviour of WC–NiCr coatings. *Surface and Coatings Technology*, *202*(20), 4893–4900. doi:10.1016/j.surfcoat.2008.03.014

Luo, F., Cockburn, A., Sparkes, M., Lupoi, R., Chen, Z., O'Neill, W., Yao, J., & Liu, R. (2015). Performance characterization of Ni60-WC coating on steel processed with supersonic laser deposition. *Defence Technology*, *11*(1), 35–47. doi:10.1016/j.dt.2014.09.003

Maev, R. G., & Leshchynsky, V. (2006). Air gas dynamic spraying of powder mixtures: Theory and application. *Journal of Thermal Spray Technology*, *15*(2), 198–205. doi:10.1361/105996306X108048

Mathapati, M., Ramesh, M. R., & Doddamani, M. (2017). High temperature erosion behavior of plasma sprayed NiCrAlY/WC-Co/cenosphere coating. *Surface and Coatings Technology*, *325*, 98–106. doi:10.1016/j.surfcoat.2017.06.033

Mayrhofer, E., Janka, L., Mayr, W. P., Norpoth, J., Ripoll, M. R., & Gröschl, M. (2015). Cracking resistance of Cr_3C_2–NiCr and WC–Cr_3C_2–Ni thermally sprayed coatings under tensile bending stress. *Surface and Coatings Technology*, *281*, 169–175. doi:10.1016/j.surfcoat.2015.09.002

Meekhanthong, K., & Wirojanupatump, S. (2014). Characterization and comparison of thermally sprayed hard coatings as alternative to hard chrome plating. *Advanced Materials Research*, *974*, 183–187. doi:10.4028/www.scientific.net/AMR.974.183

Mi, P., Zhao, H., Wang, T., & Ye, F. (2018). Sliding wear behavior of HVOF sprayed WC-(nano-WC-Co) coating at elevated temperatures. *Materials Chemistry and Physics*, *206*, 1–6. doi:10.1016/j.matchemphys.2017.09.066

Mostajeran, A., Shoja-Razavi, R., Hadi, M., Erfanmanesh, M., Barekat, M., & Firouzabadi, M. S. (2020). Evaluation of the mechanical properties of WC-FeAl composite coating fabricated by laser cladding method. *International Journal of Refractory Metals & Hard Materials*, *88*, 105199. doi:10.1016/j.ijrmhm.2020.105199

Murthy, J. K. N., & Venkataraman, B. (2006). Abrasive wear behaviour of WC–CoCr and Cr_3C_2–20 (NiCr) deposited by HVOF and detonation spray processes. *Surface and Coatings Technology*, *200*(8), 2642–2652. doi:10.1016/j.surfcoat.2004.10.136

Myalska, H., Lusvarghi, L., Bolelli, G., Sassatelli, P., & Moskal, G. (2019). Tribological behavior of WC-Co HVAF-sprayed composite coatings modified by nano-sized TiC addition. *Surface and Coatings Technology*, *371*, 401–416. doi:10.1016/j.surfcoat.2018.09.017

Nieto, A., Kim, J., Penkov, O. V., Kim, D.-E., & Schoenung, J. M. (2017). Elevated temperature wear behavior of thermally sprayed WC-Co/nanodiamond composite coatings. *Surface and Coatings Technology*, *315*, 283–293. doi:10.1016/j.surfcoat.2017.02.048

Ozgurluk, Y., Doleker, K. M., & Karaoglanli, A. C. (2018). Hot corrosion behavior of YSZ, $Gd_2Zr_2O_7$ and YSZ/$Gd_2Zr_2O_7$ thermal barrier coatings exposed to molten sulfate and vanadate salt. *Applied Surface Science*, *438*, 96–113. doi:10.1016/j.apsusc.2017.09.047

Pakseresht, A. H., Rahimipour, M. R., Vaezi, M. R., & Salehi, M. (2015). Effect of splat morphology on the microstructure and dielectric properties of plasma sprayed barium titanate films. *Applied Surface Science*, *324*, 797–806. doi:10.1016/j.apsusc.2014.11.041

Pakseresht, A. H., Rahimipour, M. R., Vaezi, M. R., & Salehi, M. (2016). Thermal plasma spheroidization and spray deposition of barium titanate powder and characterization of the plasma sprayable powder. *Materials Chemistry and Physics*, *173*, 395–403. doi:10.1016/j.matchemphys.2016.02.028

Pakseresht, A. H., Saremi, M., Omidvar, H., & Alizadeh, M. (2019). Micro-structural study and wear resistance of thermal barrier coating reinforced by alumina whisker. *Surface and Coatings Technology*, *366*, 338–348. doi:10.1016/j.surfcoat.2019.03.059

Papyrin, A. (2001). Cold spray technology. *Advanced Materials & Processes*, *159*(9), 49.

Podgornik, B., Vižintin, J., Jacobson, S., & Hogmark, S. (2004). Tribological behaviour of WC/C coatings operating under different lubrication regimes. *Surface and Coatings Technology*, *177–178*, 558–565. doi:10.1016/S0257-8972(03)00927-7

Quint, M., & Kopech, H. (1999). High Energy Plasma Ceramic Coating Optimization. In *23rd Annual Conference on Composites, Advanced Ceramics, Materials, and Structures: B: Ceramic Engineering and Science Proceedings* (pp. 335–345). Academic Press.

Rathod, S. W., Khanna, A. S., Karthikeyan, J., & Rathod, C. R. (2014). Effect of N_2 and He Carrier Gases on Oxidation Behavior of Cold Sprayed CoNiCrAlY Powder to Deposit Bond Coats. *Transactions of the Indian Institute of Metals*, *67*(2), 247–262. doi:10.100712666-013-0344-9

Rathod, W. S., Khanna, A. S., Rathod, R. C., & Sapate, S. G. (2014). Wear and Corrosion Behavior of CoNiCrAlY Bond Coats. *Journal of The Institution of Engineers (India): Series C, 95*(3), 261–271.

Rovatti, L., Lecis, N., Dellasega, D., Russo, V., & Gariboldi, E. (2018). Influence of aging in the temperature range 250–350° C on the tribological performance of a WC-CoCr coating produced by HVOF. *International Journal of Refractory Metals & Hard Materials*, *75*, 218–224. doi:10.1016/j.ijrmhm.2018.04.017

Santecchia, E., Hamouda, A. M. S., Musharavati, F., Zalnezhad, E., Cabibbo, M., & Spigarelli, S. (2015). Wear resistance investigation of titanium nitride-based coatings. *Ceramics International*, *41*(9), 10349–10379. doi:10.1016/j.ceramint.2015.04.152

Sapate, S. G., & Manish, R. (2015). Solid Particle Erosion of Thermal Sprayed Coatings. In M. Roy & J. P. Davim (Eds.), *Thermal Sprayed Coatings and their Tribological Performances* (pp. 193–226). IGI Global. doi:10.4018/978-1-4666-7489-9.ch007

Suresh Babu, P., Chanikya Rao, P., Jyothirmayi, A., Sudharshan Phani, P., Rama Krishna, L., & Srinivasa Rao, D. (2018). Evaluation of microstructure, property and performance of detonation sprayed WC-(W,Cr)2C-Ni coatings. *Surface and Coatings Technology*, *335*, 345–354. doi:10.1016/j.surfcoat.2017.12.055

Suresh Babu, P., Srinivasa Rao, D., Rama Krishna, L., & Sundararajan, G. (2017). Weibull analysis of hardness distribution in detonation sprayed nano-structured WC-12Co coatings. *Surface and Coatings Technology*, *319*, 394–402. doi:10.1016/j.surfcoat.2017.04.028

Tyagi, A., Walia, R. S., Murtaza, Q., Pandey, S. M., Tyagi, P. K., & Bajaj, B. (2019). A critical review of diamond like carbon coating for wear resistance applications. *International Journal of Refractory Metals & Hard Materials*, *78*, 107–122. doi:10.1016/j.ijrmhm.2018.09.006

Ulianitsky, V., Batraev, I., Dudina, D., & Smurov, I. (2017). Enhancing the properties of WC/Co detonation coatings using two-component fuels. *Surface and Coatings Technology*, *318*, 244–249. doi:10.1016/j.surfcoat.2016.08.008

Usmani, S., Sampath, S., Houck, D. L., & Lee, D. (1997). Effect of carbide grain size on the sliding and abrasive wear behavior of thermally sprayed WC-Co coatings. *Tribology Transactions*, *40*(3), 470–478. doi:10.1080/10402009708983682

Van Steenkiste, T. H., Smith, J. R., Teets, R. E., Moleski, J. J., Gorkiewicz, D. W., Tison, R. P., Marantz, D. R., Kowalsky, K. A., Riggs, W. L. II, Zajchowski, P. H., Pilsner, B., McCune, R. C., & Barnett, K. J. (1999). Kinetic spray coatings. *Surface and Coatings Technology*, *111*(1), 62–71. doi:10.1016/S0257-8972(98)00709-9

Vardelle, A., Moreau, C., Akedo, J., Ashrafizadeh, H., Berndt, C. C., Berghaus, J. O., Boulos, M., Brogan, J., Bourtsalas, A. C., Dolatabadi, A., Dorfman, M., Eden, T. J., Fauchais, P., Fisher, G., Gaertner, F., Gindrat, M., Henne, R., Hyland, M., Irissou, E., ... Vuoristo, P. (2016). The 2016 Thermal Spray Roadmap. *Journal of Thermal Spray Technology*, *25*(8), 1376–1440. doi:10.100711666-016-0473-x

Vashishtha, N., Khatirkar, R. K., & Sapate, S. G. (2017). Tribological behaviour of HVOF sprayed WC-12Co, WC-10Co-4Cr and Cr3C2–25NiCr coatings. *Tribology International*, *105*, 55–68. doi:10.1016/j.triboint.2016.09.025

Verstak, A., Andrew, A., & Baranovski. (2004). Deposition of carbides by Activated Combustion HVAF spraying. *Proceedings of the International Thermal Spray Conference*, 551–555.

Verstak, A., & Baranovski, V. (2003). Activated Combustion HVAF Coatings for Protection against Wear and High Temperature Corrosion. In B. R. Marple & C. Moreau (Eds.), *ASM International, Volunme 1* (pp. 535–541).

Vicenzi, J., Villanova, D. L., Lima, M. D., Takimi, A. S., Marques, C. M., & Bergmann, C. P. (2006). HVOF-coatings against high temperature erosion (~300 °C) by coal fly ash in thermoelectric power plant. *Materials & Design*, *27*(3), 236–242. doi:10.1016/j.matdes.2004.10.008

Vinayo, M. E., Kassabji, F., Guyonnet, J., & Fauchais, P. (1985). Plasma sprayed WC–Co coatings: Influence of spray conditions (atmospheric and low pressure plasma spraying) on the crystal structure, porosity, and hardness. *Journal of Vacuum Science & Technology. A, Vacuum, Surfaces, and Films*, *3*(6), 2483–2489. doi:10.1116/1.572863

Vuoristo, P. (2014). Comprehensive materials processing, 1st edition Volume 4: Coatings and films. In D. Cameron (Ed.), Comprehensive materials processing, Volume 4: Coatings and films (4th ed., pp. 229–276). Academic Press.

Wang, C. B., Wang, D. L., Chen, W. X., & Wang, Y. Y. (2002). Tribological properties of nanostructured WC/CoNi and WC/CoNiP coatings produced by electro-deposition. *Wear*, *253*(5–6), 563–571. doi:10.1016/S0043-1648(02)00173-4

Wang, T., & Ye, F. (2018). The elevated-temperature wear behavior evolution of HVOF sprayed tungsten carbide coatings: Respond to heat treatment. *International Journal of Refractory Metals & Hard Materials*, *71*, 92–100. doi:10.1016/j.ijrmhm.2017.11.007

Wang, X., Guo, L., Peng, H., Zheng, L., Guo, H., & Gong, S. (2015). Hot-corrosion behavior of a $La_2Ce_2O_7$/YSZ thermal barrier coating exposed to Na_2SO_{4+} V_2O_5 or V_2O_5 salt at 900° C. *Ceramics International*, *41*(5), 6604–6609. doi:10.1016/j.ceramint.2015.01.107

Wänstrand, O., Larsson, M., & Hedenqvist, P. (1999). Mechanical and tribological evaluation of PVD WC/C coatings. *Surface and Coatings Technology*, *111*(2–3), 247–254. doi:10.1016/S0257-8972(98)00821-4

Wu, D., Liu, Y., Li, D., Zhao, X., & Liu, Y. (2016). Tribo-corrosion properties of WC-10Co-4Cr coating in natural silt-laden waters when sliding against Si3N4. *International Journal of Refractory Metals & Hard Materials*, *58*, 143–151. doi:10.1016/j.ijrmhm.2016.04.019

Xie, M., Zhang, S., & Li, M. (2013). Comparative investigation on HVOF sprayed carbide-based coatings. *Applied Surface Science*, *273*, 799–805. doi:10.1016/j.apsusc.2013.03.010

Yang, Q., Senda, T., & Hirose, A. (2006). Sliding wear behavior of WC–12% Co coatings at elevated temperatures. *Surface and Coatings Technology*, *200*(14–15), 4208–4212. doi:10.1016/j.surfcoat.2004.12.032

Yao, J., Zhang, J., Wu, G., Wang, L., Zhang, Q., & Liu, R. (2018). Microstructure and wear resistance of laser cladded composite coatings prepared from pre-alloyed WC-NiCrMo powder with different laser spots. *Optics & Laser Technology*, *101*, 520–530. doi:10.1016/j.optlastec.2017.12.007

Young, E. J., Mateeva, E., Moore, J. J., Mishra, B., & Loch, M. (2000). Low pressure plasma spray coatings. *Thin Solid Films*, *377–378*, 788–792. doi:10.1016/S0040-6090(00)01452-8

Yuan, K., Peng, R. L., Li, X.-H., Talus, A., Johansson, S., & Wang, Y.-D. (2015). Hot corrosion of MCrAlY coatings in sulphate and SO2 environment at 900 C: Is SO2 necessarily bad? *Surface and Coatings Technology*, *261*, 41–53. doi:10.1016/j.surfcoat.2014.11.065

Zhang, W., Liu, L., Zhang, M., Huang, G., Liang, J., Xian, L. I., & Zhang, L. (2015). Comparison between WC–10Co–4Cr and Cr_3C_2–25NiCr coatings sprayed on H13 steel by HVOF. *Transactions of Nonferrous Metals Society of China*, *25*(11), 3700–3707. doi:10.1016/S1003-6326(15)64011-0

Zhao, X.-Q., Zhou, H.-D., & Chen, J.-M. (2006). Comparative study of the friction and wear behavior of plasma sprayed conventional and nanostructured WC–12% Co coatings on stainless steel. *Materials Science and Engineering A*, *431*(1–2), 290–297. doi:10.1016/j.msea.2006.06.009

Zheng, Y. (2013). Plating hard chrome plating alternative technologies-HVOF tungsten carbide coating. *Advanced Materials Research*, *712*, 395–398. doi:10.4028/www.scientific.net/AMR.712-715.395

KEY TERMS AND DEFINITIONS

Cold Spray Coatings: The protective layer of powder material is deposited on the substrate material under kinetic energy causing powder particles to undergo plastic deformation.

Corrosion Resistance: The ability of a material to resist the loss of material from the surface under the chemical and electrochemical reactions taking place in actual working environments.

Erosion: Erosion is the process of material removal from the surface due to the striking of inertia of small solid particles.

Surface Engineering: The modification techniques used in improvement in surface properties of metallic and non-metallic components and their effects in minimisation of failures.

Thermal Spray Coatings: The protective layer of powder material is deposited on the substrate material by virtue of thermal energy causing powder particles to melt fully or partially.

Tribology: The science related to friction, wear, and lubrication collectively termed tribology.

Wear Resistance: The ability of a material to resist the loss of material from the surface under the various mechanisms taking place in actual working environments.

Chapter 5
Recent Advances in Design and Fabrication of Wear Resistant Materials and Coatings:
Surface Modification Techniques

Santosh Kumar
https://orcid.org/0000-0003-4414-3305
Department of Mechanical Engineering, Chandigarh Group of Colleges, Mohali, India

Rakesh Kumar
Department of Regulatory Affair, Auxein Medical Private Limited, Sonipat, India

ABSTRACT

In recent years, the demand of wear resistance material and coating is increasing very rapidly as it reduces substantial energy losses resulting from wear and friction. To overcome these energy losses, surface engineering is employed. Surface engineering is the process of coating or modifying the surface of part to minimize wear, friction, corrosion, as well as to enhance the lifespan of machine components and reduce the manufacturing cost. Recently, numerous coating methods are available for distinct material (pure metals to alloys, carbides, composites, and ceramics) applications. Hence, this chapter provides an overview on the prevention of tribo-surfaces through distinct methods of surface modification such as thermal, physical, and chemical methods of coating. Further, distinct coating properties, applications, future scope, and challenges are described.

INTRODUCTION

Tribology is the study of wear, friction, and the lubrication of interacting surfaces. Various techniques such as laser processing, mechanical processing, plasma processing, electro processing, chemical processing, thermal spraying, ion implantation, and gas nitriding are employed to enhance the resistance against the wear of components. However, coatings obtained by thermal spraying is gaining more in-

DOI: 10.4018/978-1-7998-9683-8.ch005

terest. These coatings processes provide various benefits by combining versatility, cost effectiveness, and the ability to coat wide range of coating materials on different substrate. In addition, wide range of coating thickness (from tens to hundreds of microns) can be achieved on parts having complex geometry by using thermal sprayed processes (Joshi, 2019). Thermal spraying is a group of coating deposition techniques that includes flame spray (FS), high-velocity oxy or air-fuel spray (HVOF, HVAF), vacuum and atmospheric plasma spray (VPS, APS), electric arc spray also known as twin wire arc spray, wire or powder flame spray, detonation gun spray (D-Gun), high velocity oxygen liquid fuel (HVOLF), cold gas dynamic spray (CS), suspension and solution spraying etc. These coating deposition methods aim to enhance the desired surface performance/life by improving the distinct properties including tribological and mechanical properties. Then, these techniques are highly used for retaining superior mechanical properties, resistance against corrosion, resistance against wear, oxidation resistance, repair of machine parts, obtaining high thickness of the coating, retain bond strength with a greater coefficient of thermal expansion, It also exhibits electrical conductivity, low friction, sacrificial wear, insulation, lubricity, chemical resistance, and many other desirable surface properties (Garcia et al., 2018; Liu et al., 2019). The wide variety of non-metallic or metallic materials utilization which are extremely high melting point, recoat worn part without changing surface properties/size, enhance aesthetic properties etc. are main key merits of thermal spray processes (Joshi, 2019). Consequently, thermal spray technologies are commonly used in distinct industrial areas such as aeronautics and defense, paper production, power generation, automotive, marine and mining, electronics and semiconductor, steel and iron, textile, medical among others.

Despite that in the starting of the twenty century the Max Ulrich Schoop engineer write a patent that involves the thermal spray, thermal spraying process s development started in the decade of 1980s with the flame spray as the base of other techniques. In 1882 the Oerlikon published a first patient which consists of spraying lead powder against a fixed surface. Thereafter, in 1909 Max. Ulrich Schoop work on a patent called metal casting. In 1912 practical flame spray (wire gun) was introduced and in plasma spray coating method with hybrid feedstock was developed in 2019.Schoop's group also developed the electric arc spray technique using a metal with a low melting point as feedstock material for the coating deposition to protect metal surfaces from corrosion..During the same decade the 3rd family of TS techniques known as cold spray (solid-state) was introduced. that does not melt the feedstock material (Kumar et al.,2020(a); Kumar et al., 2020 (b); Kumar et al. (2020 (c)). Although, the temperature of the cold spray process is much lower than other thermal spray processes, which can effectively avoid phase transition, thermal cracking, oxidation, and other problems caused by the high temperature (**Figure 1**).

Figure 1. Flame temperature and particle velocity profile of cold spray and other thermal spray technology

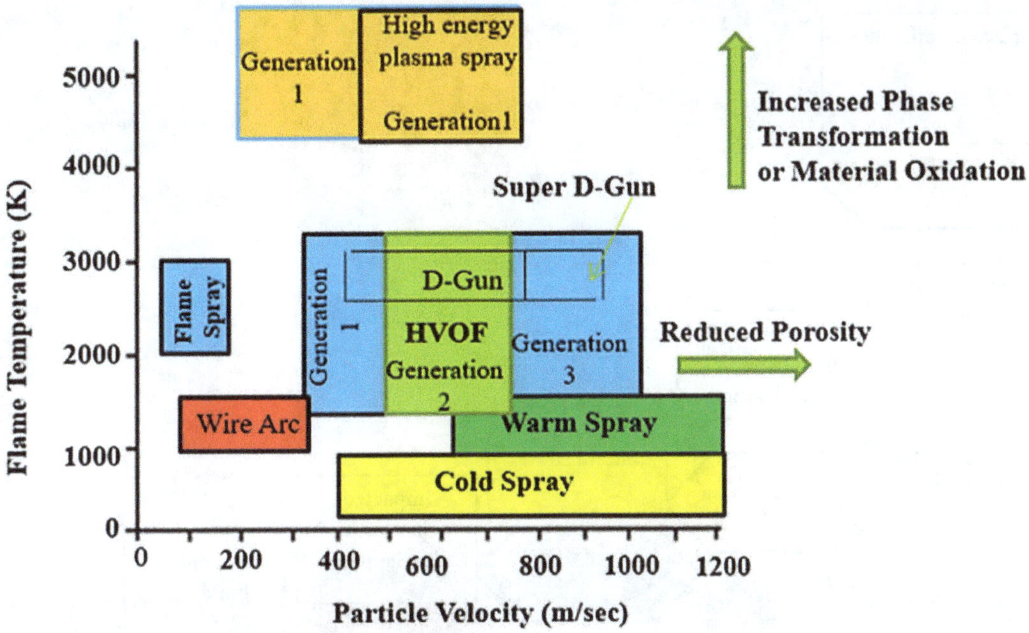

"During CS process, the flying particles impact the surface of the substrate at high speed ranging from 300 to 1200m/s. Thereafter, undergo severe plastic deformation to develop a coating at a completely solid state. These thermal spray processes uses distinct types of spray materials (pure metal, metals alloys, carbides, ceramics, polymers etc. Further, these processes are suitable for many industrial applications as it increases wear resistance, adhesion strength, hardness, corrosion resistance etc. These processes offers high production rate. All these processes, utilize powder, suspensions, wire, rod shaped coating materials for the development of coating. The selection of coating materials is mainly depends upon the equipment of particular thermal spray process (Kumar et al., 2020(c)). Although the coating material has a low coefficient of efficiency, dust/fume developed during the coating process causes an adverse effect on human health.

The principle work of thermal spraying is the projection of particles with enhanced thermal and kinetic energy to be deposited and attached to the desired surface. The overlapping of these particles built up the coatings (Kumar et al., 2021). The common step of the thermal spraying process is illustrated in **Figure 2.**

Figure 2. Thermal spray coatings steps

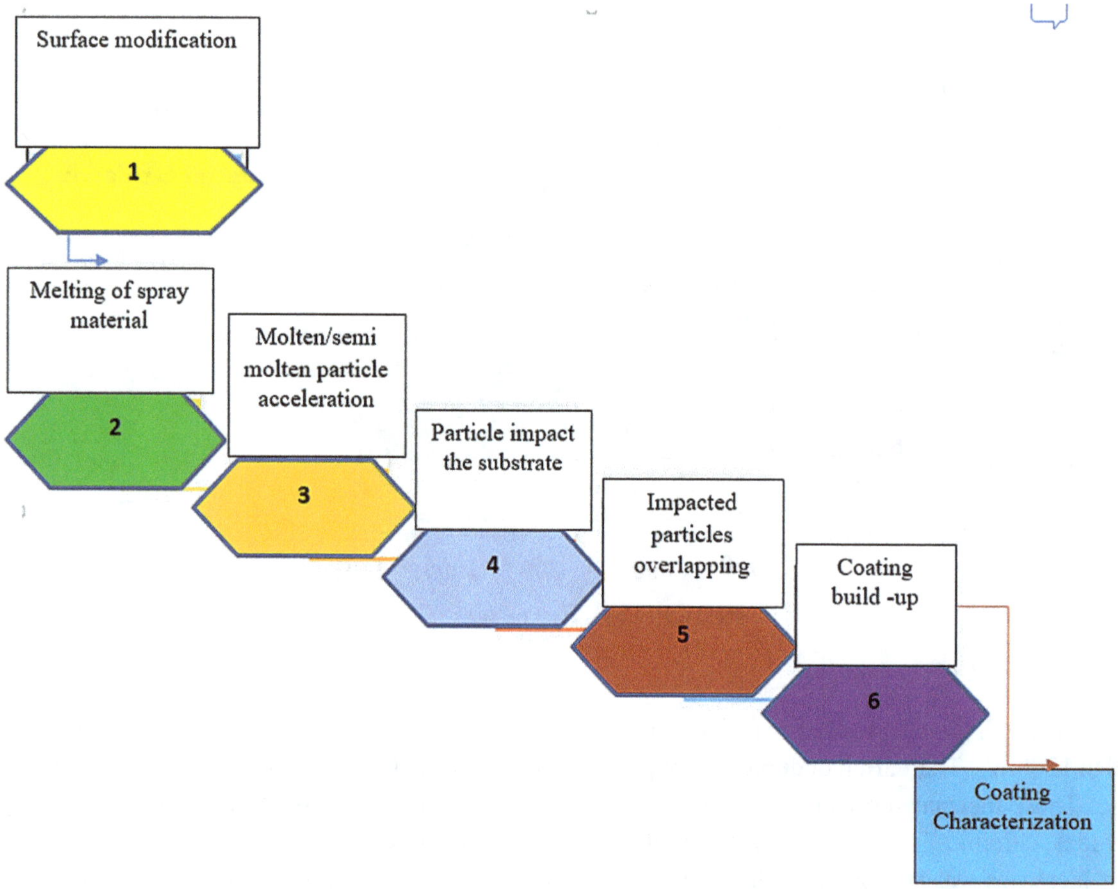

The prospects shows that coating fabrication by thermal spray methods has come a long way and has yet much more to exhibit over the coming fifteen years. Nowdays, the reason for increasing the coating demand in a material is related to the use of materials on corrosive atmospheres, metal removal difficulties, requirement of biological properties, abrasive wear, the cost factor and high temperatures application among others(Kumar, et al., 2021). Overall, the aim of this chapter is to provide an overview of distinct surface modification techniques (thermal spray coating techniques, including physical & chemical methods of coatings surface). Further, the application of surface modification methods in distinct sector is described.

BACKGROUND

Coating is covering/protection that is applied to the material surface so that the part functionally performs as needed. In addition, coatings protect the material/part by enhancing its lifespan or to make it decorative (Howarth, et al., 1997). An important consideration for coating deposition processes is to control its thickness. Based on this, coatings deposition processes can be classified on the obtained thickness as shown in **Figure 3**

Figure 3. Types of coating layers

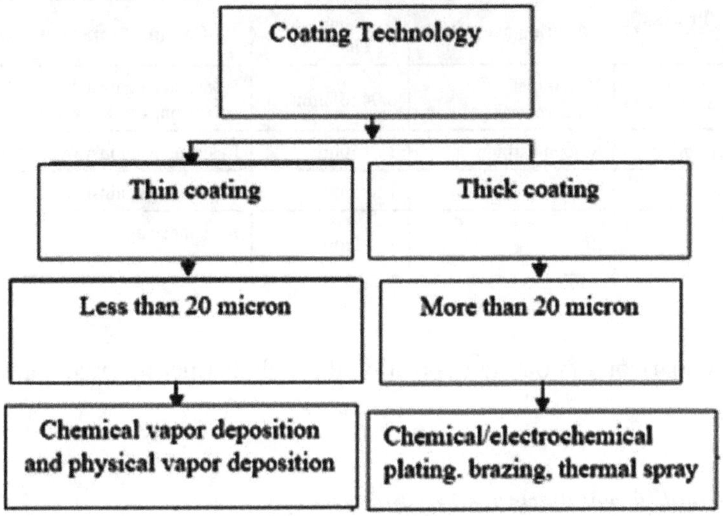

"The coating technologies can be categorized into thin films and thick films. The thin films (thickness <20μm) provide excellent improvement of surface characteristics. The examples of thin film coatings are chemical vapour deposition and physical vapour deposition processes. Both process exhibits surfaces with unparalleled corrosion resistance or hardness. Although, most of the thin film technologies need a decreased pressure environment &, hence are more costly. Thus, impose a limit on the shape and size of the substrate. On the other hands, thick films have a thickness of 30μm to several millimetres. Thick films are needed when the functional performance is based upon the layer thickness. For examples in thermal barrier coatings, when strong oxidation, corrosion, erosion conditions results in wear & the part life based upon the layer thickness (Fauchais et al., 2014).

.However, when the coating layer is applied less from the recommended thickness range the following issue can be faced such as sample visibility, pinpoint rusting, cracking, or brittleness of coating (decreased coating flexibility result in decreased coating layer cohesive strength which in turn create a cause of delamination or cracking. In recent days, many types of thermal spray processes and feedstock materials are available. However, the most important difference in the properties, temperature and thickness has been achieved. These parameters help the researcher for the selection of coating as per the field of application (Fauchais et al., 2014)..

Comparison of District Coating Properties

Recently several types of coating processes and materials are applied surfaces for its protection. Nonetheless the selection of a proper combination as per the field of application is the major threat for researchs, because it requires a considerable literature study & the knowledge and specialists are mainly required *(Billard, et al. 2018)*. The example of some coating processes, thickness, the material used, and properties are represented in **Table 1.**

Table 1. Example of distinct coating methods, materials, and their properties

S. N.	Type of Coating Process Method	Coating Material	Coating Thickness	Coating Properties	Reference
1	Thermal Spray	Metallic alloy & ceramics	0.04 to 3mm	Resistance against corrosion, good aesthetics	*(Pratap, et al., 2015)*
2	Chemical vapor deposition	Silicon carbide	1 to 50μm	Resistance against wear	
3	Physical vapor deposition	Titanium (C, N)	1 to 5μm	Resistance against wear	
4	Galvanize	Zn	1 to 5μm	Resistance against corrosion	

In addition, the comparison of coating deposition through distinct thermal spray process method is illustrated in **Table 2.**

Table 2. Comparison of different thermal spray processes

Parameters		**Thermal Spray Processes**						**References**
		Flame Spray	**Wire Arc Spray**	**D-Gun Spray**	**HVOF Spray**	**Plasma Spray**	**Cold Spray**	
Materials used	-	-	Metals/ cermets (cored wires)		Low melt ceramic& Metals/cermets	Metals, Ceramics	Metals	*(Pratap et al., 2015, Kahar et al., 2020; Smith, 2007)*
Coating parameters	Porosity (%)	10 -15	10 -20	< 2	< 2	5 – 10	< 5	
	Thickness (mm)	0.1-2.5	0.1-2.0	0.1-0.4	0.05-2.0	0.1-1.5	0.25-0.6	
	Spray rate (kg/hour)	7.0	16	1	14	5	8	
	Relative Cost	3	1	10	5	4.5	-	

Finally, the comparison of distinct coating processes coating thickness Vs coating temperature is shown in **Figure 4.**

Figure 4. Ccomparison of coating process in terms of coating thickness w.r.t. coating temperature of different coating processes

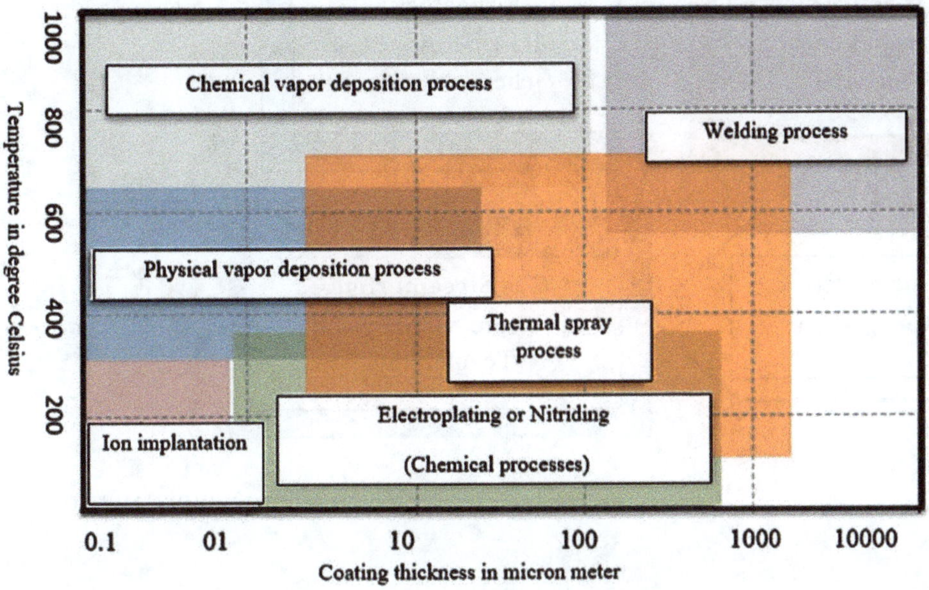

The above figure indicates the comparison between distinct coating processes in terms of coating thickness (0.1µm to 10nm) and substrate temperature (0°C to 1000°C). However, in case of physical vapour deposition process, the primary route for innovation is the development of new power supplies leading to metallic vapour with a high ionization rate that result in specific characteristics. For example the control of interfaces and microstructure. However, in case of chemical vapour deposition the appropriate selection of metalorganic precursors permits us 1st to decrease the deposition temperatures, making the process compatible with heat sensitive substrate, & 2nd to access the total diversity of chemistry in terms of compounds and compositions. Moreover, in case of thermal spray technologies, it is very much clear that the access to innovate & on demand powders constitutes one of the primary vectors of technological rupture. It is for example, the case with the possibility to develop gradient coatings or via a controlled implementation of nanopowders utilize through the form of aggregates or suspension (Billard, et al. 2018).

THERMAL SPRAY COATINGS TECHNIQUES

It is a group of surface modifications or distinct coating process methods where metals, cermet, ceramic and polymeric materials are fed in the form of wire, powder or rod form to a torch where the materials are melted or heated, & accelerated in the gas stream towards a substrate where they form a splat. Thus coating by thermal spray methods has gained extreme interest because of some major key advantages such as less operating cost, versatility, and being able to apply many types of material on any surface/ workpiece (Robert, et al., 1994; Pintilei, *2008; (Kumar, et al. 2019)*

The schematics illustration of the thermal spray process concept or principle is shown in **Figure 5.**

Figure 5. General principle of thermal spray coating processes

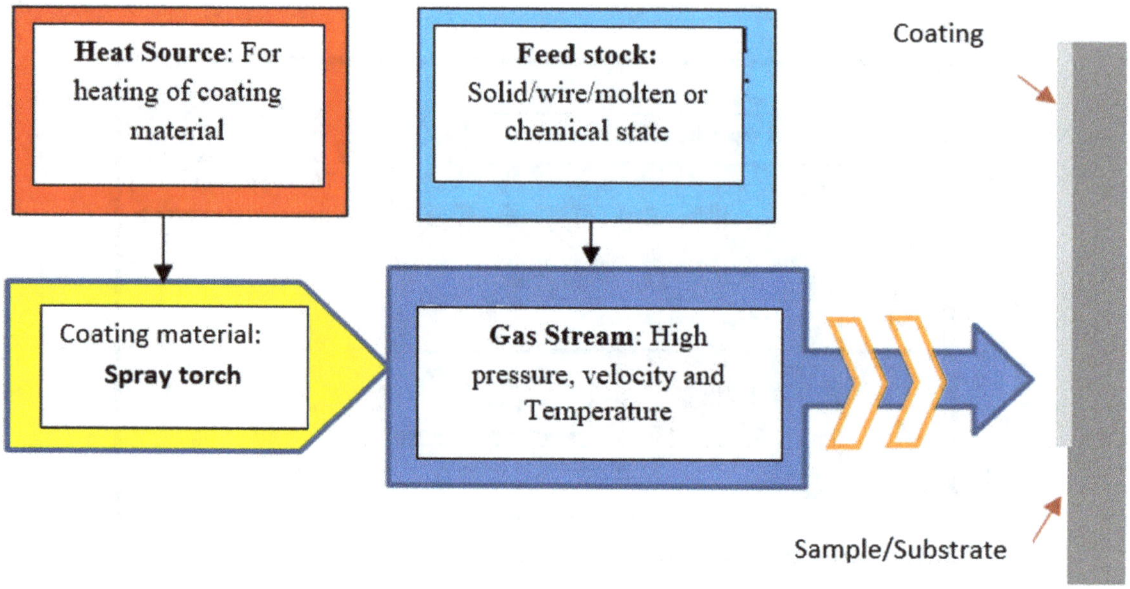

However, based on the nature of the applied energy, the thermal spray technology is classified into the following categories as shown in Figure 6 *(Ang et al., 2013).*

Figure 6. Classification of thermal spray technology

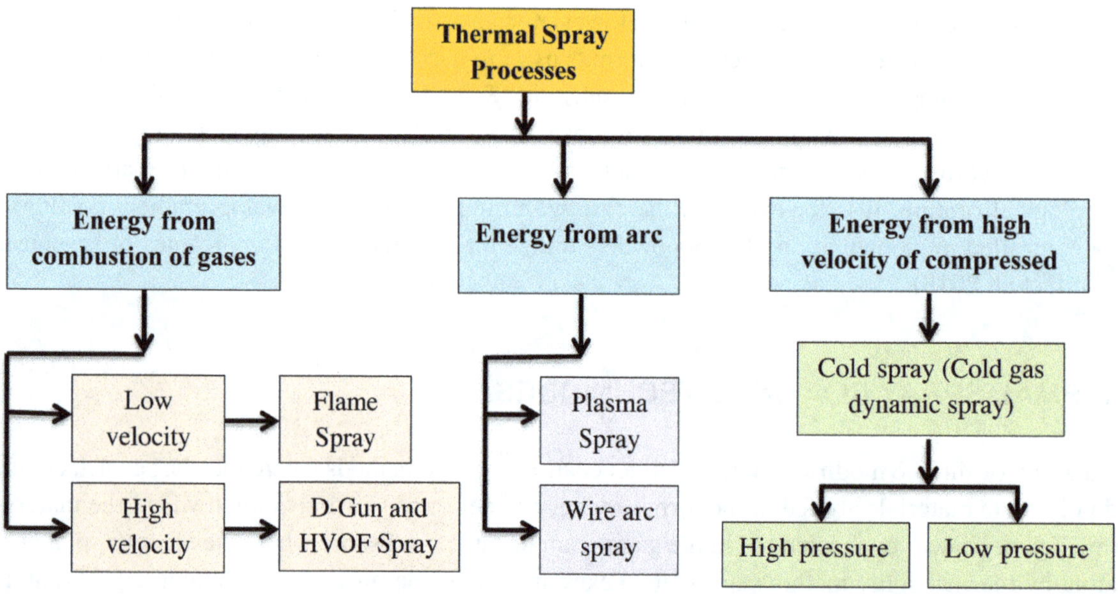

Coatings obtained by thermal spray can exhibit a thick feature onto the substrate area with high deposition rate than other coating methods. In addition, the porosity measurement helps to measure the coating quality. In the comparison study of spray method, some process have exhibited higher bond strength (Plasma, HVOF, Wire arc spray, D-gun), and some of them extremely low (flame spray). The oxide content, resistance against corrosion, resistance against wear, porosity level, etc. are depended upon the spray process, parameters used during coating fabrication leading to improve the coating properties (Talib et al, 2003).

In addition, the major key points of coating requirement after coating deposited by TS is that no/ very less machining needed for coated components (Thakur and Vasudev, 2021). Further, coatings are required for 3 main reasons (a) Cost mitigration by enhancing functionality (b) to enhance functional performance and (c) for enhancement of component value by corrosion, wear reduction.

The distinct thermal spray processes are described in the section below.

Flame Spray

In 1917, the flame spray or oxy/acetylene combustion spray process method is first invented by Guenther and Schoop. The flame spray process used combustible gas for melting of coating material that is available either rod form or wire form or powder form (Oksa, et al., 2011). . Most flame spray gun can be adapted to utilize numerous combinations of gases to balance coating characteristics and operating cost. Propane, acetylene, hydrogen along with O_2 are generally used flame spray gases. This coating method has less cost, ease to use, coating possibility of complicated shape, less maintenance, less equipment cost, and high coating deposition rates than other thermal spray processes. However, it has some limitations like higher coating porosity, low bond strengths. The main use of coatings deposited by FS are related to application of this coating for corrosion protection in case of structure, self-fluxing hard facing alloys, and to increase the protection against molten glasses at 700°C. The merits of the spray process lie in its portability, fewer costs, and versatility characteristics w.r.t. coating materials, temperature control (Singh, et al., 2021; Amin & Panchal, 2016; Thorpe & Merle, 1993). . During FS, the molten material is propelled from the nozzle tip onto the substrate surface by the application of compressed air injection. In other words, during FS process a mixture of O_2 and fuel (propane, acetylene or hydrogen) is injected. Further, the gas mixture is combusted in the front of the nozzle to develop a flame external to the nozzle. Thus, based upon the ratio of O_2 to fuel, the temperature of the flame varies from 3000°C to 3300°C. The coating properties are find out by, rate of air and gas flow, the temperature of flame, feed rate, the pressure of the gas etc. In addition the deposition rate of the coating is mainly depends upon the material used as well as coating thickness varies from 50μm to 1000 μm. In addition, any material that can be melted at the combustion temperature can be deposited by FS process (Bose, 2007).

Wire Flame Spray Method

This process used wire as a coating material (metals and alloys) which is melted with the application of flame/electric arc. The flame temperature and velocity of the particles are 3000°C and less than 300m/ sec respectively to deposit a material on to the surface of metal. The main examples of distinct coating material include the material having low melting point & Stainless steel (SS), molybdenum (Mo) etc. Characteristics of wire flame spraying process coatings (Lombardi, et al., 2013) . The image of flame spray process is presented in **Figure 7.**

Figure 7. Schematic illustration of wire flame spray

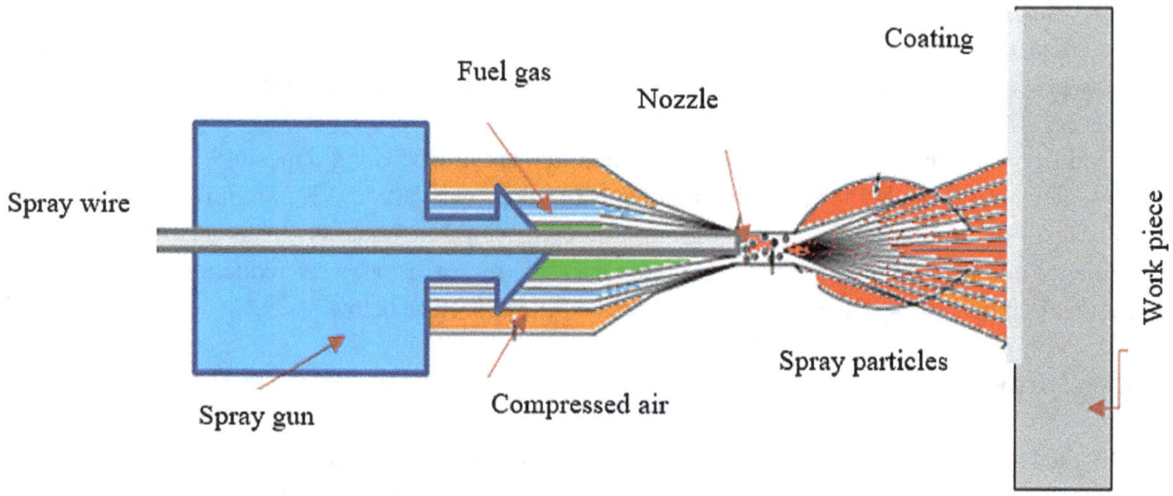

The main benefits of wire flame spray are: (a) Process is simple and has low cost than other thermal spray processes. (b) The wire used in the process is generally cheaper than powder. (c) The equipment of wire flame spray is portable. This process is used in distinct sectors: aerospace, automotive, civil, electrical and electronic etc.

Powder Flame Spray

This working process of wire flame spray process is same as that of powder flame spray, but only difference is in the feed stock material i.e. wire is replaced with powder. In this process the feed stock material is fed axial to the torch whereas, hot gases flows axially for the deposition of coating on to the material surface. The bond strength and cohesive strength is also generally lower. The schematic illustration of powder flame spraying method is depicted in **Figure 8**. The wide /distinct range of materials utilization i.e. Nickel (Ni), ceramic can processed in easy way into powder form giving a larger choice of coatings. The main reason of powder flame process development is owing to complication in the production in wire form (Grum & Bergant, 2008; Musztyfaga-Staszuk, et al., 2018; Sharma, 2012)

Figure 8. Schematic illustration of powder flame spray

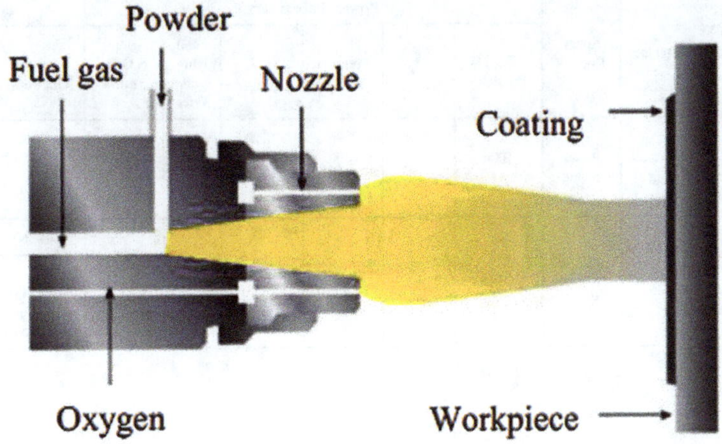

In solidifying and fusing process, the coating deposits a solid layer of carbides and borides, which add excellent resistance against wear to the coating. Also, these coating has excellent resistance to corrosion of most chemical solutions, cavitation erosion, hot hardness and erosion properties etc. Due to these properties, this process finds wide applications in distinct industries. The process parameters also affect on properties, therefore, some important process parameters that mainly effect on mechanical properties in case of powder flame sprayed coating method are represented in the **Table 3.**

Table 3. Examples of the influence of flame spray set of parameters on coatings mechanical properties

Author/ Year	Base Matel	Coating Material	Spray Process	Process Parameters						Mechanical Properties		
				Size of Particles	O_2 Pressure	Pressure of Air	C_2H_2 pressure	Feed Rate of Wire	S.O.D. (mm)	Coating Thickness	Roughness (Ra), µm	Micro Hardness
Staszuk, M.M. et al. (2018)	Unalloyed structural steel (grade S235JR)	1000 (Al)	Powder Flame Spray	131.64 µm	4 bar	4 bar	0.7 bar	-	~250	125 µm	29.05 µm	32 HV
		1000 (Sn)		16.79 µm	4 bar	4 bar	0.7 bar	-	~250	292 µm	11.28 µm	12 HV
		Polymer- yellow (PA12)		166.14 µm	4 bar	4 bar	0.7 bar	-	Greater than 250	1.43 µm	1.55 µm	9HV
Sharma, S. et al.(2012)	Carbon Steel	NiCrFeSiB	Power Flame spray	-	0.3MPa	-	0.12MPa	2g/ min.	22mm	-	4-6 µm	$346{\pm}8Hv_5$

Continued on following page

Table 3. Continued

<table>
<tr><th rowspan="2">Author/ Year</th><th rowspan="2">Base Matel</th><th rowspan="2">Coating Material</th><th rowspan="2">Spray Process</th><th colspan="6">Process Parameters</th><th colspan="3">Mechanical Properties</th></tr>
<tr><th>Size of Particles</th><th>O_2 Pressure</th><th>Pressure of Air</th><th>C_2H_2 pressure</th><th>Feed Rate of Wire</th><th>S.O.D. (mm)</th><th>Coating Thickness</th><th>Roughness (Ra), μm</th><th>Micro Hardness</th></tr>
<tr><td rowspan="2">Uyulgan, B. et al.(2007)</td><td rowspan="2">Plain carbon steel</td><td>FeCr</td><td rowspan="2">Power Flame</td><td rowspan="2">100μ M</td><td rowspan="2">3 bar</td><td rowspan="2">0.4 bar</td><td rowspan="2">0.7 bar</td><td rowspan="2">-</td><td rowspan="2">150mm</td><td>865 μm</td><td rowspan="2">~ 1.5μm</td><td rowspan="2">800HV</td></tr>
<tr><td>Ni-based bond coatings</td><td>182 μm</td></tr>
<tr><td rowspan="2">Hidalgo, HV et al. (2001)</td><td rowspan="2">AISI 304 austenitic stainless steel</td><td>NiCrAlTi</td><td rowspan="2">Powder Flame Spray</td><td rowspan="2">-</td><td rowspan="2">-</td><td rowspan="2">-</td><td rowspan="2">-</td><td rowspan="2">-</td><td rowspan="2">150mm</td><td>540-560</td><td rowspan="2">-</td><td rowspan="2">301Hv-395Hv</td></tr>
<tr><td>NiCrAlTi+NiMoAl (Bond coat)</td><td>425-540+(100-110 μm Bond coat)</td></tr>
</table>

Plasma Spray

It is a versatile commercial spray coating process method that utilizes a high energy heat source to melt & accelerate fine particles onto the surface of substrate. After impact, these molten droplets cool down & solidify instantly by transferring the heat to the underlying base material (substrate). This process is mainly used for distinct purposes such as thermal barrier, resistance against wear, biocompatibility & bioactivity etc. Gases such as argon, helium, hydrogen and nitrogen are mainly used as the arc gases. The plasma flame temperature varies from 10000°C to 15000°C. Due to this features, almost any ceramic or metal including refractory oxides or metals can be melted & deposited to develop a coating by using this process (Uyulgan, et al., 2007). This process is used in mechanical engineering, marine engineering, automotive engineering, aerospace, biomedical, chemical & nuclear industries etc. The common application of this process includes: zirconia based thermal barrier coating on turbine combustion chambers, seal ring grooves in the compressor area of aero-engine turbines, molybdenum alloys on to diesel engine piston rings and biocompatible coatings of hydroxyapatite for medical and dental implants, etc. The schematic illustration of a plasma spray process is represented in **Figure 9**. Further, this process may be classified in two categories termed as atmospheric and vacuum plasma spray "The atmospheric plasma spraying is performed under atmospheric conditions in a spray booth with exhaust system to protect from the emissions of the process. The spray material must be flowable and introduced in the form of solutions of nitrates/ organometallic compounds. Then, after injection into the plasma, droplet break up, sintering, pyrolysis and melting take place to develop splats of desired spray material" (Uyulgan, et al., 2007). On the other side, "vacuum plasma spraying uses argon (3 to 7 kPa) to protect spray materials which are susceptible to oxidation. In addition to this, a transferred arc can be applied to heat up and activate the surface prior to spraying Beyond vacuum plasma spray, lower pressure of 50 to 200Pascal are usually applied to attain dense and thin metallic and ceramic coatings" (Higuera, et al., 2001; Mishra & Prakash, 2015, Fauchais, et al., 1992; Pfender, 1994; Sanjay, et al., 2017; Bedi, et al., 2019)].

Figure 9. Schematics illustration of plasma spray process

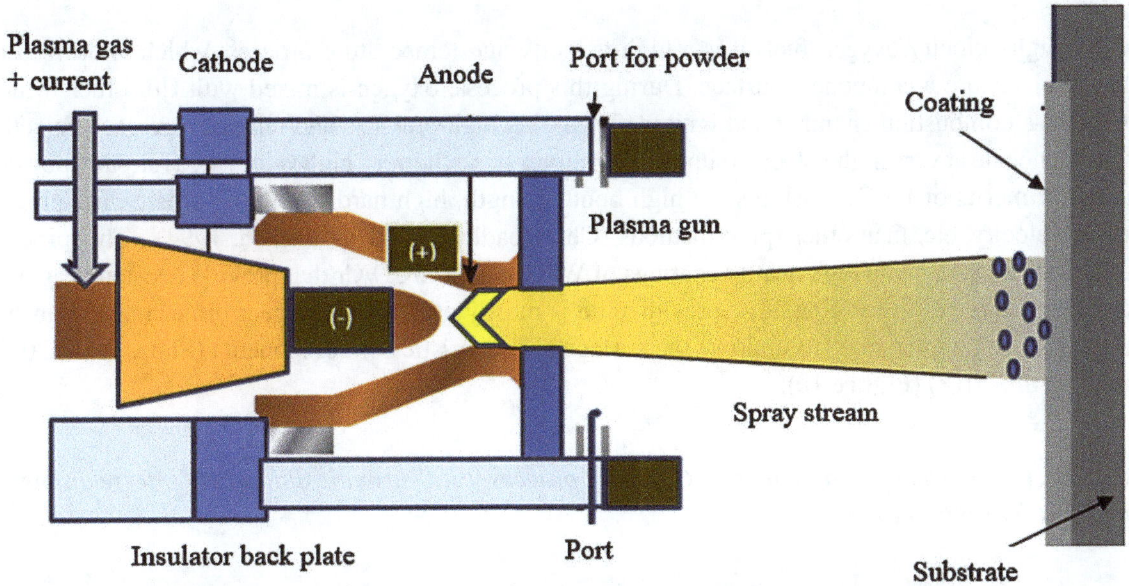

However, in order to understand the observation on the deformation behavior of the coatings, it is essential to consider the microstructure and mechanical properties of the coatings. There are various process parameters such as size of spray particles, stand of distance, nozzle diameter, pressure of plasma gas etc. on which the coating performance depends. From literature survey, it has been observed that stand off distance (SOD) has no significant effect on the bending modulus and bending yield strength.The SOD mainly control the cohesion b/w splats. This may be due to the reason that velocity and temperature of the particles in the plasma flame significantly change with change in SOD. The effect of distinct process parameters on the mechanical characteristics of plasma sprayed coatings is summarized in **Table 4.**

Table 4. Examples of the influence of plasma spray set of parameters on coatings mechanical properties

Author/ Year	Base Metal	Coating Material	Process Parameters						Mechanical Properties	
			Size of Spray Particles	Input Current (A)	Stand Off Distance (mm)	Diameter of Nozzle (mm)	Pressure of Plasma Gas (MPa)	Pressure of Carrier Gas (MPa)	Coating Thickness s (μm)	Micro Hardness ($HV_{0.2}$)
Sundaresan et al., (2020)	T91	NiCoCrAlY	22-45μm	500	100 mm	-	-	1bar	---	μ= 335 with σ = 57
Singh et al., (2006)	Superni 600	NiCrAlY	-45±10μm	700	90-110mm	-	59Psi	40Psi	Bond coat=228μm	390Hv
		stellite-6	-180±53μm			-			Bond coat=162μm, outer coat= 365μm	487Hv
		Ni-20Cr	-45±5μm			-			Bond coat=155μm, outer coat= 211μm	320Hv
Singh et al. (2005)	SF 800H	NiCrAlY	-45+10 μm	700	90-110mm	--	59Psi	40Psi	Bond coat=150μm, -outer coat= 200 μm	-
		Ni_3Al	Ni: 74 μm and Al: fine powder			-				-
		stellite-6	-180+53 μm			-				-

HVOF Spray

HVOF (high-velocity oxygen fuel) it is a high-velocity and temperature process, which is utilized to enhance or restore a component surface. During this process, oxygen is mixed with fluid fuel, which is fed into a combustion chamber and ignited. Then, this high pressure and temperature gas is ejected through a nozzle at very high velocity (supersonic velocity). To the very high velocity gas stream, powder The major merits of HVOF coatings are high bond strength, high hardness, less porosity and greater particle velocity, etc. than other spray methods (Chattopadhyay, 2004; Bunshah, 1994) . This process mostly utilized to develop high quality coatings of WC/Co & WC/CO/Cr that provide resistance against corrosion, wear, etc. In addition, this spray method is mostly used across a spectrum of manufacturing and engineering applications to improve the surface characteristics of components (Staia, et al., 2020; Kumar, et al., 2018) (**Figure 10**).

Figure 10. (a): Schematic diagram of HVOF spray process (b): Schematic diagram of interrelationship between HVOF parameters

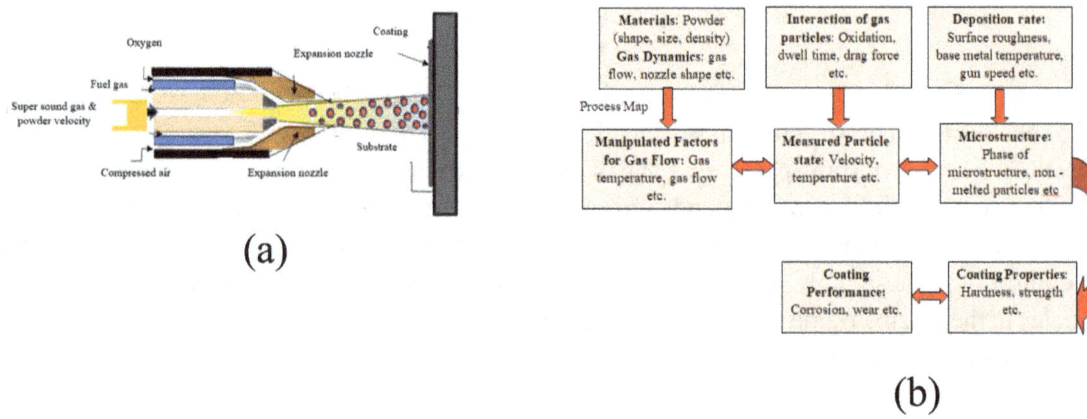

The coating performance is depends upon the HVOF process parameters (powder feedstock, gas flow, fuel gas, stand of distance etc.) During this process, due to high temperature, the evaporation, dissolution and phase transformation of the powder occurs *(Sreenivasulu & Manikandan, 2019; Kaur, et al. 2012; Sidhu, et al. 2006 (a); Sidhu, et. al., 2006(b).* Thus, in order to attain coating with desired properties, the optimization and control of process parameters is very important (**Table 5**).

Table 5. The effect of spray parameters on the mechanical characteristics of HVOF coatings

Author/ Year	Base Metal	Coating Material	Process Parameters							Mechanical Properties		
			Sizeof Particle	Oxygen Flow Rate	Fuel Flow Rate	Oxygen Pressure (bar)	Propane Pressure (bar)	Stand of Distance	Powder Feed Rate	Coating Thickness (µm)	Roughness (Ra), µm	Micro Hardness ($HV_{0.2}$)
Tailor, S. et al. 2019	T91	Cr3 C2-25%NiCr	15-45 µm	260LPM	65 LPM	10 Kg/ cm2	7 Kg/ cm2	20cm	60 gm/min.	350-400µm	5.36± 1.5µm	755VHN
Chatha et al., (2012)	T91	Ni-20Cr	-53+15 µm	250L/ min	60L/ min	-	7.5 kg/ cm^2	180mm	-	260-310 µm	9.21 ± 1.05 µm	270Hv
Sidhu et al., (2006)	Superni 600	NiCrBSi	±45 µm	250L/ min	60 L/ min	6kg/cm2	6kg/cm2	200mm	-	250 µm		810-850Hv
		Stellite-6	-53±45 µm							265 µm		850-900Hv
Sidhu et al.,(2006)	Superni 601	NiCrBSi	-45µm	200L/ min	50 L/ min	8kg/cm2	6kg/cm2	200mm	-	250-300 µm		Approx. 700 Hv

Cold Spray Method

Cold spray or cold-gas dynamic-spray is a solid-state process that is developed by Russia. It utilizes the material in powder form and converging-diverging nozzle to spray the metal parts to create a layer with supersonic velocity onto the surface of the substrate material. During this process, a nitrogen or helium gas is compressed to 3.5MPa & heated to 873k by a gas heater. Then, this high pressure and temperature gas expands in converging-diverging nozzle and reaches to supersonic velocity. The material (metal, ceramics, polymer) in powder form is injected into the throat of the nozzle & the powder particles are accelerated to higher velocities & heated to temp. well beneath the melting temperature. To enhance the gas flow velocities preheated compressed gas is used. The key merits of this spraying process are that it provides thick, excellent bonding characteristics, less heat delivered on base material during coating, uses corrosion resistance material, and provides more dense coating than other thermal spray process methods. In addition, the cold spray method is widely used in aerospace for component repair and biotechnology sectors etc.(Kumar, R. &Kumar, S., 2018; Bala, et al., 2013; Sudharshan, et al., 2007; Kumar, S., & Kumar, R. 2021). The schematic image of the cold spray method including its types, parameters, merits and demerits (Hamid, et al., 2016; Koivuluoto & Vuoristo 2017; Kumar, R. & Kumar, S., 2018 (a)) is shown in Figure 11.

Figure 11. Schematics diagram of cold spray process, parameters, merits and demerits

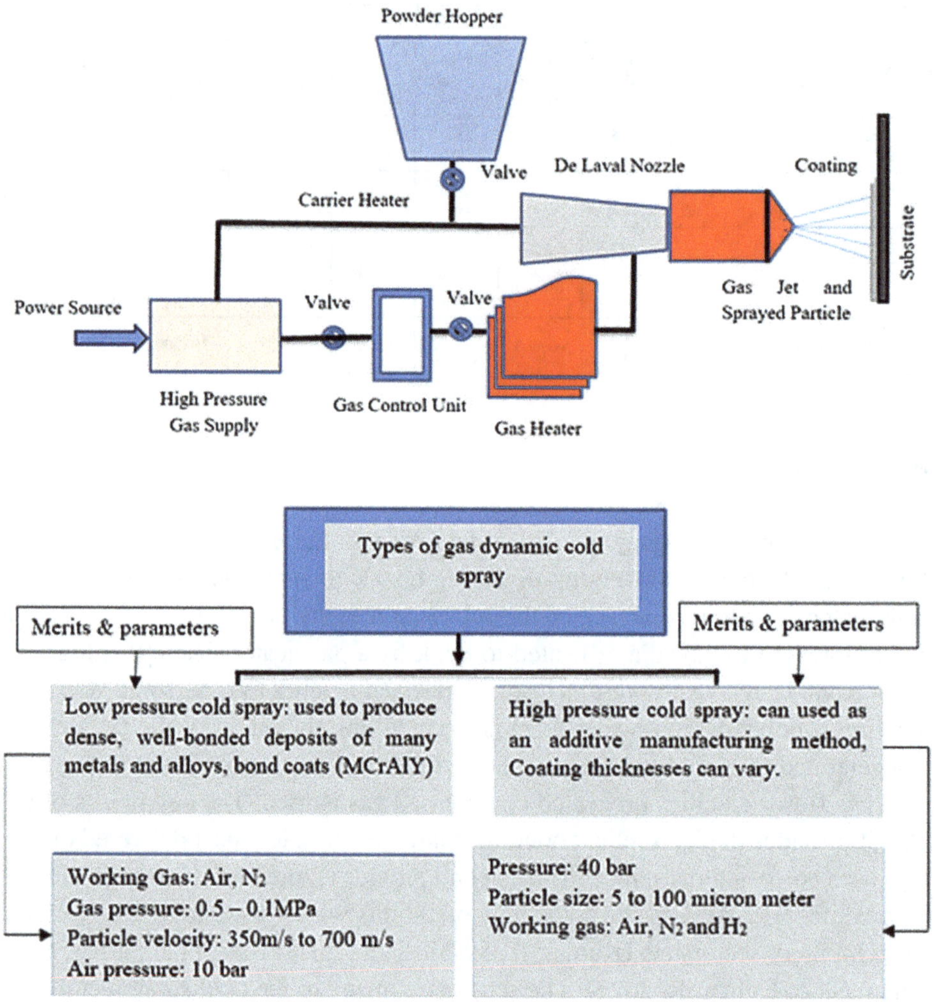

This process has been used as a coating technology in a wide range of industries: aerospace, automotive, medical, energy, etc. This process offers effective protection against high temperature oxidation, corrosion, erosion and chemicals.

Detonation Gun Spray

D-Gun is a versatile spray method that was introduced firstly in 1955 by H.B. Sargent. In this spray method the coating powder particles &acetylene gas + oxygen blend is supplied through tube-shaped barrel, thereafter these heated powder particles are sprayed onto the substrate material with a high velocity (1200 m/s). The gas mixture inside the chamber is ignited by a simple spark plug. The combustion of the gas mixture generates high pressure shock waves (detonation wave), which then propagate through the gas stream. Depending upon the ratio of the combustion gases, the temperature of the hot gas stream can go up to 4000 deg C and the velocity of the shock wave can reach 3500m/sec (Singh, et al. 2012). The main advantages of this spray method is to providea good quality coating surface, wear resistance,

resistance to corrosion, dense coatings, and high adhesive strength. In addition the major scope or field of application in power plant, turbine blade, aviation engine, aircraft ndustry, etc.(Du, et al. 2015; Kumar, et al., 2018 (b); Amin, et al.,2016; Cashon, 1975). The schematic illustration of D-Gun spray method is represented in **Figure 12.**

Figure 12. Schematic image of detonation gun spray method

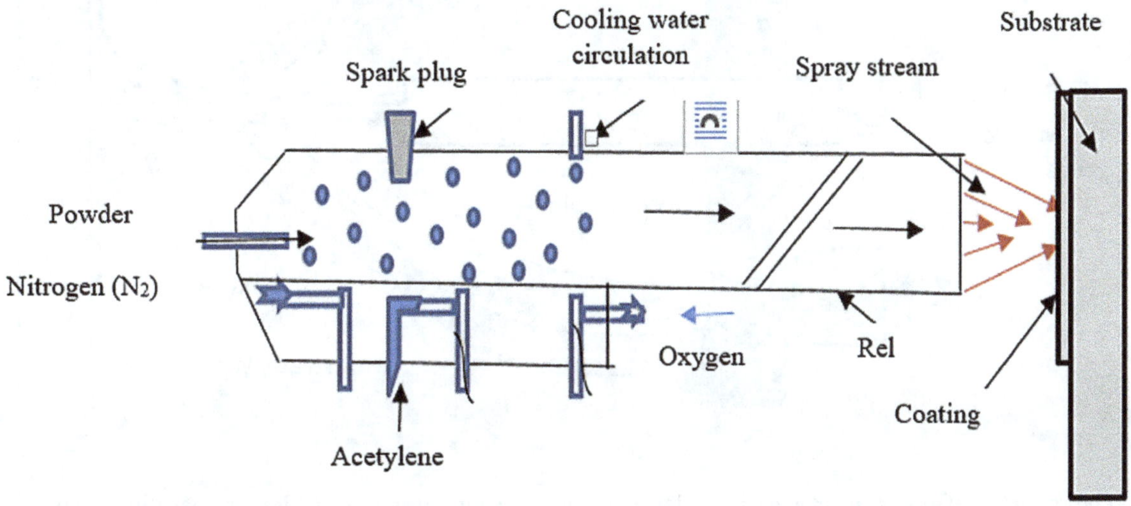

Wire Arc Spray Method

Wire arc spray or electric arc spray is an economical thermal spray coating method that uses conductive two consumable wire electrodes (joined with D.C. power source) which are fed through the gun and then melt it with the use of electric arc. Thereafter the air of compressed nature is used that help to deposit the coating material with greater velocity onto the surface of the base material. In addition, the high bond strength, more spray rate, dense coating on-site application, ease to handle, etc. are the major merits of this spray method. The electric arc spray also considered as energy-efficient process owing to all the input energy are utilized to melt the coating material such as Zinc (Zn), copper-base/ Iron-based alloy are sprayed at a feed rate of 11Kg and 4.5Kg respectively. The application of this thermal spray process is for corrosion resistance in steam-generating tubes, transport pipes, large infrastructures coatings, etc. (Kumar, et al., 2019; Malek, et al., 2014; Thorpe, 1993; Kant, et al.. 2020). The schematics illustration of the wire arc spray process is shown in **Figure 13**

Figure 13. The schematics diagram of wire arc/electric arc spray method

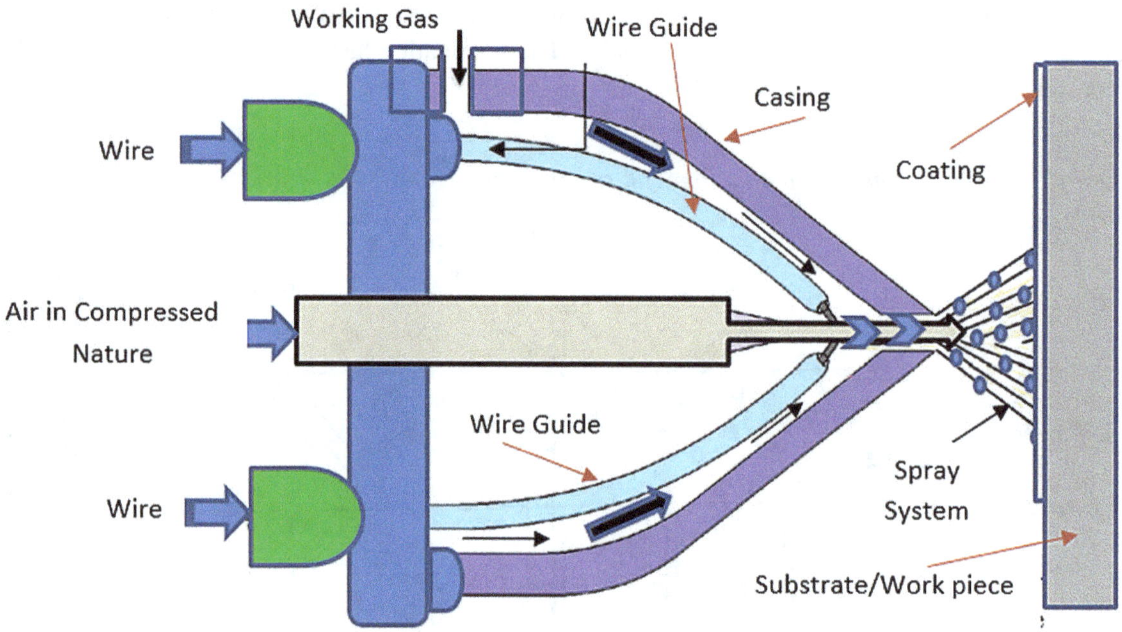

The main benefits of this process includes: can coat a variety of materials, lower operating costs, high material output per hour etc. Due to these benefits, this process find application in shaft repair, boiler tubes, bearing fit rebuilding etc.

CHEMICAL, PHYSICAL AND MECHANICAL METHODS OF COATING DEPOSITION

The several distinct coatings process methods, materials, and process parameters are recently used to prevent the corrosion, abrasion, and wear resistance of the base materials. Examples of distinct corrosion prevention techniques include thermal spray coatings, Chemical spray, Physical and Mechanical spray methods. In the case of chemical spray, deposition methods vacuum chamber is always needed, whereas in most of the thermal spray deposition methods are carried out in atmospheric environment and large heat content is used to melt a wide range of materials from metal to ceramics.

Chemical Vapor Deposition (CVD)

It is a versatile, high-quality thin layer/coating development method where the dissociation or chemical reactions of gaseous reactants in a activated condition, is followed by the development of a stable solid product. Nowdays, CVD coating fabrication method is commonly used in industries owing to the application of adopting new materials. The coating deposition rate mainly depends upon base metal temperature, the kinetics of the reaction and decomposition product crystallization etc.(Mikell, 2010).In the 1970s, this coating method already provides significant success in the production of electronic semiconductors as well as protective coatings for electronic circuits. In the recent day, the CVD technique used for several advanced products, materials, uniformly coats the shape of complex parts, coatings, and composites. It applies hard coatings to the surfaces of the cutting tool that help to enhance resistance against wear large bandgap preparation application in the material of semiconducting nature. In recent days the CVD application are grouped into four distinct sections (a) surface modification coatings (b) pyro metallurgy & extraction (c) electronic and optoelectronic materials and (d) ceramic matrix composites & ceramic fibers (Gill & Tucker, 1986; Choy, 2003).

CVD has some merits and demerits the major key characteristics /merits of CVD technology are; It develops uniform films/layers, pure and highly dense coating development capabilities, control crystal structure, orientation, flexibility and reasonable processing cost, adjustable deposition rate, etc. Thus CVD exhibits clear merits by relying on chemical reactions; it enables tunable deposition rates and superior quality coatings with good conformity. On the other hand, CVD coating has many demerits such as the explosive precursors, toxic, flammable, and corrosive nature, cost of fabrication, high owing to the use of plasma and photo-assisted CVD, difficulty in multi-component material deposition, more temperature are needed for film deposition (Xiu-Tian & Yongdong, 2010; Sun, et al., 2021; Kalita & Tanemura 2017). The schematic diagram of the chemical vapor deposition process is shown in **Figure 14.**

Figure 14. Schematic image of CVD technique

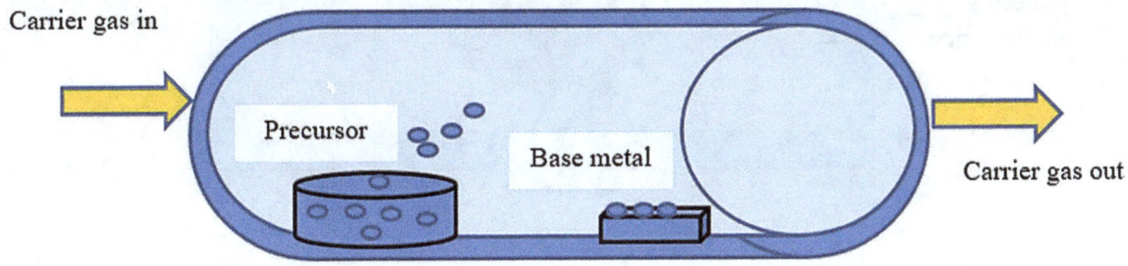

During this technique, a solid material is deposited from a vapor by some chemical reaction occurring on or in the vicinity of a normally heated surface of the base material (substrate) (Mittal, et al., 2021).

This process can also be categorized on the basis of physical characteristics of vapor, base material heating (cold & hot wall CVD) and operating environment.

Physical Vapor Deposition (PVD)

PVD is, vacuum, thin coating deposition technique that produce inert atmosphere, thus no chemical reactions are shown. Recently, the coating demand is increasing more rapidly in distinct sector applications (biomedical, metalworking, electrical, or optical industries). This is because PVD may improve several coating properties (provide high resistance against corrosion, improve hardness, resist wear, improve the electrical property, help to make the part smooth/decorative, etc. (Kalita, et al., 2017).

The key merits of PVD technique are utilize inorganic/organic material, alloys, compounds, metals and melt in same shape and size. In addition, it produce graded composition deposits, uniform coating, produce multilayer coating, and superior adhesion strength with the base metal. Although it has some drawbacks such as it produces a coating of very thin, less coating material deposition rate up to nanometer. This process utilizes complex equipment and has very high cost. In addition, the production speed of physical vapor deposition coatings is slow than other coating deposition methods. ((Mittal, et al. 2021; Helmersson, et al., 2006; Ferrari, & basko (2013); Chi, 2010; Verlinden, 2005). The schematic diagram of PVD process technique is represented in **Figure 15**.

Figure 15. Schematic illustration of PVD process

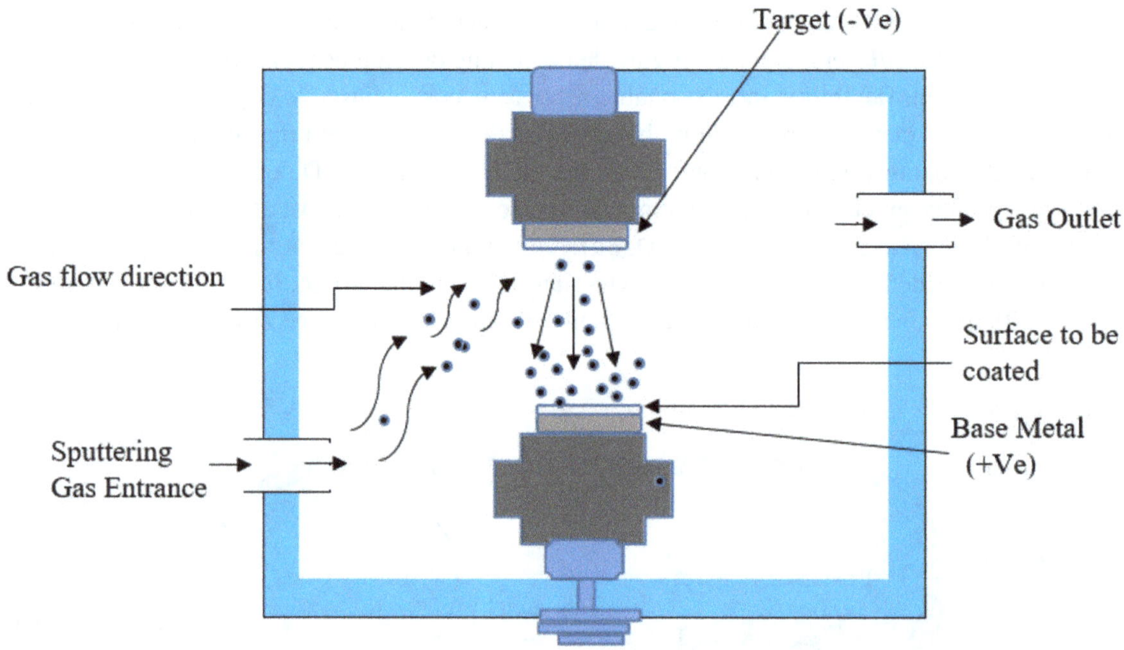

Anodizing

Anodizing coating deposition is an electrolytic passivation process that exhibits the oxide layer formation onto the surface of a metal part such as (Ti) using electric current. The major function of anodizing is to provide finishing, hardening & coloring the metal surface without changing the mechanical properties of the products. In the case of the coloring of Titanium, this process is used for medical applications such as orthopedic implant. This process has some benefits (i) Nontoxic as well as biocompatible (ii) corrosion-resistant (formation of oxide layer on to the surface of product). (iii) Provide resistance against wear etc. Anodizing process utilized distinct chemicals such as for etching: (Water hydrofluoric acid+ Ammonium B1 Fluoride+ Nitric acid), For Preparing Electrolyte (Type 1 Coloring): water is added to ortho- phosphoric acid for obtaining seven different types of colors requirement. Similarly chemicals for a gray color called Type 2 color chemical used: Orthophosphoric acid+ sulfuric acid. The important point must focus on periodical measurement of water quality, proper application of machine parameters (current, voltage, immerge time, electrolyte temperature during the process, chemical concentration) that help to make the process effective and avoid the color defect. The colors of the oxide layers were changed with the change in the applied voltage &the slope b/w applied current density and time of anodization was rapidly fallen and after a particular value, it becomes a straight line approximately (Sharma, 1992; Manjaiah, & Laubscher 2017, Williamson et al. 2013; Indira, et al., 2015; Kumar, 2020).

Sol-Gel Coating Process Method

Solgel is a simpler method to develop thin films. In this process an organic solvent solution is developed by the elements of the desired compounds & then the solution is polymerized to develop a gel. To enhance the resistance against oxidation, to improve the part properties of the surface, and to enhance the resistance against sol-gel derived films is deposited onto the surface.. The main advantages of this method are: low processing temperature, high purity, ease of manufacture of homogenous stoichiometric coatings. The major limitations of this coating method are (a) crack ability (b) complicated part shape coat difficulty (d) thickness limits etc.(Guglielmi, 1997) . In addition, the cost of raw materials i.e. chemicals may be high (Kallawar & Bhanvase, 2021; Zhong, & Li, 2010; Lamaka, et al., 2008; Innocenzi, et al., 1992; Maggio, et al., 1990; Izumi, et al., 1989; Wang & Bierwagen, 2009; Akpan, & Hameed, 2010; Kumar, et al., 2021).

SURFACE TEXTURING

It involves the controlled improvement of topography to develop functional surfaces (Rosenkranz, et al. 2020). In addition, "to minimize the wear in piston system, bearing, seals and in medical devices surface texturing play a significant role. The various mechanism of surface texturing for the enhancement of tribological properties are: (a) use of additional lubricant over the actual area by the inlet suction effect (b) enhancement of hydrodynamic pressure over the surface texture to increase the load-carrying capacity (c) decrease the actual contact area (d) lubricant storage by the reservoir effect (e) capture of wear particles by the debris trapping effect"Saeidi,, et al., 2021; Klimczak, et al., 1994; Lu,, et al., 2020; Boidi, et al., 2021) . The examples of some commonly used techniques for the fabrication of surface texture are electrochemical processing, electro discharge machining (EDM), jet machining, ultrasonic-assisted machining, beam and laser machining, etc. (Baharin, et al., 2016).

SURFACE HARDENING

This method involves heating or utilizing mechanical techniques on the metal material to enhance its surface hardness. This results in enhancement against fatigue and wear. Archards law is used.to calculate the resistance against wear of surface. According to this law, V=KlF/H, where K=wear coefficient, l=sliding distance, F=load applied, H= hardness, and V= wear volume.

Some examples of traditional coatings obtained byhardening technique include nitriding, carburizing, boriding among others. These conventional techniques comprise large surface hardened thickness due to a phase transformation in the under-layer microstructure. However,some advanced surface hardening techniques are used to improve the resistance against wear and induce fine microstructure. These advanced surface hardening techniques include duplex hardening, plasma surface diffusion, microwave heating, plasma surface diffusion, friction stir processing (FSP) and high current pulsed electron beam processing etc"(Xing, et al., 2017; Steve, et al., 2012; Kumar, et al., 2018; Vasudev, et al., 2019; Vipin, et al., 2015; Gao,, 2011).

APPLICATION OF SURFACE MODIFICATION METHODS USED IN DIFFERENT SECTORS

Medical Applications

Nowdays, the surface properties of the materials i.e. biocompatibility, adhesion strength etc. are not sufficient; therefore the modification in these material properties is very essential before use or coating. In addition, both the chemical structure and morphological characteristics should be modified in order to attain a desired surface finish (Mozetič et al., 2019). .The distinct sectors such as biomedical (implant) automotive, aerospace etc. the superior chemical as well physical properties of the materials is highly important. For example,titanium shows an excellent biocompatible behaviour and provides resistance against corrosion (López, et al., 2010) . Mainly surface properties especially in clinical is more important so, the examples of some surface modification methods utilized for titanium (Ti) and its alloys i.e. Ti 6Al 4V for implants (Liu, et al., 2004; Barshilia, 2021).

The major reason on the use of the distinct surface modification process methods for biomedical purposes are due to the enhancement in bioactivity, the hardness, reduce the wear rate, improve cell growth, develop and enhance fatigue strength of implants.

Aerospace and Engineering Applications

The material degradation owing to corrosion and wear is estimated to be very high. Therefore surface engineering plays a vital role to enhance the performance of the distinct material. Currently, surface modification techniques are used such as spraying based, chemical-based, mechanical based, etc. There is a high demand for coatings obtained by economical and eco-friendly coating fabrication processes. The surface modification is being used in aerospace, energy, automobile, chemical, societal, textiles, engineering sectors for providing functional characteristics to the engineering materials at reduced cost compare with other techniques. In addition, further development in the nanostructured coatings or nanosurfaces is expected to carve a niche in several industrial uses. Now Distinct types of materials are used in aerospace sector include titanium (Ti), Al alloys etc. (Barshilia, et al., 2010; Selva, et al., 2010; Ma, P-C, et al., 2010; Siwal, et al.,2021).

FUTURE CHALLENGES

Surface treatments or interactions are the threats ahead for the researchers in relation to the development of advanced, sophisticated micro-fluidic devices. Polydimethylsiloxane (PDMS) will be a famous material for future microfluidic devices. In addition, some challenges found with the recently used treatments, (hydrophobic recovery of PDMS surfaces) need to be adequately addressed. The scaling up of these coatings is extremely difficult to retain nanostructure under harsh conditions. Nevertheless, for aerospace and automobile sectors, self –cleaning coatings can be developed for drag reduction & as rain repellant coatings on wind shields. By reducing the drag (5 to 10%) in the aircraft, large amount of fuel can be saved along with less pollution to the environment. However, the challenges like abrasion resistance, thermal stability, and weather stability faced during sol gel process should be addressed.

Overall, there exist a large number of challenges and opportunities in the area of surface engineering, which will enable us to exploit the vast potential of these technologies at affordable cost in the near future.

CONCLUSION

Coating plays an important role in different field of applications, such as medical, agriculture, aerospace, research, and development, engineering, automotive, where there is a requirement of a highly smooth surface. The HVOF and Plasma spray is most widely used thermal spray techniques to develop anti wear and corrosion resistant coating with distinct, types of materials (metals, alloys, composites, ceramics) etc. Although, the wire arc spray provide better outcomes (resistance against corrosion, wear, oxidation and erosion) due to attractive properties such as low cost, high spray rate, on site application, etc. of the spray method. However, the coating deposited by cold spray has considerably less residual stress, thermal stress and other defects. In addition, the experimental study by the distinct researcher indicate that Ni based coating exhibit good result in term of corrosion resistance, oxidation etc. This improves the part performance and service life. Overall, the Surface Engineering plays a major role in distinct sectors, because it provides effective solution to engineering applications.

REFERENCES

Akpan, U. G., & Hameed, B. H. (2010). The advancements in sol–gel method of doped-TiO_2 photocatalysts. *Applied Catalysis A, General, 375*(1), 1–11. doi:10.1016/j.apcata.2009.12.023

Amin, S., & Panchal, H. (2016). A Review on Thermal Spray Coating Processes. *International Journal of Current Trends in Engineering & Research, 2*, 556 – 563.

Amin, S., & Panchal, H. A. (2016). Review on thermal spray coating process. *International Journal of Current Trends in Engineering & Research, 2*(4), 556–563.

Ang, A. S. M., & Sanpo, N. (2013). Thermal Spray Maps: Material Genomics of Processing Technologies. Journal of Thermal Spray. *Technology, 22*(7), 1170–1183. doi:10.100711666-013-9970-3

Baharin, A. F. S., Ghazali, M. J., & Wahab, J. A. (2016). Laser surface texturing and its contribution to friction and wear reduction: A brief review. *Industrial Lubrication and Tribology, 68*(1), 57–66. doi:10.1108/ILT-05-2015-0067

Bala, N., Singh, H., Karthikeyan, J., & Prakash, S. (2013). Cold spray coating process for corrosion protection: A review. *Surface Engineering, 30*(6), 414–421. doi:10.1179/1743294413Y.0000000148

Barshilia, H. C. (2021). Surface modification technologies for aerospace and engineering applications: Current Trends. *Challenges and Future Prospects. Trans Indian Natl. Acad. Eng., 6*(2), 173–188. doi:10.100741403-021-00208-z

Barshilia, H. C., Ghosh, M., Shashidhara, Ramakrishna, R., & Rajam, K. S. (2010). Deposition and characterization of TiAlSiN nanocomposite coatings prepared by reactive pulsed direct current unbalanced magnetron sputtering. *Applied Surface Science, 256*(21), 6420–6426. doi:10.1016/j.apsusc.2010.04.028

Bedi, T. S., Kumar, S., & Kumar, R. (2019). Corrosion performance of hydroxyapatite and hydroxyapatite/titanium bond coating for biomedical applications. *Materials Research Express*, *7*(1), 015402. doi:10.1088/2053-1591/ab5cc5

Billard, A., Maury, F., & Aubry, P. (2018). Emerging processes for metallurgical coatings and thin films. Comptes Rendus Physique. *Elsevier Masson*, *19*(8), 755–768. doi:10.1016/j.crhy.2018.10.005ff

Boidi, G., Grützmacher, P., & Varga, M. (2021). *Tribological Performance of Random Sinter Pores vs. Deterministic Laser Surface Textures: An Experimental and Machine Learning Approach*. Friction. doi:10.5772/intechopen.100245

Bose, S. (2007). High temperature coatings. Elsevier. doi:10.1016/B978-075068252-7/50007-X

Bunshah & Rointan. (1994). *Handbook of deposition technologies for films and coatings science, technology and applications* (2nd ed.). Noyes Publications.

Cashon, E. P. (1975). Wear resistant coatings applied by the detonation gun. *Tribology International*, *8*(3), 111–115. doi:10.1016/0301-679X(75)90028-6

Chattopadhyay, R. (2004). Advanced thermally assisted surface engineering processes. Kluwer Academic Publishers.

Chi, L. (2010). Nanotechnology: Vol. 8. *Nanostructured Surfaces*. John Wiley & Sons.

Choy, K. L. (2003). Chemical vapour deposition of coatings. *Progress in Materials Science*, *48*(2), 57–170. doi:10.1016/S0079-6425(01)00009-3

Di Maggio, R., Scardi, P., & Tomasi, A. (1990). Characterization of Ceria Stabilized Zirconia Coatings on Metal Substrates Materials Research Society symposia proceedings. *Materials Research Society*, *180*, 481. doi:10.1557/PROC-180-481

Du, H., Hau, W., & Liu, J. (2015). Influence of process variable on the quality of detonation gun sprayed WC-Co coatings. *Materials Science and Engineering*, *408*(1-2), 202–210. doi:10.1016/j.msea.2005.08.008

Fauchais, P., Coudert, J. F., Vardelle, M., Vardelle, A., & Denoirjean, A. (1992). Diagnostics of thermal spraying plasma jets. *J. Thermal Spray Technology*, *1*(2), 117–128. doi:10.1007/BF02659011

Fauchais, P. L., Heberlein, J. V. R., & Boulos, M. I. (2014). Introduction. In *Thermal Spray Fundamentals*. Springer. doi:10.1007/978-0-387-68991-3_1

Ferrari, A. C., & Basko, D. M. (2013). Raman spectroscopy as a versatile tool for studying the properties of graphene. *Nature Nanotechnology*, *8*(4), 235–246. doi:10.1038/nnano.2013.46 PMID:23552117

Gao, Y. (2011). Surface modification of TA2 pure titanium by low energy high current pulsed electron beam treatments. *Applied Surface Science*, *257*(17), 7455–7460. doi:10.1016/j.apsusc.2011.03.005

Garcia, E., Lee, H., & Sampath, S. (2018). Phase and microstructure evolution in plasma sprayed Yb-2Si2O7 coatings. *Journal of the European Ceramic Society*, *39*(4), 1477–1486. doi:10.1016/j.jeurceramsoc.2018.11.018

Gill, B. J., & Tucker, R. C. (1986). Plasma spray coating processes. *Materials Science and Technology, 2*(3), 207–213. doi:10.1179/mst.1986.2.3.207

Grum, J., & Bergant, Z. (2008). The optimization of powder flame-spraying parameters using a Taguchi method. *International Journal of Microstructure & Materials Properties, 3*(4/5), 4–5. doi:10.1504/IJMMP.2008.022043

Guglielmi, M. (1997). Sol-Gel Coatings on Metals. *Journal of Sol-Gel Science and Technology, 8*(1-3), 443–449. doi:10.1007/BF02436880

Hamid, A., Kreye, G. F., & Klassen, T. (2016). Cold spraying: A material perspective. *Acta Materialia, 116*, 382-407. doi:10.1016/j.actamat.2016.06.034

Helmersson, U., Lattemann, M., Böhlmark, J., Ehiasarian, A. P., & Gudmundsson, J. T. (2006). Ionized physical vapor deposition (IPVD): A review of technology and applications. *Thin Solid Films, 513*(1-2), 1–24. doi:10.1016/j.tsf.2006.03.033

Higuera, H. V., Belzunce, V. J., Menéndez, C. A., & Martinez, P. S. (2001). High temperature erosion wear of flame and plasma-sprayed nickel–chromium coatings under simulated coal-fired boiler atmospheres. *Wear, 247*(2), 214–222. doi:10.1016/S0043-1648(00)00540-8

Howarth, G. A., & Manock, H. L. (1997). Water-borne polyurethane dispersions and their use in functional coatings. *Surface Coatings International, 80*(7), 324–328. doi:10.1007/BF02692680

Indira, K., Mudali, U. K., Nishimura, T., & Rajendran, N. (2015). A Review on TiO2 Nanotubes: Influence of Anodization Parameters, Formation Mechanism, Properties, Corrosion Behavior, and Biomedical Applications. *Journal of Bio- and Tribo-Corrosion, 1*(4), 28. doi:10.100740735-015-0024-x

Innocenzi, P. C., Guglielmi, M., & Gobin, M. (1992). Coating of metals by the sol-gel dip-coating method. *J. Europ. Ceram. Soc., 10*, 431. (92) 90018-9 doi:10.1016/0955-2219

Izumi, K., Murakami, M., & Deguchi, T. (1989). Zirconia Coating on Stainless Steel Sheets from Organozirconium Compounds. *Journal of American Ceramic Society, 72*(8), 1465–1468. doi:10.1111/j.1151-2916.1989.tb07677

Joshi, S., & Nylen, P. (2019). Advanced Coatings by Thermal Spray Processes. *Technologies, 7*(79), 1–14. doi:10.3390/technologies7040079

Kahar, S., Singh, A., Vala, U., Desai, A., & Desai, S. (2020). Thermal sprayed coating using zinc: A review. *International Research Journal of Engineering and Technology, 6*, 6497–6505.

Kalita, G., & Tanemura, M. (2017). Fundamentals of Chemical Vapor Deposited Graphene and Emerging Applications. In *Graphene Materials - Advanced Applications*. IntechOpen. Available from: https://www.intechopen.com/chapters/54395 doi:10.5772/67514

Kallawar, G. A., & Bhanvase, B. A. (2021). Nanomaterial-based photocatalytic membrane for organic pollutants removal. Handbook of Nanomaterials for Wastewater Treatment, 699–737. doi:10.1016/B978-0-12-821496-1.00007-6

Kant, S., Kumar, M., Chawla, V., & Singh, S. (2020). Study of high temperature oxidation behavior of wire arc sprayed coatings. *Materials Today: Proceedings*, *21*(4), 1741–1748. doi:10.1016/j.matpr.2020.01.226

Kaur, M., Singh, H., & Prakash, S. (2012). High-temperature behaviour of a high-velocity oxy-fuel sprayed Cr3C2-NiCr coating. *Metallurgical and Materials Transactions. A, Physical Metallurgy and Materials Science*, *43*(8), 2979–2993. doi:10.100711661-012-1118-4

Klimczak, T., & Jonasson, M. (1994). Analysis of real contact area and change of surface texture on deep drawn steel sheets. *Wear*, *179*(1-2), 129–135. doi:10.1016/0043-1648(94)90230-5

Koivuluoto, H., & Vuoristo, P. (2017). Structure and corrosion properties of cold sprayed coatings: A review. *Surface Engineering*, *30*(6), 404–413. doi:10.1179/1743294413Y.0000000201

Kumar, A. (2020). Anodization of Titanium Alloy (Grade 5) to Obtain Nanoporous Surface Using Sulfuric Acid Electrolyte. *Journal of the Institution of Electronics and Telecommunication Engineers*, 1–7. Advance online publication. doi:10.1080/03772063.2020.1780958

Kumar, R., Kumar, R. and Kumar, S. (2018). Erosion corrosion study of HVOF sprayed thermal sprayed coating on boiler tubes: A Review. *IJSMS*, *1*(3), 1–6. doi:10.51386/25815946/ijsms-v1i3p101

Kumar, R., & Kumar, S. (2018). Thermal Spray Coating Process: A Study. *International Journal of Engineering Sciences & Research Technology*, *7*(3), 610–617. doi:10.5281/zenodo.1207005

Kumar, R., & Kumar, S. (2018). Comparative Parabolic Rate Constant and Coating Properties of Nickel, Cobalt, Iron and Metal Oxide Based Coating: A Review. *I-Manager's Journal on Material Science*, *6*(1), 45–56. doi:10.26634/jms.6.1.14379

Kumar, R., Singh, R., & Kumar, S. (2018). Erosion and hot corrosion phenomena in thermal power plant and their preventive methods: A study. *Asian Journal of Mechanical Engineering*, *7*(1), 38–45.

Kumar, S., Handa, A., Chawla, V., & Kumar, R. (2021). Performance of thermal-sprayed coatings to combat hot corrosion of coal-fired boiler tube and effects of process parameters and post coating heat treatment on coating performance: A review. *Surface Engineering*, *37*(5), 1–28. doi:10.1080/02670844.2021.1924506

Kumar, S., Handa, A., & Kumar, R. (2019). Overview of wire arc spray process: a review. *Journal of Composition Theory, 12*(7), 900-907. http://jctjournal.com/gallery/106-july2019.pdf

Kumar, S., Kumar, M., & Handa, A. (2018). Combating Hot Corrosion of Boiler Tubes- A Study. *Journal of Engineering Failure Analysis*, *94*, 379–395. doi:10.1016/j.engfailanal.2018.08.004

Kumar, S., Kumar, M., & Handa, A. (2019). Comparative Study of High Temperature Oxidation Behavior of Wire Arc Sprayed Ni-Cr and Ni-Al Coatings. *Engineering Failure Analysis*, *106*, 104173–104189. doi:10.1016/j.engfailanal.2019.104173

Kumar, S., Kumar, M., & Handa, A. (2020). Erosion corrosion behavior and mechanical property of wire arc sprayed Ni-Cr and Ni-Al coating on boiler steels in actual boiler environment. *Materials at High Temperatures*, *37*(6), 1–15. doi:10.1080/09603409.2020.1810922

Kumar, S., Kumar, M., & Jindal, N. (2020). Overview of Cold Spray Coatings Applications and Comparisons: A Critical Review. *World Journal of Engineering*, *17*(1), 27–51. doi:10.1108/WJE-01-2019-0021

Kumar, S., & Kumar, R. (2021). Influence of processing conditions on the properties of thermal sprayed coating: A review. *Surface Engineering*, *37*(11), 1339–1372. doi:10.1080/02670844.2021.1967024

Kumar, S., & Kumar, R. (2021). Gas Dynamic Cold Spraying: A Review on Materials, Parameters, Applications and Challenges. *I-manager's Journal on Future Engineering and Technology*, *16*(2), 43–56. doi:10.26634/jfet.16.2.17624

Kumar, S., Kumar, R., & Singh, S. (2020). The role of thermal spray coating to combat hot corrosion of boiler tubes: A study. *Journal of Xidian University*, *14*(5), 229–239. doi:10.37896/jxu14.5/024

Lamaka, S. V., Montemor, M. F., Galio, A. F., Zheludkevich, M. L., Trindade, C., Dick, L. F., & Ferreira, M. G. S. (2008). Novel hybrid sol–gel coatings for corrosion protection of AZ31B magnesium alloy. *Electrochimica Acta*, *53*(14), 4773–4783. doi:10.1016/j.electacta.2008.02.015

Liu, X., Chu, P. K., & Ding, C. (2004). Surface modification of titanium, titanium alloys, and related materials for biomedical applications. *Materials Science and Engineering R Reports*, *47*(3-4), 9–12. doi:10.1016/j.mser.2004.11.001

Liu, X., He, D., Zhou, Z., Wang, G., Wang, Z., & Guo, X. (2019). Effect of post-heat treatment on the microstructure of micro-plasma sprayed hydroxyapatite coatings. *Surface and Coatings Technology*, *367*, 225–230. doi:10.1016/j.surfcoat.2019.03.056

Lombardi, A. N., Casteletti, L. C., & Totten, G. E. (2013). Thermal spray technologies: an overview. In Q. J. Wang & Y. W. Chung (Eds.), *Encyclopedia of Tribology*. Springer. doi:10.1007/978-0-387-92897-5_684

López, M. F., Gutiérrez, A., Jiménez, J. A., Martinesi, M., Stio, M., & Treves, C. (2010). Thermal oxidation of vanadium-free Ti alloys: An X-ray photoelectron spectroscopy study. *Materials Science and Engineering C*, *30*(3), 465–471. doi:10.1016/j.msec.2010.01.004

Lu, P., & Wood, R. J. K. (2020). Tribological performance of surface texturing in mechanical applications—A review. *Surface Topography: Metrology and Properties*, *8*(4), 043001. Advance online publication. doi:10.1088/2051-672X/abb6d0

Ma, P.-C., Siddiqui, N. A., Marom, G., & Kim, J.-K. (2010). Dispersion and functionalization of carbon nanotubes for polymer-based nanocomposites: A review. *Composites. Part A, Applied Science and Manufacturing*, *41*(10), 1345–1367. doi:10.1016/j.compositesa.2010.07.003

Malek, M. H. A., Saad, N. H., Abas, S. K., & Shah, N. B. M. (2014). Critical process and performance parameters of thermal arc spray coating. *International Journal of Materials Engineering Innovation*, *5*(1), 12–27. doi:10.1504/IJMATEI.2014.059489

Manjaiah, M., & Laubscher, R. F. (2017). Effect of anodizing on surface integrity of Grade 4 titanium for biomedical applications. *Surface and Coatings Technology*, *310*, 263–272. doi:10.1016/j.surfcoat.2016.12.038

Mikell, P. G. (2010). Fundamentals of modern manufacturing materials, processes, and systems. John Wiley and Sons, Inc.

Mishra, S. B., & Prakash, S. (2015). Erosion –corrosion behaviour of Ni-20Cr plasma coationg in actual boiler environment. *Surface Engineering, 31*(1), 29–38. doi:10.1179/1743294414Y.0000000338

Mittal, M., Sardar, S., & Janta, A. (2021). Nanofabrication techniques for semiconductor chemical sensors. Handbook of Nanomaterials for Sensing Applications, Micro and Nano Technology, 119-137. doi:10.1016/B978-0-12-820783-3.00023-3

Mozetič, M. (2019). Surface Modification to Improve Properties of Materials. *Materials (Basel, Switzerland), 12*(3), 441. doi:10.3390/ma12030441 PMID:30709009

Musztyfaga-Staszuk, M. Czupryński, A., & Kciuk, M. (2018). Investigation of mechanical and anti-corrosion properties of flame sprayed coatings. *Advances in Materials Science, 18*(4), 42-53. . doi:10.1515/adms-2017-0049

Oksa, M., Turunen, E., Suhonen, T., Varis, T., & Hannula, S. P. (2011). Optimization and Characterization of High Velocity Oxy-fuel Sprayed Coatings: Techniques, Materials, and Applications. *Coatings, 1*(1), 17–52. doi:10.3390/coatings1010017

Pfender, E. (1994). Plasma jet behavior and modeling associated with the plasma spray process. *Thin Solid Films, 238*(2), 228–241. doi:10.1016/0040-6090(94)90060-4

Pintilei, L. (2008). *The science and engineering of thermal spray coatings* (2nd ed.). Wiley. doi:10.1002/9780470754085

Pratap, B., Bhatt, V., & Chaudhary, V. (2015). A review on thermal spray coating. *International Journal of Scientific and Engineering Research, 6*(5), 53–59.

Rosenkranz, A., Costa, H. L., Baykara, M. Z., & Martini, A. (2020). Synergetic Effects of Surface Texturing and Solid Lubricants to Tailor Friction and Wear - A Review. *Tribology International, 155*, 106792. doi:10.1016/j.triboint.2020.106792

Saeidi, F., Parlinska-Wojtan, M., & Hoffmann, P. (2021). *Effects of laser surface texturing on the wear and failure mechanism of grey cast iron reciprocating against steel under starved lubrication conditions.* . doi:10.1016/j.wear.2017.05.015

Sanjay, L. H., Bhamre, V. G., & Londhe, B. C. (2017). A Review on plasma spray coatings and its characterization. *IRJET, 4*(7), 1627–1631.

Selva, K. N., Krupanidhi, S. B., & Barshilia, H. C. (2014). Carbon nano tubes based tandem absorber with tunable spectral selectivity: Transition from near-perfect blackbody absorber to solar selective absorber. *Advanced Materials, 26*(16), 2552–2557. doi:10.1002/adma.201305070 PMID:24474148

Sharma, A. K. (1992). Anodizing titanium for space applications. *Thin Solid Films, 208*(1), 48–54. doi:10.1016/0040-6090(92)90946-9

Sharma, S. (2012). High Temperature Erosive Wear Study of NiCrFeSiB Flame Sprayed Coatings. *Journal of The Institution of Engineers (India): Series D, 93*(1), 7–12. . doi:10.1007/s40033-012-0006-9

Sidhu, B.S., & Prakash, S. (2006). Performance of Ni-CrAlY, Ni-Cr, stellite –6, and Ni3 Al coating in Na2SO4-60%V_2O_5 environment at 900°C under cyclic conditions. *Surface and Coating Technology, 201*(3), 1643–1654. doi:10.1016/j.surfcoat.2006.02.035

Sidhu, T. S., Prakash, S., & Agrawal, R. D. (2006a). Hot corrosion resistance of high-velocity oxy-fuel sprayed coatings on a nickel-base super alloy in molten salt environment. *Thermal Spray Technology.*, *15*(3), 387–399. doi:10.1361/105996306X124392

Sidhu, T. S., Prakash, S., & Agrawal, R. D. (2006b). Hot corrosion behaviour of HVOF sprayed NiCrBSi coatings on Ni- and Fe- base super alloys in Na2SO4-60%V2O5 environment at 9000C. *Acta Materialia (Elsevier)*, *54*(3), 773–784. doi:10.1016/j.actamat.2005.10.009

Singh, H., Kumar, S., & Kumar, R. (2021). Overview of Corrosion and its Control: A Critical Review. *Proceedings on Engineering Sciences.*, *3*(1), 13–24. Advance online publication. doi:10.24874/PES03.01.002

Singh, H., Prakash, S., Puri, D., & Phase, D. M. (2006). Cyclic oxidation behavior of some plasma-sprayed coatings in Na2SO4-60%V2O5 environment. *Materials Engineering and Performance.*, *15*(6), 729–741. doi:10.1361/105994906X150858

Singh, L., Chawla, V., & Grewal, J. S. (2012). A review on detonation gun spray coating. *Journal of Minerals & Materials Characterization & Engineering*, *11*(03), 243–265. doi:10.4236/jmmce.2012.113019

Siwal, S. S., Saini, A. K., Rarotra, S., Zhang, Q., & Thakur, V. K. (2021). Recent advancements in transparent carbon nanotube films: Chemistry and imminent challenges. *Journal of Nanostructure in Chemistry*, *11*(1), 93–130. doi:10.100740097-020-00378-2

Smith, M. F. (2007). Comparing cold spray with thermal spray coating technologies. *The Cold Spray Materials Deposition Process*, 43–61. doi:10.1533/9781845693787.1.4

Sreenivasulu, V., & Manikandan, M. (2019). Hot corrosion studies of HVOF sprayed carbide and metallic powder coatings on alloy 80A at 900ºC. *Materials Research Express*, *5*, 42–53. doi:10.1088/2053-1591/aaf65d

Staia, M. H., Mejias, A., La Barbera, G., Puchi-Cabrera, E. S., Villalobos-Gutierrez, C., Santana, Y. Y., Montagne, A., Iost, A., & Rodriguez, M. A. (2020). Mechanical properties of WC/Co-CNT HVOF sprayed coatings. *Surface Engineering*, *36*(11), 1156–1164. doi:10.1080/02670844.2018.1529285

Steve, O., & Bhadeshia, H. K. D. H. (2012). Duplex hardening of steels for aero engine Bearings. *ISIJ International*, *52*(11), 1927–1934. doi:10.2355/isijinternational.52.1927

Sudharshan Phani, P., Srinivasa Rao, D., Joshi, S., & Sundararajan, G. (2007). Effect of process parameters and heat treatments on properties of cold sprayed copper coatings. *Journal of Thermal Spray Technology*, *16*(3), 425–434. doi:10.100711666-007-9048-1

Sun, L., Yuan, G., Gao, L., Yang, J., Chhowalla, M., Gharahcheshmeh, M. H., & Liu, Z. (2021). Chemical vapour deposition. *Nature Reviews Methods Primers*, *1*(5), 5. Advance online publication. doi:10.103843586-020-00005-y

Sundaresan, C., Rajasekaran, B., & Rao, D. S. (2020). Hot corrosion behavior of plasma and D-Gun sprayed coatings on T91 steel used in boiler applications. *IOP Conference Series. Materials Science and Engineering*, *872*(1), 45–56. doi:10.1088/1757-899X/872/1/012092

Talib, R. J., Saad, S., & Toff, M. R. M. (2003). Thermal spray coating technology a review. *Solid State Science and Technology*, *11*, 109–117.

Tang, W., Zhou, Y., Zhu, H., & Yang, H. (2013). The effect of surface texturing on reducing the friction and wear of steel under lubricated sliding contact. *Applied Surface Science*, *273*, 199–204. doi:10.1016/j.apsusc.2013.02.013

Thakur, L., & Vasudev, L. (2021). *Thermal Spray Coatings* (1st ed.). CRC Press. doi:10.1201/9781003213185

Thorpe, M. L. (1993). Thermal Spray: Industry in Transition. *Advanced Materials & Processes*, *143*(5), 50–56.

Thorpe & Merle. (1993). Thermal spray: Industry in transition. *Advanced Materials & Processes*, *143*(5), 50–56.

Tucker. (1994). Thermal Spray Coatings. ASM Handbook, Journal of Surface Engineering, 5, 497-509.

Uyulgan, B., Dokumaci, E., Celik, E., Kayatekin, I., Ak Azem, N. F., Ozdemir, I., & Toparli, M. (2007). Wear behavior of thermal flame sprayed FeCr coatings on plain carbon steel substrate. *Journal of Materials Processing Technology*, *190*(1-3), 204–210. doi:10.1016/j.jmatprotec.2007.02.044

Vasudev, H., Singh, G., Bansal, A., Vardhan, S., & Thakur, L. (2019). Microwave heating and its applications in surface engineering: A review. *Materials Research Express*, *6*(10), 52–59. doi:10.1088/2053-1591/ab3674

Verlinden, B. (2005). Severe Plastic Deformation, Metalurgija-. *Journal of Metallurgy*, *11*(3), 1–17. doi:10.30544/380

Vipin, S., Ujjwal, P., & Manoj, K. B. V. (2015). Surface composites by friction stir processing: A review. *Journal of Materials Processing Technology*, *224*, 117–134. doi:10.1016/j.jmatprotec.2015.04.019

Wang, D., & Bierwagen, G. P. (2009). Sol–gel coatings on metals for corrosion protection. *Progress in Organic Coatings*, *64*(4), 327–338. doi:10.1016/j.porgcoat.2008.08.010

Williamson, R. S., Disegi, J., Griggs, J. A., & Roach, M. D. (2013). Nanopore formation on the surface oxide of commercially pure titanium grade 4 using a pulsed anodization method in sulfuric acid. *Journal of Materials Science. Materials in Medicine*, *24*(10), 2327–2335. doi:10.100710856-013-4985-3 PMID:23807314

Xing, Y. Z., Wang, G., Zhang, Y., Chen, Y.-N., & Dargusch, M. (2017). Development in plasma surface diffusion techniques of Ti-6Al-4V alloy: A review. *International Journal of Advanced Manufacturing Technology*, *92*(5-8), 1901–1912. doi:10.100700170-017-0302-5

Xiu-Tian, Y., & Yongdong, X. (2010) *Chemical vapour deposition an integrated engineering design for advanced materials*. Springer-Verlag London Limited.

Zhong, X., Li, Q., Hu, J., Zhang, S., Chen, B., Xu, S., & Luo, F. (2010). A novel approach to heal the sol–gel coating system on magnesium alloy for corrosion protection. *Electrochimica Acta*, *55*(7), 2424–2429. doi:10.1016/j.electacta.2009.11.063

KEY TERMS AND DEFINITIONS

Anodizing: It is a coating deposition electrolytic passivation process which is utilized to form oxide layer onto the surface of a metal parts.

Chemical Vapor Deposition (CVD): It is a versatile process mainly employed to synthesize of coatings, fibers, powders, and nanotubes. In addition, CVD can be utilized to produce most metals, nonmetallic elements including carbon, silicon, and a large number of compounds.

Functionally Graded Materials (FGMs): are novel materials whose properties change gradually with regards to their dimensions. These materials may be characterized by variation in composition and structure gradually over volume. This results in corresponding change in the characteristics of the material.

Physical Vapor Deposition (PVD): It is a vacuum coating deposition process in which the solid coating material vaporizes and then vapor is deposited onto the prepared substrate.

Thermal Spray: Thermal spraying methods are the group of coating processes in which the material (metals, ceramics, cermet, and polymer) are fed in powder or rod or wire form to a torch where the materials are heated/melted, and accelerated towards a prepared substrate surface in the gas stream to develop a splat.

Tribology: The term tribology is derived from the Greek word called tribos meaning rubbing, so that the literal meaning would be 'the science of rubbing'. It is the study of friction, wear, and the lubrication of interacting surfaces.

Chapter 6
Cold Spray Method for Wear-Resistant Surface Coating:
Supersonic Jet Structure and Its Impact on the Particle Deposition Process

Srikrishnan A. R.
Department of Aerospace Engineering, Amrita School of Engineering, Amrita Vishwa Vidyapeetham, India

ABSTRACT

This chapter is focused on cold spray deposition of particles for surface modification. The method, which has been recently proven to have wide applicability in the domain of tribology and wear-resistant coatings, relies on supersonic gaseous jets to deposit the particle without phase change. The chapter aims at examining the influence of the unique gas dynamic characteristics of the high-speed jets on the deposition process. The general structure of the supersonic jets, including the velocity field, pressure gradients, and the impingement behaviour, is discussed with specific attention to the requirements of the sprays for tribological coatings. Results of detailed numerical simulation of the impingement process are made use of to demonstrate the parametric influence of the supersonic jet structure on critical spray characteristics, like the particle velocity. The study also examines various aspects of the energy conversion as applied to the basic nature of the supersonic jet as well as its interaction with the microparticles.

INTRODUCTION

Cold spray deposition of micro-particles is a relatively new technique of surface coating, which has recently found extensive applicability in improving the wear resistance of surfaces that are subjected to mechanical contact and friction (He & Hassani, 2020). In this method, the micro-particles (typically in the diameter range of 5 to 50 microns) are carried by a driver gas which is pressurized and accelerated to supersonic velocity by an appropriately contoured nozzle and this high velocity stream is made to

DOI: 10.4018/978-1-7998-9683-8.ch006

impinge on the substrate, the surface to be modified. The high kinetic energy imparted to the particles by the supersonic stream of the driver gas leads to plastic deformation during the impact and thereby results in their deposition on the surface. The cold spray approach differs from the conventional thermal spray deposition process in two critical aspects: (i) As the name indicates, the method is characterized by a relatively "cold" system: neither the particles nor the substrate are heated to their melting temperature – Hence it is a completely solid-state process (ii) The driver gas is accelerated (by way of an appropriately designed nozzle) to a high velocity, exceeding the velocity of sound in the gas ('supersonic' velocity).

Figure 1 provides a schematic representation of the cold-flow setup. Since both the particle material and the substrate material are maintained at temperatures significantly lower than their melting temperatures, the microstructure of the coating will be superior to what would result in a conventional thermal spray process where the temperatures are high. In conventional thermal spray deposition the particles are melted and accelerated through either a flame or thermal-plasma arc. The molten or semi-molten materials solidify on impingement and get deposited as a thin film (Herman, Sampath & McCune, 2000). In cold spray method, since there is no phase change involved the process, the deposition does not entail tensile residual stresses and the concomitant distortion (Wang & Zhang, 2012), which are some of the disadvantages of thermal spray deposition. Also, the lower temperature levels ensure that surface deposition of materials like Aluminium, Copper and Titanium can be carried out without the risk of them getting oxidized (Ghelichi & Guagliano, 2009). From a metallurgical perspective, grain growth phase transformation is mostly avoided or reduced in comparison to the traditional thermal spray processes (Chenxi et al, 2019 and Karthikeyan, 2007). Substrates which are sensitive to temperature can also be deposited on by this process (Karthikeyan, 2007).

Figure 1. A Schematic representation of Cold Spray particle deposition system

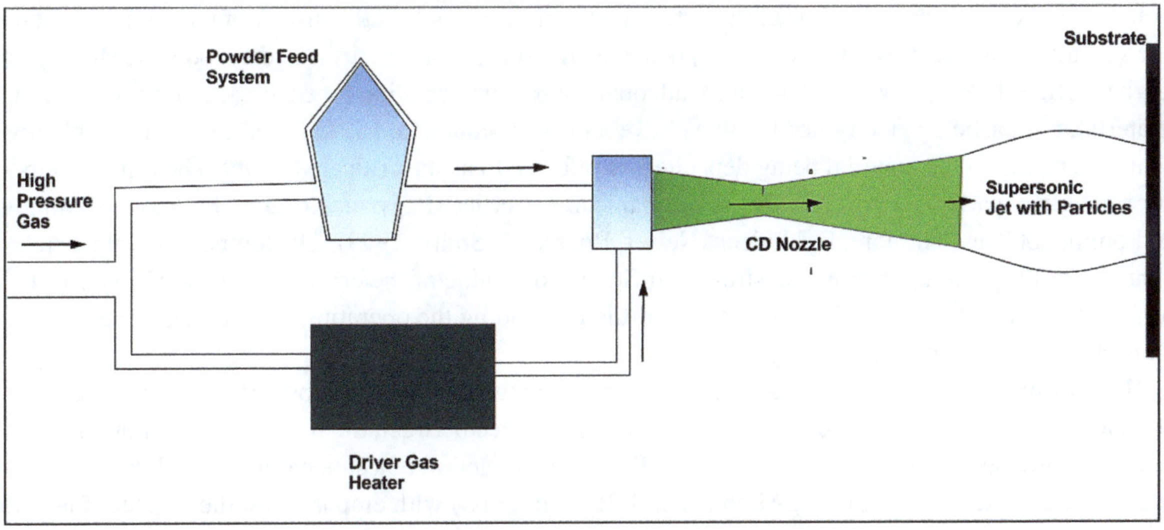

Specific to the applications in tribology, the capability of cold spray process in providing wear resistant coatings has attracted research interest in the recent years. While the initial applications were constrained to ductile materials, more recent investigations have shown that with adequate pressurization of the driver gas the cold spray method can also be developed to provide wear-resistant coatings (Ghelichi & Guagliano, 2009 and Chromik, Alidokht, Shockley & Zhang, 2018) which opens up a wider arena of applicability of this technique in tribology. Effectiveness of this method in depositing metal composites with embedded ceramic materials is also shown to be promising in very recent reviews of the process (He & Hassani, 2020). Studies also highlight the benefits of combining the process advantage of cold spray deposition with the material property of ceramics specific to tribological applications like gears (Loganathan, Narasimalu, Kannappan & Sekar, 2017).

While the lower temperature levels make cold spray an attractive alternative for the conventional thermal spray techniques, the most challenging aspect of the process is the supersonic jet that acts as the depositing gas stream and the complex fluid dynamics associated with its structure. As the high kinetic energy required to deposit the particles without phase change derives from the velocity of the driver gas, the gas needs to be accelerated to a velocity that is significantly higher than the speed of sound. The structure of a supersonic jet is characteristically different from that of a subsonic jet (Witze, 1974 and Subramanian, Nagarjun, Satish Kumar, Srikanth & Srikrishnan, 2018). The key difference is the presence of sharp local gradients in pressure and velocity –the "shocks"- which form along the free jet stream with a recurring pattern. While there are numerous studies that address the jet structure and shock pattern in supersonic jet, most of them in the past were inspired by the conventional aerospace engineering applications of such jets. Since the performance goals and parameters that drive cold spray surface deposition are unique, studies focussed on those aspects are required for optimizing the process. Ever since the potential of supersonic sprays was identified in surface coating, research on the dynamics of particle laden supersonic jets for deposition has also been gaining momentum.

The velocity with which the particles impinge on the substrate surface is important in the process of deposition (Dykhuizen et al, 1998). Since the deposition takes place with plastic deformation caused by the high kinetic energy of the spray particles, the deposit bonding is most significantly influenced by the velocity of the particles at the time of impact. Velocity of the particle at the instant of its impingement on the substrate determines whether it gets deposited on the surface or the surface gets eroded due to impact of the particle-laden spray. In order to have adequate plastic deformation for successful deposition of the material, the particle velocity needs to be above a critical value (Li, Li, & Liao, 2006). This velocity, which depends on the material being deposited, is referred to as the critical velocity. The critical velocity for a given particle is governed by its mechanical properties (Papyrin, Kosarev, Klinkov, Alkhimov & Fomin, 2007 and Gilmore, Dykhuizen, Neiser, Roemer & Smith, 1999). The temperature of particles (and hence the jet temperature) can strongly influence the value of the critical velocity (Gilomore et al 1999). The critical velocity plays an important role in deciding the operating conditions, which in turn, impact structure of the supersonic jet.

In cold spray, the "free jet" zone terminates with impingement on the substrate. The impact characteristics are significantly influenced by the stand-off distance between the nozzle and the impingement surface (Pattison, Celotto, Khan & O'Neill, 2008). The interaction of a supersonic jet with the substrate was studied by Kosarev, Klinkov, Alkimov, and Papyrin (2003) with emphasis on the impact of nozzle geometry on parameters like the stagnation temperature distribution within the impingement region. The observations of this study are important as many of the jet characteristics that can strongly influence the deposition performance are governed by the geometry of the nozzle. The deposition angle also can have considerable impact on the particle deposition efficiency (Klinkov, Kosarev & Shikalov, 2019 and Klinkov & Kosarev, 2016).

The brief survey presented above brings out two salient aspects that are key to the topic of discussion in this chapter:

1. Cold spray methods and tribological applications: As a surface modification method, the cold spray deposition process has several advantages over the conventional thermal spray techniques specific to the improvement of wear resistance of the surfaces. The research thrust that this method is receiving in tribology in recent years is reflective of the important benefits of it in this specific domain.
2. Impact of the structure and dynamics of the supersonic jet on the performance characteristics of the cold spray deposition system: Almost all the deposition characteristics are strongly influenced by the structure of the supersonic jet which is inherently complex.

This chapter is focused on the discussion of how the shock structure and the related properties of the supersonic jet influence the deposition characteristics. Wherever relevant, the parameters which are of specific interest to surface modification for tribological/wear-resistance applications are addressed. The emerging role of cold spray process in tribology is examined briefly first, followed by an overview of the key characteristics of the supersonic jets that are an integral part of any cold spray deposition system. Subsequently, the influence of these parameters on surface coating characteristics will be discussed. In view of the higher kinetic energy requirement of hard particles that are important for wear-resistant coatings, the energy issues related to the driver gas and the particles will be considered with emphasis on the deposition performance. This also requires consideration of the momentum and heat transfer between the driver gas and the particles. Key results from an ongoing study by the author will be provided at appropriate places to highlight the influence of some of the parameters.

COLD SPRAY PROCESS FOR SURFACE COATINGS IN TRIBOLOGY

Metallic contacts with relative motion are characterized by high friction and the concomitant wear. Modification of surface properties by providing surface coating is one of the most effective methods to mitigate the possibility of wear in applications where reliability and durability are important. Modifying the wear-resistance of the surface by applying coatings of materials with high hardness surface can improve the friction resistance; applying a solid lubricant coating can reduce friction – Both of these methods are widely used in tribology (Erdemir and Voevodin 2010).

In cold spray deposition method, the surface coating is accomplished by the adhesion of the coating particles to the surface of the substrate by way of plastic deformation caused by the high kinetic energy of the jet. In addition to the velocity of particles at the time of impact, the temperature and pressure of the driver gas can also have strong influence on the deposition process. Another parameter of importance is the spray distance which is the distance between the nozzle of the spray gun and the surface for deposition. This is the path that the particles travel after leaving the nozzle till the collide with the substrate.

Since wear-resistant coatings require the deposition of hard materials, the design of the cold spray system for tribological applications should be tailored to meet the requirements of materials which are characterized by low ductility. Hence each of the operational parameters should be evaluated in view of the specific requirements of this domain of application. Recent advances in cold spray research have shown that the process can be effectively used for the deposition of wear-resistant materials including metal matrix composites (MMC) (Chromik et al, 2018, He and Hasani 2020 and Xie et al, 2019), with appropriate selection of the operating conditions. The overall design of the cold spray system for any such application has to be made with due consideration of the parameters that influence the gas dynamic structure of the driver gas jet.

SUPERSONIC JET IMPINGEMENT: SHOCK STRUCTURE AND ITS IMPACT ON THE PROCESS

Cold spray fundamentally consists of a particle-laden jet at supersonic velocity, impinging directly on a surface. Since both the supersonic jet structure and the unique gas dynamic features of the impingement are critical to the deposition process, these two should be individually examined to gain insight to the parameters that influence the relevant characteristics of the coatings. With this perspective, the jet structure and the impingement zone are analysed separately in the following discussion.

Structure of a Supersonic Jet

The most important phenomenon that characterizes the supersonic flow field in general and the supersonic jet structure in particular is the formation of *shock waves* and *expansion waves* (pressure waves that lead to compression and expansion respectively) (Anderson, 2003). All practical applications of supersonic flows will encounter the formation of these waves. A shock wave is highly dissipative in nature, spans over a very thin region (thickness being of the order of a few microns at the most) and manifests as a pressure discontinuity which leads to a rise in static pressure and drop in velocity and stagnation pressure.

In the schematic diagram shown in Figure 1, the jet is shown to be issued from a nozzle which has a "convergent-divergent" (CD) design for the flow passage. This geometry (usually referred to as the *de Laval nozzle*) is required to accelerate the driver gas to supersonic velocity. The velocity field is most often characterized by the non-dimensional parameter *Mach number* defined as $M = \frac{V}{\sqrt{\gamma RT}}$ where, γ is the ratio of specific heats (C_p/C_v) of the driver gas, R is the specific gas constant (J/kg-K) and T is the local static temperature in Kelvin. Mach number is the ratio of the local gas velocity to the velocity of sound through the gas at the local temperature. The divergent section of the nozzle is designed based on the gas dynamic formulations that relate the geometry to the exit Mach number.

The nozzle performance and the jet characteristics are strongly influenced by one of the most important parameters related to the nozzle operation, the nozzle pressure ratio (NPR). NPR is defined as the ratio of the stagnation pressure (often assumed to be constant within the nozzle) and the ambient pressure at the exit of the nozzle (Anderson, 2003). When NPR is different from the value that corresponds to 'ideal' expansion, a series of shock waves and expansion waves are formed in the jet field (downstream of the nozzle). If the flow static pressure at the exit is lower than the ambient pressure, the jet is said to be "over-expanded" and if it is higher than the ambient pressure, the jet is "under-expanded". The ideal expansion corresponds to the condition where the static pressure of the gas at nozzle exit equals the ambient pressure. Both under-expansion and over-expansion mark deviation from the ideal expansion. These two non-ideal conditions are common in real applications and they lead to the formation of the compression and expansion waves downstream of the nozzle exit. In a free jet, the compression waves incident on the jet boundary reflect as expansion waves; expansion waves reflect from the jet boundary as compression waves (Anderson, 2003). Figure 2 schematically shows the flow field of an under-expanded supersonic jet. Such a flow pattern is quite common in cold spray applications.

Figure 2. Schematic diagram of the characteristic features of a supersonic free jet

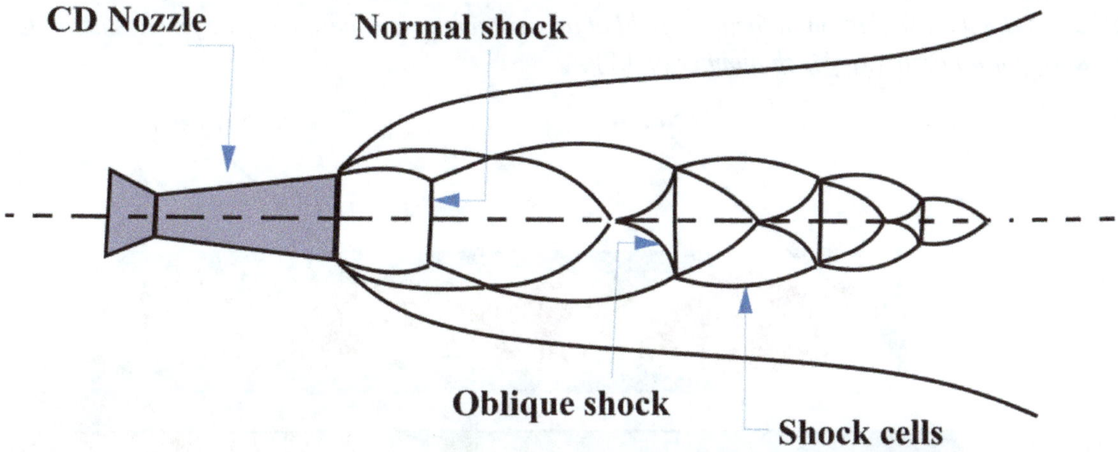

While the pressure gradients caused by the shocks and the expansion waves dominate the flow field, the following key elements of the jet structure are important and each of them is marked in the Figure 2: (i) The mixing layer that marks the outer boundary of the jet where the jet mixes with the ambient air (ii) The *Mach disk* which refers to the strong shocks near the axis of the jet where the wave angle is normal to the axis (iii) The *oblique* shock waves which are also compression waves, but inclined at an angle with the jet axis (iv) the expansion fans which are standing pressure waves that lead to a drop in static pressure and increase in velocity.

With reference to the schematic representation in Figure 2, the repeating pattern of shocks is usually referred to as "shock diamonds" (due to their appearance) or "shock cells" (due to their periodic occurrence) – Numerical correlations for the length of shock cells are provided by Murugesan et al (2020). The development of the unconfined jet is governed by the pressure equilibrium between the shock-infested region and the ambient air. Along the axis of the jet, the static pressure may be so low (due to the preceding expansions) that it will require recompression through the formation of a strong *normal* shock. This normal shock (marked in Figure 2) is often referred to as *shock disk* or *Mach disk*. The flow just down downstream of the normal shock is subsonic, whereas the outer region that undergoes oblique shock will continue to be supersonic. A *slip line* (a line of velocity gradient) separates the core subsonic region and the outer supersonic region. The subsonic region accelerates along and becomes supersonic before the next cell. Further decay of the jet almost follows this pattern as long as the jet remains supersonic (the normal shock may be replaced by oblique shock depending on the pressure gradient). Detailed studies on free jet shock structure can be seen in Zapryagaev, Kiselev, Gubanov (2018) and Zapryagaev et al (2017). Figure 3 shows a typical jet structure for an underexpanded jet, based on numerical simulations of jet with a nozzle exit Mach number of 3.7. The velocity field shown in the figure illustrates the recurring pattern of the shock structure along the jet – For clarity, only the near-field of the nozzle is shown in the figure. [The simulation results presented in this chapter are based on numerical modelling carried out using the commercially available computational fluid dynamics (CFD) tool, ANYS Fluent. Validation of the simulation process with experimental data can be found in Sudhan, Prasad, Kothurkar, & Srikrishnan, 2020]. It may be noted that repeating pattern of acceleration and deceleration along the axis of the jet is reflective of the formation of shocks and expansion waves in the stream-wise direction.

Figure 3. Velocity distribution in the near-field of an axisymmetric supersonic free jet (Results of CFD simulation for a nozzle exit Mach number of 3.7)

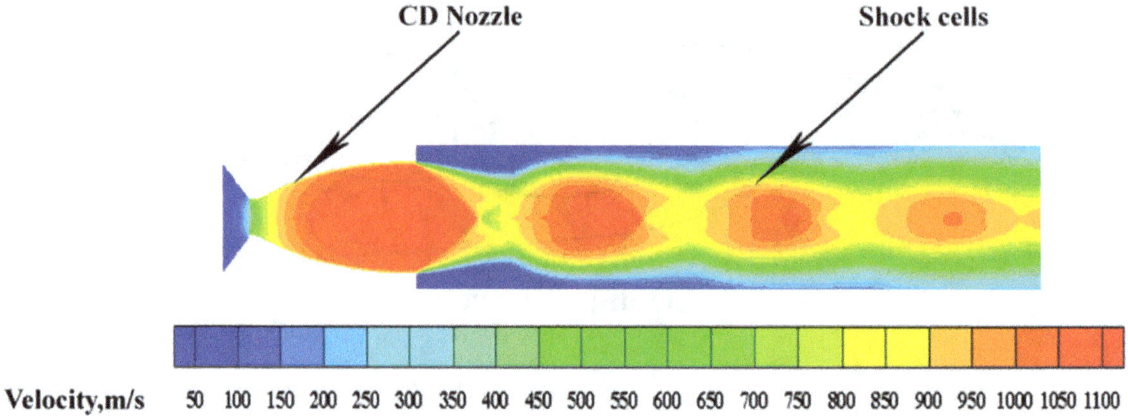

In typical applications of cold spray process, flow field of the driver gas will be dominated by the jet structure described above. This shock pattern will also have a strong influence on the deposition process. From the perspective of the concerned application, the most important question in this context is: How does the shock pattern in the driver gas stream affect the deposition ? This can be best answered by considering the particle velocity achieved. Since the driver gas velocity is what imparts the high kinetic energy required for the particles to undergo plastic deformation upon impact, the gas velocity and the particle impact velocity are strongly coupled. Figure 4 compares the particle velocities (based on numerical simulations) for five values of storage tank pressure (P_o). Each data point in the plot represents a specific case of simulation. It may be noted that all the plots in Figure 4 correspond to the same vale of Mach number at the nozzle exit. The gas velocity distributions along the jet (for these conditions) also differ both qualitatively and quantitatively. This is one of the unique characteristics of supersonic jets and is important in their application to cold spray.

Figure 4. The variation of particle velocity with supply pressure of the driver gas (Results of CFD simulations for a particle material density of 2700 kg/m³ and air as the driver gas)

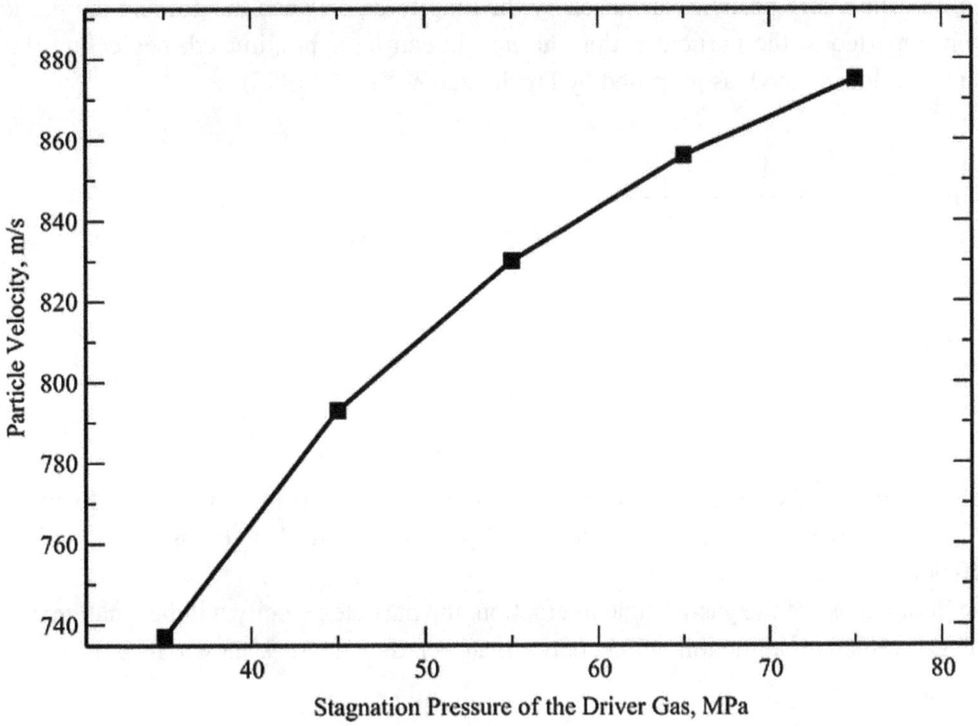

Since tribological coatings require particles that have characteristically higher critical velocities, the dependence of velocity field on the operating conditions is particularly important in such systems.

Particle-Laden Spray: The Jet Structure and Energy Considerations

Fluid dynamically, the most important aspect of the technological process of cold spray deposition is the energy transfer between the driver gas and the micro particles. Plastic deformation of the particles, which leads to the bonding with the substrate surface, is caused by the high-velocity impact which in turn is caused by the kinetic energy gained from the driver gas. Hence the influence of the jet structure on the kinetic energy of the jet and the factors that influence the transfer of energy from the jet to the micro-sized particles are important in the analysis and design of the cold spray system.

Since the process relies on the energy exchange between the gas and the solid particles the following two aspects of the process are important:

1. Interaction between the gas (the continuum phase) and the particles (which are dispersed in the fluid stream)
2. Structure of the jet and the resulting velocity field that accelerates the particles.

Particle Acceleration by the Gas

Acceleration of the micro-particles is caused by the drag force between the gas and the particles. The acceleration imparted to the particle within the nozzle can be approximated, neglecting the particle interaction (for dilute sprays), as proposed by Dykhuizen & Smith (1998).

$$m\frac{dV_p}{dt} = mV_p\frac{dV_p}{dx} = \frac{C_D A_p \rho \left(V - V_p\right)^2}{2} \tag{1}$$

(as $V_p = \frac{dx}{dt}$)

In (1) the subscript p indicates particle properties. A, V, m and ρ denote the area, velocity, mass and density respectively. The spatial and temporal parameters are denoted as *x* and *t* and C_D is the coefficient of drag.

During the initial phase of the gas-particle interaction, the particle velocity will be small in comparison to the gas velocity and the following approximate relations hold as shown in [21]:

$$V_p \propto \sqrt{\frac{\rho_g}{m}}, or\, V_p \propto \sqrt{\frac{\rho_g}{\rho_p D_p}} \tag{2}$$

It may be noted that the parameter on the RHS of relation (2) characterizes the susceptibility of the solid particle to be accelerated by the gas. In (2) the subscript *g* denotes gas properties. A higher value for the local density of the gas (ρ_g) leads to higher particle velocity. Since the density is strongly influenced by pressure, this shows the importance of the gas pressure as well. Assadi, Schmidt & Richter (2011) have proposed the following relation for the maximum particle velocity that can be achieved in a nozzle for a given value of stagnation temperature (which in turn can be approximated as the temperature in the storage tank of the driver gas):

$$v_p = f \cdot \left(\left[\frac{1}{v_{max}} + \frac{1}{v_1}\right]^{-1}\right) \tag{3}$$

where *f* is a coefficient of correlation, approximately equal to 1.

Based on an analytical model of the cold spray system, the authors the study cited above showed that for typical applications the following relation can be used to estimate the highest particle velocity achievable under a set of operating conditions, particularly when the storage pressure (p_o) and temperature are high:

$$a_{ref}^{2} = \frac{C_d L_d p_o}{\rho_p d_p} \quad (4)$$

Correlations that can be used to calculate the drag coefficients for spherical particles over a wide range of Reynolds numbers is given by Morsi and Alexander (1972).

Value of the drag coefficient in equation (4) will be influenced by the compressibility of flow and the Mach number. For supersonic flows correlations have been proposed based on empirical formulations, as summarized in Chenxi et al (2019).

The formulations for particle velocity and acceleration discussed above are valid only within the nozzle; as the particles leave the nozzle they pass through the shock-infested flow field of the jet (Morsi and Alexander, 1972). Deceleration of the gas due to the shock formation in the jet will also have an adverse impact on the magnitude of particle velocity. Particularly, In the near field of the substrate a normal shock is formed (often referred to as the *"bow"* shock due to its shape) which significantly decelerates the gas and hence the impingement velocity of the particle can also be affected to some extent. Hence the flow physics of the impingement region needs to be separately considered in view of the importance of impact velocity of the particles.

Structure of the Jet Impingement Region

As stated earlier, the velocity of particles at the time of impact needs to be higher than the critical velocity for the plastic deformation for a given material to be sufficient for bonding on to the substrate. Hence the jet flow field near the substrate deserves particular attention. The impingement of a supersonic jet on a surface is characterized by the formation of a *detached* shock – a normal shock formed at a stand-off distance from the surface. Concomitant to the formation of this shock, the flow abruptly changes to subsonic just upstream of the stagnation region. This results in an abrupt increase in the density, static pressure and static temperature locally, along with a significant reduction in the gas velocity downstream.

Due to the structural changes in the flow field at the location of the impingement, the resulting flow field (adjacent to the impingement surface) can be considered to be consisting of three different regions with distinct characteristics: (i) The free jet zone that terminate with the formation of the normal shock at a stand-off distance from the impingement surface (ii) The zone of impingement where the fluid-surface interaction can be considered to be practically of inviscid nature and (iii) The radial flow region where the wall-bound jet is directed outward after impingement (Donaldson et al, 1971).

Figure 5 shows a schematic representation of the impingement region, highlighting the various phenomena adjacent to it. The recirculation zone close to the axis is to be noted. The impingement does not *always* lead to recirculation near the wall – Depending on operating conditions, a normal flow field without the recirculation is also possible (Donaldson et al, 1971). The static pressure distribution along the substrate wall subjected to the impingement of supersonic jet is shown in Figure 6 for three values of upstream stagnation pressure (CFD simulations with a Mach 3.7 jet). The dip in static pressure observed at the central location of the wall is attributed to the formation of the recirculation zone as pointed out by Xu, Lin, Liu & Mo, (2010). The wall temperature distribution for two operating conditions are shown in Figure 7 – as expected the central region has the highest temperature owing to the normal impingement.

Figure 5. Impingement of supersonic jet on a straight wall – Schematic representation of the salient aspects of the flow field. The impingement region encloses a low velocity zone characterized by recirculation

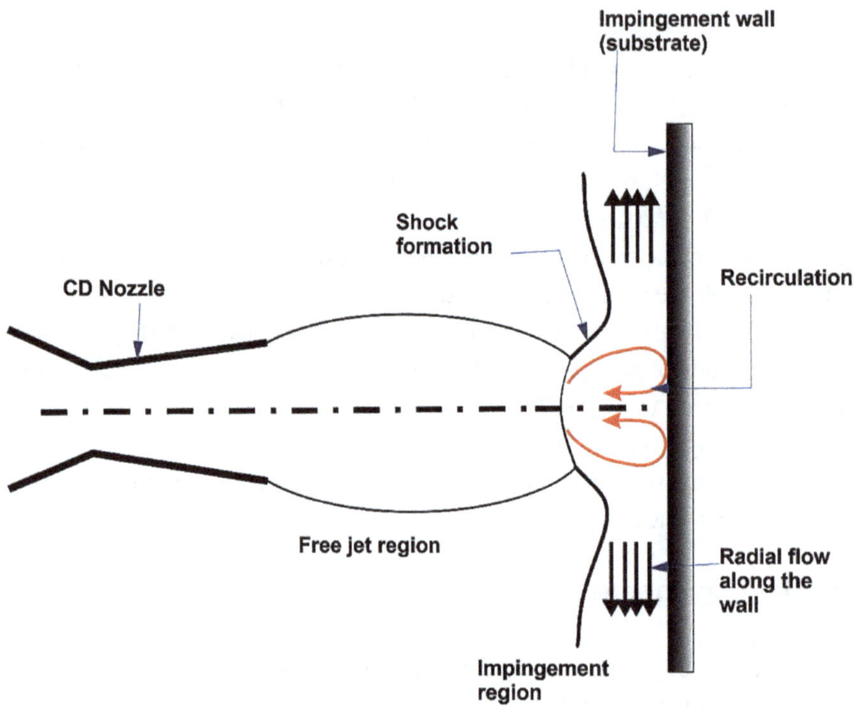

Figure 6. Static pressure along the wall of impingement (the substrate)

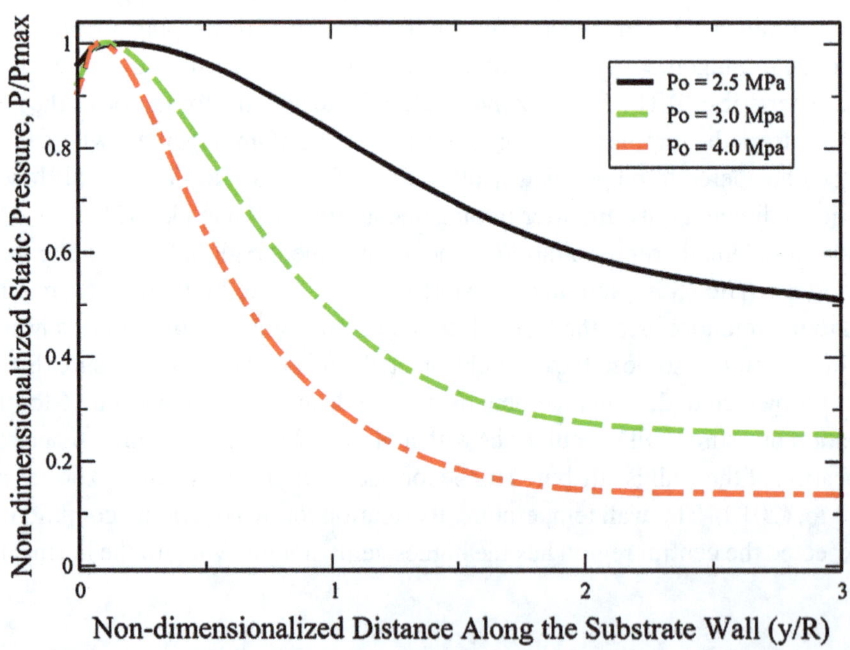

Figure 7. Static temperature along the wall of impingement (the substrate)

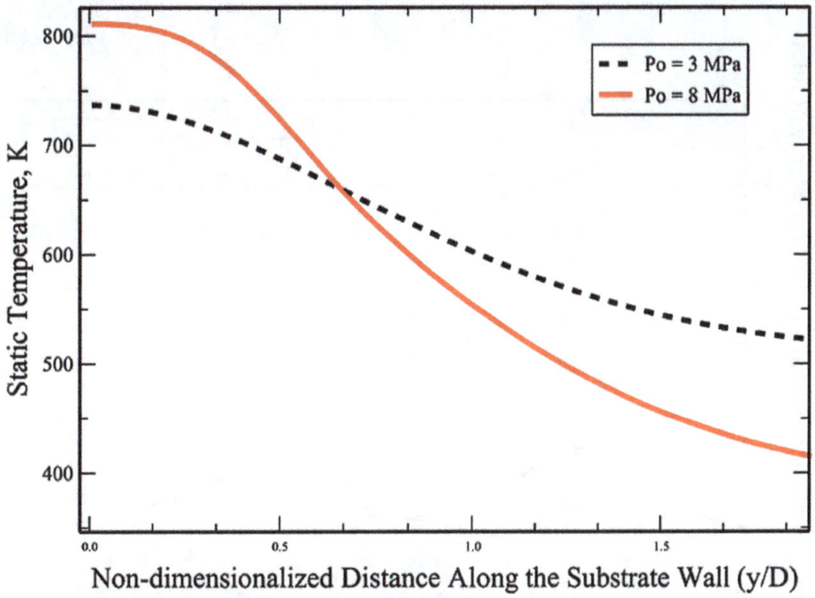

The shock formation near the wall and the ensuing flow pattern in the stagnation zone will have an impact on the deposition efficiency as well. Akhtar_and Ragab (2014) point out that the velocity of particles with larger diameters will be less influenced by the shock, but the deposition efficiency may be lower because they get accelerated to a lower velocity range. These observations point at the possibility of optimizing the particle diameter in view of improving the rate of deposition.

Since tribological applications typically require higher velocity of impact due to the hard nature of the depositing material, the particle-size optimization mentioned here has even more practical relevance in such applications.

Shock Structure and the Energy Conversion

Shock structure of the jet has an important concomitant effect on the energy of the jet: (i) Each occurrence of the shock wave (normal or oblique) results in an irreversible conversion of part of the kinetic energy of the driver gas to its internal energy. As a result of this the jet velocity reduces and the local static temperature of the jet increases. Figure 8 shows the variation of internal energy and kinetic energy along the axis of a supersonic jet (Results of axisymmetric CFD simulations done for a CD nozzle that operates with an exit Mach number of 3.7 and an NPR = 30; The gas (air in this case) is assumed to be calorically perfect). The variations in kinetic energy and internal energy are reflective of the nature of the shock structure along the axis. Across a shock the kinetic energy drops due to its dissipation as internal energy. This leads to a rise in the internal energy. This pattern repeats across each shock cell. The axial variations of the two of forms of energy as reflected by the respective curves in the figure are due to this repeating pattern of energy conversion. (As stated earlier, the deceleration of the driver gas caused by the shock structure may adversely impact the particle velocity as well and this can eventually affect the deposition efficiency).

Figure 8. Variation of internal energy and kinetic energy along the axis of a supersonic jet with a nozzle exit Mach number of 3.7

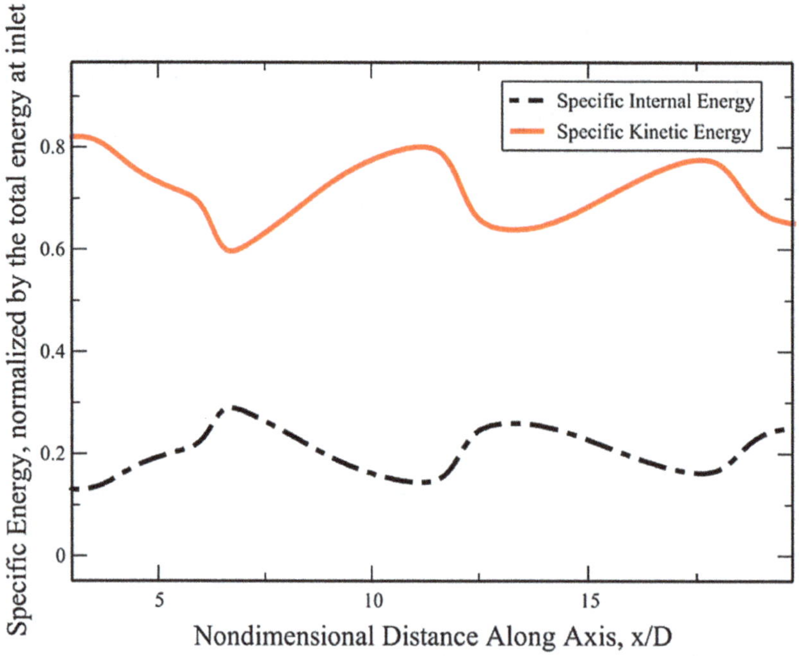

In the context of analysis of this conversion between kinetic and internal energies in a cold spray system, the non-dimensional parameter Mach number is of particular importance. While the most obvious definition of Mach number is as the ratio of the local velocity of the gas (V) and the velocity of sound through it (a) corresponding to the local static temperature (T), it can be easily shown that the square of Mach number is also proportional to the ratio of the of kinetic energy to the internal energy for a given gas (Anderson, 2003). This shows that from the perspective of the kinetic energy requirement for deposition, a high Mach number (at any location along the jet) is desirable as it implies that the kinetic energy of the gas is not significantly dissipated into thermal energy, which in turn facilitates the particles to get accelerated to the critical velocity required for deposition. Since the Mach number variation along the jet is governed by the shock cells their structure should be given due attention while designing the deposition system.

The operating conditions of the jet facility is important in view of the energy considerations highlighted above. Several studies like those of Huang & Fukanuma (2009) and Sudhan et al (2020) have examined the impact of parameters such as the temperature and composition of driver gas on the deposition performance of the system.

The Influence of Gas Properties on Jet Characteristics

High-speed flows, in general, are impacted by the *compressibility* of the gas. In supersonic flows the gas properties strongly influence the flow field due to the compressibility effects. In cold spray applications, the choice of the driver gas is critically important from this perspective. The shock formation, the pressure gradient across the shocks and the resulting velocity field are all functions of the compressibility of the driver gas. For example, Table 1 shows the variation of nozzle exit velocity for a fixed geometry nozzle that has a throat-to-exit area ratio corresponding to a Mach number of 2 at the nozzle exit. The Table shows, under identical operating conditions, the resulting gas velocities at the nozzle exit for various driver gases (these calculations are based on standard compressible flow relations under the assumption of perfect gas; the calculations are done for a static temperature of 600K. Details of the related theory can be found in Anderson (2003).

Table 1. The velocity at nozzle exit corresponding to a Mach number of 2.0 for various gases, based on perfect gas assumption

Gas	Molecular Mass	Velocity at Nozzle Exit, M/S
Helium	1	216
Argon	40	34
Nitrogen	28	40
Air	29	41

Studies on the impact of gas properties on cold spray process have been reported by Yue, Wong, Irissou, and Legoux (2011) with emphasis on the deposition of Titanium. In applications where a higher critical velocity is to achieved (as is the case with coatings for wear-resistance) a lighter gas like Helium will be preferable from the gas dynamics perspective.

Heat Transfer Between the Gas and the Particles

The transfer of thermal energy by way of convective heat transfer between the pre-heated high velocity gas and the solid particles is also important in the deposition process. An experimental correlation based on extensive measurements (Assadi, H., Gärtner, T., Stoltenhoff, H. K., 2003) provides a formulation for the critical velocity required for deposition, in terms of some of the process parameters and material properties as follows:

$$\overline{V}_C = 667 - 14\overline{\rho}_g + 0.08\overline{T}_M + 0.1\overline{\sigma}_u - 0.4\overline{T}_i \quad (5)$$

(In (5) ρ is density in g/cm3· TM is the melting temperature in °C, σu_is the ultimate strength in MPa and Ti_is the initial particle temperature in °C).

This relation shows that a high value of particle impact temperature (Ti) can reduce the critical velocity required for deposition, thereby improving the deposition efficiency. This reveals the operational importance of the gas-particle heat transfer. The heat transfer rate can be related to the temperature difference between the particles and the gas through the non-dimensional thermophysical parameters Prandtl number (Pr), Nusselt number (Nu) and Reynolds number (Re) as follows:

$$\frac{dT_p}{dt} = \left(\frac{N_u}{\Pr \mathrm{Re}_p}\right)\left(\frac{\mu A_p}{d_p m_p}\right)\left(T_g - T_p\right) \tag{6}$$

(T_g and T_p denote the static temperature of the gas and the particles respectively; μ denotes dynamic viscosity of the gas). When the non-dimensional parameter Biot Number, defined as the ratio of the volumetric thermal resistance of the particle to the surface thermal resistance, is less than 0.1 the following relation can be used to estimate the particle temperature:

$$m_d C_d \frac{dT_d}{dt} = hS\left(T_c - T_d\right) \tag{7}$$

The heat transfer coefficient h can be determined using the Ranz-Marshal equation (Fan, & Zhu, 1998), Crowe, Sommerfeld & Tsuji, 1998 and Jadidi, Moghtadernejad & Dolatabadi, 2015):

$$Nu = \frac{hd}{k_c} = 2 + 0.6\,\mathrm{Re}_p^{1/2}\,\Pr^{1/3} \tag{8}$$

It is important to note that the shock formation in each cell results in an increase in the local static temperature of the gas, which in turn, can cause an increase in the particle temperature as well. Hence the particle temperature is also strongly coupled to the shock structure in the flow field.

TRIBOLOGICAL APPLICATIONS AND JET DYNAMICS

The recent upsurge in the application of cold spray for tribological/wear-resistant coatings and the allied research primarily relies on the deposition of Metal Matrix Composites (MMC) (Chromik et al, 2018 and Xie et al, 2019) with ceramic reinforcement and mechanical blend of ceramic and metallic powders. Specific to such applications, the operating conditions under which the supersonic jets drive the particles are found to influence the deposition efficiency through the impact velocity as well as the resulting particle temperature. The particle temperature induces thermal softening of the metallic particles and also improve the retention of the reinforcement on the coated layer. The metallic particles in the spray undergo enhanced plastic deformation at higher temperatures, which favours the retention and the bonding (Dosta, Cinca, Vilardell, Cano & Guilemany, 2017 and Lewei & Hassani, 2020).

The driver gas pressure needs to be in the higher range for the deposition of MMCs for wear-resistant coatings. For example, in Chen et al (2019), a high injection pressure of 4-5 MPa WC helped enhance the reinforcement of WC (Tungsten Carbide) in maraging steel 300 composites, thereby significantly improving the wear resistance as well as mechanical strength. The study also showed that better metallurgical bonding and lower interior defects could be achieved by designing the cold spray system to operate with a higher level of particle impact velocity.

In summary, tribological coatings will require typically higher supply pressure for the driver gas and a jet structure that favours higher impact velocity as well as a higher particle temperature. A higher particle density, as in the case of particles like Tungsten carbide (WC), is likely to result in lower particle velocity (for a given set of supply conditions) (Yue et al, 2011). The operating conditions of the jet should be tailored to provide the required impact velocity for the particle material. The particle velocities achieved under identical operating conditions (P_0 = 35 MPa) are compared in figure 9 for three particle materials of particular interest in tribological coatings, namely, tungsten carbide (WC), silicon carbide (SiC) and aluminium oxide (Al_2O_3). (These results are based on discrete particle modelling simulations of the process). Tungsten carbide being a denser material achieves a lower velocity in comparison to the particles of the other two materials. These observations are in line with the findings of Steenkiste (2007) that heavier particles accelerate to lower ranges of velocity in cold spray systems.

Figure 9. Particle velocities achieved under identical operating conditions by three wear-resistant coating particles (Tungsten carbide, Silicon carbide and Aluminium oxide) which differ in density

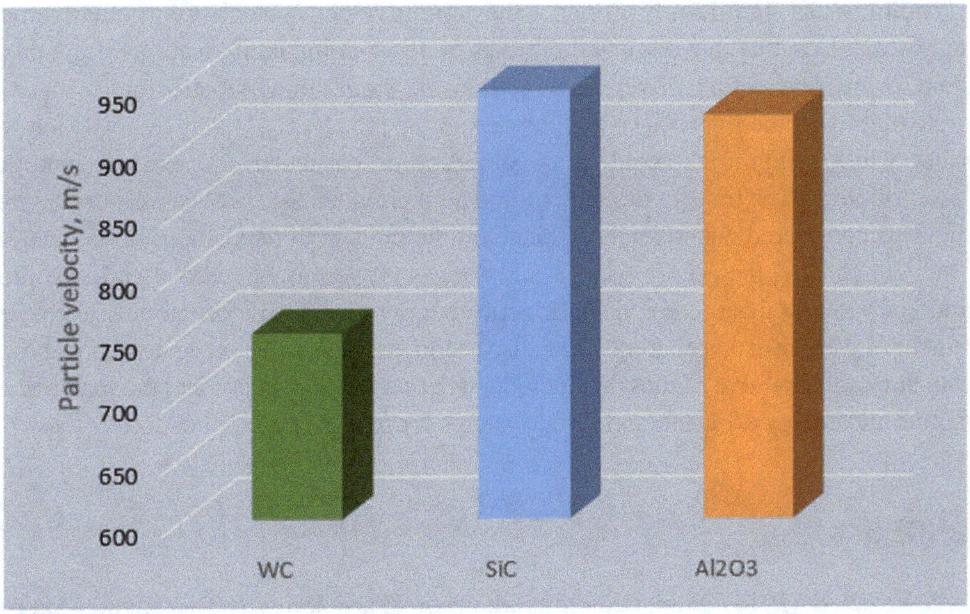

FUTURE DIRECTIONS: THE NEED FOR COLLABORATIVE RESEARCH

As cold spray deposition has been established as a viable, economic and sustainable technique for surface coating with ample potential in the domain of wear-resistant coating, further research should be attuned to tap the full benefits of this method in tribology. It may be noted that this technology has a prominent interdisciplinary character: The material characteristics (of the substrate, the particles and the coating) are all related to material science and chemical engineering, while the spray dynamics which significantly impacts the process fundamentally belongs to a branch of high-speed fluid dynamics/gas dynamics. The coupling between these two domains is so strong in cold spray deposition systems that collaborative research of teams with focused expertise in each of these is required for advancing the technology.

Computational modeling can be of immense value in the analysis of free jets, impinging jets and particle laden jets as has already been shown by several studies. Experiments in supersonic flows are particularly expensive and time-consuming and hence computational modeling can be of great assistance in exploratory studies in this domain. Sustainability and energy efficiency are two important parameters to be considered in the advancement of the cold spray technology for wear-resistant coatings. Optimization studies based on numerical modeling can be carried out to refine the design with adequate focus on these aspects.

CONCLUSION

There is an upsurge in applied research aimed at the development of supersonic cold spray deposition systems that can provide wear-resistant surface coating for tribological applications. This article discusses the impact of the unique fluid dynamic characteristics of supersonic jets on the cold spray process. The jet structure is analysed with emphasis on the parameters that directly influence the deposition process. Operating conditions that significantly influence the particle velocity, which is particularly important in wear-resistant surface coating, are discussed. Studies of an on-going numerical investigation are made use to explain the influence of the driver gas conditions on the resulting jet-structure.

The article highlights the importance of considering the processes of energy conversion within the supersonic jet, while designing the cold spray configuration particularly for surface modification in the domain of tribology. The transfer of momentum and thermal energy between the driver gas and the mirco particles is considered due to its importance on the process of deposition. In the analysis of the impact of the jet structure, the energy transfer and the gas properties, specific attention is provided to the deposition of particles that require higher kinetic energy for plastic deformation – as they represent the requirement of wear-resistant coatings. The discussion and the results presented here show that it is imperative to have collaborative studies that include the gas dynamics of jets and the material aspects to device effective and energy efficient cold spray systems for tribological applications.

REFERENCES

Akhtar, K., & Ragab, S. (2014). Numerical Simulations of a Particle-Laden Rectangular Supersonic Jet Impinging on a Solid Wall. *AIAA Science Tech. Forum, 13-17 January 2014, National Harbor, Maryland, 52nd Aerospace Sciences Meeting.*

Anderson, J. D. (2003). *Modern compressible flow: with historical perspective*. McGraw-Hill.

Assadi, H., Gärtner, T., Stoltenhoff, H. K., & Kreye, H. (2003). Bonding mechanism in cold gas spraying. *Acta Materialia, 51*(15), 4379–4394. doi:10.1016/S1359-6454(03)00274-X

Assadi, H., Schmidt, T., Richter, H., Kliemann, J.-O., Binder, K., Gärtner, F., Klassen, T., & Kreye, H. (2011). On Parameter Selection in Cold Spraying. *Journal of Thermal Spray Technology., 20*(6), 1161–1176. doi:10.100711666-011-9662-9

Chen, C., Xie, Y., Yan, X., Huang, R., Kuang, M., Ma, W., Zhao, R., Wang, J., Liu, M., Ren, R., & Liao, H. (2019). Cold Sprayed WC Reinforced Maraging Steel 300 Composites: Microstructure Characterization and Mechanical Properties. *Journal of Alloys and Compounds, 785*, 499–511. doi:10.1016/j.jallcom.2019.01.135

Chenxi, L., Narendra, S., Austin, A., Bernard, A., Thomas, E., & Christopher, J. (2019). Tribological properties of wear-resistant coatings obtained by cold gas dynamic spray. *International Journal of Heat and Mass Transfer*, *129*, 1161–1171.

Chenxi, L., Singh, N., Andrews, A., Olson, B., Schwartzentruber, T., & Hogan, C. Jr. (2019). Mass, momentum, and energy transfer in supersonic aerosol deposition processes. *International Journal of Heat and Mass Transfer*, *129*, 1161–1171. doi:10.1016/j.ijheatmasstransfer.2018.10.028

Chromik, R. R., Alidokht, S. A., Shockley, J. M., & Zhang, Y. (2018). *Tribological coatings prepared by cold spray*. Cold-Spray Coatings. doi:10.1007/978-3-319-67183-3_11

Crowe, C., Sommerfeld, M., & Tsuji, Y. (1998). *Multiphase Flows with Droplets and Particles*. CRC Press.

Donaldson, C., & Snedeker, R. S. (1971). A Study of Free Jet Impingement, Part 1: Mean Properties of Free and Impinging Jets. *Journal of Fluid Mechanics*, *45*(2), 281–319. doi:10.1017/S0022112071000053

Dosta, S., Cinca, N., Vilardell, A., Cano, I., & Guilemany, J. M. (2017). Cold Spray Coatings for Biomedical Applications. In Cold Spray Coatings: Recent Trends and Future Perspectives. Springer International Publishing.

Dykhuizen, D. C., & Smith, M. F. (1998). Gas Dynamic Principles of Cold Spray. *Journal of Thermal Spray Technology*, *7*(2), 205–212. doi:10.1361/105996398770350945

Dykhuizen, D. C., Smith, M. F., Neiser, R. A., Gilmore, D. L., Jiang, X., & Sampath, S. (1998). Impact of High Velocity Cold Spray Particles. *Journal of Thermal Spray Technology.*, *8*(4), 559–564. doi:10.1361/105996399770350250

Erdemir, A., & Voevodin, A. A. (2010). Nanocomposite coatings for severe applications. In Handbook of deposition technologies for films and coatings. William Andrew Publishing.

Fan, L. S., & Zhu, C. (1998). *Principles of Gas-Solid Flows*. Cambridge University Press. doi:10.1017/CBO9780511530142

Ghelichi, R., & Guagliano, M. (2009). Coating by the Cold Spray Process: A State of the Art. *Frattura Ed Integrità Strutturale*, *3*(8), 30–44. doi:10.3221/IGF-ESIS.08.03

Gilmore, D. L., Dykhuizen, R. C., Neiser, R. A., Roemer, T. J., & Smith, M. F. (1999). Particle velocity and deposition efficiency in the cold spray process. *Journal of Thermal Spray Technology*, *8*(4), 576–582. doi:10.1361/105996399770350278

He, L., & Hassani, M. A. (2020). Review of the Mechanical and Tribological Behavior of Cold Spray Metal Matrix Composites. *Journal of Thermal Spray Technology*, *29*(7), 1565–1608. doi:10.100711666-020-01091-w

Herman, H., Sampath, S., & McCune, R. (2000). Thermal Spray: Current Status and Future Trends. *MRS Bulletin*, *25*(7), 17–25. doi:10.1557/mrs2000.119

Huang, R. Z., & Fukanuma, H. (2009) The Influence of Spray Conditions on Deposition Characteristics of Aluminum Coatings in Cold Spraying. *Thermal Spray 2009: Proceedings of the International Thermal Spray Conference*.

Jadidi, M., Moghtadernejad, S., & Dolatabadi, A. (2015). A comprehensive review on fluid dynamics and transport of suspension/liquid droplets and particles in high-velocity oxygen-fuel (HVOF) thermal spray. *Coatings*, *5*(4), 576–645. doi:10.3390/coatings5040576

Karthikeyan, J. (2007). The advantages and disadvantages of the cold spray coating process. In V. K. Champagne (Ed.), *Woodhead Publishing Series in Metals and Surface Engineering, The Cold Spray Materials Deposition Process* (pp. 62–71). Woodhead Publishing.

Klinkov, S., & Kosarev, V. (2016). Fundamentals of cold spraying. *AIP Conference Proceedings*, *1770*. Advance online publication. doi:10.1063/1.4963926

Klinkov, S. V., Kosarev, V. F., & Shikalov, V. S. (2019). Control of cold spray process by changing of nozzle setting angle. *AIP Conference Proceedings,* 21-25. doi:10.1063/1.5117382

Kosarev, V. F., Klinkov, S. V., Alkimov, A. P., & Papyrin, A. N. (2003). On Some Aspects of Gas Dynamics of the Cold Spray Process. *Journal of Thermal Spray Technology.*, *12*(2), 265–281.

Lewei, H., & Hassani, M. (2020). A Review of the Mechanical and Tribological Behavior of Cold Spray Metal Matrix Composites. *Journal of Thermal Spray Technology*, *29*, 1565–1608. https://doi.org/10.1007/s11666-020-01091-w

Li, C. J., Li, W. Y., & Liao, H. (2006). Examination of the Critical Velocity for deposition of Particles in Cold Spraying. *Journal of Thermal Spray Technology*, *15*(2), 212–222.

Loganathan, P. S., Narasimalu, S., Kannappan, L., & Sekar, M. (2017). Supersonic particle deposition coatings for improved Tribological performance of off-shore wind turbine gears. *Proceedings of Asian Conference on Energy, Power and Transportation Electrification (ACEPT).*

Morsi, S., & Alexander, A. (1972). An Investigation of Particle Trajectories in Two-Phase Flow Systems. *Journal of Fluid Mechanics*, *55*(2), 193-208.

Murugesan, P., Kumar, A. B., Kambhampati, A. T., Pillai, S., Chandrasekar, G. C., Srikrishnan, A. R., & Velamati, R. K. (2020). Numerical Study of Characteristics of Underexpanded Supersonic Jet. *Journal of Aerospace Technology and Management*, 12.

Papyrin, A. N., Kosarev, V. F., Klinkov, S. V., Alkhimov, A. P., & Fomin, V. (2007). *Cold Spray Technology*. Elsevier.

Pattison, J., Celotto, S., Khan, A., & O'Neill, W. (2008). Standoff distance and bow shock phenomena in the cold spray process. *Surface and Coatings Technology*, *202*(8), 1443–1454.

Steenkiste, T.V. (2007). The role of particle temperature and velocity in cold spray coating Formation. In The Cold Spray Materials Deposition Process: Fundamentals and Applications. Woodhead Publishing

Subramanian, H., Nagarjun, C.V.S., Satish Kumar, K.V., Srikanth, V., & Srikrishnan, A.R. (2018). *Mixing enhancement using chevron nozzle: studies on free jets and confined jets.* Academic Press.

Sudhan, K., Prasad, G., Kothurkar, N., & Srikrishnan, A. R. (2020). Studies on supersonic cold spray deposition of microparticles using a bell-type nozzle. *Surface and Coatings Technology*, *383*, 125–244.

Wang, Q., & Zhang, M. X. (2012). Review on Recent Research and Development of Cold Spray Technologies. *Key Engineering Materials*, *533*, 1–52. https://doi.org/10.4028/www.scientific.net/kem.533.1

Witze, P. (1974). Centerline Velocity Decay of Compressible Free Jets. *AIAA Journal*, *12*(4).

Xie, X., Maa, Y., Chen, C., Ji, G., Verdy, C., Wu, H., Chen, Z., Yuan, S., Normand, B., Yin, S., & Liao, H. (2019). Cold Spray Additive Manufacturing of Metal Matrix Composites (MMCs) Using a Novel Nano-TiB2-Reinforced 7075Al Powder. *Journal of Alloys and Compounds, 819.*

Xu, J., Lin, C., Liu, K., & Mo, J. (2010). PIV Study and Numerical Simulation of the Unsteady Flowfield of Overexpanded Supersonic Impinging Jet at Ma=3. *46th AIAA/ASME/SAE/ASEE Joint Propulsion Conference & Exhibit.*

Yue, S., Wong, W., Irissou, E., & Legoux, J. G. (2011). Influence of Helium and Nitrogen Gases on the Properties of Cold Gas Dynamic Sprayed Pure Titanium Coatings. *Journal of Thermal Spray Technology*, *20*, 213–226.

Zapryagaev, V., Gubanov, D., Kavun, I., Kiselev, N., Kundasev, S., & Pivovarov, A. (2017). Investigation of supersonic jets shock-waveStructure. *AIP Conference Proceedings*. doi:10.1063/1.5007516

Zapryagaev, V., Kiselev, N., & Gubanov, D. (2018). Shock-Wave Structure of Supersonic Jet Flows. *Aerospace*, *5*(2), 60.

KEY TERMS AND DEFINITIONS

Cold Spray: A process of providing surface coating by way of deposition of micro-sized particles that are carried by a gas jet at supersonic velocity.

Jet Dynamics: The study of fluid jets considering the property gradients and their impact on the flow field.

Particles: Small spherical grains of deposition materials which have diameters of the order of microns (10^{-6} m).

Shock Structure: The pattern of abrupt pressure variations along a supersonic jet characterized by sharp gradients in properties.

Supersonic Jet: A stream of gas moving with a velocity which exceeds the velocity of sound in that gas.

Tribology: The study of relative motion between solid surfaces in contact along with the consideration of the resulting wear and the methods to minimize the friction.

Wear Resistance: The ability of the surface of material to resist the loss of material by way of friction with another solid surface.

Section 2

Chapter 7
Tribological Behaviour of Plasma-Nitrided Die Steels

Manpreet Kaur
Baba Banda Singh Bahadur Engineering College, India

Ashish Kumar
Baba Banda Singh Bahadur Engineering College, India

Pankaj Chhabra
Baba Banda Singh Bahadur Engineering College, India

Alphonsa Joseph
Institute for Plasma Research, Gandhinagar, India

Ghanshyam Jhala
Institute for Plasma Research, Gandhinagar, India

ABSTRACT

Wear, plastic deformation, and mechanical fatigue of dies are the most common failures found during hot forming operations at elevated temperatures. The change in frictional behaviour also happens. The performance of the forming operation is affected. To ensure the quality of the end products and productivity, it has become very important to control the wear and friction of die materials. Surface treatment techniques with superior wear properties and good performance can enhance the life and functionality of dies. Plasma nitriding is the most rapidly developing technique for hot forming dies. It is a cost-effective technique and improves the mechanical properties of the die surfaces. This chapter explains the tribology of hot forming dies, the plasma nitriding technique, and the procedures to develop plasma nitriding on the die steels. Thereafter, the tribological behaviour of AISI H11 and AISI H13 plasma nitrided die steels has been reported. Plasma nitriding was found to be most promising and effective in reducing wear and friction at elevated temperatures.

DOI: 10.4018/978-1-7998-9683-8.ch007

INTRODUCTION TO TRIBOLOGY

Tribology, or the study of friction, wear and lubrication, is the branch of science that deals with the study of interacting surfaces in relative motion (Hutchings, 1992). It is a complex science with theoretical calculations of few possibilities of wear and friction in the materials in contact and sensitivity of the system to operating conditions and environment. A tribological system consists of four main elements that control the overall behaviour. The first two elements are the materials of the surfaces close to each other. The third element is the interfacial medium, which can be a lubricant such as oil, gas or another form of an intermediate layer. Finally, the fourth element is the surrounding environment which is usually the surrounding air (of varying humidity), other gases, fluids or even a vacuum in the case of mechanisms utilized in space. All these elements together determine the friction and wear characteristics of the particular system. In other words, friction and wear properties are not intrinsic or inherent properties of the material but are highly dependent on the tribological system (Hardell et al., 2007; Kumar et al., 2017).

Friction is "*the resistance against sliding between two contacting bodies*" (Hardell et al., 2007). In the metal forming process, the existence of friction is necessary. It exists in terms of reaction force. In practice, it is the force that opposes motion between two contacting surfaces. Thus, friction is a part of the energy dissipative process. Sliding friction occurs mainly through contributions from two components, namely the adhesion component and ploughing component. Adhesive friction originates when atomic junctions form between the two surfaces in contact. The higher affinity of the materials will result in increased friction, for example, two steel surfaces in dry sliding will form atomic bonds if the oxide layer on the surfaces is ruptured. Ploughing friction occurs when one of the surfaces is harder than the mating surface, and the asperities of the harder one plough through the softer surface. It may also be caused due to the ploughing action of entrapped hard particles between the bodies. Johnson (2003) mentioned the relation of friction coefficient (μ) by the law of friction as (equation 1):

$$\mu = \frac{F_T}{F_N} \quad (1)$$

The friction coefficient is further divided into two parts, that is, adhesive coefficient (μ_a) and ploughing coefficient (μ_p) as shown in equation 2.

$$\mu = \mu_a + \mu_b \quad (2)$$

The μ_a represents the part of the surfaces which are in contact with each other. The function of μ_a is to control adhesive force between the surfaces in contact. During sliding, the inter surface bonds are formed between the contacting surfaces and are further broken by the adhesive forces. The adhesive force is required, and its magnitude depends upon the surface roughness of the contacting bodies. The μ_p represents the deformation force that results in the ploughing of the weakest part of the surface sliding against the hard surface. It may also be caused due to the ploughing action of entrapped hard particles between the two surfaces in relative motion (Hutchings, 1992; Johnson, 2003; Jacobson and Hogmark, 2005).

Wear is "*a loss of material from the surface*" (Hardell et al., 2007). It is a system parameter that arises when two surfaces are in contact with one another. Wear occurs due to the friction force. The wear mechanisms such as abrasive, adhesive, fatigue and chemical wear occur in hot forming dies and hot rolling rolls. The proper selection of the material and effective lubricants can lower the wear. Four main wear types are involved in wear processes:

1. **Adhesive wear**: It occurs when the surfaces of two bodies are in shear contact during relative motion. Strong bonds form between the contacting surfaces as the properties change. The formation and breaking of atomic junctions take place between mating surfaces. It results in the removal or transfer of material and is called sliding wear. The tangential relative motion results in splitting of the boundary of the contacting bodies due to which wear have occurred (Hutchings, 1992; Johnson, 2003; Jacobson and Hogmark, 2005)
2. **Abrasive wear**: It occurs when the asperities of a hard surface plough and scratch the softer mating surface resulting in 2-body abrasion. It also occurs when hard particles plough and scratch one or both interacting surfaces resulting in 3-body abrasion. The particles of hard materials are forced into soft materials due to tangential relative motion. Hence it results in abrasive wear (Hutchings, 1992; Johnson, 2003; Jacobson and Hogmark, 2005).
3. **Fatigue wear**: It is caused by tribological stress cycles that result in fatigue and subsurface crack formation eventually leading to material removal. The chemical reactions will consume the material of the surfaces.
4. **Fretting wear**: It occurs when sliding surfaces are under high loads. There is an oscillatory movement with small amplitude. The generation of wear particles takes place. The particles adhere to the contacting surface and give rise to more wear.
5. **Erosive wear:** It occurs on the solid surface under the impinging action of the solids, liquids, and gases or a combination of these. The material removal takes place.

Lubrication is the method to decrease wear and friction between two moving surfaces in contact with each other. It is achieved by introducing an easily sheared material (liquid, solid or gas) between the two surfaces. A liquid lubricant such as oil (mineral or synthetic) consists of a base fluid and different additives. Grease is another common lubricant and is considered semi-solid and consists of oil mixed with a thickener. Greases are also commonly used where oils would not stay in place and can also provide sealing against contaminants. The lubricants may be solid like graphite. In lubrication, it is common to classify the contact situation between the lubricated surfaces into different regimes (Hardell et al., 2007).

HIGH TEMPERATURE TRIBOLOGY

In recent years, the need for mechanical systems working under an extreme range of conditions i.e. high loads, speeds, and temperatures are increasing. Some examples of applications that require the use of elevated temperature are hot forming and hot forging, aerospace systems, advanced combustion engines, mining, and metalworking processes.

The dividing line between what is low or high-temperature is rather diffusing and greatly depends on the materials and applications involved. For example, while a temperature of 2000°C is already above the melting point for most materials, it is observed that certain ceramics can still offer good mechanical properties despite exceeding the threshold temperatures (Eakins et al., 2011). For this reason, a term commonly referred to by metallurgists called homologous temperature, T_{hom}, is used to categorize both the cold and hot forming operations. T_{hom} can be defined as the working temperature of the material normalized by its melting point in absolute temperature scale (K). The creep does not occur when the resultant fraction of T_{hom} is less than 0.4 and the operation is called cold working (Tisza, 2002). In tribology, high or elevated temperature applications are the ones where the use of conventional lubricants i.e. oils and greases are no longer effective since they rapidly decompose usually at around 300°C. In the case of metals and alloys, the introduction of elevated temperatures into a tribological system involves several chemical and physical phenomena occurring simultaneously.

Several problems occur in hot-metal forming of high-strength steels like oxidation of tool and workpiece surfaces, increased wear of the tools, and scaling of the workpiece. Hot forming operations at elevated temperatures significantly influence the frictional behavior and can result in the unexpected or undesirable performance of the tribological system (Hardell and Prakash, 2010). Moreover, the materials and manufacturing costs are very high as the industries desire more wear-resistant components (Prchlik and Sampath, 2007). A survey was carried out in various countries. About 1-2% of GDP approximately was lost due to friction and wear. Rs. 20,000 crores approximately were lost in India due to this problem (MRS Bulletin, 2009).

TRIBOLOGY OF HOT FORMING DIES

The die surface and near-surface region are subjected to the most severe conditions during forging. The main failure mechanisms that occur on dies are: wear, thermally and mechanically induced strain, thermal fatigue, cracks, micro-cracks, and tempering. The tempering is loss of hardness with increasing temperatures. It is influenced by the absolute temperature that causes spalling and dies deformation. The cracks and microcracks occur due to thermal fatigue, suffering the strong influence of the temperature cycle. The cracks also occur due to mechanical loading (Lenhard et al., 2006).

Oxide layers are present in almost all metals due to exposure to oxygen in the air. The oxidation rate is governed by the interaction of oxygen and metal ions that is in turn controlled by the temperature. Chemical reaction rates generally increase at elevated temperatures. Over the years, oxidation wear has been analyzed by several authors (Archard and Hirst, 1956; Quinn, 1971; Quinn 1992; Quinn 1994; Stott et al., 1985). Quinn (1994), developed a general theory of oxidation wear that not only took into account the oxidation caused by frictional heating through asperity contacts, but also the one occurring when the sliding surfaces are exposed to the surrounding temperature. In his work, it was shown that for oxidation, the tribological activation energy is approximately half the value of the static oxidation. In other words, oxidation under the sliding conditions can occur more easily than in the static oxidation, exposed to the same ambient temperature. This might be attributed to a higher diffusion rate of ions through a growing oxide layer which is generally replete with defects due to mechanical perturbations (Bhushan, 2002).

The presence of oxides at the tribo contact interface does not necessarily need to be beneficial. The oxides can be beneficial or detrimental for the component. For example, ductile oxide reduces wear and friction. On the other hand, the brittle oxide is very hard. It detaches from the surface and may result in increased wear. The properties of the oxides like strength, adherence to the substrate, capability to get sintered, and forming glaze layers, all depend on the sliding and surrounding conditions and the materials in contact. In static conditions, alloying elements less noble than iron such as aluminium, chromium, or silicon tend to form passive oxide layers, protecting the steels against corrosion (Chang and Wei, 1989). However, during sliding conditions, the formed layers are prone to break and further promote wear by abrasion. In other words, an oxide behaves differently under tribological conditions. When oxides such as BaO, B_2O_3, CoO, PbO, MoO_3, and ZnO were found in the tribo layers, the lower friction and wear values were observed (Amateau and Glaeser, 1964).

Three different iron oxides are formed depending on temperature, exposure time, and surrounding atmosphere (Chang and Wei, 1989). Between the 200°C to 570°C temperature range, the scale consists of an outer layer of hematite (Fe_2O_3) and an inner layer of magnetite (Fe_3O_4). Above 570°C temperature, an extra layer of wustite (FeO) appears on the substrate. The mechanical properties such as hardness and the crystalline structure of these oxides affect the tribological behavior. The rhombohedral ferric oxide (Fe_2O_3) is known to present an abrasive behavior due to its remarkably higher hardness at room temperature (Takeda et al., 2009; Vergne et al., 2006). Fe_3O_4 and FeO show a cubic crystalline system and are more ductile (Vergne et al., 2006). Several authors suggested that the formation of Fe_2O_3 or/and Fe_3O_4 oxide layers on ferrous materials during sliding is advantageous in reducing the friction and wear (Quinn, 1967; Kong et al, 1995; Nakamuraa et al, 2002). Kabaya et al. (1981) credited the presence of Fe_3O_4 over Fe_2O_3 within the oxide layer for reducing friction and wear at elevated temperatures.

Contrary to these observations, Wang et al. (2010) observed that the formation of Fe_3O_4 over Fe_2O_3 did not involve the reduction of wear when investigating the effects of oxides and matrix on the tribological behavior of different cast steels. According to their results, under certain conditions, the formation of iron-based tribo oxides contributed to a further increase in wear. They suggested analyzing the strength of the matrix. If it is high enough to support the oxides, it can protect against wear. On the other hand, if the matrix is soft, the presence of tribo oxides does not reduce wear and, they may even act as three-body abrasives. Interestingly, other authors have found that the decrease in hardness due to thermal softening leads to the development of protective wear layers (Hernandez et al., 2014; Stott and Jordan, 2001).

In unidirectional and reciprocating tribological systems, the wear debris is accumulated at the interface of the sliding bodies. The development of protective layers was commonly found in closed test configurations (Straffelini et al., 2001). However, Inman et al. (2005) have shown that the formation of such layers can be achieved even under limited debris retention conditions. The establishment of oxide layers can provide better friction control and wear protection, but their use is limited depending on whether or not they form under specific tribological conditions. The knowledge and understanding on assisting their formation and control their stability is essential.

The formation of protective layers may be beneficial in industrial processes involving high temperatures and pressures. The hot stamping process is being used to manufacture automotive components. In this process, a steel sheet (commonly made out of boron steel) is heated and transferred to a forming section where it is formed and quenched. Various authors investigated the sliding wear behavior at elevated temperatures of different tool steel grades sliding against coated and uncoated boron steel (Hardell et al., 2009; Pelcastre et al., 2012; Mozgovoy et al., 2014; Geiger et al., 2008). The main wear mechanisms were found as adhesive and abrasive wear.

It becomes significant to understand the mechanism of abrasive wear at elevated temperatures. It affects the tribological behavior of metals. Venkatesan et al. (1997) studied the high-temperature three-body abrasive behavior of medium carbon low alloyed steel with different surface treatments. According to their results, at relatively low temperatures, the wear resistance depended on the strength and ductility of the modified surface. But at higher temperatures, however, oxidation resistance and hot hardness were parameters to control wear. Varga et al. (2013) studied the wear-reducing effects of the formation of tribo layers in an abrasive environment at elevated temperatures. They observed that the thickness and morphology of the formed tribo layers at high temperatures rely upon the microstructure, size, and structure of the hard phases of the metal investigated. Similarly, some other authors have found the occurrence of tribo layers to decrease the wear rate of the tested materials especially at higher temperatures (Berns and Franc, 1997; Antonov et al., 2010; Badisch et al., 2010; Jones et al., 2009; Winkelmann et al., 2010).

PREVENTIVE MEASURES TO CONTROL HIGH TEMPERATURE TRIBOLOGY

Some of the commonly employed methods for wear prevention at elevated temperatures are:

- Select suitable wear-resistant materials keeping in mind the economic and technological factors
- As wear is a surface phenomenon, the use of protective coatings/surface treatments on the surfaces in contact can reduce the wear
- By providing proper lubrication between the two mating surfaces
- Ensuring that the contact pressures should not exceed beyond the allowable contact pressures for the material. It may be different for various materials and applications
- Try to avoid misalignment, and, if it is a must, use design or materials which can accommodate the same
- Use proper maintenance schedules, procedures, methods to minimize wear and friction
- Check if wear testing is required

According to Deshpande and Altan (2011), the life of hot forming die steels has been improved by various surface engineering processes like surface welding, surface coatings, such as chemical and physical vapor deposition of heat resistant ceramic materials, weld overlays, and surface treatments such as nitriding. Plasma Nitriding (PN) is the most commonly used surface treatment for hot forging dies (Deshpande and Altan, 2011). It is a thermochemical heat treatment process used to modify the surface of hot working and gearing steels (Karami, 1991; Karami and Staines, 1989). Literature reveals that the improved wear resistance is imparted by plasma nitriding to hot and cold working tool steels and stainless steels (Sun et al., 1994; Staines and Bell, 1979; Edenhofer, 1973).

PLASMA NITRIDING

Plasma nitriding is a method of surface hardening using glow discharge technology. Nascent (elemental) nitrogen is added to the surface of a metal part for subsequent diffusion into the material. In a vacuum, high-voltage electrical energy is there to form plasma, through which nitrogen ions accelerate to strike the workpiece. That, in return, heats the workpiece, cleans the surface, and provides active nitrogen. Ion nitriding gives better control of case chemistry and uniformity and has other advantages, such as lower part distortion than conventional (gas) nitriding. A key difference between gas and ion nitriding is the mechanism used to generate nascent nitrogen at the surface of the work (ASM Handbook, 1990; Li et al., 2020; Phulera et al., 2021). It has been widely used in the industry over the past decades to improve the hardness, wear, corrosion resistance, and fatigue strength of various metallic materials such as low alloy steels, titanium alloys, and hot working tool steels (Forati et al., 2011; Kumar et al., 2020; Kovac and Secer., 2020; Fernandes et al., 2020). Three reactions occurred during plasma nitriding. Sputtering occurs during the first reaction, that is, the removal of iron and other particles from the surface of the steel with the help of hydrogen gas. After sputtering, the surface is ready for the diffusion of the nitrogen ions (Pye, 2003). After the first reaction, iron nitrides form with the help of nitrogen ions.

$$Fe + N \rightarrow FeN \tag{3}$$

In the second and third reaction, FeN breaks down into ε phase followed by the γ´ phase and then followed by the Iron/nitrogen compound zone due to continuous sputtering from plasma. This process is explained as follows:

$$2FeN \rightarrow Fe_2N + N \tag{4}$$

$$3Fe_2N \rightarrow 2Fe_3N + N(\varepsilon \text{Phase}) \tag{5}$$

$$4Fe_3N \rightarrow 3Fe_4N + N(\gamma' \text{Phase}) \tag{6}$$

$$Fe_4N \rightarrow 4Fe + N(\text{Iron / nitrogen compound zone}) \tag{7}$$

Plasma Nitriding Benefits

Technology has not gained much popularity in many industries. However, the equipment cost is less. A few benefits of the technology mentioned by Pye (2003) are:

- The processed gases that are not harmful such as nitrogen, hydrogen, and argon are in use
- The oxygen level is less due to the vacuum, and the fire risk is less. Furthermore, it will avoid the reaction of carbon with oxygen, and thus the decarburization of the steel surface will be less
- The deposition rate of the nitrogen inside the steel surface gets increased by the process properties. The ion bombardment and existence of highly active species play a big role. The cycle time gets shortened by the lower heating time
- The operator requirement is less. The problems during production become less, the reliability increases, and the process repeatability is allowed at a metallurgical level. Equipment maintenance costs and operation costs are low.
- Gas consumption is less, and this further results in low operating costs.
- The technique allows the nitriding of various materials while other conventional processes do not.
- The selected areas on the surface can be nitrided.

TRIBOLOGICAL BEHAVIOUR OF PLASMA NITRIDED HOT FORMING DIE STEELS

Materials to Investigate

Among different hot forming die steels, two substrate materials namely, AISI H11 and AISI H13 were chosen as hot forming tool materials. The materials are widely used in the manufacturing of hot working dies. They were bought after consulting the actual hot press forging/forming industry. Cylindrical samples (pins) with dimensions 50 mm length × 8 mm diameter were made from the two tool materials.

The workpiece material DIN 20MnCr5 (selected as the disc material) was taken in consultation with the industry. The steel DIN 20MnCr5 is as per the European DIN standard. Table 1 shows the chemical composition of the AISI H11 and AISI H13 steels. Here 20 refers to 0.20% carbon, and 20MnCr5 indicates that this steel contains Mn and Cr as the main alloying elements, and the summation of all the other alloying elements present in the steel is close to 5 wt. The discs of 125 mm diameter × 6 mm thickness were made.

Table 1. Chemical composition of the tool and disc material

Tool Steels	Elemental Composition (wt %)										
	Cr	Mo	Si	S	C	Mn	V	Cu	P	Ni	Fe
H11	4.75-5.50	1.10-1.60	0.80-1.20	<.03	0.33-0.43	0.20-0.50	0.3-0.6	<0.25	<.03	<0.3	Bal.
H13	4.75-5.50	1.10-1.75	0.80-1.20	<.03	0.32-0.45	0.20-0.50	0.8-1.2	<0.25	<.03	<0.3	Bal.
Disc Material DIN 20MnCr5	C	Mn	P	S	Si	Cr	Fe	-	-	-	-
	0.17-0.22	1.1-1.4	Max 0.025	Max 0.03	Max 0.4	1.0-1.3	Bal	-	-	-	-

PLASMA NITRIDING DEVELOPMENT

Heat Treatment of Materials

In the first step, the die steel specimens were heat treated. Annealing was done in the furnace above the austenitizing temperature in three steps. After heating, the specimens were quenched in oil and then cooled to room temperature in air. It resulted in the formation of the martensitic microstructure. Furthermore, to relieve the stresses induced, high-temperature tempering was done. The tempering was repeated three times to remove all the stresses. The heat treatment process before the PN was done to increase the performance of the tool steels. The disc specimens were also heat treated and plasma nitrided to increase their hardness. Steel discs were surface hardened. They were pack carburized by heating to a temperature of 920°C in the presence of coal and then allowed to cool slowly in the furnace. In the next step, the discs were again heated to 820°C (just above recrystallization temperature) and then quenched in oil. Finally, discs specimens were heated again for grain refining and internal stress removal. The final hardness attained by pack carburizing is about 60 HRC. After this, the plasma nitriding was done to raise hardness in the range of 60-80 HRC.

Plasma Nitriding of Tool Steels

Plasma nitriding was developed on the H11 and H13 tool steel specimens in collaboration with Facilitation Center for Industrial Plasma Technologies (FCIPT), Institute for Plasma Research (IPR), Gandhinagar, Gujarat, India. Figure 1 shows the apparatus used for plasma nitriding. Table 2 shows the specifications of the plasma nitriding hot-wall system. Steps involved in plasma nitriding of the AISI H11 and H13 specimens are as follows:-

1. The specimens were placed inside the vacuum chamber on a specimen holder that acted as a cathode, while vacuum chamber walls acted as an anode (Figure 2(a)). The specimens were performed with acetone to remove any foreign particles.
2. The mechanical rotary pump was used to develop the vacuum pressure of 8 Pa in the chamber. Then, sputter cleaning of the specimens was done using H_2 gas at a 100 Pa pressure and 250°C temperature. The sputtering removed the surface oxide layers on the specimens.
3. After this, the specimens were plasma nitrided with a gas mixture of N_2 and H_2. The ratio of the selected gases was 20:80, 50:50, and 80:20. The process was performed at a constant pressure of 500 Pa. During plasma nitriding, a glow occurred around the specimens (Figure 2 (b)). This was due to the impinging of the ionized nitrogen and hydrogen gases accelerated under voltage bias. On striking, the kinetic energy of ionized particles gets converted into heat energy. That led to a rise in the temperature of the specimens. Further, the voltage bias was adjusted to 450-500 V to attain the specimen's temperature of 500ºC. This temperature was selected in the current study and is less than the tempering temperature by 50ºC. The process was for a constant time of 24 hours for all three gases ratios.
4. After 24 hours of process, the voltage bias was removed, and the specimens cooled down to 180°C. The cooling was performed in the presence of nitrogen gas to avoid oxidation.

Figure 1. Plasma Nitriding Set-up

Figure 2. (a) Camera pictures of substrate specimens kept inside (b) Plasma-nitriding of the samples inside chamber.

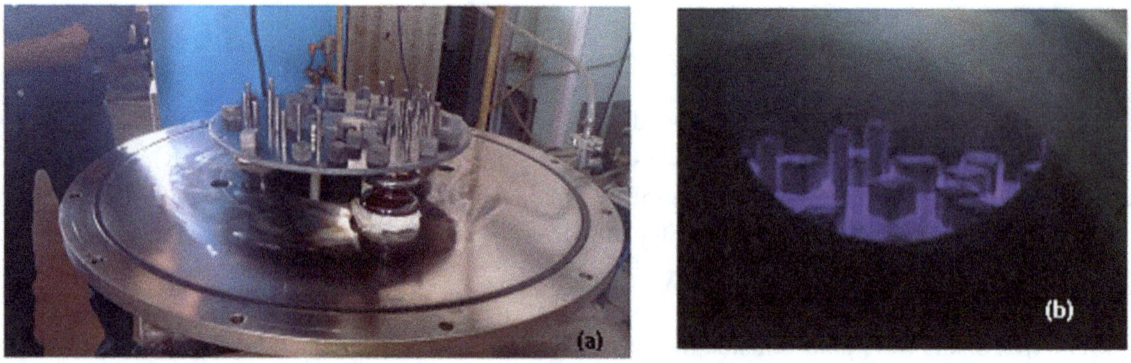

Table 2. Plasma nitriding system specifications

Nitriding temperature	450°C, 500°C
Diameter	500mm
Height	500mm
Working diameter	400mm
Working height	300mm
Working capacity	50kg
DC pulse power supply	20KW
Heater power capacity	5KW
Cooling	only for O rings

Characterization of Plasma Nitrided Specimens

The following procedure was followed for the characterization of the nitrided specimens:

1. The surface roughness of the plasma nitrided specimens was recorded at different places on the specimens after heat treatment. Before plasma nitriding, the heat-treated specimens were mirror polished and reduced to 0.05 microns.
2. The microhardness of the plasma nitrided specimens was evaluated along the etched cross-sections. The mean values of the microhardness of three observations are plotted as a function of distance from the external edge of the plasma nitrided layer to the internal core of the specimen.
3. The phases present in the nitride layer of the plasma nitrided samples were evaluated using XRD analysis. The intensities were recorded at a scanning speed of 1Kcps in the 2θ range of 30° to 90°.
4. The microstructure and elemental compositions of the nitride layer were investigated with the help of FE-SEM attached with EDS Genesis software. The investigation was performed for the surface, as well as, along the cross-section.

ROOM TEMPERATURE AND HIGH TEMPERATURE WEAR AND FRICTION STUDIES

Experimental Setup

Friction and wear investigations were performed following the ASTM G99-04 wear testing standard on the high-temperature pin-on-disc tribometer. Figure 3 shows the equipment photograph. The tests were performed under dry and unidirectional sliding conditions. The pin was loaded against the selected disc material DIN 20MnCr5 (hardened up to 60-80 HRC). The objective was to simulate the actual hot forming operation. In actual practice, the workpiece is heated to the required temperatures and then transferred between the die surfaces for forming. In the present work, the disc was heated to the required temperature and the pin was loaded against the disc thereafter. The heating was done using an induction heater controlled by the automated system to maintain the testing temperature. The friction and wear tests were conducted under the two load conditions (25N and 50N), sliding speed 0.5 m/s, and sliding distance of 1500 m. The testing temperatures range from room temperature to 600°C. The pin-on-disc machine was equipped with a strain gauge force transducer and a computerized data acquisition and control system to measure the value of frictional force. The wear rate was calculated by evaluating the loss in mass of pin specimens.

Analysis of Wear and Friction

1. The wear weight loss of the pin specimens was evaluated by measuring the weight change in pin specimens before and after the wear tests. The pins were weighed using an electronic microbalance with a precision of 0.1 mg.
2. 2. The wear volume loss (mm^3) was calculated for each test. The coefficient of friction was determined from the values of frictional force using the relation:

$$\mu = \text{frictional force (F)/applied load (N)} \quad (8)$$

3. Average coefficients of friction values were evaluated at various conditions for the untreated and plasma-nitrided specimens.
4. The exposed specimens were further analyzed to see the morphology and surface topography by using SEM and XRD techniques.

Figure 3. Camera picture of high temperature pin-on-disc tribometer

Characterization Results of the Plasma Nitrided Die Steels

The results of characterization carried out after the development of Plasma Nitriding are explained as follows:

- **Surface Roughness:** Before PN, the surface roughness (Ra) value was evaluated as 0.05 microns. After PN, the values found were different for the 20:80, 50:50, and 80:20 (N_2:H_2) gas ratios. For the H13 specimens the Ra value was found as 0.44, 0.35, and 0.33 μm when treated with 20:80, 50:50, 80:20 (N_2:H_2) gas ratios respectively and for the H11 specimens, the value was found as 0.52, 0.41, and 0.37 μm when treated with 20:80, 50:50, 80:20 (N_2:H_2) gas ratios respectively. The Ra values increased after plasma nitriding.
- **Microhardness:** The surface microhardness values for the H11 steel were observed to be in the range of 209-287 HV and for the H13 steel were in the range 190-257 HV. After heat treatment, the microhardness values increased to 550±25 HV and 620±25 HV for the H11 and H13 steels respectively. A microhardness tester was used to measure the values across the cross-section. Microhardness depth profiles for both the nitrided steels are shown in Figure 4 and Figure 5. The values were measured after every 100 μm from the top surface to the core region. Surface micro-hardness of H13 specimens with $20N_2$:$80H_2$ ratio was found as 1045±52 HV; with $50N_2$:$50H_2$

ratio it was found as 1214±60 HV and with $80N_2$:$20H_2$ ratio it was found as 1219±61 $HV_{0.1}$ (Figure 4). Surface microhardness of H11 specimens with $20N_2$:$80H_2$ ratio was found as 1089±54 HV, with $50N_2$:$50H_2$ ratio it was found as 1102±55 HV and with $80N_2$:$20H_2$ ratio it was found as 1216±61 HV (Figure 5). The hardness of the steels increased two times after the plasma nitriding. This indicated that the steels can do better in hot forming applications after the nitriding.

- **The Thickness of the Nitriding Layer:** The thickness of the nitriding layer was calculated from the point where the case hardness value reaches the substrate hardness plus 50 HV (Franc and Berns, 1997) Keeping this in mind, the thickness of the nitriding layer for AISI H13 was evaluated as ~290, ~407, and ~698 μm for the N_2:H_2 as 20:80, 50:50, and 80:20 respectively. The nitriding layer thickness for AISI H11 was evaluated as ~150, ~180, and ~205 μm for the N_2:H_2 as 20:80, 50:50, and 80:20 respectively. The thickness varied with the change in gas ratios.
- **X-Ray Diffraction (XRD) Analysis:** On the plasma nitrided specimens, the specimens nitrided with N_2:H_2 as 20:80 showed Fe_3N phase on the high-intensity peaks and Fe_3N-Fe_4N phase on the medium intensity peaks. The specimens nitrided with N_2:H_2 as 50:50, showed the CrN peaks in addition to the other iron nitrides. It was observed that with the increase in N_2 concentration, the intensity of the iron nitride peaks increased. The analysis showed that the microstructure of the nitrided layer in the specimen can be specifically controlled by adjusting the N_2 concentration.
- **SEM/EDS Analysis:** The SEM morphologies on both the PN steels showed many microparticles on the surfaces (Figure 6 and Figure 7). These microparticles were assumed to be nitrides. The EDS analysis endorsed these observations. The presence of Fe, N, Cr, and C elements was found on the top nitride layer. Figure 8 and Figure 9 show the cut sections of the nitrided specimens. The PN layer was found uniform. It depicted the martensite matrix. Nitrogen in the top layer confirmed the possibility of the formation of nitrides. The content of iron was more in the matrix.

Figure 4. Microhardness depth profiles of AISI H13 PN samples

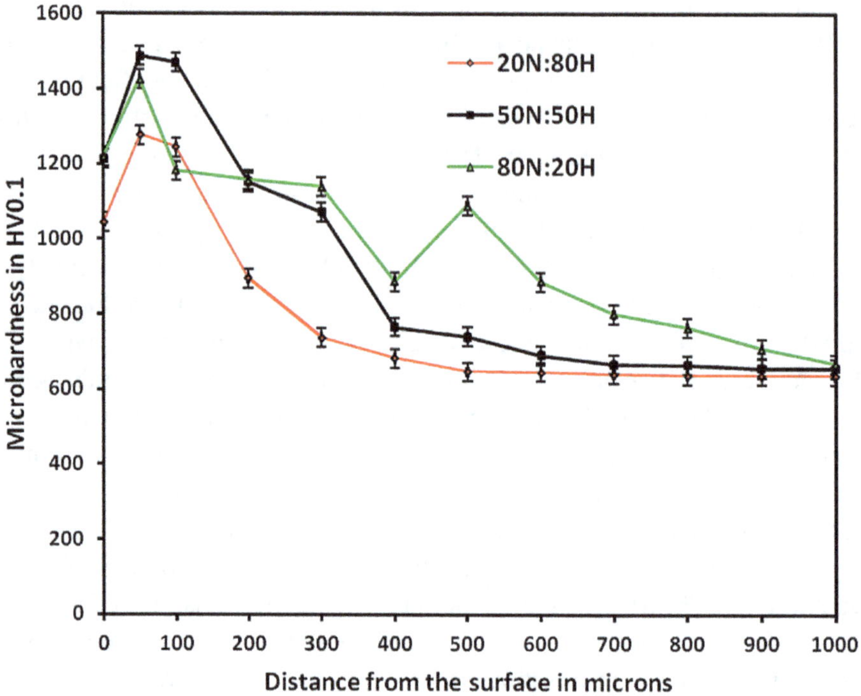

Figure 5. Microhardness depth profiles of AISI H11 PN samples

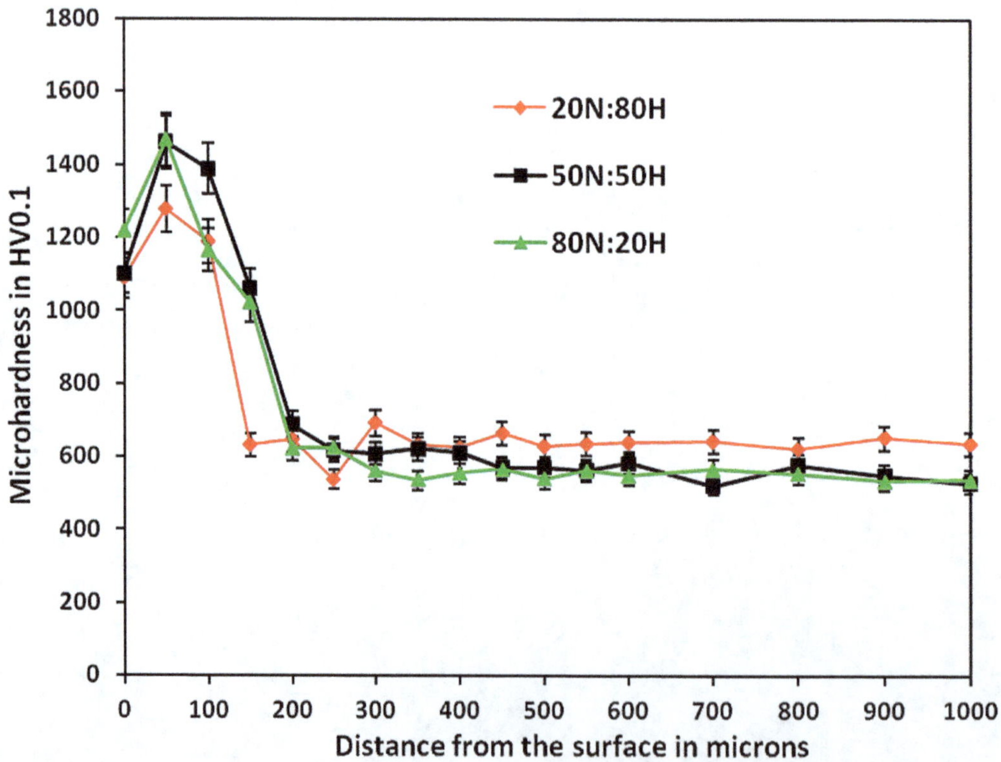

Figure 6. SEM/EDS analysis of AISI H13 PN at 500°C with N_2:H_2 ratio as (a) 20:80 (b) 50:50 (c) 80:20 and 24 h nitriding time

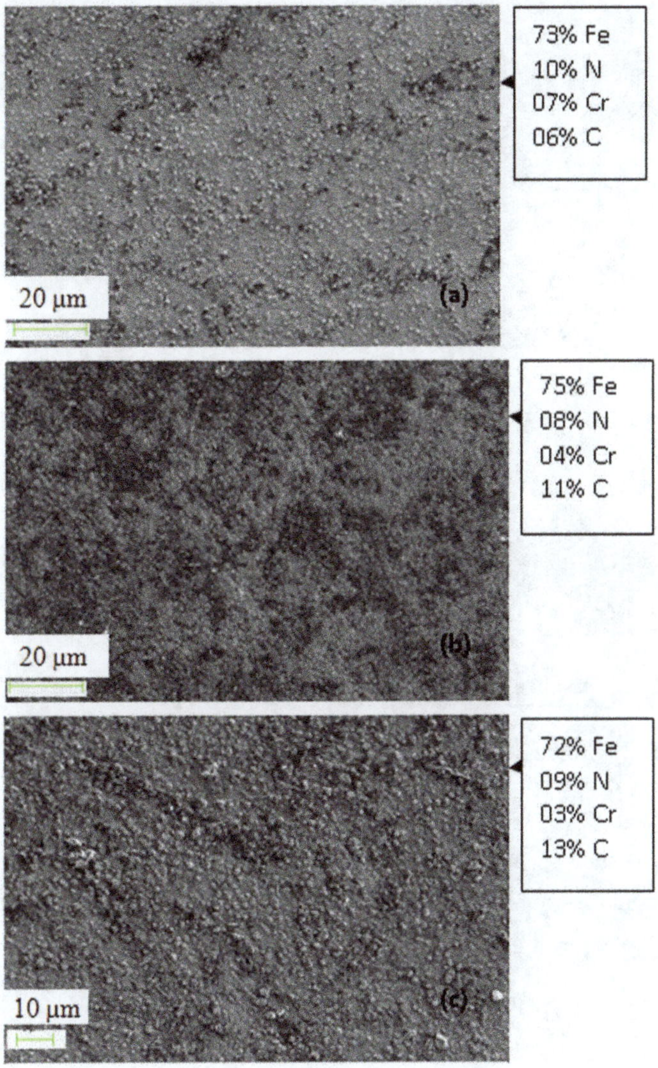

Figure 7. SEM/EDS analysis of AISI H11 PN at 500°C with N_2:H_2 ratio as (a) 20:80 (b) 50:50 (c) 80:20 and 24 h nitriding time

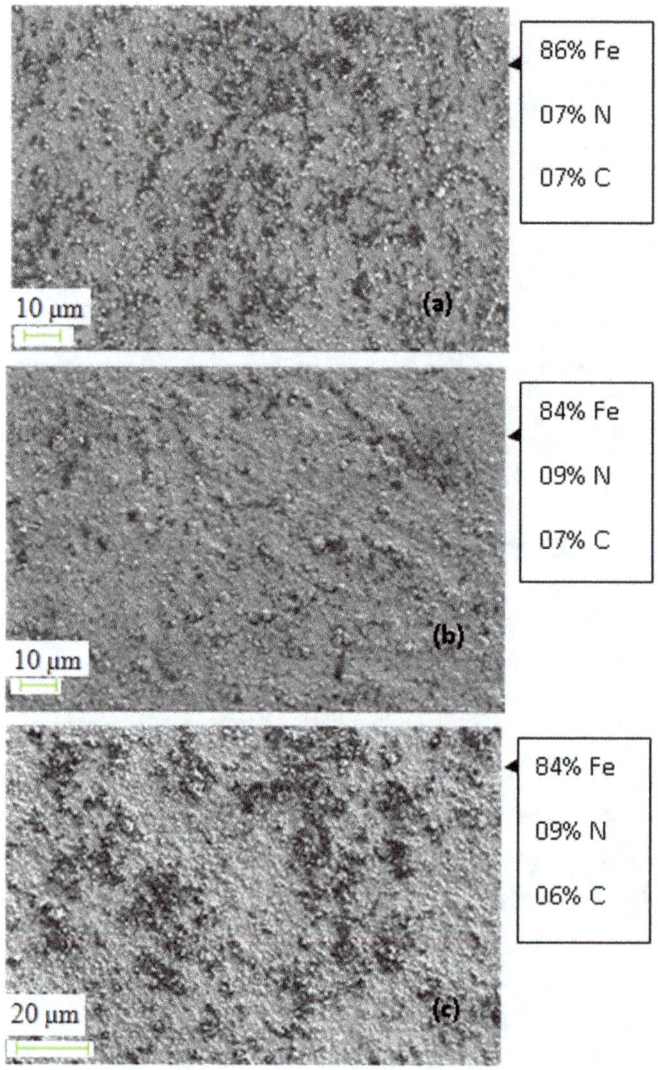

Figure 8. Cut section surface morphology of PN AISI H13 specimens using (a) $20N_2$:$80H_2$ (b) $50N_2$:$50H_2$ and (c) $80N_2$:$20H_2$ showing the case depth thickness

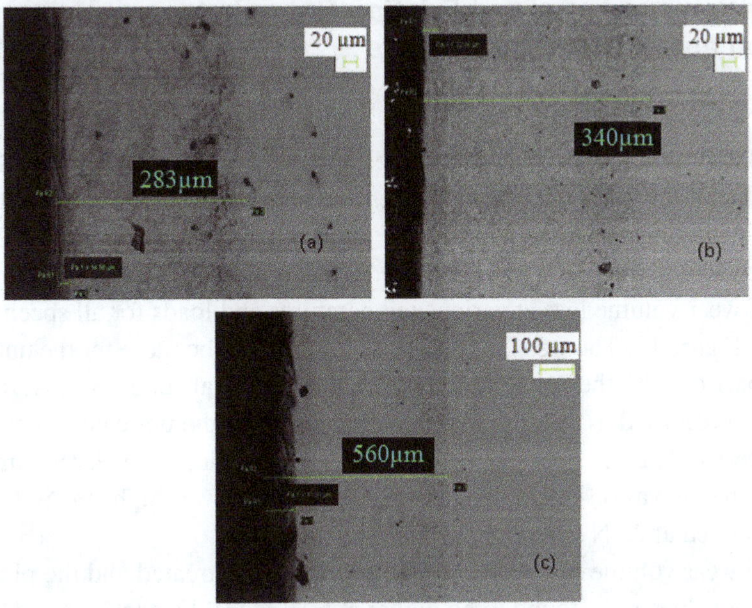

Figure 9. Cut section surface morphology of PN AISI H11 specimens using (a) $20N_2$:$80H_2$ (b) $50N_2$:$50H_2$ and (c) $80N_2$:$20H_2$ showing the case depth thickness

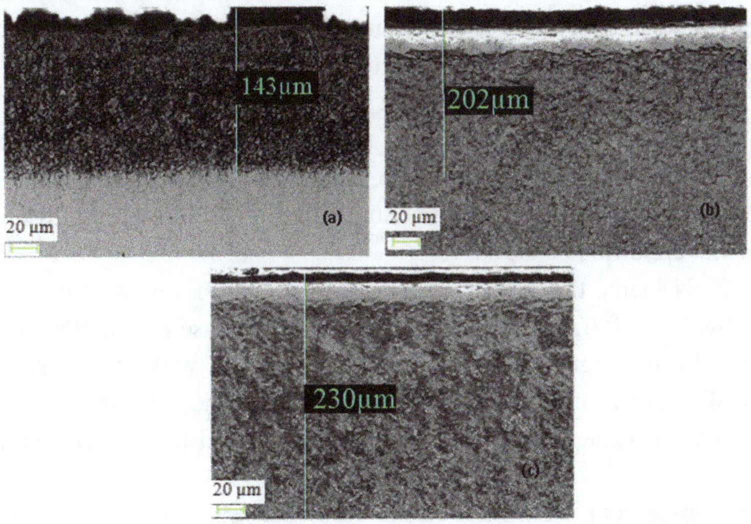

RESULTS OF THE TRIBOLOGICAL STUDIES OF THE WORN-OUT SPECIMENS

Visual Examination and Wear and Friction Behaviour of the Untreated and Nitrided Specimens

- Figure 10 and Figure 11 show the worn surfaces of the untreated and nitrided H13 steel specimens after wear testing at four different temperatures and both the 25 N and 50 N loads. All specimens showed wear marks on the surface. After the 50 minutes experiment at 400°C and 600°C temperature, the oxidation of the top layer occurred along with the adhesive and abrasive wear marks.
- The results of wear volume loss at various temperatures and loads for all specimens are shown in Figure 12 and Figure 13. The wear volume loss was found to be more for the untreated H13 specimens in comparison with the PN specimens at all experimental conditions. At room temperatures, the values were recorded as ~2.40 mm^3 and ~4.12 mm^3 for the untreated steel at 25 N and 50 N loads, respectively. The PN specimens showed the lesser values of wear volume loss at both the test loads. Further, it was observed that the wear volume loss was higher at 50 N loads in comparison to that obtained at 25 N loads.
- At 200°C, the wear volume loss values decreased for the untreated and the plasma nitrided steel specimens. The values were found to be higher at 50N loads. Further at 400°C, the wear volume loss values again decreased for the untreated and the plasma nitrided steel specimens at both the loads. The decrease was noticeable. The wear volume loss data endorse the average coefficient of friction values data. The least values of the average coefficient of friction were also observed at 400°C. The wear volume loss values increased at 600°C at both loads. Among the PN specimens, the specimens nitrided with N_2:H_2 as 20:80 showed lower wear volume loss at both loads. The least value of wear volume loss as 0.40 mm^3 was observed at 400°C and 25N loads.
- The average coefficients of friction (CoF) value obtained for all the untreated and treated specimens are shown graphically in Figure 14 and Figure 15. At room temperatures, the values observed for the untreated H13 specimens were ~0.68 and ~0.58 at 25 N and 50 N loads, respectively. After PN at three different parameters, the CoF values decreased. The average coefficient of friction values was found as ~0.36, ~0.56, ~0.58 at 25N loads respectively. Further, it was observed that the average coefficients of friction values at 50N load were lower for all specimens at all experimental conditions.
- At 200°C and 25 N loads, the average coefficient of friction values remained the same for the untreated H13 steel as (~0.67). The value was found to increase for the PN specimens with N_2:H_2 as 20:80 (Figure 14 and Figure 15). The specimens nitrided with other gas ratios showed a decrease in the CoF values. The reduction was marginal. Further, it was observed that the average coefficients of friction values at 50N load were lower for all the specimens at all experimental conditions
- At 400°C, the average coefficient of friction values got reduced. The untreated H13 specimens showed the CoF values as ~0.56 and ~0.48 at 25 N load and 50 N loads respectively. The average coefficient of friction values for all the PN specimens nitrided with N_2:H_2 as 20:80, 50:50, and 80:20 was found as ~0.36, ~0.48, ~0.49 at 25N loads respectively. The average coefficient of friction values with N_2:H_2 as 20:80, 50:50, and 80:20 were found as ~0.31, ~0.40, ~0.41 at 50N loads respectively. The values for the plasma nitrided specimens were observed to be lower than the

untreated steel. Again at elevated temperatures of 600°C, the increase in the average CoF values was seen (Figure 14 and Figure 15).

- Among all the PN specimens, the one plasma nitrided with $20N_2$:$80H_2$ gas ratio gave the lower values of average coefficients of friction at both test loads and all temperatures. Overall the values were found to be less at 400°C. The least average COF value ~0.314 was observed for the specimen PN with $20N_2$:$80H_2$ at 400°C and 50N loads.

Figure 10. Macrographs of worn surfaces of untreated H13 specimens at (a) RT (b) 200°C (c) 400°C, (d) 600°C; plasma nitrided with $20N_2$:$80H_2$ at (e) RT (f) 200°C (g) 400°C (h) 600°C; plasma nitrided with $50N_2$:$50H_2$ at (i) RT (j) 200°C (k) 400°C (l) 600°C; and plasma nitrided with $80N_2$:$20H_2$ at (m) RT (n) 200°C (o) 400°C, (p) 600°C after sliding wear tests on High temperature Tribometer under constant load of 25N for 50 minutes run

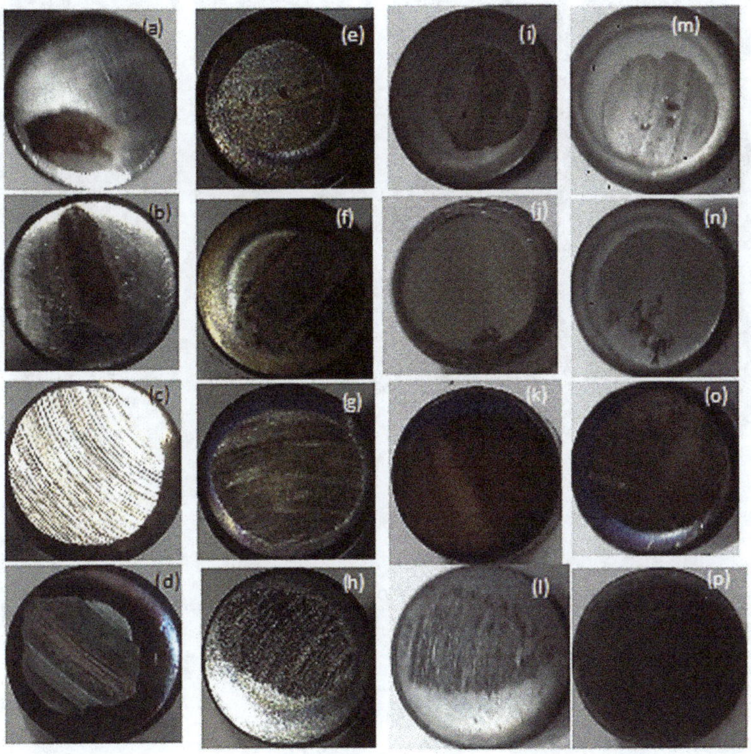

Figure 11. Macrographs of worn surfaces of untreated H13 specimens at (a) RT (b) 200°C (c) 400°C, (d) 600°C; plasma nitrided with $20N_2$:$80H_2$ at (e) RT (f) 200°C (g) 400°C (h) 600°C; plasma nitrided with $50N_2$:$50H_2$ at (i) RT (j) 200°C (k) 400°C (l) 600°C; and plasma nitrided with $80N_2$:$20H_2$ at (m) RT (n) 200°C (o) 400°C, (p) 600°C after sliding wear tests on High temperature Tribometer under constant load of 50N for 50 minutes run

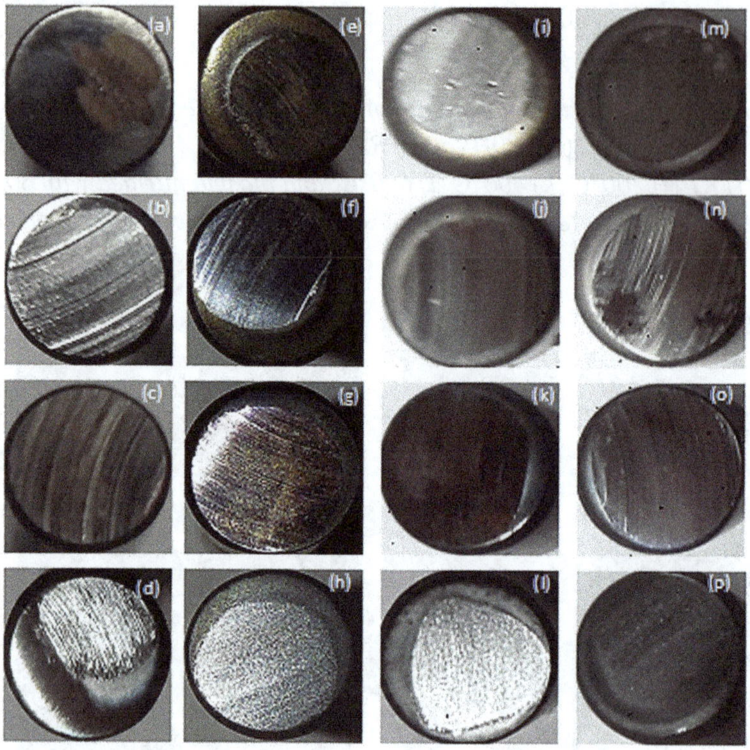

Figure 12. The wear volume loss of untreated and PN AISI H13 steels after wear tests at 25 N loads

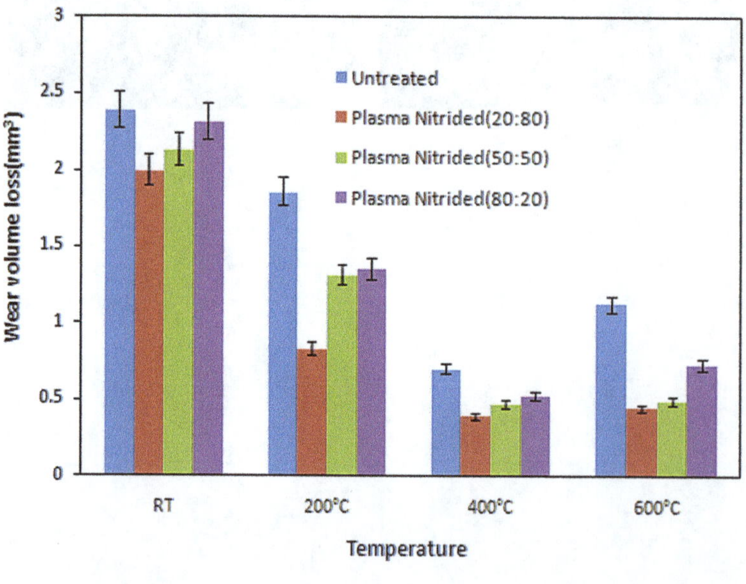

Figure 13. The wear volume loss of untreated and PN AISI H13 steels after wear tests at 50 N loads

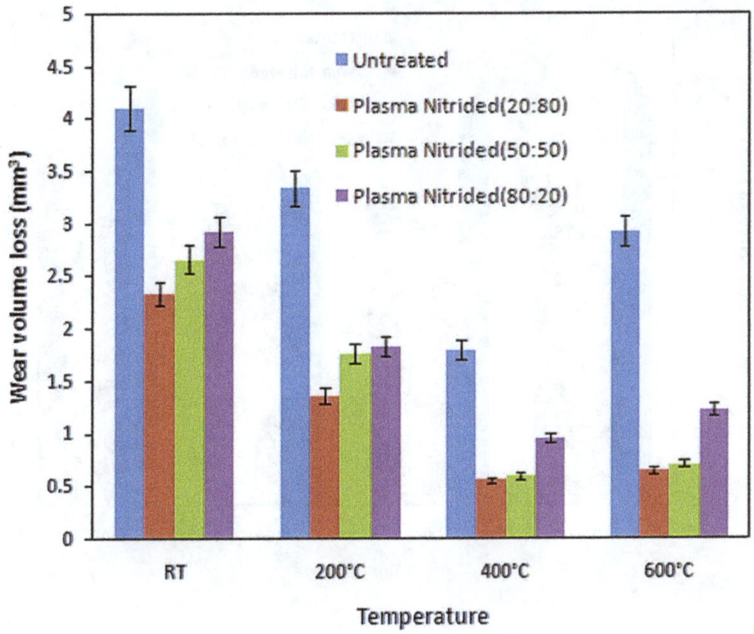

Figure 14. The average COF of untreated and PN AISI H13 steels after wear tests at 25 N loads

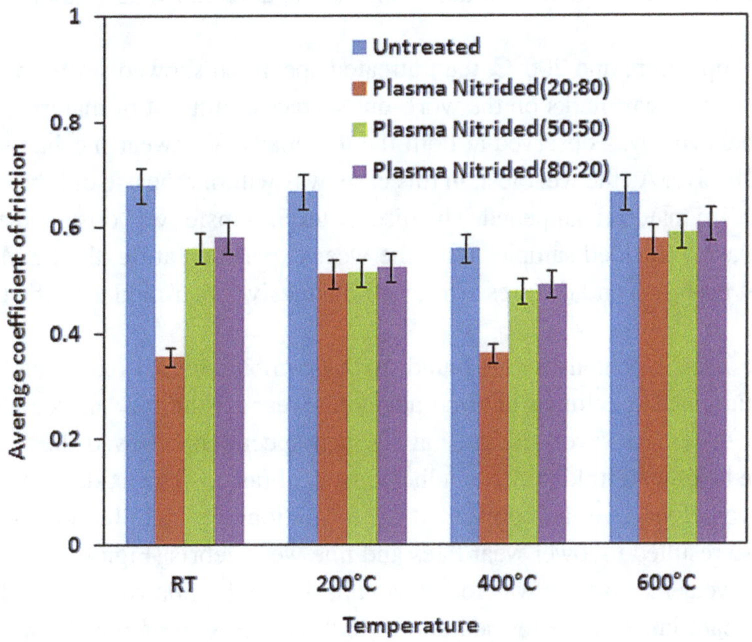

Figure 15. The average COF of untreated and PN AISI H13 steels after wear tests at 50 N loads

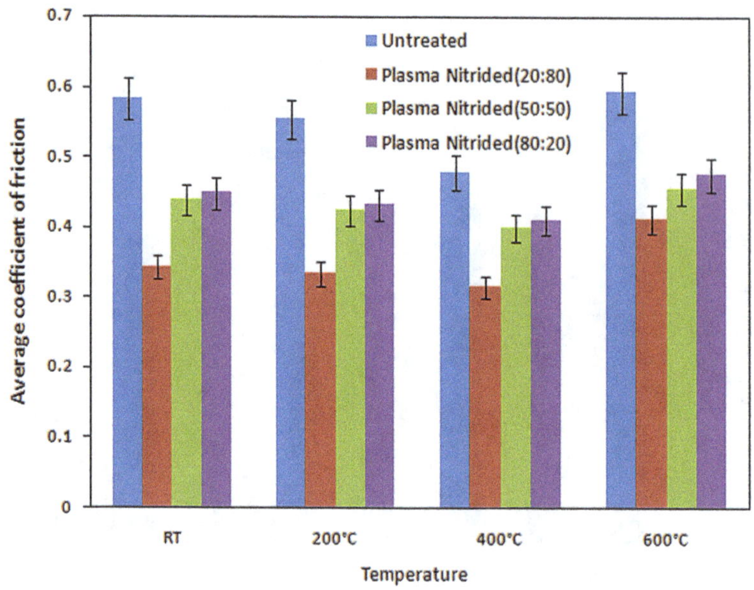

SEM Examination of Worn Surfaces

- Figures 16 to 20 show the SEM images of the specimens tested at two loads and four temperatures. The SEM morphologies explain the existing wear mechanisms. The morphologies showed the presence of pull out of materials, adhesion marks, abrasion wear marks, and oxidation wear marks.
- At the room temperature and 200°C, the untreated specimen showed severe plastic deformations and larger adhesive wear marks on the worn-out surfaces. Pull-out of materials was also noticed. This type of behavior was observed at both the test loads. The wear mechanism was found predominantly adhesive. As the wear test, in this case, was without the use of lubricants, the adhesion of the pin and disc material happened. The disc material transfer was observed onto the surface of the pin. For plasma nitrided samples after the wear experimentation, the SEM images indicated lesser material pull-outs and the presence of mild adhesive wear marks on the top (Figure 16 and Figure 17).
- At 400°C, the wear mechanism was found as the combination of oxidation, abrasion, and adhesion. The untreated specimens showed adhesion wear on the top surface alongwith the solid compact oxide layer. However, the plasma nitrided specimens showed the formation of a solid compact oxide layer with mild traces of adhesion and abrasion. The oxide film on the two surfaces acted as the protective cover and prevented the formation of a metallic bond between the sliding surfaces. These resulted in lower wear rates and fine wear debris (Figure 18).
- At 600°C, the wear mechanism was found as adhesive and oxidative (Figure 19 and Figure 20). The solid compact layers were formed due to high compressive forces between the two sliding components. These stable oxide layers might have acted as solid lubricants and reduced the friction between the two mating components. The lesser wear was observed at 400°C than at 600°C.

XRD Analysis of Worn Surfaces

The XRD diffractograms for the worn-out untreated specimen subjected to wear tests at 400°C and 25 N loads revealed the formation of Fe on the strong and medium intensity peaks. Further, on the worn surface of the PN specimen (nitrided with $20N_2$:$80H_2$), the CrO phase was identified on the strong intensity peak. Fe_2N, Fe_3N, CrN, Cr_2N phases were identified on the medium and weak intensity peaks. On the PN specimen (nitrided with $50N_2$:$50H_2$), again the CrO phase was identified on the strong intensity peak. Fe_2N, Fe_3N, Cr_2N, Fe_2O_3 phases were identified on the medium and weak intensity peaks. In the case of the PN specimen (nitrided with $80N_2$:$20H_2$), Fe_4N and Fe_2N were observed on the strong intensity peak. Fe_4N and Fe_2O_3 phases were identified on the medium and weak intensity peaks.

Figure 16. SEM micrographs of H13 steels after wear tests at RT and 50N load (a) Untreated specimen; (b) PN with N_2:H_2 as 20:80 (c) PN with N_2:H_2 as 50:50 and (d) PN with N_2:H_2 as 80:20

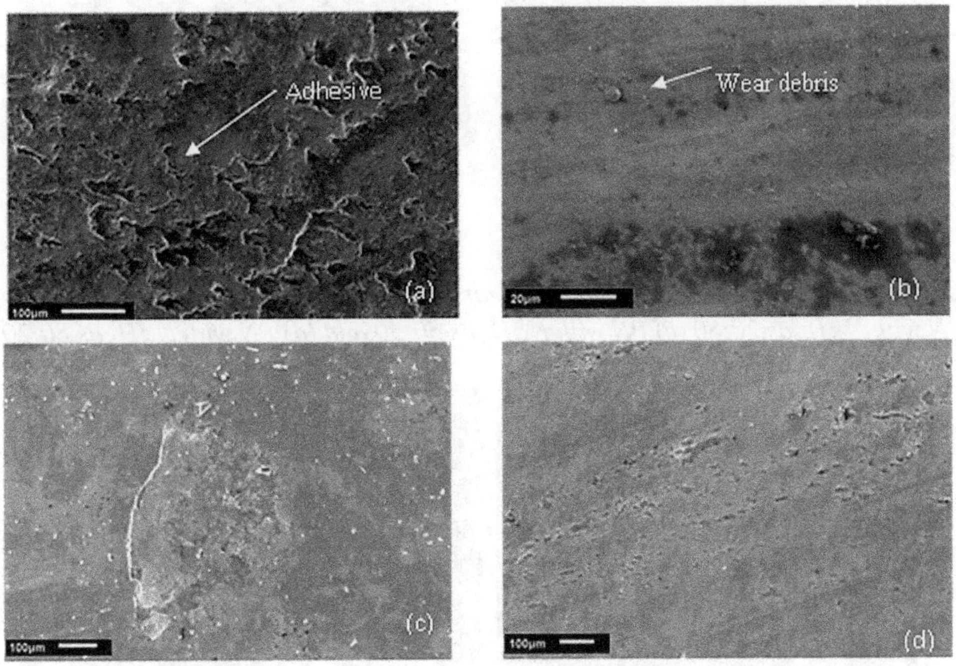

Figure 17. SEM micrographs of H13 steels after wear tests at 200°C and 25N load (a) Untreated specimen; (b) PN with N_2:H_2 as 20:80 (c) PN with N_2:H_2 as 50:50 and (d) PN with N_2:H_2 as 80:20

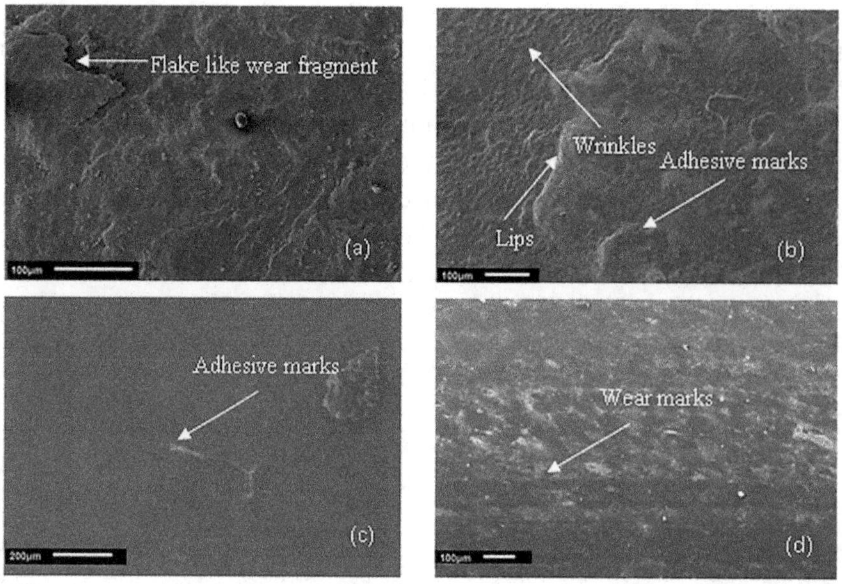

Figure 18. SEM micrographs of H13 steels after wear tests at 400°C and 25N load (a) Untreated specimen; (b) PN with N_2:H_2 as 20:80 (c) PN with N_2:H_2 as 50:50 and (d) PN with N_2:H_2 as 80:20

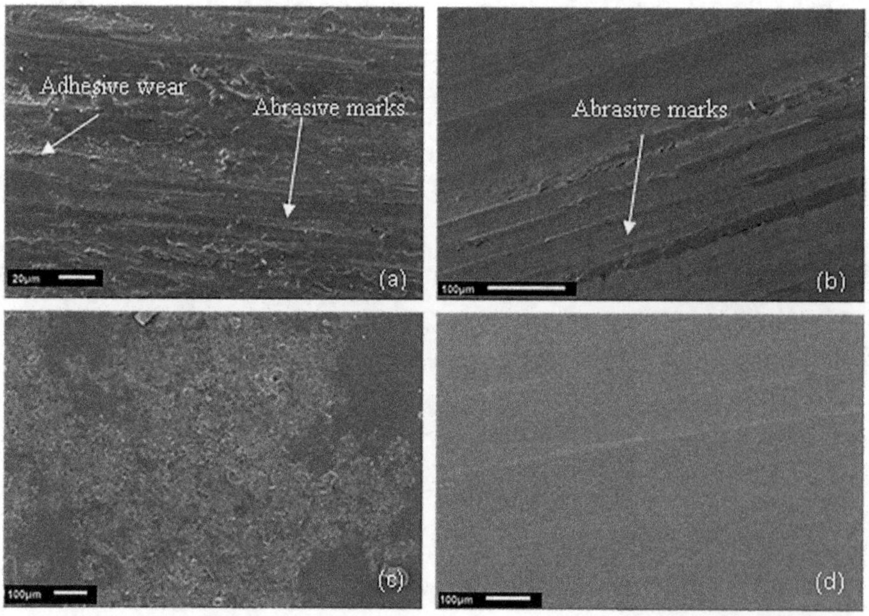

Figure 19. SEM micrographs of H13 steels after wear tests at 600°C and 25N load (a) Untreated specimen; (b) PN with N_2:H_2 as 20:80 (c) PN with N_2:H_2 as 50:50 and (d) PN with N_2:H_2 as 80:20

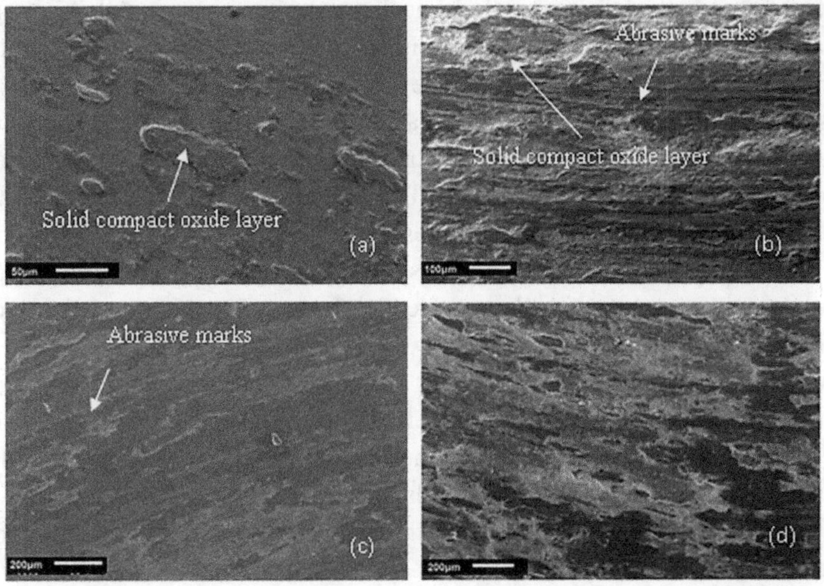

Figure 20. SEM micrographs of H13 steels after wear tests at 600°C and 50N load (a) Untreated specimen; (b) PN with N_2:H_2 as 20:80 (c) PN with N_2:H_2 as 50:50 and (d) PN with N_2:H_2 as 80:20

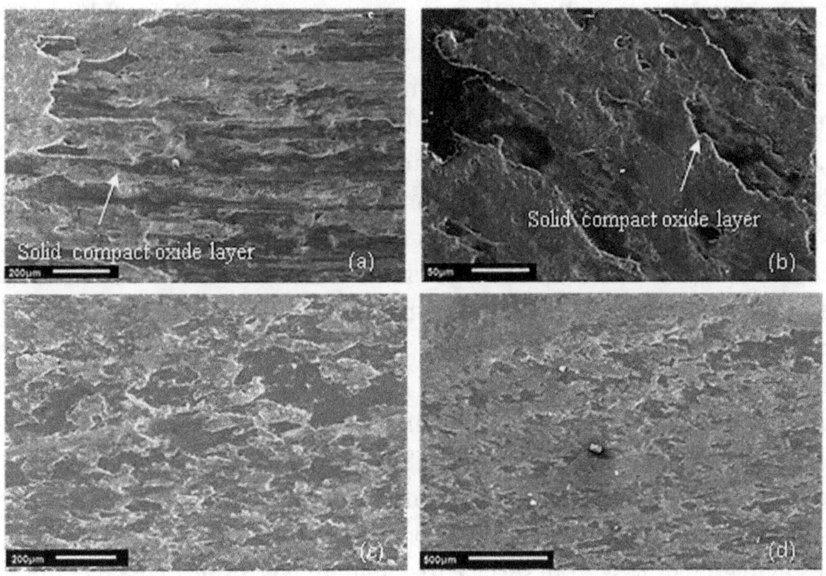

DISCUSSION

Hot-work tool steels are usually delivered in the annealed state. These are usually high-alloy steels. The steels contain carbon in lower amounts and chromium in higher amounts. Therefore, these steels are called as high chromium tool steels. These steels are not adequately used in the hot sheet metal forming operations. This is because different damage mechanisms act at the same time to generate damage to the tool thus causing the increase in the deviations from the original tool dimensions due to wear, heat-checking, micro-chipping, and breaking of the tool (Fnides et al., 2008; Klobcar et al., 2008). Die failure due to wear is mostly noticed. The material of the dies must be improved. Therefore, it becomes essential to heat treat the dies before using them. The hardness and toughness of dies are very important factors. The heat treatment processes can provide the required hardness and toughness to the steel (Pellizzari et al., 2007; Barrau et al., 2003; Nolan et al., 2006).

Further, some surface treatment techniques must be explored and developed to increase the wear resistance of the steels. In the current investigation, before the development of plasma nitriding, the steels were heat-treated to obtain the required hardness. The heat treatment process consisted of three steps i.e. austenitizing, quenching, and tempering to achieve the martensitic phase transformations which increase wear and thermal fatigue resistance. Further, with the heat treatment process, the better dissolution of vanadium-rich carbides could be obtained (Ochoa et al., 2009). Similar procedures were also adopted by other researchers (Tulsyan et al., 1993). After heat treatment, the microstructure of the steels changed to martensite. The martensitic structure mainly consists of ferrite and cementite. It gives good strength to the steel at room temperatures (Mebarki et al., 2002).

To increase the performance of the tool steels, it was decided to explore the potential of the plasma nitriding technique on the selected specimens. Plasma Nitriding is a surface treatment technique that is used to improve the mechanical properties of the steels thereby enhancing their wear resistance. After plasma nitriding the pin and disc samples in the present study, the top surface of the specimens was grey. The surface was smooth without the presence of any type of cracks. Visuttipilukul et al., (2006) carried out an in-depth study of the PN AISI H13 steels. Similar microstructures were observed by the authors.

In the current study, it was observed that the Ra values increased with the increase in the percentage of H_2. This was because $20N_2:80H_2$ gas composition caused more sputtering. The formation of iron nitrides was observed on surfaces. The EDS analysis confirmed the same. Capa et al. (2000) and Visuttipitukul et al. (2006) also recorded the same observations. It is evident from the literature that surface hardness is an important factor together with the microstructure. It was observed that the surface hardness values increased after PN. Further, it was noticed that the hardness value increased when the gas ratio $N_2:H_2$ was taken as 80:20. So more amounts of nitrogen led to more iron nitrides and thus increased the hardness of the steel surface.

The literature revealed that precipitation of the nitrides and the supersaturation with the nitrogen of the bcc matrix in the diffusion zone led to more surface hardness (Visuttipitukul et al, 2006; Pellizzari et al, 2003; Birol, 2012). The EDS analysis confirmed the presence of nitrogen along with Fe and Cr on the surface (Figures 6 and 7). Furthermore, Zhong et al., (2016) explained in their studies that the decrease in values in microhardness from surface to core region was due to the reduction in the percentage of nitrogen from surface to the core region during the diffusion process.

Figures 8 and 9 showed an increase in the thickness of the nitride layer with an increase in the nitrogen gas. The thickness of the layer was more with when the gas ratio $N_2:H_2$ was taken as 80:20. The diffusion of N_2 increased towards the depth with this gas ratio thus giving more layer thickness. Gammer et al.,

(2004); Corengia et al., (2005); Alsaran., (2003) reported similar results. Further, the figures show that the nitride layer was divided into two zones that is the white layer and the diffusion zone. By varying the gas mixtures, the white zone varied. The thickness was lowest for the gas ratio N_2:H_2 as 20:80 for the H13 tool steel. The high-intensity iron nitride peaks in the specimens nitrided with N_2:H_2 as 80:20 might be responsible for the thick white layer. A similar trend was observed in H11 tool steel. Further, in the present investigation, it was observed that the nitride layer thickness was different in both the tool steels. It was found more in the H13 steel. This might be due to the presence of higher amounts of alloying elements in the H13 tool steel. The higher percentage of Molybdenum has resulted in higher case depths. Ravindra et.al (2018) has also reported higher case depths with the presence of higher alloying elements.

The XRD analysis was carried out to identify new phase formation on the surface after the plasma nitriding process. The intensity of iron nitride peaks was more in H11 steel as compared to that observed in H13 steels indicating more formation of nitrides, however, peaks of CrN were lower than other peaks as the content of chromium is very less. Furthermore, in comparison with the case depth thickness values with varying percentages of nitrogen and hydrogen, the value came out to be more with 80% Nitrogen and 20% hydrogen. Therefore the results supported the XRD analysis. Karamis et al. (2012) also observed similar phases in a nitrided specimen. Visuttipilukul et al. (2006) analyzed the XRD profiles and observed that the nitriding layers of all specimens consisted of Fe_3N and Fe_4N, further the carbides and nitrides were seen by the EDS analysis. Kumar et al. (2012) also observed the cross-section of plasma nitrided samples under an optical microscope. The author observed approximately 200 μm diffusion zones. The results were similar to our experimentation. The observed XRD phases Fe_2N, Fe_3N, and Fe_4N were similar to the present study.

Further from the wear and friction experiments, it was observed that at room temperatures the average CoF values decreased in the case of the PN H13 steels. The decrease in the values was likely due to an increase in the hardness of the surfaces after PN. The presence of nitrides in the layer was also likely to be responsible for this observation. The PN H11 steel also showed similar behavior. CoF values were found lowest when the specimen was nitrided with N_2:H_2 as 20:80. Barrau et al. (2003); Podgornik et al. (2016); Rad et al. (2011) recorded similar values on their untreated and treated specimens. At 200°C, the specimens nitrided with a lesser percentage of nitrogen showed an increase in average CoF values. Other specimens with N_2:H_2 as 50:50 and 80:20 showed a reduction in the friction values. The reduction was nominal. Ebrahimzadeh et al. (2014) observed similar behaviors in their study on H13 steels. Furthermore, the decrease in friction values for all specimens was noticed at 400°C. Yang et al. (2011) reported this behavior as a result of the experiments conducted in the open atmospheres. The oxidation of the tool steel took place due to higher hardness. A reduction in the contact area was observed. These all parameters led to a reduction in wear and friction values. The oxide layers acted as a solid lubricant (Yang et al, 2011; Liu and Yan, 2010). At 600°C, softening of the surface layer occurred and led to the increase in wear and CoF values of all specimens. The contact area increased and gave rise to more sites of adhesive junctions. The formation of metal oxide and more wear at higher temperatures was observed by many researchers (Kchaou et al., 2013; Hardell et al., 2008; Wang et al., 1987).

The COF values for all the specimens were lower at 50 N loads in comparison to those obtained at 25 N loads at all temperatures. More amount of wear debris and higher Ra values at 50 N loads might attribute to the lower friction values (Bhushan, 1996; Blau, 1992; Chowdhury and Nuruzzamanb, 2013). For better wear resistance properties the white layer should be thin and uniform as mentioned by Das et al., (2016). In the present study, the specimen PN with N_2:H_2 as 20:80 gas ratio showed better results likely because of the formation of a thin white layer.

From the wear volume loss graphs, the maximum wear was seen for the untreated steels at all test parameters. The wear mechanism was mainly adhesive. During sliding, plastic deformation occurred and led to material detachment and debris creation (Vardavoulias et al., 1996). The values were observed to be more under 50N load at all temperatures than that obtained at 25N load. Further, in the case of plasma nitrided specimens, the wear volume loss was lesser in comparison to the untreated counterparts. The plasma nitrided specimens showed higher wear volume loss at room temperatures at both the testing loads. At 200°C and 400°C, the values decreased and again at 600°C, the increase in the wear volume loss was noticed. The wear rate was lowest at 400°C due to the formation of an oxidative film on the surface. It served as the lubricant to decrease the friction between the two mating components. It was observed that the wear volume loss of specimens treated with N_2:H_2 as 20:80 was found to be lesser for both the die steels. These specimens enhanced the wear resistance from 16% to 60% with respect to the wear volume loss of untreated specimens at all test temperatures respectively. Similarly, at 50 N load, PN H13 steel specimens nitrided with N_2:H_2 as 20:80 gas mixture enhanced the wear resistance from 43% to 78% at all test temperatures.

The SEM images indicated the wear mechanism as the blend of oxidation, abrasion and adhesion at 400°C. At higher temperatures of 600°C, the mode of wear was oxidative and adhesive. Scott, (2002) and Gaard et al. (2010) also observed the adhesive wear mechanism on the tool steels at various testing temperatures. XRD analysis of a few worn specimens verified the presence of oxides on the surface. Fe_2N, Fe_3N, Fe_4N, CrN, Cr_2N, CrO, and Fe_2O_3 phases were identified on the PN specimens tested at 400°C and at 25N loads. It is evident from the observations that specimen PN with N_2:H_2 gas ratio as 20:80 showed maximum wear resistance.

CONCLUSION

Plasma Nitriding is a rapidly developing surface treatment technique used to increase the wear resistance of industrial components especially hot forming/forging dies. The technique is being adopted by many industries. The authors developed the PN on the AISI H11 and AISI H13 die steels and analysed the wear and friction behaviour of the untreated and PN steels. Salient conclusions from the study are as below:

1. The plasma nitriding with different N_2:H_2 gas ratios were successfully developed on the selected tool steels. The nitride layer was free from any cracks, was dense, uniform and intact with the substrate steel.
2. The thickness of the nitride layer on plasma nitrided H13 steel was more than that on the plasma nitrided H11 steel due to the presence of more alloying elements in the H13 steel. The thickness value (including the white layer) on the H13 steel plasma nitrided with $20N_2$:$80H_2$, $50N_2$:$50H_2$ and $80N_2$:$20H_2$ was observed to be 283, 340 and 560 µm respectively. The thickness value (including the white layer) on the H11 steel plasma nitrided with $20N_2$:$80H_2$, $50N_2$:$50H_2$ and $80N_2$:$20H_2$ was observed to be 143, 202 and 230 µm respectively. The thickness of the nitride layer was found to be more on the specimens with the gas mixture as $80N_2$: $20H_2$.
3. The surface roughness values (Ra) on the plasma nitriding H13 specimens nitrided with N_2:H_2 gas ratios as 20:80, 50:50 and 80:20 were measured to be 0.44, 0.35, and 0.33 mm respectively. The surface roughness values (Ra) on the plasma nitriding H11 specimens nitrided with N_2:H_2 gas ratios as 20:80, 50:50, and 80:20 were measured to be 0.52, 0.41, and 0.37 mm respectively. The

Ra values were observed to be higher after plasma nitriding. The values were found to more with the increase in the percentage of hydrogen gas. This was because the gas composition consisting of a higher amount of hydrogen gas caused more sputtering.

4. The surface microhardness of the AISI H13 plasma nitrided specimens nitrided with $20N_2:80H_2$, $50N_2:50H_2$ and $80N_2:20H_2$ was found to be 1045±52 HV, 1214±60 HV and 1219±61 $HV_{0.1}$ respectively. The surface microhardness of the AISI H11 plasma nitrided specimens nitrided with $20N_2:80H_2$, $50N_2:50H_2$ and $80N_2:20H_2$ were found to be 1089±54 HV, 1102±55 HV and 1216±61 $HV_{0.1}$ respectively. The plasma nitrided steels showed higher hardness with an increase by about a factor of two as compared to the heat-treated die steels which indicate that the steels can perform better for hot forming applications.
5. The SEM images showed the presence of microparticles on the PN specimens. Iron, nitrogen, chromium, and carbon in the top surface layer were confirmed by the EDS analysis. This further indicated the formation of iron and chromium nitrides in the form of microparticles.
6. The plasma nitrided specimens showed enhanced wear resistance at all test parameters. The untreated H13 specimens showed more wear volume loss values than the plasma nitrided specimens.
7. The average COF values and the wear volume loss were observed to be lower at 400°C. This was mainly due to the formation of protective oxide film formed at this temperature that served as the lubricant and reduce the friction between the two mating components.
8. The XRD analysis after wear experiments indicated the formation of oxides. The enhanced wear resistance of the PN specimen has been attributed to the formation of these protective phases.
9. The wear mechanism for the untreated plasma nitrided specimens at RT and 200°C was predominantly adhesive. At 400°C, the wear mechanism was the blend of oxidative, adhesive and abrasive wear. At 600°C, the mode of wear was observed as oxidative and adhesive.

ACKNOWLEDGMENT

The authors very gratefully acknowledge the grant from Department of Science and Technology, New Delhi, India for conducting the current research. The authors are highly grateful to Dr. Harpreet Singh, Professor and Dean- ICSRII, Indian Institute of Technology (IIT), Ropar for providing the necessary facilities in conducting the analysis at his institute. The author also expresses his sincere thanks to Prof. Satya Prakash, Emeritus Professor, Materials and Metallurgy Department, Indian Institute of Technology, Roorkee, for his invaluable guidance.

REFERENCES

Alsaran, A. (2003). Determination of Tribological Properties of Ion-nitrided AISI 5140 Steel. *Materials Characterization*, *49*(2), 171–176. doi:10.1016/S1044-5803(03)00008-1

Amateau, M. F., & Glaeser, W. A. (1964). Survey of Materials for High Temperature Bearing and Sliding Applications. *Wear*, *7*(5), 385–418. doi:10.1016/0043-1648(64)90134-6

Antonov, M., Hussainova, I., Veintha, R., & Pirso, J. (2010). Effect of Temperature and Load on Three Body Abrasion of Cermets and Steel. *Tribology International*, *46*(1), 261–268. doi:10.1016/j.triboint.2011.06.029

Archard, J. F., & Hirst, W. (1956). The Wear of Metals Under Unlubricated Conditions. *Proc. R. Soc. Lond. Series A, 236*, 397-410.

Badisch, E., Katsich, C., Winkelmann, H., Franek, F., & Roy, M. (2010). Wear Behaviour of Hardfaced FeCrC Alloy And Austenitic Steel Under 2-Body and 3-Body Conditions at Elevated Temperature. *Tribology International*, *43*(7), 1234–1244. doi:10.1016/j.triboint.2010.01.008

Barrau, O., Boher, C., Gras, R., & Rezai-Aria, F. (2003). Analysis of the Friction and Wear Behaviour of Hot Work Tool Steel for Forging. *Wear*, *255*(7-12), 1444–1454. doi:10.1016/S0043-1648(03)00280-1

Berns, H., & Franc, S. (1997). Effect of Coarse Hard Particles on High Temperature Sliding Abrasion of New Metal Matrix Composites. *Wear*, *203-204*, 608–614. doi:10.1016/S0043-1648(96)07427-3

Bhushan, B. (1996). *Tribology and Mechanics of Magnetic Storage Devices* (2nd ed.). Springer-Verlag. doi:10.1007/978-1-4612-2364-1

Bhushan, B. (2002). Introduction to Tribology. John Wiley & Sons Inc.

Birol, Y. (2010). Response to Thermal Cycling of Plasma Nitrided Hot Work Tool Steel at Elevated Temperatures. *Surface and Coatings Technology*, *205*(2), 597–602. doi:10.1016/j.surfcoat.2010.07.035

Blau, P. J. (1992). Scale Effects in Sliding Friction: An Experimental Study, in Fundamentals of Friction: Macroscopic and Microscopic Processes. Kluwer Academic.

Capa, M., Tamer, M., Gulmez, T., & Bodur, C. T. (2000). Life Enhancement of Hot-Forging Dies by Plasma-Nitriding. *Turk J Engin Environ Sci.*, *24*, 111–117.

Chang, Y. N., & Wei, F. I. (1989). High temperature Oxidation of Low Alloy Steels. *Journal of Materials Science*, *24*(1), 14–22. doi:10.1007/BF00660927

Chowdhurya, M. A., & Nuruzzamanb, D. M. (2013). Experimental Investigation on Friction and Wear Properties of Different Steel Materials. *Tribology in Industry.*, *35*, 42–50.

Corengia, P., Ybarra, G., & Moina, C. (2005). Microstructural AISI 4140 Low-alloy Steel. *Surface and Coatings Technology*, *200*, 2391–2397. doi:10.1016/j.surfcoat.2005.01.060

Das, K., Joseph, A., Ghosh, M., & Mukherjee, S. (2016). Microstructure and Wear Behaviour of Pulsed Plasma Nitrided AISI H13 Tool Steel. *Canad. Meta. Quar.*, *55*(4), 402–408.

Deshpande, M., & Altan, T. (2011). Selection of Die Materials and Surface Treatments for Increasing Die Life in Hot and Warm Forging. FIA tech Conference, 1-23.

Eakins, E., Jayaseelan, D. D., & Lee, W. E. (2011). Toward Oxidation Resistant Zrb_2-SiC Ultra High Temperature Ceramics. *Metallurgical and Materials Transactions. A, Physical Metallurgy and Materials Science*, *42A*(4), 878–887. doi:10.100711661-010-0540-8

Ebrahimzadeh, I., & Ashrafizadeh, F. (2014). High Temperature Wear and Frictional Properties of Duplex-Treated Tool Steel Sliding Against a Two Phase Brass. *Ceramics International*, *40*(10), 16429–16439. doi:10.1016/j.ceramint.2014.07.151

Edenhofer, B. (1973). Part 2. Industrial Applications of the Process (conclusion). *Trait. Therm.*, *73*, 31–43.

Fernandes, F. A. P., Heck, S. C., Picone, C. A., & Casteletti, L. C. (2020). On the wear and corrosion of plasma nitrided AISI H13. *Surface and Coatings Technology, 381*.

Fnides, B., Yallese, M. A., & Aouici, H. (2008). Hard Turning of Hot Work Steel AISI H11: Evaluation of Cutting Pressures, Resulting Force and Temperature. *Mechanic.*, *72*(4), 59–63.

Forati, H. R., Amadeh, A., & Moradi, H. (2011). Wear Assessment of Plasma Nitrided AISI H11 steel. *Materials & Design*, *32*(5), 2635–2643. doi:10.1016/j.matdes.2011.01.027

Gaard, A., Hallback, N., Krakhmalev, P., & Bergström, J. (2010). Temperature Effects on Adhesive Wear in Dry Sliding Contacts. *Wear*, *268*(7-8), 968–975. doi:10.1016/j.wear.2009.12.007

Gammer, K., Stoiber, M., Wagner, J., Hutter, H., Kullmer, R., & Mitterer, C. (2004). Investigations on the Effects of Plasma-Assisted Pre-Treatment for Plasma-Assisted Chemical Vapour Deposition Tin Coatings on Tool Steel. *Thin Solid Films*, *461*(2), 277–281. doi:10.1016/j.tsf.2004.02.013

Geiger, M., Merklein, M., & Lechler, J. (2008). Determination of Tribological Conditions within Hot Stamping. *Production Engineering*, *2*(3), 269–276. doi:10.100711740-008-0110-8

Hardell, J., Kassfeldt, E., & Prakash, B. (2007). Friction and Wear Behaviour of High Strength Boron Steel at Elevated Temperatures of up to 800°C. *Wear*, *264*(9-10), 788–799. doi:10.1016/j.wear.2006.12.077

Hardell, J., & Prakash, B. (2008). High-Temperature Friction and Wear Behaviour of Different Tool Steels During Sliding Against Al-Si Coated High-Strength Steel. *Tribology International*, *41*(7), 663–671. doi:10.1016/j.triboint.2007.07.013

Hardell, J., & Prakash, B. (2010). Tribological Performance of Surface Engineered Tool Steel at Elevated Temperatures. *International Journal of Refractory Metals & Hard Materials*, *28*(1), 106–114. doi:10.1016/j.ijrmhm.2009.07.009

Hardell, J., Prakash, B., & Steinhoff, K. (2009). High Temperature Tribological Studies on Surface Engineered Tool Steel and High Strength Boron Steel. *Steel Research International*, *80*, 665–670.

Hernandez, S., Hardell, J., Courbon, C., Winkelmann, H., & Prakash, B. (2014). High Temperature Friction and Wear Mechanism Map For Tool Steel and Boron Steel Tribopair. *Tribology-Materials. Surfaces and Interfaces*, *82*(2), 74–84. doi:10.1179/1751584X13Y.0000000049

Hutchings, I. M. (1992). *Tribology-Friction and Wear of Engineering Materials*. Butterworth-Heinemann. doi:10.1016/0261-3069(92)90241-9

Inman, I. A., Datta, P. K. M., Du, H. L., Gray, J. S. B., Pierzgalski, S., & Luo, Q. (2005). Studies of High Temperature Sliding Wear of Metallic Dissimilar Interfaces. *Tribology International*, *38*(9), 812–823. doi:10.1016/j.triboint.2005.02.007

International, A. S. M. (1990). Author. *ASM Metal Handbook*, *1*, 439–444.

Jacobson, S., & Hogmark, S., (2005). *Tribologi-friction notning och smorjning. andra upplagan*. Academic Press.

Johnson, K. L. (2003). *Contact Mechanics* (204th ed.). Cambridge University Press.

Jones, L. C., & Llewellyn, R. J. (2009). Sliding Abrasion Resistance Assessment of Metallic Materials for Elevated Temperature Mineral Processing Conditions. *Wear, 267*(11), 2010–2017. doi:10.1016/j.wear.2009.06.023

Kabaya, T., & Iwabuchi, A. (1981). The Fretting Wear of 0.45% C Steel and Austenitic Stainless Steel from 20 to 650 °C in Air. *Wear, 74*(2), 229–245. doi:10.1016/0043-1648(81)90165-4

Karami, M. (1991). An Investigation of the Properties and Wear Behaviour of Plasma Nitrided Hot-Working Steel (H13). *Wear, 150*(1-2), 331–342. doi:10.1016/0043-1648(91)90327-Q

Karami, M., & Staines, A. (1989). An Evaluation of the Response of 722M24 Material to High Temperature Plasma Nitriding Treatments. *Heat Treatment Metals., 3*, 79–82.

Karamis, M. B., Baki, M., & Aydın, G. C. (2012). Sliding/rolling Wear Performance of Plasma Nitrided H11 Hot Working Steel. *Tribology International, 51*, 18–24. doi:10.1016/j.triboint.2012.02.005

Kchaou, M., Alimi, A., Elleuch, R., & Desplanques, Y. (2013). Characterisation of Oxidation and Wear Oxidized Surfaces of H13 Steel/ Brass in Dry Sliding Conditions. *International Journal of Microstructural and Material Properties., 8*(4/5), 373–384. doi:10.1504/IJMMP.2013.057073

Klobcar, D., Tusek, J., & Taljat, B. (2008). Thermal Fatigue of Materials for Die-Casting Tooling. Materials Science and Engineering, 472(A), 198-207.

Kong, H., Yoon, E., & Kwon, O. (1995). Self-Formation of Protective Oxide Films at Dry Sliding Mild Steel Surfaces under a Medium Vacuum. *Wear, 181*, 183325–183333. doi:10.1016/0043-1648(94)07042-3

Kovac, H., & Secer, Y. (2020). Improved tribological performance of AISI 316L stainless steel by a combined surface treatment: Surface texturing by selective laser melting and plasma nitriding. *Surface and Coatings Technology, 400.*

Kumar, A., Kaur, M., Joseph, A., & Jhala, G. (2020). Surface engineering analysis of plasma-nitrided die steels. *Proceedings of the Institution of Mechanical Engineers. Part J, Journal of Engineering Tribology, 234*(6), 917–931. doi:10.1177/1350650119873237

Kumar, A., Kaur, M., Singh, S., Joseph, A., Jhala, G., & Bhandari, S. (2018). High-temperature tribological studies of plasma-nitrided tool steels. *Surface Engineering, 34*(8), 620–633. doi:10.1080/02670844.2017.1341223

Kumar, R., Bhardwaj, D., & Sharma, Y. C. (2018). A Review on Plasma Ion Nitriding (PIN) Process. *Journal of Materials Science, 6*(1), 31–44. doi:10.26634/jms.6.1.14000

Kumar, R., Prakash, R., Jospeh, A. J., Jain, J., Pareek, A., Rayjada, P. A., Raole, P. M., & Mukherjee, S. (2012). Impact of Forging Conditions on Plasma Nitrided Hot-Forging Dies and Punches. *Journal of Material Science and Research., 1*(4), 11–18. doi:10.5539/jmsr.v1n4p11

Kumar, S., Maity, S. R., & Patnaik, L. (2020). Friction and tribological behavior of bare nitrided, TiAlN and AlCrN coated MDC-K hot work tool steel. Ceramics International, 46(11A), 17280-17294.

Lenhard, A., Damasio, S., Milke, A., & Schaeffer, L. (2006). Method to Estimate Workpiece-Die Heat Transfer Coefficient on Precision Warm Forging Process. *4th JSTP International Seminar on Precision Forging*.

Li, G., Liang, Y., Sun, H., & Cao, Y. (2020). Nitriding behavior and mechanical properties of carburizing and nitriding duplex treated M50NiL steel. *Surface and Coatings Technology, 384*.

Liu, R. L., & Yan, M. F. (2010). Improvement of Wear and Corrosion Resistances of 17-4PH Stainless Steel by Plasma Nitro-carburizing. *Materials & Design, 31*(5), 2355–2359. doi:10.1016/j.matdes.2009.11.069

Mebarki, N., Delagnes, D., Lamesle, P., Delmas, F., & Levaillant, C. (2002). Relationship between Microstructure and Mechanical Properties of a 5% Cr hot work tool steel. *Proceedings of the Sixth International Conference on Tooling*.

Mozgovoy, S., Hardell, J., Deng, L., Oldenburg, M., & Prakash, B. (2014). Effect of Temperature on Friction and Wear of Pre Hardened Tool Steel during Sliding Against 22MnB5 Steel. *Tribology-Materials. Surfaces and Interfaces, 82*, 65–73. doi:10.1179/1751584X13Y.0000000056

MRS Bulletin. (2009). *Technology Advances, 34*(792).

Nakamuraa, M., Hirao, K., Yamauchi, Y., & Kanzaki, S. (2002). Tribological Behaviour of Unidirectional Aligned Silicon Nitride against Steel. *Wear, 252*(5-6), 484–490. doi:10.1016/S0043-1648(02)00005-4

Nolan, D., Leskovsek, V., & Jenko, M. (2006). Estimation of Fracture Toughness of Nitride Compound Layers on Tool Steel by Application of Vickers Indentation Method. *Surface and Coatings Technology, 201*(1-2), 182–188. doi:10.1016/j.surfcoat.2005.11.077

Ochoa, E. A., Wisnivesky, D., Minea, I., Ganciu, M., Tauziede, C., Chapon, P., & Alvarez, F. (2009). Microstructure and Properties of the Compound Layer Obtained by Pulsed Plasma Nitriding in Steel Gears. *Surface and Coatings Technology, 203*(10-11), 1457–1461. doi:10.1016/j.surfcoat.2008.11.025

Pelcastre, L., Hardell, J., Herrera, N., & Prakash, B. (2012). Investigations into the Damage Mechanisms of Form Fixture Hardening Tools. *Engineering Failure Analysis, 25*, 219–226. doi:10.1016/j.engfailanal.2012.05.014

Pellizzari, M., Zadra, M., & Molinari, A. (2007). Tribological Properties of Surface Engineered Hot Work Tool Steel for Aluminum Extrusion Dies. *Surface Engineering, 23*(3), 165–168. doi:10.1179/174329406X150477

Phulera, N., Kaushik, S., Kshetri, R., Kanojia, N., Rawat, K., & Uniyal, V. (2021). Effects of plasma nitriding on mechanical, tribological and corrosion properties of friction stir welded joints of Al 2024. Materials Today: Proceedings, 4(15), 6726-6732.

Podgornik, B., Zuzek, B., Kafexhiu, F., & Leskovšek, V. (2016). Effect of Si Content on Wear Performance of Hot Work Tool Steel. *Tribology Letters, 63*(5), 5. Advance online publication. doi:10.100711249-016-0695-6

Prchlik, L., & Sampath, S. (2007). Effect of the Microstructure of Thermally Sprayed Coatings on Friction and Wear Response under Lubricated and Dry Sliding Conditions. *Wear*, *262*(1-2), 11–23. doi:10.1016/j.wear.2006.03.042

Pye, D. (2003). *Practical Nitriding and Ferritic Nitrocarburizing*. ASM International. doi:10.31399/asm.tb.pnfn.9781627083508

Quinn, T. F. J. (1967). The Effect of Hot Spot Temperatures on the Un-lubricated Wear of Steels. *Trans-asle.*, *10*, 158–168.

Quinn, T. F. J. (1971). Oxidational Wear. *Wear*, *18*(5), 413–419. doi:10.1016/0043-1648(71)90005-6

Quinn, T. F. J. (1992). Oxidational Wear Modelling: I. *Wear*, *153*(1), 179–200. doi:10.1016/0043-1648(92)90269-E

Quinn, T. F. J. (1994). Oxidational Wear Modelling: Part II. The General Theory of Oxidational Wear. *Wear*, *175*(1-2), 199–208. doi:10.1016/0043-1648(94)90183-X

Rad, H. F., Amadeh, A., & Moradi, H. (2011). Wear Assessment of Plasma Nitrided AISI H11 Steel. *Materials & Design*, *32*(5), 2635–2643. doi:10.1016/j.matdes.2011.01.027

Staines, A., & Bell, T. (1979). Plasma Nitriding of High Alloy Steels. In *Proceedings of the Conference on Heat Treatment-Methods and Media*. Institute of Met.

Stott, F. H. (2002). High Temperature Sliding Wear of Metals. *Tribology International*, *35*(8), 489–495. doi:10.1016/S0301-679X(02)00041-5

Stott, F. H., Glascott, J., & Wood, G. C. (1985). Models for the Generation of Oxides during Sliding Wear. *Proc. R. Soc. Lond. A*, 167-186.

Stott, F. H., & Jordan, M. P. (2001). The Effects of Load and Substrate Hardness on the Development and Maintenance of Wear Protective Layers During Sliding at Elevated Temperatures. *Wear*, *250*(1-12), 391–400. doi:10.1016/S0043-1648(01)00601-9

Straffelini, G., Trabucco, D., & Molina, A. (2001). Oxidative Wear of Heat Treated Steels. *Wear*, *250*(1-12), 485–491. doi:10.1016/S0043-1648(01)00661-5

Sun, Y., Bell, T., & Wood, G. (1994). Wear Behaviour of Plasma Nitrided Martensitic Stainless Steel. *Wear*, *178*(1-2), 131–138. doi:10.1016/0043-1648(94)90138-4

Takeda, M., Onishi, T., Nakakubo, S., & Fujimoto, S. (2009). Physical Properties of Iron Oxide Scales on Si Containing Steels at High Temperature. *Materials Transactions*, *50*(9), 2242–2246. doi:10.2320/matertrans.M2009097

Tisza, M. (2002). *Physical Metallurgy for Engineers*. ASM International Materials Park and Freund Publishing House Ltd.

Tulsyan, R., Shivpuri, R., & Altan, T. (1993). *Investigation of Die Wear in Extrusion and Forging of Exhaust Valves*. ERC for Net Shape Manufacturing, Report No. B-93-28. The Ohio State University.

Vardavoulias, M., Jeandin, M., Velasco, F., & Torralba, J. M. (1996). Dry Sliding Wear Mechanism for P/M Austenitic Stainless Steels and Their Composites Containing Al_2O_3 and Y_2O_3 Particles. *Tribology International*, *29*(6), 499–506. doi:10.1016/0301-679X(95)00110-P

Varga, M., Rojacz, H., Winkelman, H., Mayer, H., & Badisch, E. (2013). Wear Reducing Effects and Temperature Dependence of Tribo Layers Formation in Harsh Environment. *Tribology International*, *65*, 190–199. doi:10.1016/j.triboint.2013.03.003

Venkatesan, K., Subramanian, C., & Surnmerville, E. (1997). Three Body Abrasion of Surface Engineered Die Steel at Elevated Temperatures. *Wear*, *203-204*, 129–138. doi:10.1016/S0043-1648(96)07442-X

Vergne, C., Boher, C., Gras, R., & Levaillant, C. (2006). Influence of Oxides on Friction in Hot Rolling: Experimental Investigations and Tribological Modelling. *Wear*, *260*(9-10), 957–975. doi:10.1016/j.wear.2005.06.005

Visuttipitukul, P. C., & Kuwahara, H. (2006). Characterization of Plasma Nitrided AISI H13 Tool Steel. *Acta Metall Slov.*, *12*, 264–274.

Wang, D. Y., Shu, D. L., & Guo, X. C. (1987). Effect of Microstructure and Properties on High Temperature Wear Characteristics of $3Cr_2W_8V$ (H21) steel. *Wear*, *119*(1), 101–117. doi:10.1016/0043-1648(87)90101-3

Wang, S. Q., Wei, M. X., & Zhao, Y. T. (2010). Effects of the Tribo Oxide and Matrix on Dry Sliding Wear Characteristics and Mechanisms of A Cast Steel. *Wear*, *269*(5-6), 424–434. doi:10.1016/j.wear.2010.04.028

Winkelmann, H., Badisch, E., Varga, M., & Danninger, H. (2010). Wear Mechanisms at High Temperatures. Part 3: Changes of the Wear Mechanism in the Continuous Impact Abrasion Test with Increasing Testing Temperature. *Tribology Letters*, *37*(2), 37419–37429. doi:10.100711249-009-9534-3

Yang, J., Liu, Y., Ye, Z., Yang, D., & He, S. (2011). Microstructural and Tribological Characterization of Plasma and Gas-Nitrided 2Cr13 Steel in Vacuum. *Materials & Design*, *32*(2), 808–814. doi:10.1016/j.matdes.2010.07.022

Zhong, H., Dai, L., Yue, Y., Wang, B., Zhang, X., Tan, C., Ma, M., & Liu, R. (2016). Tribological Properties of Plasma-Nitrided AISI 4340 Steel In Vacuum. *Materials Science and Technology*, *32*(4), 275–281. doi:10.1080/02670836.2015.1121341

Chapter 8
Recent Advancements in Wear-Resistant Coatings Prepared by PVD Methods

Kamalan Kirubaharan Amirtharaj Mosas
https://orcid.org/0000-0002-6514-9063
FunGlass – Centre for Functional and Surface Functionalized Glass, Alexander Dubček University of Trenčín, Slovakia

Dinesh Kumar Devarajan
Sathyabama Institute of Science and Technology, India

Subhenjit Hazra
Sathyabama Institute of Science and Technology, India

Gobi Saravanan Kaliaraj
Sathyabama Institute of Science and Technology, India

ABSTRACT

Physical vapor deposition (PVD) technologies are widely used to produce wear and corrosion resistant coatings for a variety of industrial applications. In recent years, there has been remarkable interest in the development of novel wear resistant coatings prepared through PVD methods, which helps to reduce friction and wear, as a result of recovering energy losses up to 30% due to friction and economy loss due to wear. This chapter provides comprehensive data of recent progress in wear resistant coatings prepared using PVD methods, starting with the introduction of it needs, significance, physiochemical properties, and the selection criteria of wear resistant coatings. The applications, physical, and chemical properties of superhard materials such as diamond like carbon (DLC), titanium nitride (TiN), chromium nitride (CrN), and tantalum nitride (TaN) are also presented.

DOI: 10.4018/978-1-7998-9683-8.ch008

INTRODUCTION - NEED FOR WEAR RESISTANT COATINGS

Wear is a material degradation process due to the continuous tribological interactions. Surface modification on alloys/metals are often applied to modify the surface properties, which are generally used in applications having aggressive environments, show explicit performance at their surfaces, such as excellent corrosion and wear resistance. Wear resistant coatings can be implemented, where the use of liquid lubricants (grease, oil, etc.) proves unsuccessful due to excessive load, migration, extreme temperature or neglected surfacing. The tribological, corrosion and mechanical properties of any alloy/metals can be enhanced by deposition of thin/thick hard coating onto the surface of a metal to elevate the overall component performance.

Generally organic (paints, polymers) and inorganic (electroplating, electrophoretic, conversion coatings, and galvanizing) based coatings are applied onto metals, but many of these types of coatings suffer from several glitches including limitations towards high temperature application or lack of wear resistance. When a unique combination of high temperature, wear, and corrosion resistance is needed, coating processes such as electroless plating and electroplating are used, but these coatings often face environmental problems associated with the coatings themselves. Often, it might be hard to find the perfect balance between corrosion and wear resistance among these. Transition metal carbides and nitrides based hard coatings are extensively used nowadays to protect the materials against corrosion and wear.

BACKGROUND

Every industrial production, mechanical tools are the key components. Therefore, it is inevitable to enhance the tool lifetime and productivity. In the last decade, ceramic and composite materials have also been implemented other than metals. Wear resistance is a common term and there are many types of wear, which are encountered such as abrasion, impact, fretting, and frictional sliding and so on. Surface coatings find usage in drills, cutters, taps and milling tools. Wear resistance coatings on interacting surfaces are important in industrial corrosion protection, which demands more innovative ideas. They are essential and requires scheduled maintenance to make sure good working condition. Excessive wear of components during the production resulting in immediate replacement of component or maintenance certainly occurs. This is a major cause of concern in terms of maintenance, depreciation, costs due to delay in production.

Lightweight alloys and metals have become essential components in automotive, transportation, aerospace, and many other industries. There is a huge demand for their wide range of applications, these lightweight metals and alloys have met a wide range of surface solutions. The feasibility of cost to weight ratio, they can contribute to improve the performance and efficiency in these industries. In spite of having improved weight-to-strength ratios, lightweight alloys show less hardness and have reduced resistance to wear.

One possible solution to address these stability related issues and also getting superior performance is to deposit wear resistant coatings onto the components that are prone to wear and damage. This outstanding wear resistance can be achieved through an optimum combination of fracture toughness and hardness of the coatings. Enhanced hardness protects the overlay coating from impact, cutting and fragmentation. In addition, ductility, increased toughness and compressive stresses of the coating reduce surface fatigue, micro cracking and failure.

PROPERTIES OF HARD WEAR RESISTANT COATINGS

The properties of hard wear resistant coatings can be categorized into three ways:

1. Structural and microstructural properties (roughness, thickness, crystal structure, morphology, texture, chemical composition, and surface topography).
2. Physical and chemical properties (density, magnetic, electrical, optical and thermal properties, oxidation and corrosion resistance).
3. Mechanical properties (adhesion, stresses, hardness, wear, fracture toughness, and friction).

However, for tribological applications, the essential properties of hard wear resistant coatings are microstructure, hardness, Young's modulus, thickness, surface roughness, corrosion and erosion resistance, fracture toughness, reduced internal stress and enhanced adhesion to the substrate or subsequent layers (Bull 2006). Hardness term is referred to as the resistance of a material against permanent deformation by the influence of another interacting hard surface. For a tribological point of view, the hardness of the substrate is very crucial. If a thin layer of material deposited onto the substrate is not sufficient, then it may fail at low contact stresses, because it cannot follow substrate deformation.

Coating hardness is used to evaluate coating performance, and to assess coating quality in numerous tribological applications. However, the intrinsic hardness of a coating should not be overstated. Resistance to fatigue and chemical stability are generally show a larger influence on wear resistance than coating hardness. Adhesion is referred to the interaction between the adjacent surface of two bodies (for example substrate and film). These two surfaces are bound together through either valence forces, mechanical interlocking or the combination of both. Young's modulus (also called modulus of elasticity) may difficult to determine for thin films. But, if the thickness of the coating is sufficient, then it can be determined with the help of nanoindentation technique. This Young's modulus show a strong influence on delamination of the coating, contact stresses, fracture toughness, residual stresses within the coating. Generally, hard coatings have higher Young's modulus than pure substrate and also yield low thermal expansion coefficient (TEC). An increase in surface temperature through frictional force may create the introduction of tensile stress within the coating.

Coatings prepared through PVD methods are extensively used in many tools and machine components. The requirements of coating for cutting tools have accelerated for the design and improvement of advanced hard coatings in this era. The fundamental requirements of hard coating, the coating material should have the following properties (Hoornaert, Hua, and Zhang 2009).

1. High hardness
2. High fracture toughness
3. High wear resistance
4. Excellent adhesion to the substrate
5. Low heat transfer coefficient
6. Low friction coefficient
7. Chemically inert
8. Fine grain and crystalline microstructure
9. Smooth morphology

Bulk materials cannot satisfy all the requirements as mentioned above. Hence, PVD based coatings have gained more attention, particularly in machine sliding and tool components. The coating properties can be often explained based on their structure during thin film growth. To achieve high wear resistance coatings, several approaches have been included such as; a) doping of multiple elements which improves the hardness of the coating and coating performance (Dai et al. 2018), b) fabrication of compositionally or functionally graded nanocomposite coatings to enhance the overall coating adhesion, stress accumulation at the coating interface, coefficient of thermal expansion, and lattice misfit (Yuan et al. 2020), and c) ion bombardment treatment to improve the adhesion and wear resistance (Zhai et al. 2021). During film deposition, the defects formed can act as an obstacle for dislocations. This is one of the major strengthening mechanism of PVD based coatings (Bhushan 2000). Other strengthening mechanisms are through secondary phase particles, line defects, internal boundaries, solutes, etc.

SELECTION CRITERIA OF WEAR RESISTANT PVD COATINGS

At first, it is significant to know that wear resistance is a common term and there are different types of wear such as abrasion, frictional sliding, impact, etc. Earlier, it was assumed that higher hardness (H) of materials surface would yield better wear resistance. It may partly true but different types of wear mechanism have not been taken into account. A more useful indicator for a hard wear resistant coating is the ratio of hardness (or tensile strength) with respect to its stiffness (Young's Modulus, E). Coatings showed high H/E ratio would exhibit better toughness and able to withstand deformation of the substrate without yielding (Krella 2020). Such properties offer wear resistance of the substrate more consistently than hardness only.

Secondly, low frictional force acts between the wear counterpart and PVD coating. Friction is defined by a renowned macroscopic formula called $F = \mu N$. Upon decreasing the frictional coefficient, μ, the transfer of energy between the sliding surfaces reduces and hence damage potential decreases. This could be achieved to some extent by reducing the roughness of the surface of the substrate through polishing. Alternatively, an introduction of a lubricating layer between the surfaces will diminish the transfer of energy and hence damage reduces. These lubricating layers could be any one of the liquids, gases or solids (e.g. MoS_2 or graphite), providing lamella microstructures showing weakly bonded layers, which can be easily decoupled through shear forces. This kind of approach works well where low coefficient of friction (CoF) is the main requirement, and not useful where transferring energy between the surfaces is a major concern.

In the search of hard wear resistance coatings, the PVD coating should have load-bearing capabilities, particularly if the coating has higher strength than that of substrate. This can happen with the substrates having lightweight, where low yield strength may not support Hertzian stresses delivered by wear partner. In such cases two approaches can be done by a) depositing a thick layer of coating that can be able to withstand higher load without fracture or yield stress, or b) increasing the base metal strength via diffusion of secondary elements to its lattice system.

For other applications, a hard coating may able to withstand variable wear, environmental shock or sliding conditions, but resistance to abrasion is also required. The major requirement for modern hard wear resistant coatings is to reduce wear and frictional coefficient in order to increase the tool component lifetime. In this case, we must know which wear mechanism has the highest contribution to wear (adhesion, abrasion, corrosion, high temperature, etc) in order to select the optimum coating for the desired application. The possibility of unique combinations of different materials properties unlocks the feasibility of advanced coatings having low CoF, high temperature stability, enabling lubrication without the use of liquid lubricants (Fotovvati, Namdari, and Dehghanghadikolaei 2019).

Though the single layer coatings are finding wide applications in many sectors of engineering and technology, there are an increasing number of requirements where the properties of a single layers are not sufficient (Bull and Jones 1996). One way to conquer this problem is to use a multilayer coating (**Figure 1**) that combines the attractive properties of several materials. The multilayer concepts are more promising with respect to their properties and performance in many industrial applications. It meets many requirements like multi-functional character, good adherence to alloy substrates, moderate residual stresses, low friction coefficients, proper hardness to toughness ratio, etc. (Holleck and Schier 1995a). Metastable and multi-component materials can be arranged in a graded multilayer arrangement. In multilayer PVD coatings, one can realize the combination of single layer materials properties and different concepts can be combined in a coating, tailored for specific requirements.

Figure 1. Different types of coating prepared through PVD methods

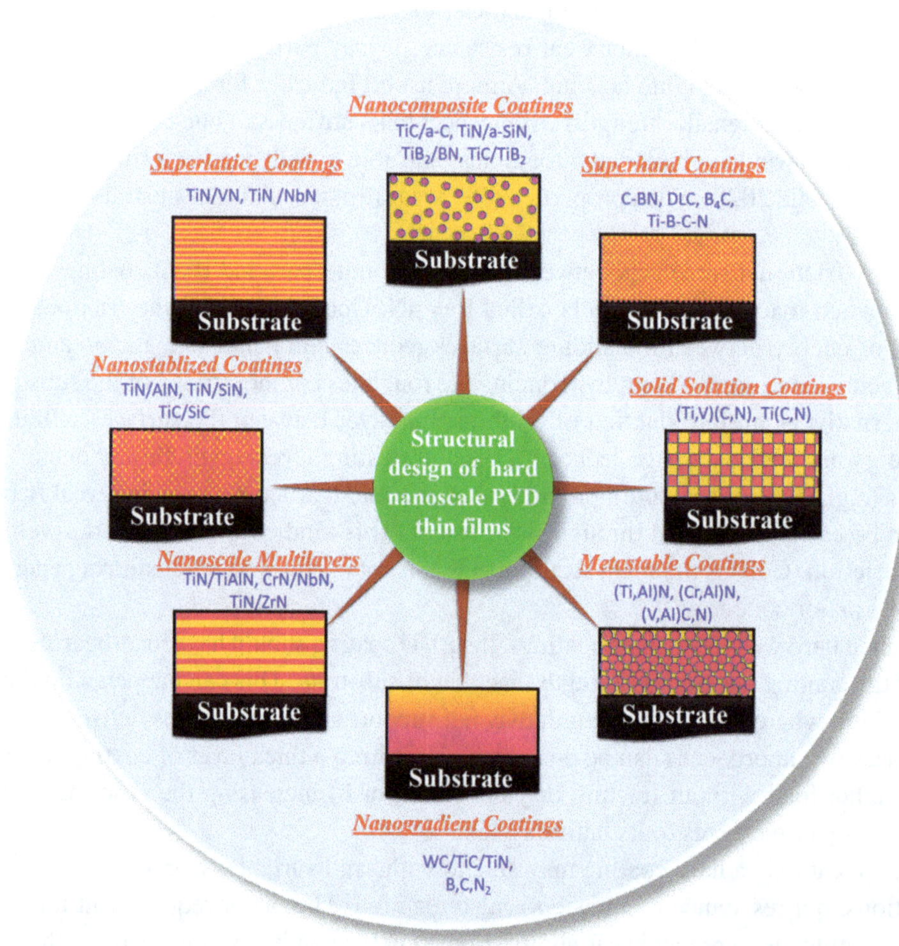

MULTILAYER FILMS

Recent advancements in multi-layered PVD coatings have led to improved performance of mechanical component. To reduce the CoF and improved wear resistance of a metallic component, the substrate can be deposited with a thin layer of materials with appropriate chemical and physical properties. The limitations of single component and monolayer films such as very high residual stresses, lattice mismatch, lack of multi-functional character, and adhesion strength to the substrate have promoted interest to use multi-layered films. Multilayer and multicomponent structures provide the feasibility to improve the shortcomings of the individual coating by the addition of properties of every individual layer. The higher wear resistance combined with low CoF and chemical inertness makes the multilayer films very attractive for many industrial applications.

Multilayer films are the combinations of alternate arrangements of two or more materials, having a thickness of few nanometres to microns. Multilayer films offer a combination of superior properties from individual materials and increased tribological performance than monolithic films. The multilayer coatings use interlayers which improves the adhesion of the films with the subsequent layers or substrate by reducing the thermal expansion mismatch, lattice misfit, residual stress and other necessary properties. The stacking of alternate layers may increase the fracture toughness through introducing a layer to deflect the cracks or to arrest the propagation of cracks.

Generally, multilayer films are differed from sandwich layers because of some complementary properties. The unique combination of several component microstructures and their compositions within one coating provides the overall combination of individual component properties, reduction of thermal and lattice mismatch between the substrate and the film, reduced residual stresses associated within the film, ability to hold crack propagation under service conditions, and enhanced fracture toughness. The flexibility in PVD technology and the development of combinatorial processes led to substantial developments in multicomponent and multi-layered thin films, which has found major industrial applications.

NANOCOMPOSITE COATINGS

Nanocomposite coatings have recently attracted attention as they promote hardness, improved wear resistance, oxidation resistance, and other physiochemical properties (Patscheider 2003). Nanocomposite coatings consist of two or more phases and are composed of nanocrystalline grains of transition-metal nitrides, carbides or oxides surrounded by an amorphous hard matrix. Such coating shows enhanced hardness mainly due to nanostructured grain size of materials and higher resistance to movement of dislocations (Pogrebnjak and Beresnev 2012). Other remarkable properties such as high fracture toughness, thermal stability and interface toughness are key parameters for their performance as protective coatings.

Advanced surface related properties such as magnetic (Redolfi Riva et al. 2017), optical (Keskin et al. 2017), electronic (Xue et al. 2017), mechanical (G. Zhang et al. 2015), chemical (Saikia, Sahariah, and Pandey 2016), catalytic (Bazyari et al. 2016), and tribological properties (Vepřek and Reiprich 1995; S. Veprek et al. 2000; Voevodin, O'Neill, and Zabinski 1999; Musil, Karvánková, and Kasl 2001) can be obtained through advanced nanocomposite coatings. These properties make nanocomposite coatings suitable for industrial applications in optical, tooling, and magnetic storage devices. Nanostructured coatings are divided into sub categories such as nanocomposite coatings, nano-scale multilayer coating, superlattice coating, nano-graded coatings, etc (Musil 2000; Holubar, Jilek, and Sima 2000; Stan

Veprek and Argon 2001; Musil, Karvánková, and Kasl 2001; Voevodin, O'Neill, and Zabinski 1999). The superior performance of nanocomposite coatings significantly depends on material selection, proper choice of deposition method, and processing parameters (S. Zhang et al. 2003). Nanocomposite yttria stabilized zirconia (YSZ) coatings with Ag and Mo produced by PVD process showed better wear resistance properties at higher temperatures (Hu, Muratore, and Voevodin 2007).

The desired properties of nanocomposite films can be modified by tailoring the crystallite size and volume fractions of nanocrystalline phases embedded in an amorphous matrix and the distribution of amorphous matrix among the nano crystallites. The amorphous nanocrystalline composite films result in strength, due to restriction in crystallites dislocations, which was surrounded by an amorphous matrix. The amorphous phase surrounding the crystalline phase helps to reduce the lattice mismatch, because the dislocation cannot propagate into the amorphous phase.

DEPOSITION METHODS

The design and development of coatings for wear resistant applications have started in the mid of 1960s using chemical vapor deposition (CVD), physical vapor deposition (PVD), and pulsed laser deposition (PLD) methods. Coatings can be deposited by means of a variety of deposition techniques with each one having its own advantages and disadvantages. Coating techniques can be broadly classified into three categories namely; solution methods, CVD and PVD methods. Each process can again be sub-classified into types based on the source of energy used for the deposition of coatings. Some of the coating techniques used for the deposition of coatings are listed below.

1. Powder Spray deposition methods
    a. Thermal spray, Plasma spray
    b. High velocity oxy fuel (HVOF)
2. Chemical vapor deposition (CVD) methods
    a. Low pressure and high pressure (LPCVD/HPCVD)
    b. Plasma enhanced (PECVD)
    c. Metal-Organic (MOCVD)
3. Physical vapor deposition (PVD) methods
    a. Thermal evaporation
    b. Electron beam PVD
    c. Sputtering
        i. Direct current diode sputtering
        ii. Radio frequency sputtering
        iii. Balanced and unbalanced magnetron sputtering
    d. Cathodic arc
    e. Ion plating

PVD method produces hard surface coatings that provide excellent wear and corrosion resistance. These developments are very important in many industrial applications providing drastic improvement in terms of efficiency, productivity, and lifetime and therefore providing cost reductions. Such types of coatings are commonly used for industrial coating including automotive parts, engines, and machines, as they provide excellent wear and abrasion resistance. PVD consists of a variety of vacuum deposition techniques, which will useful to fabricate coatings and thin films. In general, PVD describes as a process in which the target or source material transform from a condensed phase to a vaporized phase followed by the formation of thin film. The most familiar PVD techniques are sputtering and evaporation.

Atmospheric Plasma Spray (APS) Method

Plasma spraying is a deposition technique in which a direct current electric arc field will create between two electrodes, while a required (mixed gases) gases passes through the arc. The high potential arc will turn the gases into an ionized state at high temperatures along with the high speed of up to 400 m/s (Greene 2017). APS process operates at high temperature and velocity of the plasma medium to heat the coating powder feedstock to molten or semi-molten droplets. The feedstock material has to be sprayed by injecting into the plasma plume, where it gets melted and accelerated towards the substrate surface with the help of career gas (**Figure 2**). The molten particle strike the substrate surface at a very high velocity (~150 to 200 m/s). When the molten droplets strike the substrate surface, it gets anchored through mechanical interlocking mechanism to form a lamella structured strong adherent coating.

Figure 2. Schematic diagram of atmospheric plasma spray method

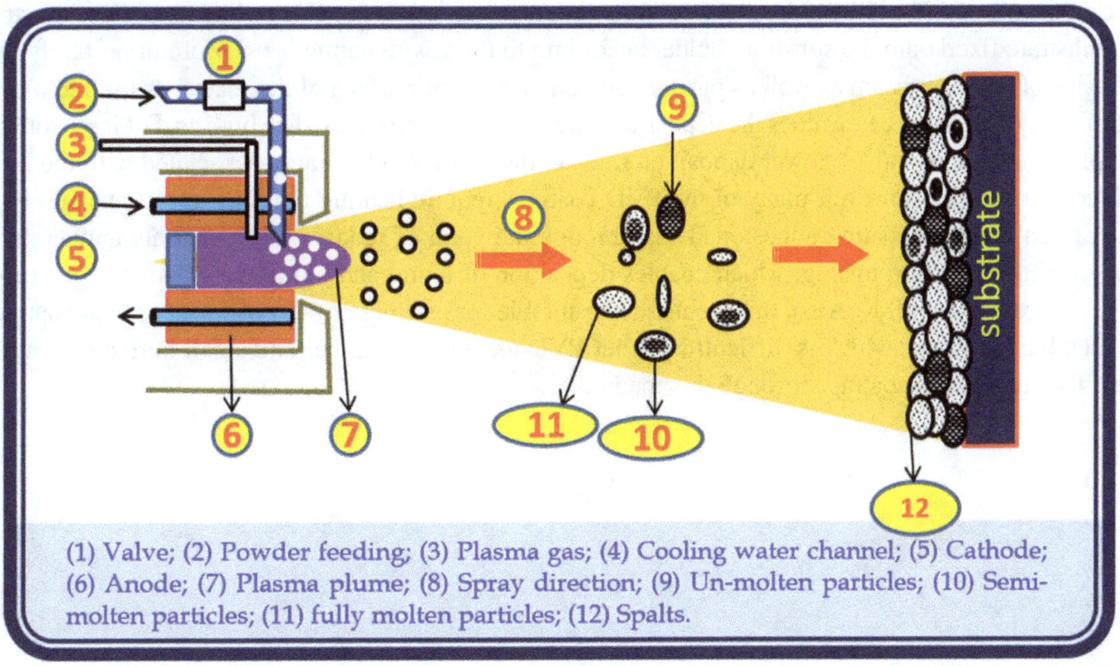

This method is widely used in developing protective coatings and thermal barrier coatings onto metals and alloy substrates for corrosion resistance, erosion and abrasion resistant coating applications. The main advantages of plasma spray process are; a) the substrate can be kept at lower temperatures, b) larger components can be coated, c) relatively fast, d) thickness of the coating can vary from few μm to mm, and f) strong adhesion through mechanical as well as metallurgical bonding can be obtained. Different substrate materials such as ceramics, metals, glass, silicon wafer or even organic materials that are placed in plasma, will then be deposited with a layer of few microns from the source materials (Lugscheider and Weber 1990). This method is widely used to produce wear resistant coatings on automotive parts such as piston rings, synchronizer rings, and cylinder bores due to its high deposition rate. Since this is a line-of-sight process, coating on smaller diameter and internal bores is still challenging.

Pulsed Laser Deposition (PLD) Process

Pulsed laser deposition (PLD) is another type of PVD technique that has gained a great attention for the past few years due to its ease of handling and fabricating the thin film in a successive way. Unlike other methods, complex stoichiometry films can also be developed through this technique. Many materials and compounds that are normally very difficult to deposit by other techniques, especially multi-element oxides, can be done by PLD method. The formation of thin film in PLD process can be divided into different stages (**Figure 3**). At first, the high energetic laser beam will be focused onto target materials surface. At adequately high density flux and sufficient laser pulse duration, all elements present in the target materials are readily heated up to their evaporation temperature. Then, the materials are ablated out from the target surface without losing the stoichiometry as in the target materials (Michelmore 2016). During the ablation process, several parameters will influence the density of the plume (complex physical phenomena) such as electronic excitation, collision, exfoliation, thermal and hydrodynamics. During the second stage, the ablated materials by high energetic laser source will tend to move towards the substrate fixed onto the substrate holder according to the gas-dynamic laws. More importantly, the spot size of the laser beam as well as plasma temperature on the surface of the materials used to play a significant role, which determines the deposition rate, and film uniformity. The high cost of laser source, sample coating area and the lower deposition rate are the main disadvantages associated with the laser ablation technique. However, many of the PVD coatings require heating the substrate above 500 K in order to ensure high coating adhesion. Thus, through the point of reducing the substrate temperature, PLD method seems promising, which enables deposition of mono and multilayer coating even at low temperature below 100 K. Also, this technique is suitable for producing wear resistant carbon coatings on metallic substrates which is difficult by other PVD techniques. Larger and small bore components are difficult to apply coatings through this method.

Figure 3. Schematic diagram of pulsed laser deposition system

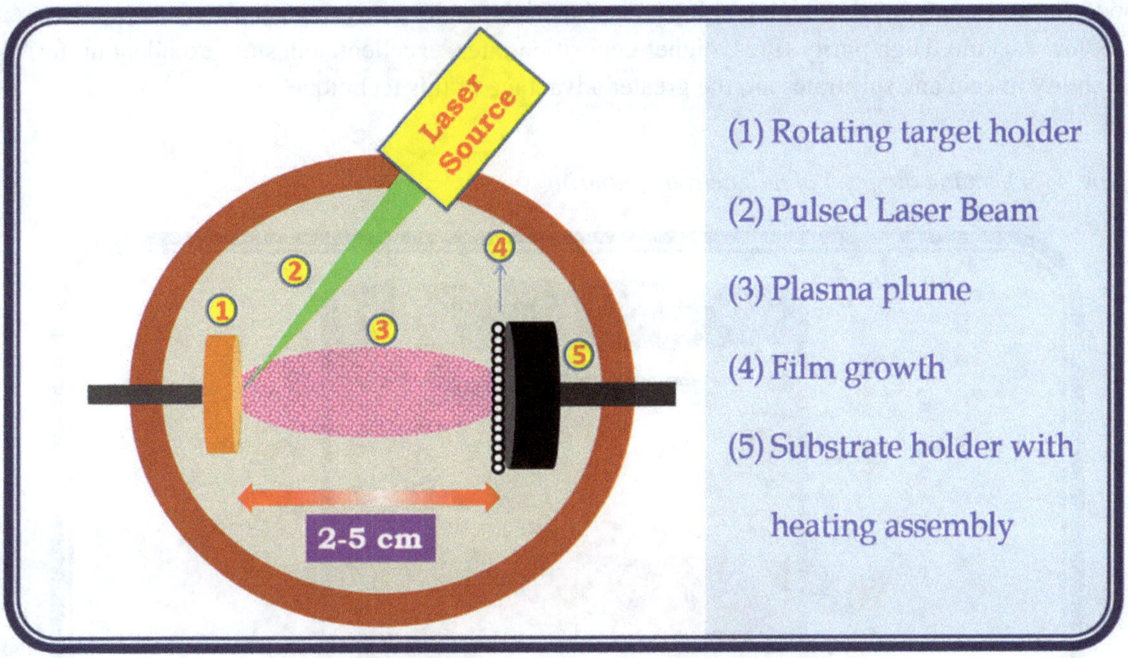

Magnetron Sputtering Process

Sputtering method involves the impingement of bulk materials (target) with high energetic sputtering gases such as argon ions that will be deposited onto a desired substrates like metals, silicon wafer, and glass substrates. The desired substrates, which has to be coated, are placed in a high vacuum chamber and generally argon gas is used as a sputtering gas (**Figure 4**). Moreover, during deposition, the negative electric field will place onto magnetron gun as target material which has to be bombarded with argon (Ar) gas and subsequent Ar atoms get ionized resulting plasma in the chamber (Premphet et al. 2017). The formed argon ions in the chamber will get high kinetic energy and will bombard onto the target surface with high kinetic mobility of Ar ions and the ejected atoms are deposited to form a thin film. Apart from the magnetron sputtering method, several other methods of plasma-assisted vapor deposition coating systems are also widely used, such as reactive sputtering with oxygen, nitrogen and methane environment, ion-assisted sputtering and ion beam magnetron sputtering. In general, the magnetron sputtering was divided into 2 major types, namely direct current (DC) and radio frequency (RF) magnetron sputtering methods.

DC sputtering technique is used to deposit the conducting bulk materials. If the target to be deposited is non-conducting, then the positive charge accumulation onto target material during deposition is formed and then the sputtering yield will be very low and even zero. In RF magnetron sputtering, both non conducting and conducting target materials can be sputtered to fabricate the thin films. In general, higher sputter rate will occur at the lower pressure due to the presence of high quantity of argon in the chamber. However, the minimal mean free path will affect the rate of sputtering owing to the purging of excess of carrier and sputtering gases in inside the chamber (Baudet et al. 2017). Low sputtering will be observed, if the substrate and target distance is larger.

Sputtering is a better alternative for evaporation techniques to produce better surface morphology and good adhesion strength. Also, high melting point materials can be easily deposited and the thickness can be as low as 5 nm. High purity films, higher deposition rates, excellent adhesion, excellent uniformity and ability to coat any substrates are the greater advantage of this technique.

Figure 4. Schematic diagram of magnetron sputtering system

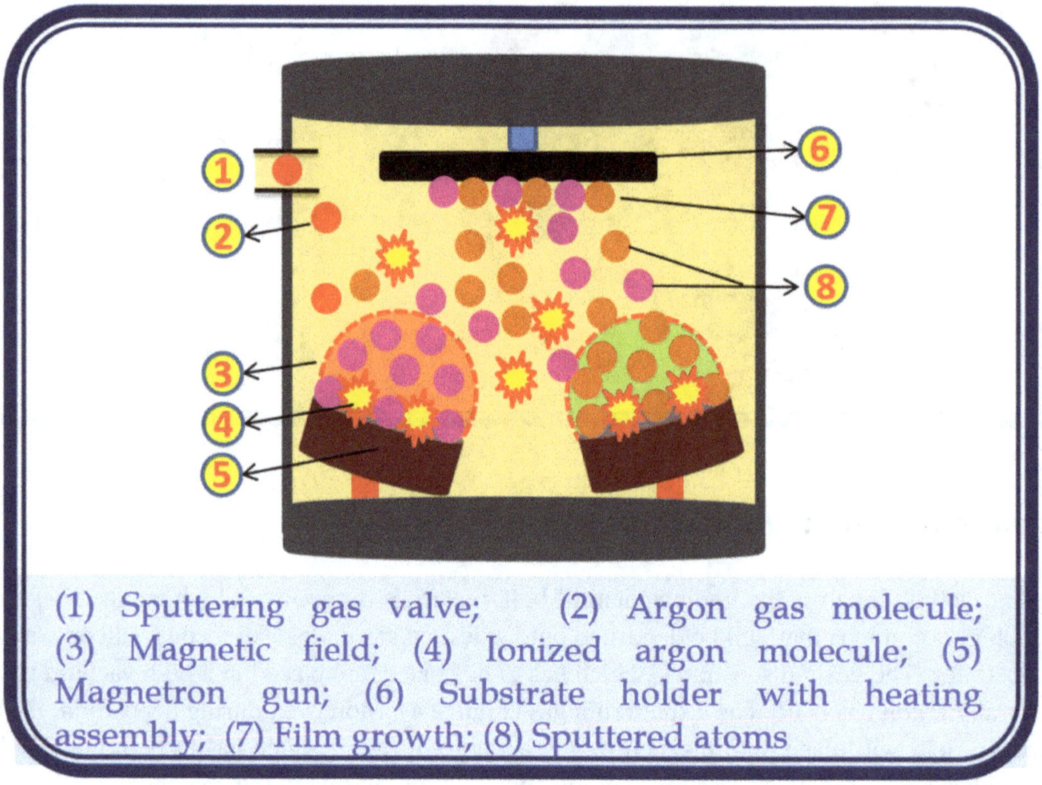

Electron Beam Physical Vapor Deposition (EBPVD) Technique

EBPVD is a type of physical vapor deposition in which the source will be bombarded with a high energetic electron beam, which come from a charged tungsten filament that to be used for evaporation. The common system consists of a vacuum chamber and an electron beam power source, target material to be evaporated and substrate holder. A charged tungsten filament is used to produce a bunch of electrons and is directed to the source material that to be evaporated. The generated electron beam is directed to the target material with the help of magnets. Upon interaction with the target material, the kinetic energy of the electron is transferred into heat energy and heating up the material. Once the melting point of the material reached, the material gets vaporized (evaporated) and then condensed onto the substrate surface and forms a thin film. The advantage of this method is higher deposition rate, uniform deposition, strain tolerant microstructure and wide range of materials starting from metals to ceramic can be developed through this technique. Only the material evaporates without decomposition can be deposited using this technique. Generally, metals, ceramics and a combination of both can be evaporated using this technique. This technique provides columnar microstructure which is a greater advantage for wear and erosion resistant coating applications. **Table 1** provides a brief comparison of various coating processes.

Table 1. Comparison between PVD, PLD, Spray process and EBPVD techniques (Fauchais and Vardelle 2000; Singh and Wolfe 2005)

Coating Process	Deposition Rate	Surface Roughness	Microstructure	Substrate Temperature
Powder Spray	>100 μm/minute	Very high	Lamellar	RT to 1073 K
PLD	0.02-0.05 nm per laser shot	high	Layer-by-layer growth, epitaxy	100 K to 973 K
PVD-Sputtering	< 5 μm/hr	Low	Columnar	Less than 973 K
EBPVD	10nm to 100 μm/minute	Low	Columnar, nanograined, equiaxed	RT to 1073 K

TITANIUM NITRIDE (TiN) BASED PVD COATINGS

Transition metal nitrides are also known as refractory metal nitrides, show unique combination of chemical and physical properties, thus make them more attractive from both fundamental and technological point of view. They usually have high melting points, high hardness, good thermal and electrical conductivity, and good corrosion resistance. Their unique combination of properties made possible for a variety of applications including, wear resistant coatings for cutting tool components, diffusion barriers, microelectronics applications, and a corrosion and abrasion resistant layers on optical and mechanical components.

TiN coating exhibits low coefficient of friction (CoF), excellent corrosion resistance and good hardness which makes it a stronger candidate for wear resistant applications. TiN coating developed through PVD methods have shown a wide range of applications in cutting, metal forming, and tool components. PVD TiN coating reduces the wear of tool components and increase the life of the component by tenfold. When the TiN coating thickness is about 2-3 μm, it provides optimal tool life (Posti and Nieminen 1989). Another crucial parameter is the substrate temperature. When the substrate is heated above 550°C, the hardness and dimension stability of the TiN coating are reduced (Singer, Fayeulle, and Ehni 1991). The advantage of sputtered PVD TiN coating is the ability to deposit below 200 °C. At lower substrate temperature, the adatom (adsorbed atoms) mobility is low and hence the coating density is low with more columnar structure with reduced internal stresses. In general, reduced compressive stress resulted in lower tendency of spallation. Thus lower substrate temperature provide better wear resistance, toughness and hardness through the formation of fine grained microstructure (Wu et al. 1998).

Deposition of single layered TiN coating by varying the DC power values between 200 W to 350 W on brass, aluminium, and mild steel substrates showed decrease in wear rate and CoF compared to uncoated substrate (Chavda et al. 2016). Depending on grain size distribution in TiN coating and the type of substrates, the CoF decreased up to 25% and resistance to wear increased from ~ 30% to 700%. It is also noticed that, the smaller the grain size, lesser the CoF and wear rate. TiN coating exhibited crystalline in nature at lesser power and become amorphous structure at 350 W. Also, the lower value of CoF and wear was obtained at lower power of 200 W. The wear rate of TiN coating (4 μm thickness) decreased by increasing the load and sliding speed. Less wear rate was observed in TiN coating with steel ball having low hardness and elastic modulus and the wear rate is highest for Al_2O_3 ball possessing high hardness and elastic modulus (Yan et al. 2018). The reason could be due to the contact pressure range produced on the TiN coating surface. For Al_2O_3 ball, the contact pressure is higher than steel ball. In addition, sliding speed and load also influencing the CoF values. The results suggest that CoF does not alter the wear behavior of TiN coating and the degradation of the coating caused due to abrasive and adhesive wear.

Ternary compounds of transition metal nitrides (e.g. TiAlN) system show better properties than binary compounds (e.g. TiN). The introduction of third element in binary compound disrupts the lattice system and showed remarkable properties. By the addition of Al into the TiN matrix, the properties of TiAlN coating showed improved thermal stability and diffusion barrier and high hardness than TiN coating. Single layered TiN and TiAlN coating may not be adequate to provide superior wear resistance. Compared to single layer of TiAlN coating, multilayer TiAlMoN coating with alternative Mo and Ti enriched nitride coating exhibited better wear resistance (Sergevnin et al. 2018). Also, layers of TiN/TiC, TiN and TiC coating showed less CoF and increased life of the component by two-fold as compared to monolayer coatings (Holleck and Schier 1995b). Another fact is that by increasing the thickness of the bilayer system can cause decrease in the coating adhesion. But this was not observed in the case of TiN/TiAlSiN system. High adhesion strength and better wear resistance was observed in the case of multilayer TiN/TiAlSiN/TiN system (Dobrzański, Pakuła, and Hajduczek 2004).

Multilayer coating layer thickness influences the wear resistance of the coatings. When the layer thickness decreases, wear rate decreases. The wear resistance of TiN/Ti coating consists of each layer thickness of about 20 nm was compared with TiN/Ti coating with an individual layer thickness of 100 nm. The study showed that a decrease in layer thickness to 20 nm showed 50% lesser wear than 100 nm thicker layers (Srinivasan, Kulkarni, and Anand 2007). In TiN/Ti coating, a decrease in wear resistance is associated with the presence of soft metallic phase in ceramic phase. An increase in the soft metallic phase caused decreased in elastic modulus and hardness of the coating and hence wear rate increases (Cheng et al. 2010).

Addition of Cu and Ag into TiAlN coating (thickness of about 2 μm) alters the hardness of the system ranging from 27 GPa to 6.7 GPa. TiAlN coating containing 11 at. % of Cu-Ag content shows 15.2 GPa hardness and having wear volume than 20 atomic percent Cu-Ag content which had 6.7 GPa. The results showed that the increased hardness of the coating would not decrease the rate of wear (Mejía V., Perea, and G. 2020). By the addition of Cu-Ag system to TiAlN coating, the decreased coating hardness, residual stress, elastic modulus, CoF, wear rate, and thus improved wear resistance were obtained. However, the individual constituent of the total system played a major role in wear resistance. For example, TiN, TiAlSiN and TiCrAlSiN coating deposited on 316 SS substrates show a hardness value of 26.4 GPa, 30.4 GPa and 31.6 GPa. The wear rate is decreased by increasing the coating hardness, thus wear resistance increases (Khanchaiyaphum et al. 2017).

CHROMIUM NITRIDE (CrN) BASED PVD COATINGS

Coatings based on chromium nitride (CrN) prepared through PVD method have shown extensive range of applications as protective coatings. The addition of elements such as Si, C, Zr, Ni, Nb, Al, etc. to CrN have shown enhanced friction, adhesion strength, wear, oxidation and corrosion resistance, which is attributed to their hardness, frictional coefficient and microstructure. A recent report suggested that the incorporation of Ni to CrN impose over their structure, adhesion, tribological performance and mechanical behaviour of the coatings (Akhter et al. 2021). CrN coating having thickness of 200 nm showed low CoF than that of thick W-C:H coating (Beer et al. 1999). Despite smaller thickness of CrN coating, lower abrasive wear and CoF is obtained than thicker coatings.

In the case of CrAlN coatings, an increase in frictional force followed by decrement in wear rate was observed (Ding and Zeng 2005). Investigation on CrN and AlCrN coating showed no correlation between wear rate and CoF (Mo and Zhu 2009). The wear rate of AlCrN coating having high hardness (30.6 GPa) was different when subjected to the different ball counter bodies used for wear testing (Antonov et al. 2018). In the case of ZrO_2 ball, higher wear rate was observed due to lower hardness and low thermal conductivity. But, CoF is lower by use of Al_2O_3 ball, which showed much lesser wear rate. Also, increase in the frictional force did not increase the wear rate. The increasing sliding temperature alters the CoF and wear depending on the indenter used for testing. Thus, these changes were related to thermal conductivity and hardness of the balls used for wear testing.

The addition of carbon in the multilayer CrN coating showed lower CoF than monolayer CrN coating. CrN/CrCN coating showed the lowest CoF and wear rate by the presence of carbon in the multilayer (Kong et al. 2018). Better wear resistance was observed for multilayer coating than single layer of CrN coating (Pougoum et al. 2018). In addition, multilayer PVD coatings wear resistance depends on the coating microstructure. The wear rate and wear depth of CrCN single layer coating was about five-fold higher than that of CrCN/CrN multilayer coating tested in the same condition (Gilewicz and Warcholinski 2014). The improvement in wear resistance could be connected to the adhesion of the coating to the substrate. With the addition of the number of bilayers, the adhesion of the coating improved. However, increasing the CrN layer coating thickness in the bilayer system caused less adhesion with the substrate. In spite of the thickness of individual layer, bilayer thicknesses also play a huge role in wear rate of multilayer system. CrN/CrCN system having thinner bilayer showed better wear resistance.

The decrease in CoF occurs due to the formation of craters and removal of droplets. Polishing provides the removal of micro droplets from the surface of the coating. The reduction in roughness of the surface by about 60 percent was accompanied by decrease in CoF to about 50 percent. It must be pointed out that lower roughness of the surface always does not mean lower CoF. Irrespective of coating wear test conditions, AlCrN coating exhibited lower CoF than CrWN coating, despite high roughness (Mozgovoy et al. 2019).

TANTALUM NITRIDE (TaN) BASED PVD COATINGS

TaN has gained interest due to its excellent physical and chemical properties and also exhibits good biocompatibility. TaN is widely used in producing diffusion barrier layers, hard wear resistant coatings, thin film resistors and masking layers. TaN coating deposited through magnetron sputtering results in a variety of crystal systems of TaN (body-centered cubic, face-centered cubic, hexagonal) with different physical, chemical and mechanical properties (Zaman and Meletis 2017).

Recent studies on TaN coatings prepared through PVD method showed that hard coating does not always improve the wear resistance of the components. TaN coating showing lower hardness than WC-Co substrate showed reasonable wear resistance than that of substrate. TaN coatings having higher hardness than WC-Co showed increased wear resistance (Tan et al. 2019). Also, TaN coating with higher hardness showed lower wear resistance than the coating having low hardness than TaN coating. Hence, an increase in hardness of the coating did not increase the wear resistance of coatings. Studies of TaN coatings also exhibited that the wear rate of steel ball was about 7-fold higher than TaN coatings as compared with uncoated WC-Co substrate despite hardness of the coatings.

DIAMOND LIKE CARBON (DLC) BASED PVD COATINGS

DLC is a metastable amorphous phase carbon film that contains a three-fold coordinated sp^2 sites presented in graphite-like carbon and four–fold coordinated sp^3 sites presented in diamond like carbon atoms, showed excellent properties such as high wear resistance, high hardness, ultra-low CoF, chemically inert and excellent biocompatibility, etc. Diamond having band gap of ~ 5.5 eV exhibit hard but electrically insulative, whereas graphite show zero band gap conducts electricity and soft as lubricant. Robertson (Robertson 2002) proposed that a cluster model of amorphous carbon film consists of a three-dimensional network in which sp^2 islands are surrounded. The presence of graphite clusters (sp^2 hybridization) determine the electrical and optical properties of the DLC film, while the sp^3 hybridized carbon atom mainly controls the mechanical behavior (Donnet 1998). The ratio of sp^3/sp^2 in DLC films varies depending upon the deposition methods and conditions, concentration of hydrogen and the presence of other elements. Higher sp^2 content in DLC film tend to have a lower CoF in dry conditions (a-C:H), but show higher wear rate than a film with higher sp^3 content and lower hydrogen content (ta-C) (Erdemir and Donnet 2006).

Though tetrahedral amorphous carbon (ta-C) and hydrogenated tetrahedral amorphous carbon (ta-C:H) films show high residual stress and high hardness, it will easy delamination from the substrate due to poor adhesion. In contrast, hydrogenated (a-C: H or DLC) films exhibit comparatively low residual stress than ta-C and ta-C:H films, and shows good adhesion due to the presence of hydrogen (Charitidis 2010). DLC films are often used as a solid lubricating material for its high hardness, high wear resistance and low CoF properties (Yang et al. 2002; Moseler et al. 2005). Presently, the recent development of hard (20 GPa), super-hard (40 GPa) and ultra-hard (80 GPa) coatings are found to be a remarkable research topic owing to their wide range of industrial applications such as automobile parts, cutting tools, magnetic storage media and glasses (Dwivedi, Kumar, and Malik 2012). Besides the film properties, it is important that DLC films can be deposited at relatively low film process temperature (less than 200 °C). A major problem with this DLC is poor adhesion to the substrates. Films having high sp^3/sp^2 ratios and 90 percent sp^3 content have been reported and show residual compressive stresses close to 10 GPa (Dearnley et al. 2010). Whereas coatings of a-C: H containing high amounts of H and sp^2 bonded carbon structures with very few to none amount of sp^3 structures typically show very low residual stress (less than 1 GPa). As we know, coatings with higher residual stress are prone to delamination due to thermal expansion mismatch and lattice misfit leading to catastrophic failure.

Earlier reports reveal that the solution of this problem by doping other elements into the DLC film reduces the high residual stress and enhancing the properties. In particular, Ti, Cu, and WC doped DLC films showed significant interest due to their mechanical properties and reduces intrinsic stresses. DLC coating can be deposited through several methods including radio frequency (RF) (Chowdhury and Laugier 2004) Direct current (DC) (Tomala, Pauschitz, and Roy 2013) magnetron sputtering, PECVD (Choi et al. 2005), arc ion plating (AIP) (Zou et al. 2004), Filtered Cathodic vacuum arc (FCVA) (Chen et al. 2013) and PLD (Shum, Zhou, and Li 2013). However, magnetron sputtering is still widely used due to their advantages like simple process, low cost, film homogeneity and control of the process. There are several parameters for the deposition of DLC thin films through magnetron sputtering including chamber vacuum, gas pressure, Ar and reactive gas flow rate, chamber pressure, substrate temperature, distance between substrate and target, etc.

FUTURE RESEARCH DIRECTIONS

Hard wear resistant coatings used for high temperature sliding applications suffer from wear damage than expected. The microstructural changes of the coating have to be studied in-situ in order to understand the wear behaviour with respect to microstructure. This includes dislocations, grain boundary sliding, deformation mechanism, twinning, etc. These atomistic level understanding allow us to realize the key component for wear rate and thereby steers us to develop advanced materials for wear resistant applications. The introduction of super-lubricity into the multilayer coating minimize the effect of CoF and wear rate. Hence, the focus on the fabrication of these materials will be useful to develop more effective hard coatings. In addition, the development of large-scale nanostructured coatings with stability and reproducibility are critical factor in the design and development of larger components. The selection of proper choice of methods and materials would solve the developmental issues in the upcoming days. It is believed that the recent advancements in the coating characterization for microstructure and morphology make a significant impact on the fundamental understanding in materials science would provide advanced wear resistant coatings based on PVD techniques in a wide range of industrial applications.

SUMMARY AND CONCLUSION

PVD coatings provide a good platform to tailor the physical and chemical properties of the hard wear resistant coatings. The wear and corrosion performance of the coatings were tested in several conditions, which are widely used in many industries. The deformation of coating depends on the hardness and elasticity. Their resistance to plastic deformation and erosion rate, PVD coating shows better resistance to corrosion and wear. In the case of higher resistance to plastic deformation, the change in ductile to brittle behaviour and wear rate increment is observed. The lower wear rate of the coatings depends on the microstructure and growth morphology, which affect the residual stress distribution and crack propagation. Multilayer coatings generally show better wear resistant than monolayer coatings. For example, multilayer Ti/TiN coating showed less tensile stress than TiN monolayer. With the addition of metallic interlayer, there is a decrement in interfacial stress. This could be, the metallic interlayer absorbs more load bearing energy on elastic deformation and improve the resistance.

Surface roughness of the coating affects the CoF of PVD coatings. An increase in CoF caused by the increase in surface roughness of the coating. Nevertheless, lower surface roughness does not provide lower CoF. For example, AlCrN coating showed lower CoF than CrWN despite of higher surface roughness. Moreover, there is no correlation between wear rate and CoF. Therefore, lower CoF does not mean that higher wear resistance. Many investigations showed that lower the CoF and lower wear rate was due to the lower surface roughness. There is still a need to establish the relationship between the coating properties and its wear resistance.

ACKNOWLEDGMENT

This work was supported by the VEGA grant no. 1/0171/21. This work is a part of dissemination activities of project FunGlass. This project has received funding from the European Union´s Horizon 2020 research and innovation programme under grant agreement No 739566. This work was also created in the frame of the project Centre for Functional and Surface Functionalized Glass (CEGLASS), ITMS code is 313011R453, operational program Research and innovation, co-funded from European Regional Development Fund.

REFERENCES

Akhter, R., Zhou, Z., Xie, Z., & Munroe, P. (2021). Enhancing the Adhesion Strength and Wear Resistance of Nanostructured NiCrN Coatings. *Applied Surface Science*, *541*(March), 148533. doi:10.1016/j.apsusc.2020.148533

Antonov, M., Afshari, H., Baronins, J., Adoberg, E., Raadik, T., & Hussainova, I. (2018). The Effect of Temperature and Sliding Speed on Friction and Wear of Si3N4, Al2O3, and ZrO2 Balls Tested against AlCrN PVD Coating. *Tribology International*, *118*(February), 500–514. doi:10.1016/j.triboint.2017.05.035

Baudet, E., Sergent, M., Němec, P., Cardinaud, C., Rinnert, E., Michel, K., Jouany, L., Bureau, B., & Nazabal, V. (2017). Experimental Design Approach for Deposition Optimization of RF Sputtered Chalcogenide Thin Films Devoted to Environmental Optical Sensors. *Scientific Reports*, *7*(1), 3500. doi:10.103841598-017-03678-w PMID:28615650

Bazyari, A., Khodadadi, A. A., Mamaghani, A. H., Beheshtian, J., Thompson, L. T., & Mortazavi, Y. (2016). Microporous Titania–Silica Nanocomposite Catalyst-Adsorbent for Ultra-Deep Oxidative Desulfurization. *Applied Catalysis B: Environmental*, *180*(January), 65–77. doi:10.1016/j.apcatb.2015.06.011

Beer, P., Djouadi, M. A., Marchal, R., Sokolowska, A., Lambertin, M., Czyzniewski, A., & Precht, W. (1999). Antiabrasive Coatings in a New Application-Wood Rotary Peeling Process. *Vacuum*, *53*(1–2), 363–366. doi:10.1016/S0042-207X(99)00110-4

Bhushan, B. (2000). *Modern Tribology Handbook, Two Volume Set. Modern Tribology Handbook, Two Volume Set*. CRC Press. doi:10.1201/9780849377877

Bull, S. J. (2006). Physical Vapour Deposition Methods for Protection against Wear. *Surface Coatings for Protection Against Wear*, (January), 146–183. doi:10.1533/9781845691561.146

Bull, S. J., & Jones, A. M. (1996, January). Multilayer Coatings for Improved Performance. *Surface and Coatings Technology*, *78*(1-3), 173–184. doi:10.1016/0257-8972(94)02407-3

Charitidis, C. A. (2010). Nanomechanical and Nanotribological Properties of Carbon-Based Thin Films: A Review. *International Journal of Refractory Metals & Hard Materials*, *28*(1), 51–70. doi:10.1016/j.ijrmhm.2009.08.003

Chavda, M. R., Dave, D. P., Chauhan, K. V., & Rawal, S. K. (2016). Tribological Characterization of TiN Coatings Prepared by Sputtering. *Procedia Technology*, *23*(January), 36–41. doi:10.1016/j.protcy.2016.03.070

Chen, L., Yang, M. C., Song, C. F., Yu, B. J., & Qian, L. M. (2013). Is 2 Nm DLC Coating Enough to Resist the Nanowear of Silicon. *Wear*, *302*(1–2), 909–917. doi:10.1016/j.wear.2013.01.088

Cheng, Y. H., Browne, T., Heckerman, B., Bowman, C., Gorokhovsky, V., & Meletis, E. I. (2010). Mechanical and Tribological Properties of TiN/Ti Multilayer Coating. *Surface and Coatings Technology*, *205*(1), 146–151. doi:10.1016/j.surfcoat.2010.06.023

Choi, W. S., Heo, J., Chung, I., & Hong, B. (2005). The Effect of RF Power on Tribological Properties of the Diamond-like Carbon Films. *Thin Solid Films*, *475*(1–2), 287–290. doi:10.1016/j.tsf.2004.07.033

Chowdhury, S., & Laugier, M. T. (2004). Non-Contact AFM with a Nanoindentation Technique for Measuring the Mechanical of Thin Films. *Nanotechnology*, *15*(8), 1017–1022. doi:10.1088/0957-4484/15/8/027

Dai, W., Kwon, S. H., Wang, Q., & Liu, J. (2018). Influence of Frequency and C2H2 Flow on Growth Properties of Diamond-like Carbon Coatings with AlCrSi Co-Doping Deposited Using a Reactive High Power Impulse Magnetron Sputtering. *Thin Solid Films*, *647*(February), 26–32. doi:10.1016/j.tsf.2017.12.016

Dearnley, P. A., Neville, A., Turner, S., Scheibe, H. J., Tietema, R., Tap, R., Stüber, M., Hovsepian, P., Layyous, A., & Stenbom, B. (2010). Coatings Tribology Drivers for High Density Plasma Technologies. *Surface Engineering*, *26*(1–2), 80–96. doi:10.1179/174329409X451218

Ding, X. Z., & Zeng, X. T. (2005). Structural, Mechanical and Tribological Properties of CrAlN Coatings Deposited by Reactive Unbalanced Magnetron Sputtering. *Surface and Coatings Technology*, *200*(5–6), 1372–1376. doi:10.1016/j.surfcoat.2005.08.072

Dobrzański, L. A., Pakuła, D., & Hajduczek, E. (2004). Structure and Properties of the Multi-Component TiAlSiN Coatings Obtained in the PVD Process in the Nitride Tool Ceramics. *Journal of Materials Processing Technology*, *157–158*, 331–340. doi:10.1016/j.jmatprotec.2004.09.052

Donnet, C. (1998). *Recent Progress on the Tribology of Doped Diamond-like and Carbon Alloy Coatings: A Review. In Surface and Coatings Technology*. Elsevier. doi:10.1016/S0257-8972(97)00611-7

Dwivedi, N., Kumar, S., & Malik, H. K. (2012). Superhard Behaviour, Low Residual Stress, and Unique Structure in Diamond-like Carbon Films by Simple Bilayer Approach. *Journal of Applied Physics, 112*(2), 023518. Advance online publication. doi:10.1063/1.4739287

Erdemir, A., & Donnet, C. (2006). *Tribology of Diamond-like Carbon Films: Recent Progress and Future Prospects. Journal of Physics D: Applied Physics*. doi:10.1088/0022-3727/39/18/R01

Fauchais, P., & Vardelle, A. (2000). Heat, Mass and Momentum Transfer in Coating Formation by Plasma Spraying. *International Journal of Thermal Sciences, 39*(9–11), 852–870. doi:10.1016/S1290-0729(00)01195-9

Fotovvati, Namdari, & Dehghanghadikolaei. (2019). On Coating Techniques for Surface Protection: A Review. *Journal of Manufacturing and Materials Processing, 28*(3), 28. . doi:10.3390/jmmp3010028

Gilewicz, A., & Warcholinski, B. (2014). Tribological Properties of CrCN/CrN Multilayer Coatings. *Tribology International, 80*, 34–40. doi:10.1016/j.triboint.2014.06.012

Greene, J. E. (2017). Review Article: Tracing the Recorded History of Thin-Film Sputter Deposition: From the 1800s to 2017. *Journal of Vacuum Science & Technology A: Vacuum, Surfaces, and Films, 35*(5), 05C204. . doi:10.1116/1.4998940

Holleck, H., & Schier, V. (1995a). Multilayer PVD Coatings for Wear Protection. *Surface and Coatings Technology, 76–77*(1–3 pt 1), 328–336. doi:10.1016/0257-8972(95)02555-3

Holleck, H., & Schier, V. (1995b). Multilayer PVD Coatings for Wear Protection. *Surface and Coatings Technology, 76–77*(November), 328–336. doi:10.1016/0257-8972(95)02555-3

Holubar, P., Jilek, M., & Sima, M. (2000). Present and Possible Future Applications of Superhard Nanocomposite Coatings. *Surface and Coatings Technology, 133–134*(November), 145–151. doi:10.1016/S0257-8972(00)00956-7

Hoornaert, T., Hua, Z. K., & Zhang, J. H. (2009). Hard Wear-Resistant Coatings: A Review. *Advances in Tribology*, 774–779. doi:10.1007/978-3-642-03653-8_257

Hu, J. J., Muratore, C., & Voevodin, A. A. (2007). Silver Diffusion and High-Temperature Lubrication Mechanisms of YSZ–Ag–Mo Based Nanocomposite Coatings. *Composites Science and Technology, 67*(3–4), 336–347. doi:10.1016/j.compscitech.2006.09.008

Keskin, İ. Ç., Türemiş, M., Katı, M. İ., Kibar, R., Şirin, K., Çipiloğlu, M. A., Kuş, M., Büyükçelebi, S., & Çetin, A. (2017). The Radioluminescence and Optical Behaviour of Nanocomposites with CdSeS Quantum Dot. *Journal of Luminescence, 185*(May), 48–54. doi:10.1016/j.jlumin.2016.12.048

Khanchaiyaphum, S., Saikaew, C., Wisitsoraat, A., & Surinphong, S. (2017). Wear Behaviours of Filtered Cathodic Arc Deposited TiN, TiAlSiN and TiCrAlSiN Coatings on AISI 316 Stainless Steel Fishing Net-Weaving Machine Components under Dry Soft-Sliding against Nylon Fibres. *Wear, 390–391*, 146–154. doi:10.1016/j.wear.2017.07.018

Kong, Y., Tian, X., Gong, C., & Chu, P. K. (2018). Enhancement of Toughness and Wear Resistance by CrN/CrCN Multilayered Coatings for Wood Processing. *Surface and Coatings Technology, 344*(June), 204–213. doi:10.1016/j.surfcoat.2018.03.027

Krella, A. (2020). Resistance of PVD Coatings to Erosive and Wear Processes: A Review. *Coatings,* 10(10), 921. . doi:10.3390/coatings10100921

Lugscheider, E., & Weber, T. (1990). Plasma Spraying—An Innovative Coating Technique: Process Variants and Applications. *IEEE Transactions on Plasma Science*, *18*(6), 968–973. doi:10.1109/27.61511

Mejía, V., Hernán, D., Perea, D., & Gilberto Bejarano, G. (2020). Development and Characterization of TiAlN (Ag, Cu) Nanocomposite Coatings Deposited by DC Magnetron Sputtering for Tribological Applications. *Surface and Coatings Technology*, *381*(January), 125095. Advance online publication. doi:10.1016/j.surfcoat.2019.125095

Michelmore, A. (2016). Thin Film Growth on Biomaterial Surfaces. *Thin Film Coatings for Biomaterials and Biomedical Applications*, (January), 29–47. doi:10.1016/B978-1-78242-453-6.00002-X

Mo, J. L., & Zhu, M. H. (2009). Tribological Oxidation Behaviour of PVD Hard Coatings. *Tribology International*, *42*(11–12), 1758–1764. doi:10.1016/j.triboint.2009.04.026

Moseler, M., Cumbsch, P., Casiraghi, C., Ferrari, A. C., & Robertson, J. (2005). The Ultrasmoothness of Diamond-like Carbon Surfaces. *Science*, *309*(5740), 1545–1548. doi:10.1126cience.1114577 PMID:16141070

Mozgovoy, S., Alik, L., Hardell, J., & Prakash, B. (2019). Material Transfer during High Temperature Sliding of Al-Si Coated 22MnB5 Steel against PVD Coatings with and without Aluminium. *Wear*, *426–427*(April), 401–411. doi:10.1016/j.wear.2018.12.042

Musil, J. (2000). Hard and Superhard Nanocomposite Coatings. *Surface and Coatings Technology*, *125*(1–3), 322–330. doi:10.1016/S0257-8972(99)00586-1

Musil, J., Karvánková, P., & Kasl, J. (2001). Hard and Superhard Zr–Ni–N Nanocomposite Films. *Surface and Coatings Technology*, *139*(1), 101–109. doi:10.1016/S0257-8972(01)00989-6

Patscheider, J. (2003). Nanocomposite Hard Coatings for Wear Protection. *MRS Bulletin*, *28*(03), 180–183. doi:10.1557/mrs2003.59

Pogrebnjak, A. D., & Beresnev, V. M. (2012). *Hard Nanocomposite Coatings, Their Structure and Properties. Nanocomposites - New Trends and Developments* (Vol. 1). InTech Publishers.

Posti, E., & Nieminen, I. (1989). Influence of Coating Thickness on the Life of TiN-Coated High Speed Steel Cutting Tools. *Wear*, *129*(2), 273–283. doi:10.1016/0043-1648(89)90264-0

Pougoum, F., Qian, J., Laberge, M., Martinu, L., Klemberg-Sapieha, J., Zhou, Z., Li, K. Y., Savoie, S., & Schulz, R. (2018). Investigation of Fe3Al-Based PVD/HVOF Duplex Coatings to Protect Stainless Steel from Sliding Wear against Alumina. *Surface and Coatings Technology*, *350*(September), 699–711. doi:10.1016/j.surfcoat.2018.07.070

Premphet, P., Prasoetsri, M., Boonyawan, D., Supruangnet, R., Udomsom, S., & Leksakul, K. (2017). Optimization of DC Magnetron Sputtering Deposition Process and Surface Properties of HA-TiO2 Film. *Materials Today: Proceedings*, *4*(5), 6372–6380. doi:10.1016/j.matpr.2017.06.141

Riva, R., Eugenio, I. P.-S., Lak, A., Pellegrino, T., Pérez-Juste, J., & Mattoli, V. (2017). Plasmonic/Magnetic Nanocomposites: Gold Nanorods-Functionalized Silica Coated Magnetic Nanoparticles. *Journal of Colloid and Interface Science*, *502*(September), 201–209. doi:10.1016/j.jcis.2017.04.089 PMID:28486141

Robertson, J. (2002). Diamond-like Amorphous Carbon. *Materials Science and Engineering R Reports*, *37*(4–6), 129–281. doi:10.1016/S0927-796X(02)00005-0

Saikia, U., Sahariah, M. B., & Pandey, R. (2016). Stability of Cu–Nb Layered Nanocomposite from Chemical Bonding. *Chemical Physics Letters*, *655–656*(July), 59–65. doi:10.1016/j.cplett.2016.05.022

Sergevnin, V. S., Blinkov, I. V., Belov, D. S., Smirnov, N. I., Volkhonskii, A. O., & Kuptsov, K. A. (2018). Wear and Erosion of Arc-PVD Multilayer Ti-Al-Mo-N Coatings under Various Conditions of Friction and Loading. *International Journal of Advanced Manufacturing Technology*, *98*(1–4), 593–601. doi:10.100700170-018-2235-z

Shum, P. W., Zhou, Z. F., & Li, K. Y. (2013). To Increase the Hydrophobicity and Wear Resistance of Diamond-like Carbon Coatings by Surface Texturing Using Laser Ablation Process. In *Thin Solid Films* (Vol. 544, pp. 472–476). Elsevier. doi:10.1016/j.tsf.2013.02.075

Singer, I. L., Fayeulle, S., & Ehni, P. D. (1991). Friction and Wear Behavior of TiN in Air: The Chemistry of Transfer Films and Debris Formation. *Wear*, *149*(1–2), 375–394. doi:10.1016/0043-1648(91)90386-9

Singh, J., & Wolfe, D. E. (2005). Review Nano and Macro-Structured Component Fabrication by Electron Beam-Physical Vapor Deposition (EB-PVD). *Journal of Materials Science*, *40*(1), 1–26. doi:10.100710853-005-5682-5 PMID:15754137

Srinivasan, Kulkarni, & Anand. (2007). Thermal Stability and High-Temperature Wear of Ti-TiN and TiN-CrN Nanomultilayer Coatings under Self-Mated Conditions. *Tribology International*, *40*(2), 266–77. . doi:10.1016/j.triboint.2005.09.035

Tan, P., Fu, L., Teng, J., Zhu, J., Yang, W., Li, D., & Zhou, L. (2019). Effect of Texture on Wear Resistance of Tantalum Nitride Film. *Tribology International*, *133*(May), 126–135. doi:10.1016/j.triboint.2019.01.001

Tomala, A., Pauschitz, A., & Roy, M. (2013). Nanotribology of Pulsed Direct Current Magnetron Sputtered Diamond like Carbon Films. *Surface Science*, *616*(October), 60–70. doi:10.1016/j.susc.2013.06.008

Veprek, S., & Argon, A. S. (2001). Mechanical Properties of Superhard Nanocomposites. *Surface and Coatings Technology*, *146–147*(September), 175–182. doi:10.1016/S0257-8972(01)01467-0

Veprek, S., Niederhofer, A., Moto, K., Bolom, T., Männling, H.-D., Nesladek, P., Dollinger, G., & Bergmaier, A. (2000). Composition, Nanostructure and Origin of the Ultrahardness in Nc-TiN/a-Si3N4/a- and Nc-TiSi2 Nanocomposites with HV=80 to ≥105 GPa. *Surface and Coatings Technology*, *133–134*(November), 152–159. doi:10.1016/S0257-8972(00)00957-9

Vepřek, S., & Reiprich, S. (1995). A Concept for the Design of Novel Superhard Coatings. *Thin Solid Films*, *268*(1–2), 64–71. doi:10.1016/0040-6090(95)06695-0

Voevodin, A. A., O'Neill, J. P., & Zabinski, J. S. (1999). Nanocomposite Tribological Coatings for Aerospace Applications. *Surface and Coatings Technology*, *116–119*(September), 36–45. doi:10.1016/S0257-8972(99)00228-5

Wu, K. R., Bayer, R. G., Engel, P. A., & Sun, D. C. (1998). Wear of Physical Vapor Deposition Tin Coatings Sliding against Cr-Steel and WC Counterbodies. *Journal of Tribology*, *120*(3), 482–488. doi:10.1115/1.2834576

Xue, L., Wang, W., Guo, Y., Liu, G., & Wan, P. (2017). Flexible Polyaniline/Carbon Nanotube Nanocomposite Film-Based Electronic Gas Sensors. *Sensors and Actuators. B, Chemical*, *244*(June), 47–53. doi:10.1016/j.snb.2016.12.064

Yan, Z., Jiang, D., Gao, X., Hu, M., Wang, D., Fu, Y., Sun, J., Feng, D., & Weng, L. (2018). Friction and Wear Behavior of TiN Films against Ceramic and Steel Balls. *Tribology International*, *124*(August), 61–69. doi:10.1016/j.triboint.2018.03.031

Yang, S. H., Kong, H., Lee, K. R., Park, S., & Kim, D. E. (2002). Effect of Environment on the Tribological Behavior of Si-Incorporated Diamond-like Carbon Films. *Wear*, *252*(1–2), 70–79. doi:10.1016/S0043-1648(01)00856-0

Yuan, S., Lin, N., Zou, J., Lin, X., Liu, Z., Yu, Y., Wang, Z., Zeng, Q., Chen, W., Tian, L., Qin, L., Xie, R., Li, B., Zhang, H., Wang, Z., Tang, B., & Wu, Y. (2020). In-Situ Fabrication of Gradient Titanium Oxide Ceramic Coating on Laser Surface Textured Ti6Al4V Alloy with Improved Mechanical Property and Wear Performance. *Vacuum*, *176*(June), 109327. doi:10.1016/j.vacuum.2020.109327

Zaman, A., & Meletis, E. I. (2017). Microstructure and Mechanical Properties of TaN Thin Films Prepared by Reactive Magnetron Sputtering. *Coatings*, *7*(12), 209. doi:10.3390/coatings7120209

Zhai, W., Bai, L., Zhou, R., Fan, X., Kang, G., Liu, Y., & Zhou, K. (2021). *Recent Progress on Wear-Resistant Materials: Designs, Properties, and Applications. In Advanced Science*. John Wiley and Sons Inc. doi:10.1002/advs.202003739

Zhang, G., Ke, Y., He, J., Qin, M., Shen, H., Lu, S., & Xu, J. (2015). Effects of Organo-Modified Montmorillonite on the Tribology Performance of Bismaleimide-Based Nanocomposites. *Materials & Design*, *86*(December), 138–145. doi:10.1016/j.matdes.2015.07.090

Zhang, S., Sun, D., Fu, Y., & Du, H. (2003). Recent Advances of Superhard Nanocomposite Coatings: A Review. *Surface and Coatings Technology*, *167*(2–3), 113–119. doi:10.1016/S0257-8972(02)00903-9

Zou, Y. S., Wang, W., Song, G. H., Du, H., Gong, J., Huang, R. F., & Wen, L. S. (2004). Influence of the Gas Atmosphere on the Microstructure and Mechanical Properties of Diamond-like Carbon Films by Arc Ion Plating. *Materials Letters*, *58*(26), 3271–3275. doi:10.1016/j.matlet.2004.06.017

KEY TERMS AND DEFINITIONS

Adatom: An atom that lies on the substrate surface.

Columnar: The perpendicular grain growth of coating morphology with respect to substrate.

Epitaxy: The well-defined orientation of coating growth with respect to the substrate orientation.

Erosion: The process of removal of coating materials by means of abrading substances.

Lamellar: A type of microstructure composed of fine and alternative layers in the form of lamellae.

Nanocomposite: The composite which contains at least one phase in nanoscale morphology (nanoparticles, nanotubes, nanorods, etc.).

Nanocrystalline: The average grain size is less than 100 nm.

Chapter 9
Tribological Performance of Coatings Obtained by PVD Techniques:
From Industrial to Biological Applications

Mihaela Dinu
National Institute of R&D for Optoelectronics, Romania

Iulian Pana
https://orcid.org/0000-0003-1368-6219
National Institute of R&D for Optoelectronics, Romania

Anca C. Parau
National Institute of R&D for Optoelectronics, Romania

Alina Vladescu
https://orcid.org/0000-0001-5770-4541
National Institute of R&D for Optoelectronics, Romania

ABSTRACT

The chapter presented a short review related to the factors that dictate the wear and friction behavior of various coatings obtained by PVD techniques used for various applications. An important parameter with high impact on the final properties of the coatings prepared by cathodic arc evaporation is the reactive atmosphere. The friction and wear performance of the carbide coatings were strongly dependent on the carbon content which can be controlled by varying the C2H2 mass flow rate and the arc current on the cathode. Regarding the carbonitrides coatings, the ratio of C/N plays an important role; the coating with a low C/N ratio showed reduced friction coefficients, while for the coatings with a high C/N ratio the wear was improved. For biomedical applications, the magnetron sputtering deposition technique was employed to enhance the tribological performance of Ti6Al4V alloy in Ringer solution using various types of coatings such as carbonitrides, carbide, or hydroxyapatite.

DOI: 10.4018/978-1-7998-9683-8.ch009

INTRODUCTION

The present chapter highlights the tribological results obtained by the authors during the last decade and it takes into consideration the friction and wear behavior of various coatings obtained by Physical Vapour Deposition (PVD) techniques such as magnetron sputtering and cathodic arc evaporation. It should be noted that within the next few years the tribological properties of functional surfaces are set to become a vital key factor in several interdisciplinary fields of physics, engineering, biology, chemistry, and materials science. Also, the tribological knowledge advancements are receiving an increasing attention in the new context of the accelerated global warming, electronics components shortage crisis or to the increasing number of patients with medical devices. In addition, the past decade has seen a renewed importance in prolonging the service life of nanomaterials as the Fourth Industrial Revolution is ongoing and the opening of new areas of applications is mandatory. In order to address the above-mentioned specific issues, the researcher's worldwide efforts were focused on developing eco-friendly technologies (no harmful substances are used or result from) which are capable to produce cost-efficient coatings with enhanced tribological properties. This approach was meant to have a social, economic and ecological impact by introducing a new generation of wear resistant coatings, complying with all the environmental and health regulations, with increased life quality outcomes for the citizens. By carefully controlling the PVD deposition processes one may obtain coatings with different component elements in mono- and multilayer structures depending on their final applications. The high deposition rate and reproducibility are crucial assets of PVD processes that are currently used in industry for large area coaters. Considering the increasing need for developing wear resistant coatings, the PVD techniques represent a suitable solution which is able to address all market demands. Through selective modification of deposition parameters, it is possible to achieve a high tunability of tribological and mechanical properties. Therefore, the phase structure of nitride coatings can be changed by varying the reactive gas flow rate, bias voltage or substrate temperature (Kazmanli et al., 2003). Also, similar effects with an additional increasing of wear resistance at elevated temperatures can be obtained by controlling the doping element within the binary or ternary coatings (Tillmann & Dildrop, 2017). Besides, by using a high plasma density PVD technology such as High Power Impulse Magnetron Sputtering (HiPIMS), high performance protective films with good mechanical properties ($H \geq 20$ GPa, $E \geq 300$ GPa) were produced (Zin et al., 2018). Superior tribological behavior of quintenary duplex coatings as compared to non-duplex or nitrided layers can be achieved by combining surface treatments such as plasma nitriding with PVD deposition techniques such as cathodic arc evaporation (M. Braic, Braic, Balaceanu, Vladescu, et al., 2011). The aim of this chapter is to extend current knowledge of wear resistant coatings developed for industrial, severe service and biological applications. The information included in this chapter can be beneficial for the scientific community in order to obtain an optimal surface finish for tools working in dry or wet environment, at room temperature or in severe conditions, or even in the human body.

BACKGROUND

In broad terms, tribology can be defined as the science and technology of rubbing surfaces in contact and relative motion and entails the study and application of friction, lubrication and wear principles. There is a considerable amount of available literature on the outstanding mechanical properties of carbides, nitrides or carbonitrides which were found to be suitable for a large variety of tribological applications

(Aissani et al., 2019; Yao et al., 2017). In recent years, various deposition techniques were employed to obtain enhanced tribological properties such as magnetron sputtering, pulsed laser deposition, plasma enhanced chemical vapor deposition or cathodic arc evaporation. The reported results showed excellent wear and friction reduction, while microscale surface texturing proved to have also a beneficial effect on the tribological performances (Meng et al., 2018). The high entropy alloys represent a recently novel category which also showed superior mechanical properties as compared to binary or ternary coatings.

The tribological tests performed on different coatings were carried out according to DIN 50324 and ASTM G99. A pin-on-disk tribometer (CSM Instruments, load range from 0.5 N to 30 N, sapphire or steel ball, linear speed from 1 to 50 cm/s) and a universal mechanical tester (TriboLab UMT Bruker, EN1071-3:2005 standard) were used for this purpose.

INDUSTRIAL APPLICATIONS

Cathodic Arc Evaporation

The tribological properties of various coatings obtained by cathodic arc evaporation in dry environment with industrial applications were further presented and the influencing factors which dictate their wear and friction behavior were listed in Table 1.

Table 1. Tribological performance of different coatings deposited by cathodic arc evaporation (dry testing conditions)

Cathodic Arc Evaporation					
Material	Varying Factor	μ	k (x 10^{-6} $mm^3N^{-1}m^{-1}$)	Testing Conditions	Ref.
C45 steel (substrate)	N/A	0.61	22.00	400 m sliding distance, 5 N load, 0.15 m/s sliding speed, sapphire ball counterpart	(M. Braic et al., 2020)
TiSiC–SS (A1)	C_2H_2 reactive gas; C/(metal + Si) <1, A1-4:100 A arc current, 75-155 sccm flow rates, 4-8*10^{-2}Pa deposition pressure	0.45	7.89		
TiSiC–SS (A2)		0.27	6.97		
TiSiC–SS (A3)		0.27	5.20		
TiSiC–SS (A4)		0.23	3.10		
TiSiC–SS (B1)	C_2H_2 reactive gas; C/(metal + Si) >1, B1-3: 70-90 A arc current, 155 sccm flow rates, 8*10^{-2}Pa deposition pressure	0.21	2.79		
TiSiC–SS (B2)		0.17	1.87		
TiSiC–SS (B3)		0.15	1.07		
C45 steel (substrate)	N/A	0.56	22.00	300 m sliding distance, 5 N load, 0.15 m/s sliding speed, sapphire ball counterpart	(M. Braic et al., 2017)
TiSiC-Cr (M-90)	M: CH_4 reactive gas; 90,110 and 130 sccm gas flow rates	0.40	4.80		
TiSiC-Cr (M-110)		0.48	3.50		
TiSiC-Cr (M-130)		0.43	2.20		
TiSiC-Cr (A-90)	A: C_2H_2 reactive gas; 90,110 and 130 sccm gas flow rates	0.22	3.02		
TiSiC-Cr (A-110)		0.21	2.10		
TiSiC-Cr (A-130)		0.11	1.44		

Continued on following page

Table 1. Continued

Cathodic Arc Evaporation					
Material	**Varying Factor**	**μ**	**k (x 10^{-6} $mm^3N^{-1}m^{-1}$)**	**Testing Conditions**	**Ref.**
C45 steel (substrate)	N/A	0.61	25.00	400 m sliding distance, 5 N load, 0.15 m/s sliding speed, sapphire ball counterpart	(Vitelaru et al., 2014)
TiSiC	CH_4 reactive gas; Cr and Zr alloying element	0.37	7.00		
TiSiC-Cr		0.72	16.00		
TiSiC-Zr		0.32	2.00		
C45 steel (substrate)	N/A	0.56	N/A	400 m sliding distance, 5 N load, 0.15 m/s sliding speed, sapphire ball counterpart	(Balaceanu et al., 2015)
TiSiC	CH_4 reactive gas; 1, 2, 3: 3.2, 6.8 and 16.2 at.% Ni content, respectively	0.49	7.00		
TiSiC–Ni-1		0.33	3.16		
TiSiC–Ni-2		0.37	3.66		
TiSiC–Ni-3		0.59	7.33		
TiSiN	N_2 reactive gas; 26.7, 14.5 and 7.3 nm bilayer periods, 200, 400, 800: the number of the individual layers	0.44	5.20	400 m sliding distance, 3 N load, 0.2 m/s sliding speed sapphire ball counterpart	(Vladescu et al., 2013)
TiSiN/Ni200		0.36	3.20		
TiSiN/Ni400		0.35	2.60		
TiSiN/Ni800		0.38	3.80		
C45 steel (substrate)	N/A	0.93	N/A	400 m sliding distance, 3 N load, 0.2 m/s sliding speed, sapphire ball counterpart	(M. Braic, Braic, Balaceanu, Zoita, et al., 2011)
ZrCN-1	$(CH_4 + N_2)$ reactive gas mixture; C/(C +N) = 0.72 (series 1), C/(C +N) = 0.24 (series 2), 25, 13, 6.3 and 4.4 nm bilayer periods	0.30	5.60		
ZrCN-2		0.45	9.20		
Zr/ZrCN-1 (Λ= 25 nm)		0.27	2.90		
Zr/ZrCN-2 (Λ= 25 nm)		0.38	4.10		
Zr/ZrCN-1 (Λ= 13 nm)		0.21	2.30		
Zr/ZrCN-2 (Λ= 13 nm)		0.25	2.80		
Zr/ZrCN-1 (Λ= 6.3nm)		0.24	7.10		
Zr/ZrCN-2 (Λ= 6.3 nm)		0.30	9.80		
Zr/ZrCN-1 (Λ= 4.4 nm)		0.29	6.30		
Zr/ZrCN-2 (Λ= 4.4 nm)		0.35	9.20		

An important parameter with high impact on their final properties is the reactive atmosphere. Thus, by varying the C_2H_2 mass flow rate and the arc current on the cathode it has been shown that friction and wear performance of the carbide coatings were strongly dependent on the carbon content in the case of alloyed TiSi-stainless steel carbide coatings (M. Braic et al., 2020). In this recent study two series of coatings were obtained (3.6–3.8 μm thickness) such as the carbon content to be varied and the composition to be changed from understoichiometric to overstoichiometric. It was showed that a higher acetylene flow rate and lower arc current led to the formation of smaller droplets with a decreased density on the coatings surface, result confirmed by other research report (Polcar et al., 2010). In the present study, the stainless steel was selected as cathode due to strong carbide-forming metals, namely Cr. Moreover, Fe and Ni have a contribution in an amorphous free carbon phase formation with great impact on tribological performance (Yeh et al., 2004). When friction coefficients were investigated, it was found that coatings from series B ((C/(Me + Si) > 1) outperformed those with C/(Me + Si) < 1 (series A). It can be noticed a more stable evolution of the friction coefficient over the testing distance in the case of overstoichiometric coatings. This result can be ascribed to a free-carbon phase formation on the counterpart in the form of a thin solid lubricant present at the interface, resulting in superior friction performance, as indicated in the literature (Guruvenket et al., 2009; Lindquist et al., 2009; Strauss et al., 2011; Wei, 2008). The carbon content had also a contribution on the wear rate, leading to lower amount of debris and less severe adhesive wear. According to their composition, the adhesive and oxidative wear mechanism changed towards a mild polishing wear, as indicated by SEM and EDS results. The lowest wear rate (1.1 x 10^{-6} $mm^3N^{-1}m^{-1}$) and friction coefficient (0.15) was demonstrated by the coating with the highest carbon content (~ 64 at. %). Therefore, the above mentioned (TiSi-SS) C coatings which were investigated in dry conditions can be a start point for industrial application by coating C45 steel tools parts.

A comparison between two series of TiSiC-Cr coatings deposited on C45 steel by using either CH_4 or C_2H_2 as precursor gases was further made (by varying the gas flow rate: 90, 110 and 130 sccm) (coatings with 3.6–3.9 μm thickness) (M. Braic et al., 2017). As indicated by the SEM cross-section images, when C_2H_2 was used, a featureless microstructure was obtained in the case of A–90 coating, in contrast with the rough cross-section morphology showed by M–90 coating. Based on the XPS results, this difference can be explained by the formation of higher free carbon phase content. Also, a decreased crystallite size was observed in this case due to higher proportion of a–C:H phase present in TiSiC-Cr (A-series) distributed at the grain boundaries with a hindering effect on the crystal growth. The results also showed that when C_2H_2 was used, better friction and wear performance was obtained. Friction coefficients values were lower for A-series coatings and a decreasing tendency with increasing gas flow rate was observed. Considering the higher amount of a–C:H phase for TiSiC-Cr (A-series), as already mentioned, it is reported in various studies the formation of a graphite like transfer film at the coating-counterpart interface with beneficial effect on the tribological properties (Jiang et al., 2010; Mitterer et al., 2011). The values of friction coefficients and wear rates for the series obtained by using C_2H_2 showed lower values in the range 0.1–0.2 and 1.4–3.0 × 10^{-6} mm^3 N^{-1} m^{-1}, respectively. Also, the wear tracks for these samples exhibited a mild polishing wear, while adhesive and oxidative wear mechanism was observed for the coatings obtained in CH_4 reactive atmosphere.

TiSiC-Zr and TiSiC-Cr coatings were also comparatively investigated and the effect of the metallic alloying element into TiSiC structure was considered (Vitelaru et al., 2014). The mentioned coatings were deposited on C45 steel using CH_4 as reactive atmosphere, resulting in coatings with 3.6-3.8 μm thickness. Intended for protection of irrigation water pump parts, coatings with high carbon content (over 50 at.%) were targeted, since it was reported that the formation of a free carbon phase showed an improvement on tribological properties (Jiang et al., 2010; Lindquist et al., 2009). Both TiSiC-Cr and TiSiC-Zr led to smaller values of grain size and residual stress, as compared with TiSiC reference. However, TiSiC-Zr showed the highest hardness among all the investigated coatings, leading also to the lowest worn volume, according to the Archard's law (Archard, 1953). Indeed, Zr addition led to the best tribological properties among the investigated coatings, with a friction coefficient of ~0.3 and a wear rate ~2 x 10^{-6} mm^3 N^{-1} m^{-1} (400 m sliding distance, 5 N load, 0.15 m/s sliding speed). Moreover, combined adhesive and oxidative wear were the main mechanisms present in this study.

Another alloying element was considered and TiSiC-Ni coatings were prepared on C45 steel substrates by adding different amounts of Ni into TiSiC structure (Balaceanu et al., 2015). Thus, by controlling the arc current on Ni cathode (70, 90, and 110 A), coatings with 3.2, 6.8 and 16.2 at.% Ni content were obtained (2.8–3.1 μm thickness), being labelled as TiSiC–Ni-1, TiSiC–Ni-2 and TiSiC–Ni-3, respectively. It was reported a decrease in residual stress with Ni addition and also a hardness dependence according to its content. When it comes to friction and wear, similar behavior was found, the ball-on disc tribological investigation showed better results when Ni content did not exceed 7 at.% (μ~ 0.32 and k ~3.16 - 3.66 x 10^{-6} mm^3 N^{-1} m^{-1}). This result can be explained by the presence of an amorphous graphite-like carbon, reported also in the case of addition of a weak carbide-forming metal into TiC structure (Jansson et al., 2011; Jansson & Lewin, 2013). On the other hand, for TiSiC–Ni-3 coating, a higher amount of metallic Ni was found, according to XPS and XRD investigation, explaining the poor tribological properties in this case.

When it comes to nitride/metal multilayers, TiSiN/Ni coatings (7 to 27 nm bilayer periods) were deposited on C45 steel in order to investigate their tribological behavior and the results were analyzed in comparison with those obtained for TiSiN monolayer (Vladescu et al., 2013). It was concluded that the proposed multilayer structures led to an improvement of friction coefficient (values in 0.34 - 0.38 range) as compared with the monolayer, but without a dependence of the coating type. Similar behavior was observed for the wear rate, being explained by the higher film adhesion obtained in this case. TiSiN-400 coating (with Λ ~15 nm) showed the best wear resistance, proving the same dependence on the bilayer period as the hardness and indicating that abrasive wear was the dominant wear mechanism (De Pellegrin et al., 2009; Peters & Nastasi, 2002).

Zr/ZrCN coatings (4.4 to 70 nm bilayer periods) deposited also on C45 steel were obtained in a CH_4 +N_2 reactive atmosphere (M. Braic, Braic, Balaceanu, Zoita, et al., 2011). ZrCN monolayer was also prepared as reference. Unlike previous studies (where C/(C +N) ratio was ~ 0.4) (Balaceanu et al., 2005, 2008), in this case the C/(C +N) ratio was set at ~0.7 for series 1 and ~0.2 for series 2 (for ZrCN sub-layer composition). For tribological evaluation only coatings with superior hardness and adhesion were selected: monolayers and multilayers with Λ= 4.4–25 nm (3.2 μm total thickness). Results obtained following the microchemical, microstructural, mechanical and tribological investigations showed a strong dependence of the coatings properties on C/N ratio and bilayer period. Moreover, all Zr/ZrCN multilayers proved superior behavior as compared with ZrCN monolayers.

MAGNETRON SPUTTERING

When it comes to coatings obtained by magnetron sputtering method, the tribological behavior investigated by the authors in dry environment is presented in the following section. The values of friction coefficient, wear rate along with their testing conditions are presented in Table 2.

Table 2. Tribological performance of different coatings deposited by magnetron sputtering (dry testing conditions)

Magnetron Sputtering					
Material	**Varying Factor**	**μ**	**k ($\times 10^{-6}$ $mm^3N^{-1}m^{-1}$)**	**Testing Conditions**	**Ref.**
ZrCN-1	(CH_4+N_2+Ar) reactive gas mixture; (C+N)/(Me_1+Me_2) ~1 (series 1); (C+N)/(Me_1+Me_2) ~3 (series 2)	0.32	N/A	350 m sliding distance, 3 N load, 0.2 m/s sliding speed, sapphire ball counterpart	(M. Braic, Balaceanu, et al., 2011)
ZrCN-2		0.30			
(Zr,Ti)CN-1		0.38			
(Zr,Ti)CN-2		0.28			
(Zr,Hf)CN-1		0.51			
(Zr,Hf)CN-2		0.24			
(Zr,Nb)CN-1		0.32			
(Zr,Nb)CN-2		0.27			
C45 steel (substrate)	N/A	0.61	N/A	400 m sliding distance, 5 N load, 0.15 m/s sliding speed, sapphire ball counterpart	(V. Braic et al., 2014)
(CrCuNbTiY)C-1-80	(Ar+CH_4) reactive gas mixture; C/Me ~1.1 (series 1); C/Me ~2.7 (series 2); 80, 300, 500 and 650°C substrate temperature	0.53	64.00		
(CrCuNbTiY)C-2-80		0.26	4.90		
(CrCuNbTiY)C-1-300		0.18	7.60		
(CrCuNbTiY)C-2-300		0.22	4.00		
(CrCuNbTiY)C-1-500		0.20	3.90		
(CrCuNbTiY)C-2-500		0.20	0.70		
(CrCuNbTiY)C-1-650		0.23	0.90		
(CrCuNbTiY)C-2-650		0.17	5.60		
TiAlCrNbY-1	(Ar+CH_4) reactive gas mixture; 1-5 series: 0, 0.10, 0.17, 0.26, 0.33 CH_4/(CH_4+Ar) flow rate ratio, respectively	0.86	50.00	100 m sliding distance, 1 N load, 0.05 m/s sliding speed, sapphire ball counterpart	(M. Braic et al., 2010)
(TiAlCrNbY)C-2		0.24	8.10		
(TiAlCrNbY)C-3		0.24	4.40		
(TiAlCrNbY)C-4		0.12	0.70		
(TiAlCrNbY)C-5		0.40	9.20		
D3 steel (substrate)	N/A	0.93	890.00	400 m sliding distance, 3 N load, 0.2 m/s sliding speed, sapphire ball counterpart	(M. Braic et al., 2013)
(TiZr)C-1	(Ar+CH_4) reactive gas mixture; 1-3 series: ~ 0.25; 0.35; 0.50 CH_4/(CH_4+Ar) flow rate ratio, respectively	0.72	74.70		
(TiZr)C-2		0.23	11.15		
(TiZr)C-3		0.17	2.87		
(CuSiTiYZr)C-1		0.44	1.24		
(CuSiTiYZr)C-2		0.18	4.40		
(CuSiTiYZr)C-3		0.15	1.89		
TiN	used as reference	0.83	4.18	400 m sliding distance, 3 N load, 0.2 m/s sliding speed, sapphire ball counterpart	(V. Braic, Vladescu, et al., 2012)
TiC		0.17	9.35		
TiZrNbHfTa		0.87	16.70		
(TiZrNbHfTa)N	(Ar+N_2), (Ar+CH_4) reactive gas mixture	0.96	3.10		
(TiZrNbHfTa)C		0.15	0.83		

The addition of Ti, Hf and Nb to ZrCN structure was investigated and the friction coefficient was assessed in dry environment at room temperature (M. Braic, Balaceanu, et al., 2011). The coatings were deposited in a (CH_4+N_2+Ar) reactive gas mixture and (Zr,Ti)CN, (Zr,Hf)CN and (Zr,Nb)CN coatings were obtained in different deposition conditions in order to obtain (C+N)/(Me_1+Me_2)~1 (for series 1) and (C+N)/(Me_1+Me_2) ~3 (for series 2). The fluctuation of the friction coefficients over the sliding distance can be explained by the unstable tribo-layers formed/destroyed at the sliding contact. Moreover, analyzing the surface profiles of the wear tracks, the coating removal is dominated by a grooving process. It was found that the friction coefficients were in the 0.2-0.5 range and the coatings with high non-metal content showed better behavior. As indicated by Raman analysis results, there was an amorphous carbon phase formation which acts as solid lubricant (Balaceanu et al., 2009; Jang et al., 2005), more evidenced in the case of (Zr,Hf)CN-2 coating, which showed the lowest value of friction coefficient (μ=0.24).

The influence of carbon content was also analyzed in the case of (CrCuNbTiY)C (thickness ~2 μm) multi-component coatings deposited on C45 steel as protective films for tribological applications (V. Braic et al., 2014). The coatings were obtained in (Ar+CH_4) reactive gas mixture and the substrate temperature varied between 80, 300, 500 and 650°C. Different carbon contents were examined by setting the carbon/metal ratios values at ~1.1 (series 1) and ~2.7 (series 2). The metallic constituents were chosen based on their different atomic radii with beneficial effect on nano-crystalline structures or amorphous phases formation (Guo & Liu, 2011). As demonstrated from XPS, an increase in free carbon phase was found for overstoichiometric coatings, as also reported in literature (Jang et al., 2005; Mitterer et al., 2011; Strauss et al., 2011). The same higher carbon content in coatings from series 2 had a positive influence on their morphology, which improved according to temperature substrate, the reason being the higher atomic species mobility (Liang et al., 2011). Moreover, the carbon content can led to lower crystallite sizes and reduced columns width (Johnson et al., 2010). However, when using a temperature of 650 °C, large columns were observed analyzing the surface topography, being linked with thickness increase of coatings deposited by magnetron sputtering technique (Pei et al., 2008), more evident in the case of (CrCuNbTiY)C-2-650 coating. From tribological point of view, as expected, the overstoichiometric coatings showed improved properties as compared with the coatings with from series 1 (carbon/metal ~1.1).

In the case of (TiAlCrNbY)C, deposited also in Ar+CH_4 reactive atmosphere, the CH_4/(CH_4+Ar) flow rate ratio (F_R) was varied in 0 - 0.33 range (M. Braic et al., 2010). The friction results showed low coefficients (0.05–0.25) dependent of carbon content in each coating, which has a contribution in the surface lubricant layer. The abrasive wear mechanism was found to be dominant; the surface wear followed the hardness behavior, being also dependent on the carbon content.

(CuSiTiYZr)C coatings were comparatively investigated with respect to ternary (TiZr)C used as reference (M. Braic et al., 2013). Moreover, their CH_4/(CH_4 + Ar) ratio (0.25; 0.35; 0.50) was used as weighting factor. The constituent elements were selected based on their different atomic radii, thus the structure proved to be amorphous, according to XRD analysis. It was reported that in these conditions a crystalline structure is difficult to form, due to high lattice distortion energy (Guo & Liu, 2011; Yang & Zhang, 2012; Zhang et al., 2008). The friction behavior was superior in the case of MPE coatings as compared with ternary structures, although as carbon content increase similar values were found. The same results were showed by the wear rates, for this behavior their high hardness and low surface roughness being considered to be responsible. The optimum value proved to be at a carbon/metal ratio of about 1.3 which was obtained at CH_4/(CH_4+Ar) flow rate ratio of 0.50.

Both (TiZrNbHfTa)N and (TiZrNbHfTa)C multi-element coatings were obtained in Ar+N_2 and Ar+CH_4 reactive gas mixtures, respectively, and compared with TiN, TiC and TiZrNbHfTa coatings (V.

Braic, Vladescu, et al., 2012). Considering friction coefficients, MPE coatings performed better than Ti based nitride and the metallic coating. However, TiC reached also a lower friction coefficient (0.18), characteristic to carbide structures, slightly higher as compared with (TiZrHfNbTa)C (0.15). The wear rate showed similar behavior, the carbon based MPE showing $k = 0.8 \times 10^{-6}$ mm^3 N^{-1} m^{-1}, attributed to hardness and resistance to plastic deformation improvement.

SEVERE SERVICE APPLICATIONS (CORROSIVE ENVIRONMENT/ HIGH TEMPERATURE

Cathodic Arc Evaporation

This section presents the wear and friction behavior of certain coatings deposited by cathodic arc, starting from a mild corrosive environment based on saline solution (0.9% NaCl) and moving forward to a higher concentration (3.5% NaCl) in order to underline the tribological properties in a more aggressive medium. Figure 1 presents the friction coefficients and wear rates for 316L and various coated samples immersed in 0.9% NaCl solution at room temperature. The effect of wear and corrosion was previously investigated and it was demonstrated that when it comes to a corrosive environment, the tribological properties are influenced by the liquid layer between contact surfaces (Jun & Fengyuan, 2013; Kuptsov et al., 2013; Vitu et al., 2008). Moreover, the results obtained during an investigation performed at 23 °C and 250 °C were comparatively presented, since there are recent studies that showed a lower friction coefficient at high temperature for uncoated samples, related probably with oxide layers formation at the sample surface, whereas the high temperature had a negative impact on the coatings tribological properties (Polcar et al., 2006; Rester et al., 2006).

TiSiC coatings alloyed with Zr and Cr were deposited on 316 L (in a CH_4 reactive atmosphere) and friction and wear behaviour was evaluated in 0.9% NaCl solution (Parau et al., 2016). The coated samples showed an unstable evolution of coefficient of friction over 400 m sliding distance (at 5 N load and 0.15 m/s sliding speed). This result was ascribed to the destruction of the film formed on the sample surface and its passivation over different periods of time or the presence of a third body at the sliding contact (Landolt et al., 2004; Matthews et al., 2007). When it comes to the measured wear rate, TiSiC-Cr coating demonstrated the best behaviour among the investigated coatings ($k = 3.2 \times 10^{-6}$ $mm^3N^{-1}m^{-1}$). The study emphasized the effect of different coating types on the wear process involved and the environment used, since in dry testing conditions TiSiC-Zr showed the best tribological behaviour (Vitelaru et al., 2014).

Considering also a more complex structure, multilayered TiSiC/NiC coatings with different bilayer periods (Λ ranging from 2.5 to 19.8 nm) (1.8 μm total thickness) were deposited on 316 L steel and were investigated for friction and wear behaviour in 0.9% NaCl corrosive environment (M. Braic et al., 2015). The testing conditions in this study were set as follows: 300 m sliding distance, 3 N applied load and 0.1 m/s sliding speed. When compared with TiSiC monolayer, the friction coefficient decreased from 0.38 to 0.16 (the lowest value being assigned to TiSiC/NiC multilayer with $\Lambda = 4.8$ nm). Moreover, the mentioned coating proved to have also the lowest wear rate, reaching a value of $k = 0.7 \times 10^{-6}$ mm^3 N^{-1} m^{-1}). It was showed in this study that by controlling the thicknesses of the individual layers, one can improve the tribological properties of tools working in corrosive environment. The obtained results can be explained by the fact that when it comes to nitride/metal systems (where component layers have different structures) the dislocation motion is blocked at the interfaces (Yashar & Sproul, 1999). The fact that among all investigated multilayers, TiSiC/NiC-720 (Λ= 4.8 nm) proved to have superior friction properties, is due to its higher demonstrated adhesion to 316L steel (Lc~ 57 N). As it was stated before, friction between two sliding surfaces is controlled also by adhesion (Ma et al., 2003). In the present study it was pointed out that adhesion has an important role in friction and wear behaviour, since a decrease of film spallation or delamination can lead also to a low amount of debris. The SEM image of the wear track, for the mentioned multilayer, showed a smaller cross-sectional area as compared with TiSiC monolayer, the processes involved in the tribological evaluation are of adhesive, corrosive, oxidative and third-body nature.

Figure 1. a) Friction coefficients and b) wear rates for 316L and coated samples in 0.9% NaCl solution at room temperature

(Ref. (M. Braic et al., 2015; Parau et al., 2016))

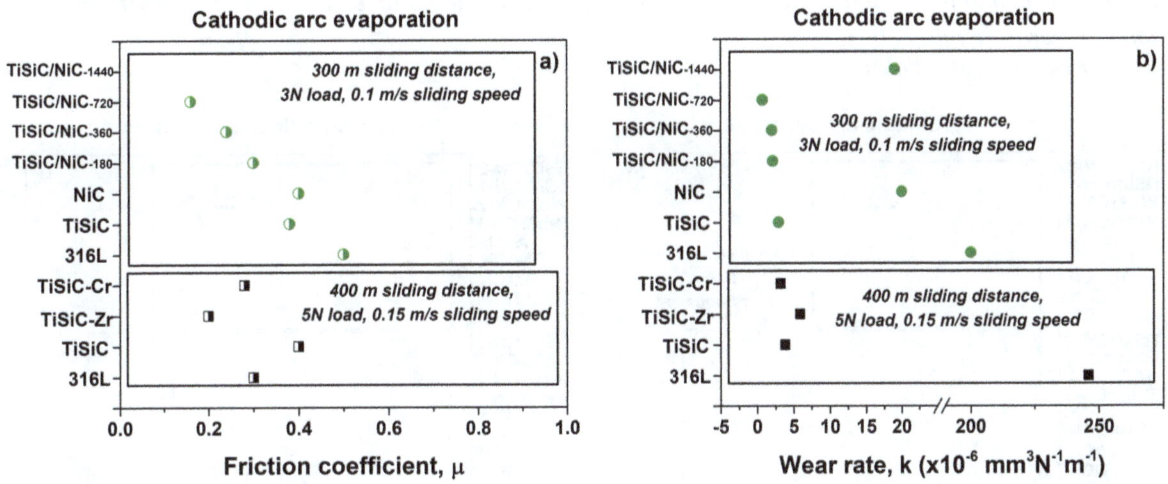

Severe service applications were also explored and the corrosion resistance was assessed in a more aggressive environment (3.5% NaCl). Moreover, friction and wear were evaluated in dry environment, both at 23 ° C and 250 ° C. Figure 2 presents the friction coefficients and wear rates for 316L and various coated samples in dry environment at room temperature and 250 °C. Recent papers reported the addition of Zr, Nb and Si into TiCN structure and the resulted ternary, quaternary and quinary systems were evaluated from tribological point of view (Pruncu et al., 2017, 2019). For this purpose, a reciprocating friction rig was used, the tests being conducted under 5 N normal load, 2 Hz frequency and 12.4 mm stroke length, in dry environment at both room temperature and 250 °C. In the case of ternary and quaternary carbo-nitrides, coating with C/N ratio of about 2 was prepared. It can be observed that while at high temperature the friction coefficient of the uncoated sample was lower as compared with the results obtained at room temperature (probably due to oxide layer formed (V. Braic et al., 2014; Matthews et al., 2007; Polcar et al., 2006), the temperature had a higher impact on the coatings, with severe damage. Coating deposition proved to be a better solution against friction in all tested conditions, and among TiCN, TiNbCN, TiZrCN and TiSiCN coatings, TiNbCN had the lowest wear rate at 23 °C, while TiCN coating performed best at 250 °C. Moreover, at the end of the tribological test, all the coatings showed abrasion and oxidation as the main processes involved. Moving forward to more complex structures,

TiNbZrCN and TiZrSiCN coatings were deposited in order to improve thermal stability, wear and friction of 316 L steel. This time, the C content was changed for each coating combination, resulting in series 1- where C/N ratio ~ 0.4 and series 2- with a C/N ratio of 2.5. One may observe that at 23 °C coatings with a low C/N ratio showed reduced friction coefficients, while for the coatings with a high C/N ratio the wear was improved. Both TiNbZrCN and TiZrSiCN coatings were found to be a proper solution for parts that work in severe corrosion environments, such as the ones in nuclear power plants. The results showed the major role of carbon content with high impact on the tribological properties, which results in an amorphous free carbon phase according to Ref. (Polcar et al., 2010). Moreover, it was showed also to have a contribution in improving the corrosion resistance by increasing grain boundaries length (He et al., 2001; Schwarzer & Richter, 2006; Scully et al., 2007). Thus, by changing the C/N ratio their properties can be tailored (Jansson & Lewin, 2013; Lackner et al., 2004; Lengauer, 2000).

Figure 2. a) Friction coefficients and b) wear rates for 316L and coated samples in dry environment at room temperature and 250 °C
(Ref. (Pruncu et al., 2017, 2019))

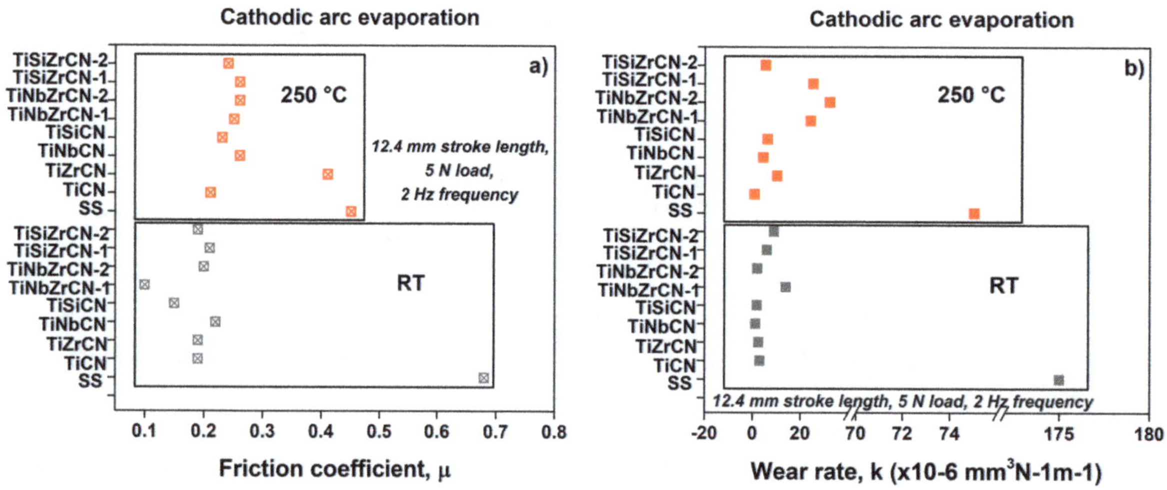

BIOLOGICAL APPLICATIONS

Within this section biological applications are explored. Different coatings deposited by cathodic arc evaporation and magnetron sputtering on different types of substrates (316L, Ti6Al4V, CoCr, AZ31 alloy) were reported. Starting from 0.9% NaCl solution used as corrosive medium up to 3.5% NaCl, Ringer solution or simulated body fluid, at room or body temperature, their friction and wear behaviour was assessed and the results are further presented.

Cathodic Arc Evaporation

When it comes to cathodic arc evaporation, recent studies investigated the tribological properties of TiSiON, TiC, ZrC, TiNbC, ZrCN and ZrCrSiCN deposited on CoCr and 316L in saline environment (Constantin et al., 2019; Dinu et al., 2015; Pana et al., 2020). Corresponding friction coefficients and wear rates for the mentioned substrates and coated samples immersed in saline solution are presented in Figure 3. In the case of TiSiON coatings, with a thickness of about 3 μm, the effect of the substrate bias (−50 V, −100 V, −150 V and −200 V) was explored, in order to enhance the CoCr substrate properties from tribological point of view. It was reported the importance of the synovial fluid on improving friction when it comes to sliding surfaces (Dinu et al., 2015) and the high amount of NaCl present in the human body environment. Thus, intended to be used for load bearings applications, 0.9% NaCl was used as a medium for the experimental setup. The measurements were carried out over 400 m sliding distance, under 3 N applied load and 15 cm/s a sliding speed, at room temperature. The obtained results show that by choosing certain deposition parameters the final properties can be tailored, since the coatings obtained at a higher substrate bias (−200 V) presented a shallow wear track, the lowest friction coefficient and better wear results. As shown also by previous authors, when it comes to cathodic arc evaporation, the increase in atom mobility is linked with the higher energy of the ions bombarding the substrate (Manova et al., 2010) and by increasing the substrate bias, their properties can be enhanced (Aouadi et al., 2019). The results were also linked with surface roughness, since there is a tendency of friction coefficient increase with the real contact area (Yan et al., 2012) and also with the higher Ti content exhibited by the specimen deposited at higher substrate bias. Thus, the friction coefficient for TiSiON (-200 V) reached at the end of the test a value of 0.1 and the wear rate was 6.29×10^{-6} mm^3 N^{-1} m^{-1}.

To simulate the temperature conditions found in the human body for load bearing implants, where a low friction coefficient and high corrosion resistance are required, the temperature of the 0.9% NaCl solution was set at 37 °C. This study reported the enhancement of 316L stainless steel by covering with binary and ternary systems (Pana et al., 2020). Having a thickness of about 1.3 μm, the TiC, ZrC, and TiNbC stoichiometric coatings were investigated from tribological point of view after 250 m sliding distance, 5 N applied load at a speed of 15 cm/s. As indicated by Ducheyne a material with a stoichiometric structure is more appropriate for human body conditions, being more stable and having a longer service life (Ducheyne, 2017). Among all the tested materials, TiNbC proved to have superior tribocorrosive properties with the lowest μ (1.6) and k ($0.99 \times 10^{-5} mm^3 N^{-1} m^{-1}$) values as compared with ZrC and TiC coatings. As other results in the literature, ternary coatings seemed to have better properties as compared with binary ones (Mandes et al., 2018). The higher corrosion resistance of TiNbC coating was ascribed to its crystalline structure, namely the low crystalline size and hence the paths surrounding the grains with an effect on electrolyte penetration. Moreover, the study reports the formation of Nb_2O_5 oxide with high chemical stability and the best adhesion to the metallic substrate showed by TiNbC.

More complex carbonitrides based on an early transition metal to which metallic and/or non-metallic elements were added was the goal of the next study (Constantin et al., 2019). For this purpose, a harsher environment was used (3.5% NaCl) and ZrCrSiCN was compared with ZrCN coating and 316L stainless steel substrate used as reference. After a sliding distance of 200 m, under an applied load of 3N at 0.15 m/s, it was found that the addition of Cr and Si to the base structure led to a decrease of the wear rate of about 60%. This result can be ascribed to a more amorphous phase content, as indicated by XPS. Similar results were obtained in literature when the effect of the carbon content on the tribological behaviour was investigated, a carbon rich surface acting as a solid lubricant (Nguyen et al., 2013). Moreover, the measurements were influenced by both corrosion and mechanical wear, ZrCrSiCN proved to have a finer morphology and a carbon richer phase. It was found that the addition of Cr and Si can have a great impact on corrosion resistance, wear and friction properties (Jahodova et al., 2013; Jansson & Lewin, 2013; Shtansky et al., 2011).

Figure 3. a) Friction coefficients and b) wear rates for CoCr, 316L substrates and coated samples in saline solution
(Ref. (Constantin et al., 2019; Dinu et al., 2015; Pana et al., 2020))

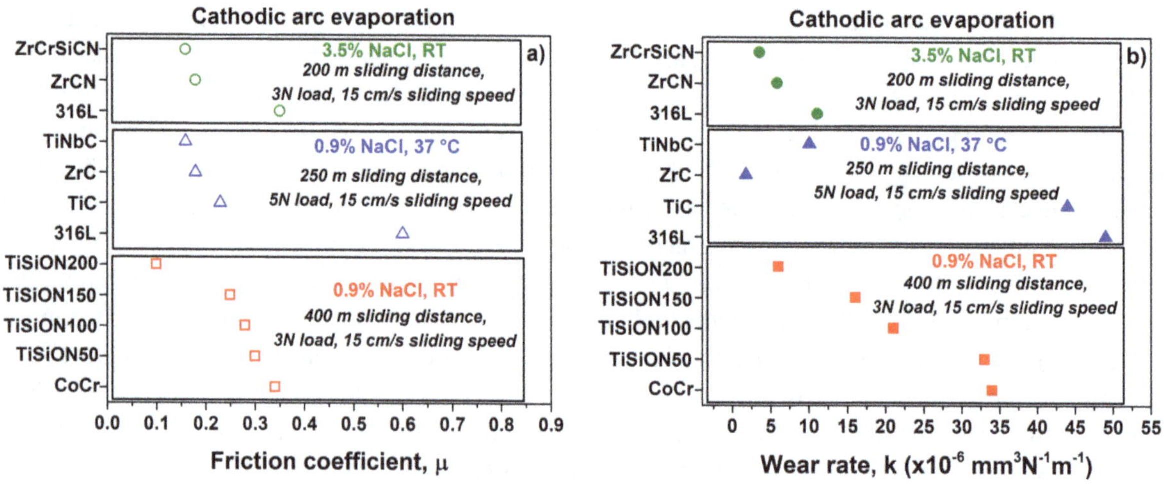

MAGNETRON SPUTTERING

Friction coefficients and wear rates for Ti6Al4V, AZ31 substrates and coated samples in Ringer and SBF solution were further presented. Figure 4 presents the friction coefficients and wear rates for the mentioned substrates and various coated samples immersed in Ringer and SBF solution. Magnetron sputtering deposition technique was employed to enhance the tribological performance of Ti6Al4V alloy in Ringer solution at room temperature. Using a reactive atmosphere formed from a mixture of Ar, N_2 and CH_4 gases, Hf containing ZrCN protective layers were obtain (~2 μm thickness) with two overstoichiometric compositions (~ 2.0 and 3.5, respectively) (Cotrut et al., 2013). The tribological characteristics were assessed under 3 N load, 0.15m/s sliding speed and an overall sliding distance of 600 m. Significantly lower values of friction coefficient were obtained in the case of the coated parts (0.19-0.29) as compared with the bare substrate (~0.5). The results were explained by the carbon rich layer formed at the contact surface, present also in other types of carbonitrides (Cotrut et al., 2012). As the carbon content increased, ZrHfCN-2 showed even better friction properties, reaching a 0.19 coefficient. It was acknowledged also the effect of hardness on the wear behaviour, since a reduced contact area will lower the possibility of adhesive contact. When a corrosion environment is used, synergistic effects are expected (Xue et al., 2020).

Multi-principal-element (TiZrNbHfTa)N and (TiZrNbHfTa)C coatings were further investigated (V. Braic, Balaceanu, et al., 2012). Also, in this case Ti6Al4V alloy was used as substrate and the results showed an improvement of tribological performance in SBF solution (measurements performed at room temperature) when the mentioned coatings were used. With near equiatomic concentrations of the metals (obtained by modifying the power applied to each cathode) coatings with ~2 μm thickness were obtained. By modifying the carbon content, two series of carbide coatings were deposited: (TiZrNbHfTa)C-1 and (TiZrNbHfTa)C-2 with low and high C/metal ratios, respectively. High hardness was found for the carbide coating (with high C content) which exhibited the best values of friction coefficient ($\mu = 0.12$) and wear rate ($k = 0.2 \times 10^{-6}$ mm^3 N^{-1} m^{-1}) (after 800 m sliding distance, 0.2 m/s sliding speed at 3 N load).

Hydroxyapatite coatings (with 1.83 to 1.97 Ca/P ratio) deposited by magnetron sputtering were also proposed for biological applications (Dinu et al., 2017). Recent reports showed that AZ31 magnesium alloy, used as substrate, has a high degradation rate in contact with human fluids (Kirkland et al., 2010; Witte, 2015). Therefore, by deposition of hydroxyapatite coatings and combining their bioactive and osseoconductive properties, one can improve the dissolution rate of magnesium (Vladescu et al., 2015; Vranceanu et al., 2016). The main focus of the study was to investigate the tribological properties of the hydroxyapatite coatings according to their position relative to the target erosion zone. All the tribological investigations were performed in a simulated body fluid (SBF) medium (pH=7.4), with 1 N applied normal load, 15 cm/s sliding speed and the sliding distance of 300 m. It was reported that the apatite formation in SBF is similar with the one formed in the human body conditions, promoting the formation of the bonds with the hard tissue (Kokubo & Takadama, 2006). The lowest friction coefficient (μ= 0.184) and also the lowest wear rate (4.83×10^{-5} mm^3 N^{-1} m^{-1}) was shown by the coating located under target erosion zone, these results being a first step in tailoring the performance of the Mg alloy. Apart from the sample position relative to target erosion zone, other parameters were investigated: deposition temperature (Bramowicz et al., 2016; Vladescu et al., 2015), bias voltage (Surmeneva et al., 2013, 2015) and deposition time (Surmenev et al., 2011).

Figure 4. a) Friction coefficients and b) wear rates for Ti6Al4V, AZ31 substrates and coated samples in Ringer and SBF solution
(Ref. (V. Braic, Balaceanu, et al., 2012; Cotrut et al., 2013; Dinu et al., 2017))

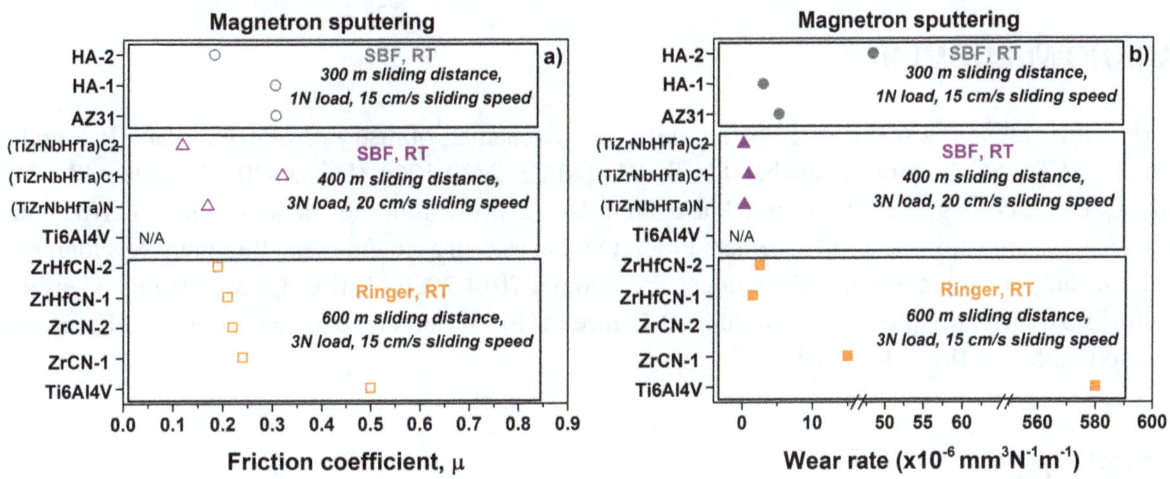

FUTURE RESEARCH DIRECTIONS

The coatings described in this chapter need further investigations related to reactive environmental conditions and deposition parameters in order to enhance their tribological behaviour correlated with their future applications. Looking towards their approach in real conditions, since their performance in laboratory conditions may not be similar with the ones obtained in the real life, thus implementation issue might arise.

CONCLUSION

The authors' tribological findings over the last decade are highlighted in this chapter, which considers the friction and wear behaviour of various coatings obtained by using either magnetron sputtering or cathodic arc evaporation. The tribological properties of coatings can be successfully tailored as a function of deposition parameters (working gases, base pressure, substrate bias voltage, intentional heating) as well as ex-situ surface treatments (e.g. plasma nitridation or thermal treatment under different atmospheres). As it was observed, better tribological properties were achieved when using acetylene working gas, as compared with methane reactive atmosphere. Moreover, increasing the gas flow rate led to the formation of a free-carbon phase which acted as a thin solid lubricant present at the sliding interface, improving the friction performance. Also, higher substrate temperature proved to increase atomic species mobility with further effect on the coating structure and friction and wear properties. Substrate bias represents another deposition parameter with impact on tribological properties, accounted for providing higher energy of the ions bombarding the substrate. Thus, the investigated coatings showed lower friction coefficients and better wear results when increasing the negative bias voltage. Moreover, further surface treatments can be applied in order to obtain coatings with improved tribological properties. Depending on the envisaged applications, new mono- and multilayer structures can expand the current knowledge and will serve as input in achieving an optimal surface finish for industrial applications, to ensure a prolonged lifetime of tools that work in a dry or wet environment, at room temperature or in extreme conditions, or even in the human body.

ACKNOWLEDGMENT

Part of this research was supported by a grant of the Romanian Ministry of Education and Research, CNCS - UEFISCDI, project number PN-III-P4-ID-PCE-2020-1264 (PCE95/2021), within PNCDI III. A. Vladescu is grateful to Tomsk Polytechnic University within the frame a Tomsk Polytechnic University-Competitiveness Enhancement Program grant as well as to European Regional Development Fund through Competitiveness Operational Programme 2014-2020, Action 1.1.3 Creating synergies with H2020 Program; project H2020 Support Centre for European project management and European promotion, MYSMIS code 107874.

REFERENCES

Aissani, L., Fellah, M., Radjehi, L., Nouveau, C., Montagne, A., & Alhussein, A. (2019). Effect of annealing treatment on the microstructure, mechanical and tribological properties of chromium carbonitride coatings. *Surface and Coatings Technology*, *359*, 403–413. doi:10.1016/j.surfcoat.2018.12.099

Aouadi, K., Tlili, B., Nouveau, C., Besnard, A., Chafra, M., & Souli, R. (2019). Influence of substrate bias voltage on corrosion and wear behavior of physical vapor deposition CrN coatings. *Journal of Materials Engineering and Performance*, *28*(5), 2881–2891. doi:10.100711665-019-04033-y

Archard, J. F. (1953). Contact and rubbing of flat surfaces. *Journal of Applied Physics*, *24*(8), 981–988. doi:10.1063/1.1721448

Balaceanu, M., Braic, M., Braic, V., & Pavelescu, G. (2005). Properties of arc plasma deposited TiCN/ZrCN superlattice coatings. *Surface and Coatings Technology*, *200*(1–4), 1084–1087. doi:10.1016/j.surfcoat.2005.01.077

Balaceanu, M., Braic, V., Braic, M., Vladescu, A., Zoita, C. N., Grigorescu, C. E. A., Grigore, E., & Ripeanu, R. (2009). Characteristics of Ti-Nb, Ti-Zr and Ti-Al containing hydrogenated carbon nitride films. *Solid State Sciences*, *11*(10), 1773–1777. doi:10.1016/j.solidstatesciences.2008.12.001

Balaceanu, M., Braic, V., Kiss, A., Zoita, C. N., Vladescu, A., Braic, M., Tudor, I., Popescu, A., Ripeanu, R., Logofatu, C., & Negrila, C. C. (2008). Characteristics of arc plasma deposited TiAlZrCN coatings. *Surface and Coatings Technology*, *202*(16), 3981–3987. doi:10.1016/j.surfcoat.2008.02.005

Balaceanu, M., Parau, A. C., Braic, M., Vladescu, A., Luculescu, C. R., Logofatu, C., & Braic, V. (2015). Growth and characterization of arc evaporated TiSiC-Ni coatings. *Tribology Letters*, *58*(43), 1–9. doi:10.100711249-015-0521-6

Braic, M., Balaceanu, M., Parau, A. C., Dinu, M., & Vladescu, A. (2015). Investigation of multilayered TiSiC/NiC protective coatings. *Vacuum*, *120*(A), 60–66. doi:10.1016/j.vacuum.2015.06.019

Braic, M., Balaceanu, M., Vladescu, A., Zoita, C. N., & Braic, V. (2011). Study of (Zr,Ti)CN, (Zr,Hf)CN and (Zr,Nb)CN films prepared by reactive magnetron sputtering. *Thin Solid Films*, *519*(12), 4092–4096. doi:10.1016/j.tsf.2011.01.375

Braic, M., Balaceanu, M., Vladescu, A., Zoita, C. N., & Braic, V. (2013). Deposition and characterization of multi-principal-element (CuSiTiYZr)C coatings. *Applied Surface Science*, *284*, 671–678. doi:10.1016/j.apsusc.2013.07.152

Braic, M., Braic, V., Balaceanu, M., Vladescu, A., Zoita, C. N., Lungu, C. P., Grigorescu, C. E. A., Grigore, E., & Logoftu, C. (2011). (Ti,Cr,Nb)CN coatings deposited on nitrided high-speed steel by cathodic arc method. *Surface and Coatings Technology*, *205*, S209–S213. doi:10.1016/j.surfcoat.2011.03.030

Braic, M., Braic, V., Balaceanu, M., Zoita, C. N., Kiss, A., Vladescu, A., Popescu, A., & Ripeanu, R. (2011). Structure and properties of Zr/ZrCN coatings deposited by cathodic arc method. *Materials Chemistry and Physics*, *126*(3), 818–825. doi:10.1016/j.matchemphys.2010.12.036

Braic, M., Braic, V., Balaceanu, M., Zoita, C. N., Vladescu, A., & Grigore, E. (2010). Characteristics of (TiAlCrNbY)C films deposited by reactive magnetron sputtering. *Surface and Coatings Technology*, *204*(12–13), 2010–2014. doi:10.1016/j.surfcoat.2009.10.049

Braic, M., Vladescu, A., Balaceanu, M., Luculescu, C., Padmanabhan, S. C., Constantin, L., Morris, M. A., Braic, V., Ana Grigorescu, C. E., Ionescu, P., Dracea, M. D., & Logofatu, C. (2017). A comparative study of the structural, mechanical and tribological characteristics of TiSiC-Cr coatings prepared in CH4 and C2H2 reactive atmosphere by cathodic vacuum arc. *Applied Surface Science*, *400*, 318–328. doi:10.1016/j.apsusc.2016.12.160

Braic, M., Vladescu, A., Parau, A. C., Pruncu, C. I., & Braic, V. (2020). Tribological properties of alloyed TiSi-stainless steel carbide coatings deposited by reactive cathodic arc method. *Wear*, *460–461*, 203456–203466. doi:10.1016/j.wear.2020.203456

Braic, V., Balaceanu, M., Braic, M., Vladescu, A., Panseri, S., & Russo, A. (2012). Characterization of multi-principal-element (TiZrNbHfTa)N and (TiZrNbHfTa)C coatings for biomedical applications. *Journal of the Mechanical Behavior of Biomedical Materials, 10*, 197–205. doi:10.1016/j.jmbbm.2012.02.020

Braic, V., Parau, A. C., Pana, I., Braic, M., & Balaceanu, M. (2014). Effects of substrate temperature and carbon content on the structure and properties of (CrCuNbTiY)C multicomponent coatings. *Surface and Coatings Technology, 258*, 996–1005. doi:10.1016/j.surfcoat.2014.07.044

Braic, V., Vladescu, A., Balaceanu, M., Luculescu, C. R., & Braic, M. (2012). Nanostructured multi-element (TiZrNbHfTa)N and (TiZrNbHfTa)C hard coatings. *Surface and Coatings Technology, 211*, 117–121. doi:10.1016/j.surfcoat.2011.09.033

Bramowicz, M., Braic, L., Ak, F., Kulesza, S., Birlik, I., & Vladescu, A. (2016). Mechanical properties and fractal analysis of the surface texture of sputtered hydroxyapatite coatings. *Applied Surface Science, 379*, 338–346. doi:10.1016/j.apsusc.2016.04.077

Constantin, L. R., Parau, A. C., Balaceanu, M., Dinu, M., & Vladescu, A. (2019). Corrosion and tribological behaviour in a 3.5% NaCl solution of vacuum arc deposited ZrCN and Zr–Cr–Si–C–N coatings. *Proceedings of the Institution of Mechanical Engineers. Part J, Journal of Engineering Tribology, 233*(1), 158–169. doi:10.1177/1350650118774132

Cotrut, C. M., Balaceanu, M., Titorencu, I., Braic, V., & Braic, M. (2012). ZrNbCN thin films as protective layers in biomedical applications. *Surface and Coatings Technology, 211*, 57–61. doi:10.1016/j.surfcoat.2011.08.016

Cotrut, C. M., Braic, V., Balaceanu, M., Titorencu, I., Braic, M., & Parau, A. C. (2013). Corrosion resistance, mechanical properties and biocompatibility of Hf-containing ZrCN coatings. *Thin Solid Films, 538*, 48–55. doi:10.1016/j.tsf.2012.12.100

De Pellegrin, D. V., Torrance, A. A., & Haran, E. (2009). Wear mechanisms and scale effects in two-body abrasion. *Wear, 266*(1–2), 13–20. doi:10.1016/j.wear.2008.05.015

Dinu, M., Cojocaru, M., Braic, V., Tarcolea, M., Braic, M., Miculescu, F., Vladescu, A., & Cotrut, C. M. (2015). Improvement of the tribological performance in corrosive environment of CoCr alloy by TiSiON coatings. *Applied Surface Science, 332*, 295–299. doi:10.1016/j.apsusc.2015.01.221

Dinu, M., Ivanova, A. A., Surmeneva, M. A., Braic, M., Tyurin, A. I., Braic, V., Surmenev, R. A., & Vladescu, A. (2017). Tribological behaviour of RF-magnetron sputter deposited hydroxyapatite coatings in physiological solution. *Ceramics International, 43*(9), 6858–6867. doi:10.1016/j.ceramint.2017.02.106

Ducheyne, P. (2017). *Comprehensive Biomaterials* (2nd ed.). Elsevier.

Guo, S., & Liu, C. T. (2011). Phase stability in high entropy alloys: Formation of solid-solution phase or amorphous phase. *Progress in Natural Science, 21*(6), 433–446. doi:10.1016/S1002-0071(12)60080-X

Guruvenket, S., Li, D., Klemberg-Sapieha, J. E., Martinu, L., & Szpunar, J. (2009). Mechanical and tribological properties of duplex treated TiN, nc-TiN/a-SiNx and nc-TiCN/a-SiCN coatings deposited on 410 low alloy stainless steel. *Surface and Coatings Technology, 203*(19), 2905–2911. doi:10.1016/j.surfcoat.2009.03.009

He, G., Bian, Z., & Chen, G. (2001). Corrosion behavior of a Zr-base bulk glassy alloy and its crystallized counterparts. *Materials Transactions*, *42*(6), 1109–1111. doi:10.2320/matertrans.42.1109

Jahodova, V., Ding, X. Z., Seng, D. H. L., Gulbinski, W., & Louda, P. (2013). Mechanical, tribological and corrosion properties of CrBN films deposited by combined direct current and radio frequency magnetron sputtering. *Thin Solid Films*, *544*, 335–340. doi:10.1016/j.tsf.2013.02.103

Jang, C. S., Jeon, J. H., Song, P. K., Kang, M. C., & Kim, K. H. (2005). Synthesis and mechanical properties of TiAlCxN1-x coatings deposited by arc ion plating. *Surface and Coatings Technology*, *200*(5–6), 1501–1506. doi:10.1016/j.surfcoat.2005.08.065

Jansson, U., & Lewin, E. (2013). Sputter deposition of transition-metal carbide films - A critical review from a chemical perspective. *Thin Solid Films*, *536*, 1–24. doi:10.1016/j.tsf.2013.02.019

Jansson, U., Lewin, E., Råsander, M., Eriksson, O., André, B., & Wiklund, U. (2011). Design of carbide-based nanocomposite thin films by selective alloying. *Surface and Coatings Technology*, *206*(4), 583–590. doi:10.1016/j.surfcoat.2010.06.017

Jiang, J., Hao, J., Pang, X., Wang, P., & Liu, W. (2010). Structure and characteristics of amorphous (Ti,Si)-C:H films deposited by reactive magnetron sputtering. *Diamond and Related Materials*, *19*(10), 1172–1177. doi:10.1016/j.diamond.2010.05.005

Johnson, L. J. S., Rogström, L., Johansson, M. P., Odén, M., & Hultman, L. (2010). Microstructure evolution and age hardening in (Ti,Si)(C,N) thin films deposited by cathodic arc evaporation. *Thin Solid Films*, *519*(4), 1397–1403. doi:10.1016/j.tsf.2010.08.150

Jun, C., & Fengyuan, Y. (2013). Corrosive wear performance of Monel K500 alloy in artificial seawater. *Tribology Transactions*, *56*(5), 848–856. doi:10.1080/10402004.2013.804967

Kazmanli, M. K., Ürgen, M., & Cakir, A. F. (2003). Effect of nitrogen pressure, bias voltage and substrate temperature on the phase structure of Mo–N coatings produced by cathodic arc PVD. *Surface and Coatings Technology*, *167*(1), 77–82. doi:10.1016/S0257-8972(02)00866-6

Kirkland, N. T., Lespagnol, J., Birbilis, N., & Staiger, M. P. (2010). A survey of bio-corrosion rates of magnesium alloys. *Corrosion Science*, *52*(2), 287–291. doi:10.1016/j.corsci.2009.09.033

Kokubo, T., & Takadama, H. (2006). How useful is SBF in predicting in vivo bone bioactivity? *Biomaterials*, *27*(15), 2907–2915. doi:10.1016/j.biomaterials.2006.01.017

Kuptsov, K. A., Kiryukhantsev-Korneev, P. V., Sheveyko, A. N., & Shtansky, D. V. (2013). Comparative study of electrochemical and impact wear behavior of TiCN, TiSiCN, TiCrSiCN, and TiAlSiCN coatings. *Surface and Coatings Technology*, *216*, 273–281. doi:10.1016/j.surfcoat.2012.11.058

Lackner, J. M., Waldhauser, W., Ebner, R., Bakker, R. J., & Scho, T. (2004). Room temperature pulsed laser deposited (Ti,Al)Cx N1-x coatings - chemical, structural, mechanical and tribological properties. *Thin Solid Films*, *468*(1–2), 125–133. doi:10.1016/j.tsf.2004.05.089

Landolt, D., Mischler, S., Stemp, M., & Barril, S. (2004). Third body effects and material fluxes in tribocorrosion systems involving a sliding contact. *Wear*, *256*(5), 517–524. doi:10.1016/S0043-1648(03)00561-1

Lengauer, W. (2000). Transition metal carbides, nitrides and carbonitrides. In *Handbook of Ceramic Hard Materials Edited*. WILEY-VCH Verlag GmbH. doi:10.1002/9783527618217.ch7

Liang, S., Chang, Z., Tsai, D., Lin, Y., Sung, H., Deng, M.-J., & Shieu, F.-S. (2011). Effects of substrate temperature on the structure and mechanical properties of (TiVCrZrHf)N coatings. *Applied Surface Science*, *257*(17), 7709–7713. doi:10.1016/j.apsusc.2011.04.014

Lindquist, M., Wilhelmsson, O., Jansson, U., & Wiklund, U. (2009). Tribofilm formation and tribological properties of TiC and nanocomposite TiAlC coatings. *Wear*, *266*(3–4), 379–387. doi:10.1016/j.wear.2008.04.046

Ma, X. G., Komvopoulos, K., Wan, D., Bogy, D. B., & Kim, Y. S. (2003). Effects of film thickness and contact load on nanotribological properties of sputtered amorphous carbon thin films. *Wear*, *254*(10), 1010–1018. doi:10.1016/S0043-1648(03)00307-7

Mandes, A., Vladoiu, R., Prodan, G., Dinca, V., Porosnicu, C., & Dinca, P. (2018). The properties of binary and ternary Ti based coatings produced by Thermionic Vacuum Arc (TVA) technology. *Coatings*, *8*(3), 114–125. doi:10.3390/coatings8030114

Manova, D., Gerlach, J. W., & Mändl, S. (2010). Thin film deposition using energetic ions. *Materials (Basel)*, *3*(8), 4109–4141. doi:10.3390/ma3084109

Matthews, A., Franklin, S., & Holmberg, K. (2007). Tribological coatings: Contact mechanisms and selection. *Journal of Physics. D, Applied Physics*, *40*(18), 5463–5475. doi:10.1088/0022-3727/40/18/S07

Meng, R., Deng, J., Liu, Y., Duan, R., & Zhang, G. (2018). Improving tribological performance of cemented carbides by combining laser surface texturing and W-S-C solid lubricant coating. *International Journal of Refractory Metals & Hard Materials*, *72*, 163–171. doi:10.1016/j.ijrmhm.2017.12.024

Mitterer, C., Fateh, N., & Munnik, F. (2011). Microstructure-property relations of reactively magnetron sputtered VCxNy films. *Surface and Coatings Technology*, *205*(13–14), 3805–3809. doi:10.1016/j.surfcoat.2011.01.037

Nguyen, X. H., Kim, I. K., Jang, B. K., & Oh, Y. S. (2013). Effect of carbon content on the tribological behavior of TiCxN1-x films prepared by arc-vapor deposition. *Journal of the Ceramic Society of Japan*, *121*(1420), 961–967. doi:10.2109/jcersj2.121.961

Pana, I., Vladescu, A., Constantin, L. R., Sandu, I. G., Dinu, M., & Cotrut, C. M. (2020). In vitro corrosion and tribocorrosion performance of biocompatible carbide coatings. *Coatings*, *10*(7), 654–670. doi:10.3390/coatings10070654

Parau, A. C., Vitelaru, C., Balaceanu, M., Braic, V., Constantin, L. R., Braic, M., & Vladescu, A. (2016). TiSiC, TiSiC-Zr, and TiSiC-Cr coatings—Corrosion resistance and tribological performance in saline solution. *Tribology Transactions*, *59*(1), 72–79. doi:10.1080/10402004.2015.1077406

Pei, Y. T., Chen, C. Q., Shaha, K. P., De Hosson, J. T. M., Bradley, J. W., Voronin, S. A., & Čada, M. (2008). Microstructural control of TiC/a-C nanocomposite coatings with pulsed magnetron sputtering. *Acta Materialia*, *56*(4), 696–709. doi:10.1016/j.actamat.2007.10.025

Peters, A. M., & Nastasi, M. (2002). Effect of carrier gas on the deposition of titanium carbo-nitride coatings by a novel organo-metallic plasma immersion ion processing technique. *Vacuum*, *67*(2), 169–175. doi:10.1016/S0042-207X(02)00108-2

Polcar, T., Novák, R., & Široký, P. (2006). The tribological characteristics of TiCN coating at elevated temperatures. *Wear*, *260*(1–2), 40–49. doi:10.1016/j.wear.2004.12.031

Polcar, T., Vitu, T., Cvrcek, L., Vyskocil, J., & Cavaleiro, A. (2010). Effects of carbon content on the high temperature friction and wear of chromium carbonitride coatings. *Tribology International*, *43*(7), 1228–1233. doi:10.1016/j.triboint.2009.12.010

Pruncu, C. I., Braic, M., Dearn, K. D., Farcau, C., Watson, R., Constantin, L. R., Balaceanu, M., Braic, V., & Vladescu, A. (2017). Corrosion and tribological performance of quasi-stoichiometric titanium containing carbo-nitride coatings. *Arabian Journal of Chemistry*, *10*(7), 1015–1028. doi:10.1016/j.arabjc.2016.09.009

Pruncu, C. I., Vladescu, A., Parau, A. C., Braic, M., Dearn, K. D., Constantin, L. R., & Braic, V. (2019). Multifunctional Ti based carbonitride coatings for applications in severe environments. *Thin Solid Films*, *682*, 63–75. doi:10.1016/j.tsf.2019.04.052

Rester, M., Neidhardt, J., Eklund, P., Emmerlich, J., Ljungcrantz, H., Hultman, L., & Mitterer, C. (2006). Annealing studies of nanocomposite Ti-Si-C thin films with respect to phase stability and tribological performance. *Materials Science and Engineering A*, *429*(1–2), 90–95. doi:10.1016/j.msea.2006.05.053

Schwarzer, N., & Richter, F. (2006). On the determination of film stress from substrate bending: STONEY's formula and its limits. *Materials Science*, *1*, 1–17.

Scully, J., Gebert, A., & Payer, J. (2007). Corrosion and related mechanical properties of bulk metallic glasses. *Journal of Materials Research*, *22*(2), 302–313. doi:10.1557/jmr.2007.0051

Shtansky, D. V., Kuptsov, K. A., Kiryukhantsev-Korneev, P. V., Sheveiko, A. N., Fernandez, A., & Petrzhik, M. I. (2011). Comparative investigation of Al- and Cr-doped TiSiCN coatings. *Surface and Coatings Technology*, *205*(19), 4640–4648. doi:10.1016/j.surfcoat.2011.04.012

Strauss, H. W., Chromik, R. R., Hassani, S., & Klemberg-Sapieha, J. E. (2011). In situ tribology of nanocomposite Ti-Si-C-H coatings prepared by PE-CVD. *Wear*, *272*(1), 133–148. doi:10.1016/j.wear.2011.08.001

Surmenev, R. A., Surmeneva, M. A., Evdokimov, K. E., Pichugin, V. F., Peitsch, T., & Epple, M. (2011). The influence of the deposition parameters on the properties of an rf-magnetron-deposited nanostructured calcium phosphate coating and a possible growth mechanism. *Surface and Coatings Technology*, *205*(12), 3600–3606. doi:10.1016/j.surfcoat.2010.12.039

Surmeneva, M. A., Chaikina, M. V., Zaikovskiy, V. I., Pichugin, V. F., Buck, V., Prymak, O., Epple, M., & Surmenev, R. A. (2013). The structure of an RF-magnetron sputter-deposited silicate-containing hydroxyapatite-based coating investigated by high-resolution techniques. *Surface and Coatings Technology*, *218*, 39–46. doi:10.1016/j.surfcoat.2012.12.023

Surmeneva, M. A., Tyurin, A. I., Mukhametkaliyev, T. M., Pirozhkova, T. S., Shuvarin, I. A., Syrtanov, M. S., & Surmenev, R. A. (2015). Enhancement of the mechanical properties of AZ31 magnesium alloy via nanostructured hydroxyapatite thin fi lms fabricated via radio-frequency magnetron sputtering. *Journal of the Mechanical Behavior of Biomedical Materials*, *46*, 127–136. doi:10.1016/j.jmbbm.2015.02.025

Tillmann, W., & Dildrop, M. (2017). Influence of Si content on mechanical and tribological properties of TiAlSiN PVD coatings at elevated temperatures. *Surface and Coatings Technology*, *321*, 448–454. doi:10.1016/j.surfcoat.2017.05.014

Vitelaru, C., Balaceanu, M., Parau, A., Luculescu, C. R., & Vladescu, A. (2014). Investigation of nanostructured TiSiC-Zr and TiSiC-Cr hard coatings for industrial applications. *Surface and Coatings Technology*, *251*, 21–28. doi:10.1016/j.surfcoat.2014.04.001

Vitu, T., Polcar, T., Cvrcek, L., Novak, R., Macak, J., Vyskocil, J., & Cavaleiro, A. (2008). Structure and tribology of biocompatible Ti-C:H coatings. *Surface and Coatings Technology*, *202*(22–23), 5790–5793. doi:10.1016/j.surfcoat.2008.06.040

Vladescu, A., Braic, M., Azem, F. A., Titorencu, I., Braic, V., Pruna, V., Kiss, A., Parau, A. C., & Birlik, I. (2015). Effect of the deposition temperature on corrosion resistance and biocompatibility of the hydroxyapatite coatings. *Applied Surface Science, 354*(B), 373–379. doi:10.1016/j.apsusc.2015.05.059

Vladescu, A., Braic, V., Braic, M., & Balaceanu, M. (2013). Arc plasma deposition of TiSiN/Ni nanoscale multilayered coatings. *Materials Chemistry and Physics*, *138*(2–3), 500–506. doi:10.1016/j.matchemphys.2012.12.010

Vranceanu, D. M., Cotrut, C. M., Bramowicz, M., Titorencu, I., Kulesza, S., Kiss, A., Berbecaru, A., Pruna, V., Branzei, M., & Vladescu, A. (2016). Osseointegration of sputtered SiC-added hydroxyapatite for orthopaedic applications. *Ceramics International*, *42*(8), 10085–10093. doi:10.1016/j.ceramint.2016.03.114

Wei, R. (2008). Plasma enhanced magnetron sputter deposition of Ti-Si-C-N based nanocomposite coatings. *Surface and Coatings Technology*, *203*(5–7), 538–544. doi:10.1016/j.surfcoat.2008.05.019

Witte, F. (2015). Reprint of: The history of biodegradable magnesium implants: A review. *Acta Biomaterialia, 23*(S), S28–S40. doi:10.1016/j.actbio.2015.07.017

Xue, C., Zhang, P., Wei, D., Hu, H., Li, F., & Yang, K. (2020). Corrosion and tribocorrosion behaviors for TA3 in Ringer's solution after implantation of Nb ions. *Applied Sciences (Basel, Switzerland)*, *10*(23), 8329–8339. doi:10.3390/app10238329

Yan, P., Deng, J., Wu, Z., Li, S., Xing, Y., & Zhao, J. (2012). Friction and wear behavior of the PVD (Zr,Ti)N coated cemented carbide against 40Cr hardened steel. *International Journal of Refractory Metals & Hard Materials*, *35*, 213–220. doi:10.1016/j.ijrmhm.2012.06.003

Yang, X., & Zhang, Y. (2012). Prediction of high-entropy stabilized solid-solution in multi-component alloys. *Materials Chemistry and Physics*, *132*(2–3), 233–238. doi:10.1016/j.matchemphys.2011.11.021

Yao, S. H., Su, Y. L., & Lai, Y. C. (2017). Antibacterial and tribological performance of carbonitride coatings doped with W, Ti, Zr, or Cr deposited on AISI 316L stainless steel. *Materials (Basel)*, *10*(10), 1189–1205. doi:10.3390/ma10101189

Yashar, P. C., & Sproul, W. D. (1999). Nanometer scale multilayered hard coatings. *Vacuum*, *55*(3), 179–190. doi:10.1016/S0042-207X(99)00148-7

Yeh, J. W., Chen, S. K., Lin, S. J., Gan, J. Y., Chin, T. S., Shun, T. T., Tsau, C. H., & Chang, S. Y. (2004). Nanostructured high-entropy alloys with multiple principal elements: Novel alloy design concepts and outcomes. *Advanced Engineering Materials*, *6*(5), 299–303. doi:10.1002/adem.200300567

Zhang, B. Y., Zhou, Y. J., Lin, J. P., Chen, G. L., & Liaw, P. K. (2008). Solid-solution phase formation rules for multi-component alloys. *Advanced Engineering Materials*, *6*(6), 534–538. doi:10.1002/adem.200700240

Zin, V., Miorin, E., Deambrosis, S. M., Montagner, F., & Fabrizio, M. (2018). Mechanical properties and tribological behaviour of Mo-N coatings deposited via high power impulse magnetron sputtering on temperature sensitive substrates. *Tribology International*, *119*, 372–380. doi:10.1016/j.triboint.2017.11.007

KEY TERMS AND DEFINITIONS

Carbide: A compound formed based on carbon and a metal.

Cathodic Arc Evaporation: A Physical Vapour Deposition technique where a material target is vaporised using an electric arc in order to condensate on the substrate and form a thin film.

Friction: The resistance of motion that occurs during sliding of two objects.

Lubrication: The process of reducing friction and wear during sliding of two objects.

Magnetron Sputtering: A Physical Vapour Deposition technique where a material target is sputtered by energetic particles bombardment in order to condensate on the substrate and form a thin film.

Nitride: A compound formed based on nitrogen and a metal.

Tibocorrosion: The combined degradation process that occurs during sliding of two objects in a corrosive environment.

Wear: The process of removal and deformation of a material that occurs during sliding of two objects.

Chapter 10
Investigation on the Wear Resistance of Ni-B-TiO2 Composite Coatings for Dry Crushing Application

Mouna Kallel
National Engineering School of Sfax, Tunisia

Amir Bahri
National Engineering School of Sfax, Tunisia

Khaled Elleuch
National Engineering School of Sfax, Tunisia

ABSTRACT

To achieve a more important service life of hammers, used in crushing process, a Ni-B-TiO2 composite coating was electrodeposited on heat treated AISI P20 using conventional and novel methods. The prepared coatings underwent different tribological tests to quantify the coating that offers the best resistance against wear. For this reason, abrasive wear tests such as pin-on-disk test and multi-pass scratch test were performed to evaluate the abrasive wear resistance of the coatings under a round counterbody (alumina ball) and a sharper contrerbody (sphero-conical indenter), respectively. In addition, the impact-sliding test was also performed to assess the impact resistance of the composite coatings. The obtained results showed that the novel method promotes the best mechanical and tribological properties of the elaborated Ni-B-TiO2 composite coating. This is attributed to the fact of adding TiO2 sol into Ni-B electrolyte which enhances the dispersive strength of the formed TiO2 nanoparticles, contrary to adding solid TiO2 nanoparticles into the electroplating bath.

DOI: 10.4018/978-1-7998-9683-8.ch010

INTRODUCTION

The wear represents a serious issue in many industries in various fields, especially the one using crusher machines (food industries, mineral industries, automotive and aerospace companies). Different components of crushers are exposed to severe erosion. For example, in the olive oil extraction industry, the hammers of the crusher, made of 304L stainless steel, underwent premature damage in three weeks. Based on expert study, the researchers have found that the wear phenomenon originated from the acidity and the continuous impact of olive sides on hammers, leading mainly to a corrosive and erosive wear (Ben Saada & Elleuch, 2015; Bahri et al., 2015) respectively. Furthermore, during the coke process in steel manufacturing, the repetitive impact of coke particles provokes a serious wear of hammers. The latter are made of austenitic manganese steel and have lost a weight of 5Kg within only 8.5 hours (Kallel et al., 2017). To reduce this damage, multiple attempts and solutions have been proposed and studied. Some industries have chosen to change either the material or the design of the hammer while others decided to keep the base material of the hammer, focusing their studies on enhancing its mechanical properties. This could be done by strengthening its working surfaces through a heat treatment technique or by a coating deposition process.

Metal alloy coatings were intensively elaborated to confer on the surface of a sample some required properties such as decorative aspects, corrosion protection and engineering properties. In some cases, this coating does not satisfy the properties in question, which is why composite metallic coatings are performed to enhance the performance of pure films. This includes higher physical properties (thermal and electrical conductivity), better corrosion resistance and particularly significant mechanical properties. The combination between ceramic particles (Al_2O_3, TiO_2, ZrO_2 SiC, WC, etc.) and a metal or polymer matrix favors a synergy impact on the surface of the target material. In this respect, the use of composite coatings has progressively interested academics and professionals in medical, engineering and industrial applications.

Furthermore, many researchers have proven that introducing particles, especially nano-particles, in a metallic matrix enhances the mechanical properties, the wear performance and the self-lubricating characteristics of the target material (Shen et al., 2013). As tribology science is interested in interacting surfaces in relative contact, nanoparticles act as a lubricant to release the contact between the surfaces in relative motion. This, in turn, favors a higher wear resistance of the material. For instance, in automotive applications such piston rings and cylinder nanoparticles prove their central role in raising the lubricant behavior by the formation of tribofilms, which consequently reduces friction and wear (Singh et al., 2018). It has been reported that the beneficial effects of nanoparticles on metallic matrices are attributed to three important reasons. First, the nanoparticles can prevent the migration as well as dislocation of the grain boundaries. Second, introducing nanoparticles in a metallic coating decreases its porosity, leading to a more compact and dense structure with less cracks. Third, the crystalline size of the metallic coating can be refined thanks to the code-position of nanoparticles (Singh et al., 2018).

As regards this work, mineral industries suffer from wear occurring in various machines and more particularly in hammer crushers. In this context, an industry operating on crushing mineral rocks encountered premature wear in hammers of barite rock crusher (Kallel et al., 2017). Therefore, the dysfunction of the crusher causes a larger downtime in the industry, resulting in an important financial loss. The brittleness of the hammer material (High chromium cast iron 'HCCI') and the sharpness of the barite rocks are responsible for the hammer premature wear according to an expert study. As a matter of fact, the aim of this work is to find a solution to lengthen the service life of hammers. The substitution of the hammer material with AISI P20 tool steel reduces its wear but it remains insufficient.

In order to achieve an important service life for the hammer, a Ni-B-TiO_2 composite coating is electrodeposited on AISI P20 after undergoing a heat treatment. The electrodeposition is a simple, easy to control and a cheaper process that offers great mechanical and tribological properties. The introduction of TiO_2 nanoparticles is conducted through two techniques: the conventional process and the sol-enhanced one. The conventional method consists in adding a quantity of crystalline TiO_2 nanoparticles (in form of powder) into the Ni-B electroplating solution. With regard to the sol-enhanced method, this process has been recently created, combining the sol-gel route and the electrodeposition process. It consists in dropping a small amount of transparent TiO_2 sol solution, prepared via sol-gel route, into a Ni-B electroplating bath under vigorous mechanical stirring. As a result, amorphous TiO_2 nanoparticles formed in-situ through hydrolysis reaction and condensation process (Chen and Gao, 2011).

The conventional method should give rise to a composite coating with better tribological properties than that of the pure Ni-B coating. Meanwhile, the preliminary tribological tests showed that the results are underestimated. Thus, it is crucial to search for more useful composite coating to fulfill the requested goal. Based on the literature, the positive outcomes of the composite coating, deriving from the conventional way, is still under demand due to the nanoparticle agglomeration. Indeed, solid nanoparticles have a higher surface energy, which generates attractive forces leading to the nanoparticle agglomeration in the deposit. To overcome this serious obstacle, many surveys have been conducted to diminish the agglomeration phenomenon of the nanoparticles with chemical surfactant, ultrasonic agitation and mechanical stirring. The novel method, according to various publications, has successfully reduced the agglomeration of nanoparticles while the mechanical performance and the corrosion resistance of the deposit showed great improvement.

After the deposition of the Ni-B-TiO_2 composite coating by using both methods, the elaborated coatings have undergone a series of different tests to characterize their morphology, structure, chemical composition, mechanical properties (microhardness and Young's modulus) and tribological performance. It is important to note that the pure Ni-B coating was also elaborated for comparison purposes. The test aimed at determining the best coating and the most convenient process to carry out a composite coating particularly, in terms of the tribological performance, which not only ensures a better resistance to wear but also increases the service life of the hammer.

MATERIALS AND METHODS

Materials

Nickel sulphate hexahydrate 98% purity, nickel chloride hexahydrate 98% purity and boric acid were bought from different suppliers. Titanium dioxide (TiO_2) nanopowder with a primary particle size of about 21 nm (TEM) and purity of 99.5% was imported from Sigma Aldrich. TiO_2 sol solution was prepared in Advanced Material Laboratory at Sfax National School of Engineering, Tunisia.

AISI P20 mold steel was bought from Böhler, Germany.

Experimental Approach

The electrodeposition of Ni-B and Ni-B-TiO_2 composite coatings was conducted on the heat-treated mold steel AISI P20 substrate. Firstly, the mold steel AISI P20 bar was cut through robot-wire machine to obtain specimen with dimensions of 40*40*4 mm^3. The specimens were quenched in oil and then were tempered before electrodeposition. The substrates were mechanically polished in order to obtain a mirror-like surface with silicon carbide abrasive papers of grit size 180, 240, 320, 400, 600, 800 and 1000. The water is used as a lubricant during polishing process. After grinding, the specimens were cleaned with acetone. Prior to electrodeposition, the substrates were activated in 5% H_2SO_4 solution, rinsed in distilled water, and finally set in the electroplating bath.

The DC power supply is used throughout the electrodeposition process. The substrate, forming a cathode, was connected to the negative terminal of the DC power supply. The nickel sheet (99% purity), as an anode, was connected to the positive terminal of the power supply. The schematic representation of the experimental setup of the electrodeposition process is drawn in Figure 1. Ni-B and Ni-B-TiO_2 composite coatings were developed at 37 °C ± 2. The current density was kept at 1A/dm^2 and the pH was 3.5.The time of the electrodeposition is an hour from the start of the DC power supply. A modified Ni-B solution was prepared as electroplating solution for all the coatings. The bath composition is listed in Table 1. The electroplating bath was kept agitated throughout the entire electrodeposition process to guarantee a uniform distribution of the reinforcing nanoparticles into the Ni-B matrix. The Ni-B-TiO_2 powder was elaborated using TiO_2 powder with a concentration of 5g/L. The electroplating bath with TiO_2 powder was agitated during 24 hours and then treated with by ultrasound for 30 min, before starting the electrodeposition, to reduce the agglomeration of TiO_2 nanoparticles into the bath. The Ni-B-TiO_2 sol coating was produced by adding 25ml/L TiO_2 sol into the electroplating bath just before initiating the electrodepositing process. The preparation of the TiO_2 sol solution was described elsewhere (Masseoud et al., 2010).

Figure 1. Schematic representation of the setup of the electrodeposition process to develop Ni-B-TiO_2 composite coatings using (a) conventional method and (b) sol-enhanced method

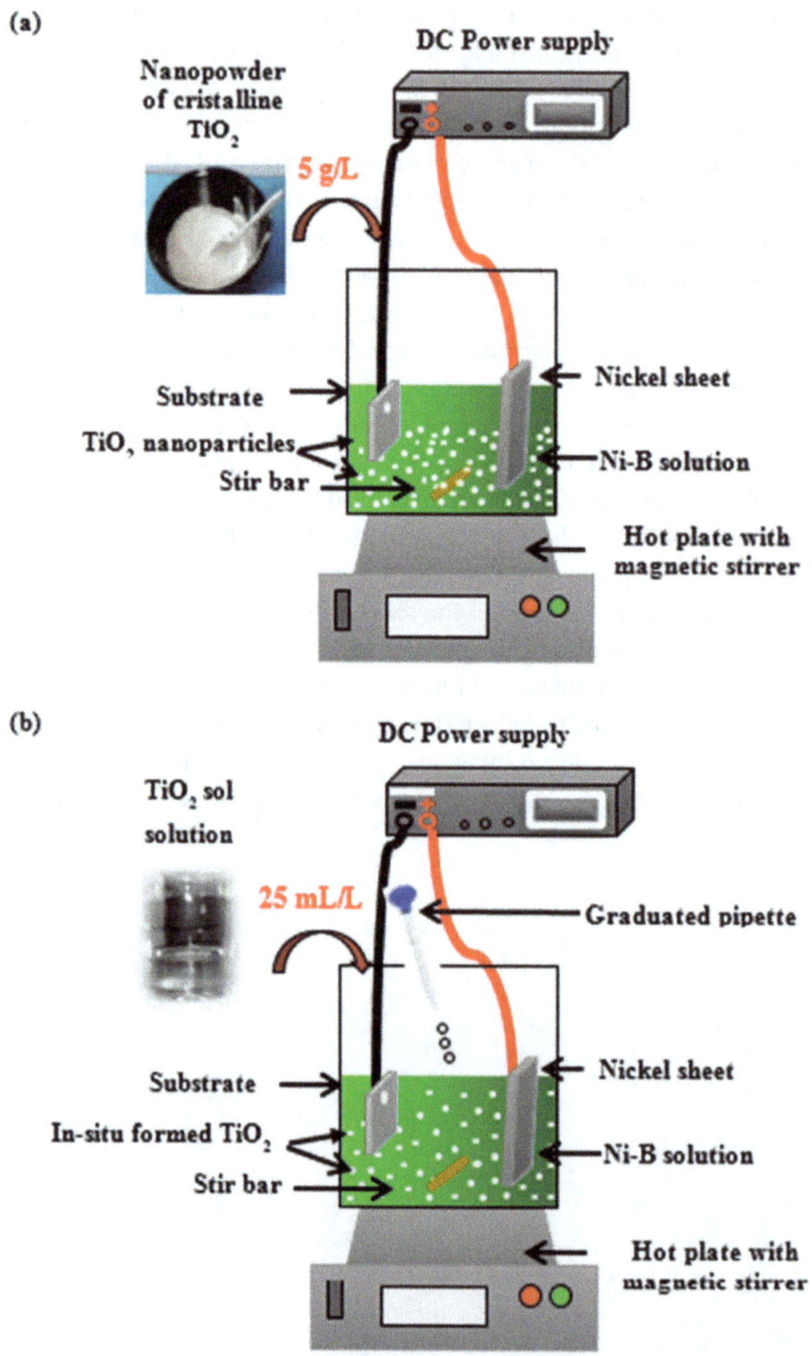

Table 1. The bath composition

Bath Composition	Concentration
Nickel sulfate $NiSO_4.6H_2O$ (98%)	280 g/L
Nickel chloride $NiCl_2$.6H2O	40 g/L
Boric acid H_3BO_3 (99,8%)	38 g/L
Trimethylamine borane (TMAB)	3 g/L
Anionic Surfactant	Small quantity
TiO_2 powder	5g/L
TiO_2 sol	25 ml/L

Coating Characterization

Field Emission Gun Scanning Electron Microscopy analysis (SEM) was performed to examine the morphology of each coating. The SEM is coupled with Energy Dispersive Spectroscopy (EDS) to identify the chemical composition of the elaborated coatings. Moreover, the cross-section of the studied coatings was carried out to investigate the microstructure of the bottom region of each coating, demonstrating the distribution of TiO_2 nanoparticles in the Ni-B matrix and quantifying the thickness of the coatings. The roughness of the coatings was measured using 3D profilometer (TALYSURF CLI 2000, Taylor Hobson).

Concerning the mechanical properties, a micro indentation test (Micro Combi tester MCT^3) was performed to determine the microhardness and Elastic modulus of the coatings.. The test was performed using a Vickers indent under a 0.5N load at a rate of 100 N/min. To obtain an average value as the final property, ten measurements were conducted at different locations for each coating and the average was quoted. The Oliver and Phar method was used to extract the mechanical properties of the coating by analyzing the indentation curves (Olivier &Pharr, 1992).

Based on an expert study, the hammers are subject to sliding and impact solicitations. For this reason, the wear resistance of the coatings was explored through repetitive sliding and impact-sliding tests. Concerning the repetitive sliding test, the pin-on-disk wear test (pin-on-disk tribometer, LGME MZ03) was first employed to assess the wear behavior of each coating and to get a primary idea about the wear endurance of the coating. A 6 mm diameter alumina ball is used as a counter-body by rotating over a 6mm diameter wear track. The conditions of this test were selected to simulate the operative parameters of hammers by considering the maximum parameters at which the tribometer can operate in the laboratory (Normal load and rotating speed). The wear endurance of each coating is evaluated with a test duration ranging from 30 minutes to 4 hours. Second, as the barite rocks are characterized by sharp edges, the wear resistance of the coatings was evaluated against sharper counterparts through a multi-pass scratch test (Micro-Combi, CSM Instruments, Peseaux, Switzerland). A rounded conical indenter (Rockwell type-c diamond indenter) with a 200 µm radius tip was employed. The number of passes was also changed to assess the wear resistance of the coating. The normal load is selected after preliminary tests. The wear tests were repeated at least three times in each sliding duration for each coating and the obtained measurements were averaged. All of the wear tests were conducted in air at room temperature (25°C) under non-lubricated conditions. The wear tracks of the two tests are then analyzed by optic microscopy observation (Leica). The depth and volume of the wear in the coating after the pin-on-disk test are quantified through a 2D profilometer device (Surtronic 25-Taylor Hobson). The wear rate is calculated thanks to the volume of the wear as shown in the following equation:

$$K = \frac{V}{d.F} \tag{1}$$

K represents the wear rate (mm^3/N.m); d is the sliding distance (m) and F corresponds to the applied normal load (N).

The 3D profilometer (TalySurf CLI 2000, Taylor Hobson) is used on the wear scratch to determine the wear mechanism of each elaborated coating.

The impact-sliding test was performed on the elaborated coatings using two impact angles (90°and 45°). An alumina ball with a 10 mm diameter was employed as a counterbody. The choice of these impact angles aimed at, firstly, establishing a complete normal impact load at an impact angle of 90°. Moreover, it is important to evaluate the wear resistance of the coating when the sliding motion is much more important than the load impact by using an impact angle of 45°. The sliding-impact test was conducted at the same impact energy and frequency of 4 mJ and 16 Hz, respectively, for all experiments. The number of impacts is kept at 10000 for all tests.

The generated wear traces of each coating were characterized through two instruments: a Scanning Electron Microscopy (SEM) to analyze the worn surface and a 2D profilometry to draw the wear track profile and to measure the wear depth. The impact load curves were also presented.

RESULTS AND DISCUSSIONS

Coating Morphology

Figure 2 displays the surface morphology of the elaborated coatings. The Ni-B coating surface appears uniform, and is characterized by a dense nodular structure. Each nodule is formed by numerous smaller nodules with a cauliflower appearance (Figure 2.a). The figures 2b and c reveal the influence of TiO_2 powder and TiO_2 sol, respectively, on the surface morphology of the Ni-B matrix. It was seen that the co-deposition of TiO_2 powder into the Ni-B matrix completely changes its typical nodular morphology to a cauliflower-like morphology (Yang et al., 2010). Furthermore, Ni-B-TiO_2 powder morphology presents a porous structure with random grains.

However, the Ni-B-TiO_2 sol shows a surface morphology close to that of the Ni-B coating. However, some clusters made of fine nano-sized TiO_2 nanoparticles take place on the coating surface. No cracks were observed on the entire surface of each coating. Besides, it seems that the codeposition of TiO_2 nanoparticles by the sol-enhanced method results in a decrease of the Ni grain size of the coating. Wang et al. (Wang et al., 2014) reported similar results. They added different concentrations of TiO_2 sol in the Ni-B solution and demonstrated that the grain size reduces with the TiO_2 sol amount.

It is worthy to note that the Ni-B-TiO_2 powder coating exhibits the highest roughness of approximately 0.2 μm, whereas the pure Ni-B coating and Ni-B-TiO_2 sol are smoother with a roughness of 0.11 μm and 0.12 μm, respectively. The roughness values of the coatings match the surface morphology of the coating.

The EDS plots indicate the presence of a peak representing Ni elements with higher intensity for the elaborated coatings. An attentive observation is essential for the Ti element in the composite coatings to confirm the codeposition of TiO_2 nanoparticles in the Ni-B matrix. It was found that the Ti peak is more pronounced in the Ni-B-TiO_2 powder coating while that of the Ni-B-TiO_2 sol is nearly indefinite.

Therefore, the physical state of the TiO_2 nanoparticles added into the Ni-B solution (solid powder or sol-gel solution) plays a central role in the morphology surface of Ni-B coating. A few grams of TiO_2 nanoparticles powder can result in the destruction of the uniformity of Ni-B coating by the creation of large pores. This could be ascribed to the fact that the distribution of TiO_2 powder into the grain boundaries forms a physical barrier, preventing the growth of the Ni grains (Mohajeri et al., 2016). So, the Nickel grains grow randomly to the point of the formation of a variable collection of different sizes (Mohajeri et al., 2016). However, a little quantity of TiO_2 sol does not have a significant influence on the surface morphology of the Ni-B coating since it does not affect the nickel grain growth.

Figure 2. Surface morphology of the elaborated coatings with their corresponding EDS plots: (a) Ni-B, (b) Ni-B-TiO_2 solid and (c) Ni-B-TiO_2 sol

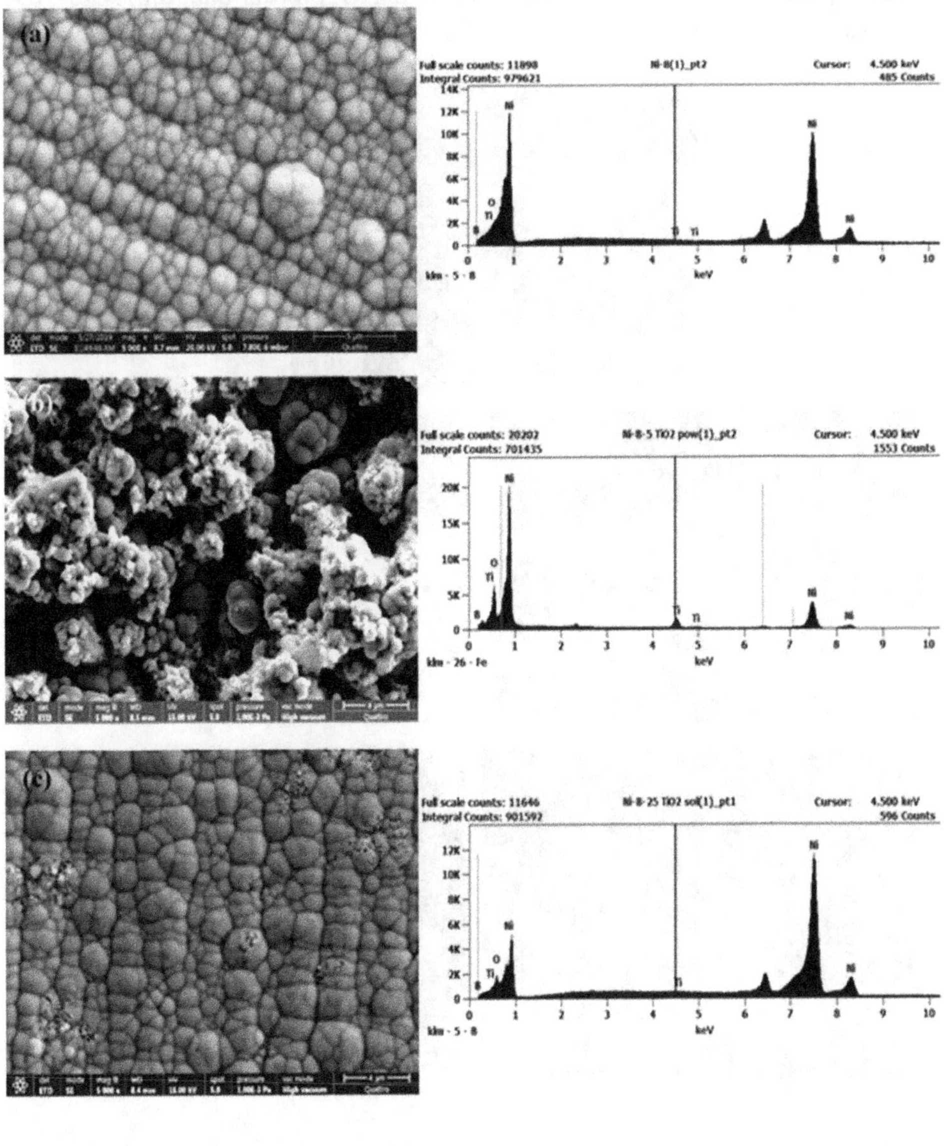

The cross-section of the studied coatings were analyzed using SEM images to examine their bulk morphology and to focus on the distribution of TiO_2 nanoparticles in the Ni-B matrix (Figure 2). The interface quality and the thickness of each coating are investigated as well.

The Ni-B coating was uniformly and continuously deposited on the AISI P20 substrate with a thickness of about 12 μm (Figure 3.a). Figure 3.b reveals that the bottom surface of the Ni-B-TiO_2 powder is compact and the top surface seems very rough. The thickness of the Ni-B-TiO_2 powder coating is averaged between 14 and 16 μm. However, the local area of the cross section of the coating exposes the appearance of the reinforcing TiO_2 nanoparticles in the Ni-B matrix (Figure 3.c). The nanometer-sized TiO_2 nanoparticles are presented as an aggregated area with an average size ranged between 75 nm and 405 nm. The solid TiO_2 nanoparticles tend to agglomerate to minimize their surface energy (Ogihara et al., 2014). Contrary to the top surface, an absence of porosity was noted in the bottom of the Ni-B-TiO_2 powder coating. Concerning the Ni-B-TiO_2 sol coating, the SEM cross section image does not show the TiO_2 nanoparticles dispersed in the Ni-B matrix. This could be attributed to the uniform dispersion of TiO_2 nanoparticles throughout the coating as well as their very small size and relatively low amount.

Figure 3. SEM cross section of (a) Ni-B coating, (b,c) Ni-B-TiO_2 powder and (d,e) Ni-B-TiO_2 sol nanocomposite coatings

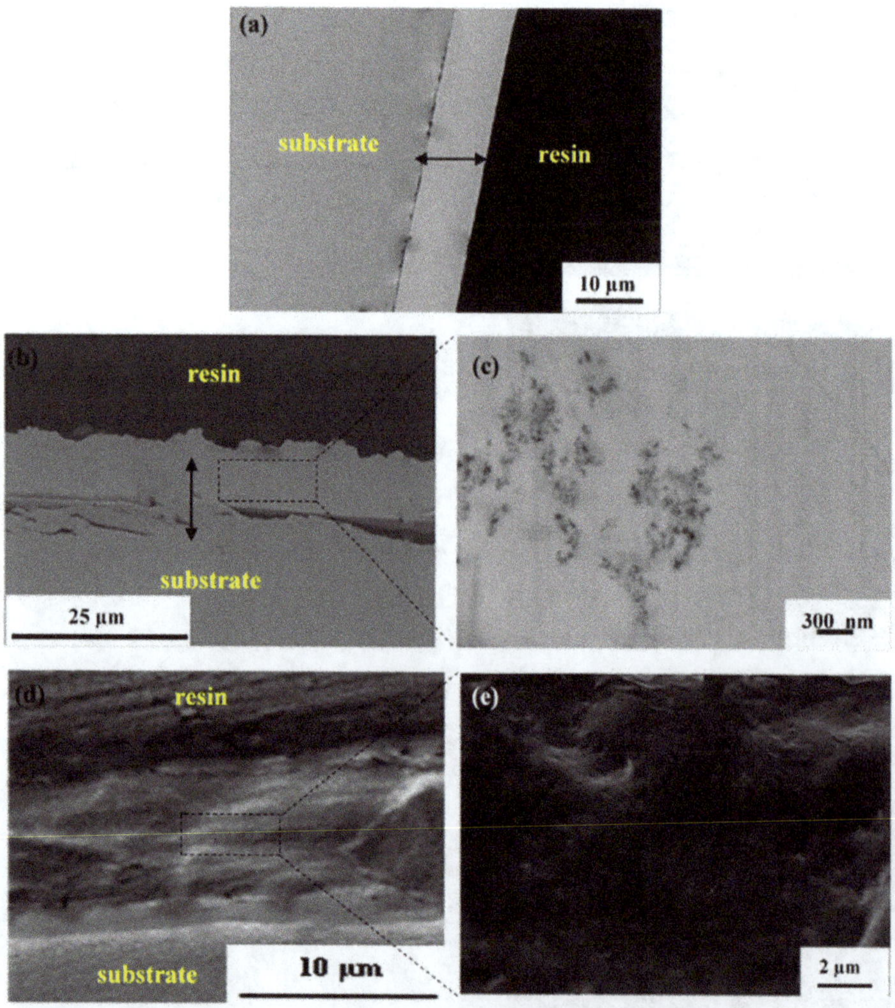

Mechanical Characterization

In order to have insight into the influence of the TiO_2 nanoparticles, with both structures (amorphous and crystalline), on the mechanical properties of the Ni-B matrix; micro-indentation experiments were carried out on Ni-B, Ni-B-TiO_2 powder and Ni-B-TiO_2 sol nanocomposite coatings.

The Load–displacement curves of the studied coatings are presented in Figure 4. These curves can be divided into 2 parts: loading and unloading parts. During loading, the displacement of the indenter increases with load. The maximum penetration depth at the applied load of 0.5 is less than 10% of the coating roughness and much higher than the surface roughness of the coatings, excluding any effect of the substrate on the results. A comparative investigation of loading and unloading curves shows that the indentation depth of the pure Ni-B coating is the smallest (1.8 µm) compared to the two Ni-B-TiO_2 nanocomposite coatings. This result indicates that Ni-B has the highest hardness. The incorporation of TiO_2 nanoparticles in the Ni-B coating through the sol-enhanced method slightly increases the maximum penetration depth to 1.4 µm. Obviously, the microhardness of the Ni-B matrix decreases slightly. The highest maximum penetration depth is recorded at 1.9 µm for the Ni-B-TiO_2 powder composite coating suggesting a remarkable decrease in the Ni-B microhardness. Thus, the codeposition of TiO_2 nanoparticles with different methods play a central role in the mechanical response of the Ni-B coating.

Figure 4. The load–displacement microindentation curves corresponding to the pure Ni-B coating, Ni-B-TiO_2 powder and Ni-B-TiO_2 sol nanocomposite coatings

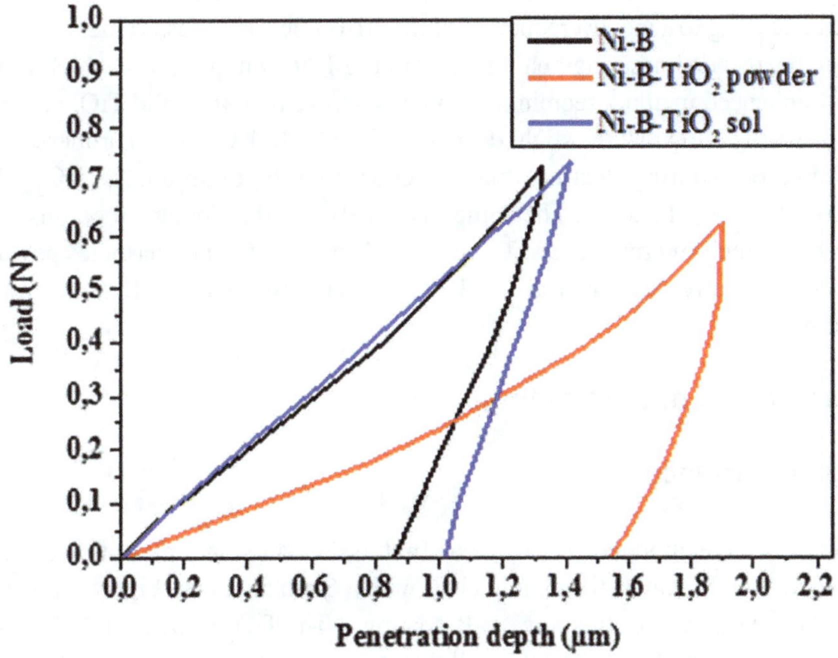

Table 2. Micro-hardness and Young's Modulus of the elaborated coatings

	Micro-Hardness (GPa)	Young's Modulus (GPa)
Pure Ni-B	13.7±1.5	262±23
Ni-B-TiO_2 powder	7.7±0.4	203±12
Ni-B-TiO_2 sol	12.6±1.2	264±20

For further quantitative analysis of the micro-indentation results, the micro-Hardness (H) and the Elastic Modulus of the studied coating are calculated following the Oliver and Pharr procedure as listed in Table 2. The results have the same tendency as the load-displacement microindentation curve shown in Figure 4.

The microhardness of the Ni-B coating presents the highest value of approximately 13.7 GPa. This value is consistent with the typical values of a pure Ni-B metal. The addition of TiO_2 nanoparticles either in the form of powder or sol-gel solution decreases the microhardness of the Ni-B coating. Moreover, it is essential to note that the microhardness of the Ni-B coating dropped to 7.7 GPa when TiO_2 powder was co-deposited in the coating whereas introducing TiO_2 sol causes a slight decrease of the Ni-B coating microhardness, reaching 12.6 GPa. These results are evident and could be accounted for by the morphology of both composite coatings. Indeed, the porosity of the coating lowers its bond strength by a significant amount and reduces its compact structure. Besides, the solid TiO_2 nanoparticles agglomeration also deteriorates the strength of the Ni-B matrix, explaining the low microhardness of the Ni-B-TiO_2 powder coating. In contrast, the Ni-B-TiO_2 sol composite coating exhibits a nearly similar morphology to the Ni-B coating leading to a modest reduction of the Ni-B microhardness value. In addition, the latter can be ascribed to the dispersion strengthening effect of the TiO_2 nanoparticles, which were codeposited following the sol-enhanced method, seemingly more dispersive than the solid TiO_2 nanoparticles in the Ni-B matrix. It is worth noting that the slight decrease of the Ni-B-TiO_2 microhardness can be explained by the release of hydrogen during electroplating as a consequence of introducing TiO_2 sol into the Ni-B bath (Beltowska-Lehman et al., 2015). The comparison between the Young's modulus values of all the coatings match that of their microhardness. The codeposition of TiO_2 nanoparticles powder in the Ni-B coating decreases its rigidity. Nonetheless, the TiO_2 sol virtually does not affect the young's modulus of the Ni-B coating.

The Wear Resistance of the Coatings

Abrasive Wear Resistance

The wear behavior of the elaborated coatings was first assessed using a pin-on-disc tribometer tester, under a normal load of 10N against the alumina ball with a 6mm diameter. Figure 5 presents the optical micrographs of the wear tracks of the pure Ni-B coating, Ni-B-TiO_2 powder and Ni-B-TiO_2 sol nanocomposite coatings after a continuous rotating sliding process lasting 30 minutes, 2 hours and 4 hours. The majority of the wear tracks show a grey appearance with some white areas. Continuous grooves parallel to the direction of rotating motion can also be perceived, representing the important aspect of the abrasive wear mode (Gyftou et al., 2004). The depth and width of the grooves are largely in control of the appearance of the wear track. Indeed, the pure Ni-B coating exhibits continuous and deep grooves delaminated by ridges for a sliding duration of 30 minutes. However, when the sliding duration increases, the wear tracks of the coating contain some areas exempt from the coating material. The removed coating forms wear debris. The Ni-B-TiO_2 powder coating presents a similar behavior to the Ni-B coating. Hence, introducing TiO_2 powder into the Ni-B coating does not enhance the wear resistance of the Ni-B matrix. Here, the main wear mechanism includes material removal combined with severe microcutting. Regarding the Ni-B-TiO_2 sol coating, only grooves take place on the wear track along the sliding direction for all sliding durations. Here, the predominant wear mechanism is grooving.

Based on the above results, the Ni-B-TiO_2 sol promotes the best wear resistance due to the addition of TiO_2 sol into the Ni-B solution. The formation of TiO_2 nanoparticles in-situ strengthens the wear resistance of the Ni-B coating. In contrast, the solid TiO_2 nanoparticles show no influence on the improvement of the wear resistance of the Ni-B coating.

Figure 5. The worn surfaces of pure Ni-B coating, Ni-B-TiO_2 powder and Ni-B-TiO_2 after continuous rotating sliding during 30 minutes, 2 hours and 4 hours

Figure 6 comparatively presents the friction coefficient of different elaborated coatings under dry sliding conditions. The pure Ni-B coating and Ni-B-TiO_2 sol coating show a similar friction coefficient trend. Indeed, at the beginning the friction coefficient curves increase, raching high values to later on decrease significantly. The curves are then stabilized by a slow increase during the remaining of the test until a constant value is obtained. The friction coefficient of the Ni-B-TiO_2 powder greatly increases in the start of the test, then decreases up to a certain value and is kept nearly stable until the end of the test. The rise of friction coefficient at the beginning can be explained by the coating roughness, which could be suppressed at the first sliding contact between the coating surface and the alumina ball contact area. The friction coefficient of the Ni-B coating is stable between 0.4 and 0.42. The codeposition of TiO_2 powder into the Ni-B coating slows down its friction coefficient and holds the lowest value reaching 0.35. When examining the friction coefficient of the Ni-B matrix, an increase can be detected when the TiO_2 sol is added in the Ni-B solution, ranging from 0.46 and 0.49. Some studies referred the evolution of the friction coefficient to the Young's modulus of the rubbed coatings (Gomez-Navarro et al., 2008; Poot & Zant, 2008; Frank et al., 2007). They reported that the high Young's modulus of the reinforcement material prevent it from behaving as a lubricant agent in the matrix, especially when the amount of this material is too small. This clearly explains the higher friction coefficient of Ni-B-TiO_2 sol coating as it holds the highest young's modulus (table 2). In contrast, the TiO_2 powder reduced the young's modulus of the Ni-B matrix, thus, allowing the formation of a lubrication layer, soothing the contact between the coating surface and the conterbody. Hence, the friction coefficient of the Ni-B-TiO_2 powder is reduced.

Figure 6. Friction coefficient of pure Ni-B coating, Ni-B-TiO_2 powder and Ni-B-TiO_2 sol after continuous rotating sliding during 4 hours

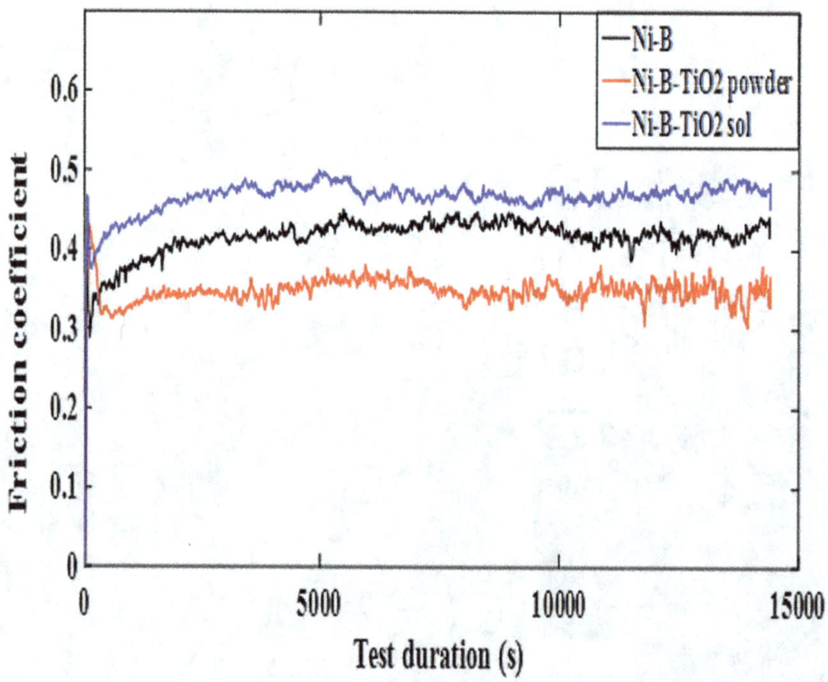

Optic microscopy images of wear tracks of different coatings are an intuitive way to have a preliminary idea about their wear resistance. In order to obtain a clear comparison between the wear resistance of different coatings, the evolution of the wear depth and the wear rate with a specific sliding duration are presented in figure 7 a and figure 7b., respectively. The wear depth of pure Ni-B and of Ni-B-TiO_2

powder increases with the sliding duration. The Ni-B coating exhibits the highest wear depth (around 13μm) after a sliding distance of 4 hours, exceeding the coating thickness. For the Ni-B-TiO_2 sol, the wear depth is lower and is distinguished by a nearly steady-state wear depth, especially when the sliding duration ranges between 1 and 3 hours. A wear depth of 3.8 μm is the highest value recorded after using a sliding duration of 4 hours for the Ni-B-TiO_2 sol composite coating.

The evolution of the wear rate of the different coatings differs from the wear depth. Indeed, at the beginning, the wear rate of the pure Ni-B is noticeably high ($9.2.10^{-6}$ mm^3/N.m), then it progressively decreases up to a sliding duration of 2 hours, and finally increases slightly to 5.10^{-6} mm^3/N.m at a sliding duration of 4 hours. The Ni-B-TiO_2 powder achieves a wear rate more important than that of the pure Ni-B coating at a sliding duration ranging from 1 to 3 hours. Comparing the two previous coatings, the Ni-B-TiO_2 sol coating has much lower wear rate values for all sliding durations. At a sliding duration of 4 hours, the wear rate reaches 1.10^{-6} mm^3/N.m.

Based on the above mentioned results, it is clear that the co-deposition of TiO_2 nanoparticles into the Ni-B matrix by the sol-enhanced method remarkably intensifies the wear endurance. The Ni-B-TiO_2 sol nanocomposite coating is much denser and contains finer grain size as reported in the previous investigation (Kallel et al., 2021a). As a result, the wear resistance of the Ni-B coating was considerably improved. When introducing TiO_2 powder in the Ni-B matrix, the wear resistance of the Ni-B coating was mediocre. This could be correlated to the weak structure of the Ni-B-TiO_2 powder originating from its higher porosity. Furthermore, the low microhardness of the Ni-B-TiO_2 powder could be explained by its higher grain size compared to the Ni-B coating and the Ni-B-TiO_2 sol composite coating (Kallel et al., 2021 a). It was also reported that the roughness of the coating intervenes in its wear endurance. The smoother surface contributes in the reduction of the stress concentration between the friction materials, resulting in a wear resistance enhancement (Hamid & Ghayad, 2002). Therefore, the wear resistance of the Ni-B-TiO_2 composite coating depends not only on the incorporation of the TiO_2 nanoparticles, but also on the nature of their structure (crystalline or amorphous).

Note that the decrease of the wear rate with the sliding duration could be attributed to the formation of an oxidation layer between the coating surface and the alumina ball during the sliding friction. This phenomenon reduces the contact between the materials in relative motion, ending in a reduction of the wear rate.

Figure 7. Evolution of (a) wear depth and (b) wear rate of the pure Ni-B coatings, Ni-B-TiO_2 powder and Ni-B-TiO_2 sol nanocomposite coatings

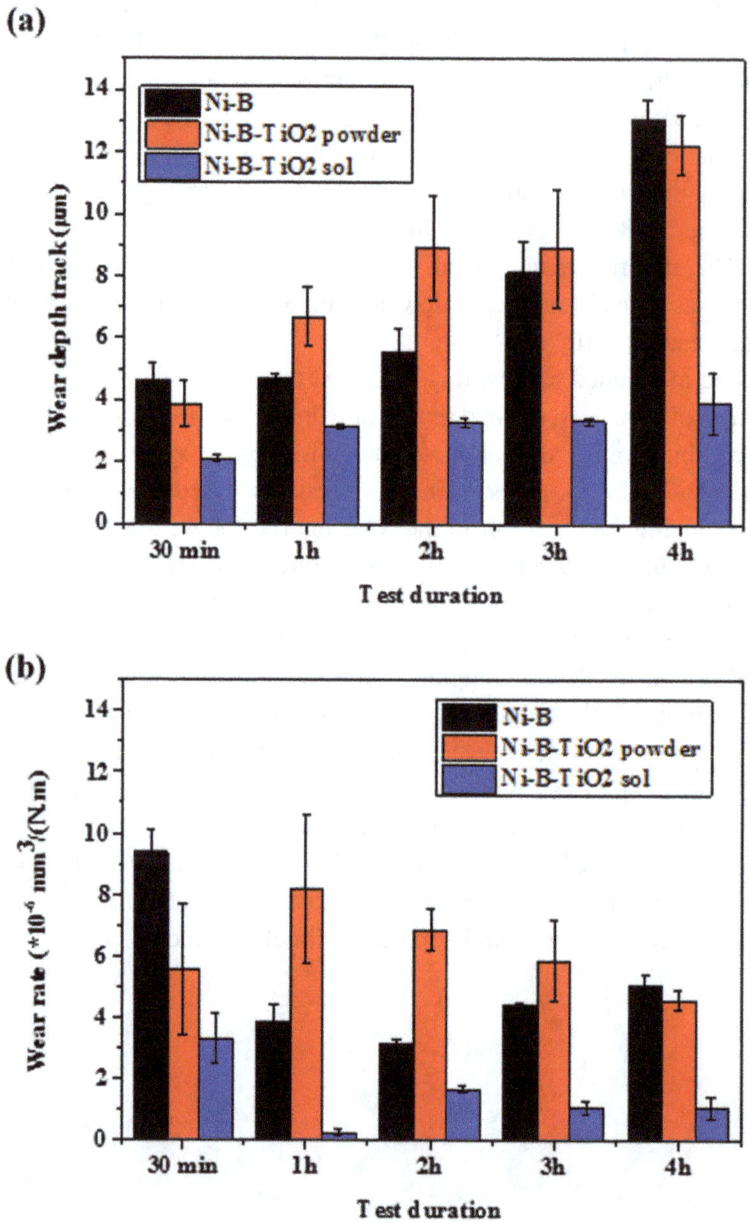

Based on the findings of a previous paper characterizing the barite rocks shape (Kallel et al., 2017), it is crucial to assess the resistance of the coatings against a sharper counter-body (Rockwell type-C indenter with a tip radius of 200 µm). For this reason, a multi-pass scratch test was performed on the Ni-B-TiO_2 composite coatings under a 20N load for multiple passes (5, 20 and 100). It is worth noting that the multi-pass scratch test permits an evaluation of the adhesion strength of the coating. This test is a suitable way to estimate the fatigue endurance of the coatings under continuous friction. Pure Ni-B has not been evaluated by this test due to the preliminary results showing the spallation of the coating at the first pass (Kallel et al., 2021 b).

The worn surfaces of the composite coatings, obtained at 5, 20 and 100 passes, are firstly investigated by microscopic optical images (Figure 8). The 3D profilometry images are also performed for the same scratch traces to further quantify the damage on the coating surface (Figure 9). The variations of number of passes allow a progressive detection of damage under repetitive sliding motions. The wear rate is also calculated after a multi-pass scratch test (Figure 10).

During the multi-pass scratch test, the applied load is kept constant leading to a constant wear track width. According to the microscopic observation images, an absence of failure was observed at the sides of the scratch traces during the multi-pass scratch test. For the Ni-B-TiO_2 powder composite coating, the optic microscopy images of scratch grooves obtained at 5 and 20 passes showed a total fragmented surface. Meanwhile, the 3D profilometry images of both scratch traces presented a deep groove with a little plastic deformation at their ridges, as a result of a considerable wear rate, reaching about $1.75.10^{-3}$ $mm^3/m.N$. At the beginning of the test, the surface roughness of the Ni-B-TiO_2 powder coating is higher, which facilitates the wear of the coating. The porosity of the surface coating weakens its bond strength. Thus, it could be a relevant reason for the formation of cracks and the large wear rate of the composite coating as well. A further increase in the number of passes will result in a dramatic damage of the coating as it will undergo a gross spallation, reflecting coating detachment. The spalling phenomenon is a consequence of the repetitive and continuous loadings, leading to the accumulation of cracks and a fracture of the coating material once the stress concentration reaches a critical value. The material removed forms wear debris (Figure 8.c). The wear rate decreases with the number of passes in spite of the severe failure occurring at a higher number of passes. This result indicates that the TiO_2 nanoparticles form a protective layer on the coating surface which consequently reduces the contact between the materials in motion. Therefore, the incorporation of TiO_2 nanoparticles in the Ni-B coating achieves an interesting improvement on the Ni-B wear resistance as more durability and strength in the Ni-B matrix is achieved.

The microscopic observation of the scratch grooves of the Ni-B-TiO_2 sol only shows a plastic deformation within the grooves scratch for all the number of passes. Furthermore, the 3D profilometry of the residual scratch groove after 100 passes exposes a pile-up with the greatest height at one side of the residual scratch groove. This reflects the development of a buckle parallel to the scratch groove, separating the coating surface and the substrate underneath without delamination. This interfacial adhesion failure is considered evidence of an adequate adhesion of the coating to the substrate material by several researchers as mentioned in their previous work (Bull & Berasetegui, 2006). In addition, the same investigations have ascribed the creation of a buckle failure to the non-uniformity of the scratch indenter tip, resulting in the formation of a compressive stress at the side of the scratch groove following the continuous repetitive scratches (Bull & Berasetegui, 2006). It was seen that the wear rate of the Ni-B-TiO_2 sol composite coating is lower than that of the Ni-B-TiO_2 powder composite coating for all of the number of passes. It varies between $0.5.10^{-3}$ $mm^3/m.N$ and $0.05.10^{-3}$ $mm^3/m.N$ corresponding to the numbers of passes of 5 and 100, respectively. Thus, it is clear that the codeposition of TiO_2 nanoparticles in the Ni-B matrix through the sol-enhanced technique strongly enhances its resistance against continuous rubbing by a sharper indenter when compared to the same nanoparticles codeposited via the conventional method. Moreover, according to the optical microscopy images, the Ni-B-TiO_2 powder coating shows very critical aspects of failure such as coating detachment. The latter was not observed on all the scratch traces of the Ni-B-TiO_2 sol composite coating. This is especially ascribed to the crucial role of the sol-enhanced method as a most convenient way to strengthen the bond within the

Ni-B matrix microstructure in order to easily support higher stress concentrations. In addition, the big difference between the microhardness of both composite coatings reflects the dissimilarities in their wear endurance and wear mechanism (Table 2). Indeed, the predominant wear mechanism of the Ni-B-TiO_2 powder coating is material removal while the Ni-B-TiO_2 sol coating mostly shows plastic deformation which could be considered as the dominant wear mechanism.

Figure 8. Optical microscopy image of the scratch grooves of Ni-B-TiO_2 composite coatings coatings after multi-pass scratch test

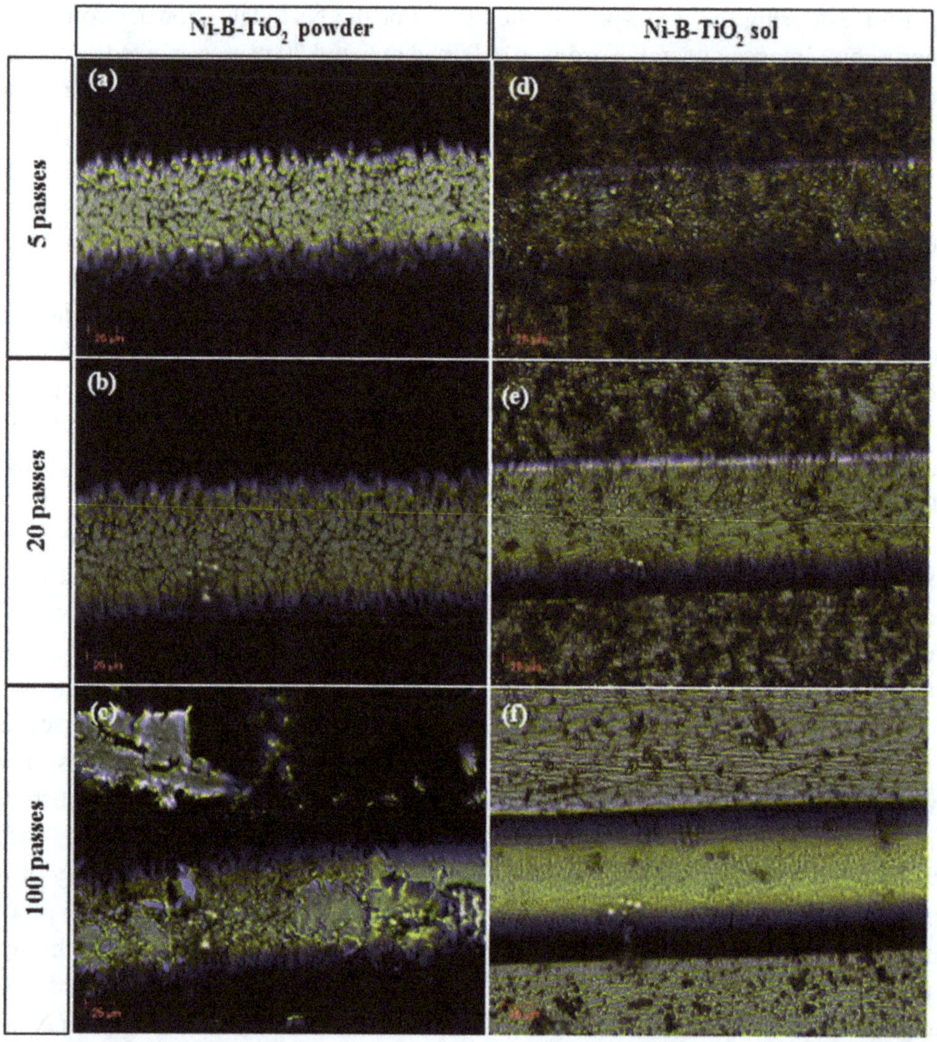

Figure 9. 3D profilometry images of the scratch grooves of (a) Ni-B-TiO_2 powder and (b) Ni-B-TiO_2 sol nanocomposite coatings after multi-pass sliding test

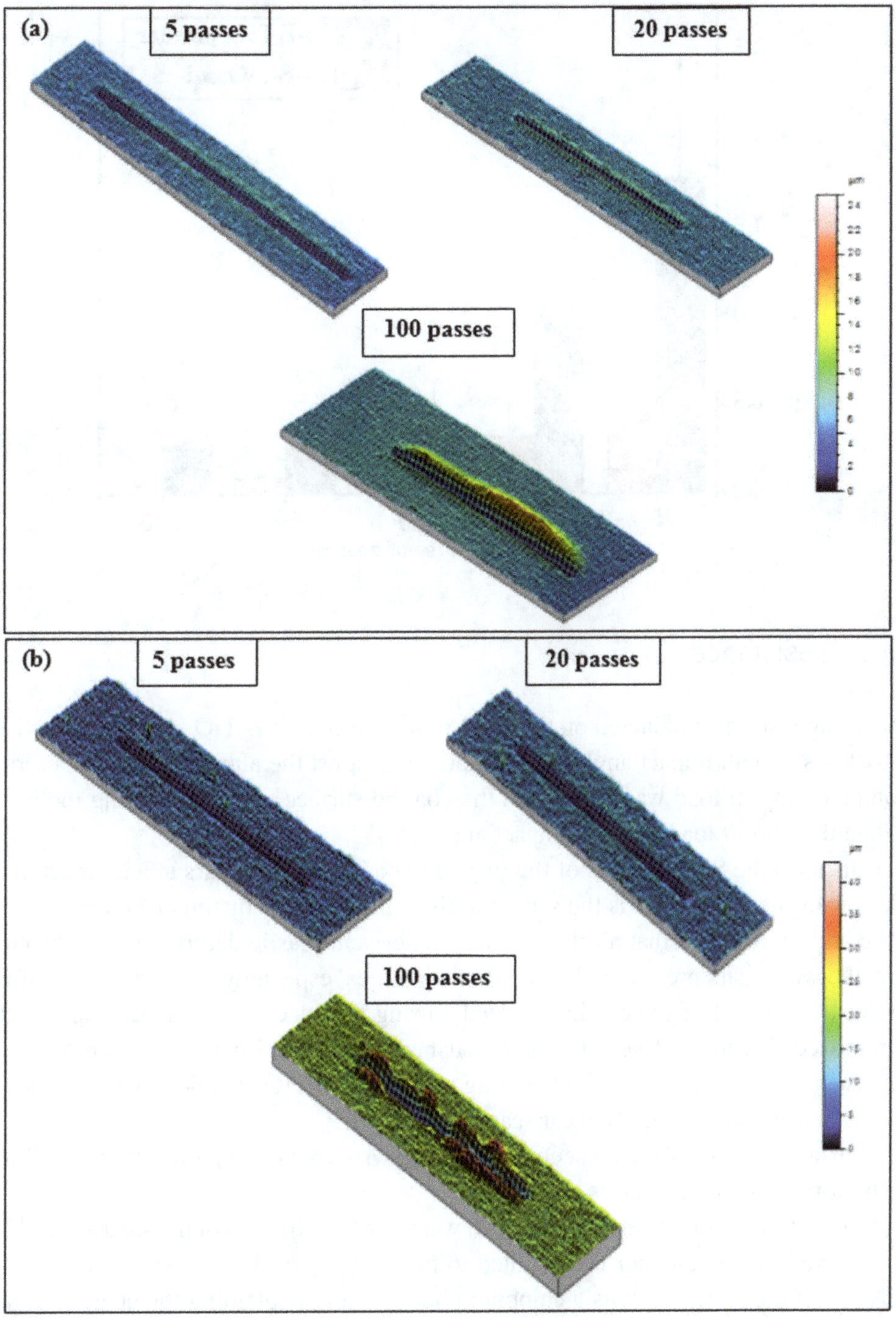

Figure 10. Wear rate of Ni-B-TiO_2 powder and Ni-B-TiO_2 sol nanocomposite coatings during multi-pass scratch test

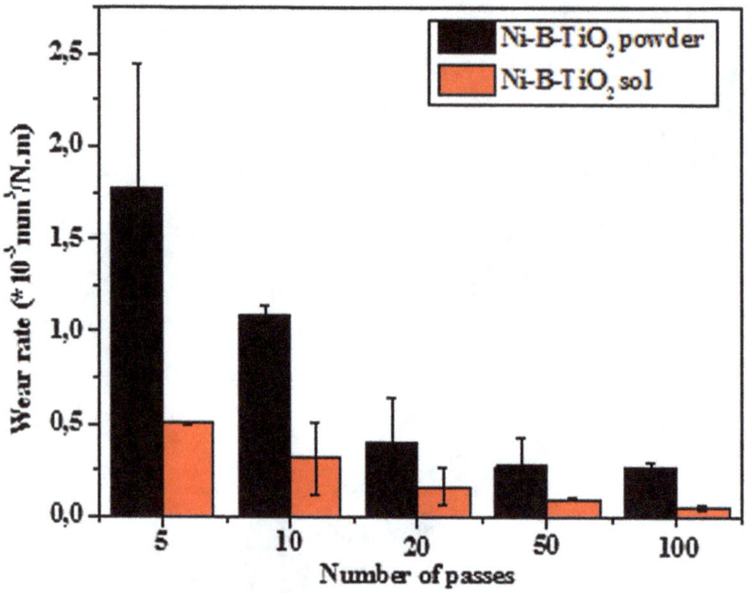

Impact Wear Resistance

The impact-sliding test was conducted on the Ni-B pure coating, Ni-B-TiO_2 powder and Ni-B-TiO_2 sol composite coatings at both impact angles of 90° and 45° against the alumina ball. For an impact angle of 90°, a complete impact load was applied on the coating surface while the sliding motion was more significant than the impact loading at an impact angle of 45°.

Figure 11 displays the SEM images of the impacted coatings wear scars at an impact angle of 90°. The overall view of the worn scars is the same for all coatings and is distinguished by a circular shape (Figure 11a, c and e). It seems that all the coatings underwent plastic deformation at the contact area. Furthermore, the wear scars present dark zones at their edges, especially in the direction of the impact, indicating a local oxidation of the coating. Indeed, during the successive normal impacts, the coating was locally removed. The formed wear debris was pushed towards the impact direction and then expelled from the wear scars. For instance, the Ni-B coating and the Ni-B-TiO_2 powder composite coating figure out wear debris inside and outside the wear scars.

The worn surface of Ni-B-TiO_2 sol seems to be very wrinkly and exposes an oxidized wear debris down at the bottom of the wear scar.

Besides, figure 10.b highlights the accumulated wear debris at the end of the wear scar which appears to be totally cracked. Such behavior is attributed to the fragility of the formed wear debris. The wear scar of the Ni-B-TiO_2 powder exhibits a chipping phenomenon, explaining the coating removal under the continuous impacts and the higher amount of wear debris (Figure 11.d).

A higher magnification observation of the dark area of the Ni-B-TiO_2 sol wear scar indicates the presence of very fine and short cracks, which took place in the oxidized surface (Figure 11.f).

Figure 11. SEM observation of the worn surface of (a,b) pure Ni-B coating, (c,d) Ni-B-TiO_2 powder coating and (e,f) Ni-B-TiO_2 sol coating after the impact sliding test at an impact angle of 90°

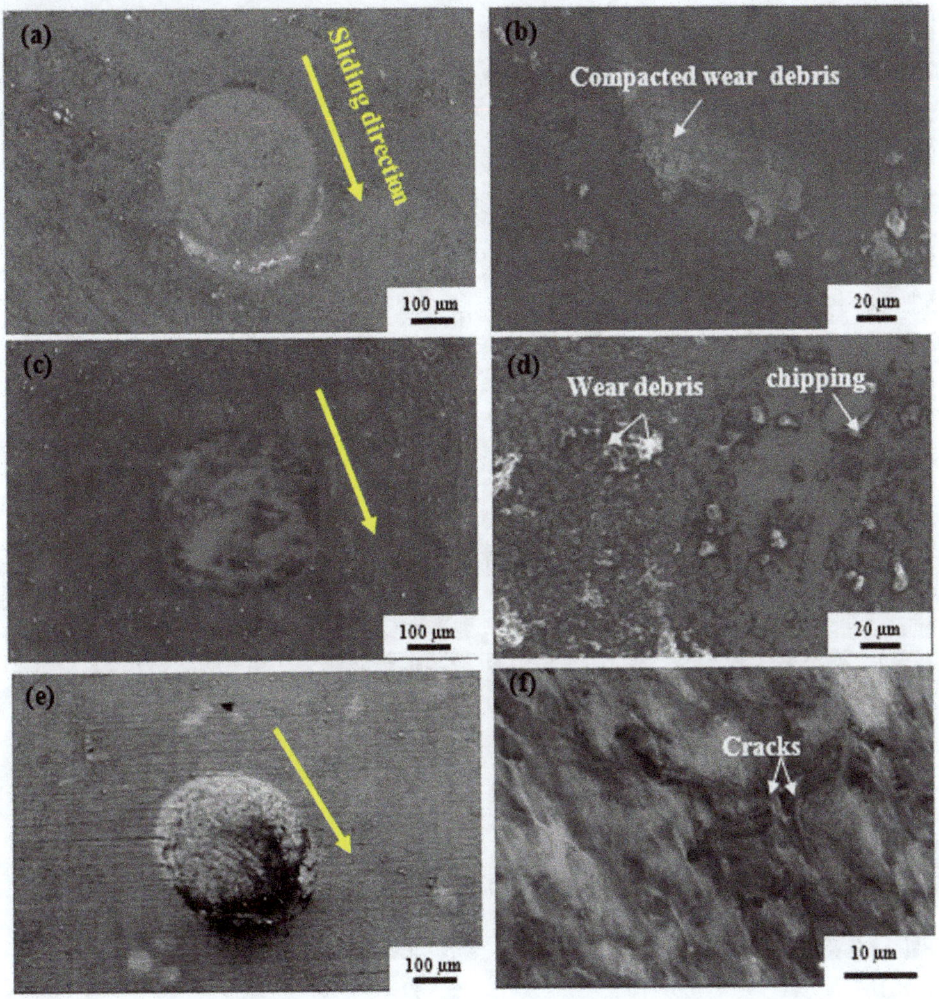

Figure 12 reports the SEM images of the impacted coatings wear scars at an impact angle of 45°. Opposite to the wear scars obtained at an impact angle of 90° on the surface coating, the overall view of the worn scars changed and the indentation size increased. This clearly indicates two distinct zones: the impact zone and the sliding zone (Figure 12.a, c and e). Indeed, the circular shape generated from the impact motion is followed by a continuous path with a width inferior to the diameter of the impacted area. It is noticeable that the impacted area and the end of the rectangular path are surrounded by compacted wear debris. A higher magnification of the wear scars of pure Ni-B and that of the Ni-B-TiO_2 powder composite coating highlights the cracks covering the totality of the compacted area (Figure 12.b and d). The formation of the wear debris is more visible on the worn surface of the Ni-B-TiO_2 powder coating compared to the other worn scars based on the compacted area. Similar to the wear scar obtained at an impact angle of 90°, the worn surface of the Ni-B-TiO_2 sol showed an oxidized area containing very fine cracks. Furthermore, this oxidized area contains a detached material.

Figure 12. SEM observation of the worn surface of (a, b) pure Ni-B coating, (c, d) Ni-B-TiO_2 powder coating and (e, f) Ni-B-TiO_2 sol coating after an impact sliding test at an impact angle of 45°

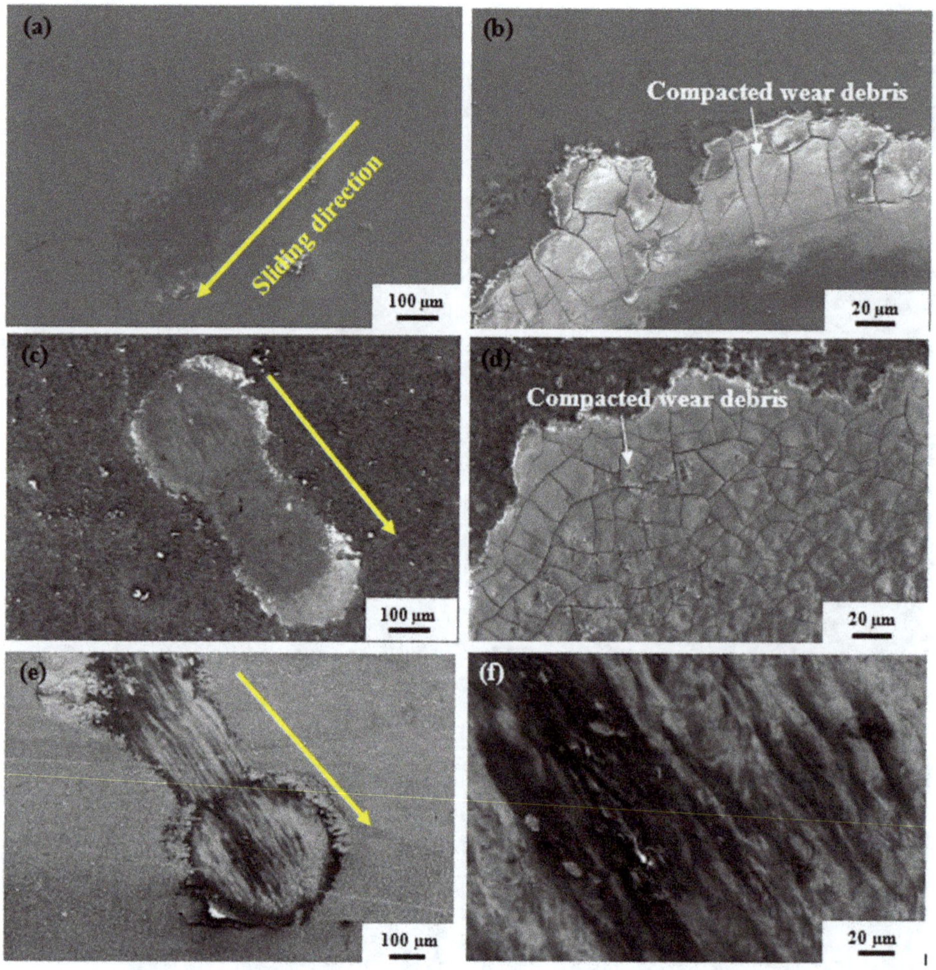

Figure 13 shows the variation of the impact load against the number of cycles for two cases, namely under 45° (figure 13a.) and 90° (figure 13b.) impact angles. These curves describe the responses of all considered coatings. It was found that pure Ni-B and Ni-B-TiO_2 sol present similar impact loads under a 45° impact angle. However, the Ni-B-TiO_2 powder demonstrates lower impact load values which could be attributed to its lower hardness. All these coatings present a similar variation according to the number of cycles i.e. there is a beginning step in which there is an impact load increase followed by a steady step where the impact load has a constant value. Such behavior could be explained by a structure modification in the first step induced by a plastic deformation, especially encouraged by the sliding loading. This led to the wear of a material and the establishment of a complete contact area.

However, under a 90° impact angle where the impact is normal on the contact surface, there is no sliding loading. The recorded curves demonstrate important impact load values compared to the ones obtained under the 45° impact load. The impact load at an impact angle of 90° is almost two times greater than that at an impact angle of 45°. Indeed, the evolution of the impact load under the 90° impact angle presents an oscillating evolution against the number of cycles. It varies from 240 to 260 N. This behavior could be explained by a pure elastic impact on a material with high stiffness. Such a trend is more significant for the coating with TiO_2 particles.

Figure 13. Impact force evolution during impact-sliding test of Ni-B coating, Ni-B-TiO_2 powder and Ni-B-TiO_2 sol nanocomposite coatings at contact angle of (a) 45 and (b) 90

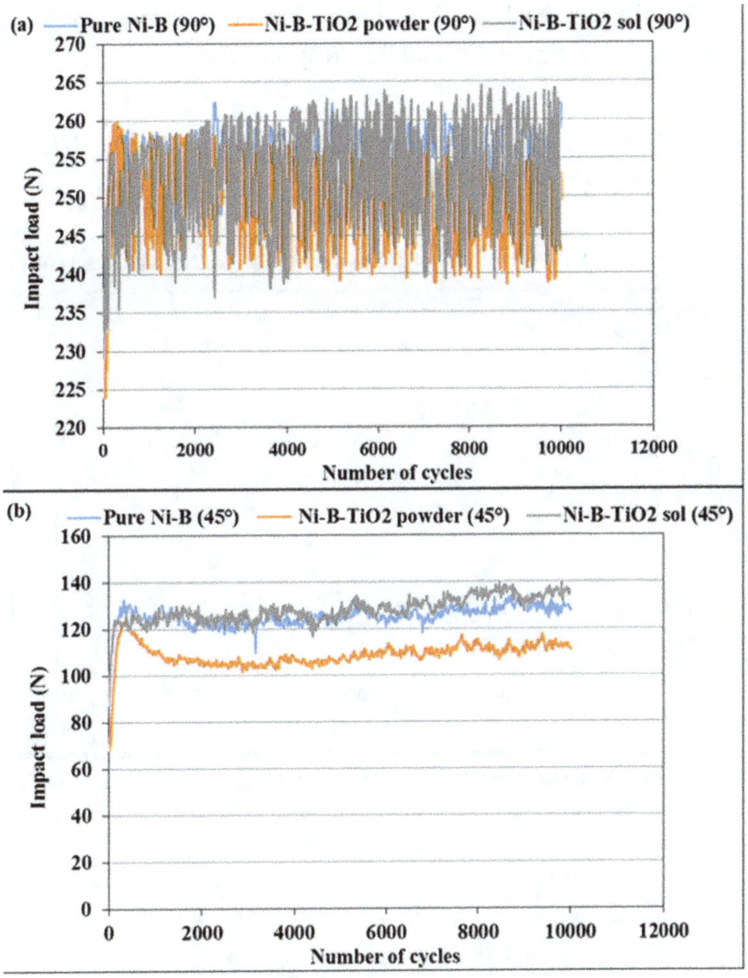

The wear scars profiles of a worn surface generated on the Ni-B coating, Ni-B-TiO_2 powder and Ni-B-TiO_2 sol nanocomposite coatings at the end of the sliding-impact test at an impact angle of 45° and 90° are presented in Figure 14a and Figure 14b, respectively. The analysis of the interferometer of each impact trace is crucial to quantify the endurance of each coating to resist impact loading by quantifying the wear depth. The wear scars profiles also permit the detection of some aspects of damage like the presence of groves, craters, bulges and material accumulation on worn traces. It is noticeable that the generated cicatrices at impact angle of 45° present more bulges than those generated at an impact angle of 90°.

Regardless of the impact angle, the Ni-B-TiO_2 sol nanocomposite coating exhibits the lowest wear depth as opposed to the Ni-B-TiO_2 powder, which shows the highest wear depth. Indeed, the maximum wear depth of Ni-B-TiO_2 sol reached 2 µm at both impact angles, whereas the maximum wear depth of the Ni-B-TiO_2 powder ranged between 6.2 µm and 17 µm at an impact angle of 45° and 90°, respectively. For the impact angle of 90°, the wear scars profile of the Ni-B-TiO_2 powder is characterized by a large diameter in comparison with the other two deposits. Furthermore, the wear scar profiles of this composite coating are distinguished by apparent edges at their sides due to the wear debris accumulation (figure 11.c and d). With regard to the pure Ni-B coating, the maximum wear depth shows very close values of about 5 µm and 6 µm at impact angles of 45° and 90°, respectively.

Figure 14. Wear track profiles of pure Ni-B coating, Ni-B-TiO_2 powder and Ni-B-TiO_2 sol nanocomposite coatings at contact angle of (a) 45 and (b) 90

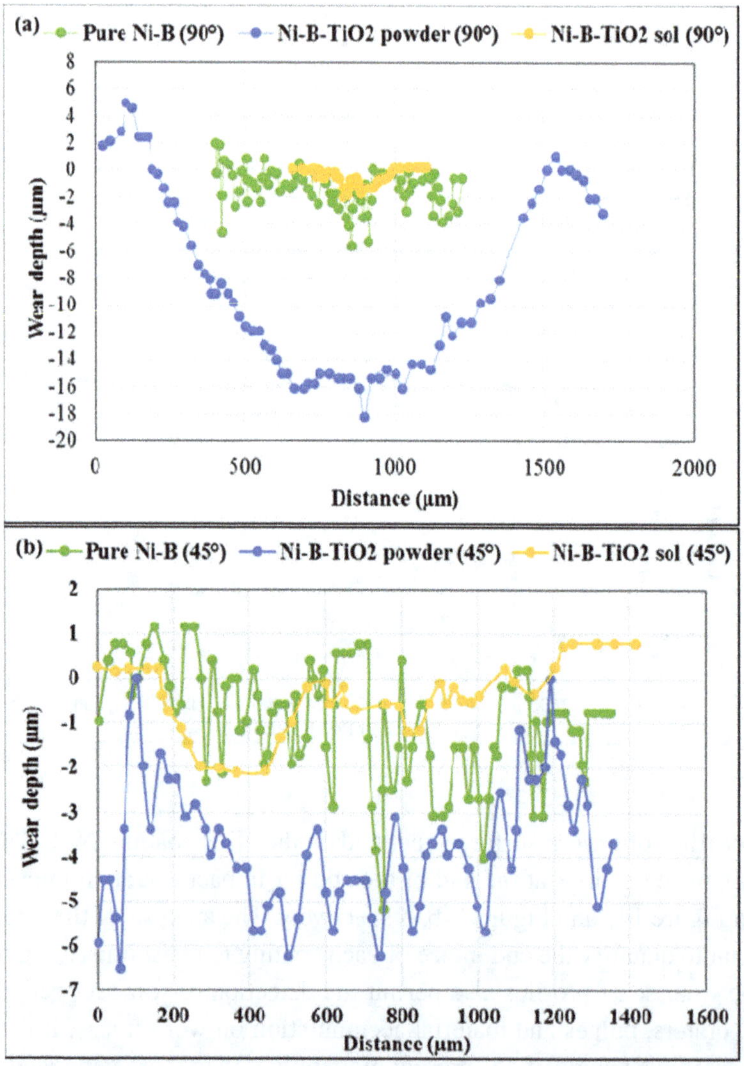

Based on the above-mentioned results, it seems that the wear impact resistance of the pure Ni-B coating and that of the Ni-B-TiO_2 sol nanocomposite coating are not mainly influenced by the impact angle as they showed similar wear depths. This is attributed to their higher microhardness as well as their greater stiffness, leading to a mostly elastic behavior of both coatings with a very little plastic deformation, indifferently reacting against impact angles (table 2). On the contrary, an interesting plastic deformation took place on the impacted surfaces of the Ni-B-TiO_2 powder coating owing it to its ductility, which comes from its lower microhardness. In addition to that, the stiffness of this deposit is lower than that of the Ni-B and Ni-B-TiO_2 coatings (table 2). As a result, the Ni-B-TiO_2 powder behaves plastically against repeated impact loadings. Therefore, the impact angle intervenes in the impact wear resistance of this coating, hence the relevant difference between both wear depths obtained at different impact angles. The findings of the wear track profiles of the studied coatings are in agreement with the evolution of the impact load curves at both impact angles.

CONCLUSION AND FUTURE ASPECTS

In this paper, the codeposition of TiO_2 nanoparticles into the Ni-B matrix using two different methods (conventional method and sol-enhanced method) was successfully performed on a heat-treated P20 tool steel substrate. The effect of the incorporation of TiO_2 nanoparticles in the Ni-B matrix on the morphology, composition and mechanical properties as well as the abrasive wear resistance and impact wear resistance of the elaborated nanocomposite coatings were largely discussed.

The surface morphology of Ni-B coating changes from a nodular and dense structure to a cauliflower-like and porous structure via the incorporation of solid TiO_2 nanoparticles through the conventional method. Meanwhile, the surface morphology of the Ni-B-TiO_2 sol composite coating preserves almost the same morphology of the Ni-B coating due to the sol-enhanced method. The incorporated TiO_2 nanoparticles by the conventional method is clearly visible in the EDS plot of the Ni-B-TiO_2 powder coating as well as in the cross–section where the TiO_2 nanoparticles appear in the form of aggregates. In contrast, the TiO_2 nanoparticles are barely detected in the EDS plot of the Ni-B-TiO_2 sol composite coating while they are completely absent in the cross-section of the Ni-B-TiO_2 sol composite coating. This proves that the higher distribution of TiO_2 nanoparticles resulted from the introduction of TiO_2 sol in the Ni-B electrolyte. On the contrary, adding solid TiO_2 nanoparticles directly into the Ni-B bath hinders the dispersive strength of the coating. The microindentation findings clearly demonstrate the relation between the mechanical properties of the studied coatings and their morphology. The pure Ni-B and Ni-B-TiO_2 sol composite coatings provide a much higher microhardness and an important Young's modulus thanks to their compact structure, offering a promising bond strength between the grain boundaries of the coatings. However, the porous structure as well as the TiO_2 agglomeration into the Ni-B-TiO_2 powder nanocomposite coating provokes a lower microhardness and stiffness.

The abrasive wear resistance of the pure Ni-B coating, Ni-B-TiO_2 powder and Ni-B-TiO_2 sol nanocomposite coatings was explored using a pin-on-disk tribomoter test to evaluate the evolution of the wear endurance of the coatings under spherical counterbody (alumina ball) through time. The multi-pass scratch test was also carried out to assess the coating wear resistance against a much sharper indenter (Rockwell Type-C indenter). The results obtained from the pin-on-disk test reflects the outstanding abrasive wear resistance of the composite coating elaborated with the sol-enhanced method. This novel process greatly improves the strength endurance of the Ni-B matrix by preventing the detachment of the coating from the P20 substrate, unlike both pure Ni-B coating and Ni-B-TiO_2 powder nanocomposite coating, showing coating removal even in an earlier sliding time. However, the use of the conventional process in developing the Ni-B-TiO_2 composite coating decreases the friction coefficient of the Ni-B matrix contrary to the use of the novel process. These findings are consistent with the multi-pass scratch test results. The codeposition of the TiO_2 nanoparticles with the traditional method in the Ni-B coating causes a total deterioration as the scratch trace of the Ni-B-TiO_2 powder exhibits spallation failure within the groove at a number of passes of 100. Under the same conditions, only plastic deformation takes place along the scratch trace of the Ni-B-TiO_2 sol nanocomposite coating, while buckling failure appears at the one side of the scratch trace, indicating a separation between the coating and the substrate. This also proves the higher strength of the composite coating, capable of withstanding the sliding motion against a sharper indenter without destruction. Such a response reflects directly the adhesive strength of the coating. The resulted wear rate during both abrasive wear tests decreases with the sliding time. This originated from the formation of a tribo-oxide layer between the materials in relative motion. Furthermore, the wear rate evolution allows the distinction between the susceptibility of each codeposition process to strengthen the wear resistance of the Ni-B matrix.

The impact-sliding test at the impact angles of 45° and 90° elucidates a close relationship between the impact response and the mechanical behavior of each studied coating. On one hand, the higher stiffness of the pure Ni-B coating and the Ni-B-TiO_2 sol composite coating leads to a predominant elastic response against the impact loadings. As a result, the wear depth is independent from the impact loadings and reaches a lower maximum value, which appears to be similar for both impact angles. On the other hand, the Ni-B-TiO_2 powder composite coating exhibits a sufficient ductility, suitable for a plastic impact response, depending on the impact loading level and the impact angle. Moreover, regardless of the tested coating, the impact angle of 45° creates a more important wear area, composed of circular zones followed by groove traces. Meanwhile, only circular areas take place on the impacted surface at the impact angle of 90°.

To sum up all of the above-mentioned results, the main concept to retain is, firstly, the difference between the form and the way of introducing TiO_2 sol in the Ni-B solution shows a total dissimilarity in the surface morphology, the mechanical properties and the wear response of the Ni-B matrix

Second, it has been proven that the use of the sol-enhanced method associated with the electrodeposition process results in a crucial improvement of the tribological behavior of the Ni-B matrix. Therefore, the introduction of a solution of TiO_2 sol into the electrolyte bath is more convenient than adding solid TiO_2 nanoparticles in the same bath. Besides, this new process is more suitable for the industry as it is more cost effective: it is cheaper than the conventional method and easier to manipulate.

This study is considered as a first step to elucidate and clarify the advantages of the sol-enhanced method in comparison to the conventional one, with respect to the dispersive strength of the incorporated TiO_2 nanoparticles in the Ni-B matrix and the evolution of the tribological response of the composite coatings. Further work could be interesting so as to develop a numerical model for a better clarification of the experimental test results, for either the abrasive or impact wear test.

REFERENCES

Bahri, A., Guermazi, N., Elleuch, K., & Ürgen, M. (2015). Tribological performance of TiN coatings deposited on 304L stainless steel used for olive-oil extraction. *Wear*, *342–343*, 77–84.

Beltowska-Lehman, E., Indyka, P., Bigos, A., Szczerba, M. J., & Kot, M. (2015). Ni–W/ZrO_2 nanocomposites obtained by ultrasonic DC electrodeposition. *Materials & Design*, *80*, 1–11.

Ben Saada, F., & Elleuch, K. (2015). Damage of stainless steel components by olive paste. *Tribology Transactions*, *59*, 856–864.

Bull, S.J., & Berasetegui E. (2006). An overview of the potential of quantitative coating adhesion measurement by scratch testing. *Tribology Interface Engineering*, 99-114.

Chen,, W.W., & Gao, W. (2011). Microstructures and properties of sol-enhanced nanostructured metal-oxide composite coatings. *Progress in Material Science*, *21*, 355–362.

Frank, I. W., Tanenbaum, P. M., & van der Zande, A. M. (2007). Mechanical properties of suspended graphene sheets. *Journal of Vacuum Science & Technology B. Microelectronics and Nanometer Structures Processing*, *25*, 2558–25610.

Gomez-Navarro, C., Burgard, M., & Ken, K. (2008). Elastic properties of chemically derived single graphene sheets. *Nano Letters*, *8*(7), 2045–2049.

Gyftou, P., Stroumbouli, M., Pavlatou, E. A., Asimidis, P., & Spyrellis, N. (2004). Tribological study of Ni matrix composite coatings containing nano and micro SiC particles. *Electrochimica Acta*, *50*(23), 4544–4550.

Hamid, Z. A., & Ghayad, I. M. (2002). Characteristics of electrodeposition of Ni-polyethylene composite coatings. *Materials Letters*, *53*(4-5), 238–243.

Kallel, M., Zouch, F., Antar, Z., Bahri, A., & Elleuch, K. (2017). Hammer premature wear in mineral crushing process. *Tribology International*, *115*, 493–505.

Kallel, M., Antar, Z., Masseoud, M., Vesco, S., Barletta, M., & Elleuch, K. (2021). Comparative investigation of scratch resistance and tribological performance of Ni–B–TiO2 composite coatings prepared by conventional and novel processing methods. *Ceramics International, 47*(10) Part A, 14438-14454. doi:10.1016/j.ceramint.2021.02.023

Kallel, M., Masseoud, M., Antar, Z., Fridrici, V., Barletta, M., & Elleuch, K. (2021). Tribological properties of Ni-B-TiO_2 sol composite coating elaborated by sol-enhanced process: abrasive wear and impact wear. *Journal of Materials Research and Technology, 1*(3), 857 – 871.

Messaoud, M., Houmard, M., Briche, S., Roussel, F., & Langlet, M. (2010). Hydrophobic functionalization of cotton-based textile fabrics through a non-fluorinated sol–gel route. *Journal of Sol-Gel Science and Technology*, *55*, 243–254.

Mohajeri, S., Dolati, A., & Ghorbani, M. (2016). An investigation on the electrodeposition mechanism of Ni-TiO_2 nanocomposite coatings. *Journal of Ultrafine Grained Nanostructure Materials*, *49*, 51–63.

Ogihara, H., Wang, H., & Saji, T. (2014). Electrodeposition of Ni–B/SiC composite films with high hardness and wear resistance. *Applied Surface Science*, *296*, 108–113.

Oliver, W. C., & Pharr, G. M. (1992). An improved technique for determining hardness and elastic modulus using load and displacement sensing indentation experiments. *Materials Research Society. Journal of Materials Research, 7*(6). Advance online publication. doi:10.1557/JMR.1992.1564

Poot, M., & van der Zant, H. S. J. (2008). Nanomechanical properties of few-layer graphene membranes. *Applied Physics Letters, 92*(6), 063111.

Shen, J., Yin, W., Wei, Q., Li, Y., Liu, J., & An, L. (2013). Effect of ceramic nanoparticle reinforcements on the quasistatic and dynamic mechanical properties of magnesium-based metal matrix composites. *Journal of Materials Research, 28*, 1835–1852.

Singh, S. K., Chattopadhyaya, S., Pramanik, A., Kumar, S., & Gupta, N. (2018). Influence of Nanoparticle on the Wear behaviour of Thin Film Coatings (A Review). *International Journal of Applied Engineering Research., 13*(6), 4053–4058.

Wang, Y., Wang, S. J., Shu, X., Gao, W., & Lu, W. (2014). Preparation and property of sol-enhanced Ni–B–TiO_2 nano-composite coatings. *Journal of Alloys and Compounds, 617*, 472–478.

Yang, X. K., Li, Q., Zhang, S. Y., Zhong, X. K., Dai, Y., & Luo, F. (2010). Electrochemical corrosion behaviors and protective properties of Ni–Co–TiO_2 composite coating prepared on sintered NdFeB magnet. *Materials and Corrosion, 61*(7), 618–625.

KEY TERMS AND DEFINITIONS

Crushing: Is a mineral process aimed at reducing the size of rocks to be further processed.

Electroplating: A process which consists of depositing a metallic coating on a substrate through electrochemical reaction. The metal ion, through a solution, migrate from positive electrode (anode) to negative one (cathode).

Friction: A force resists to the motion (sliding or rolling) of two surface of solids in contact.

Plastic Deformation: An irreversible deformation takes place in a solid and changes its shape without fracture under the action of applied stress exceeding the yield strength of the material.

Sol-Enhanced Coating: A composite metallic coating reinforced by ceramic nanoparticles that formed in-situ into the bath electroplating.

Spalling: Failure of metallic surface in which the surface of metal breaks into small pieces because of different reasons like an excessive sliding or impact loadings.

Wear: Is a damage of a solid surface caused by friction or excessive use.

Chapter 11
Tribo-Corrosion Behaviour and Characterization of Biocompatible Coatings

Amol Bajarang Chavan
Sanjeevan Engineering and Technology Institute, Panhala, India

Sanjaykumar S. Gawade
Rajarambapu Institute of Technology, Islampur, India

Digvijay G. Bhosale
Dr. D. Y. Patil Institute of Technology, Pimpri, India

ABSTRACT

Commercially available metallic orthopaedic implant materials cause major problems like stress shielding and the release of harmful ions due to corrosion and wear. Also, the secondary operation is a must for the implant removal. Therefore, the biodegradable and biocompatible magnesium (Mg) implant materials have been investigated. Mg shows favorable biological properties and matching mechanical properties with the natural bone. Mg alloys rapidly corrode in the physiological environment, which cause failure of the implant before completing the expected function. Surface coating is the most effective method for improving the corrosion performance of Mg and its alloys. Hydroxyapatite (HA), being the most stable phase of calcium phosphates in physiological conditions, is preferred as a coating material. The chapter focuses on the tribo-corrosion and characterization of HA coatings prepared by electrodeposition process on Mg alloys. The results are useful for the designer community in the selection of biocompatible coatings and process parameters to maximize the life of bio-implants.

DOI: 10.4018/978-1-7998-9683-8.ch011

INTRODUCTION

Biomaterials interact and coexist with biological systems to repair or replace or augment the diseased tissue, organ or function of the body. The biomaterials must be biocompatible, i.e. they should not produce any harmful or undesirable effect. The biomaterials include metallic, ceramic, polymeric and composite materials (Ramakrishna et al., 2010; Park & Kim, 2003). Biomaterials that are used in orthopaedic applications are generally called as 'orthopedic biomaterials' or 'implants'. Orthopedic biomaterials are mainly employed in osteosynthesis, arthroscopy, joint replacement, dental implant reconstruction and spine stabilization. This chapter focuses on the implants used in osteosynthesis. Osteosynthesis is fixation and repair of fractured bone (Hallab & Jacobs, 2020; Rodríguez-González, 2009). A bone fracture is break or disruption in normal architecture of bone. Whenever bone fractures, it needs a surgery to implant extra material, which can support the fractured bone and helps to heal it. The biomaterials used as orthopaedic implants include bone plates, rods, screws, pins, wires, etc. They are used to serve the purpose of bone support, fracture fixation or total joint replacement.

The biocompatibility and mechanical properties are two most important factors for selection of a material as orthopaedic implant. The outstanding mechanical properties and acceptable biocompatibility of metallic materials make them a potential candidate for orthopaedic application (Niinomi, 2002). The traditional metallic implants used to support fractured bone during its healing process are stainless steel, titanium and cobalt-chromium alloys. Though these materials are having excellent mechanical properties, they induce some problems in bone healing process. In comparison with natural bone, they have higher elastic modulus. Due to this, the implant material carries more external load than the bone, which leads to loss of density of healing bone and bone becomes weaker. This undesirable effect is called as 'stress shielding effect' (Nagels et al., 2003). Also, the corrosion and wear of these metallic implants in physiological environment, releases harmful ions which cause inflammation and detrimental tissue reactions causing tissue damage (Jacobs et al., 1998; Lhotka et al., 2003). Besides this, these implants are biocompatible but not biodegradable. They provide only temporary support; therefore, they must be removed by secondary surgery after bone healing process is over (Park & Kim, 2003). This causes loss of financial resources and may additionally induce post operative infections. Therefore, in order to reduce such complications, the use of degradable and biocompatible metallic implants has been investigated.

Magnesium (Mg), which is biocompatible and biodegradable, has been studied extensively to replace traditional metallic implants. Magnesium is degraded gradually in the physiological environment and is replaced by the healing tissue. Mg^{2+} ions, released during the degradation process, are harmless to the human body. Also, the biodegradability of Mg eliminates the need for secondary surgery. But the corrosion rate of Mg in the body is very high in the physiological environment, which may lead to loss of mechanical integrity of Mg-based implant before the completion of the bone healing process. Surface coating is a preferential method used to delay the corrosion of Mg-based implants. It is mandatory that, the coating material should also be biocompatible and biodegradable. The HA is a natural element of human bone and teeth. But HA cannot be used for load-bearing applications due to its intensive brittleness. Therefore, it is used as a coating material on Mg-based substrate. This chapter focuses on the electrodeposition of hydroxyapatite (HA) on the surface of Mg-based implants to delay its corrosion in a physiological environment.

MAGNESIUM AS ORTHOPAEDIC MATERIAL

Mg is lightweight alkaline earth metal. It is biocompatible and biodegradable. Mg is an essential nutrient required to carry out many biological functions in the human body. The mechanical and biological properties of Mg can be utilised to replace traditional orthopaedic implants.

Mechanical Properties of Mg

In comparison with natural bone, traditional orthopaedic implants have high modulus and density. Whereas, Mg is a lightweight metal with a hexagonal packed structure with a density of 1.738 g/cm^3. It has a high strength to weight ratio with good ductility and castability (Baker & Avedesian, 1999; Emley, 1966). As compared to traditional metallic implants, the density, elastic modulus and yield strength of Mg-alloys are closer to bone tissue. The elastic modulus of commercially available Mg alloys is 41-45 GPa, which is nearer to the bone. This controls the stress shielding effect which is a limitation of traditional metallic implants. Mg shows the matching mechanical properties with natural bone, which distinguishes it from traditional metallic implants. Thus the matching mechanical properties and strength to sustain applied external loads make the Mg and its alloys a potential candidate for biomedical and orthopaedic applications. Table 1 shows the comparison of the mechanical properties of metallic implants with natural bone. It illustrates that, as compared to traditional metallic implants, the properties of Mg are nearer to the bone. Thus, the Mg can overcome the limitations of traditional metallic implants used for orthopaedic applications.

Table 1. Properties of metallic implant materials in comparison to natural bone (Staiger et al., 2006; Guan et al., 2012)

Material	Density (kg/m^3)	Elastic Modulus (GPa)	Yield Stress (MPa)	Compressive Yield Stress CYS (MPa)
Natural bone	1800-2100	3-20	104.9–114.3	130-180
Magnesium alloys	1740-2000	41-45	90-170	65-100
Titanium alloys	4400-4500	110-117	760–880	758-1117
Cobalt chromium alloys	8300-9200	230	310	450-1000
Stainless steel alloys	7900-8100	189-205	200–300	170-310

Biological Properties of Mg

Mg is one of the most important nutrient elements present in the human body. An average 70 kg human being has 21-28 gm of Mg present in the body. The daily allowable intake of Mg is 320-420 mg. It has been reported that Mg is required for normal muscle contraction, healthy immune system, steady heart rhythm, strong teeth, bone and muscles (Gröber et al., 2015). It regulates blood pressure, blood sugar levels and insulin metabolism (Saris et al., 2000). Thus it prevents and manages diabetes and cardiovascular diseases. It is an essential element in more than 300 enzymatic reactions. It is required for

DNA and RNA synthesis as well as in protein synthesis (Hartwig, 2001). Mostly, the Mg available in the body is found in bones and teeth. Mg plays important role in bone remodelling. It can accelerate the growth of new bone tissue. The Mg^{2+} ions present in the physiological environment helps to accelerate the bone healing process by promoting Calcium deposition. Almost there is no evidence indicates the toxicity of Mg (Saris et al., 2000).

Corrosion Behaviour of Mg

The corrosion behaviour of implants in physiological conditions plays a vital role in maintaining their mechanical strength. From the fracture of a bone, it needs 12-16 weeks for healing and restoring most of the bone to its original strength. It means that an orthopaedic implant should maintain its mechanical property at least for 3 months to avoid failure of the implant due to rapid loss of integrity of the Mg implant. Therefore, it becomes necessary to study the corrosion behaviour of the implant material. The corrosion of Mg alloys in aqueous environment proceeds in the following reaction produces H_2 gas, OH^- ions and magnesium hydroxide (Zhou & Lee, 2011).

$$Mg \rightarrow Mg^{2+} + 2e^- \quad (1)$$

$$2H_2O + 2e^- \rightarrow H_2 + 2(OH)^- \quad (2)$$

$$Mg^{2+} + 2(OH)^- \rightarrow Mg(OH)_2 \quad (3)$$

The $Mg(OH)_2$ formed on the surface of Mg, acts as a protective layer (Shaw, 2003). But this layer is not so strong to protect the Mg surface from corrosion. The human body fluids contain high levels of Cl^- ions, which adversely affects the corrosion protection of Mg-based alloys. $Mg(OH)_2$ layer loses stability and it is converted into soluble MgCl when the chloride concentration is above 30 mmol/L (Shaw, 2003). The presence of Chloride ions is detrimental to the corrosion resistance of Mg alloys. The following chemical equations represent the corrosion of Mg in a physiological environment

$$Mg(OH)_2 + 2Cl^- \rightarrow MgCl_2 \quad (4)$$

$$Mg + 2H_2O \rightarrow Mg(OH)_2 + H_2 \quad (5)$$

$$Mg + 2Cl^- \rightarrow MgCl_2 \quad (6)$$

The $MgCl_2$ layer is formed in a chloride-containing environment. This layer is reported to be non-toxic, biocompatible and moderately soluble. In a physiological environment, the corrosion process occurs rapidly. This faster degradation leads to the excessive release of H_2 gas (Song & Atrens, 2000). This slows down the healing process and causes the embrittlement of the metal. Further, the bone is poorly vascularised and transports excessive hydrogen gas. This results in the formation of H_2 gas pockets near the implant. Rapid corrosion causes local alkalisation as OH^- ions increase pH which hampers biocompatibility. The higher corrosion rates lead to a reduction in mechanical strength and thereby causing failure of the implant. Therefore, to achieve the desired performance of Mg implants for an expected period, the enhancement of corrosion resistance is strongly needed.

To understand and evaluate the corrosion performance of Mg-based biodegradable implants various in-vivo and in-vitro corrosion tests are performed. For in-vitro corrosion analysis, EIS (Electrochemical Impedance Spectroscopy), immersion and H_2 evolution tests are carried out. The artificial body fluids are used for in-vitro tests. The extensively used artificial body fluids are Simulated Body Fluids (SBF), Artificial Plasma (AP), Hank's solution, Dulbecco's Modified Eagle's Medium (DMEM) etc.

CORROSION CONTROL METHODS

For biodegradable implants, the corrosion rate should be such that, the implant retains its strength till the bone healing process is over. The rapid corrosion can cause weakening of the implant and further its failure may occur. As discussed earlier, the implant should maintain its mechanical integrity at least for three months. Pure Mg corrodes rapidly in physiological conditions. Besides the loss of strength, it also suffers from the high amount of H_2 gas evolution, local alkalisation and high levels of Mg^{2+} ions in body fluids. Therefore, an effective strategy must be utilized to enhance the corrosion resistance of Mg-based alloys. Figure 1 shows the methods of improving corrosion resistance of Mg-based alloys. The improvement in the corrosion can be significantly achieved by alloying and surface coating. The alloying elements influence the mechanical properties and corrosion resistance of Mg-based alloys. Surface coatings can be an effective strategy to delay the initial corrosion of Mg-based implants.

Figure 1. Methods of improving corrosion resistance of Mg based biodegradable implants

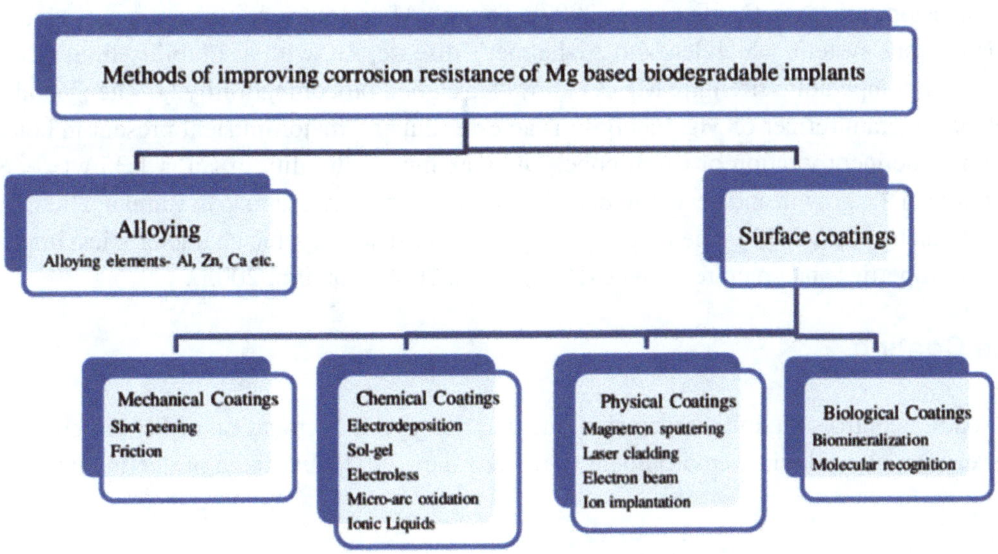

Alloying Elements of Mg

Alloying improves the mechanical and corrosion properties of Mg-based implants (Atrens, et al., 2022; Witte et al., 2005; Kannan & Raman, 2008). As the degradation is going to occur in the human body, the basic requirement of alloying elements is that they should produce no toxicity. Therefore, Mg alloys used as biomedical implants are alloyed with nutrient elements (Seiler et al., 1988). These elements improve the mechanical and corrosion properties by solid solution strengthening and grain refinement of base Mg metal. Aluminium (Al), zinc (Zn), manganese (Mn), calcium (Ca), strontium (Sr), tin (Sn), silicon (Si) and most rare earth elements (REE) have proven to increase corrosion resistance. For alloys, the solubility limit is the maximum concentration of the solute atoms that can dissolve within the solvent without the formation of a second phase rich in solute (Mouritz, 2012, p. 89). Table 2 indicates the maximum allowable daily intake and solubility limits of the most investigated alloying elements of Mg alloys for biomedical application.

Table 2. Solubility limits and Maximum Daily allowable dosage of the alloying elements in magnesium matrix (Radha & Sreekanth, 2017; Kirkland et al., 2011)

Element	Solubility Limits (wt. %)	Maximum Daily Allowable Dosage(mg)
Zn	6.2	15
Ca	1.35	1400
Mn	2.2	2.3-2.6
Si	0.003	20-25
Al	12.70	14
Sr	0.11	5
Y	12.5	0.016

AZ (Al-Zn) alloys have attracted the attention of many researchers due to their commercial availability. AZ31, AZ61, AZ91 are common Mg-Al-Zn alloys studied for biomedical application. Many researchers have claimed that Al, with a maximum solubility of 12.7 wt. %, aids in improving mechanical properties and corrosion resistance (Müller et al., 2007). Though Al is a nutrient element, excess intake of Al harms the nervous system, which leads to Alzheimer's disease. Zn, with a solubility limit of 6.7 wt. % improves room temperature strength. Mg-Zn alloys have good biocompatibility (J. Zheng et al., 2010). Ca, an effective grain refiner of Mg-Ca alloys is an essential and major nutrient present in bone. Mg is necessary for the incorporation of Ca in bone. The maximum solubility of Ca is 1.35 wt. %. Si plays important role in the growth and development of the immune system (Kannan & Raman, 2008; E. Zhang et al., 2010; Zhang et al., 2011). The REE, which exhibit anti-carcinogenic characteristics, improves the mechanical properties and creep resistance (Rettig et al., 2008; Rokhlin, 2003).

Surface Coating

Surface coating controls the initial corrosion rate and delays the degradation of Mg-based implants in a highly reactive physiological environment (Gray & Luan, 2002). In this, a protective layer of metal, polymer or ceramic is deposited on the Mg substrate. The coating helps to enhance the chemical, biological, morphological and surface engineering properties of the substrate. Coating characteristics, such as adhesion, thickness and corrosion behaviour must be considered for the selection of coat material and deposition technique.

The most important coating characteristic is the adhesion of the coating material with the substrate. This can be measured by using a universal testing machine (UTM) or by scratch test. The coating material should adhere to the substrate; it is a basic requirement of coating. The coating formed can be dense or porous; it can be amorphous or crystalline. In porous coatings, the body fluids may infiltrate through the coat and corrode the substrate material. There is no such infiltration found in dense coatings. Amorphous coatings generally have a high dissolution rate whereas crystalline coatings provide a low dissolution rate therefore, they can provide long term stability. The thickness of coating plays an important role in delaying initial corrosion. The coating thickness may vary from millimetres to a few nanometres. But higher thickness can cause detachment of outer layers leading to poor adhesion. The parameters of the deposition process decide the stability and performance of coatings. If the deposition parameters are not selected with care, it may lead to several undesirable issues like the possibility of separation of coat layer, fatigue failure of coat and premature coating dissolution which ultimately cause coating failure. Therefore, proper care should be taken while selecting process parameters of coat deposition.

CaP (Calcium Phosphates), a well-known family of bioceramics, have received the attention of many researchers for biomedical applications (Shahmohammadi & Khazaei, 2021; Narayanan et al., 2008). This chapter focuses on the coating of hydroxyapatite (HA), which is the most stable phase of CaP ceramic in physiological pH, on Mg alloy substrate.

HA COATING

HA is a naturally occurring mineral present in the human body with the chemical formula $Ca_5(PO_4)_3(OH)$ / $Ca_{10}(PO_4)_6(OH)_2$. Being the most stable phase of CaP in physiological pH (7.2-7.5) HA is mostly the preferred coat material for biomedical implants (Posner & Betts, 1975). HA has already proven biomedical applications such as bone tissue engineering, orthopaedic, dentistry and drug delivery systems (Narayanan et al., 2008). The properties of HA and coat deposition methods are discussed in the following sections.

Biological and Mechanical Properties of HA

HA has a similar chemical composition as that of the bone. It is the main component and essential element of bone and teeth, which provides rigidity. It shows the ability to bind chemically with bone. When implanted, newly formed bone binds directly to HA through a carbonated calcium-deficient apatite layer at the bone-implant interface (Dorozhkin, 2009; LeGeros, 2008). HA surface supports osteoblastic cell adhesion, growth, and differentiation, and new bone is deposited by the creeping substitution from the adjacent living bone (Kattimani et al., 2016). It is osteoinductive and osteoconductive. Osteoinduction is part of normal bone healing and is responsible for the majority of newly formed bone, e.g. after a fracture or the insertion of an implant (Albrektsson & Johansson, 2001). Osteoinductivity, which is normally used to demonstrate the ability of the implant to induce bone growth in an ectopic site (Lu et al., 2018). Osteoconduction is a phenomenon where afer bioactive materials are implanted into the bone environment, bone tissue will grow along the surface or internal pore of the implant (Lu et al.,

2018). HA is biocompatible, bioactive and biodegradable CaP. Ca and P are the main components. Stoichiometric HA has a monoclinic crystal system with a (Ca/P) ratio of 1.67. But it is found difficult to achieve an exact stoichiometric ratio, based on this, various phases of HA include, Ca-deficient-HA, Oxy-HA and carbonated-HA. As the Ca/P of bone nearly equals 1.5, Ca-def-HA has received more attention because its Ca/P ranges between 1.5-1.67. Ca-def-HA plays a vital role in new bone formation and bone remodelling. HA has a density of 3.16 g/cm^3. It cannot be used for load-bearing applications because of its poor mechanical properties (Zhou & Lee, 2011). It is very brittle in nature. Therefore, it is used as a coat rather than as an implant.

HA Deposition Methods

To obtain a coating with desirable characteristics, the selection of an appropriate deposition method is very important. The features such as substrate material, coat material, design and geometry of the component and required coating thickness decide the method to be selected. The currently investigated deposition methods are plasma spraying, ion implantation, physical vapour deposition, sputtering sol-gel coating, chemical deposition and electrodeposition, etc (Abdullah et al., 2013; Surmenev, 2012; Liu et al., 2004; Makwana & Bhingole, 2018; Zheng-Zheng et al., 2020; Rahman et al., 2020).

Methods involving high temperatures cannot be used with Mg substrate as its melting point is low and they are poor heat resistant. Besides this, HA may decompose at higher temperatures. Therefore, chemical and electrochemical methods are generally preferred because they are carried out at room or moderately high temperatures. The electrodeposition of HA on Mg substrate has been discussed in detail in subsequent sections.

ELECTRODEPOSITION OF HA ON MAGNESIUM SUBSTRATE

The overview of elctrodeposition process and steps followed for the deposition of HA on Mg-based substrate material has been discussed in this section.

Overview of HA Electrodeposition

Electrodeposition method is based on principle of electrolysis. It is used to obtain metallic coatings by the action of an electric current on a conductive material immersed in a solution containing a salt of the metal to be deposited (Basile et al., 2010). Electrodeposition is popular method for coating because of its simplicity and low cost. It is generally carried out at ambient temperature, so even substrates with low melting points can be coated suitably with this method. It can be used with any shape of the component. The thickness and chemical composition of coating can be easily controlled by choosing suitable process parameters (Narayanan et al., 2006). The important parameters of electrodeposition are current, voltage, electrolyte concentration, pH and temperature. The modes of electrodeposition are potentiostatic, galvanostatic and pulsed reverse current mode. Figure 2 depicts the HA deposition procedure on Mg substrate.

Figure 2. Electrodeposition of HA on Mg substrate

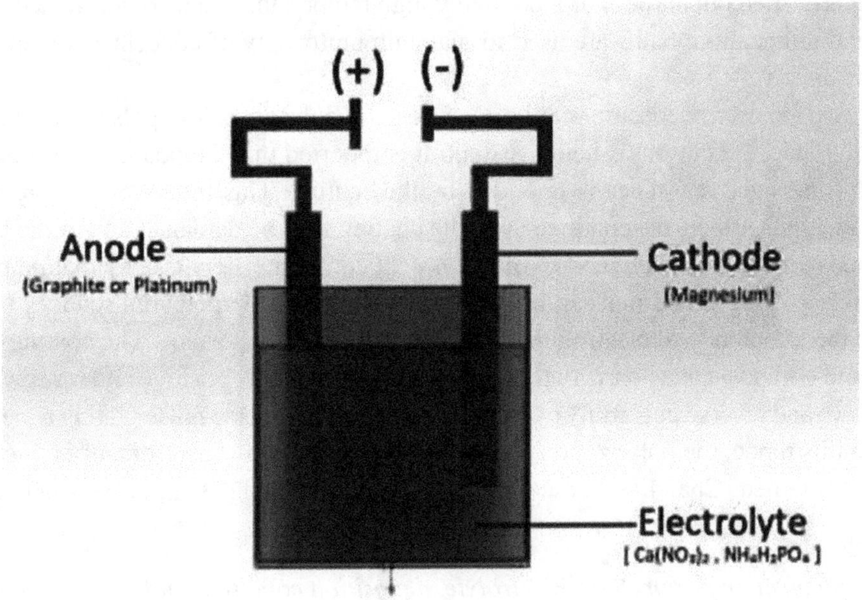

In electrodeposition of HA on Mg substrate, Mg alloys act as a cathode and as an anode graphite or platinum material is used. The electrolytes used in various researches mainly consisted of calcium nitrate, $Ca(NO_3)_2$ and ammonium dihydrogen phosphate, $NH_4H_2PO_4$. The electrolyte parameters that influence the coating characteristics are adjusted accordingly. These parameters mainly include pH, concentration and temperature of the electrolyte. The values of voltage, current are maintained as per the requirement.

Major Steps of Electrodeposition

The major steps for obtaining HA coatings on the Mg-based substrate are sample preparation, pre-treatment, electrodeposition and post-treatment. The deposition steps are followed sequentially. First of all sample preparation is carried out. The sample is cut into a suitable shape, thickness and size. Further, it is ground to obtain uniform roughness. The ultrasonic cleaning is carried out for degreasing the sample. Then, it is activated with HF solution and finally dried by drying apparatus. Some researchers pre-treated the sample in NaOH solution for 1-2 hours at 80 °C (Kannan & Orr, 2011; Guan et al., 2012).

The actual electrodeposition is carried out after the pre-treatment process is over. Table 3 lists an overview of electrodeposition parameters used in various researches. It indicates the electrolyte compositions, deposition conditions as well as microstructure of HA coatings obtained in various experiments. The electrodeposition is carried out in two-electrode cells or three-electrode cells. In the two-electrode cell, graphite or platinum material is used as the anode while Mg alloy substrate acts as the cathode. In three-electrode cells, graphite or platinum electrode acts as a counter electrode, saturated calomel electrode (SCE) acts as reference electrode and Mg serves as working electrode. The electrolyte used in the process mainly contains $Ca(NO_3)_2$, $NH_4H_2PO_4$ and H_2O_2. The pH of electrolyte is obtained between the range 4-6. The pH of the electrolyte is adjusted by dilute HNO_3, NH_4OH and $(CH_2OH)_3CNH_2$. A suitable deposition time is selected, in most of the researches, it ranges between 30 minutes to 2 hours. The deposition is carried out at room or moderately high temperatures. The other devices such as magnetic stirrers and temperature baths are used to maintain uniformity of concentration and temperature of electrolyte respectively.

The modes used for electrodeposition can be potentiostatic, galvanostatic or pulse reverse current (PRC). In galvanostatic mode, the current is held constant. It is reported that the potential of the Mg electrode decreases with time since resistance is provided by the coating. This reduction in potential increases the H_2 evolution which affects the coating. Even the coating may be damaged due to the bursting of H_2 bubbles. Whereas, in potentiostatic mode, voltage is kept constant. Therefore, by maintaining optimum voltage, desirable coating properties can be achieved. But this static potential reduces the rate of ion diffusion from the main body of electrolyte solution to the substrate surface. The coatings obtained are loose, porous and with low adhesion with the substrate. In PRC mode, positive and reverse pulse current densities, positive and reverse pulse duty cycles and positive and reverse pulse plating times are adjusted accordingly. In this mode, the polarization of electrolytes is reduced. Therefore, effective and adherent coatings can be obtained (Chandrasekar & Pushpavanam, 2008; Peng et al., 2006; Lin et al., 2003).

Table 3. Electrodeposition parameters- Electrolyte, deposition condition and microstructure

Alloy With Dimension of Deposition	Electrolyte	Deposition Conditions	Post Treatment of As-deposited Samples	Microstructure and Crystal Structure	Ca/P Ratio	Reference
AZ91D 30×25×5 mm^3	0.1 M $Ca(NO_3)_2$, 0.06 M $NH4H_2PO_4$, H_2O_2 10 ml/l	pH 4.3, 4 V, 2 h, room temperature	1 M NaOH solution for 2 h at 80 °C	*As-deposited* two kinds of crystal characteristics. One appeared regular flake-like structure diverging from centre toward periphery other appeared irregular flake-like structure with different dimension. *Post treatment*- uniform flake-like morphology HA not compact	NA	(Y. W. Song et al., 2008)
AZ31 30×10×8 mm^3	0.042 mol/L $Ca(NO_3)_2$, 0.025 mol/L $NH4H_2PO_4$ and 0.1 mol/L NaNO3	pH 5, current density 0.4 mA/cm^2, 85 °C, 1 hr	0.2 M NaOH solution for 4 h at 80 °C	*As-deposited* DCPD, radiated plates-like phase, *Post-treated*HA single phase completely, needle-like phase with needle-like particles 1000 nm in length and 35 nm in diameter *Both*-chrysanthemum flowers aggregated on the surface Uneven thickness and aggregation	As-deposited 1.40 Post-treated 1.50	(Wen et al., 2009)
Mg–Zn–Ca alloy 25×10×5 mm^3	0.042 mol/l $Ca(NO_3)_2$, 0.025 mol/l $NH4H_2PO_4$ and 0.1 mol/l NaNO3	pH 5.0, 30 min, 85 °C, the positive and reverse pulse-\current densities are 10 and 20 mA/cm^2, duty cycles 0.1 and 0.5, and plating times 10 and 2 ms, respectively	…	*Two different morphologies observed*: a *tabular region* composed of an irregular flake-like structure approximately 100 nm in thickness and *a fluffy region* composed of a fine acicular structure less than 100 nm in diameter, which partly covers the plate-like coating. Crystallite size 40 nm, No cracks observed	1.33	(Wang et al., 2010)
Mg-Zn, 3 coatings obtained DCPD, HA and FHA.	0.042 mol l^{-1} $Ca(NO_3)$_$4H_2O$and 0.025 mol l^{-1} $NH4H_2PO_4$ in deionized water. *DCPD*- adding 10 ml l^{-1} H_2O_2 into the electrolyte for 4 h. *FHA*-The FHA coating was directly electrodeposited by adding NaNO3 (0.1 mol l/1) and NaF (2 mmol l 1) at 60 ± 1 °C into the electrolyte.	pH 4.4 *DCPD*- 5 mA/cm2, *HA*- 80 °C for 4 h. *FHA* 0.5 mA/cm2,, 60 ± 1 °C into the electrolyte	$Ca(OH)_2$ and 0.1 mol/l NaOH solution at 80 °C	*DCPD*a flake-like structure, flakes are 5–20 µm in size, and the thickness of the coating is about 20 µm *HA* inherits the flake-like feature of DCPD . The large crystal layer is loosely packed and small plates disperse among these large flakes. The thickness of the HA coating is about 5–10 µm. *The FHA* two layers The inner layer closely packed columnar crystals, hexagonal cross-section width 100–200 nm. the external crystals are irregularly arranged with spheres, branches or even broccoli-like structures. The prepared FHA coating is no more than 3 µm thick.	DCPD- slightly higher than 1.00 HA- 1.54 FHA- 1.48	(Song et al., 2010)

Continued on following page

Table 3. Continued

Alloy With Dimension of Deposition	Electrolyte	Deposition Conditions	Post Treatment of As-deposited Samples	Microstructure and Crystal Structure	Ca/P Ratio	Reference
AZ91 30×30×5 mm³	0.1M $Ca(NO_3)_2$ and 0.06M $NH4H_2PO_4$	−2, −3 and −4 V w.r.t. SCE	1 M NaOH solution	The coating after alkaline treatment at both voltages (i.e. −2 and −3 V) showed rod-like shaped particles; however, a few clustered particles were also observed on the sample coated at −3 V.	NA	(Kannan & Orr, 2011)
Mg–Zn–Ca 25×10×4 mm³	0.15 mol/L NaNO3, 0.025 mol/L $NH_4H_2PO_4$, 0.042 mol/L $Ca(NO_3)_2$ and 20 mL/L H_2O_2. For FHA coatings, 2 mmol/L NaF	65 °C, Avg. current density 1 mA/cm², pH 5.0, room temperature	--	*TED* coating comprised columnar crystals, and the average diameter of columnar crystals was about 0.5μm. *PRC* closely packed rod-like nano-crystals structure. The diameters of the rods ranged from 60 to 120 nm	TED- 1.30 PRC 1.44	(Meng et al., 2011)
Mg-4.0Zn-1.0Ca-0.6Zr 20×20×1 mm³	0.05 mol/L $Ca(NO_3)_2$, and 0.1 mol/L $NH4H_2PO_4$	pH 4.5 to 4.6 at 25 C, 2 h Three temperatures 25, 50, 75 °C Three voltages 5, 10, 15 V	250 mL of 0.5 mol L 1 NaOH, and the temperature was maintained at 80 °C, 2 hr	*Preteatment-* Loosely packed spherical particles *50°C, 10V* posttreated- dense, uniform coating without cracks coating at *5V and 15V* showed less desirable microstructures compared to the coating deposited at the voltage of 10 V, when the temperature was set at 50°C. *50°C, 5V-* surface cracks and loose bonding. Particle size increased when the deposition temperature increased to 75 °C.	NA	(Guan et al., 2012)
AZ91D 45×16×5 mm³	0.1mol/L $Ca(NO_3)_2$ and 0.06mol/L $NH4H_2PO_4$. 10mL/L H_2O_2	ph- 4.3, current density 2-2.5 mA/cm², 2 hr, room temperature	HA- 1 M NaOH solution for 2 h at 80 °C, SAHA- 0.1mol/L stearic acid solution with ethanol as solvent at 50 °C for 2h	Microstructure of HA and SAHA both are similar. Composed of porous black surface. *SAHA* more compact and uniform	posttreated- 50 °C, 10 V 1.77. 50 °C, 5 V 1.30 50 °C, 15 V 1.22	(Zhang et al., 2012)
CP-Mg	0.1 M $Ca(NO_3)_2$, 0.06 M $(NH_4)_3PO_4$, 10 mL/L of 30 vol.% H_2O_2	pH: 4.0, temperature:27 °C, 100 mV vs saturated calomel electrode (SCE) for 1 h	1 M NaOH solution for 2 h at 80 °C	*As-deposited* coating regular flake-like structure diverging from center toward periphery and an irregular flake-like structure with different dimensions. *Posttrated* uniform flake-like morphology	NA	(Jamesh et al., 2012)
AZ31	3mM $Ca(H_2PO_4)_2$ + 7mM $CaCl_2$ solution	Ph 6 for 30 min. 37 °C The electrodeposition was performed at −1.4, −1.6, and −1.8V	--	Flake like particles observed on all. Crystallite becomes finer with increasing cathodic potential to more negative side.	.	(Salman et al., 2013)
Mg-Zn-Ca	0.042 mol·L-1 $Ca(NO_3)_2$, 0.025 mol·L⁻¹ $NH4H_2PO_4$ and 0.1 mol·L⁻¹ $NaNO_3$,	pH 5, positive and reverse pulse-current densities 10 and 20 mA/cm², duty cycles 0.1 and 0.5, plating times 10 and 2 ms, The deposition process lasted for different time (0-30 min) at 80±1 °C.	--	*For 30 min* deposition, the acicular structure has gradually transformed into flake one, and adjacent curled flakes lap over on one joint to construct the macroporous structure, fluffy precipitations composed of fine acicular structure are observed on the coating, considered as the just formed crystals.	1.33	(Wang et al., 2014)
Mg-1.2Ca-4.5Zn 15×10×10 mm³	$Ca(NO_3)_2$ 0.042 mol/L), $NH4H_2PO_4$ (0.025 mol/L), $NaNO_3$ (0.1 mol/L) and H_2O_2 (10 mol/L)	*PVD* room temperature sputtering pressure of 0.24 Pa, RF sputtering current of 200 W, 90 min, and bias voltage of -150 V. *ED* pH 5.0 by adding HNO_3 and $(CH2OH)_3CNH_2$ at room temperature, *FHA* 60 °C current 4 mA/cm² 1 h. The addition of NaNO3 leads to an enhancement of the ionic strength. FHAp coating was soaked in 100 ppm AgNO3, room temperature for 30 min for Ag+ deposition	--	*Nano-SiO_2* coating possesses a uniform surface within which nanometer-sized SiO_2 particles are dispersed along with bubble-like structure that was formed within the surface. Porosity observed in the SiO_2 layer, The cracks are more than 1 μm long, deep. *Ag-doped FHAp coating* as top layer - large plate-like crystals grew perpendicularly to the substrate, presence of small rounded Ag nano-particles collected in the form of agglomerates which are generally close to each other which are precipitated within the lamellar plates. *nano-SiO_2 coating* has a much smoother surface compared to the Ag/FHAp coating	NA	(Bakhsheshi-Rad et al., 2016)
Mg-2.1Zn-0.22Ca, 5 mm in height and 12 mm in diameter	calcium nitrate ($Ca(NO_3)_2$.4H_2O, 0.042 mol/L), ammonium dihydrogen phosphat ($NH4H_2PO_4$, 0.025 mol/L) and sodium nitrate (NaNO3, 0.1 mol/L) in deionized water.	pH-5, at 80 °C for 20 min. *TED* process - cathodic current density 1 mA/cm² for 20 min	--	*Fluoride coating* is uniform and compact, and no microcrack is detected, *TED-HA/MgF_2* coating is uneven. micro-flake-like structure, and many agglomerated parts with pores in the center. *PRC-HA/MgF_2* compact nano-rod-like structure and covers the fluoride coating surface completely. The diameter of rod-like crystals is ranging from 50 to 80 nm.	TED-HA/ MgF_2 1.21 to 1.29 PRC-HA/MgF_2 1.50	(Feng et al., 2017)

The post-treatment of the as-deposited Mg substrate is the final step for obtaining HA on the surface. The potentiostatically and galvanostatically deposition coatings mainly consist of dicalcium phosphate dihydrate (DCPD) and β-tricalcium Phosphate (β-TCP) Ca-P phases. Post-treatment is carried out to convert these phases into HA. The NaOH solution is used to serve this purpose. Mostly, it is carried out by immersing as-deposited samples in NaOH at 80 °C for 2-4 hrs. It means that a two-step method is used to obtain HA on Mg substrate potentiostatically or galvanostatically. Whereas, PRC mode is a single-step deposition method. When a high cathode current density is applied, the cathodic polarization of the Mg alloy leads to an increase in the pH at the interface between the alloy and the electrolyte due to the formation of OH^- ions. This sudden increase in pH triggers crystal nucleation and initiates crystal growth of the desired CaP phase directly on the substrate surface (Wang et al., 2010, pp. 1746). Therefore, post-treatment is not needed, in the case of PRC mode.

The electrodeposited HA on Mg-based implants show outstanding biocompatibility, osteoinductivity and osteoconductivity which enhance the healing rate of affected bone tissue. The coatings show long term stability as compared to bare/uncoated Mg.

CHARACTERIZATION, CORROSION BEHAVIOR, CYTOTOXITY AND MECHANICAL PROPERTIES OF HA COATINGS

This section discusses the microstructure, composition and crystal structure of as-deposited and post-treated HA coatings. The in-vitro and in-vivo degradation behaviour and the improvement in corrosion resistance are illustrated. The bioactivity, cytocompatibility and mechanical properties of HA coatings are discussed. The effect of composite coats is finally discussed.

Microstructure and Characterization of HA Coatings

The composition, crystal structure and morphology influence the stability of coatings in human body fluids. The instruments like scanning electron microscope (SEM), X-ray diffractometer (XRD), Energy dispersive spectroscopy (EDS) are utilized to understand microscopic properties and composition of the coatings obtained by the electrodeposition process. The galvanostatic and potentiostatic deposition processes are commonly refered as traditional electrodeposition (TED) process. The TED process mainly deposits DCPD, TCP and octacalcium phosphate (OCP) crystals on Mg substrate (Song et al., 2010). The main constituents of as-deposited coatings are highly soluble in the severe physical environment. Therefore, the alkaline treatment is carried out to convert them into HA. The post-treated coatings contain HA (Wen et al., 2009). HA is the most stable phase amongst all CaPs, in a physiological environment. The alkaline treatment is accompanied by a reduction in weight, volume and thickness of coatings (Song et al., 2010; Jamesh et al., 2012).

As-deposited coatings developed in the TED process mainly contains flake-like structures diverging from the centre towards the periphery and irregular flake-like structures with different dimensions (Jamesh et al., 2012). The post-treated coatings show uniform flake-like or needle-like morphology. In TED, loose, porous and low adhesive coatings were developed. The obtained HA coatings were not compact. Also, uneven thickness and aggregation were observed (Wen et al., 2009). The defects observed in the coating deposited by TED are mainly due to two reasons. First reason is polarization in concentrations of electrolyte is formed, since the speed of ion diffusion from the main body of the solution to the surface of the metallic substrate is too slow. The other reason is due to H_2 produced at the cathode due to reduction of H_2O (Wang et al., 2010, pp. 1743-1744). These limitations can be overcome by utilizing the PRC mode of deposition. The coatings obtained with PRC mode were having closely packed rod-like nanostructures. The coating was compact, uniform and with no microcracks. Compared to flake-like structure, rod-like structures provides much better corrosion protection due to dense and uniform structure and provide favourable conditions for deposition of apatite owing to increased contact area with surrounding fluid after implantation (Feng et al., 2017). In addition, Lu et al. (2008) reported that the nanosized HA is more beneficial in bone cell adhesion and proliferation.

Ca/P ratio of coatings plays an important role in osteoinduction and bone regeneration. The stoichiometric Ca/P ratio of HA is 1.67 and that of the natural bone is nearly 1.5. The HA coatings with Ca/P ratio for coatings between 1.33 to 1.65 is considered as Calcium deficient HA (Ca-def-HA) coatings. Dorozhin (2002) reported that Ca-def-HA could be more conducive to inducing the formation of new bone. Ca-def-HA is most soluble and induces bone regeneration (Wang et al., 2011). The Ca-def HA meets the requirements for biodegradable implants, because many studies indicated that, the dissolution of HA in the human body after implantation is too slow to achieve optimal results, while Ca-def-HA seems to be more soluble and may induce precipitation of new bone-like apatite after implantation (Dorozhkin, 2002; Monteiro et al., 2003).

Degradation Behavior

To understand degradation behavior and corrosion resistance of coatings Electrochemical impedance spectroscopy (EIS), immersion and H_2 evolution tests are performed in simulated body fluids (SBF).

EIS

The EIS test is performed with classical three electrode cells. In this, platinum electrode acts as counter electrode, SCE as refernce electrode and Mg substrate acts as working electrode. The test solution used is SBF at human body conditions i.e. pH = 7.4, temperature = 37 °C. the samples are immesed in solution prior to test to establish open circuit potential (OCP). The test results indicate the degradation performance of the HA coated Mg based substrates in SBF.

Wen et al. (2009) prepared HA coatings on AZ91 alloy substrate by catodic electrodeposition method. The deposition parameters and morphology are as stated in table 3. The electrochemical tests indicated that E_{corr} of the substrate increased significantly from -1.62 to -1.42 V after the surface modification with HA. The polarization curves showed that the I_{corr} of substrate reduced from 2.51 x 10^{-5} to 2.51 x 10^{-8} A. this indicated that the alkali treatment could help to improve the corrosion resistance of the substrate. This improvement in corrosion resistance shows that the coatings with alkali treatment have better stabilty. Similar results were obtained by Wang et al. (2011).

Wang et al. (2011) deposited Ca-def-HA coating on Mg-Zn-Ca alloy substrate by pulse electrodeposition. The potentiodynamic polarization tests were performed in Kokubo's SBF. The results indicated that the corrosion potential of Mg alloy substrate increased from -1645 to -1414 mV, whereas the corrosion current density decreased from 110 to 25 μA/cm^2. This proves there is significant increase in the corrosion resistance of the substrate due to obtained HA coating.

Meng et al. (2011) obtained fluorine-doped hydroxyapatite (FHA) coating by TED and PRC deposition methods. They compared the corrosion protection capacities of TED and PRC coated Mg alloy substrates. The E_{corr} of Mg alloy substrate increased from -1.67 to -1.58 V and I_{corr} value reduced from 126.89 to 6.31 μA/cm^2. Similarly, E_{corr} of Mg alloy substrate deposited by PRC method was increased to -1.51 V and I_{corr} value reduced from 2.51 μA/cm^2. From the results, it can be stated that, the corrosion rate of PRC coatings was lowest amongst three. Similar results were obtained by Feng et al. (2017).

According to the various EIS tests performed in the literature, it can be stated that the HA coated Mg alloys show improvement in corrosion resitance. Also, the alkali treatment of as-deposited coatings enhance the degradation performance in SBF. Further, as compared to TED, PRC mode of deposition is more attractive method for HA deposition (Wang et al., 2011; Feng et al., 2017).

Immersion Test

It is reported that the Mg alloys suffer severe corrosion attacks in Cl^- ions containing fluids. The immersion test is performed in SBF on HA-coated Mg alloy substrate to evaluate the corrosion protection capacity and bioactivity of the coating. It was observed that the degradation of the Mg alloy substrate occurs in a cyclic process: pitting degradation – pit enlargement and surface flakes break-off – then pitting degradation restarted again at newly exposed surfaces (Guan et al., 2012).

Y. W. Song et al. (2008) conducted an immersion test in SBF for 48 hrs. the uncoated samples suffered a severe corrosion attack after immersion in SBF. The surface was covered with plenty of white corrosion products and numerous cracks were observed on the surface. Whereas HA-coated samples suffered the corrosion attack to some extent. The flake-like HA coating was dissolved in SBF to some extent. But the coating was intact and corrosion could not penetrate the coating. Wen et al. (2009) and Wang et al. (2011) obtained similar results.

Guan et al. (2012) produced a novel biodegradable Mg-Zn-Ca-Zr alloy by series of metallurgical processes. Further, they electrodeposited HA coating on the alloy by electrodeposition process. The immersion tests were carried out in SBF. The samples were characterized after 3, 7 and 15 days of immersion. As soon as the uncoated alloy samples were immersed in SBF, they showed visible gas bubbles on the surface. After 5 hrs of immersion, the surface turned black and white degradation pits were observed on the surface. After 3 days of immersion in SBF, a thin layer of degradation products was observed on the surface of uncoated samples. Degradation pits of an average of 1μm were found on the surface, but still, some intact areas were left. After 7 days, white degradation products covered the entire surface and flakes started to break off. The degradation pits were further enlarged and a few intact areas were left on the surface. Finally, after 15 days, pits became much larger and more interconnected. Whereas for HA-coated all samples, after 3 days, degradation pits of 6 μm average diameter appeared on the surface, but still, some intact areas could be observed on the surface. After 7 days, the samples were severely corroded in SBF. The almost entire surface was covered by the enlarged and interconnected pits. In deep degradation pits, the white degradation products were observed. After 15 days of immersion, the HA-coated samples were severely corroded. Also, the further pit enlargement caused a break-off of degradation products in the pits. This corrosion behaviour of HA was contradictory to earlier researches. Here, the degradation of HA coatings was faster than the substrate. But, at the early stages of biodegradation (< 24 hrs), the HA-coated alloy samples showed a slower degradation rate than uncoated alloy samples. It is well known that the rapid initial degradation of the surface of the implant promotes osteolysis and bone tissue regeneration adversely affected. Therefore the HA-coated samples provide favourable conditions for bone regeneration as compared to uncoated samples.

Thus the HA coatings control and decrease the initial degradation rate which is a crucial requirement of the Mg alloy substrate to be used for biodegradable implants application.

H_2 Evolution

The H_2 evolution tests are performed to evaluate the degradation performance of biodegradable implants. The test is carried out in SBF maintained as per physiological conditions. The H_2 evolution occurs as a result of cathodic reduction of H_2O. the bubbles formed at substrate may hamper the coating characteristics. Even bursting of H_2 bubbles occurs so the coating is damaged. The H_2 evolution also increases the pH of the medium.

Song et al. (2010) fabricated three kinds of CaP coatings, namely, DCPD, HA and FHA on Mg-Zn alloy by electrodeposition process. The H_2 evolution test was carried out in m-SBF. Due to fast degradation, the H_2 evolution of uncoated samples was observed to be highest than the three coatings. DCPD coated samples showed a sharp acceleration in H_2 release from the 6th to 9th day of immersion, as the substrate was exposed to the solution. For the first week of immersion, as compared to DCPD, the HA-coated samples showed faster H_2 evolution but the rate was relatively stable. Overall, FHA coated sample showed the lowest degradation rate. During the first week of immersion, the DCPD samples were having the lowest increase in pH. But the DCPD coatings are not suitable for the long term. For long term applications, FHA coatings are showed better performance. After two weeks of immersion, the FHA coated samples indicated long term corrosion resistance. Also, the increase in pH value was lowest than uncoated, DCPD, HA-coated samples.

Meng et al. (2011) compared pH variation in TED and PRC coated samples. During the early stages of immersion, both the coatings showed a rapid increment in pH. But, as compared to PRC, TED coatings showed a higher increase in pH. The corrosion in SBF resulted in a higher pH rise and more H_2 evolution. But compared with pH value of uncoated Mg alloy substrate in reference (Yang & Zhang, 2009), both, TED and PRC coated samples showed lesser increase in pH. Thus, it is clear that HA coating on Mg alloys is beneficial to retard the initial degradation. Also, they impart structural stability in long term applications.

In-Vivo and In-Vitro Biocompatibility Test

The implant materials must be harmless and should produce no toxicity. The biocompatibility and cytotoxicity tests evaluate whether the implant is safe for its use in the human body. In-vitro tests are carried out in artificial physiological conditions. Whereas, in in-vivo tests, the material is directly implanted into a living organism.

Biomineralization

For orthopaedic implants, during bone remoulding, a stable and favourable bone-implant interface is very important (Franz et al., 2011). Rapid degradation of the implant during bone remoulding causes the precipitated CaP mineral to break apart. Thus the surface bioactivity is hampered. The osteoconductive minerals needed to be formed on the implant surface to obtain osseointegrated interface (Song et al., 2010). The interface activity between the implant surface and body fluids is important to induce bone formation.

Song et al. (2010) performed mineralization experiments on DCPD, HA, and FHA coated Mg-Zn alloy substrate in m-SBF. After one week of immersion, DCPD coated implants showed precipitates with cotton-like structures distributed on the coating. The Ca/P molar ratio was found to be 1.09 to 1.31. For HA-coated samples, the ball like precipitates with Ca/P molar ratio 1.29 to 1.40 was observed. Whereas FHA coated samples showed in-situ growth and hemispherical crystals, with an average Ca/P molar ratio of 1.57, spring up from inner columnar crystals. After the first week, the HA and FHA coated samples provided the nucleation sites for apatite/ β-TCP. The HA and FHA coatings have the higher Ca/P ratio of the precipitates. This indicates that HA and FHA coated samples improve surface bioactivity.

In-vitro Cytotoxicity

The in-vitro tests are performed in simulated physiological environment outside the body of living organism. These tests are used to determine the cytoxicity of biomaterials before actually implanting the biomaterial in the body of living organism.

Guan et al. (2012) performed cytotoxicity and hemolysis tests to understand the biocompatibility of the HA-coated Mg-Zn-Ca-Zr alloy. Cytotoxicity tests were performed on uncoated and HA-coated Mg alloy. The L-929 fibroblasts cultured with the extracts of the HA-coated and uncoated Mg-4.0Zn-1.0Ca-0.6Zr (wt %) alloy samples for 1, 2, and 3 days. No significant difference in fibroblast morphology was observed between the HA-coated sample group and the uncoated sample group. Based on the thorough visual examinations of cell morphologies, the cytotoxicity score of the HA-coated and uncoated Mg-4.0Zn-1.0Ca-0.6Zr (wt %) alloy samples was assessed. All samples had a cytotoxicity score within acceptable limit after 1–3 days of culture. Also, the evaluated hemolysis rate for both the samples was less than 5%, thus the requirement as an implant material is achieved.

Li (2011) evaluated in-vitro indirect cytotoxicity of the ions released including F ions by extracts using 3-(4,5)-dimethylthiahiazo(-z-y1)-3,5-di-phenytetrazoliumromide (MTT) test and morphologies' observation on human bone marrow stromal cells. The study was carried out after 2, 4 and 7 days of culture. After the first 2 days' and 4 days' culture, there was no significant difference among the cells in the dilutions and the negative controls. But when the 7th day arrived, higher cells viabilities appeared compared with the control group. According to the absorbance value, the appearance of the higher cells viabilities was not because of the nature reduction of the cells in normal cells culture. On the 2nd and 7th day normal spindle morphology for the negative control and dilution groups was observed. While a little numbers of twisted ones appeared for the toxic positive controls. As a result, the ions released during the degradation exhibited no toxicity to hBMSCs from both the quantitative and visual way. The results indicated that the ions released were beneficial for the cells proliferation avoiding the pH factors.

In-vivo Test

These tests are performed by actually implanting the biodegradable material into a living organism. The results obtained are more realistic as the implant is exposed to real physiological conditions.

Li et al. (2011) prepared FHA coating on Mg-Zn alloy by electrodeposition. The coating thickness was 2-8 μm. The comparative study was carried out by performing in-vivo animal tests for coated and uncoated implant samples. The samples were implanted into the femoral condyle of an adult Newzealand rabbit. The artificial bone defect of diameter 2.5 mm was created and the cuspidal material of diameter 2.4 mm was implanted into the defect. Thus the gap of 0.1 mm was maintained to evaluate interface bioactivity. The implants were removed after one month of implantation. The corrosion morphology was studied with the help of a micro-CT scan. The SEM images of the implants were taken after the removal of the implant from the condyle. As compared to uncoated implants, more direct contacts with the surrounding tissues were found for FHA coated implants. Thus the enhanced interface bioactivity was observed with coated implants. The results concluded that the coating can be suitably used for interface improvement.

The in-vivo and in-vitro test results obtained by many researchers show that the HA coatings are biologically safe and no pieces of evidence of toxicity are found so far. HA-coated implants are biocompatible and can serve the purpose of bone healing without inducing any toxicity in the body.

Mechanical Properties of HA Coatings

The mechanical integrity of biodegradable implants plays a crucial role in the structural stability of adverse physiological conditions. The Cl^- ions cause rapid degradation of the Mg-based implants and cause a loss in mechanical strength. This leads to the failure of the implants before serving the purpose they are meant for. Therefore, it is necessary to study and understand the changes in mechanical properties of Mg-based implants.

Wang et al. (2010) performed slow strain rate tensile (SSRT) tests to evaluate mechanical properties of Ca-def-HA coated Mg-Zn-Ca alloy. The tests were performed in m-SBF at 36.5 ± 0.5 °C. Cylindrical tensile specimens with gauge dimensions of 10 mm in length and 4 mm in diameter were used. These samples were pulled at a strain rate of 2.16×10^{-5} mm/s until fracture. For comparison, SSRT of Mg–Zn–Ca alloy was also carried out in air at the same strain rate. The data for the ultimate tensile strength (UTS) and time of fracture (TOF) extracted from SSRT were used to evaluate the mechanical properties of the Mg–Zn–Ca alloy. In comparison with uncoated samples, the Ca-def-HA coated samples showed a 5.6% increment in UTS and a 16.6% increment in TOF when tested in SBF. The results indicated that the higher values of coated implants are beneficial for maintaining the mechanical strength augment the bone healing process. Further, the shear strength tests were performed to evaluate the bonding strength between the coat and Mg alloy substrate. The values of bonding strength were found to be 41.8 ± 2.7 MPa, which was greater than HA coatings obtained by the TED process. This value is even greater than the shear strength of natural cortical bone (Wang et al., 2011). Therefore, the pulsed deposition improves the mechanical properties and can be used for high load-bearing applications. The possible reasons for the improvement in bonding strength between HA coating and the substrate are when the high current density is applied, the Mg^{2+} is produced from the cathode i.e. Mg alloy substrate. They are incorporated in the coatings. This plays a vital role in increasing the adhesion strength. Also, when the reversed pulse current is applied for a short duration, the loose HA particles are resolved and only adherent HA particles continue to grow.

Kannan and Orr (2011) coated HA on AZ91 alloy at different cathodic voltages. The samples were exposed to SBF for five days. After the exposure to SBF, to obtain UTS values, the tensile tests were performed on the samples. Also, the fractured analysis of the surfaces was carried out to understand the modes of failure. The UTS values for uncoated, unexposed samples were 165 Mpa, whereas for exposed samples it was 95 Mpa. This indicates that the uncoated samples could maintain 60% of UTS in comparison with the uncoated, unexposed sample. Interestingly the HA-coated coated samples were nearly having 15% higher UTS than the uncoated samples. Fracture analysis tests indicated the modes of failure in uncoated and coated samples before and after the exposure to SBF. The uncoated, unexposed Mg alloy showed ductile tearing i.e. the ductile mode of the mode of failure was observed. A higher magnification analysis of fractures area indicated brittle mode of failure for exposed samples. For HA-coated samples, a mixed mode of failure i.e. ductile tearing and brittle features were observed. This means that even after exposure to SBF, the HA-coated alloy showed better UTS values than the uncoated.

Thus, the HA-coated samples could significantly maintain their mechanical properties after exposing them to SBF. As compared to uncoated samples, the values of mechanical properties of SBF exposed coated samples show that they can be used for long term implant application.

TRIBOLOGICAL CHARACTERISTICS OF ELECTRODEPOSITED COATINGS

Electrodeposition provides a handy and precise route to engineering coatings in tribology. Tailor-made characteristics such as wear-resistant, corrosion-resistant, magnetic etc can be achieved by electrodeposited coatings. Studies have proven that these coatings can extend the applicability of every substrate by enhancement in desirable properties such as biocompatibility, hardness, tribo-corrosion behaviour, hydrophobicity etc. The H/E ratio is one of the significantly affecting parameters that will affect tribological properties of electrodeposited coatings (Ghosh et al., 2007). Similarly, the consequences of grain measurement variant and copper content on the H/E ratio in pulse electroplated coatings were investigated (Ghosh, et al., 2010). The reduction in coefficient of friction and improvement in wear resistance is possible by the development of coatings on Ti6Al4V alloy, which will increase the applicability of alloys for its tribological usage (Pawlak & Wendler, 2009).

The thickness of coatings also affects the tribological properties. The tribological properties improve with an increase in coating thickness. The reason is contact pressure at friction surfaces had the longest duration of fidelity at some point of the wear test. Wear mechanism modified in the multilayer coating with skinny layers main to minimization of friction and coating destruction (Tokarz et al., 2008). The use of Fe-W alloy help to improve tribo-properties of coatings by a reduction in the coefficient of friction. The addition of Fe-W acts as a self lubricant at the contact surfaces during the rubbing. With electrodeposition beneath one-of-a-kind modern densities, these coatings confirmed greater wear depth (Tsyntsaru et al., 2010).

The formation of layered coatings will improve tribological characteristics of coatings by the reduction in the coefficient of friction. The coating layer will act as a barrier, which could reduce the stress range at the interface between coatings and substrate. The delamination and crack propagation can also be limited by the structure of the coating. Initial investigations of electro-deposits were frequently inspired by the possibility of creating complex tensile strength and ductility; current endeavours tend to emphasise the potential of particular optical, electronic properties, modified tribological surfaces or better corrosion resistance.

SUMMARY

HA coatings can effectively improve the corrosion resistance performance of Mg-based biodegradable implants. The electrodeposition method is the most favourable method for HA deposition as coating characteristics can be controlled easily by optimising deposition parameters. HA coatings improvise degradation performance and maintain the stability of coating for a desirable duration. HA coatings are biocompatible, osteoconductive and osteoinductive. The mechanical integrity of the substrate is maintained by HA coating. Therefore, for further development of biodegradable implants, HA coatings are needed to be studied extensively.

REFERENCES

Abdullah, C. W. (2013). Hydroxyapatite-Coated Magnesium-Based Biodegradable Alloy: Cold Spray Deposition and Simulated Body Fluid Studies. *Journal of Materials Engineering and Performance*, *22*(10), 2997–3004. doi:10.100711665-013-0589-9

Albrektsson, T., & Johansson, C. (2001). Osteoinduction, osteoconduction and osseointegration. *European Spine Journal*, *10*(0), S96–S101. doi:10.1007005860100282 PMID:11716023

Atrens, A., Shi, Z., Mehreen, S. U., Chen, X., Johnston, S., Song, G.-L., Chen, X., & Pan, F. (2022). Corrosion of Mg Alloys. Encyclopedia of Materials: Metals and Allloys, 1, 46-74.

Baker, H., & Avedesian, M. M. (1999). *Magnesium and Magnesium Alloys*. ASM International.

Bakhsheshi-Rad, H. R., Hamzah, E., Ismail, A. F., Kasiri-Asgarani, M., Daroonparvar, M., Parham, S., Iqbal, N., & Medraj, M. (2016). Novel bilayered nanostructured SiO2/Ag-FHAp coating on biodegradable magnesium alloy for biomedical applications. *Ceramics International*, *42*(10), 11941–11950. doi:10.1016/j.ceramint.2016.04.119

Basile, F., Benito, P., Fornasari, G., Monti, M., Scavetta, E., Tonelli, D., & Vaccari, A. (2010). A novel electrochemical route for the catalytic coating of metallic supports. *Studies in Surface Science and Catalysis*, *175*, 51–58. doi:10.1016/S0167-2991(10)75007-2

Chandrasekar, M. S., & Pushpavanam, M. (2008). Pulse and pulse reverse plating – conceptual, advantages and applications. *Electrochimica Acta*, *53*(8), 3313–3322. doi:10.1016/j.electacta.2007.11.054

Dorozhkin, S. V. (2002). A review on the dissolution of calcium apatites. *Progress in Crystal Growth and Characterization of Materials*, *44*(1), 45–61. doi:10.1016/S0960-8974(02)00004-9

Dorozhkin, S. V. (2009). Calcium Orthophosphates in Nature, Biology and Medicine. *Materials (Basel)*, *2*(2), 399–498. doi:10.3390/ma2020399

Emley, E. F. (1966). *Principles of Magnesium Technology*. Pergamon Press.

Feng, Y., Zhu, S., Wang, L., Chang, L., Yan, B., Song, X., & Guan, S. (2017). Characterization and corrosion property of nano-rod-like HA on fluoride coating supported on Mg-Zn-Ca alloy. *Bioactive Materials*, *2*(2), 63–70. doi:10.1016/j.bioactmat.2017.05.001 PMID:29744413

Franz, S., Rammelt, S., Scharnweber, D., & Simon, J. C. (2011). Immune responses to implants – A review of the implications for the design of immunomodulatory biomaterials. *Biomaterials*, *32*(28), 6692–6709. doi:10.1016/j.biomaterials.2011.05.078 PMID:21715002

Ghosh, S. K., Limaye, P. K., Bhattacharya, S., Soni, N. L., & Grover, A. K. (2007). Effect of Ni sublayer thickness on sliding wear characteristics of electrodeposited Ni/Cu multilayer coatings. *Surface and Coatings Technology, 201*(16-17), 7441–7448. doi:10.1016/j.surfcoat.2007.02.014

Ghosh, S. K., Limaye, P. K., Srivastava, C., & Tewari, R. (2010). Comparison of sliding wear behaviour of pulse electrodeposited Ni–Cu nanocrystalline alloys and Ni–Cu/Cu multilayers. *Transactions of the IMF, 88,* 158-162. 10.1179/174591910X12692576434653

Gray, J. E., & Luan, B. (2002). Protective coatings on magnesium and its alloys-a critical review. *Journal of Alloys and Compounds, 336*(1-2), 1–2, 88–113. doi:10.1016/S0925-8388(01)01899-0

Gröber, U., Schmidt, J., & Kisters, K. (2015). Magnesium in Prevention and Therapy. *Nutrients, 7*(9), 8199–8226. doi:10.3390/nu7095388 PMID:26404370

Guan, R.-G., Johnson, I., Cui, T., Zhao, T., Zhao, Z.-Y., Li, X., & Liu, H. (2012). Electrodeposition of hydroxyapatite coating on Mg-4.0Zn-1.0Ca-0.6Zr alloy and in vitro evaluation of degradation, hemolysis, and cytotoxicity. *Journal of Biomedical Research Part A, 100*(4), 999–1015. doi:10.1002/jbm.a.34042 PMID:22307984

Hallab & Jacobs. (2020). Orthopedic Applications. *Biomaterials Science,* 1079–1118.

Hartwig, A. (2001). Role of magnesium in genomic stability. *Mutation Research, 475*(1-2), 113–121. doi:10.1016/S0027-5107(01)00074-4 PMID:11295157

Jacobs, J. J., Skipor, A. K., Patterson, L. M., Hallab, N. J., Paprosky, W. G., Black, J., & Galante, J. O. (1998). Metal release in patients who have had a primary total hip arthroplasty. *Journal of Bone & Joint Surgery, 80*(10), 1447–1458. doi:10.2106/00004623-199810000-00006 PMID:9801213

Jamesh, M., Kumar, S., & Sankara Narayanan, T. S. N. (2012). Electrodeposition of hydroxyapatite coating on magnesium for biomedical applications. *Journal of Coatings Technology and Research, 9*(4), 495–502. doi:10.100711998-011-9382-6

Kannan & Orr. (2011). In vitro mechanical integrity of hydroxyapatite coated magnesium alloy. *Biomedical Materials, 6*(4), 11.

Kannan, M. B., & Singh Raman, R. K. (2008). In vitro degradation and mechanical integrity of calcium-containing; magnesium alloys in modified-simulated body fluid. *Biomaterials, 29*(15), 2306–2314. doi:10.1016/j.biomaterials.2008.02.003 PMID:18313746

Kattimani, Kondaka, & Lingamaneni. (2016). Hydroxyapatite—Past, Present, and Future in Bone Regeneration. *Bone and Tissue Regeneration Insights, 7,* 9-19.

Kirkland, N. T., Staiger, M. P., Nisbet, D., Davies, C. H. J., & Birbilis, N. (2011). Performance-driven design of Biocompatible Mg alloys. *Journal of the Minerals Metals & Materials Society, 63*(6), 28–34. doi:10.100711837-011-0089-z

Lamaka, S. V., Montemor, M. F., Galio, A. F., Zheludkevich, M. L., Trindade, C., Dick, L. F., & Ferreira, M. G. S. (2008). Novel hybrid sol–gel coatings for corrosion protection of AZ31B magnesium alloy. *Electrochimica Acta, 53*(14), 4773–4783. doi:10.1016/j.electacta.2008.02.015

LeGeros, R. Z. (2008). Calcium Phosphate-Based Osteoinductive Materials. *Chemical Reviews*, *108*(11), 4742–4753. doi:10.1021/cr800427g PMID:19006399

Lhotka, C., Szekeres, T., Steffan, I., Zhuber, K., & Zweymüller, K. (2003). Four-year study of cobalt and chromium blood levels in patients managed with two different metal-on-metal total hip replacements. *Journal of Orthopaedic Research*, *21*(2), 189–195. doi:10.1016/S0736-0266(02)00152-3 PMID:12568948

Li, J., Han, P., Ji, W., Song, Y., Zhang, S., Chen, Y., Zhao, C., Zhang, F., Zhang, X., & Jiang, Y. (2011). The in vitro indirect cytotoxicity test and in vivo interface bioactivity evaluation of biodegradable FHA coated Mg–Zn alloys. *Materials Science and Engineering B*, *176*(20), 1785–1788. doi:10.1016/j.mseb.2011.05.029

Lin, S., LeGeros, R. Z., & LeGeros, J. P. (2003). Adherent octacalciumphosphate coating on titanium alloy using modulated electrochemical deposition method. *Journal of Biomedical Materials Research. Part A*, *66*(4), 819–828. doi:10.1002/jbm.a.10072 PMID:12926034

Liu, X., Chu, P. K., & Ding, C. (2004). Surface modification of titanium, titanium alloys, and related materials for biomedical applications. *Materials Science and Engineering*, *47*(3–4), 49–121. doi:10.1016/j.mser.2004.11.001

Lu, J., Yu, H., & Chen, C. (2018). Biological properties of calcium phosphate biomaterials for bone repair: A review. *RSC Advances*, *8*(4), 2015–2033. doi:10.1039/C7RA11278E

Lu, Y. P., Chen, Y. M., Li, S. T., & Wang, J. H. (2008). Surface nanocrystallization of hydroxyapatite Coating. *Acta Biomaterialia*, *4*(6), 1865–1872. doi:10.1016/j.actbio.2008.05.016 PMID:18567551

Makwana & Bhingole. (2018). Electrochemical and Plasma Surface Modification of Magnesium and its Alloy: Review. *Materials Today: Proceedings*, *5*(9), 18260-18267.

Meng, E. C., Guan, S. K., Wang, H. X., Wang, L. G., Zhu, S. J., Hu, J. H., Ren, C. X., Gao, J. H., & Feng, Y. S. (2011). Effect of electrodeposition modes on surface characteristics and corrosion properties of fluorine-doped hydroxyapatite coatings on Mg–Zn–Ca alloy. *Applied Surface Science*, *257*(11), 4811–4816. doi:10.1016/j.apsusc.2010.12.073

Monteiro, M. M., Campos da Rocha, N. C., Rossi, A. M., & Gloria de Almeida, S. (2003). Dissolution properties of calcium phosphate granules with different compositions in simulated body fluid. *Journal of Biomedical Materials Research. Part A*, *65A*(2), 299–305. doi:10.1002/jbm.a.10479 PMID:12734825

Mouritz. (Ed.). (2012). Introduction to Aerospace Materials. Academic Press.

Müller, W. D., Nascimento, M. L., Zeddies, M., Córsico, M., Gassa, L. M., Fernández, M. A., & de Mele, L. (2007). Magnesium and its alloys as degradable biomaterials. Corrosion studies using potentiodynamics and EIS electrochemical techniques. *Journal of Materials Research*, *10*(1), 5–10. doi:10.1590/S1516-14392007000100003

Nagels, J., Stokdijk, M., & Rozing, P. M. (2003). Stress shielding and bone resorption in shoulder arthroplasty. *Journal of Shoulder and Elbow Surgery*, *12*(1), 35–39. doi:10.1067/mse.2003.22 PMID:12610484

Narayanan, R., Dutta, S., & Seshadri, S. K. (2006). Hydroxy apatite coatings on Ti-6Al-4V from seashell. *Surface and Coatings Technology*, *200*(16–17), 4720–4730. doi:10.1016/j.surfcoat.2005.04.040

Narayanan, R., Seshadri, S. K., Kwon, T. Y., & Kim, K. H. (2008). Calcium phosphate-based coatings on titanium and its alloys. *Journal of Biomedical Materials Research. Part B, Applied Biomaterials*, *85*(1), 297–299. doi:10.1002/jbm.b.30932 PMID:17853421

Niinomi, M. (2002). Recent metallic materials for biomedical applications. *Metallurgical and Materials Transactions. A, Physical Metallurgy and Materials Science*, *33*(3), 477–486. doi:10.100711661-002-0109-2

Park, J. B., & Kim, Y. K. (2003). *Metallic biomaterials. Biomaterials principles and application*. CRC Press.

Pawlak, W., & Wendler, B. (2009). Multilayer, hybrid PVD coatings on Ti6Al4V titanium alloy. *The Journal of Achievements in Materials and Manufacturing Engineering*, *37*(2), 660–667.

Peng, P., Kumar, S., Voelcker, N. H., Szili, E., Smart, R. S. C., & Griesser, H. J. (2006). Thin calcium phosphate coatings on titanium by electrochemical deposition in modified simulated body fluid. *Journal of Biomedical Materials Research*, *76*(2), 347–355. doi:10.1002/jbm.a.30514 PMID:16270340

Posner, A. S., & Betts, F. (1975). *Synthetic Amorphous Calcium Phosphate and Its Relation to Bone Mineral Structure* (Vol. 8). Bone Mineral Structure.

Radha, R., & Sreekanth, D. (2017). Insight of magnesium alloys and composites for orthopedic implant applications – a review. *Journal of Magnesium and Alloys*, *5*(3), 286–312. doi:10.1016/j.jma.2017.08.003

Rahman, M., Li, Y., & Wen, C. (2020). HA coating on Mg alloys for biomedical applications: A review. *Journal of Magnesium and Alloys*, *8*(3), 929–943. doi:10.1016/j.jma.2020.05.003

Ramakrishna, S., Ramalingam, M., Kumar, T. S. S., & Soboyejo, W. O. (2010). *Biomaterials: A Nano Approach. CRC Press*. Taylor & Francis Group.

Rettig, R., & Virtanen, S. (2008). Time-dependent electrochemical characterization of the corrosion of a magnesium rare-earth alloy in simulated body fluids. *Journal of Biomedical Materials Research. Part A*, *85*(1), 167–175. doi:10.1002/jbm.a.31550 PMID:17688266

Rodríguez-González. (2009). *Biomaterials in Orthopaedic Surgery*. ASM International.

Rokhlin, L. L. (2003). *Magnesium Alloys Containing Rare Earth Metals: Structure and Properties*. Taylor & Francis. doi:10.1201/9781482265163

Salman, S. A., Kuroda, K., & Okido, M. (2013). Preparation and Characterization of Hydroxyapatite Coating on AZ31 Mg Alloy for Implant Applications. *Hindawi Publishing Corporation, Bioinorganic Chemistry and Applications. Article ID, 175756*, 6.

Saris, N. E., Mervaala, E., Karppanen, H., Khawaja, J. A., & Lewenstam, A. (2000). Magnesium: An update on physiological, clinical and analytical aspects. *Clinica Chimica Acta*, *394*(1-2), 1–26. doi:10.1016/S0009-8981(99)00258-2 PMID:10727669

Seiler, H. G., Sigel, H., & Sigel, A. (1988). *Handbook on Toxicity of Inorganic Compounds*. Marcel Dekker.

Shahmohammadi, P., & Khazaei, B. A. (2021). Characterization of Zn/Mg-enriched calcium phosphate coating produced by the two-step pulsed electrodeposition method on titanium substrate. *Surfaces and Interfaces*, *22*, 100819. doi:10.1016/j.surfin.2020.100819

Shaw. (2003). Corrosion resistance of magnesium alloys. *ASM handbook volume 13a: Corrosion: Fundamentals, testing and protection.*

Song, G. L., & Atrens, A. (2000). Corrosion mechanisms of magnesium alloys. *Advanced Engineering Materials*, 11–33.

Song, Y., Zhang, S., Li, J., Zhao, C., & Zhang, X. (2010). Electrodeposition of Ca–P coatings on biodegradable Mg alloy: In vitro biomineralization behavior. *Acta Biomaterialia*, *6*(5), 1736–1742. doi:10.1016/j.actbio.2009.12.020 PMID:20018262

Song, Y. W., Shan, D. Y., & Han, E. H. (2008). Electrodeposition of hydroxyapatite coating on AZ91D magnesium. *Materials Letters*, *62*(17–18), 3276–3279. doi:10.1016/j.matlet.2008.02.048

Staiger, M. P., Pietak, A. M., Huadmai, J., & Dias, G. (2006). Magnesium and its alloys as orthopedic biomaterials: A review. *Biomaterials*, *27*(9), 1728–1734. doi:10.1016/j.biomaterials.2005.10.003 PMID:16246414

Surmenev, R. A. (2012). A review of plasma-assisted methods for calcium phosphate-based coatings fabrication. *Surface and Coatings Technology*, *206*(8–9), 2035–2056. doi:10.1016/j.surfcoat.2011.11.002

Tokarz, A., Wieczorek, P., Fraczek, T., & Nitkiewicz, Z. (2008). Preparation, structural and mechanical properties of electrodeposited Co/Cu multilayers. *Physica Status Solidi*, *5*(c), 3526–3529. doi:10.1002/pssc.200779431

Tsyntsaru, N. I., Bobanova, Z. I., Kroitoru, D. M., Cheban, V. F., Poshtaru, G. I., & Dikusar, A. I. (2010). Effect of a multilayer structure and lubrication on the tribological properties of coatings of Fe-W alloys. *Surface Engineering and Applied Electrochemistry*, *46*(6), 538–546. doi:10.3103/S1068375510060025

Wang, H., Guan, S., Wang, Y., Liu, H., Wang, H., Wang, L., Ren, C., Zhua, S., & Chend, K. (2011). In vivo degradation behavior of Ca-deficient hydroxyapatite coated Mg–Zn–Ca alloy for bone implant application. *Colloids and Surfaces. B, Biointerfaces*, *88*(1), 254–259. doi:10.1016/j.colsurfb.2011.06.040 PMID:21783346

Wang, H., Zhu, S., Wang, L., Feng, Y., Ma, X., & Guan, S. (2014). Formation mechanism of Ca-deficient hydroxyapatite coating on Mg-Zn-Ca alloy for orthopaedic implant. *Applied Surface Science*, *307*, 92–100. doi:10.1016/j.apsusc.2014.03.172

Wang, H. X., Guan, S. K., Wang, X., Ren, C. X., & Wang, L. G. (2010). In vitro degradation and mechanical integrity of Mg–Zn–Ca alloy coated with Ca-deficient hydroxyapatite by the pulse electrodeposition process. *Acta Biomaterialia*, *6*(5), 1743–1748. doi:10.1016/j.actbio.2009.12.009 PMID:20004746

Wen, C., Guan, S., Peng, L., Ren, C., Wang, X., & Hu, Z. (2009). Characterization and degradation behavior of AZ31 alloy surface modified by bone-like hydroxyapatite for implant applications. *Applied Surface Science*, *255*(13-14), 6433–6438. doi:10.1016/j.apsusc.2008.09.078

Witte, F., Kaese, V., Haferkamp, H., Switzer, E., Meyer-Lindenberg, A., Wirth, C. J., & Windhagen, H. (2005). In vivo corrosion of four magnesium alloys and the associated bone response. *Biomaterials, 26*(17), 3557–3563. doi:10.1016/j.biomaterials.2004.09.049 PMID:15621246

Yang, L., & Zhang, E. (2009). Biocorrosion behavior of magnesium alloy in different simulated fluids for biomedical application. *Materials Science and Engineering C, 29*(5), 1691–1696. doi:10.1016/j.msec.2009.01.014

Yin, Z.-Z., Qi, W.-C., Rong-Chang, Z. X.-B. C., Gu, C.-D., Guan, S.-K., & Zheng, Y.-F. (2020). Advances in coatings on biodegradable magnesium alloys. *Journal of Magnesium and Alloys, 8*(1), 42–65. doi:10.1016/j.jma.2019.09.008

Zhang B. P., Wang Y. & Geng L. (2011). Research on Mg-Zn-Ca Alloy as Degradable Biomaterial. *Biomaterials-Physics and Chemistry*, 183-204.

Zhang, E., Yang, L., Xu, J., & Chen, H. (2010). Microstructure, mechanical properties and bio-corrosion properties of Mg-Si(-Ca, Zn) alloy for biomedical application. *Acta Biomaterialia, 6*(5), 1756–1762. doi:10.1016/j.actbio.2009.11.024 PMID:19941979

Zhang, X., Li, Q., Li, L., Zhang, P., Wang, Z., & Chen, F. (2012). Fabrication of hydroxyapatite/stearic acid composite coating and corrosion behavior of coated magnesium alloy. *Materials Letters, 88*, 76–78. doi:10.1016/j.matlet.2012.08.011

Zheng, J., Wang, Q. D., Jin, Z. L., & Peng, T. (2010). Effect of Sm on the microstructure, mechanical properties and creep behavior of Mg-0.5Zn-0.4Zr based alloys. *Materials Science and Engineering A, 527*(7–8), 1677–1685. doi:10.1016/j.msea.2009.10.067

Zhou, H., & Lee, J. (2011). Nanoscale hydroxyapatite particles for bone tissue engineering. *Acta Biomaterialia, 7*(7), 2769–2781. doi:10.1016/j.actbio.2011.03.019 PMID:21440094

KEY TERMS AND DEFINITIONS

Biocompatibility: It is property of material, that produces no undesirable or harmful or toxic effect while coexisting or interacting with living system.

Biodegradable: It is a material that can be broken down or decomposed naturally by living organisms.

Biomaterials: Biomaterials are natural or synthetic substances which interact with living organisms used to augment or repair or replace natural functions of diseased biological systems.

Cytotoxicity: Cytotoxicity is the toxicity caused by a material or process which leads to cell damage or cell death.

Implant: It is a man-made biocompatible material implanted in human body used to repair or replace diseased or damaged or missing part.

In-Vitro Tests: The in-vitro tests are performed in simulated physiological environment outside the body of living organism. These tests are used to determine the cytotoxicity of biomaterials before actually implanting the biomaterial in the body of living organism.

In-Vivo Tests: These tests are performed by actually implanting the biodegradable material into a living organism. The results obtained are more realistic as the implant is exposed to real physiological conditions.

Osteoconduction: It is part a of bone healing process which promotes the bone growth on a surface.

Osteoinduction: It is part a of bone healing process in which new bone formation is induced.

Chapter 12
Sol-Gel-Based Multifunctional Superhydrophobic Coatings and Its Tribological Properties:

Satish A. Mahadik
https://orcid.org/0000-0002-3072-358X
Sanjay Ghodawat University, India

F. Pedraza
Université de La Rochelle, France

Sarika S. Mahadik
Shivaji University, Kolhapur, India

ABSTRACT

This chapter summarizes key issues in the fields of multifunctional superhydrophobic coatings with an analysis of their tribological properties. In this respect, the authors explore a simple sol-gel process strategy and tribological properties controlled through a reaction of ORMOSIL-based polymers that generate multifunctional minimum free energy structures of micro- to nano-scale siloxane chains. Different compositions and dimensions of solid materials (contact angle = 150° and sliding angle = 10°) can be superhydrophobic fabricated through various deposition methods. The complete waterproof layering has been demonstrated to have excellent cost, scalability, and especially the ability to encapsulate other functional groups. The perspectives have established many significant functionalities with better tribological properties for the next generation of smart multifunctional superhydrophobic coatings.

INTRODUCTION AND BACKGROUND OF SUPERHYDROPHOBIC COATINGS

Research on the superhydrophobic surface(SHS) has been extensively developed from fundamental principles of functional investigation in recent years. In the field of practical applications such as self-cleaning, resistance to corrosion, drag reduction, antifreeze, and oil-water separation, multifunctional superhydrophobic surface applications have tremendous potential(S. A. Mahadik, Pedraza, & Mahadik,

DOI: 10.4018/978-1-7998-9683-8.ch012

2017). For SHS the existence of a micro/nanostructure is crucial. Overall, an appropriatearrangement of surface roughness witha low-energy surface produces artificial SHS. Two approaches are often used to create a superhydrophobic surface. The first technique involves generating rough surfaces, followed by covering materials with low surface energy(Satish. A. Mahadik et al., 2016). The second method is to restructure the hydrophobic material surface(S. A. Mahadik et al., 2016). Superhydrophobic coatings like smart coatings have become multifunctional, efficient, and versatile with the recent advances in coating technology and have also attracted global research interest.

Complex multi-scale structures with nanostructures on top of micro-structures efficiently enhance the texture hierarchically and decrease the interface between texture and liquid. Naturally, the results guide the creation of artificial SHSs and the design of surfaces with controlled wettability. Despite different preparation procedures, the commercialization of superhydrophobicity has been seriously affected due to their low tribologicaland chemical stability, as well as the fragility of their dual-scale roughness characteristics. There is consequently great emphasis on the mechanical strength and endurance of multifunctional superhydrophobic coatings in recent studies(S. A. Mahadik et al., 2013). Superhydrophobic coatings like smart coatings have become multifunctional, efficient, and versatile with the recent advances in coating technology and have also attracted research interest(S. A. Mahadik et al., 2012b). Due to their performance, such coatings may be applicable in all possible domestic areas. Superhydrophobic coatings are suitable for diverse applications in various fields, such as automotive windshields and photovoltaic solar panels for self-cleaning purposes in the solar power sector, marine infrastructure, aircraft, and naval ships for an anti-corrosion, and anti-icing purposes, etc.

MAIN FOCUS OF THE CHAPTER

The chapter focuses on the multifunctional and tribological properties of organically modified silica (ORMOSIL) based superhydrophobic coatings that expand every year. However, there are few relevant studies from the current literature on multifunctional superhydrophobic coatings with tribological properties in some fields. This chapter summarizes a multifunctional superhydrophobic coating with tribological performance and functionalities like self-cleaning, transparency, anti-icing, anti-corrosion, and some other functionalities.

METHODS FOR PRODUCING SUPERHYDROPHOBIC COATINGS.

Superhydrophobic surfaces fabrication strategies can be classified and shown in **figure 1**.

Figure 1. Outline of the general synthesis strategy of superhydrophobic coatings.

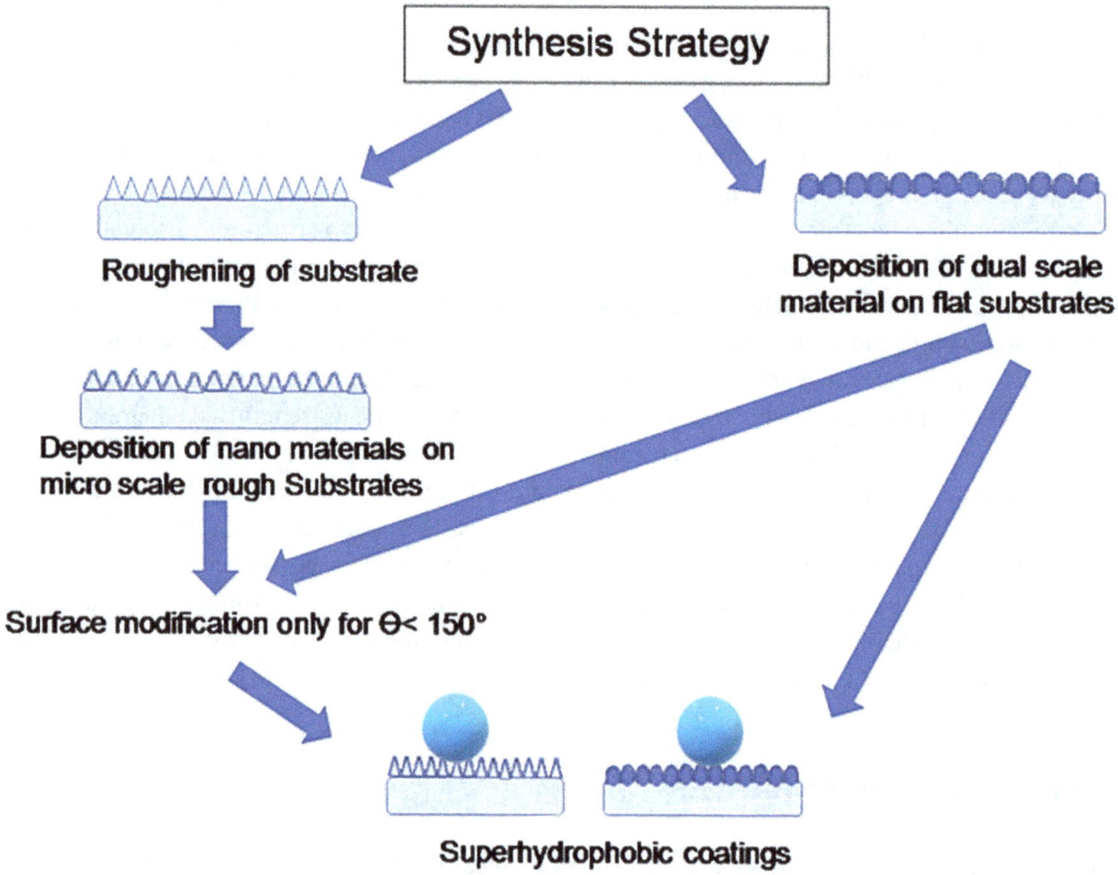

ROUGHENING OF HYDROPHOBIC MATERIALS

The superhydrophobicitycanbe created by roughening the surface without any surface functionalization. The film can be made transparent by making the size of the particle or texture on the film much smaller or larger than the light wavelength. Dual scale texture with low-energy surfaces can produce superhydrophobic surfaces. Good superhydrophobic properties can be achieved whenever of the dual scale textures are fabricatedby wet deposition methods (dipping, spinning, spray and dropping)on fine or rough polished metal substrates. For instance, Pedraza et al. (PhysChemChemPhys 2013)developed an easy, but versatile method to fabricate superhydrophobic coating by roughening thesubstrates by electropolishing. These films are superhydrophobic with multifunctional properties like higher optical transparency, reversible superoleophobicity, thermal stability, mechanical strength, and show prolonged durability(S. A. Mahadik et al., 2012, 2013).

SURFACE MODIFICATION OF ROUGH MATERIALS WITH LOW SURFACE ENERGY MATERIALS

Superhydrophobic surfaces can be achieved by surface modification and the construction of surface topographic structure. Coatings with low surface energy are popular means for surface modification and enhancement of hydrophobicity. Surface modification and the building of a topographic surface structure can be used to create superhydrophobic surfaces. Surface modification and hydrophobicity amplification are typical uses for superhydrophobic coatings with low surface free energy. In this process, hydrophobicity has to be built into organo-functionalized inorganic precursors and retained in the production by carefully controlling the sol-gel process conditions. Another phase- separation grows a three-dimensional network of hybrid silica particles network. Although starting with hydrophobic materials makes it easier to achieve superhydrophobicity, the hydrophobicity of the surface is not a necessary condition. It is also feasible to construct a meta-stable superhydrophobic state using a material that has less than 90° according to the Cassie- Baxter equation. Superhydrophobicity may be achieved with alkyl group-terminated silicon surfaces with an inherent water contact angle larger than 150°, if particular micro-textures consisting of overhang structures with well-defined geometries are used to fabricate them. It's a wonderful example of surface engineering. A more common method is to roughen the surface and then apply hydrophobic treatments to change the surface chemistry.

The dip, spin, spray, and electrodeposition methods produce a simple technique for making dual rough surfaces with superhydrophobicity. Dual-scale texture forming a superhydrophobicity can be obtained with low surface energy groups(like methyl, ethyl, chlorine, fluorine, etc.)(Satish. A. Mahadik et al., 2016). Low surface energy group-terminated silicon surfaces show a water contact angle greater than 150°. A more common technique is to roughen the surface and then employ surface modification procedures to change the surface chemistry(S. A. Mahadik et al., 2012). Surface modification of hydrophobic coatings can produce superhydrophobicity with multifunctionalityby tuning the material's roughness.

PROCESSING, FORMULATION, AND DEPOSITION OF MULTIFUNCTIONAL SUPERHYDROPHOBIC COATINGS

A superhydrophobic coating's functioning is determined by elements such as coating components, preparation processes, and coating deposition methods.Polymer matrix, curing agent, inhibitor host (micro- or nano-sized), inhibitor kind, and catalyst when necessary are among the manufacturing components of the coating solution, which should be as environmentally friendly as possible, with minimal toxicity. Smart coatings are combined with organic-inorganic components and may be produced by many techniques, with the sol-gel process shown inthe **figure 2**.

This inorganic approach involves colloidal suspension production (usually metal alkoxides) and networking via sol gelation in the continuous liquid phase (colloidal suspensions of small (1–100 nm) particles). The synthesis of sol-gel usually proceeds in four phases: (a) hydrolysis, (b) condensation and polymerization, (c) particle growth, and (d) polymer agglomeration(Brinker & Scherer, 1990). Simultaneously, they hydrolyze, and condensation processes occur once the hydrolysis reaction has started. Low molecular weight byproducts such as alcohol and water are produced as a result of hydrolysis and condensation. These smaller molecules are formed during the drying process, and the network grows strong as more condensation happens.

Figure 2. Aschematic illustration of the sol-gel coating process.

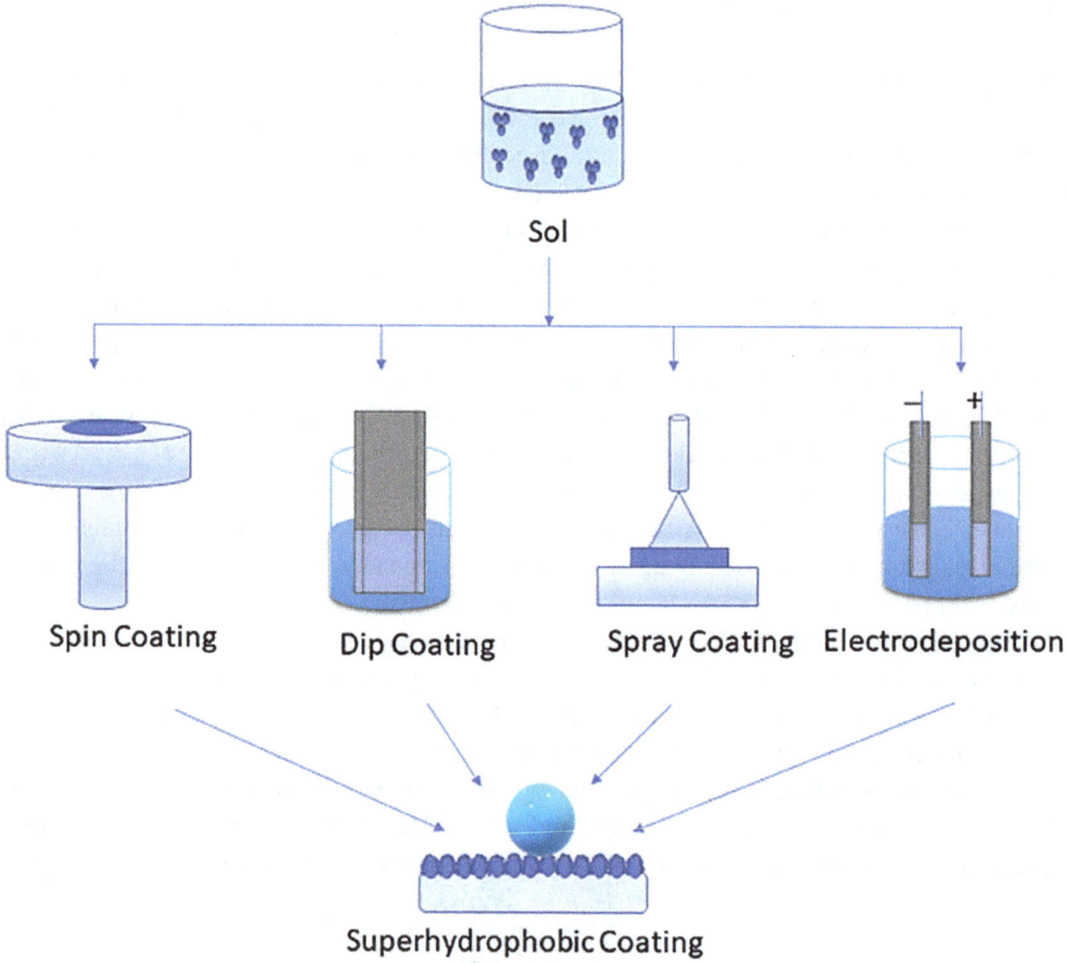

FACTORS AFFECTING THE TRIBOLOGICAL PROPERTIES OF SOL-GEL COATINGS

Organic/Inorganic Contents

Hardness and Young's modulus are systematically reduced as organic content increases in the sol-gel coatings?. It is clear from this trend that organic components are less interconnected than inorganic components (Trabelsi et al., 2011). The organic contents provide ductility and flexibility, while inorganic networking becomes less effective. Similarly, the module of Young's modulus increases progressively with the inorganic component. Some researchers constructed a transparent sol-gel coating and explored the consequence of the silica amounts on the tribological properties of coatings based on acrylate, polyurethane resin, MPTMS, and TEOS (Wouters et al., 2004). If the silica concentration increases from 0 to 18 wt %, both the Young's modulus and the hardness increased from 1.8 to 3 GPa, and from 80 to 200 MPa, respectively in relation with the good compatibility between the organic and the inorganic networks. In conclusion, it is essential to regulate the development of the organic/inorganic networks to modify the mechanical qualities of the coating.

Filler Additivities

The addition of fillers to the sol is another means of adapting the mechanical characteristics of the coating. The filler additive in the network is well known to enhance tribological properties. Silica and TiO_2arepreferably used in sol-gel components. Some results indicated that Young's modulus and hardness levels enhance with silicon concentration(L. Y. L. Wu et al., 2008). Earlier studies showed that the scratch with an 800 μm tip radius of Rockwell achieves 13.7 ± 0.7 N compared to 7.9 ± 0.6 N for pure silica coating(Dinelli et al., 2011a). From previous study claims that with 5% nanoTiO2 in a silica-based coating, the hardness of pencils may approach 6H. Increasing nanoparticles of titanium dioxide improves abrasion resistance(Hwang et al, 2003).

Silica Composites

The ZrO-TiO_2-B_2O_3-ZnO system is made with silica composite to provide additional functionality to their final coating structures. Shows improvement on the mechanical characteristics of the coating of other metallic nanoparticles added to silica-based sol. Chelation of the metal core with acetylacetone or methacrylic acid is generally requiredto regulate the hydrolysis/condensation reactivity of zirconium and titanium-based alkoxide more rapidly than silicone-based ones. Accordingly, considerable expertise is essential for both kinetic and metallic interactions of silanes to prevent the instability of sol. The type of mixed hybrid oxide affects the mechanical characteristics of the coating. The inclusion of boric acid in the silica sol forms a Si–O–B network(Mülazim et al., 2011). If the B/Si ratio increased, the hardness and abrasion resistance are improved sequentially. It indicates an improving cross-linking density after adding boron. The same scratch resistance enhancement effect observed for tetraethyl orthotitanate (TEOT) is introduced to TEOS (Dinelli et al., 2011b).

Interface

Finally, the contact between the substrate and the coating, as well as the quality of its adherence, is an important factor in determining the mechanical stability behavior of any coating. In this aspect, the surface treatment of the substrate (chemical, plasma, adhesion promoter, etc.) improves the mechanical properties. Furthermore, detailed studies of the effects of surface modifications of the coating tribological characteristics are declining in general applications. Some results estimate a transmission of around 70% after 140 rotations of steel wear abrasion tester under loading 1 kg(Tsukakoshi et al., 2011). The promotional agent employed - 3 aminopropyltrimethoxysilane (3-APMS) - has also not been considered a function of the finishing hardness. This improvement is therefore attributed entirely to improved adherence.

CURING PROCESS

Oven, UV, and microwave curing methods are commonly used for various sol-gel coatings. Oven treatment, generally 80 to 135 ^{0}C is the most conventional essential to maintain functionality below 140 ^{0}C.

Oven curing by rapid solvent evaporation accelerates the sol-gel reaction so the parameters can be easily controlled. The prolonged treatments affect the hardness like Fabri et al. demonstrated (Fabbri et al., 2009). The crosslink density is therefore associated with the surface hardness. The abrasion resistance is coupled with increases the cure time and enhances all these parameters.

Thickness and Aging

The viscosity of the sol can change from sol aging time to hydrolysis. The thickness of the coating rises equally by increasing sol viscosity, all other parameters remaining identical. Much research in the field of sol-gel reveals that the thickness and mechanical characteristics of the coating are highly correlated parameters. Microhardness (Vickers HV1) typically evolves with depth of the coating(Lin et al., 2011). For thicker coatings, the hardness values increased from 30 HV1 to 250 HV1 for thicknesses ranging from 2.2 to 8.0 μm, respectively. Lionti also underlined the strong correlation between ageing of the sol and resistance to scratch(Lionti et al., 2013). In particular, they discovered that the resistance to scratch decreases with increasing aging time in adherence, which is line with fewer bonds accessible to the substrate as the condensation occurs. As previously noted, abrasion resistance may be influenced by thickness as well as surface or contact properties.

MULTIFUNCTIONAL SUPERHYDROPHOBIC SURFACES AND SOME IMPORTANT PROPERTIES OF SUPERHYDROPHOBIC COATINGS

In this study, we demonstrate specific functions of superhydrophobic surfaces through the sol-gel process.

Self-cleaning Properties

Superhydrophobic surfaces are usually given a self-cleaning role by removing contaminated particles from water droplets hitting or rolling water, which involves external forces such as gravity (as shown in **figure 3**). Superhydrophobicity with the self-cleaning surfaces possibly applied for optoelectronics, touch screens, solar panels, non-stick cookware, and windshield glasses.

Figure 3. Mechanism of self-cleaning on a superhydrophobic surface.

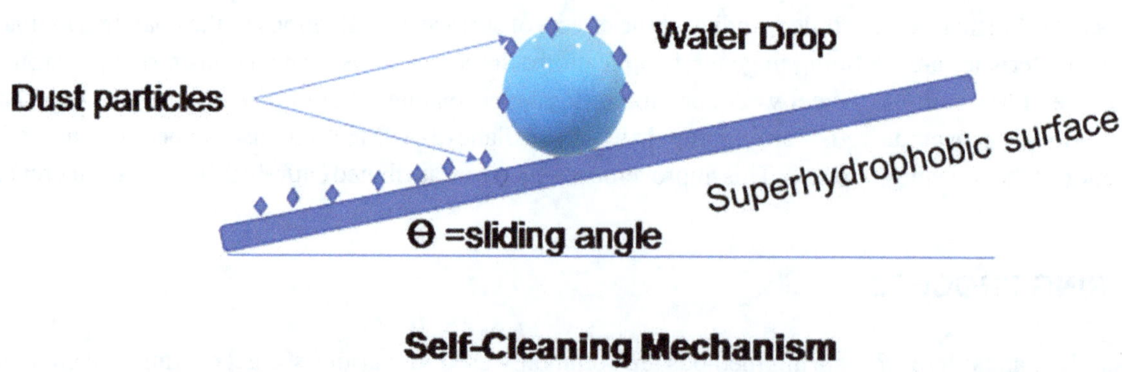

The application of ORMOSIL material through the sol-gel process gives a multifunctional film nature with a self-cleaning effect. Because of their remarkable water-repellent characteristics, superhydrophobic coatings have received a lot of research attention from the last two decades. Superhydrophobic surface areas have a WCA is exceeding 150° and an angle of sliding of less than 10°. The interaction between the metal ions and corrosive species minimizes by creating a self-cleaning effect on the superhydrophobic surfaces. Superhydrophobic coatings, the majority of which resembles the microstructure of waterproof plants or flowers and exhibit the necessary self-cleaning properties. A self-clean superhydrophobic coating with polymerization and condensation coating and utilized as quartz and glass. After several rigorous testing, the coatingshows high mechanical durability, high hardness, and self-cleaning. After more than 250 cm in a standard reciprocal abrasion machine, the materials remain superhydrophobic with a working force of 500 g. The findings show that the surface can withstand the 6H pencil hardness test due to the high hardness. Many terrestrial plants leave surfaces to repel water and have a high amount of self-cleaning capability. Some researchers have looked into the wetting and self-cleaning characteristics of superhydrophobic surfaces.

Wear strength investigation of superhydrophobic layering composed of nanoparticles and acrylic resin matrix hydrophobic silica in a glass substrate or iron-tin sheet(M. Li et al., 2019). He observed that roughening abrasive sheets quickly degraded the superhydrophobic coating, which had a hardness of up to 4H and wear resistance of up to 2000 cycles. Superhydrophobicsurfaces manufactured by electrophoresis of nanoparticles in electrodeposited DTMS sol-gel films were also shown to be water repellent and durable self-cleaning with WCA=153.1± 0.8° and CA=17.8 ± 2.0°, heat resistance up to 270 ^{0}C, and after 100 times of adhesive tape test abrasion-resistant up to 20 cycles (X.F. Zhang et al., 2017).

In some of our previous works (Mahadik et al. 2016b), the usual preparationprocedures of superhydrophobic and self-cleaning surfaces were examined to develop multifunctionalsuperhydrophobic surfaces with self-cleaning properties. Such sol-gel coatings were manufactured on glass, quartz, alumina, aluminum, copper, steel and stainless steel. The sol-gel coatings were shown to offer higher thermal and mechanical durability than organic coatings due to the high strength of the Si-O bond, ORMOSIL(S. A. Mahadik et al., 2016b) (S. A. Mahadik et al., 2012a). The surface coating is highly thermally stable up to 550 ^{0}C, static water contact angle about 167±1^0, a low sliding angles of about 2±1^0. The manufacturing approach may readily be extended to the domains of industrial development and high technology.

The polyvinyl alcohol was used to graft the silica particles' microporous network onto the glass. The water contact angle was as high as 169±1^0, and the water sliding angle was as low as 5±1^0. The silica film showed a 4 H grade pencil scratch resistance as per method ASTM D 3363. Earlier studies showed that water repellent coatings based on TiO_2 and SiO_2 using spin and spray techniques on different material substrates via a sol-gel path(S. A. Mahadik, Pedraza, & Mahadik, 2017).The superhydrophobicity derived from spray coatings has been found to maintain its self-cleaning characteristics in harsher outdoor environments than spin coatings.

In short, superhydrophobicity is easily produced with micro- or nanostructures, but water droplets on the surface are known to be difficult to roll due to the complex surface chemistry. Coatings with nanoparticles, such as silica nanoparticles are doped to increase tribological characteristics. Superhydrophobicity enhances protection against dust accumulation, and active agent encapsulation and surface finishing can provide both functions. The process of synthesizing sol-gel and utilizing silica particles provides a simple and effective path to these goals. A summary of some of the works dealing with the tribological properties of self- superhydrophobic sol-gel coatings are gathered in **Table 1.**

Table 1. Tribological properties of self-cleaning superhydrophobic sol-gel coatings.

Sr. No.	Deposition Method	Surface Properties	Tribological Properties	Percursors and References
1	Spray	CA=167.3 ± 0.9^0 SA= 1.9 ± 0.1^0	H=283MPa YM=1445MPa PH=B	PDMS- modified SiO_2 NP (Dyett& Lamb, 2016)
2	Spin coating	CA=150^0 SA= 25^0	COF=0.2	$AlCl_3$ (Qi et al., 2016)
3	Electro-deposition	CA= 159.1° SA= 1.2° Thermal stability up to 250 °C	3.2 m on 1200 grit sandpaper under 0.60 kPa.	Silica NP - polydimethylsiloxanemodified DTMS (X.-F. Zhang et al., 2017)
4	spin coating	CA=155 ± 1^0 CAH=1^0	LC= 1.97 ± 0.11 mN	VTEOS-Resin (Taurino et al., 2014)
5	dip coating	CA= 158 ± 2° SA= 4 ± 1°	abrasion against 1500 mesh sandpaper up to 80 cm	TiO_2-OTS (Pratiwi et al., 2020)
6	Spray	CA=162.5^0 SA=2^0	SR=5H(2.97 mN/m)	MTEOS and colloidal silica (Lakshmi et al., 2011)

Hardness=H, Pencil hardness=PH, Abrasion resistance =AR, Scratch resistance=SR, coefficient of friction=COF, Critical load=LC.

TRANSPARENCY

The transparency function of superhydrophobic surfaces is traditionally due to the light scattering by dual-scale roughness of a coating, which lies below or above the optical region (380-700 nm) as a forbidden roughness gap(S. A. Mahadik et al., 2010)(S. A. Mahadik et al., 2012a). The most significant part of the preparation is to provide a good balancebetween scratch-resistant and self-cleaning properties. It is quite challenging to maintain both transparent and abrasion-resistant liquid-repellent coatings simultaneously. Thisis mainly because the most transparent liquid-repellent s are manufactured via nanoparticle assembly, which have poor wear resistance in addition to having a relatively weak coating adhesion and rubbing compared to polymer-based transparent hydrophobic coatings.

Non-weatherproof coatings mechanical robustness refers to their resistance to abrasion wear. In general, a durable transparent superhydrophobic surface is created in two ways:(a) Restrict the removal of material to sustain wear of superhydrophobicity.(b) Design a material that keeps transparent superhydrophobicity as it wears off.

A wear resistance of the ORMOSIL based material tested with sandpaper for check their stability. ORMOSIL-based transparent superhydrophobic materials are long-lasting against harsh conditions(Sebastian et al., 2021). Sol-gel networks provide the necessary properties for a high steepness against low mechanical stability, low wearablity and fragile nature. Non-wettable coatings should be mechanically durable based on a conventional wear-abrasion test, but other tests such as adhesion to the substrate and scratch resistance should also be considered. It's also worth noting that most of these manufacturing techniques should be able to be used in broad regions, such as optical self-cleaning surfaces, public vehicle windows, walls, and building windows as shown in **figure 4**.

Figure 4. Applications of transparent superhydrophobic coating for different devices.

The superhydrophobic (162.1^0) coating was created using a PU-SiO_2 based three-dimensional network structure using a double-step sol-gel dip-coating process to increase the transmission of the coating to 94.38%(Luo et al., 2018). Furthermore, at the pressure of 5 KPa, the layer exhibited a strong abrasion resistance due to its good adhesive strength and structural stability to increase the anti-abrasion properties.(Righeira Carnegie et al., 2016)have designed novel multifunctional hydrophobic (SiO_2-TiO_2)-ZrO2 bilayer coatings that present high anti-reflection (AR), photocatalytic self- cleanability, high scratch, and abrasion resistance properties. Furthermore, a combination of TEOS and MPTMS improves abrasion resistance rise in Si amount. A summary of some of the works dealing with the tribological properties of transparent superhydrophobic sol-gel coatings are gathered in **Table 2.**

Table 2. Tribological properties of transparent superhydrophobic sol-gel coatings.

Sr. No.	Method	Functionality	Tribological Properties	Precursors and References
1	dip	Anti-fogging. CA=155° CA=3°. Totalscattering =0.007 %	Wear resistance= After 50 wiping cycles with a force of 1.5 N, CA changed to 141°	Al_2O_3-SiO_2 (Felde et al., 2016)
2	spray	Optically transparent CA=164o Self-cleaning=2o	Abrasion(1508MPa)up to 20cycles Hardness=0.68±0.0051 Youngs Modulus=0.7 ± 0.217	TEOS-Glymo- FAS-17 (Kumar et al., 2015)
3	spray	CA=158^0 Self-cleaning=2^0 SOP=146.7^0	Scratch resistance=9H Adhesion=5B	MTEOS-perfluoroalkylmethacrylic (Hitoki et al., 2002)

Continued on following page

Table 2. Continued

Sr. No.	Method	Functionality	Tribological Properties	Precursors and References
4	spin	Transparency=89.9 % CA=164±2 0	SHP maintains after the impact of 40 g sand impact abrasion from a 50 cm height. 100–300mm sized grains have a velocity of 8.7 km h^{-1}	TEOS-TMCS (Y. Zhang et al., 2017)
5	dip	CA=169° Self-cleaning =5° Transparency=82 %	scratch resistance=9B	17FTMS (Liu et al., 2015)
6	spray	CA=166^0 Optically transparent	Adhesion test=5B scratch=6H modulus =1.57±0.11 GPa Hardness=0.24±0.01 GPa	TEOS-GLYMO-PFOTES (X. Wu et al., 2016)
7	Spin	CA=170° Self-cleaning=3° Transparency=84%	Hardness=8H	TEOS-silica composite (Topcu et al., 2018)
8	dip	CA=160° Self-cleaning=10° Transparency=77 %	Adhesion=15B Scratch=9H	TEOS-PFOTS (Trabelsi et al., 2011)
9	Spray	CA=161.5±2 ° Self-cleaning=10.3±2° Transparency=61 %	Abrasion with knife-edge CA dropped to 145±2°.	PFOTES-SiO2 (Y. Li et al., 2015)

The transparency of coating decreased to 75%after 50 coating cycles but thethe interfacial strength per unit length increased to 61 gf/mm. Dual scale morphology of silica materials shows transparency (90.22%), thermal stability(550 ^{0}C), acid resistance, and superhydrophobic coating (WCA=172^0, SA=2^0). Surface treatment with 10%TMCS reveals superhydrophobicity in the film. A successfully manufactured MTMS based semitransparent (60%) multifunctional superhydrophobic surfaces (168±2^0) achieved with sol-gel dip coating and surface modification process(S. A. Mahadik et al., 2016). These coatings show superhydrophobic and superoleophilic nature with water and organic oils.

Dip-coating coatings are multifunctional with durability that can last up to several months, thermal stability that can last up to 550°C, acid resistance against mild acids, and the ability to easily recover from surface modifications(S. A. Mahadik, et al., 2013). A modified surface with mono-, di-, and tri-function ORMOSIL agents is used to produce transparent (>60%) superhydrophobic silica aerogel(Mahadik et al., 2016). A phenyltrimethoxysilane-based dip-coated film was fabricated by using a two-stage sol-gel method. Hydrophobic and optical transparent (82%) film was fabricated using a fixed molar ratio(silica sol (tetraethylorthosilicate): 11.03 (methanol): 0.17 (acidic water): 0.58 (base water)), where the weight of PTMS varied between 0 and 15% (Parale et al., 2013). ORMOSIL materials can improve their opitcal properties by grafting functional groups on the surface through the surface modification process. It originates from functionalities of silylating agents to regulate the hydrophobicity, andoptical transmission.

ANTI-ICING COATINGS

Nowadays, ice accumulation is amongs the biggest concerns in sectors like airplanes, ships, roads, and telecommunications equipment due to massive economic losses in these sectors. In addition, ice snow causes partial or entire loss of engine thrust. Ice build-up may cause critical damage and perhaps harm people in transmission lines, cars, and overseas platforms. A large machine and energy are needed to remove ice from the machinery. To address such difficulties, scientists have shown that superhydrophobicity can reduce ice or delay icing (**figure 5**). Superhydrophobicity may also have a slippery alternate phenomenon to the anti-icing process. It will slip off the surface of this produced ice and diminish the ice adhesion strength. An anti-icing coating helps repel the natural ice from the environment when touching the surface. It is also known as glaciophobic.

Figure 5. (A)Mechanism of anti-icing nature of the superhydrophobic surfaceand (B) Anti-icing strategies.

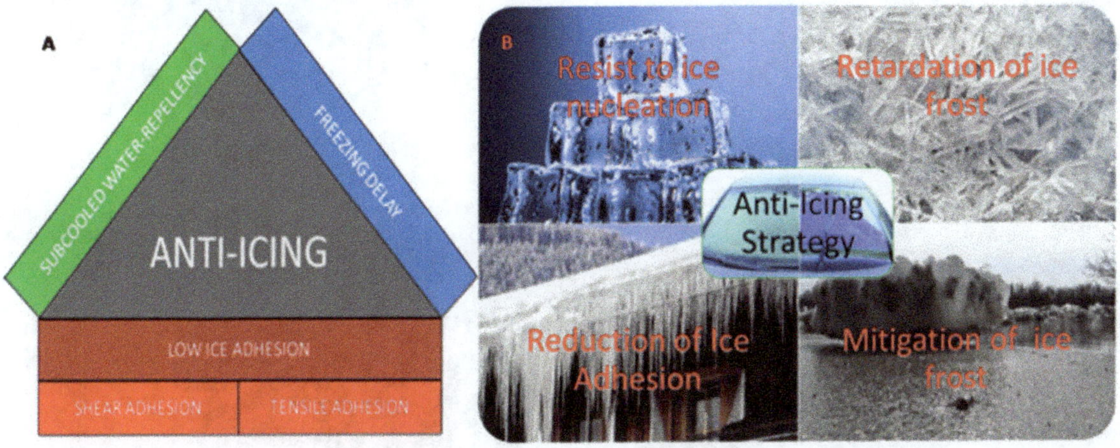

Ice accumulation may occur in a wide variety of temperatures in natural environmentsas shown in **figure 6**. The moisture is due to many situations, including freezing rain, snow, and frost. Anti-icing techniques may generally classify into active and passive ones. Actual active solutions to fight issues of glazing include heating systems, chemical deicing, and mechanical removal. On the other hand, surfaces can passively obstruct the development of ice and make the removal processeasier without additional energy.

Figure 6. Potential applications of superhydrophobic anti-icing coating.

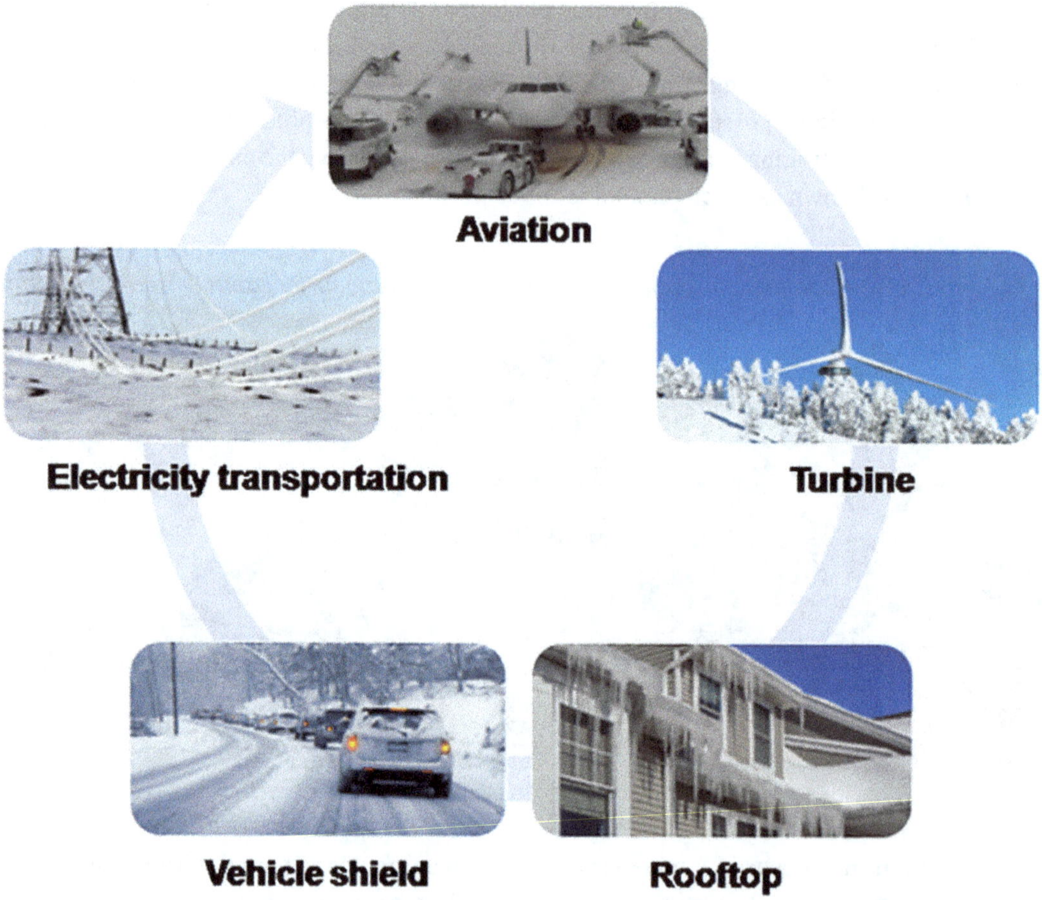

Passive techniques are more ecologically effective than active processes, which may perform using low-surface energy chemicals based upon structure-based physico-chemical processes. Superhydrophobic surfaces inspired by the idea of lotus leaves as potential anti-icing approaches are hoped because of their outstanding water repellency. Many investigations have reported effective passive anti-icing solutions using a superhydrophobic phenomenon to reduce the adhesive strength or delay freezing. In recent years, sol-gel technology has also been used to manufacture superhydrophobic coatings for aircraft and ship applications. For instance, Wu and Chen created a transparent and sol-gel-derived ice/liquid-repellent coating that delivers excellent transmitting performance (X. Wu & Chen, 2018a). Such anti-icing coating showed good ice adherence, ice formation, and self-cleaning performance compared to well-studied superhydrophobic coatings. The coating showed up to 97.8% transmission and greater mechanical strength and durability than modern slippery porous surfaces and polymer coatings saturated with liquid. The study showed that sol-gel coating with a high ratio of nanoparticles is a suitable way to achieve transparency. The particles are thoroughly mixed and ultrasonically dispersed during the sol-gel preparation process.

In the creation of a two-stage isophobic coating, a commercial polyurethane-based paint was discovered(Carreño et al., 2020). A carbon fiber-epoxy resin composite was modified by adding different

PDMS concentrations (2, 5, and 10 wt. %). The second process involves the application of a heterogeneous hybrid sol-gel coating with a drop-like structure that exposes some of the paint beneath. The results reveal that the adhesiveness strength of the ice is reduced to 80%, and breaking shear forces less than 100 kPa are achieved. Roughness, water contact angle, and ice adhesion test were all measured over the course of more than 100 shear-off/freezing ice cycles. Sol-gel coatings were shown self-cleaning performance on transmission (~ 97.8%), ice-adhesion strength, ice formation, and well-studied superhydrophobic coating(X. Wu & Chen, 2018a). The mechanical features of transparent solid coatings have been studied extensively employing nanoindentation, pencil scratch, transversal adhesion, and dolly pull-off tests.

Superhydrophobic coatings were produced using silanol and nano-silica for cement concrete(Gao, Huang, et al., 2016). The contact angle of the superhydrophobic cement reached about 160°, while the rolling angle dropped to 2.3°. During the first 100 cycles, the contact angle appeared to decrease to 60^0, and the sliding angle increased to 90^0. Hydrophobic asphalt shows significantly less snow residual than uncoated pavements. A summary of these coatings is gathered in **Table 3.**

Table 3. Tribological properties of anti-icing superhydrophobic sol-gel coatings.

Sr. No.	Method	Ice Adhesion Strength (KPa)	Tribological Properties	Percursor and References
1	Spray	CA= 166° SA=6° Freezing delay=1651 S	Pencils Hardness=3B	MTES-HFTES (Tang et al., 2015)
2	Spray	CA= 163.1° SA= 4° delay icing by 30 min	SHP maintains even after 150 cycles under 9.8 kPa, After tape-peeling under 90.5 kPa for 200 cycles.	TEOS-HDTMS-PU-HD-POS (Y. Li et al., 2018)
3	Dip	CA=164° SA= Icing delay =500 s	Adhesion=4B hardness =3H	ZnO-PTFE-SiO_2 (Saffar et al., 2021)
4	Spray	CA=117° SA=20° Ice adhesion=57–87 kPa Icing delay=349S	Hardness = 380 ± 10 MPa elastic modulus = 12.39 ± 0.46 GPa scratch resistance = 4 B	TiO_2-GLYMO- itaconicacid-TEOS (X. Wu et al., 2018)
5	Spray	CA=153° SA= Iceadhesion=216.3 kPa	The pencil hardness of < 6B and adhesion of 4B.	TEOS-PDMS (R.P.S. Chakradhar, 2012)
6	Spray	CA= >160° SA=<25° Iceadhesion=58.2 ± 1.5KPa	Hardness 270 ± 10 MPa Scratch 4H	TEOS-GLYMOL-PFOTES (X. Wu & Chen, 2018b)

The current approach of active de-icing and passive anti-icing, which uses less energy, still necessitates the development of a combined strategy.This new strategy can apply to overcome the anti-icing problem.

ANTI-CORROSION COATINGS

Superhydrophobic anti-corrosion barrier coatings avoid corrosion by using air pockets layers acting as an effective passivation layer between metal ions and corrosive species. Superhydrophobic anti-corrosive barrier coating layers are promising as a new efficient option for the anti-corrosion issue(as illustrated in **figure 7**). One approach to developing coatings with active corrosion protection is to include anti-corrosive components into sol-gel films. There have been a variety of alternative ways for introducing a pre-treatment or primer layer inhibitor. If the inhibitor coating does not interact with corrosive species, direct doping inhibitory ions or molecules into the sol-gel coating material can give good rust prevention.

Figure 7. Mechanism of anti-corrosion behavior on a superhydrophobic surface.

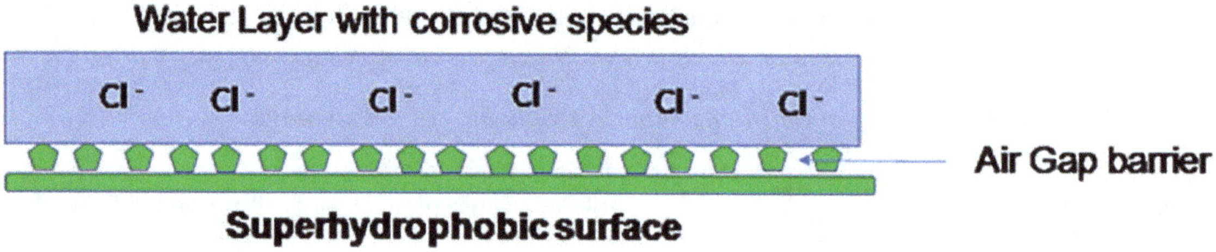

Anti-corrosion coatings are a wide set of protective sol-gel films. Their main function is to separate the layers of metal from hostile forces. The requirements for the class of layers are low porosity, strength, crack-proof, good support adhesion, and scratch resistance. Inorganic-inorganic hybrids (e.g., vinyl, epoxy, acryl groups), coatings with inhibitors (e.g., doped with permanganate ions, cerium salts (III), and mercaptobenzothiazole, etc.), and deposits containing nanoparticles are examples of sol-gel materials (e.g., montmorillonite). The inorganic component enhances scratch resistance, hardness, and substrate adhesion. The barrier of the sol-gel film may be interrupted by the chemical-modified inhibitor's inhibitory capacity being reduced or lost. The early leak of the inhibitor and the resulting reduction inefficiency are other disadvantages of this method.

Thin sol-gel anti-corrosion coatings were employed on various kinds of metal surfaces that are more resistant to corrosion than corrosive materials to (i.e., 316L) titanium alloys (e.g., Ti6Al4V) or aluminum alloys (e.g., AA2024-T3). Additionally, a superhydrophobic coating can be potentially applied to industrial infrastructures, ancient statues, cooking pans, and military vehicles as shown in the **figure 8.**

Figure 8. Potential applications of anti-corrosive nature of the superhydrophobiccoating.

An earlier report has been made of multifunctional super-hydrophobic coatings using the single sol-gel spray coating technique on germanium, brass, steel, and copper substrates(S. A. Mahadik et al., 2016). It shows good corrosion control ability for up to six months against an artificially corrosive salt solution. The produced superhydrophobicity was durable for up to 2 years, mechanical strength was sustainability under a tap water impact up to 8 hours, transparency about 83%, self-cleaning capability, acid resistance, and heat stability was above 500 ^{0}C.

Our group reported(Rao et al., 2011)the MTES based superhydrophobic coating with a contact angle to be 155° and the slide angle to be 7^{0} on the copper substrate. The superhydrophobic nature persists even after the substrate is absorbed in 50% HCl solution for up to 100 hours. The coatings remain stable and exhibit superhydrophobic with anticorrosion activity even after 90 days of anti-humidity testing. The coatingsare alsomechanically stable, and the spherical size of water droplets remains the same on the bent copper substrate above 90^{0}.

Magnesium alloys offer a variety of desired properties, including low densities (lower than Al- or Ti-based alloys), high rigidity, and recycling capability. Low resistance to corrosion ability that can be improved with sol-gel. Architecture is another possible area where the sol-gel coating is used for anti-corrosion purposes. A hot-dip galvanized method was utilized to decrease the danger of corrosion of the steel used to strengthen concrete. Hot-dipped galvanized steel, on the other hand, is very corrosive when exposed to highly alkaline environments, such as fresh concrete, where the pH is approximately 12.5. A film produced using the sol-gel technique is the best solution to this problem. As indicated in the literature, the anti-corrosion ofgalvanized steel is enhanced with organic parts containing inorganic compounds (e.g., SiO_2 epoxy group). They have suggested electrostatic spraying techniques with a superhydrophobic EP-PTFE-graphenepolydopamine (GP) -SiO_2-PFOTES coating of 156.3°±1.5° and a low of 3.5°±0.5° SA on a steel substrate. The double interfacial correction shows an increase in mechanical resistance of the coating, which combined the shear stress and pressure-flow during melting with the "glue" effect of polydopamine. The study showed numerous hydroxyl groups in the self- polymerization of dopamine on the graphene surface, which provides a secondary reaction platform.

Three-dimensional (3D) GP-SiO_2 nanoparticles were fabricated by the sol-gel method. In addition, even after the NaCl solution was scratched or immersed in 3.5 wt. % for 60 days, the produced coating maintained its strong water repellent, potential application exhibition in maritime regions. Lianget al. succeeded in creating superhydrophobic withself-cleaning, anti-icing, anticorrosion and protective clothing(Liang et al., 2018). This technique produced cylindrical silica-based particles with a hierarchical morphology, yielding contact angles of about 167° and excellent rust protection. Self-healing and superhydrophobic layers are promising ways of preventing the corroding of metallic substrates. Although various approaches are employed to produce the self-healing anti-corrosion capability of coatings, most of themare at a lab trial phase with limited handover foractual applications because of the expensive cost and the time required. Some of these coatings are gathered in **Table 4.**

Table 4. Tribological properties of anti-corrosive superhydrophobic sol-gel coatings.

Sr.No.	Substrate	Method	Anti-corrosion Properties		Tribological Properties	Precursor and Reference
1	Carbone steel	dip coating	Corrosion resistance(CR)	50 GΩ.cm^2	Durability > 18 months in saline water Self-healing Adhesion > 3.5 MPa) COF increased by 0.1-0.2 Adhesion 240 mN Wearresistance> 148 mNafter 50 cycles	PMMA-MPTS-TEOS (S. v. Harb et al., 2017)
			Testing time (T)	18 Months		
			Testing solution	3.5% NaCl		
2	Mg alloy AZ31	dip coating	CR	1 GΩ.cm^2	adhesion after 1-month immersion in NaCl it was impossible to detach from the surface unless using a sharp scalpel	Epoxy-APTES-ZnO (Brusciotti et al., 2013)
			T	30 days		
			Solution	0.05% NaCl		
3	Mg alloy	dip coating	CR	~100GΩ. cm^2	Self-healing with MBT	Epoxy-SiO_2 (Qiao et al., 2015)
			T	7 Months		
			solution	3.5% NaCl		

Continued on following page

Table 4. Continued

Sr.No.	Substrate	Method	Anti-corrosion Properties		Tribological Properties	Precursor and Reference
4	Al alloy AA2024-T3	Single blade/150	CR	~10GΩ.cm²	CA=128°	Epoxy-APTES-tetration (Ammar et al., 2016)
			T	30 days		
			solution	3% NaCl		
5	316L stainless steel	dip coating	CR	~0.01 GΩ.cm²	2.5 µm Crosshatch and Tape Pull Test have shown an excellent adhesion	PMMA-MPTS-TEOS (Hammer et al., 2010)
			T	36 days		
			solution	3.5% NaCl		
6	Al alloys AA1050	dip coating	CR	~1 GΩ.cm²	Adhesion=500 pN Young's modulus=494 MPa)	GMA-EHAG-PTMS-TEOS (Khelifa et al., 2013)
			T	21 days		
			solution	0.1 M NaCl		
7	A1010 carbon steel	Dip coating	CR	~1 GΩ.cm²	Adhesion =8 MPa	PMMA-MPTS-TEOS (Hammer et al., 2012)
			T	18 Months		
			solution	3.5% NaCl		
8	A1010 carbon steel	dip coating	CR	~10GΩ.cm²	Adhesion =8 MPa	PMMA-MPTS-TEOS-Ce(IV) (S. V. Harb et al., 2015)
			T	304/404 days		
			solution	3.5% NaCl		
9	A1020 carbon steel	dip coating	CR	~0.5GΩ.cm²	Scratch resistance= 100 mN Mechanical strength= 80 mN	PMMA-MPTS-TEOS-lignin (S. V. Harb et al., 2015)
			time	50		
			solution	3.5% NaCl		
10	A1010 carbon steel	dip coating	CR	5GΩ.cm²	Adhesion = 3.6 – 4.0 MPa, Young modulus 5.4 ±0.1GPa hardness = 311.0 ±5.0 MPa COF = 0.5 - 0.7,	PMMA-MPTS-TEOS (dos Santos et al., 2015)
			time	196 days		
			solution	3.5% NaCl		
11	Mild steel	dip coating	CR	~10GΩ.cm²	Scratch resistance Critical load=70.3 mN	PMMA-MPTS-TEOS-Ce (Mosa et al., 2016)
			time	362 days		
			solution	3.5% NaCl		
12	A1020 carbon steel	dip coating	CR	~3GΩ.cm2	Critical load=148 mN Mechanical strength= 240 mN Wear resistance COF=0.6 after 50 cycles	PMMA-MPTS-TEOS-CNTs-GO (S. v. Harb et al., 2016)
			time	211 days		
			solution	3.5% NaCL		

Continued on following page

Table 4. Continued

Sr.No.	Substrate	Method	Anti-corrosion Properties		Tribological Properties	Precursor and Reference
13	Carbon steel	spray	CR	~100GΩ.cm^2	Corrosive after 30 days	Polyurethane-ZrO_2-SiO_2(del Angel-López et al., 2013)
			Time	226 days		
			solution	3.5% NaCl		

SUPERHYDROPHOBIC SURFACES WITH OTHER PROPERTIES

In contrast to typical smooth surfaces, nanostructured superhydrophobic surfaces with high blood compatibility have no platelet adherence on the resulting superhydrophobic surfaces. Such compatible superhydrophobic coatings have many biomedical applications because of their long-term bio-stability and moderate blood compatibility as shown in **figure 9**.

TEOS-based superhydrophobic, transparent, and biocompatible coatings have been developed using the sol-gel process (Xiu et al., 2007). Preliminary investigations show the superhydrophobic coating efficiently reduces E. coli bacteria adhesion to it. Superhydrophobic biomaterial design for new applications including contact lenses, temporary implants, MEMS, precise medicine discharging coatings, medical equipment coating, etc.(Elashnikov et al., 2021); Methyltriethoxysilane-tri-methoxymethyl silane deposition provides unique characteristics in the area at various dip times, including superhydrophobicity, biocompatibility, and transparency(S. A. Mahadik, et al., 2017).

Figure 9. Application of multifunctional superhydrophobic coatings.

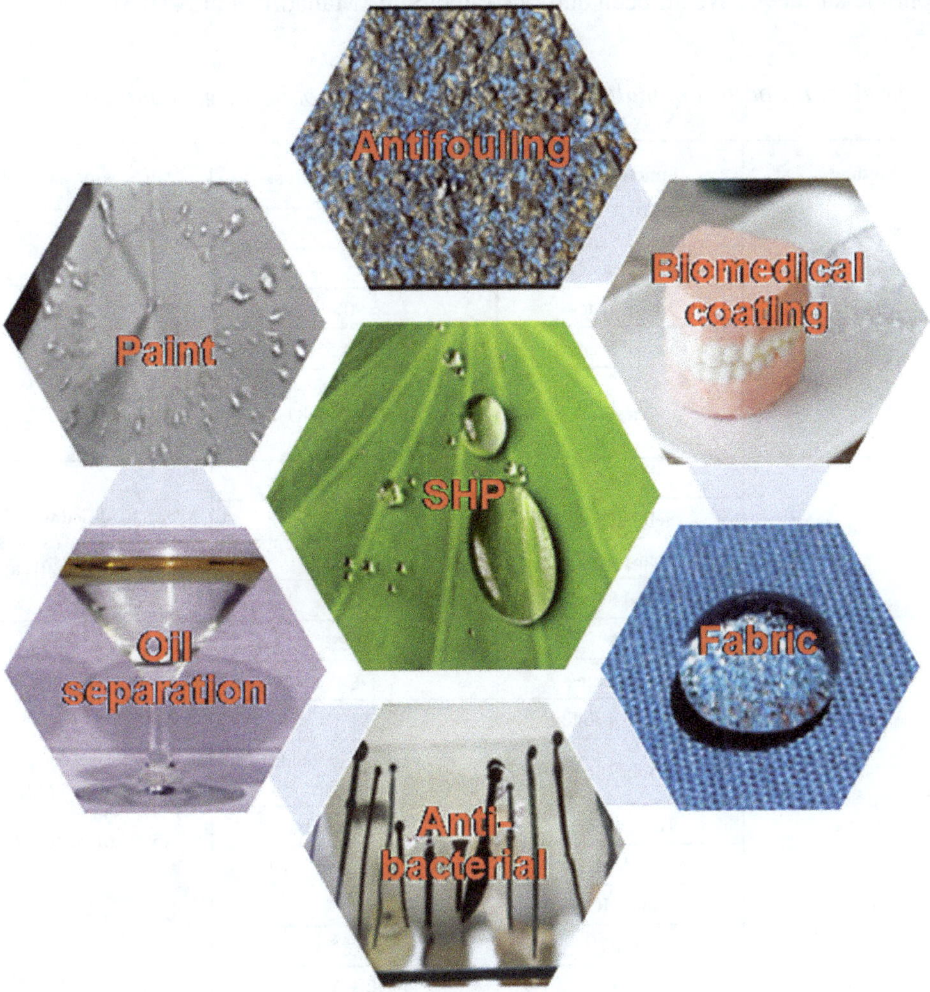

A sol-gel dip-coating method is suggested as a straightforward way to create a recoverable, superhydrophobic silica coating at a constant contact angle of 170 ± 1^0 and a water slide angle of 2 ± 1^0. Surface modification contributed to the modification of surface properties, allowing the fabrication of recoverable surfaces in superhydrophilic and superhydrophobic states. This type of coating has enormous potential uses in industry and home applications (S. A. Mahadik et al., 2012). Our group has successfully developed a superhydrophilic-superhydrophobic fabric network for oil-water separation ability. Layer-by-layer deposition of MTMS sol on cotton fabric is used to accomplish this result. The superoleophilic surface and the surface free energy cotton surface fabric (water contact angle of 167 ± 1^0 and the little sliding angle of 4 ± 1^0) were effectively achieved. The overall surface free energy is 13.23 ± 0.37 mJ.m^{-2}. Fabrics showed stable superhydrophobicity for 60 days in outdoor environments, boiling water for 5 h, and ten ultrasonic rounds of laundry with negligible changes. To conclude, superhydrophobic surfaces have been a hot topic in modern coating and surface research due to their wide variety of applications, and several successful preparation techniques have been established (**Table 5**). Although there are significant technical issues in this sector, the economic feasibility, poor tribological properties, and durability of superhydrophobic surfaces have all been demonstrated(S. A. Mahadik et al., 2016).

Table 5. Tribological properties of multifunctional superhydrophobic sol-gel coatings.

Sr. No.	Method	Functionality	Details of Functions	Precussor and References
1	Single step Sol-gel dop coating	CA	172^0	MTES-MTMS on glass substrate (Rao et al., 2011)
		Self-cleaning	2^0	
		transparency	>90%	
		Thermal stability	SHP up to 550 °C	
		Acid resistance	SHP with Conc HCL HP with Conc. HNO_3 and H_2SO_4	
2	Single-step Sol-gel Spray coating	CA	167 ± 1^0	MTES modified with TMCS on the glass substrate (S. A. Mahadik et al., 2012a)
		Self-cleaning	2 ± 1^0	
		Transparency	>78%	
		Thermal Stability	550 °C	
3	Double step Sol-Gel Dip coating	CA	168 ± 2^0	TMCS modified MTMS coating on Quartz substrate (S. A. Mahadik, et al., 2013)
		Self-cleaning	3 ± 1^0	
		Thermal stability	560 °C	
		Durability	up to six-month outdoor conditions	
		Transparency	Maximum 60%	
		Superoleophilic	Kerosene and Diesel	
		Conc HCL	155^0	
		H_2SO_4	120 ± 5o	
		HNO_3	90 ± 5°	
4	Double step sol-gel dip coating	CA	170 ± 1^0	TMCS modified MTMS coating on Glass substrate (S. A. Mahadik, et al., 2013)
		Self-cleaning	3 ± 1^0	
		Thermal Stability	550 °C	
		Durability	Hydrophobicity up 1 one year	
		Restoring	Surface modification with TMCS 10%	
		Acid resistance	Maintain SHP in 10% HNO_3 up to 1 month	

Continued on following page

Table 5. Continued

Sr. No.	Method	Functionality	Details of Functions	Precussor and References
5	Double step Sol-Gel Spray coating	CA	Steel =161 ± 2 ^{0} Germanium =163 ± 2^0 Brass =170 ± 1^0 Copper =170 ± 1^0	MTES on glass, copper, aluminum, steel, brass, germanium (S. A. Mahadik et al., 2016)
		Self-cleaning	Steel =6 ± 1^0 Germanium =5 ± 1^0 Brass= 3 ± 1^0 Copper = 4 ± 1^0	
		Durability	Maintain SHP in outdoor environments up to 24 months	
		Thermal Stability	Up to 500 ^{0}C	
		Acid resistance	Conc. HNO_3	
		Anti-corrosion	Maintain SHP after 6 months in the artificial corrosive solution.	
		Mechanical Strength	Maintain SHP underwater drop impact up to 8 H	
		Transmission	83% on Glass Substrate and the metal substrate, it is quite visible	
6	Single-step Sol-gel dip-coating process	CA	170 ± 1^0	TMCS modified MTMS on glass (S. A. Mahadik et al., 2012)
		Self-cleaning	2 ± 1^0	
		Thermal stability	550 ^{0}C	
		Reversibility	With thermal degradation and surface modification	
7	Sol-gel dip coating	CA	155^0	MTES on copper Substrate (Rao et al., 2011)
		Self-cleaning	7^0	
		Anti-corrosion & Chemical resistance	Copper maintains SHPin50% HCl for 100 h	
		Transparency	Visible transparent	
		Durability	90 days against humid conditions	
		Mechanical strength and Adhesiveness	SHP maintain even after bending with 90^0	
8	Sol-Gel dip coating	Hydrophobic	135±2^0	MTMS-TEOS on glass (Ganbavle et al., 2011)
		Self-cleaning	15°±1^0	
		Transparency	>80%	
9	Double Step Sol-gel dip coating	Hydrophobicity	120^0	PTMS modified TEOS on glass (V. G. Parale et al., 2013)
		Transparency	>80%	
		Durability	Retain HP in 85% humidity over 40 days, 30 days	
		Thermal stability	Up to 386 ^{0}C	

Continued on following page

Table 5. Continued

Sr. No.	Method	Functionality	Details of Functions	Precussor and References
10	Double Step Sol-gel dip coating	Hydrophobicity	125⁰	OTES modified TEOS on glass (V. G. Parale et al., 2013)
		Transparency	>88%	
		Durability	Retain HP in 85% humidity over 40 days, 30 days	
		Thermal stability	Up to 259 ⁰C	
11	Single-step sol-gel layer by layer process	CA	167 ± 2⁰	MTMS on cotton fabrics (S. A. Mahadik et al., 2016)
		Self-cleaning	4 ± 1⁰	
		Oil separation	96% to 90% oil separation efficiency with different oils	
		Durability	Retain SHP up to 30 days with 9.6% variation.	
		Mechanical strength	Stable up to 10 washing cycles and 5h boiled water with 2% variation.	
12	Electrodeposition	CA	151.0±1.4°	CeO_2 on Ni20Cr substrate (Pedraza et al., 2015)
		Thermal stability	1200 ⁰C	
13	Single-step sol-gel dip-coating process	CA	169 ± 1⁰	MTMS-PVA on glass (Kavale et al., 2012)
		Self-cleaning	5 ± 1°	
		Durability	Maintain SHP up to 270 days	
		Scratch resistance	Up to 4H	
		Transmission	86%	
14	CBD method	CA	152°	Zinc acetate (Gawali et al., 2017)
15	Single-step sol-gel spray and spin coating process	CA On steel	SiO_2 Spray=164⁰ TiO_2 Spray=158⁰ SiO_2 Spin=140⁰ TiO_2 Spin=130⁰	MTMS and TBT on quartz, glass, alumina, brass germanium, & steel. (S. A. Mahadik 2017)
		Self-cleaning on steel	Spray SiO_2=9⁰ Spray TiO_2=10⁰ Spin SiO_2=>36⁰ Spin TiO_2=>47⁰	
		Thermal stability	TiO_2=356⁰C MTMS=550 ⁰C	
		Durability test	6 months against outdoor conditions	
		Mechanical stability	9 hour against water drop impact	
		Switching	Thermal degradation and surface modification	
		Transparency	Spin MTMS=78% Spray MTMS=60% Spin TiO_2=80% Spray TiO_2=85%	

Continued on following page

Table 5. Continued

Sr. No.	Method	Functionality	Details of Functions	Precussor and References
16	The single-step sol-gel dip-coating process	CA	168 ±1°	MTMS-MTES on glass (S. A. Mahadik, Pedraza, Mahadik, et al., 2017)
		Self-cleaning	4 ± 1°	
		Transparency	>83%	
		Biocompatibility	Bone marrow mesenchymal stem cell (BMMSC)	
17	Single-step Sol-gel dip coating	CA	153±3°	TMCS modified MTES (S. A. Mahadik & Mahadik, 2021)

SOLUTIONS AND RECOMMENDATIONS

In further work, the best practical applications must be found forORMOSIL-based materials, and the coating characteristics optimization must be in the chosen application. In addition, the anti-icing, anti-corrosion, antifouling, antibacterial, self-cleaning, and transparency property havealso become interestingin recently, so that the surface property must be used for broader industrial applications in the next few years. In general, dual hierarchical structures are consideredamong the mostimportant superhydrophobic coatings. However,suchstructures are unstable and quickly damaged. There is a lack of standardized evaluation of tribological properties of superhydrophobic coatings. The research should focus on the improvement of durability and the establishment of better durability assessment methods. The commercial success of superhydrophobic surfaces is restricted due to cost, a complicated process, poor tribological properties, and toxicity. Research should aim at large-scale manufacturing methods using cost-effective and more environmentally friendly raw materials. The need for the synthesis, manufacturing, and application of organic-inorganic nanohybrids materials today and in the future will be the development of environmentally friendly and solvent-free materials.

Specifically, one emerging, globally important field of applications for self-healing smart superhydrophobic materials includes multiple functionalities, including antioxidation, high-temperature corrosion, salt corrosion, blood repellents, and anti-bacterial surfaces. Self-healing offers a beneficial technique that enhances the long-lasting life of the superhydrophobic surface further. The surfaces are degraded by the delicate micro-/nanoscale structure of the superhydrophobic coatings, which lowers the lifespan of the superhydrophobic coating following mechanical and chemical damage. Therefore, it is essential to make hydrophobic surfaces more durable so that they can be used in real life. However, there are still many obstacles to integrating superhydrophobic surfaces with self-healing capabilities. The regenerative properties against chemistry or physical degradation may increase the long-term sustainability of superhydrophobic surfaces. Therefore, the ideal way to ensure long-term durability while exposed to the external environment is to incorporate a superhydrophobic area with self-healing capacity. The problem of durability of superhydrophobic surfaces was dealing with mechanical stability. However, current research have revealed that highly hydrophobic self-healing surfaces are better suited to extend surface life and durability.

FUTURE RESEARCH DIRECTIONS

The industry is tremendously involved in developing a smart and multifunctional superhydrophobic coating. Failure to maintain a permanent and stable coating are preventing practical applications from implementation. The solutions for multifunctional superhydrophobic coatings are still not supported. The merging of self-healing functionalities is meaningful as it adds value and provides greater effectiveness to solve problems. Lack of stability after long-term exposure to harsh environments will destroy functional efficiency and visual impairment. Superhydrophobic coating manufacturing technology may impact its indigenous stability. Substrate adherence strength, wearing resistance, abrasion resistance, and dynamic impact resistance is variables that influence the coating's strength.

A significant element affecting the mechanical strength and stability of these structures is the interface strength within matrix components comprising nanostructures, binding materials, and substrates. From an industrial point of view, even when considering sustainability, cost-effectiveness is still a priority. The goal is to develop asmart multifunctional superhydrophobic coating with sustainable and environmentally friendly materials. This technique might provide importance by reducing the impact on the environment and the amount of toxicity.

ORMOSIL sol-gel coating with self-healing characteristics is extremely significant. Efficient models need to develop quickly to establish the threshold for self-healing capabilities of many destructive phenomena. Self-healing abilities can be used as additional functions in superhydrophobicity to solve durability problems. As a result, future research will focus on improving the mechanical robustness and long-term durability of multifunctional superhydrophobic coatings with self-healing capabilities. It will fix the functionality's functioning on its own, with no external involvement required.

CONCLUSION

This chapter reviewed on synthesis and tribological properties of multifunctional superhydrophobic organic-inorganic coating materials. ORMOSIL materials are more compatible with multifunctional superhydrophobic coatings containing multifunctional groups. Thus, future work will increase the stability of smart multifunctional coatings. Coatings with self-healing capability are expectedto restore multifunctional function trigger with external damaging environments. Tribological features are affecting their bulk and interface properties, sol-gel chemistry, thickness, and adhesion. Through the analysis of tribological performance, mechanical investigations including tribological principles can give insight into asperity failure at relevant length scales. Furthermore, the sol-gel method provides us with multiple options for meeting a variety of technical criteria. In the future, research into the development of cheaper and environmentally friendly coatings points towards the exploration of ORMOSIL based material to synthesize self-healing multifunctional superhydrophobic surfaces.

ACKNOWLEDGMENT

Authors are highly thankful to ERASMUS MUNDUS MOVER PROGRAMME.

REFERENCES

Ammar, S., Ramesh, K., Vengadaesvaran, B., Ramesh, S., & Arof, A. K. (2016). Amelioration of anticorrosion and hydrophobic properties of epoxy/PDMS composite coatings containing nano ZnO particles. *Progress in Organic Coatings*, *92*, 54–65. doi:10.1016/j.porgcoat.2015.12.007

Brusciotti, F., Snihirova, D., Xue, H., Montemor, M. F., Lamaka, S., & Ferreira, M. G. S. (2013). Hybrid epoxy–silane coatings for improved corrosion protection of Mg alloy. *Corrosion Science*, *67*, 82–90. doi:10.1016/j.corsci.2012.10.013

Carreño, F., Gude, M. R., Calvo, S., Rodriguez de la Fuente, O., & Carmona, N. (2020). Design and development of icephobic coatings based on sol-gel/modified polyurethane paints. *Materials Today. Communications*, *25*, 101616. Advance online publication. doi:10.1016/j.mtcomm.2020.101616

Chakradhar, B. T. (2012). Studies on the evaluation of ice adhesion strength of different hydrophobic and superhydrophobic coatings. *National Aerospace Laboratories, PD SE 1224*, 1–30.

del Angel-López, D., Domínguez-Crespo, M. A., Torres-Huerta, A. M., Flores-Vela, A., Andraca-Adame, J., & Dorantes-Rosales, H. (2013). Analysis of degradation process during the incorporation of ZrO2:SiO2 ceramic nanostructures into polyurethane coatings for the corrosion protection of carbon steel. *Journal of Materials Science*, *48*(3), 1067–1084. doi:10.100710853-012-6839-7

Dinelli, M., Fabbri, E., & Bondioli, F. (2011a). TiO2–SiO2 hard coating on polycarbonate substrate by microwave assisted sol–gel technique. *Journal of Sol-Gel Science and Technology*, *58*(2), 463–469. doi:10.100710971-011-2413-z

Dinelli, M., Fabbri, E., & Bondioli, F. (2011b). TiO2–SiO2 hard coating on polycarbonate substrate by microwave assisted sol–gel technique. *Journal of Sol-Gel Science and Technology*, *58*(2), 463–469. doi:10.100710971-011-2413-z

dos Santos, F. C., Harb, S., Menu, M.-J., Turq, V., Pulcinelli, S. H., Santilli, C., & Hammer, P. (2015). On the structure of high performance anticorrosive PMMA–siloxane–silica hybrid coatings. *RSC Advances*, *5*(129), 106754–106763. doi:10.1039/C5RA20885H

Dyett, B., & Lamb, R. (2016). Correlating Material Properties with the Wear Behavior of Sol–Gel Derived Superhydrophobic Films. *Advanced Materials Interfaces*, *3*(13), 1500680. doi:10.1002/admi.201500680

Elashnikov, R., Ulbrich, P., Vokatá, B., Pavlíčková, V. S., Švorčík, V., Lyutakov, O., & Rimpelová, S. (2021). Physically Switchable Antimicrobial Surfaces and Coatings: General Concept and Recent Achievements. *Nanomaterials (Basel, Switzerland)*, *11*(11), 3083. doi:10.3390/nano11113083 PMID:34835852

Fabbri, P., Messori, M., Toselli, M., Veronesi, P., Rocha, J., & Pilati, F. (2009). Enhancing the scratch resistance of polycarbonate with poly(ethylene oxide)-silica hybrid coatings. *Advances in Polymer Technology*, *27*(2), 117–126. doi:10.1002/adv.20122

Felde, N., Coriand, L., Duparré, A., & Tünnermann, A. (2016). Wear- Resistant Nanostructured Sol-Gel Coatings for Functional Applications. *Journal of Coating Science and Technology*, *3*(3), 100–108. doi:10.6000/2369-3355.2016.03.03.1

Ganbavle, V., Bangi, U. K. H., Latthe, S. S., Mahadik, S. A., & Rao, A. V. (2011). Self-cleaning silica coatings on glass by single step sol-gel route. *Surface and Coatings Technology*, *205*(23–24), 5338–5344. doi:10.1016/j.surfcoat.2011.05.055

Gao, Y., Huang, L., & Zhang, H. (2016). Study on anti-freezing functional design of phase change and temperature control composite bridge decks. *Construction & Building Materials*, *122*, 714–720. doi:10.1016/j.conbuildmat.2016.06.065

Gawali, S. A., Mahadik, S. A., Pedraza, F., Bhosale, C. H., Pathan, H. M., & Jadkar, S. R. (2017). Synthesis of zinc oxide nanorods from chemical bath deposition at different pH solutions and impact on their surface properties. *Journal of Alloys and Compounds*, *704*, 788–794. Advance online publication. doi:10.1016/j.jallcom.2017.01.228

Hammer, P., dos Santos, F. C., Cerrutti, B. M., Pulcinelli, S. H., & Santilli, C. (2012). Highly corrosion resistant siloxane-polymethyl methacrylate hybrid coatings. *Journal of Sol-Gel Science and Technology*, *63*(2), 266–274. doi:10.100710971-011-2672-8

Hammer, P., Schiavetto, M. G., dos Santos, F. C., Benedetti, A. V., Pulcinelli, S. H., & Santilli, C. V. (2010). Improvement of the corrosion resistance of polysiloxane hybrid coatings by cerium doping. *Journal of Non-Crystalline Solids*, *356*(44–49), 2606–2612. doi:10.1016/j.jnoncrysol.2010.05.013

Harb, S., dos Santos, F. C., Caetano, B. L., Pulcinelli, S. H., Santilli, C., & Hammer, P. (2015). Structural properties of cerium doped siloxane–PMMA hybrid coatings with high anticorrosive performance. *RSC Advances*, *5*(20), 15414–15424. doi:10.1039/C4RA15974H

Harb, S., Pulcinelli, S. H., Santilli, C., Knowles, K. M., & Hammer, P. (2016). A Comparative Study on Graphene Oxide and Carbon Nanotube Reinforcement of PMMA-Siloxane-Silica Anticorrosive Coatings. *ACS Applied Materials & Interfaces*, *8*(25), 16339–16350. doi:10.1021/acsami.6b04780 PMID:27266403

Harb, S. v., Trentin, A., Torrico, R. F. O., Pulcinelli, S. H., Santilli, C. v., & Hammer, P. (2017). Organic-Inorganic Hybrid Coatings for Corrosion Protection of Metallic Surfaces. In *New Technologies in Protective Coatings*. InTech. doi:10.5772/67909

Harb, S. V., Cerrutti, B. M., Pulcinelli, S. H., Santilli, C. V., & Hammer, P. (2015). Siloxane–PMMA hybrid anti-corrosion coatings reinforced by lignin. *Surface and Coatings Technology*, *275*, 9–16. doi:10.1016/j.surfcoat.2015.05.002

Hitoki, G., Takata, T., Kondo, J. N., Hara, M., Kobayashi, H., & Domen, K. (2002). An oxynitride, TaON, as an efficient water oxidation photocatalyst under visible light irradiation ($\lambda \leq 500$ nm). *Chemical Communications (Cambridge)*, *2*(16), 1698–1699. doi:10.1039/B202393H PMID:12196955

Hwang, D. K., Moon, J. H., Shul, Y. G., Jung, K. T., Kim, D. H., & Lee, D. W. (2003). Scratch resistant and transparent UV-protective coating on polycarbonate. *Journal of Sol-Gel Science and Technology*, *26*(1–3), 783–787. doi:10.1023/A:1020774927773

Kavale, M. S. M. S., Mahadik, S. A. A., Mahadik, D. B. B., Parale, V. G. G., Rao, A. V., Vhatkar, R. S. S., Wagh, P. B. B., & Gupta, S. C. C. (2012). Enrichment in hydrophobicity and scratch resistant properties of silica films on glass by grafted microporosity of the network. *Journal of Sol-Gel Science and Technology*, *64*(1), 9–16. doi:10.100710971-012-2822-7

Khelifa, F., Druart, M.-E., Habibi, Y., Bénard, F., Leclère, P., Olivier, M., & Dubois, P. (2013). Sol–gel incorporation of silica nanofillers for tuning the anti-corrosion protection of acrylate-based coatings. *Progress in Organic Coatings*, *76*(5), 900–911. doi:10.1016/j.porgcoat.2013.02.005

Kumar, D., Wu, X., Fu, Q., Ho, J. W. C., Kanhere, P. D., Li, L., & Chen, Z. (2015). Development of durable self-cleaning coatings using organic–inorganic hybrid sol–gel method. *Applied Surface Science*, *344*, 205–212. doi:10.1016/j.apsusc.2015.03.105

Lakshmi, R. V., Bharathidasan, T., & Basu, B. J. (2011). Superhydrophobic sol–gel nanocomposite coatings with enhanced hardness. *Applied Surface Science*, *257*(24), 10421–10426. doi:10.1016/j.apsusc.2011.06.122

Li, M., Li, Y., Xue, F., & Jing, X. (2019). A robust and versatile superhydrophobic coating: Wear-resistance study upon sandpaper abrasion. *Applied Surface Science*, *480*, 738–748. doi:10.1016/j.apsusc.2019.03.001

Li, Y., Li, B., Zhao, X., Tian, N., & Zhang, J. (2018). Totally Waterborne, Nonfluorinated, Mechanically Robust, and Self-Healing Superhydrophobic Coatings for Actual Anti-Icing. *ACS Applied Materials & Interfaces*, *10*(45), 39391–39399. doi:10.1021/acsami.8b15061 PMID:30351901

Li, Y., Zhang, Z., Zhu, X., Men, X., Ge, B., & Zhou, X. (2015). Fabrication of a superhydrophobic coating with high adhesive effect to substrates and tunable wettability. *Applied Surface Science*, *328*, 475–481. doi:10.1016/j.apsusc.2014.12.086

Liang, Y., Ju, J., Deng, N., Zhou, X., Yan, J., Kang, W., & Cheng, B. (2018). Super-hydrophobic self-cleaning bead-like SiO2@PTFE nanofiber membranes for waterproof-breathable applications. *Applied Surface Science*, *442*, 54–64. doi:10.1016/j.apsusc.2018.02.126

Lin, Y., Yuan, G., Sheehan, S., Zhou, S., & Wang, D. (2011). Hematite-based solar water splitting: Challenges and opportunities. *Energy & Environmental Science*, *4*(12), 4862. doi:10.1039/c1ee01850g

Liu, S., Liu, X., Latthe, S. S., Gao, L., An, S., Yoon, S. S., Liu, B., & Xing, R. (2015). Self-cleaning transparent superhydrophobic coatings through simple sol–gel processing of fluoroalkylsilane. *Applied Surface Science*, *351*, 897–903. doi:10.1016/j.apsusc.2015.06.016

Luo, G., Jin, Z., Dong, Y., Huang, J., Zhang, R., Wang, J., Li, M., Shen, Q., & Zhang, L. (2018). Preparation and performance enhancements of wear-resistant, transparent PU/SiO 2 superhydrophobic coating. *Surface Engineering*, *34*(2), 139–145. doi:10.1080/02670844.2016.1236068

Mahadik, S., Pedraza, F., Parale, V. G., & Park, H.-H. (2016). Organically modified silica aerogel with different functional silylating agents and effect on their physico-chemical properties. *Journal of Non-Crystalline Solids*, *453*, 164–171. doi:10.1016/j.jnoncrysol.2016.08.035

Mahadik, S. A., Kavale, M. S., Mukherjee, S. K., & Rao, A. V. (2010). Transparent Superhydrophobic silica coatings on glass by sol–gel method. *Applied Surface Science*, *257*(2), 333–339. doi:10.1016/j.apsusc.2010.06.062

Mahadik, S. A., Mahadik, D. B., Kavale, M. S., Parale, V. G., Wagh, P. B., Barshilia, H. C., Gupta, S. C., Hegde, N. D., & Rao, A. V. (2012b). Thermally stable and transparent superhydrophobic sol-gel coatings by spray method. *Journal of Sol-Gel Science and Technology*, *63*(3), 580–586. Advance online publication. doi:10.100710971-012-2798-3

Mahadik, S. A., Mahadik, D. B. B., Kavale, M. S. S., Parale, V. G. G., Wagh, P. B. B., Barshilia, H. C., Gupta, S. C., Hegde, N. D. D., & Rao, A. V. (2012a). Thermally stable and transparent superhydrophobic sol–gel coatings by spray method. *Journal of Sol-Gel Science and Technology*, *63*(3), 580–586. doi:10.100710971-012-2798-3

Mahadik, S. A., & Mahadik, S. S. (2021). Surface morphological and topographical analysis of multifunctional superhydrophobic sol-gel coatings. *Ceramics International*, *47*(20), 29475–29482. Advance online publication. doi:10.1016/j.ceramint.2021.07.115

Mahadik, S. A., Parale, V., Vhatkara, R. S., Mahadik, D. B., Kavale, M. S., Wagh, P. B., Gupta, S., & Gurav, J. (2013). Superhydrophobic silica coating by dip coating method. *Applied Surface Science*, *277*, 67–72. doi:10.1016/j.apsusc.2013.04.001

Mahadik, S. A., Pedraza, F., & Mahadik, S. S. (2017). Comparative studies on water repellent coatings prepared by spin coating and spray coating methods. *Progress in Organic Coatings*, *104*, 217–222. doi:10.1016/j.porgcoat.2016.11.006

Mahadik, S. A., Pedraza, F., Mahadik, S. S., Relekar, B. P., & Thorat, S. S. (2017). Biocompatible superhydrophobic coating material for biomedical applications. *Journal of Sol-Gel Science and Technology*, *81*(3), 791–796. doi:10.100710971-016-4244-4

Mahadik, S. A., Pedraza, F., & Vhatkar, R. S. (2016). Silica based superhydrophobic coating for long-term industrial and domestic applications. *Journal of Alloys and Compounds*, *663*, 487–493. Advance online publication. doi:10.1016/j.jallcom.2015.12.016

Mahadik, S. A., Pedraza, F. D., Relekar, B. P., Parale, V. G., Lohar, G. M., & Thorat, S. S. (2016). Synthesis and characterization of superhydrophobic–superoleophilic surface. *Journal of Sol-Gel Science and Technology*, *78*(3), 475–481. doi:10.100710971-016-3974-7

Mahadik, S. A. S. A., Fernando, P. D., Hegade, N. D. N. D., Wagh, P. B. P. B., & Gupta, S. C. S. C. (2013). Durability and restoring of superhydrophobic properties in silica-based coatings. *Journal of Colloid and Interface Science*, *405*, 262–268. doi:10.1016/j.jcis.2013.04.042 PMID:23746435

Mahadik, S. A. S. A., Mahadik, D. B. B., Parale, V. G. G., Wagh, P. B. B., Gupta, S., & Venkateswara Rao, A. (2012). Recoverable and thermally stable superhydrophobic silica coating. *Journal of Sol-Gel Science and Technology*, *62*(3), 490–494. doi:10.100710971-012-2753-3

Mahadik, S. A. S. A., Pedraza, F., & Vhatkar, R. S. R. S. (2016a). Silica based superhydrophobic coating for long-term industrial and domestic applications. *Journal of Alloys and Compounds*, *663*, 487–493. doi:10.1016/j.jallcom.2015.12.016

Mahadik, S. A. S. A., Pedraza, F., & Vhatkar, R. S. R. S. R. S. (2016b). Silica based superhydrophobic coating for long-term industrial and domestic applications. *Journal of Alloys and Compounds*, *663*, 487–493. doi:10.1016/j.jallcom.2015.12.016

Mosa, J., Rosero-Navarro, N. C., & Aparicio, M. (2016). Active corrosion inhibition of mild steel by environmentally-friendly Ce-doped organic–inorganic sol–gel coatings. *RSC Advances*, *6*(46), 39577–39586. doi:10.1039/C5RA26094A

Mülazim, Y., Kahraman, M. V., Apohan, N. K., Kızıltaş, S., & Güngör, A. (2011). Preparation and characterization of UV-curable, boron-containing, transparent hybrid coatings. *Journal of Applied Polymer Science, 120*(4), 2112–2121. doi:10.1002/app.33358

Parale, V. G., Mahadik, D. B., Mahadik, S. A., Kavale, M. S., Wagh, P. B., Gupta, S. C., & Rao, A. V. (2013). OTES modified transparent dip coated silica coatings. *Ceramics International, 39*(1), 835–840. Advance online publication. doi:10.1016/j.ceramint.2012.05.079

Parale, V. G. G., Mahadik, D. B. B., Kavale, M. S. S., Mahadik, S. A. A., Rao, A. V. V., & Mullens, S. (2013). Sol-gel preparation of PTMS modified hydrophobic and transparent silica coatings. *Journal of Porous Materials, 20*(4), 733–739. doi:10.100710934-012-9648-0

Pedraza, F., Mahadik, S. A., & Bouchaud, B. (2015). Synthesis of ceria based superhydrophobic coating on Ni20Cr substrate via cathodic electrodeposition. *Physical Chemistry Chemical Physics, 17*(47), 31750–31757. doi:10.1039/C5CP04723D PMID:26562006

Qi, C., Zheng, Y., Cao, L., Gao, J., & Wan, Y. (2016). Preparation and performance of sol–gel-derived alumina film modified by stearic acid. *Journal of Sol-Gel Science and Technology, 78*(3), 641–646. doi:10.100710971-016-3989-0

Qiao, Y., Li, W., Wang, G., Zhang, X., & Cao, N. (2015). Application of ordered mesoporous silica nanocontainers in an anticorrosive epoxy coating on a magnesium alloy surface. *RSC Advances, 5*(59), 47778–47787. doi:10.1039/C5RA05266A

Rao, A. V. V., Latthe, S. S. S., Mahadik, S. A. S. A., & Kappenstein, C. (2011). Mechanically stable and corrosion resistant superhydrophobic sol-gel coatings on copper substrate. *Applied Surface Science, 257*(13), 5772–5776. doi:10.1016/j.apsusc.2011.01.099

Righeira Carnegie, M., Sherine, A., Sivagami, D., & Sakthivel, S. (2016). Anti-reflection coatings with enhanced abrasion and scratch resistance properties. *Journal of Sol-Gel Science and Technology, 78*(1), 176–186. doi:10.100710971-015-3924-9

Saffar, M. A., Eshaghi, A., & Dehnavi, M. R. (2021). Fabrication of superhydrophobic, self-cleaning and anti-icing ZnO/PTFE-SiO2 nano-composite thin film. *Materials Chemistry and Physics, 259*, 124085. doi:10.1016/j.matchemphys.2020.124085

Sebastian, D., Yao, C.-W., Nipa, L., Lian, I., & Twu, G. (2021). Corrosion Behavior and Mechanical Properties of a Nanocomposite Superhydrophobic Coating. *Coatings, 11*(6), 652. doi:10.3390/coatings11060652

Tang, Y., Zhang, Q., Zhan, X., & Chen, F. (2015). Superhydrophobic and anti-icing properties at overcooled temperature of a fluorinated hybrid surface prepared via a sol–gel process. *Soft Matter, 11*(22), 4540–4550. doi:10.1039/C5SM00674K PMID:25966370

Taurino, R., Fabbri, E., Pospiech, D., Synytska, A., & Messori, M. (2014). Preparation of scratch resistant superhydrophobic hybrid coatings by sol–gel process. *Progress in Organic Coatings, 77*(11), 1635–1641. doi:10.1016/j.porgcoat.2014.05.009

Topcu, A. S. K., Erdogan, E., & Cengiz, U. (2018). Preparation of stable, transparent superhydrophobic film via one step one pot sol-gel method. *Colloid & Polymer Science*, *296*(9), 1523–1532. doi:10.100700396-018-4377-9

Trabelsi, O., Tighzert, L., Jbara, O., & Hadjadj, A. (2011). Synthesis via sol–gel process and characterization of novel organic–inorganic coatings. *Journal of Non-Crystalline Solids*, *357*(24), 3910–3916. doi:10.1016/j.jnoncrysol.2011.08.005

Tsukakoshi, S., Itatani, K., & Koda, S. (2011). Properties of silicious film on polycarbonate substrate prepared by vacuum ultraviolet irradiation: Effect of intermediate silane layer. *Journal of the European Ceramic Society*, *31*(14), 2489–2496. doi:10.1016/j.jeurceramsoc.2011.06.011

Wouters, M. E. L., Wolfs, D. P., van der Linde, M. C., Hovens, J. H. P., & Tinnemans, A. H. A. (2004). Transparent UV curable antistatic hybrid coatings on polycarbonate prepared by the sol–gel method. *Progress in Organic Coatings*, *51*(4), 312–319. doi:10.1016/j.porgcoat.2004.07.020

Wu, L. Y. L., Chwa, E., Chen, Z., & Zeng, X. T. (2008). A study towards improving mechanical properties of sol–gel coatings for polycarbonate. *Thin Solid Films*, *516*(6), 1056–1062. doi:10.1016/j.tsf.2007.06.149

Wu, X., & Chen, Z. (2018a). A mechanically robust transparent coating for anti-icing and self-cleaning applications. *Journal of Materials Chemistry. A, Materials for Energy and Sustainability*, *6*(33), 16043–16052. doi:10.1039/C8TA05692G

Wu, X., & Chen, Z. (2018b). A mechanically robust transparent coating for anti-icing and self-cleaning applications. *Journal of Materials Chemistry. A, Materials for Energy and Sustainability*, *6*(33), 16043–16052. doi:10.1039/C8TA05692G

Wu, X., Fu, Q., Kumar, D., Ho, J. W. C., Kanhere, P., Zhou, H., & Chen, Z. (2016). Mechanically robust superhydrophobic and superoleophobic coatings derived by sol–gel method. *Materials & Design*, *89*, 1302–1309. doi:10.1016/j.matdes.2015.10.053

Wu, X., Tang, Y., Silberschmidt, V. V., Wilson, P., & Chen, Z. (2018). Mechanically Robust Transparent Anti-Icing Coatings: Roles of Dispersion Status of Titanate Nanotubes. *Advanced Materials Interfaces*, *5*(18), 1800773. doi:10.1002/admi.201800773

Xiu, Y., Hess, D. W., & Wong, C. P. (2007). A Novel Method to Prepare Superhydrophobic, Self-Cleaning and Transparent Coatings for Biomedical Applications. *2007 Proceedings 57th Electronic Components and Technology Conference*, 1218–1223. 10.1109/ECTC.2007.373949

Zhang, X.-F., Zhao, J.-P., & Hu, J.-M. (2017). Abrasion-Resistant, Hot Water-Repellent and Self-Cleaning Superhydrophobic Surfaces Fabricated by Electrophoresis of Nanoparticles in Electrodeposited Sol-Gel Films. *Advanced Materials Interfaces*, *4*(13), 1700177. doi:10.1002/admi.201700177

Zhang, Y., Dong, B., Wang, S., Zhao, L., Wan, L., & Wang, E. (2017). Mechanically robust, thermally stable, highly transparent superhydrophobic coating with low-temperature sol–gel process. *RSC Advances*, *7*(75), 47357–47365. doi:10.1039/C7RA08578H

Section 3

Chapter 13
The Role of Two-Dimensional Materials in Superlubricity on Friction and Wear-Prone Surfaces

Rafael Vargas-Bernal
https://orcid.org/0000-0003-4865-4575
Instituto Tecnológico Superior de Irapuato, Mexico

ABSTRACT

Moving mechanical systems create a lot of friction; therefore, a lot of the energy produced is used to overcome it. It is vital to find unique ways to develop lubricants that allow for more effective control or decrease friction to reach a sustainable future. High friction, if not efficiently reduced or controlled, can result in higher wear losses and, as a result, shorter system life and lower reliability. Two-dimensional (2D) materials have distinct friction and wear properties from their three-dimensional (3D) counterparts. They can be used as additives in oils and composites to reduce stiction, friction, and wear, even though they are ultra-thin even with numerous layers. The role of these materials in superlubricity on surfaces prone to friction and wear is discussed in this chapter. These materials are solid two-dimensional lubricants that can address developing needs in mechanical systems in motion in current and emergent real-world applications.

INTRODUCTION

One of the issues that mankind faces is reducing energy losses due to friction and wear, which for roughly 20% of the total energy produced worldwide (Holmberg, 2019). The use of novel materials in lubricants to develop more efficient systems could cut these losses by as much as 40%. These materials should be used as low-shear-resistance coatings to reduce frictional forces and hence improve tribological performance. Furthermore, these additives must be non-toxic and environmentally beneficial. They

DOI: 10.4018/978-1-7998-9683-8.ch013

may be exposed to radiation, function at high temperatures, or work in a vacuum or under ultra-clean conditions (Rosenkranz, 2020).

Without the addition of additives, base fluids such as mineral oil and synthetic goods cannot meet the standards of a high-performance lubricant (Braun, 2017). Synthetic substances that can increase the qualities of lubricants are known as additives. They serve three purposes in base fluids: to enhance existing properties, inhibit unwanted qualities, and introduce new properties. Lubricants typically comprise at least 90% mineral oil and no more than 10% additives. Tribology divides additives into two categories: 1) those that affect the physical and chemical properties of base fluids, and 2) those that affect metal surfaces by altering their physicochemical qualities. The additives examined in this chapter are those that are designed to reduce friction, support severe pressures, and protect surfaces from wear. The lubrication system operates in a mixed friction event when two components of equipment in contact move and hydrodynamic lubrication has not been formed, or when high stress and powerful pressures are applied. Antiwear (AW) and extreme pressure (EP) additives are necessary for metalworking fluid, motor oil, hydraulic fluid, or lubricating grease in this scenario to prevent moving parts from wearing out. These additives have a polar structure that forms layers on the metal surface through adsorption or chemisorption, ensuring their instant availability through the formation of a coating. As the temperature rises, these additives react with the metal surface, generating tribochemical reaction layers (such as phosphides, sulfides, sulfates, oxides, and/or carbides) that keep the sliding metals from making direct contact. Friction modifiers (FM) are layers that are physically adsorbed and have only moderately high or poor pressure support qualities. The coefficient of friction and viscosity is reduced by these modifiers. They perform at high temperatures with low sliding speeds, substantially increased loads, and low viscosities. By lowering friction forces, these reduce stick-slip oscillations and noise.

The newly introduced additives, which are based on finely ground powders of two-dimensional materials and their dispersions, are utilized as solid additions because of their outstanding circulation capabilities and high thermal stability when the oil supply fails. Because they minimize friction and increase the anti-friction capabilities of basic oil, two-dimensional materials can be used as developing lubrication additives (Yang, 2014). These even multi-layer materials are ultra-thin and can be used as additives in oils to reduce stiction, coefficient of friction ($0.0001 < \mathrm{CoF} < 0.01$), and wear in mechanical systems with oscillating, rotating, and sliding contacts. Ultra-low shear strength, high specific surface strength, in-plane strength, poor layer-to-layer interaction, and surface chemical stability are all physical properties of two-dimensional materials that are important for superlubricity (Liu, 2019). Graphene and its derivatives were the first two-dimensional materials to be utilized as oil additives, and when applied in small amounts, they considerably reduced the friction and wear of steel parts. Other two-dimensional materials have been developed to optimize tribological performance with positive results due to their simple chemistry and the weak van der Waals bond between graphene layers. Molybdenum disulfide (MoS_2), tungsten disulfide (WS_2), phosphorene or black phosphorus (BP), hexagonal boron nitride (*h*-BN), layered double hydroxides (LDH), graphitic carbon nitride (g-C_3N_4), metal-organic frameworks (MOFs), and MXenes are some of the other solid lubricants (Liu, 2019; Rosenkranz, 2020; Wang, H., 2020). Additionally, these two-dimensional nanomaterials have a high elastic modulus, great strength, and ultra-low friction. Two-dimensional titanium carbide Ti_3C_2 can be used as a lubrication additive, either as a single layer or stacked in many layers, to reduce friction and anti-friction qualities of the base oil when increased to 1% by weight (Yang, 2014). Film-forming, self-healing, and ball-bearing lubrication mechanisms are used in these lubricants. Lubricant additives, nanoscale lubrication films, and aeronautical lubrication materials are the major uses (Liu, 2019).

Nanocomposites based on two-dimensional materials can also be created for tribological applications (Ji, 2020). Polymers, ceramics, and metals are among the matrices employed. Through effective dispersion and self-lubrication of the two-dimensional nanofillers, these materials have a synergistic effect that reduces friction, improves wear resistance, and increases lubricity and heat conductivity. The kind and content of the two-dimensional material, as well as the interfacial bond, determine the performance of these polymeric nanocomposites. It is feasible to strengthen the interfacial bond by functionalizing the two-dimensional material. To improve the lubricity qualities of ceramic matrix composites, interdiffusion and chemical interactions of both materials at high temperatures must be avoided.

The goal of this chapter is to emphasize the importance of two-dimensional materials in maximizing superlubricity on friction and wear-prone surfaces. The lubricity mechanisms linked with the use of lubricants supplemented with two-dimensional materials are investigated in this study. These lubricants are presented in terms of their current applications and solutions. Finally, future research directions are discussed to emphasize the importance of continuing to study these lubricants.

BACKGROUND

Friction and wear are two mechanical phenomena derived from the relative movement between the solid surfaces of material elements and fluid layers, sliding against each other. Friction is created by sliding, rolling, or rotating contact interfaces in any man-made, natural, or biological system. These phenomena must be controlled because they play a vital role in the reliability, efficiency, endurance, and performance of machinery (Meng, 2021). Control can be done through the meticulous design and manufacture of machinery, selection and development of specific materials to achieve superficial modification through coatings, and the use of lubrication technology. Lubrication technology involves maintenance, lubricant supply control, and the use of additives as well as fluid and solid lubricants. As time passes, multiple requirements arise which involve the design of intelligent mechanical systems that meet the new challenges that traditional passive technologies cannot solve, such as engines running at low speeds or in the start-stop cycles in hybrid cars used in the large cities so congested of vehicular traffic. For some decades, one of the technical strategies that tribologists have in a phase of technological development and research to reduce friction and wear is active control. This control from a lubrication process perspective is associated with the behavior of the adsorption film in the nanoscale (Zhang, 2015) or the use of intelligent lubrication materials (Gong, 2020).

Lubrication can be defined as a process or technique used to reduce friction and wear between one or a pair of surfaces when they are in proximity and move relatively from one to another through a lubricant between them (Zhang, 2015). The lubricant can be a gas, a liquid, a liquid-liquid dispersion, a solid-liquid dispersion, or a solid. In practice, the effects of a lubricating film are not limited to friction and wear of surfaces but produce effects between the interface of both materials such as adhesion, cohesion, electric double layer, electronic and ionic emissions, and forces and surface energy.

In general, there are two basic lubrication regimens which are fluid film lubrication and boundary film lubrication. A continuous fluid film separates solid surfaces in the fluid film lubrication regime. The pressure generated within the fluid layer fully supports the external load on the solid surfaces in this regime. The fluid film is generated and maintained in the hydrostatic mode when the lubricant is forced into the space or hole between the parts by an external pump. The fluid film is said to operate in the self-acting hydrodynamic mode when it is formed and retained by the viscous drag of the surfaces moving relative to each other with a wedge. This state is called elastohydrodynamic lubrication when the pressure in the fluid film is so high that the deformation of the solids under the pressure is not negligible, and/or the changes in the viscosity and density of the lubricant interposed within the space are no longer ignorable, such as under counter-formal conditions (or elastohydrodynamic lubrication when the deformation is plastic). Fluid mechanics and solid mechanics are the only two principles that govern fluid film lubrication. The effects of heat on fluid film lubrication can also be considered.

Unlike fluid film lubrication, the lubricant in boundary film lubrication is not fluid, regardless of where it comes from or whether it is formed by a solution, gas, or another medium, and so fluid mechanics is inadequate to describe the behavior of boundary films. The boundary film also has the property of covering the solid substrate in molecularly thin layers, ranging from a monolayer to many layers. The load is carried by the molecularly thin layer in boundary lubrication if the solid substrate is atomically smooth and/or the local pressure is insufficiently high to break down the intervening boundary films, as shown in Figure 1 (a) (Hardy, 1922). For most rough-designed surfaces, the external load is usually shared by the boundary film and the distributed contact roughness that penetrate the thin boundary layer, as shown in Figure 1 (b) (Bowden, 1945). A variety of processes rich in variation occur when one rough surface slides across another, including debris wear, tribofilm formation, material transfer, stick-slip movement, and temperature rise.

Figure 1. (a) Hardy's and (b) Bowden's frontier lubrication models are shown schematically (adapted from (Hardy, 1922; Bowden, 1945; Zhang, 2015))

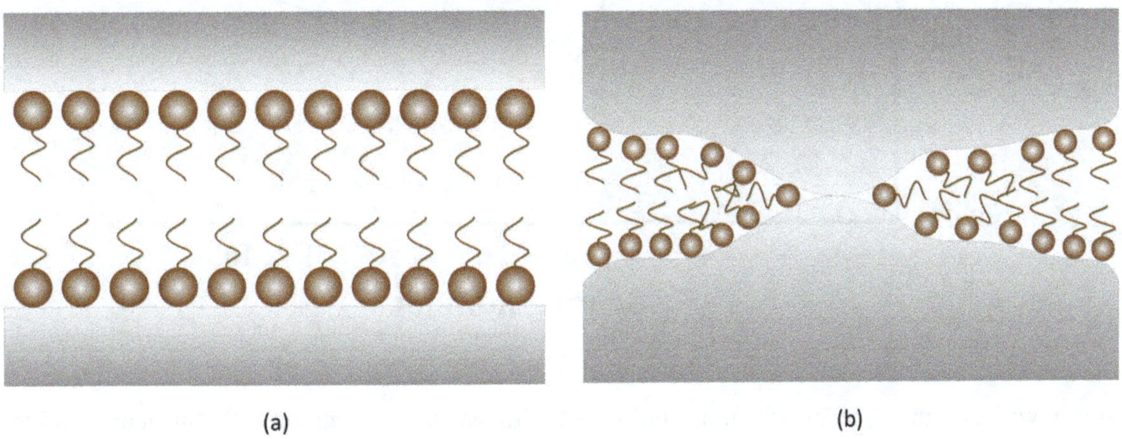

Fluid film lubrication, in which the two surfaces are completely separated from each other by the resulting fluid film, has a lower coefficient of friction; boundary lubrication, in which the surfaces are in direct contact, has a higher coefficient of friction; and mixed lubrication, which is a transitional stage between boundary lubrication and fluid film lubrication, has an intermediate coefficient of friction. Hydrodynamic lubrication (LH) and elastohydrodynamic lubrication (EH) are two sub-regimes of the fluid film regime (EHL). The difference between these two sub-regimes is that in EHL, a solid structure's deformation is of the same order of magnitude as the film thickness, which is commonly measured in nanometers to micrometers. The solid deformation in HL is substantially less than the film thickness and can be ignored. The Stribeck curve, as shown in Figure 2, is used to depict the regimes. The coefficient of friction is related to the lubricant's viscosity and the difference in speed between the contact surfaces, as well as inversely proportional to the pressure applied to the contact surfaces. The Stribeck curve, which plots the coefficient of friction versus the bearing number: $\eta\omega/p$ (lubricating viscosity, η, speed (ω), and contact pressure (p), is shown in Figure 2. Friction losses are exceedingly low in hydrodynamic lubrication, far lower than in other lubrication regimes. Due to relative motion and internal film friction, the hydrodynamic pressure created in the lubricating film is insufficient to cause surface deformation.

Figure 2. The coefficient of friction as a function of bearing number is illustrated using the Stribeck curve and the lubrication regime
(adapted from (Delprete, 2018))

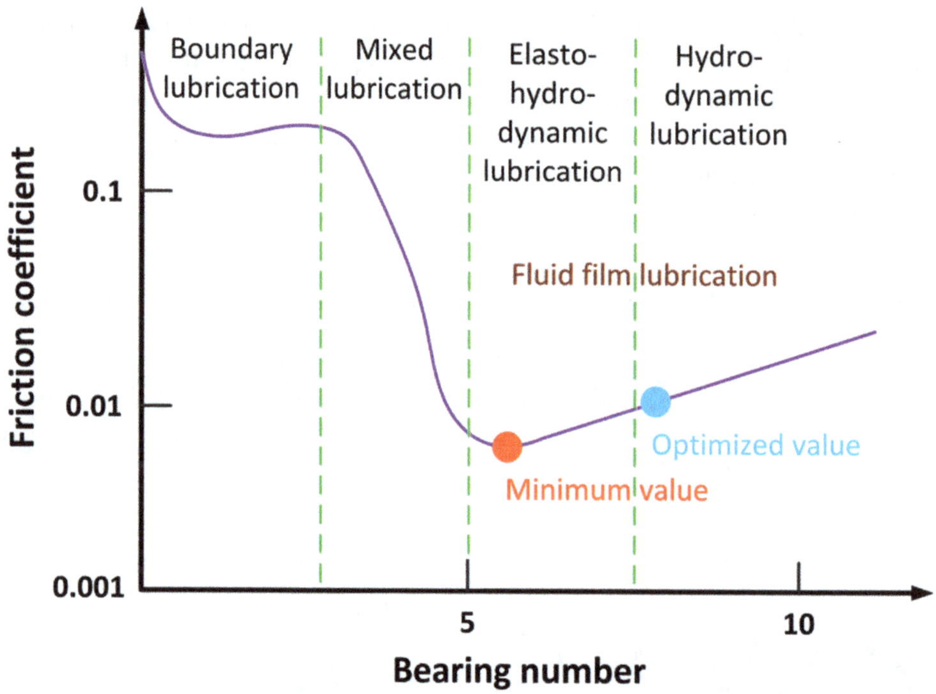

As shown in Figure 3, several friction studies including various volumetric solid lubricants and their thin coatings have been used for friction and wear reduction in previous decades. The coefficient of friction presented by solid lubricating materials in dry and wet environments is depicted in this diagram. Surprisingly, less research on the tribological characteristics of two-dimensional materials as solid lubricants at the microscale and nanoscale has been done until the previous decade.

Figure 3. Before superlubricity, friction coefficients of solid lubricants in (a) dry and (b) wet environments

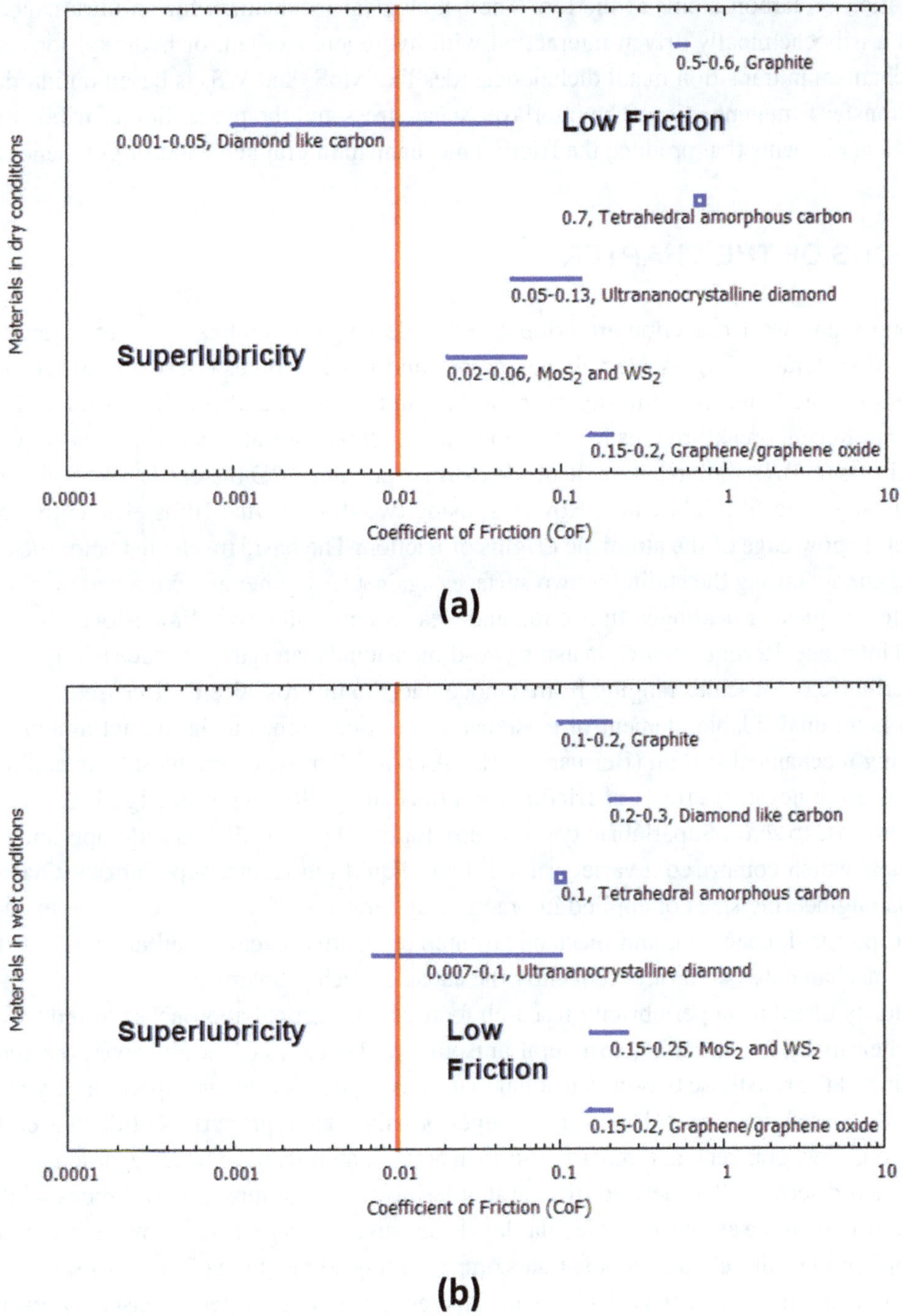

For the materials illustrated in Figure 3, the wear and/or friction mechanisms are described below. In the case of graphite, dry and wet friction is caused by interlayer shear stress and water intercalation, respectively (Berman, 2014). The dry and wet causes of friction for diamond as carbon (near-frictionless carbon) are strong chemical inertia and repulsive forces owing to hydrogen termination. The dry and wet causes of friction in amorphous tetrahedral carbon include tribochemically driven surface reaction and termination of carbon atoms at the top. The tribological mechanism for an ultra-nanocrystalline diamond is a tribochemically driven interaction with hydrogen, oxygen, or hydroxyl ions. The tribological mechanism in transition metal dichalcogenides like MoS_2 and WS_2 is based on interlayer shear stress and transfer film generation. The interlayer shear stress and the prevention of tribocorrosion are the tribological elements that produce the friction mechanism in graphene and/or graphene oxide.

MAIN FOCUS OF THE CHAPTER

Friction control and wear reduction are critical in a wide range of applications, from nanoscale electromechanical systems to large-scale vehicle engines and wind turbines (Berman, 2018). As a result, numerous efforts are being made to design materials and surfaces that provide efficient friction and handling wear. Recent breakthroughs in two-dimensional (2D) materials such as graphene (G), hexagonal boron nitride (*h*-BN), molybdenum disulfide (MoS_2), and other 2D materials have ushered in a new era of atomically solid film lubricants. However, using two-dimensional films efficiently necessitates a fundamental knowledge of the atomistic origins of friction. The basic mechanisms for the dissipation of frictional energy during the sliding of two surfaces against each other are explained in this review, as well as the techniques for dealing with friction and wear when adding two-dimensional materials at the tribological interface. Recent research in using two-dimensional materials to reduce friction to near-zero levels -superlubricity- at scales ranging from nanocontacts to macroscales is discussed.

Friction is an unavoidable element of existence since it consumes a significant amount of energy in any moving mechanical system (Berman, 2018). As a result, researchers have been exploring novel ways to reduce the negative effects of friction (and thus energy losses) to negligible levels for a long time (friction close to zero). Superlubricity is the term for this. Despite all these attempts and fundamental techniques, which comprised a variety of solid and liquid lubricants, superlubricity has only been produced on engineering sizes or applied in practical systems on a few occasions. The extraordinarily complicated physical, chemical, and mechanical interactions that occur simultaneously at the sliding interface of mechanical systems are frequently the cause of such problems.

The majority of extant superlubricity research focuses on practical approaches to reduce friction in lamellar materials by manipulating structural anisotropy (Berman, 2018). However, the fundamental understanding of the role those two-dimensional materials have in achieving superlubricity has not been extensively discussed previously. Due to their superior tribological properties of decreased friction in superlubric regimes, graphene and other two-dimensional materials have been regarded as solid lubricants since their discovery. The desired structural ordering or orientations of two-dimensional materials offer superlubricity at the atomic or molecular level. Because two-dimensional materials offer the most practical applications, this chapter concentrates on presenting some of the efforts focused on producing superlubricity using these materials. These materials restrict chemical effects caused by reactivity with the environment while controlling structural stresses.

Two topics that undoubtedly establish the general concepts involved in the chapter are solid lubricants that represent the materials that will be studied in the case of two-dimensional materials. On the other hand, the superlubricity phenomenon under which these two-dimensional materials operate is analyzed with the required intention to identify the relevance of using solid lubricants in high-value-added applications.

SOLID LUBRICANTS

A lubricant is a substance that aids in the reduction of friction, the transmission of force, the movement of foreign particles, and the cooling of moving surfaces (Bart, 2013). A solid lubricant is a powder or thin coating that decreases friction and wear between contacting surfaces in relative motion while also protecting them from degradation. Solid lubricants are used to create a continuous, adhering soft or hard layer on rubbing surfaces. Mechanical, electrochemical, and physical processes can all be used to apply these films. The influences of particle shape, size, mobility, and crystallographic characteristics of the particles must be considered in these materials.

Solid lubricants have a unique role to play in decreasing wear in conditions where liquid lubricants are either impracticable or undesirable, such as in a vacuum, space technology, or vehicular mobility (Bart, 2013). Solid lubricants are necessary for lubrication in harsh situations where tribological contact bearing surfaces must remain well separated and the lubricant must remain in place. Solid lubricant products are commonly used in applications where high specific loads are applied to sliding surfaces in the presence of boundary and mixed frictional regimes, at very low hydrodynamic effective speeds, or when the lubricant must develop over a wide temperature range or under extreme temperature conditions (for example, in aviation). In harsh environments, high-vacuum applications, nuclear reactors, and other applications where contamination by lubricating oils or greases must be avoided, dry lubrication with solid lubricants is also necessary.

Solid lubricants can be made from a variety of materials (Bart, 2013). Structural lubricants, mechanical lubricants, (reactive) soaps, and chemically active lubricants are the most common types of solid lubricants. The layered lattice structures of structural lubricants (e.g., graphite and metal dichalcogenides) account for their lubricating capabilities. Self-lubricating organic substances (such as thermoplastics and thermosets) and naturally occurring metal oxide films (typically around 10 nm thick), chemical surface coatings (produced by chemical or electrochemical action on the metal surface), and glasses are all examples of mechanical lubricants. Reactive soaps (stearic, oleic, and palmitic acid salts) in combination with a zinc phosphate coating are common solid metal-forming lubricants. The manufacture of greases is the primary function of soaps (metal salts of fatty acids) in lubrication technology. Soaps can also be generated in situ on a metal surface by a fatty acid attacking the metal chemically. Extreme pressure (EP) and anti-wear (AW) additives, as well as other compounds, are chemically active lubricants that interact with the metal surface to form a lubricating or protective layer. The lamellar solids MoS_2 and graphite, as well as poly (tetrafluoroethylene) (PTFE) (combined with organic and inorganic lubricants) and other fluorine-containing polymers, are examples of common solid lubricants. Solid lubricants such as magnesium stearate dihydrate (MgSt-D) and hydrogenated castor oil (melting point 86 °C) have also been used.

Crystal structure, thermal and oxidative stability, volatility, chemical reactivity, melting point, and hardness are all important material properties for solid lubricants (Bart, 2013). If thermal degradation of the solid lubricant is to be avoided, thermal stability (which is closely related to chemical stability) is essential. Solid lubricants have a temperature range within which they work. Graphite, for example, can resist temperatures of up to 650 °C and moderate stresses. Up to 300 degrees Celsius, starch remains stable.

Only a few solid lubricants have the necessary adhesive and cohesive qualities to form an effective lubrication layer with low coefficients of friction and long life (Bart, 2013). A transport medium, a binding agent (such as oils, greases, or water), and/or preparation of the material surface are all required for most solid lubricants. Solid lubricant compositions usually include a solid, a binder, and additives like corrosion inhibitors and solvents. Powders, dispersions and suspensions, greases and greases, pastes, and dry-film lubricants are all examples of solid lubricant products.

Solid lubricants are typically sold as powders embedded in a resin or wax film, or as oil or water dispersions (Bart, 2013). Solid lubricants in powder form must have the right qualities to form a film (cohesive properties) and adhere to surfaces (adhesive properties). Self-lubricating dry lubricants can meet these requirements, although MoS_2, graphite, and PTFE are significantly less so. To coat mass elements for hot and cold forming, solid lubricant dispersions and suspensions in water are utilized. In gear and oil lubrication systems, oil dispersions and suspensions are employed as additives. Solid lubricating chemicals (graphite, MoS_2, ZnS, *h*-BN, PTFE, ultra-high molecular weight polyethylene (UHMWPE)) can be found in high concentrations in greases. Micronized solids with an average particle size of 5 to 15 m are suitable for most applications. The addition of solid lubricants to greases improves pressure absorption capacity, wear resistance, and friction resistance. Pastes are lubricants that are solid and suspended in a carrier oil.

Tribofilms also called lubricating boundary films, tribo-boundary films or boundary films are films that reduce friction and wear in lubricated systems on tribologically stressed surfaces (Bart, 2013). These films are generally solid surface films that form when lubricating components or tribological surfaces react chemically. Complex mechanochemical interactions between surface materials and lubricants cause them to form. Reaction films can be tribochemical (lubricants containing Zinc dithiophosphate (ZDDP)) or polymeric and non-sacrificial reaction films (formed by complex esters).

Solid lubrication is an alternative method of reducing the friction that involves introducing micrometer or nanoscale particles into the contact zone (Hod, 2018). Exfoliation of consecutive layers or full collapse of onion-like formations can act as small bearings as well as a source of flake lubrication. External lubricants must be introduced into the sliding contact to reduce friction. The contacts are covered with two-dimensional materials that provide the required superlubricity to avoid large external loads and high slip speeds when the conditions are right. Additionally, the numerous asperity configurations effectively isolate the individual nanoscale connections, reducing unwanted elastic effects. However, the effective contact edge-to-surface ratio will skyrocket, resulting in unwanted edge pinning and increased friction. Deposition of two-dimensional material coatings on stiff surfaces and/or the use of multilayer stacks are two ways to lessen in-plane stretching effects. The volumetric support inhibits the tendency of the friction interface's contact layers to modify their crystal lattices, reducing the creation of locally comparable areas and favoring superlubricity.

Superlubricity

The friction regime in which friction decreases or almost vanishes is known as superlubricity (Shekhar, 2021). Under specific conditions, scientists have developed major friction reduction methods in recent years. Researchers have worked hard to find new techniques to lessen negative friction effects (and thus energy losses) while also aiming at near-zero coefficients of friction. Because it is impossible to achieve a condition of zero friction, the idea of superlubricity is defined as a regime where the coefficient of friction is decreased to the order of 0.01 or less, as shown in Figure 4. Since its inception, the concept of superlubricity has stimulated the interest of researchers in tribology, physics, chemistry, materials science, and other domains (Li, 2013). Over the last three decades, tremendous progress has been achieved in experimental studies of superlubricity. Researchers from all over the world have made significant progress in practical application, computer simulation, and theoretical concepts related to solid and liquid superlubricity. When the two friction surfaces are in direct contact with each other, traditional solid lubricants such as carbon-like diamond (DLC) films, molybdenum disulfide (MoS_2), graphite, and carbon nitride (CN_x) exhibit superlubricity. Superlubricity research is progressing more quickly thanks to advancements in nanomaterial manufacturing techniques, software, and computers, as well as ways for monitoring and characterizing material properties. With the current trend of progress, it is believed that superlubricity would be utilized in a realistic method soon, saving not only energy but also reducing pollution during the friction process. Although suitable chemistry, structural ordering, and orientations in two-dimensional materials and heterojunctions can roughly attain this regime on a molecular and/or atomic scale, changing this phenomenon on an industrial scale is difficult (Shekhar, 2021). Significant progress has been made in this domain in recent years as a result of attempts at characterizations, modeling, and simulations; the resulting mechanisms are divided into eight main categories, namely structural superlubricity, thermolubricity, liquid superlubricity, hydrogen mediated superlubricity, superlubricity by rolling effect, superlubricity by cold ion chains, superlubricity by telescopic dynamics and confined geometries, and superlubricity by self-assembly in liquid crystals (Baykara, 2018; Martin, 2018) as depicted in Figure 5.

Figure 4. Friction regimes based on the friction coefficient value
(adapted from (Wang, 2020; Shekhar, 2021))

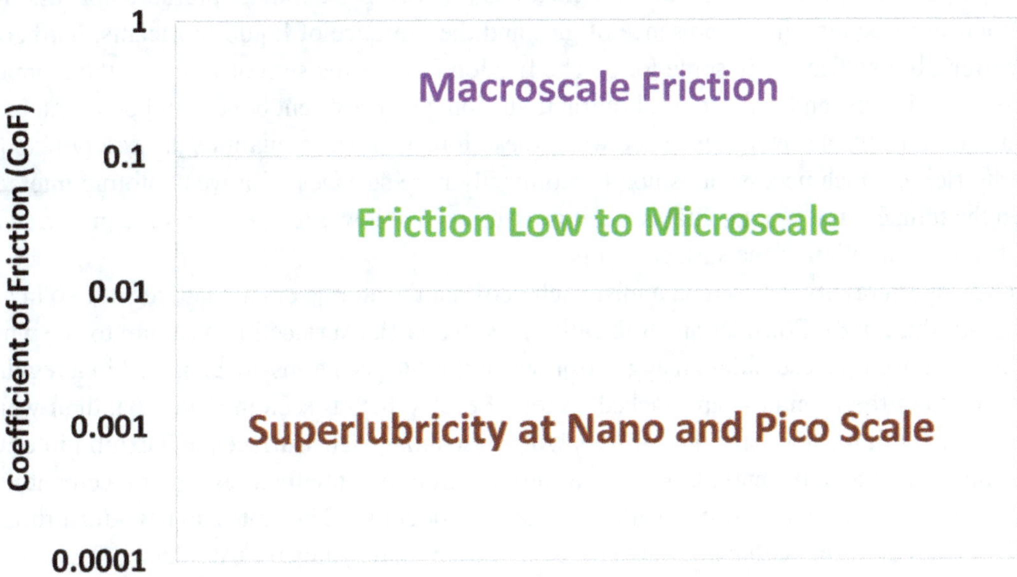

Figure 5. Possible mechanisms of superlubricity

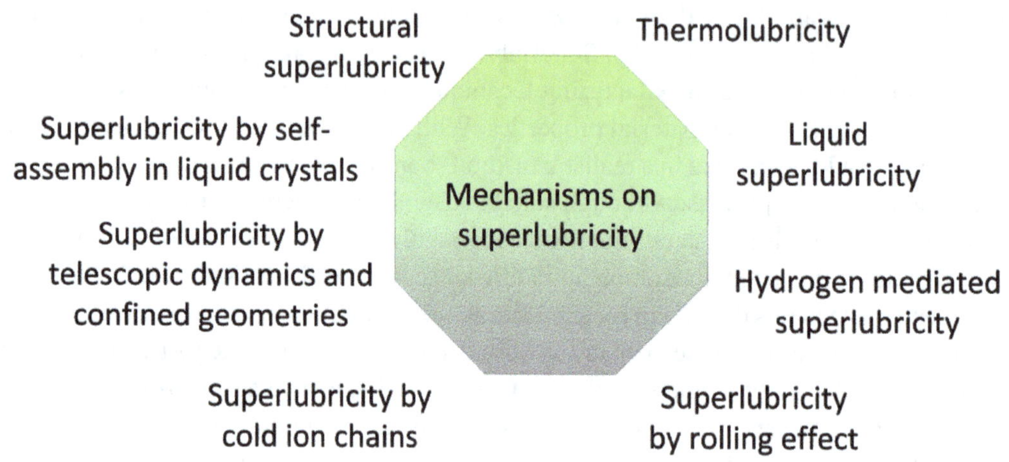

According to the materials between which the friction is made, there are two types of contacts: hard and soft (Martin, 2018). Metals, ceramics, and hard coatings with high normal loads, high sliding speeds, and peak contact pressures as high as 3 GPa are all examples of hard contacts. Soft contacts are found in polymers, elastomers, and biological tissues with peak contact pressures of less than 10 MPa. Under high vacuum, in the presence of gas, and the presence of liquid lubricants, hard contacts exhibit extremely low friction in some form. The friction force is the sum of forces at the atomic level that include Coulombic and van der Waals attractions, ionic and covalent bonds, and other interactions, as well as all conceivable interferences between these forces. Surface qualities are solely responsible for superlubricity, which necessitate smooth, atomically pure surfaces with weak atomic interactions. Although the temperature is not a vital issue, it is assumed that lower temperatures are preferred due to reduced thermal vibration of the surface atoms.

Hirano's superlubricity is based on a mismatch between the atomic crystal lattices of two in contact surfaces. (Martin, 2018). Consider an atomically smooth and flat surface in a vacuum to see why. The surface energy envelope resembles an egg carton when the surface atoms are arranged in a regular pattern. When two of these "crates" are stacked on top of each other, as seen in Figure 6a, their wells and peaks will exactly fit together, and the surfaces will resist sliding. The surfaces will exhibit intermittent stick-slip movement and, depending on the sliding direction, potentially a zig-zag movement as well. The amount of force necessary to start sliding will be considerable. The crates do not adjust further if a surface is rotated at an angle that does not conform to rotational symmetry. As seen in Figure 6 b), the mismatch of the crystal lattices cancels the lateral forces in each direction, resulting in zero net friction forces. The mismatch between the crystal lattice spacings, which causes frictional anisotropy, is another source of incommensurability, as seen in Figure 6 c).

Figure 6. Types of contacts to produce or not superlubricity: (a) without sliding or superlubricity, (b) with sliding due to relative rotation (superlubricity), and (c) with sliding due to unmatched crystalline lattice (superlubricity)

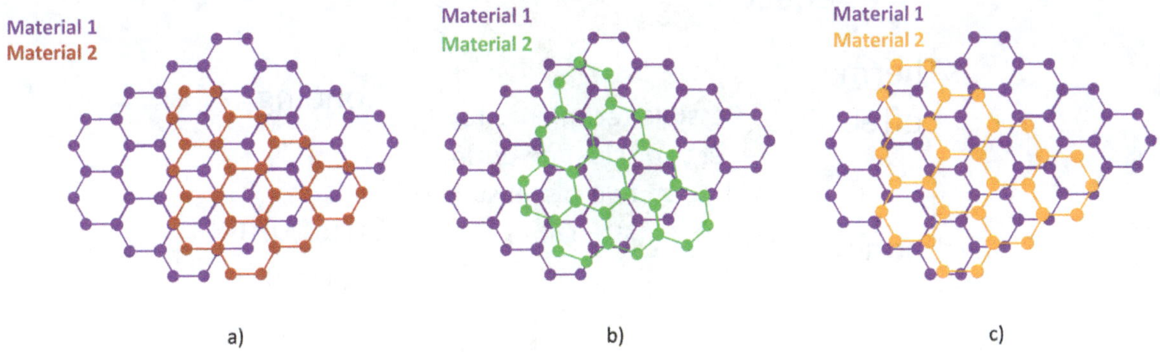

The state of superlubricity, also known as structural lubricity, occurs when two contact surfaces have no resistance to sliding (Zheng, 2014). Even after thirty years since the concept of superlubricity was proposed in 1990 (Hirano, 1990), experimental observations of superlubricity were limited to the nanoscale and in high vacuum or inert gas settings. The molecular interactions between the constituent atoms of solids provide the atomistic origin of frictional force, but not the force generated extrinsically by surface roughness, the presence of foreign atoms, and so on. There are two types of friction forces: atomistic locking and dynamic locking. When the configuration of atoms on a contact surface changes with sliding distance and interatomic potentials have arbitrary mechanical resistance, atomistic locking develops. Dynamic locking, on the other hand, occurs when the configuration changes abruptly owing to a dynamic process and the interatomic potential is greater than a predetermined value. When totally clean solid surfaces are prepared in the case of superlubricity, the friction force vanishes completely.

In general, friction between two surfaces is proportional to the normal force and independent of surface roughness, contact area, and sliding speed when macroscopic objects touch each other in a multiple roughness contact (Berman, 2018). Amontons' law gives this basic explanation of friction at the macroscopic level. Friction, on the other hand, gets much more complicated at the nanoscopic level, where several processes contribute to energy losses while sliding and therefore to friction. A description of some hypothetical frictional energy dissipation mechanisms is shown in Figure 7. Wear, bonding, structural effects, molecular deformation, environment/chemistry, phonons, thermal effects, and electronic effects are all examples of these.

Figure 7. Energy dissipation techniques when sliding on two-dimensional materials

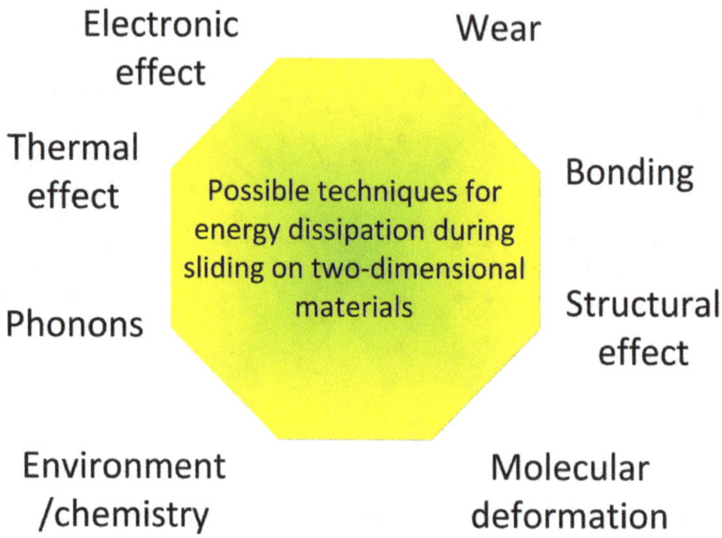

Table 1 summarizes the description of these events in two-dimensional materials that are related to energy losses owing to sliding and, as a result, lead to friction (Shekhar, 2021). The interlayer bonding forces in two-dimensional materials correspond to van der Waals interactions, therefore additional sources of friction, such as electrons, can be neglected (Zheng, 2014). As a result, the important feature of the superlubricity state is that friction seems to reduce to zero or near zero as the contact transitions from a commensurate to an incommensurate regime. When a crystal is exposed to a periodic potential, it may adjust itself to follow the periodicity of the potential under certain conditions (such as when the period of the potential is close to the crystal periodicity; the potential is strong enough, etc.), resulting in a so-called commensurate state (Woods, 2014). From magnetism and dislocations in crystals to vortices in superconductors and atomic layers adsorbed on a crystalline surface, such commensurate-incommensurate transitions are common in many aspects of condensed matter physics. The features of topological defects between the two commensurate phases, such as solitons, domain walls, and dislocation walls, may be of particular interest. One material can stretch to adjust to a slightly different periodicity (the corresponding state seen for small rotational angles) or demonstrate little change depending on the rotational angle between the two hexagonal lattices (the incommensurate state). Areas with matching lattice constants are separated by domain walls in the commensurate state, which accumulates the strain.

Table 1. Contributions to the superlubricity of two-dimensional materials

Property	Role of the Two-dimensional Materials
Wear	Corrosive, adhesive, and abrasive degradations are all inhibited by them. They enable the shear stress at the interface easier to deal with. They reduce friction, which reduces wear.
Deformation of molecules	Two-dimensional materials' high inertia and bending flexibility help them resist molecular deformation. Molecular deformation is also reduced by increasing the number of layers in two-dimensional materials.
Heat	At higher temperatures, dynamic surface corrugation improves bending. The sliding friction increases as the temperature decreases.
Static electricity	Friction at sliding contacts is dominated by the creation of an electrostatic charge. For ultrathin samples, the coefficient of friction is lowered by a factor of two. The lack of Van der Walls bonds aids in the reduction of friction.
Bonds	During sliding, bond breakdown reduces the adhesion and roughness of the contacts. By introducing functional groups between two-dimensional materials in liquids, friction can be reduced.
Phonons	Because electronic excitations (exoelectrons), optical excitations (photons), and lattice vibrations (phonons) increase friction, sliding generates heat. When compared to the roughness of the substrate, electron-phonon coupling increases friction moderately.
Environment	Depending on the type of interaction of the material's surface with the gas during sliding, the presence of gas might alter the coefficient of friction positively or negatively.
Structure	The lattice encounters jamming when the two surfaces in friction have the same rotational symmetry and matching orientation. As a result, the sliding is stopped, increasing friction, and the system becomes commensurate. To reduce friction, the opposite is desired.

The discovery of two-dimensional materials has revolutionized the science of tribology, as its low friction capabilities are owing to a unique crystalline alignment of contact materials, resulting in inter-layer sliding with a coefficient of friction near-zero (Zhang, 2019a). Superlubricity has been achieved by transitioning from the macroscopic to the nanoscale regime, thanks to advances in material discovery and manufacture (Berman, 2018). To show low friction, two-dimensional materials must have high quality and few imperfections. Controlling the optimal structure of the materials in the sliding contact at the macroscale is nearly impossible. As a result, one of the primary trends anticipated in future solid lubricants is the use of homogenous two-dimensional materials to obtain friction coefficients close to zero. The chemical functionalization of these materials' surfaces has a significant impact on their tribological properties, resulting in high friction levels. The discovery of two-dimensional materials and structures with outstanding tribological properties will continue to stimulate researchers' interest in using these materials to control friction and wear at the micro and nanoscales. An extreme slip was coined for the hypothetical geometry of two completely flat, crystalline surfaces rolling over one other in dry incommensurate contact when the word superlubricity was first coined (Boom, 2019; Zhang, 2019b). Only a few experimental systems have achieved the needed combination of flattened and crystalline perfection, such as the minuscule contact of two surfaces between two-dimensional materials that assure near-zero friction.

Nano and/or micro tribology systems and length scales range from single asperity atomic contacts to microscopic slip surfaces (Vanossi, 2020). The first is the simple asperity atomic friction found in AFM probing and manipulation, the second is the nanoscale friction found in nanoobjects such as physisorbed graphene and physisorbed islands, the third is the mesoscale friction found in ion traps and colloidal suspensions, and the fourth is the microscopic friction found in long carbon nanotubes and extended surfaces in two-dimensional systems.

Liquid and solid lubricants work together to achieve ultra-low friction (Vanossi, 2020). Liquid friction is created using lubricating additives, ionic liquids, polymer brushes, or hydration layers, and it gives interfaces exceptional durability. Coatings are used in industrial applications such as magnetic hard drives, razor blades, microelectromechanical systems, and more to generate solid friction. The capacity to forecast the physical response of a system under shear based on the geometric characteristics of the nearest interface rather than system-dependent differentiating traits is critical when dealing with well-defined bonding interfaces, both in soft and hard matter contacts. Surface crystal lattice mismatching or misalignment can cause structural incommensurability, which prevents interlocking and collective stick-slip motion of the interface asperities, resulting in minor frictional force fading.

SOLUTIONS AND RECOMMENDATIONS

Due to their desirable mechanical and physical properties, composite materials have gained tremendous popularity in the automotive, aerospace, marine, and defense industries in recent years (Sahoo, 2021). These composites, on the other hand, are limited in their applicability due to poor tribological characteristics. Researchers have proposed using solid lubricants as a reinforcing phase in the matrix phase for the fabrication of composite materials to improve the tribological properties of composites by reducing friction. Aside from various forms of solid lubricants, two-dimensional materials such as graphene, WS_2, and MoS_2 have emerged as novel solid lubricants with increased friction-reduction potential. Two-dimensional (2D) materials are crystalline materials that are made up of one or more layers of atoms (Zhang, 2019a). 2D materials have some of the lowest coefficients of friction and wear ratios, making them appealing for enhancing the efficiency, durability, and environmental friendliness of mechanical systems in the future. The following paragraphs describe solutions that have provided two-dimensional materials such as graphene, transition metal dichalcogenides, black phosphorus, hexagonal boron nitride, layered double hydroxides, metal-organic frameworks, and MXenes, to reduce friction and wear. The friction of atomically thin materials can differ significantly from ordinary friction (Boon, 2019; Zhang, 2019b). Regular 3D material surfaces introduce degrees of freedom toward interaction that are absent or at least orders of magnitude lower in 2D materials. These degrees of freedom can be adjusted, paving the way for the tribologist's ideal of tunable friction to become a reality. The creation of a dynamic, reversible, and adjustable technique to maximize the tribological performance of engineered systems with friction coefficients close to zero is one of the big issues facing the mechanics and materials field.

Graphene

Two-dimensional materials can serve as a solid or colloidal liquid lubricant thanks to their unique friction, wear, mechanical, optical, electrical, and thermal properties (Berman, 2014). The main attributes that two-dimensional materials have for impressive tribological behavior are high chemical inertness, extreme mechanical strength, and resistance to shear stress on their densely packed, atomically smooth surface. To reduce stiction, friction, and wear, ultrathin, multi-layered graphene has been applied to nanoscale and microscale systems such as microelectromechanical systems (MEMS) and nanoelectromechanical systems (NEMS) with oscillating, rotating, and sliding contacts. Liquid water reduces friction on graphene, reducing the impact of capillary forces found in humid conditions. The surface beneath the graphene layer affects the wetting angle; however, the effect of the substrate is modified by the number of layers and is minimal for numerous graphene layers. Because graphene is atomically smooth and has low surface energy, it can be utilized instead of thin solid films to reduce adhesion and friction on various surfaces. The friction force, dominated by out-of-plane deformation that reduces the area of contact with graphene, decreases as the number of layers increases. The Van der Waals forces that modify the contact area between surfaces can be tuned by functionalizing graphene with different chemical groups, increasing the friction force by two, six, and seven times when it is hydrogenated, fluorinated, and oxidized, respectively (Berman, 2014). Graphene has been demonstrated to be particularly successful in decreasing friction and wear on the nano and macro sizes, not just as a lubricant but also as an addition to oils, self-lubricating composites, and solvents. Graphene, unlike graphite, decreases friction and wear regardless of ambient conditions (wet or dry), and functions as a perfect passivation layer, limiting corrosion and wear-inducing (tribo-corrosion) in steel contacts, according to macro-scale tribological research. Graphene films significantly reduce friction and wear at sliding steel contacts, according to macro-scale experiments. In dry nitrogen and humid air settings, the friction of the steel lowers from 0.9 to 1 for steel bar to 0.15-0.16 for graphene-covered steel. When graphene is present at the sliding interfaces, the volume and wear ratios are lowered by 3-4 orders of magnitude.

To further reduce graphene's coefficient of friction, researchers have sought options to optimize its tribological properties. In the presence of graphene oxide, super high exfoliation reduced graphene oxide (SRGO) was produced and thermally reduced by KOH activation (Zhao, 2020). The nanosheets produced had a unique two-dimensional microstructure with low re-stacking and agglomeration, a high specific surface area (1665 m^2/g), and an interlayer spacing of 4.25 Å, which resulted in excellent tribological properties, such as a base oil friction coefficient reduction of 70%, a wear volume reduction of > 60%, and good self-dispersing stability and sliding orientation. These findings show that graphene microstructure may be efficiently engineered as a prospective contender for lowering frictional energy consumption and enhancing anti-wear durability. Due to Van der Waals interactions between the layers, the thinner the graphene sheet is, the easier it is for the material to penetrate the contact areas, and its adjacent layers can slide against each other with very little shear stress.

Transition Metal Dichalcogenides

Transition metal dichalcogenides, such as MoS_2, $MoSe_2$, and WS_2, are made up of hexagonal layers of transition metal atoms (M) sandwiched between two layers of chalcogen atoms (X) with a stoichiometry (MX_2) (Ji, 2020). The lubricating properties of TMDs (e.g., MoS_2, WS_2, $MoSe_2$) are connected to their layered microstructures, which result in lamella formation with low shear stress and decreased coefficient of friction due to weak van der Waals bond forces between the interlayers (Gong, 2020). TMDs are excellent lubricants in a vacuum, but their characteristics deteriorate in humid air, causing oxidation. MoS_2, which has a molybdenum atom in the middle and two layers of sulfur atoms sandwiched between them, is a typical TMD material used as a solid lubricant in vacuum or dry air, but it degrades in humid air or high temperatures due to oxidation into oxygen or water vapor. Molybdenum disulfide (MoS_2) is a good lubricant over a wide temperature range and can even sustain lubricity in composite coatings at -100 degrees Celsius.

MoS_2, a solid lubricant with excellent lubricating qualities, oxidizes spontaneously and absorbs moisture from the air, resulting in low wear resistance and short wear life (Zhang, 2021). For the first time, the additive graphitic carbon nitride (g-C_3N_4 (CN)) was successfully coupled with MoS_2 as a solid lubricant via the hydrothermal synthesis in this study. When compared to pure MoS_2 in air, the CN/MoS_2 composite has a low coefficient of friction (0.031) and a very long wear life.

Phosphorene

Research on phosphorene's possible tribological applications, such as lubrication additives and fillers in self-lubricating composites, began soon after its discovery (Wang, 2018). Micro-friction capabilities, lubrication properties as a lubrication additive based on water and oil, and friction and wear of composites based on polyvinylidene fluoride (PVDF) and phosphorene (black phosphorus) are among the advancements to be highlighted. Phosphorene as a lubricant in oil had similar lubrication qualities at low loads, but the dispersions had exceptional extreme pressure resistance and bearing capacity at high loads. When the black phosphorus content of BP/PVDF composites was greater than 10%, the improvement in friction characteristics was more noticeable, and the addition of BP improved the composite's wear resistance. For lubricating additive applications, larger flake sizes are preferred. Phosphorene interacts weakly with water.

Phosphorene has been investigated as a viable solid lubricant due to its large surface-area-to-volume ratio, which reduces the effect of surface forces (such as friction, stiction, and tiny contact wear) on parts in relative motion (Lee, 2019). The strain-induced lattice displacement of the phosphorus atoms changes the structure of their bands, and hence the electrical and optical properties of the phosphorus when a mechanical strain is applied to the phosphorene. Phosphorene has anisotropic frictional behavior because its coefficient of friction is larger in the armchair direction than in the zig-zag direction.

The use of black phosphorous flecked with silver nanoparticles as a lubricant in poly-alpha-olefin-based (PA06) oils has been proposed (Tang, 2020). The friction coefficient and wear ratio were lowered by 73.4 percent and 92 percent, respectively, with an additive of 0.075 wt percent Ag/BP. Due to the deposition of the Ag/BP composite on the rubbed surface during the sliding process, a physical protective coating is generated. Meanwhile, a carbon-based film is formed from the oil because of the tribochemical reaction at the friction interface. The exceptionally low coefficient of friction and outstanding wear resistance is due to the production of a tribofilm made up of silver nanoparticle nanosheets, phosphorene nanosheets, and amorphous carbon.

When phosphorene nanofilms were put in contact with Ti_6Al_4V (TC4)/GCr15, their tribological behavior as a lubricating additive in water was optimized (Wang, 2021). The friction coefficient and wear ratio were reduced by 32.4 percent and 61.1 percent, respectively, when 70 mg/L of additive was used. Because phosphorene nanofilms are deposited on worn surfaces via physical electrostatic interactions, their friction-reducing properties are linked to their lamellar structure. Furthermore, the adsorption and tribochemical coatings generated on the worn surfaces were the primary determinants of antiwear performance. The phosphorene nanosheets' improved tribological capabilities are due to these features, which prevented direct contact between the rubbed surfaces and considerably reduced friction and wear.

Hexagonal-Boron Nitride (h-BN)

Hexagonal boron nitride (*h*-BN) has an ultra-flat surface and a lattice mismatch with graphite of less than 2% (Cho, 2013). Because of its layered crystal structure, it has a low atomic shear plane resistance, which is related to the weak van der Waals bond between the layers, like graphite. It was also discovered that even a single layer *h*-BN could exhibit minimal friction without sliding at the layer contact.

The two-dimensional materials also serve as water-based lubricant additives, allowing water-lubricated films to maintain high thermal conductivity and bearing capacity (Liu, 2019). Because of their layered crystalline structure, superlubricity can also be seen in other two-dimensional materials such as MoS_2 and *h*-BN. To increase wear resistance and reduce friction, the lubricating effect of aqueous dispersions of *h*-BN at concentrations of 1 percent, 0.05 percent, and 0.01 percent were tested. On the sliding surface, *h*-BN is repeatedly deposited and peeled, generating a protective coating that can reduce friction and wear. The coefficient of friction increased somewhat for the compositions of 0.01 and 0.05 percent, whereas the coefficient of friction reduced slightly for the composition of 1 percent (Cho, 2013).

Layered Double Hydroxides (LDHs)

Layered double hydroxides (LDHs) are a type of inorganic mineral made up of divalent and trivalent metal cations that occur naturally (Wang, 2017). Three NiAl-LDH nanoplatelets of various sizes were produced in this study using reaction times of 6, 12, and 24 hours. Due to their favorable crystalline structure and the creation of a more protective tribofilm during the sliding process with superior mechanical properties, the big nanoplatelets (NiAl-LDH 24 h) displayed the best and most stable tribological performance. Their tribological properties in base oil were examined under pressures up to 2.16 GPa, which resulted in a 10% reduction in friction coefficient and a significant improvement in wear performance after nano-LDHs were introduced. These advantages came because of a protective tribolayer forming on the contact interface, owing to nano-LDHs' high degree of crystallinity, which makes them a promising candidate for oil-based additives in demanding industrial applications.

To test the friction capabilities of LDHs as lubricant additives, researchers produced La-doped Mg/Al double layered hydroxide and intercalated it with sodium dodecyl sulfate (SDS) (Li, 2015). After intercalation, LDHs with a La/Al molar ratio of 0.1 expanded the interlayer from 7.662 to 25.663 Å, indicating a reduction in friction between the expanded laminates, good nanoparticle dispersion in the medium, and an effective tribofilm generated on the contact surfaces. The lubricating process suggests that the interlayer anions' charge number weakened the electrostatic attraction between positively charged laminates and anions, making the layers easier to slide and reducing the frictional force given to the rubbed surface.

Metal-Organic-Frameworks (MOFs)

Metal-organic frameworks (MOFs) as nanocontainers with stimulus-responsive qualities for lubricants have been proposed as a novel class of self-lubricating nanocomposites (Zhang, 2017). 5.0 wt% Cu-BTC [copper (II)-benzene-1,3,5-tricarboxylate] catalyst in a mesoporous metal-organic framework (m-MOF) with epoxy resin and oleylamide as a lubricant are used to make these nanocomposites. When a stress stimulus is applied to the nanocomposite to generate a chemical reaction between the oleylamine released from the nanocontainer and the friction pairs, they achieve a coefficient of friction of 0.03. The mesoporous structure permits the pores to adsorb the lubricant molecules by making use of the large specific surface area and high interfacial contact, resulting in nanoparticles being packed together to create mesopores with diameters ranging from 26 to 72 nm under solvothermal conditions.

Ultrasound-assisted exfoliation in solvents produced nano-sheets with a two-dimensional metal-organic framework structure based on Zn(II)-benzimidazole-acetate (Zn (Bim) (OAc)) with a thickness of 4 nm (Wang, F.-F., 2020). The smaller the interlayer spacing, the stronger the Van der Waals forces, which means Zn (Bim) (OAc) nanosheets are less likely to reduce friction than other inorganic 2D stacked nanomaterials. As a result, we believe that the native organic interface of 2D MOFs is better suited to practical use as lubricant additives than inorganic nanomaterials. However, 2D Zn (Bim) (OAc) nanosheets currently do not meet the criteria for long-term additive stability in oil. The layer spacing of 2D MOFs is around 3.32 Å, which is smaller than 3.49 Å for transition metal dichalcogenides (e.g., MoS_2, WS_2), 3.33 Å for graphene, and 3.33 Å for hexagonal boron nitride when compared to inorganic two-dimensional layered materials.

MOFs containing zirconium functionalized with dialkyldithiophosphate (DDP) have been proposed as lubricating oil additives that can reduce friction and wear while also providing antioxidation (Wu, 2021). Self-assembly assisted by ultrasound was used to coordinatively graft DDP molecules onto MOF nanoparticles. Adding Zr-MOFs @ DDP to oil reduced both the coefficient of friction and the wear volume, and its oxidation induction time was substantially longer than that of base oil.

MXenes

Ti_3C_2 was one of the first MXenes to be studied for its tribological behavior in base oil when used as a lubricating additive (Yang, 2014). The results show that the two-dimensional layered Ti_3C_2 may significantly improve the base oil's friction reduction and anti-friction capabilities, especially when using 1.0 wt% Ti_3C_2. As a result, MXene is a promising contender for lubricating materials in practical applications.

When applied as a solid lubricant, the effect of surface finishes and sandwiched water in $Ti_3C_2T_x$ nanoparticles on friction and wear was examined (Rosenkranz, 2019). Experiments at 4 percent relative humidity demonstrated that $Ti_3C_2T_x$ had exceptional tribological properties, including a 300 percent reduction in friction and the ability to dramatically reduce adhesive, abrasive, and tribochemical wear behavior. This MXene was used as a solid lubricant under dry conditions on stainless steel surfaces. According to the findings, end groups and the amount of water intercalated can be used as defining parameters to adjust interfacial resistance and hence friction characteristics.

Few-layered $Ti_3C_2T_x$ nanosheets act as solid lubricants on SiO_2 substrates, with friction-reduction characteristics that are slightly lower than graphene and MoS_2 (Rodriguez, 2021). Since Yuri Gogotsi's group discovered MXene nanosheets in 2011, their chemical bonding and resulting mechanical properties have made them intriguing candidates in the field of tribology. These nanosheets were first employed in composite materials as lubricant additives and reinforcing phases. The first investigations on solid lubrication capacity were conducted in 2018-2019, and they investigated the feasibility of using them in composites with nanodiamonds or graphene quantum dots to create wear-free sliding conditions.

MXene's mechanical properties, such as flexibility and high adhesion, are critical in achieving low friction and low wear (Wyatt, 2021). MXenes are particularly attractive for tribological applications due to their wide range of compositions and morphologies, as well as their low shear strength between neighboring layers due to weak secondary interlayer bonding. MXenes have been used as lubricant additives, solid lubrication coatings, and reinforcement phases in composites. MXenes' frictional characteristics between distinct layers are influenced by M-X composition, stoichiometry, and surface terminations, according to density functional theory (DFT) and molecular dynamics (MD) simulations. Low sliding pathways were expected for Ti-based MXenes with friction coefficients of 0.24 to 0.27 for bar and 0.10 to

0.14 for terminated MXenes, respectively. To increase friction and wear performance, MXene multilayer particles have been applied as lubricating additives in paraffin, poly (alpha) olefins, and other base oils. For additive-free base oils with MXene concentrations of roughly 0.8 to 1.0wt percent, friction reductions of up to 54 percent and a nine-fold reduction in wear volume have been proven. The production of self-lubricating tribofilms and, as a result, load-bearing capacities were attributed to the smoother, less abraded, less deformed worn surfaces. Increasing MXene multilayer particle concentrations beyond the ideal levels tend to increase friction due to possible agglomerations, lowering dispersion and/or quality stability, and, as a result, lowering tribological characteristics. The highest friction and wear ratio reductions for $Ti_3C_2T_x$ flakes and multilayer powder, respectively, were observed at 34% and 45 percent, as well as 10% and 5%. The ability to generate MXene-rich self-lubricating tribofilms was credited with a higher rise in singular flakes than those with few layers. These results demonstrate the potential of single-flake MXene as a lubricant, and the application of higher-quality, larger flakes and more stable $Ti_3C_2T_x$ MXene could lead to even more wear reduction. MXenes produce friction coefficients of 0.06 to 0.1, however, they are still unable to match the absolute friction coefficient and relative improvement factor of other two-dimensional nanomaterials. MXenes have been utilized to improve the mechanical and tribological properties of polymer composites based on epoxy resin and ultrahigh-molecular-weight polyethylene (UHMWPE). MXene (2 wt% $Ti_3C_2T_x$) and epoxy resin composites show somewhat enhanced mechanical characteristics and a three-fold decrease in friction. When MXenes are amino-functionalized, the friction is reduced by 35% and the wear ratio is reduced by 72%. MXenes' ability to reduce friction and wear has recently been proved in hybrid coatings for solid lubrication, such as MXene/nanodiamond and MXene/nanodiamond/quantum dot. When compared to pure $Ti_3C_2T_x$ and nanodiamond coatings, this hybrid improved friction coefficients by 46 percent and 30 percent, respectively. MXenes can degrade structurally (increased defect density, sheet bending/breakage, reduced size of lateral sheets, among other things) or chemically (oxidation, modified surface terminations, among other things), both of which can have an impact on tribofilm formation.

Superlubricity in Two-Dimensional Materials (2D)

The friction coefficients of some two-dimensional materials have been reduced to values less than 0.01, which makes the friction regime be in superlubricity, as illustrated in Figure 8. When graphite slides against graphene at the nanoscale, it forms incommensurate contact under low contact pressure, resulting in structural superlubricity (Li, 2018). With pressures of 2.52 GPa, a friction coefficient of 0.0003 may be achieved for ultra-high contact pressures in micro and nanomachines, corresponding to an extraordinarily low shear resistance at a graphene/graphite interface, which is appealing for nanoscale lubrication. Superlubricity stability at ultra-high contact pressures (> 1 GPa) is still a difficulty (Ren, 2021). The friction coefficient of oxidized phosphorus floating in pure water is 0.004 when treated to this situation.

Figure 8. Coefficients of friction of two-dimensional materials exhibiting structural superlubricity

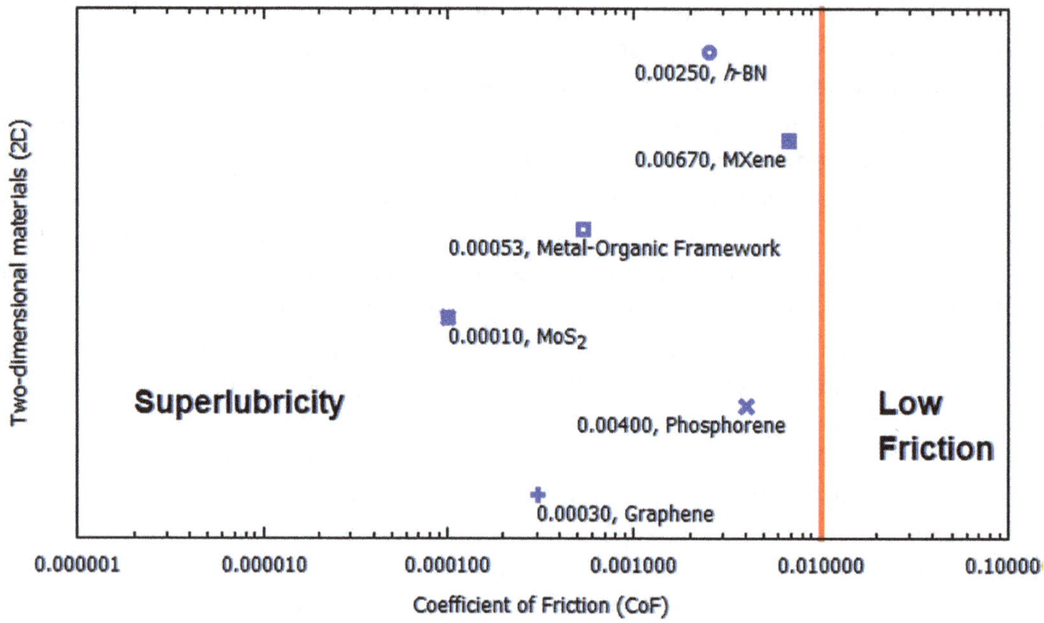

When two incommensurate monolayers are sliding tested, the ultra-low friction between atomic layers of hexagonal MoS_2, a key solid lubricant, and lubricating oil additive, exhibits a coefficient of friction of 0.0001 (Li, 2017). Organic metal frameworks can be an appropriate material for achieving solid superlubricity due to the current frictional energy dissipation mechanism. Because the influence of puckering caused by topographic considerations was minimized, they were able to obtain a friction coefficient as low as 0.00053 (Liu, 2021).

The tribological characteristics of Ti_3C_2 Mxene produced on SiO_2-covered silicon (Si) substrates subjected to sliding wear against diamond-like carbon (DLC) (Huang, 2021). In a dry nitrogen environment, the coefficient of friction was reduced to the superlubricity regime with a coefficient of friction of 0.0067. Structural superlubricity is an interesting tribological phenomenon in which lateral interactions between two incommensurate surfaces are effectively canceled, producing ultra-low sliding friction (Song, 2018). When aligned connections resist external loads under ambient conditions, interfaces between graphite and hexagonal boron nitride can withstand external loads. In multi-contact graphene/h-BN heterojunctions, the studies revealed friction coefficients as low as 0.0025.

FUTURE RESEARCH DIRECTIONS

The composites' lubrication mechanism is a tribofilm that forms on the surface of the composites during friction when it contains solid lubricating material. Self-lubricating composite materials have the potential for substantial advancements in materials, processing, and/or manufacturing. Researchers are actively exploring the introduction of newer, typically nanostructured, forms of popular lubricants in terms of materials. Different types of two-dimensional materials, their processing paths, and qualities that are used as reinforcing phases to make composites have recently aroused the interest of researchers. However, producing these two-dimensional materials with appropriate lubricating qualities poses certain obstacles. Lubricants should also be evaluated with microscopy and spectroscopic techniques in the post-processing of self-lubricating composites to quantify the state of the lubricant. In terms of manufacturing, the advanced manufacturing approach of additive manufacturing opens new possibilities for self-lubricating composites. Altogether, for most additive manufacturing procedures that focus on making bulk metal and alloy parts, there is a real prospect of tailoring the properties of composites using these approaches. Because many additive manufacturing procedures include lasers, controlling the lubricant condition in the composite could be a major concern.

For practical applications, additive dispersity and stability in oils are critical. As a result, nanosheets of two-dimensional MOFs as additives in oils were promising, and future work will focus on increasing performance and exposing the process. To promote interfacial compatibility and dispersion stability in liquid, nanomaterials must have well-designed surfaces and interfaces. Despite the advances achieved so far, future strategies must design inorganic-organic interfaces to introduce organic materials on their surfaces.

The adjustable structures of MXenes, as well as their M-X and Tx compositions, are the key difficulties and future directions for improving their tribological performance. Beyond increasing their effectiveness as lubricating additives, rational alteration of surface chemistry may be an appropriate strategy to improve their dispersion stability. Given the early stage of MXene tribological research, they may be able to cause even more favorable benefits, making them the next generation of solid lubricating systems. Understanding the MXenes' energy-dissipating friction and wear mechanisms through high-resolution chemical, structural, and morphological characterization of the tribological interface and tribofilms generated is critical for long-term low-friction and low-wear performance. To properly understand the underlying mechanisms, a complete investigation of the chemical and structural development of tribologically stressed MXenes is required due to high local temperatures and high contact pressures acting on tribological contact.

From theoretical computations and simulations to rigorous empirical trials, superlubricity has advanced at a breakneck pace during the last two decades. Superlubricity will remain a means of developing the principles of friction at the atomic scale for mechanical engineering, physics, materials science, and chemistry for the time being. Solid superlubricity can now be achieved at various sizes, however liquid superlubricity mechanisms offer greater promise due to their higher load-carrying capability. This regimen can be regarded as a powerful tool for energy issues, given the development in this field over the last three decades, with values of friction coefficients tending to 10^{-5} since laboratory research is currently underway linked to industrial applications and problems of scalability, portability, stability, and controllability, which can lead to sustainability.

The application of superlubricity to mechanical systems used in space exploration and in micro and nanoelectromechanical systems (where energy losses due to friction can be critical) emerges as a promising niche area (Baykara, 2018). Once the promise of superlubricity is realized for such niche applications, the events will likely spur research aimed at broadening the scope of technological advancements, allowing for a long-term state of excessively low friction and transformative improvements in mechanical system efficiency.

CONCLUSION

Despite all the progress made so far, one of the most important goals in the decades ahead will be to reduce friction and wear-related mechanical failures in moving mechanical systems, because friction hurts the efficiency, durability, and environmental compatibility of dynamic mechanical systems. The use of two-dimensional materials for tribological purposes is surprisingly underexplored, despite its weakly bonded multilayer structure and capacity to self-lubricate. Due to their distinct characteristics, two-dimensional nanomaterials such as graphene, GO, rGO, phosphorene, *h*-BN, g-C_3N_4, MoS_2, WS_2, $Ti_3C_2T_X$, and MOF can effectively improve the properties of composites, resulting in uniform dispersion, optimized orientation, and strong interface bonding characteristic of the van der Waals interaction to possess tribological properties such as ultra-low friction, self-lubrication action, and ultra-low wear ratio. As a result, the trying to find new materials, coatings, and lubricants (liquid and solid) that can reduce friction and wear will continue. Notwithstanding extensive research and development efforts on two-dimensional materials for a wide range of current and future applications, their tribological potential as a lubricant is mostly undeveloped. Friction can be reduced to ultra-low levels at specified loads using two-dimensional materials when sliding occurs without stick-slip. Superlubricity obtained with two-dimensional materials as solid lubricants are stable throughout a wide range of temperatures, sliding speeds, and contact pressures thus far. When moist air is present at the slip interface, however, there are still some significant constraints to obtaining near-zero friction. The coefficient of friction of 2D materials is several orders of magnitude lower than that of macroscopic materials, and it can be adjusted by manipulating energy dissipation mechanisms during sliding. Additional research is needed to understand the mechanisms underlying the lubricating properties of nanosheets of two-dimensional materials and to assess their impact on relevant applications.

ACKNOWLEDGMENT

The author is grateful for the support of the Instituto Tecnológico Superior de Irapuato (ITESI) and the National Council of Science and Technology (CONACYT) for their support for the realization of this publication related to the research topic of Advanced Materials of the Academic Group ITESI-CA-01 “Advanced Materials Applied to Engineering”. Also, he appreciates the support of his wife and son to carry out this research on nanomaterials.

REFERENCES

Bart, J. C. J., Gucciardi, E., & Cavallaro, S. (2013). Lubricants: Properties and Characteristics. In *Biolubricants: Science and Technology* (pp. 24–73). Woodhead Publishing. doi:10.1533/9780857096326.24

Baykara, M. Z., Vazirisereshk, M. R., & Martini, A. (2018). Emerging Superlubricity: A Review of the State of the Art and Perspectives on Future Research. *Applied Physics Reviews*, *5*(4), 041102. doi:10.1063/1.5051445

Berman, D., Erdemir, A., & Sumant, A. V. (2014). Graphene: A New Emerging Lubricant. *Materials Today*, *17*(1), 31–42. doi:10.1016/j.mattod.2013.12.003

Berman, D., Erdemir, A., & Sumant, A. V. (2018). Approaches for Achieving Superlubricity in Two-Dimensional Materials. *ACS Nano*, *12*(3), 2122–2137. doi:10.1021/acsnano.7b09046 PMID:29522673

Boon, D., & Frenken, J. (2019). Tunable Superlubricity of 2-Dimensional Materials. *Proceedings of the National Academy of Sciences of the United States of America*, *116*(49), 24386–24387. doi:10.1073/pnas.1918084116 PMID:31723046

Bowden, F. P., Gregory, J. N., & Tabor, D. (1945). Lubrication of Metal Surfaces by Fatty Acids. *Nature*, *156*(3952), 97–101. doi:10.1038/156097a0

Braun, J. (2017). Additives. In T. Mang & W. Dresel (Eds.), *Lubricants and Lubrication* (3rd ed., pp. 119–144). Wiley-VCH. doi:10.1002/9783527645565.ch6

Cho, D.-H., Kim, J.-S., Kwon, S.-H., Lee, C., & Lee, Y.-Z. (2013). Evaluation of Hexagonal Boron Nitride Nano-Sheets as a Lubricant Additive in Water. *Wear*, *302*(1-2), 981–986. doi:10.1016/j.wear.2012.12.059

Delprete, C., & Razavyka, A. (2018). Piston Dynamics, Lubrication and Tribological Performance Evaluation: A Review. *International Journal of Engine Research*, *21*(15), 725–741. doi:10.1177/1468087418787610

Gong, H., Yu, C., Zhang, L., Xie, G., Gui, D., & Luo, J. (2020). Intelligent Lubricating Materials: A Review. *Composites. Part B, Engineering*, *202*, 108450. doi:10.1016/j.compositesb.2020.108450

Hardy, W. B., & Doubleday, I. (1922). Boundary Lubrication – The Paraffin Series. *Proceedings of the Royal Society of London. Series A, Containing Papers of a Mathematical and Physical Character*, *100*(707), 550–574. doi:10.1098/rspa.1922.0017

Hirano, M., & Shinjo, K. (1990). Atomistic Locking and Friction. *Physical Review B: Condensed Matter*, *41*(17), 11837–11851. doi:10.1103/PhysRevB.41.11837 PMID:9993633

Hod, O., Meyer, E., Zheng, Q., & Urbakh, M. (2018). Structural Superlubricity and Ultralow Friction across the Length Scales. *Nature*, *563*(7732), 485–492. doi:10.103841586-018-0704-z PMID:30464268

Holmberg, K., & Erdemir, A. (2019). The Impact of Tribology on Energy Use and CO_2 Emission Globally and in Combustion Engine and Electric Cars. *Tribology International*, *135*, 389–396. doi:10.1016/j.triboint.2019.03.024

Huang, S., Mutyala, K. C., Sumant, A. V., & Mochalin, V. N. (2021). Achieving Superlubricity with 2D Transition Metal Carbides (MXenes) and MXene/graphene Coatings. *Materials Today Advances*, *9*, 100133. doi:10.1016/j.mtadv.2021.100133

Ji, Z., Zhang, L., Xie, G., Xu, W., Guo, D., Luo, J., & Prakash, B. (2020). Mechanical and Tribological Properties of Nanocomposites incorporated with Two-Dimensional Materials. *Friction*, *8*(5), 813–846. doi:10.100740544-020-0401-4

Lee, H. G., Yoon, H. M., & Lee, J. S. (2019). Anisotropic Nanoscale and Sub-Nanoscale Friction Behaviors between Phosphorene and Silicon Tip. *Applied Surface Science*, *481*, 1573–1584. doi:10.1016/j.apsusc.2019.01.204

Li, H., Wang, J., Gao, S., Chen, Q., Peng, L., Liu, K., & Wei, X. (2017). Superlubricity between MoS_2 Monolayers. *Advanced Materials*, *29*(27), 1701474. doi:10.1002/adma.201701474 PMID:28497859

Li, J., Li, J., & Luo, J. (2018). Superlubricity of Graphite Sliding against Graphene Nanoflake under Ultrahigh Contact Pressure. *Advancement of Science*, *5*(11), 1800810. doi:10.1002/advs.201800810 PMID:30479926

Li, J. J., & Luo, J. B. (2013). Advancements in Superlubricity. *Science China. Technological Sciences*, *56*(12), 2877–2887. doi:10.100711431-013-5387-y

Li, S., Qin, H., Zuo, R., & Bai, Z. (2015). Friction Properties of La-doped Mg/Al Layered Double Hydroxide and Intercalated Product as Lubricant Additives. *Tribology International*, *91*, 60–66. doi:10.1016/j.triboint.2015.06.012

Liu, L., Zhang, Y., Qiao, Y., Tan, S., Feng, S., Ma, J., Liu, Y., & Luo, J. (2021). 2D Metal-Organic Frameworks with Square Grid Structure. A Promising New-Generation Superlubricating Material. *Nano Today*, *40*, 101262. doi:10.1016/j.nantod.2021.101262

Liu, L., Zhou, M., Jin, L., Li, L., Mo, Y., Su, G., Li, X., Zhu, H., & Tian, Y. (2019). Recent Advances in Friction and Lubrication of Graphene and other 2D Materials: Mechanisms and Applications. *Friction*, *7*(3), 199–216. doi:10.100740544-019-0268-4

Martin, J. M., & Erdemir, A. (2018). Superlubricity: Friction's Vanishing Act. *Physics Today*, *71*(4), 40–46. doi:10.1063/PT.3.3897

Meng, Y., & Liu, C. (2021). Spatiotemporal Manipulation of Boundary Lubrication by Electro-Charging and Electrochemical Methods. In Superlubricity (2nd ed., pp. 499-516). doi:10.1016/B978-0-444-64313-1.00025-9

Ren, X., Yang, X., Xie, G., He, F., Wang, R., Zhang, C., Guo, D. & Luo, J. (2021). Superlubricity under Ultrahigh Contact Pressure Enabled by Partially Oxidized Black Phosphorus Nanosheets. *NPJ 2D Materials and Applications*, *5*, 44. doi:10.1038/s41699-021-00225-0

Rodriguez, A., Jaman, M. S., Acikgoz, O., Wang, B., Yu, J., Grützmacher, P. G., Rosenkranz, A., & Baykara, M. Z. (2021). The Potential of $Ti_3C_2T_x$ Nano-sheets (MXenes) for Nanoscale Solid Lubrication Revealed by Friction Force Microscopy. *Applied Surface Science*, *535*, 147664. doi:10.1016/j.apsusc.2020.147664

Rosenkranz, A., Grützmacher, P. G., Espinoza, R., Fuenzalida, V. M., Blanco, E., Escalona, N., Gracia, F. J., Villarroel, R., Guo, L., Kang, R., Mücklich, F., Suarez, S., & Zhang, Z. (2019). Multi-Layer $Ti_3C_2T_x$-Nanoparticles (MXenes) as Solid Lubricants – Role of Surface Terminations and Intercalated Water. *Applied Surface Science*, *494*, 13–21. doi:10.1016/j.apsusc.2019.07.171

Rosenkranz, A., Liu, Y., Yang, L., & Chen, L. (2020). 2D Nano-Materials beyond Graphene: From Synthesis to Tribological Studies. *Applied Nanoscience*, *10*(9), 3353–3388. doi:10.100713204-020-01466-z

Sahoo, S. (2021). Self-Lubricating Composites with 2D Materials as Reinforcement: A New Perspective. *Reinforced Plastics*, *65*(2), 101–103. doi:10.1016/j.repl.2020.06.007

Shekhar, H., & Dumpala, R. (2021). Overcoming Friction and Steps towards Superlubricity: A Review of Underlying Mechanisms. *Applied Surface Science Advances*, *6*, 100175. doi:10.1016/j.apsadv.2021.100175

Song, Y., Mandelli, D., Hod, O., Urbakh, M., Ma, M., & Zheng, Q. (2018). Robust Microscale Superlubricity in Graphite/Hexagonal Boron Nitride Layered Heterojunctions. *Nature Materials*, *17*(10), 894–899. doi:10.103841563-018-0144-z PMID:30061730

Tang, G., Su, F., Xu, X., & Chu, P. K. (2020). 2D Black Phosphorus Dotted with Silver Nanoparticles: An Excellent Lubricant Additive for Tribological Applications. *Chemical Engineering Journal*, *523*, 146503. doi:10.1016/j.cej.2019.123631

Vanossi, A., Bechinger, C., & Urbakh, M. (2020). Structural Lubricity in Soft and Hard Matter Systems. *Nature Communications*, *11*(1), 4657. doi:10.103841467-020-18429-1 PMID:32938930

Wang, F.-F., Liu, Z., & Cheng, Z.-L. (2020). Ultrasonic-assisted Exfoliation for 2D Zn(Bim)(OAc) Nanosheets used as an Oil-Soluble Additive in Lubricants. *Applied Organometallic Chemistry*, *34*(11), e5950. doi:10.1002/aoc.5950

Wang, H., & Liu, Y. (2020). Superlubricity achieved with Two-Dimensional Nano-Additives to Liquid Lubricants. *Friction*, *8*(6), 1007–1024. doi:10.100740544-020-0410-3

Wang, H., Liu, Y., Liu, W., Wang, R., Wen, J., Sheng, H., Peng, J., Erdemir, A., & Luo, J. (2017). Tribological Behavior of NiAl-Layered Double Hydroxide Nanoplatelets as Oil-based Lubricant Additives. *ACS Applied Materials & Interfaces*, *9*(36), 30891–30899. doi:10.1021/acsami.7b10515 PMID:28841291

Wang, Q., Hou, T., Wang, W., Zhang, G., Gao, Y., & Wang, K. (2021). *Tribological Behavior of Black Phosphorus Nanosheets as Water-based Lubrication Additives*. Friction. doi:10.100740544-020-0465-1

Wang, W., Xie, G., & Luo, J. (2018). Black Phosphorous as a New Lubricant. *Friction*, *6*(1), 116–142. doi:10.100740544-018-0204-z

Woods, C. R., Britnell, L., Eckmann, A., Ma, R. S., Lu, J. C., Guo, H. M., Lin, X., Yu, G. L., Cao, Y., Gorbachev, R. V., Kretinin, A. V., Park, J., Ponomarenko, L. A., Katsnelson, M. I., Gornostyrev, N., Watanabe, K., Tanigucho, T., Casiraghi, C., Gao, H.-J., ... Novoselov, K. S. (2014). Commensurate-Incommensurate Transition in Graphene on Hexagonal Boron Nitride. *Nature Physics*, *10*(6), 451–456. doi:10.1038/nphys2954

Wu, W., Liu, J., Li, Z., Zhao, X., Liu, G., Liu, S., Ma, S., Li, W., & Liu, W. (2021). Surface-functionalized nanoMOFs in Oil for Friction and Wear Reduction and Antioxidation. *Chemical Engineering Journal*, *410*, 128306. doi:10.1016/j.cej.2020.128306

Wyatt, B. C., Rosenkranz, A., & Anasori, B. (2021). 2D MXenes: Tunable Mechanical and Tribological Properties. *Advanced Materials*, *33*(17), 2007973. doi:10.1002/adma.202007973 PMID:33738850

Yang, J., Chen, B., Song, H., Tang, H., & Li, C. (2014). Synthesis, Characterization, and Tribological Properties of Two-Dimensional Ti_3C_2. *Crystal Research and Technology*, *49*(11), 926–932. doi:10.1002/crat.201400268

Zhang, G., Xie, G., Si, L., Wen, S., & Guo, D. (2017). Ultralow Friction Self-Lubricating Nanocomposites with Mesoporous Metal-Organic Frameworks as Smart Nanocontainers for Lubricants. *ACS Applied Materials & Interfaces*, *9*(43), 38146–38152. doi:10.1021/acsami.7b12591 PMID:28994574

Zhang, J., & Meng, Y. (2015). Boundary Lubrication by Adsorption Film. *Friction*, *3*(2), 115–147. doi:10.100740544-015-0084-4

Zhang, S., Hou, Y., Li, S., Liu, L., Zhang, Z., Feng, X.-Q., & Li, Q. (2019b). Tuning Friction to a Superlubric State via in-Plane Straining. *Proceedings of the National Academy of Sciences of the United States of America*, *116*(49), 24452–24456. doi:10.1073/pnas.1907947116 PMID:31659028

Zhang, S., Ma, T., Erdemir, A., & Li, Q. (2019a). Tribology of Two-Dimensional Materials: From Mechanisms to Modulating Strategies. *Materials Today*, *26*, 67–86. doi:10.1016/j.mattod.2018.12.002

Zhang, Y., Li, P., Ji, L., Wan, H., Chen, L., Li, H., & Jin, Z. (2021). Tribological Properties of MoS_2 Coating for Ultra-Long Wear-Life and Low Coefficient of Friction combined with Additive g-C_3N_4 in Air. *Friction*, *9*(4), 789–801. doi:10.100740544-020-0374-3

Zhao, J., Huang, Y., Li, Y., Gao, T., Dou, Z., Mao, J., Wang, H., He, Y., Li, S., & Luo, J. (2020). Superhigh-Exfoliation Graphene with a Unique Two-Dimensional (2D) Microstructure for Lubrication Application. *Applied Surface Science*, *513*, 145608. doi:10.1016/j.apsusc.2020.145608

Zheng, Q., & Liu, Z. (2014). Experimental Advances in Superlubricity. *Friction*, *2*(2), 182–192. doi:10.100740544-014-0056-0

ADDITIONAL READING

Abate, F., D'Agostino, V., Di Giuda, R., & Senatore, A. (2010). Tribological Behaviour of MoS_2 and Inorganic Fullerene-like WS_2 Nanoparticles under Boundary and Mixed Lubrication Regimes. *Tribology – Materials. Surfaces and Interfaces*, *4*(2), 91–98. doi:10.1179/175158310X12678019274282

Acikgoz, O., & Baykara, M. Z. (2020). Speed Dependence of Friction on Single-Layer and Bulk MoS_2 Measured by Atomic Force Microscopy. *Applied Physics Letters*, *116*(7), 071603. doi:10.1063/1.5142712

Fernandes, T. F. D., & Neves, B. R. A. (2016). Friction Coefficient Mapping of 2D Materials via Friction-Induced Topographic Artifact in Atomic Force Microscopy. *Journal of Advances in Nanomaterials*, *1*(2), 73–81. doi:10.22606/jan.2016.12004

Luo, T., Chen, X., Li, P., Wang, P., Li, C., Cao, B., Luo, J., & Yang, S. (2018). Laser Irradiation-induced Laminated Graphene/MoS_2 Composites with Synergistically Improved Tribological Properties. *Nanotechnology*, *29*(26), 265704. doi:10.1088/1361-6528/aabcf5 PMID:29633718

Nguyen, H. T., & Chung, K.-H. (2020). Assessment of Tribological Properties of Ti_3C_2 as a Water-based Lubricant Additive. *Materials (Basel)*, *13*(23), 5545. doi:10.3390/ma13235545 PMID:33291773

Qi, S., Wei, X., Chen, L., Geng, Z., Luo, J., Lu, Z., & Zhang, G. (2020). 3D Graphene/Hexagonal Boron Nitride Composite Nanomaterials Synergistically Reduce the Friction and Wear of Steel-DLC Contacts. *Nano Select*, *2*(4), 791–801. doi:10.1002/nano.202000233

Senatore, A., D'Agostino, V., Petrone, V., Ciambelli, P., & Sarno, M. (2013). Graphene Oxide Nanosheets as Effective Friction Modifier for Oil Lubricant: Materials, Methods, and Tribological Results. *ISRN Tribology*, *425809*, 1–9. Advance online publication. doi:10.5402/2013/425809

Taha-Tijerina, J., Peña-Paras, L., Narayanan, T. N., Garza, L., Lapray, C., Gonzalez, J., Palacios, E., Molina, D., García, A., Maldonado, D., & Ajayan, P. M. (2013). Multifunctional Nanofluids with 2D Nanosheets for Thermal and Tribological Management. *Wear*, *302*(1-2), 1241–1248. doi:10.1016/j.wear.2012.12.010

KEY TERMS AND DEFINITIONS

Friction: A force that prevents solid surfaces, fluid layers, and material elements from sliding against each other from moving in the same direction.

Lubricants: Substances that aid in the reduction of friction between surfaces in mutual contact, hence reducing the heat generated when the surfaces move.

Superlubricity: A sliding or motion regime in which friction or sliding resistance diminishes, resulting in a kinetic coefficient of friction less than 0.01.

Tribofilm: Film that is formed on tribologically stressed surfaces to reduce friction and wear in lubricated systems.

Two-Dimensional (2D) Materials: Crystalline materials are made up of single or few layers of atoms that have substantially stronger in-plane interactions than those along the stacking direction.

Wear: Damage, progressive loss, or distortion of material caused by mechanical (e.g., erosion) or chemical (e.g., corrosion) forces.

Chapter 14
Tribological Studies of Bulk Metallic Glasses:
Structure, Preparation, Surface Treatment, and Coating

Suman Gandi
National Institute of Technology, Goa, India

Venkata Satya Chidambara Swamy Vaddadi
https://orcid.org/0000-0003-4939-6360
National Institute of Technology, Goa, India

Saran Srihari Sripada Panda
National Institute of Technology, Goa, India

Saidi Reddy Parne
https://orcid.org/0000-0003-2817-4240
National Institute of Technology, Goa, India

Motilal Lakavat
https://orcid.org/0000-0002-4020-0643
National University of Science and Technology, Oman

Amiya Bhaumik
Lincoln University College, Malaysia

ABSTRACT

The materials research community has been very interested in bulk metallic glasses (BMGs) over the past two decades because of their demonstrated friction and wear properties, as well as their potential for use in a variety of important tribological applications. Because of their superior chemical, physical, and mechanical properties, BMGs are a promising candidate material for advanced engineering applications. Compared to conventional crystalline metals and alloys, BMGs have higher strength, higher elastic strain, and higher hardness, making them a promising material class for tribological applications. A unique deformation process is realized in these materials due to the lack of a crystalline structure and faults such as misalignments, which display high strength, hardness, strong wear resistance, massive plastic deformation, corrosion-resistant, and hardness. In this chapter, the authors describe the research achievements in the field of BMGs, the tribological properties, structure, and applications.

DOI: 10.4018/978-1-7998-9683-8.ch014

INTRODUCTION

Tribology is the study of friction, wear, and lubrication. Over the past few decades, there has been a steady increase in interest in alloys with tribological properties. The advent of bulk metallic glasses (BMGs) has increased demand for a better understanding of their composition, preparation, surface treatments, and coatings (Khan et al., 2018). BMGs are amorphous alloys with distinctive structure and physical characteristics that come from amorphous phases that crystallize during cooling the melt at low temperatures (<600°C). BMGs possess the advantages of high strength and good corrosion resistance. However, as a class of material, they exhibit many unique properties, which has led to tremendous interest in the area of BMGs for commercial applications such as automotive, aerospace, and biomedical (Hutchings & Shipway, 2017; Ludema & Ajayi, 2018). For example, it is used to design better bearings and lubricants. There are different types of tribological studies that can be performed to address specific needs or problems (Czichos, 2009). One type is BMGs (Khan et al., 2018) has a structure made up of many small grains with high-density amorphous regions between them (Khan et al., 2018; Wang et al., 2004). This gives BMGs their unique mechanical properties, such as hardness and low coefficient of friction (Khan et al., 2018; Suryanarayana & Inoue, 2013). BMGs have been studied to determine how they behave when exposed to different environments and conditions. Compared to equivalent alloys, they have a much greater resistance to wear and corrosion, improved ability to withstand high pressure, and the potential for electrically stimulating reversible shape memory (Chen, 2011; Löffler, 2003; Suryanarayana & Inoue, 2013; Trexler & Thadhani, 2010). In addition, BMGs can be manufactured at low temperatures in a single step from inexpensive starting materials, which makes them environmentally friendly (Johnson, 1996; Kruzic, 2016; Wang, 2009).

Furthermore, BMGs are typically composed of 60% to 90% metals, and the rest is either glass or metal oxides (Zhang & Huang, 2019). These materials have been described as "glassy metals" because they don't have the distinct crystalline structure that other metals do. This makes them super strong and durable. One of the most common causes of damage to engineering materials these days is the failure of the materials in industrial applications. In this regard, tribological features of BMGs have received substantial attention because of their mechanical performance and properties (Liao et al., 2018; Medina et al., 2020; Zhou et al., 2021).

Additionally, BMGs exhibit several features, such as amorphicity and high strength, which can be transferred to a range of other materials, such as the most common glasses, metallic glasses, metals, alloys, and different nanocomposite materials (Cornuault et al., 2020; Jiang et al., 2015; Lu & Liu, 2004; Zhou et al., 2020). However, the most distinguishing feature of BMGs, along with other properties, is the glass transition, which converts super-cooled liquids to a glassy state when cooled from a high to a low temperature and vice versa (Li et al., 2007; Löffler, 2003; Park & Kim, 2005). BMGs are extremely strong at low temperatures and plasticity at high temperatures, as amorphous alloys with a glass transition and their thermodynamic and physical characteristics change dramatically at the glass transition temperature (Tg). However, for practical reasons and to improve the reliability and performance of the material, it is desirable to lower Tg. The Klement research group was the first to successfully manufacture a glassy alloy with an $Au_{75}Si_{25}$ composition (Klement et al., 1960). On the order of 10^5-10^6 KS^{-1}, metal alloys' essential cooling rate is estimated to freeze the liquid structure. It was demonstrated experimentally in 1959 by utilizing a cool metal plate instead of water, oil, or air to demonstrate the validity of this concept. During solidification, they discovered that good contact between liquid droplets and cold metal prevented the creation of gaseous layers, which resulted in a limited amount of heat being released.

Furthermore, when the droplets collide with the solid, cold metal plate, they form a thin layer that allows quick thermal diffusion. As revealed by XRD studies, the crystalline structure of the flakes has been destroyed, and a new, noncrystalline design has emerged in the micrometer-sized fragments. The composition of most glass-forming alloys documented at the time was a factor in their capacity to produce glass in certain situations. It is common practice to restrict the search for new alloys to phase regions near deep eutectic points in order to enhance the process of discovering excellent glass formers in binary and ternary alloy systems (Chen, 1980).

The wear and friction behavior of several BMGs, such as Zr, Cu, Mg, and Ti base alloys, has been extensively studied in the literature up to this point (Eckert et al., 2007; Marimuthu et al., 2021; Pole et al., 2021; Sohrabi et al., 2021; Wang et al., 2018). The tribological properties of BMGs have been demonstrated in industrial applications in friction components, as evidenced by their low wear rate and low friction coefficient, among other characteristics. In addition, the wear properties of Co, Fe, and Ni-based BMG micro-gears were marginally superior to those of 304 stainless steel (Prakash, 2005). It will reveal that the BMGs can be used as a replacement for metal because they have higher thermal conductivity than traditional metals and maintain their structural stiffness at temperatures below those where bulk crystals form. In addition, BMGs have been shown to have lower friction than conventional metals (Khun et al., 2016). This makes them promising candidates for automotive engineering, bearings, and other tribological applications.

PREPARATION OF BMGs

Most BMGs can be obtained by various preparation techniques such as metallurgical, micro-technological, and casting (Suryanarayana & Inoue, 2017). These methods involve rapidly forming an amorphous metal by cooling at a high temperature. The resulting product is the then glassy state. For example, a BMG can be made from a single melt or multiple melts, and the metallurgical process entails the fast development of amorphous alloys. The cooling rate should be rapid enough to circumvent crystal development and straight reach the glassy state. The casting procedure creates BMGs, while the spinning and quenching method creates ribbons. Physical vapor deposition (PVD) techniques such as sputtering and evaporation make metallic glass alloys thin coatings. Due to the quick quenching rate, these techniques easily create amorphous alloy thin films of various compositions. Table 1 shows the different methods involved in preparing BMGs. This section will discuss several common BMG preparation approaches that can be addressed for practical considerations to manufacture bulk level at a lower cost.

Casting

Casting is very common in manufacturing glassware and metallic glasses (Laws et al., 2009). This casting process involves that prepares the BMG for a crystalline structure. Depending on the type of metallic glasses being cast, it can be done in one step or multiple steps. In this process, molten glass is poured into a mold and cooled until solidified (Telford, 2004). The mold can be a container of the desired shape, a pattern on its surface, or anything else that will hold the molten glass and produce the desired shape (Telford, 2004). The process involves melting raw materials to make molten glass and then pouring them into molds to create BMGs. In the casting procedure, there are various methods for creating BMGs. For example, arc melting, melt casting, and injection casting are three of the most used forms of producing

BMGs. Water-quenching, high-pressure die casting, copper mold casting, cap-cast technique, suction casting, squeeze casting, electromagnetic vibration process, and unidirectional zone melting method are some of the various ways available (Suryanarayana & Inoue, 2017). While each of these strategies has its advantages and disadvantages, they also differ in cost and complexity.

Melt Casting

Melt casting is a production process for making metal products that involve melting the metal in an enclosed space and then pouring it into a mold. It is more popular than other processes, such as die-casting or sand casting, because it does not require any external pressure to force the molten metal into the mold cavity. It can be used to cast parts that are not symmetrical or that have complicated shapes. Melt casting is popular for making automotive parts and machine components like gears and pulleys because these parts need to be durable and lightweight. The process also requires less handling and storage space than foundry casting methods like permanent mold or sand. Xiaofang Jiang and his colleagues recently developed $Cu_{36}Zr_{48}Ag_8Al_8$ BMG by using a melt-casting technique in a heated water-cooled vacuum arc melting furnace under varying concentrations of H_2SO_4 (Jiang et al., 2019). The tribological characteristics of $Cu_{36}Zr_{48}Ag_8Al_8$ BMG are outstanding in the presence of H_2SO_4. The wear rate and friction coefficient can be as low as 0.10 ± 0.01 and $(0.28 \pm 0.23) \times 10^{-6}$ $mm^3\ N^{-1}m^{-1}$ even when the material is immersed in a 15 weight percent H_2SO_4 solution. It is believed that the fundamental cause of this phenomenon is a combination of surface oxidation and passivation, as well as the formation of liquid lubricating coatings from sulfuric acid solution and dissolution reaction products. Hence, $Cu_{36}Zr_{48}Ag_8Al_8$ BMG may replace stainless steel in the sulfuric acid corrosive environment.

Arc Melting

Arc melting is how metals can be heated to their liquid state and shaped as desired. Arc melting differs from the more common induction or gas-fired furnaces in that it relies on an electric arc to heat the metal. This is done by passing an electric current through the metal, causing it to become molten instantly. The main advantages of arc melting are its high thermal efficiency and freedom from scale build-up. This technique can create alloy systems that require a low critical cooling rate for glass formation. An arc is used to melt the alloy on a copper mound. After the alloy is molten, copper acts as a heat sink, removing heat from the melt. However, arc melting has several disadvantages, such as requiring large amounts of space due to the size of the electrodes used in the process. Figure 1 shows the block diagram of arc melting used to prepare BMGs. Tariq N. H. et al. reported mechanotribo properties of Zr–Al–Ni–Cu BMGs prepared by arc melting method (Tariq et al., 2009). Wear resistance in these alloys is positively related to mechanical parameters such as fracture strength and elastic strain, elastic modulus and microhardness, and the nature and mode of shear bands formed during the deformation process. Furthermore, veins may be ruptured due to exceptionally high shear deformation in areas of significant melting during compression. As a result, veins were divided into long threads.

Figure 1. Schematic illustration of arc melting method

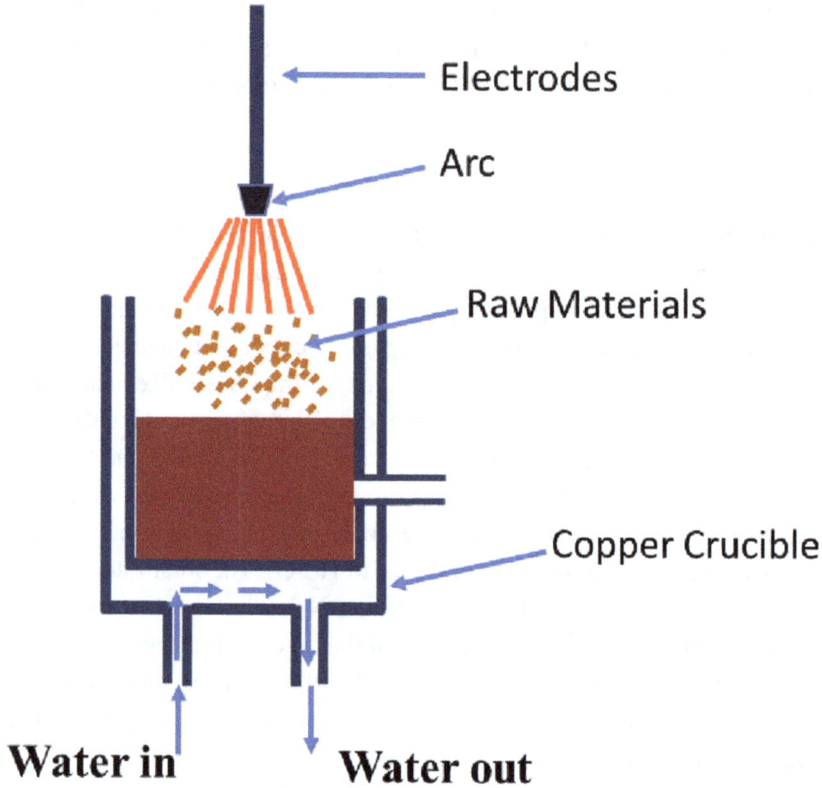

Injection Casting

Injection casting is a metalworking process that requires the use of a mold. The molten metal is injected into the mold, where it cools and hardens to produce a cast. The end product is usually an ingot or billet. Injection casting can have exceptional precision, surface finish, and mechanical properties. It's also capable of quickly producing large quantities of parts, making it ideal for mass production. It's typically used in manufacturing automotive parts, engine blocks, power plant components, and military hardware. In 2004, E. Fleury and his colleagues studied tribological properties of $Mg_{65}Cu_{15}Y_{10}Ag_{10}$, $Cu_{47}Ti_{33}Ni_6Zr_{11}Si_2Sn_1$, $Ti_{45}Ni_{15}Cu_{25}Sn_3Be_7Zr_5$, $Ti_{40}Zr_{29}Ni_7Cu_8Be_{16}$, $Ti_{40}Zr_{25}Ni_3Cu_{12}Be_{20}$, and $Ni_{59}Zr_{20}Ti_{16}Si_2Sn_3$ BMGs prepared via melt spinning and injection casting method at room temperature under dry sliding condition (Fleury et al., 2004). They paid particular attention to the effects of motion type on tribological parameters. Under identical sliding conditions, the frictional behavior of these BGMs ranged from 0.35 to 0.43. As a result, the wear performance varied significantly. The tribological properties of these BMGs can change due to surface changes in the wear tracks, which are influenced by the counterpart material and type of sliding action.

Melt Spinning

Melt spinning is a process used to create BMGs. It produces thin ribbons of molten metal drawn out into thin strands that are cooled into glassy fibers (Suryanarayana & Inoue, 2017). Melt spinning can be used to produce materials that are unusually strong, lightweight, corrosion-resistant, and thermally conductive. The process can improve many aspects of society by providing new materials for high-tech applications in aerospace engineering, telecommunications, and computer chips. The term "melt-spinning" refers to a method in which molten metal is extruded to create fine filaments and ribbons similar to those used to produce synthetic textile fibers. The molten metal is ejected through an aperture and allowed to harden either in flight or against cold during the melt-spinning process. It's known as free-flight melt spinning (FFMS) or chill-block melt spinning (CBMS), depending on whether the melt is allowed to solidify in flight or against a chill (substrate). CBMS, on the other hand, is the most often employed method by researchers. This technique drives a molten metal jet onto a cold, revolving heat sink, where the jet is reformed and allowed to harden. When the jet comes into contact with the heat sink, it creates a melted puddle with a thickness equal to and a length approximately double that of the jet itself. In conjunction with the beginning of solidification, the ribbon is released from the surface of the fast-rotating heat sink. Figure 2 depicts a simplified schematic design of the melt-spinning process.

Figure 2. Schematic illustration of the melt-spinning process

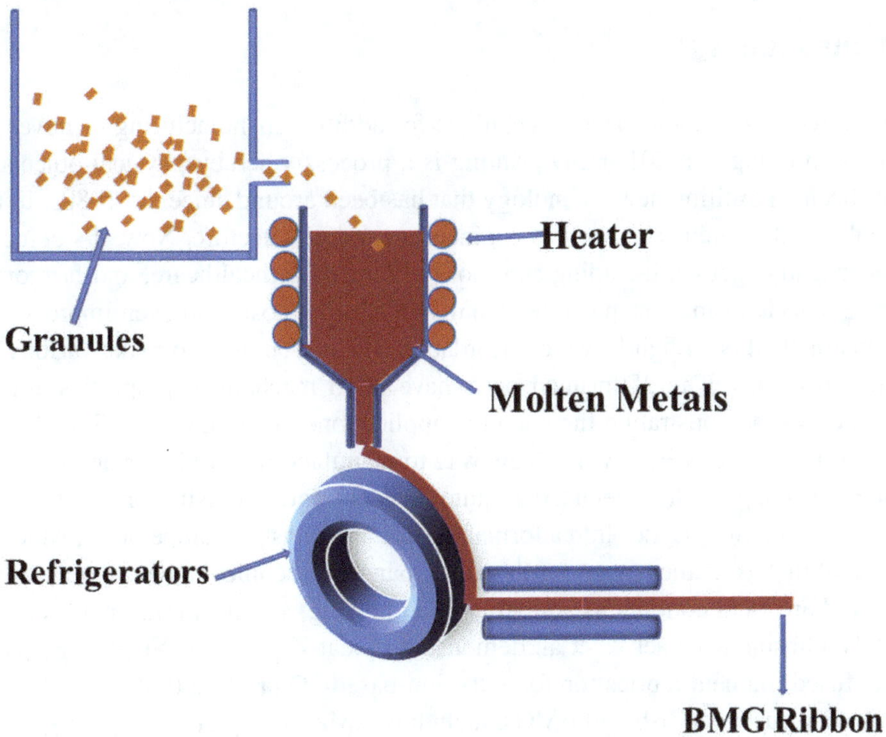

Mechanical Alloying via Ball-Milling

Mechanical ball milling is used to grind materials into superfine powder (Calka & Radlinski, 1991). The ball mill is one of the oldest industrial tools and uses balls to crush material. These can be steel or ceramic and often rotate horizontally inside the shell. Milling is done by ball mills, which are heavy rotating drums containing ceramic balls that grinding media. The drum rotates around its axis, causing the balls to collide with the material inside the drum until it becomes a fine powder. Ball mills are extensively used in mineral processing, chemical engineering, and other industrial applications where either a fine or coarse product is required. In addition, the mechanical ball milling process can enhance many physical properties, such as surface hardness and wettability. This makes it a useful tool for designing new applications for these materials to demonstrate how mechanical milling can produce uniform samples or alter crystallization pathways during processing. By using ball milling, multi-component alloy powders have also been generated from Fe, Mg, and Zr, in addition to other alloy systems (Basu & Ranganathan, 2003). Table 1 lists some of these. These particles can easily be consolidated into bulk in the super-cooled liquid area. The size restrictions imposed by solidification-processing technologies will not apply in the case of this strategy to synthesize monolithic BMGs. BMG composites, on the other hand, benefit greatly from the use of a ball mill. Ball milling has created many oxides, carbide, and other ceramic-dispersed composites. To ensure that the ceramic phase is evenly distributed, significant deformation and kneading are necessary. According to previous research, ball milling can also be used to create composites that comprise even nanometer-sized ceramic particles (Bhatt & Murty, 2008).

Additive Manufacturing

Researchers have recently demonstrated a technique for additive manufacturing of novel BMGs with three-dimensional printing (Li, 2018). 3D printing is a process whereby physical objects are created digitally. It has been an exciting new technology that has been around since the 1980s. In the past, 3D printing was only used in industries such as engineering and architecture. Now, it's being adopted by many different industry sectors, including automotive, aerospace, healthcare, and fashion. Using this process to create objects from scratch can reduce manufacturing costs and even improve efficiency in some cases. This method is straightforward, affordable, and can be used to make various BMGs with controlled microstructures. The 3D-printed parts have good mechanical properties and show good tribological properties, demonstrating the potential applications in automobiles (Sohrabi et al., 2021). Authors believe this method will provide a new way to manufacture BMGs for practical applications. Thermoplastic 3D printing, on the other hand, is quite advanced and can easily construct complex shapes. Thermal softening of thermoplastics into a formable state at a constant temperature, which is not present in conventional metals, is the reason for this discrepancy in technical advancement. BMGs show a supercooled liquid area and continuous softening upon heating, similar to thermoplastics, in contrast to typical metals. Gibson, Michael A. et al. demonstrated that $Zr_{44}Ti_{11}Cu_{10}Ni_{10}Be_{25}$ BMGs could also be employed in fused filament fabrication for extrusion-based 3D printing (Gibson et al., 2018). Many investigations have focused on Zr-based BMGs, including AMZ4 ($Zr_{59.3}Cu_{28.8}Al1_{0.4}Nb_{1.5}$), an industrial-grade BMG that has garnered the greatest attention due to its lower cost and easier availability compared to lab-grade metallic glass powders (Sohrabi et al., 2021).

Electrodeposition of BMGs

Electrodeposition is the process by which metal ions are electrochemically deposited on a cathode or anode, forming a metallic layer on their surface. The traditional method for preparing these materials involves ex-situ electrodeposition from a bath solution. The electric potential difference drives the ions to react in solution to form metallic glasses. It is often used to prepare BMGs. These BMGs can be used in fiberglass composites, plastics, paints, inks, and lubricant additives. Researchers have been proven to be more corrosion-resistant than metals, making them a good alternative for some applications. Compared to conventional melt quenching, the electrodeposition approach to generating amorphous alloys is a simple, low-cost, and adaptable alternative. However, electrodeposited metallic glasses' tribological characteristics and wear mechanisms remain unknown. Sheng-Bao investigated BMG electrodeposition by electroplating fine granular copper onto a Zr-based BMG ($Zr_{41}Ti_{14}Cu_{12.5}Ni_{10}Be_{22.5}$) as a substrate (Qiu & Yao, 2008). Cu was used as the anode, BMG as the cathode, and the electrolyte was copper sulfate/sulfuric acid. Different coating thicknesses were obtained by altering the electrodeposition duration, revealing a polycrystalline phase.

Table 1. Preparation and tribological features of BMGs

BMGs	Preparation Method	Features	Reference
$Mg_{65}Cu_{15}Y_{10}Ag_{10}$	melt spinning and injection casting	friction and wear	(Fleury et al., 2004; Park et al., 2001)
$Mg_{66}Zn_{30}Ca_{4}$	mechanical ball-milled via spark plasma sintering	mechanical, corrosion, and biocompatibility	(Li et al., 2021)
Mg_{65}-Cu_{25}-Y_{10}	electromagnetic vibration	mechanical, corrosion, and biocompatibility	(Tamura et al., 2005)
$Cu_{47}Ti_{33}Ni_{6}Zr_{11}Si_{2}Sn_{1}$	melt spinning and injection casting	friction and wear	(Fleury et al., 2004; Park et al., 2002)
$Cu_{47.5}Zr_{47.5}Al_{5}$	arc-melting	friction and wear	(Rahaman et al., 2015)
$Cu_{36}Zr_{48}Ag_{8}Al_{8}$	melt–casting	friction and wear	(Jiang et al., 2019)
$Ni_{59}Zr_{20}Ti_{16}Si_{2}Sn_{3}$	injection casting	friction and wear	(Fleury et al., 2004; Lee et al., 2004)
$(Ni_{75}Cr_{15}Si_{10})_{92}Ti_{8}$	mechanical ball-milled	friction and corrosion	(G. Guddla et al., 2021)
$[Ni\text{-}Cr\text{-}Si]_{94}$:Nb6	mechanical ball-milled	mechanical and corrosion	(G. T. Guddla et al., 2021)
$Ti_{45}Ni_{15}Cu_{25}Sn_{3}Be_{7}Zr_{5}$	injection casting	friction and wear	(Fleury et al., 2004; Kim et al., 2004)
$Ti_{40}Zr_{29}Ni_{7}Cu_{8}Be_{16}$	injection casting	friction and wear	(Fleury et al., 2004; Kim et al., 2003)
$Ti_{40}Zr_{25}Ni_{3}Cu_{12}Be_{20}$	arc-melting	friction and wear	(Rahaman et al., 2015)
(Ti, Cu, Zr)$_{92.5}Fe_{2.5}Sn_{2}Si_{1}Ag_{2}$	copper-mold casting	friction, wear, adhesion, proliferation,	(Yang et al., 2021)
$Zr_{41.3}Ti_{13.7}Ni_{10}Cu_{12.5}Be_{22.5}$	melt quenching	friction and wear	(Conner et al., 1998; Fleury et al., 2004)
$Zr_{60.6}Al_{13.6}Ni_{7.5}Cu_{18.3}$	arc-melting	friction and wear	(Tariq et al., 2009)
$Zr_{52.5}Cu_{17.9}Ni_{14.6}Al_{10}Ti_{5}$	arc-melting	friction and wear	(Siegrist et al., 2007)
$Zr_{52.5}Cu_{17.9}Ni_{14.6}Ti_{5}Al_{10}$	arc-melting	friction and wear	(Wu et al., 2013)

Continued on following page

Table 1. Continued

BMGs	Preparation Method	Features	Reference
$Zr_{52.5}Cu_{17.9}Ni_{14.6}Al_{10}Ti_5$	selective laser melting	compressive strength, plasticity	(Pauly et al., 2017)
$Zr_{61}Cu_{17.5}Ni_{10}Al_{7.5}Si_4$	magnetron sputtering	friction and wear	(Zhou et al., 2019)
$Zr_{48}Nb_8Cu_{14}Ni_{12}Be_{18}$	water quenching	friction and wear	(Zhang et al., 2003)
$Zr_{55}Cu_{30}Ni_5Al_{10}$	cap-cast	mechanical	(Yokoyama et al., 2007)
$Fe_{68.8}C_{7.0}Si_{3.5}B_{5.0}P_{9.6}Cr_{2.1}Mo_{2.0}$	inert gas atomization	friction and wear	(Yoon et al., 2011)
$Fe_{48}Cr_{15}Mo_{14}C_{15}B6Y_2$	plasma spraying	friction and wear	(Su et al., 2019)
$Fe_{43.7}Co_{7.3}Cr_{14.7}Mo_{12.6}C_{15.5}B_{4.3}Y_{1.9}$	selective laser melting	fracture toughness	(Li et al., 2018)
$Fe_{72}Al_5Ga_2C_6B_4P_{10}Si_1$	mechanical ball-milled	mechanical, thermal stability	(Chueva et al., 2007)
$Pd_{42.5}Cu_{30}Ni_{7.5}P_{20}$	water quenching	thermal stability	(Nishiyama et al., 2012)
$Hf_{48}Cu_{29}Ni_{10}Al_{13}$	arc melting and suction casting	friction and wear	(Abad & Browne, 2020)
$Mg_{65}Cu_{25}Y_{10}$	electromagnetic vibration	mechanical	(Tamura et al., 2005)

In summary, the use of the methods to prepare bulk metallic glass via casting, melt spinning, mechanical alloying, additive manufacturing, and electrodeposition has been reviewed in this section. The most common way of preparation is casting, which is used to create bulk metallic glasses-free of defects. Melt spinning can produce bulk metallic glasses that are more suitable for forming thin sheets. Mechanical alloying can create materials with various compositions. Additive manufacturing also provides a form of bulk metallic glass creation. Electrodeposition manufactures the metal oxide slurry required by using an electrolyte solution. The methods to prepare bulk metallic glasses are efficient and fast, but each has different advantages. Casting is efficient but can only be used for simple geometries. Melt spinning is efficient, but the metal qualities are not as pure as other methods. Mechanical alloying is quick and easy but takes a lot of energy. Additive manufacturing is arguably the most successful method, but it does have some limitations. Through the individual process steps, it is believed that a variety of materials can be produced with a wide range of properties. In addition to providing design flexibility, this collection of techniques allows for the production of metallic glasses. In addition, various methods have been presented here to create BMGs in multiple shapes and sizes. Since the critical cooling rate (R_c) for forming the glassy phase of BMG is comparatively low, it can be done relatively simple to obtain the crystals through various techniques.

Nevertheless, when the method includes a substrate, inevitably, the phase will eventually get heterogeneous. This is because the different methods for producing BMG yielded different results. Further, care must be taken to avoid contaminating some alloy systems, such as those based on Zr and Ti, because these alloys are extremely sensitive to impurities. Particularly considering Zr-based BMGs, excessive oxygen concentration has been found to reduce the temperature of the supercooled liquid and affect the crystallization behavior of these materials (Heinrich et al., 2012).

STRUCTURE OF BMGs

BMGs are an emerging class of materials with properties tuned over an extensive range by varying the composition, microstructure, and processing conditions. The transition from glassy to amorphous states in BMGs is based on the non-equilibrium nature of the material (Greer & Ma, 2007; Schroers, 2010). Moreover, BMGs are metastable materials with a glass-like elastic response due to their inherent low entropy. This low entropy makes them better suited for use as structural materials than traditional metals because they also have high entropy, leading to rapid relaxation after loading. In addition, structural characterization is a key to understanding BMGs and how we might use them in the future.

Microstructure

Single crystals of metals have one-dimensional atomic arrangements, which give them a uniform atomic structure. Structurally, they are highly ordered and crystalline materials. These properties illustrate the metallic nature of these materials (Figure 3). BMGs, on the other hand, have a microstructure that is a mixture of amorphous metal and crystalline metal (Figure 3). This gives them properties from both glassy phases and crystalline materials. In other words, they are neither purely glassy nor purely metallic because they have different types of structures within their microstructure. The microstructure of BMGs is essential to understanding the atomic structure and stability of these materials. The physical properties of BMGs, such as their rigidity, strength, and toughness, are highly dependent on the microstructure. The microstructure determines the properties of a BMG, which can vary wildly from one material to another. Various methods are used to determine the phases generated during the synthesis, annealing, and derivation of BMGs. One of the essential parameters that can be quantified is the morphology and size of the developed phases and the crystal structure, composition, and volume fraction of those phases. The fundamental understanding of these factors allows us to improve the alloy design process.

Additionally, these microstructural characterizations can be carried out in coincidence with tribological properties in order to investigate: i) the type of failures that occur, ii) the interaction of shear bands, and iii) the source of crack start. When looking at the surface of a high reflectivity surface, it is clear that it is amorphous. On the other hand, a dull grey surface indicates either a devitrified or crystalline nature. An examination under a microscope with magnification ranging from 100 to 1000 times reveals porosity, fissures, and gritty inclusions.

Figure 3. Schematic structural illustrations of crystalline metal and BMGs

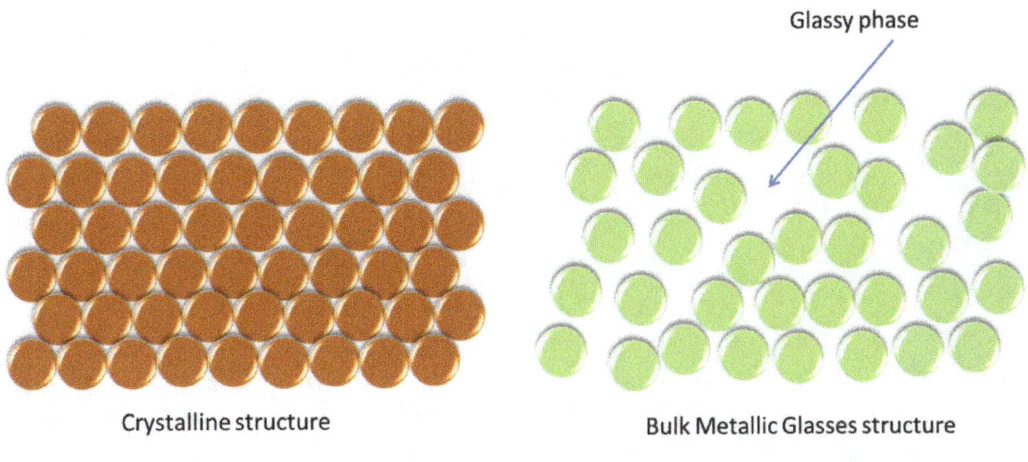

Glass Transition Temperature

The glass transition temperature (Tg) is when a material changes from an amorphous solid to a non-crystalline state. A BMG is any metallic alloy that has undergone a quench-cooling process in order to create an amorphous or non-crystalline form. While some BMGs can be cooled slowly enough for the solidification process to happen normally, it is difficult to control the rate at which other BMGs cool at room temperature. This has led scientists to believe that the "glass transition" of these alloys happens very quickly instead of gradually. Moreover, to make BMGs from molten metal, the cooling rate (R_c) must be enough to high to prevent the nucleation and development of crystalline phases in the super-cooled liquid zone between the melting temperature (T_m) and T_g during the melting process (Schneider, 2001). Using a straightforward nucleation theory, the researchers demonstrated that the reduced glass transition temperature ($T_{rg}=T_g/T_m$) is a critical factor that defines whether or not a given material's melt can solidify into a glass during cooling. They concluded that T_{rg} must be relatively high in order to produce BMGs (Turnbull, 1969).

Glass Forming Ability (GFA)

GFA is defined as the ability of a material to exhibit a glassy character, and it is a critical aspect in determining whether or not BMGs will form (Greer & Ma, 2007). The GFA of these systems has expanded from 1mm to several centimeters during the past many decades. Almost from the beginning of the development of BMGs, the GFA parameter has become a heated topic of discussion. An objective quantitative measure of GFA is the critical-cooling rate (R_c), over which crystallization is prevented during solidification. The greater the value of R_c, the greater the value of GFA. Figure 4 depicts the relationship between GFA and critical cooling rate. The T_g, the super-cooled liquid region ($\Delta T=T_x-T_g$), and the GFA parameter ($\gamma=T_x/(T_g+T_l)$) are the essential criteria for evaluating or comparing the GFA of different materials. T_x is the crystallization onset temperature, Tg, and is the glass transition temperature. The reduced glass-transition temperature, T_{rg}, is defined as the ratio of the glass-transition temperature (T_g) and the liquidus temperature (T_l) ($T_{rg}= T_g/T_l$). It is well known that the greater the value of T_{rg}, the greater the liquid stability and, thus, the greater the GFA. Table 2 lists the various compositions of BMGs and the glass transition parameters related to them (Lu, Li, et al., 2000; Lu & Liu, 2002; Lu, Tan, et al., 2000).

Figure 4. Schematic illustration of the relationship between GFA and critical cooling rate

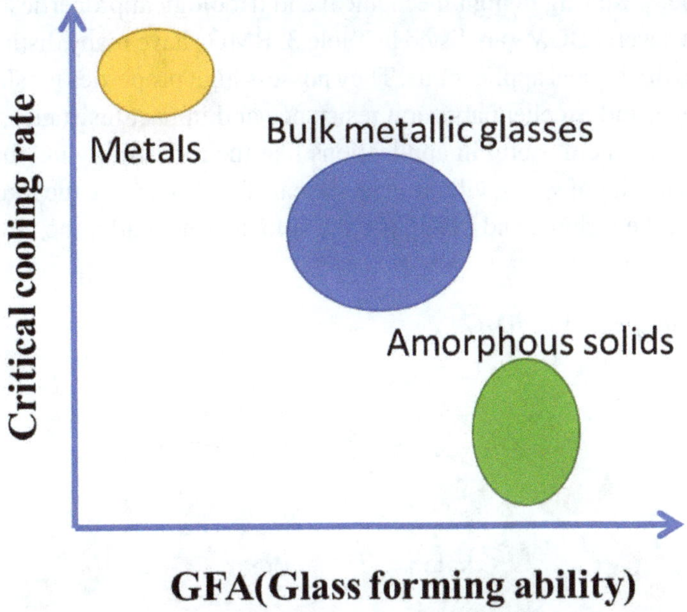

Table 2. Various BMGs compositions and their glass transition properties

BMG	R_c (K s^{-1})	T_g (K)	$\Delta T=T_x-T_g$	$\gamma=T_x/(T_g+T_l))$	$T_{rg}= T_g/T_l$
$Mg_{80}Ni_{10}Nd_{10}$	12.5×10⁴	454	16.3	0.353	0.517
$Mg_{75}Ni_{15}Nd_{10}$	46.1	450	20.4	0.379	0.57
$Zr_{66}Al_8Cu_7Ni_{19}$	22.7	662.3	58.4	0.387	0.552
$Zr_{57}Ti_5Al_{10}Cu_{20}Ni_8$	10	676.7	43.3	0.395	0.591
$La_{55}Al_{25}Ni_{20}$	67.5	490.8	64.3	0.388	0.521
$La_{66}Al_{14}Cu_{20}$	37.5	395	54	0.399	0.540
$Pd_{79.5}Cu_4Si_{16.5}$	500	635	40	0.392	0.589
$Pd_{40}Ni_{40}P_{20}$	0.167	590	63	0.409	0.585
$Ti_{34}Zr_{11}Cu_{47}Ni_8$	100	698.4	28.8	0.389	0.597

TRIBOLOGICAL STUDIES OF BMGs

Tribology studies the frictional interactions between moving surfaces (Fleury et al., 2004). Figure 5 shows the tribological properties of BMGs. It directly impacts the performance and lifespan of many mechanical parts. While aluminum, copper, and zinc elements comprise the bulk of bulk metallic glass's physical composition, most of its tribological properties are provided by its metal composition. The material has a common boundary between pure metals and the impurities such as oxygen, nitrogen, and others. A big benefit of this composition is that the alloying components are constantly interspersed, and they are very hard to distinguish even with the naked eye. This property makes it possible to process the material in a controllable manner, resulting in high mechanical and tribological properties. The mechanical and tribological features of several BGMs are listed in Table 3. BMGs have many distinctive features, which may help them in most tribological applications. They possess high properties for shearing, frictional and thermal shock resistance, and excellent abrasion resistance and impact resistance, among other things. These properties can make them useful in applications like the industrial wind turbine rotor, in which materials are subjected to a lot of stress, vibration, and wear. This is because the rotor must withstand the high rotational speed of the turbine, and the higher the velocity, the harder the materials used become.

Figure 5. Tribological properties of BMGs

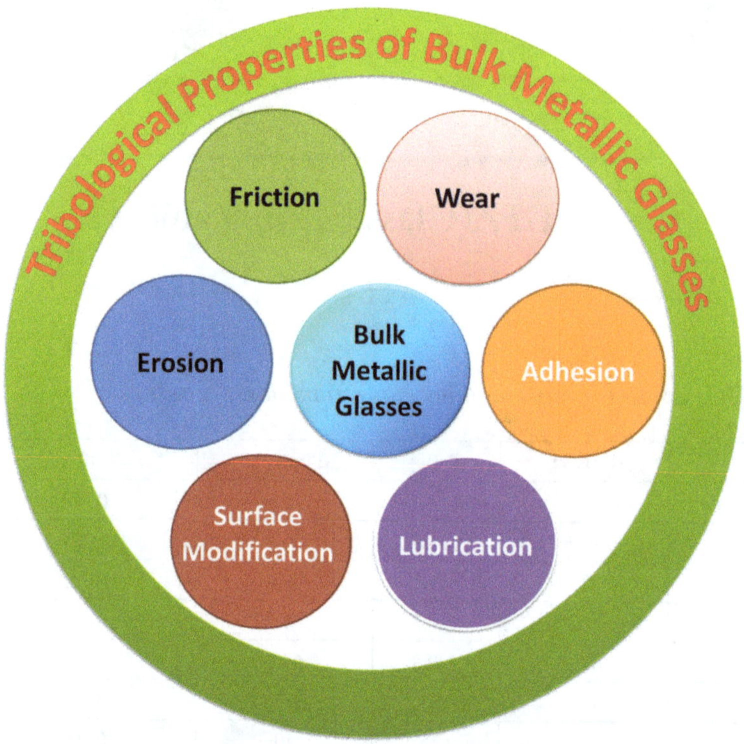

Mechanical Properties

One of the most common mechanical characteristics of metallic glasses is their hardness. This property can be measured in various ways but most commonly Vickers hardness, yield strength, and fracture strength. There is a linear relationship between the yield strength and fracture strength and between Vickers hardness and Young's modulus. The yield strength or hardness values are unaffected by structural relaxation in BMGs (Trexler & Thadhani, 2010). Because BMGs are isotropic solids, their mechanical properties are less affected by their processing history. Unlike crystalline alloys, the microstructure changes during processing, resulting in different mechanical properties. A unique deformation mechanism is realized in BMGs, which thus have high strength (~2 GPa for Cu, Ti, and Zr based alloys; ~3 GPa for Ni-based, ~4

GPa for Fe-based, and ~5 GPa for Co-based alloys) and high hardness, good wear resistance, and large elastic deformation (Trexler & Thadhani, 2010). For example, the glassy alloys produced by Amiya et al. in a cylindrical shape with a diameter of 1.8 cm have an extraordinary GFA and an extraordinary strength above 4 GPa, as indicated by their GFA measurements (Amiya & Inoue, 2008). At ambient temperature, localized shear deformation is the most common mode of elastic deformation, while tensile ductility of BMGs is only encountered in extremely small samples or at extremely high strain rates in very small pieces. The high specific strength of 270 GPa and a high specific Young's modulus was observed in $Co_{43}Fe_{20}Ta_{5.5}B_{31.5}$ BMG with an ultrahigh breakage strength above 5 GPa (Inoue et al., 2003). The hardness and specific strength results are higher than those published for any other BMG alloy.

Furthermore, for increasing the ductility of metallic glasses, the creation of heterogeneous microstructures is necessary and can be achieved in a variety of ways, including deformation-induced nanocrystallization (Das et al., 2005). Interaction with the phases embedded in the glassy matrix can block the propagation of shear bands through the sample. As in composites, where cracks are plugged by reinforcements or blunted in ductile phases or matrix, this allows for the multiplication, branching, and termination of shear bands. In $Cu_{50}Zr_{50}$ and $Zr_{65}Al_{7.5}Ni_{10}Pd_{17.5}$ glassy alloys, nanoparticles stabilized plastic deformation by blocking the crack-tip area (Hajlaoui et al., 2007). Since Pd is substituted for Cu, the primary phase shifts into a nanoscale icosahedron quasicrystalline phase, the compressive ambient temperature stress-strain curve of $Zr_{65}Al_{7.5}Ni_{10}Cu_{17.5}$ and $Zr_{65}Al_{7.5}Ni_{10}Pd_{17.5}$ BMGs have considerably altered plasticity. These nanoscale structures aid mechanical qualities such as high strength and ductility (Saida et al., 2005). In addition, there is currently a lack of understanding about the structural changes that thin film-based BMGs. Recently, Sun et al. discovered ultrastable BMGs in Zirconium-based thin films. This study looked at thin-film metallic glasses (TFMG) produced by room-temperature sputtering with a variety of $Zr_{50}Cu_{44.5}Al_{5.5}$ and $Zr_{50}Cu_{41.5}Al_{5.5}$ BMGs (Sun et al., 2020). Thermal stability and mechanical properties improved as the deposition rate was reduced, indicating a transition from bulk-like to ultrastable behavior. Lee and his colleagues used a co-sputtering method with four pure elementals (Zr–Ti–B–Si) targets to create a new TFMG with high hardness and corrosion resistance (Deng et al., 2014). Zr and B atoms differ in atomic size by 82.8 percent, and all binary atomic couples except the Zr–Ti combination have a negative heat of mixing. The formation of a thin film with a dense featureless microstructure and high hardness (more than 10 GPa) was achieved by raising the Si concentration by 12.9 percent.

Table 3. Mechanical and tribological properties of various BGMs

BMGs	Mechanical Features	Tribological Features	Applications	Ref
$Mg_{65}Zn_{30}Ca_5$	excellent hardness	excellent dry wear resistance	orthopedic, biomedical	(Hua et al., 2018)
$Zr_{44}Ti_{11}Cu_{10}Ni_{10}Be_{25}$	hardness 8.3 GPa and elastic modulus 128 GPa	wear resistance increases by 65% and does not obey Archard's wear law.	engineering applications, biomedical applications, machine tools, engine parts.	(Bajpai et al., 2021)
$Zr_{65}Cu_{15}Al_{10}Ni_{10}$	hardness 6.98 GPa elastic modulus, 115 GPa, yield strength 1.9 GPa	recovery index η = 0.71~0.86, COF=0.53	anti-wear surfaces, sports goods, structural frames, and electronic castings	(Marimuthu et al., 2021)
$Zr_{50}Ti_2Cu_{38}Al_{10}$ $(Zr_{0.5}Ti_{0.02}Cu_{0.38}Al_{0.1})_{98}Lu_2$	two at.% Lu addition enhances the critical diameter (d_c) from 5 mm to 20 mm.	improves the wear resistance	sporting goods, electromagnetic device casing, micro-geared motor parts	(Zhou et al., 2017)
$Zr_{44}Ti_{11}Cu_{10}Ni_{11}Be_{25}$	nano hardness 18.32 ± 0.21 GPa	the coefficient of friction reduced from 0.5–0.6 to 0.1–0.2; also, the wear resistance improved by more than 60 times.	orthopedic applications.	(Sawyer et al., 2020)
$Zr_{60}Cu_{10}Al_{15}Ni_{15}$	yield strength 1.6 GPa	the coefficient of friction varies from 0.39 to 0.63.	engineering applications	(Salehan et al., 2019)
$Ti_{45}Zr_{16}Cu_{10}Ni_9Be_{20}$	increased plasticity, decreased hardness after deep cryogenic-cycling treatment, and higher elastic recovery ability.	good wear resistance and anti-aging performance.	industrial	(Jia et al., 2021)
$[(Fe_{0.6}Co_{0.2}Ni_{0.2})_{0.69}B_{0.2}Si_{0.11}]_{96}Nb_4$	vickers hardness 930 HV, fracture strength above 2800 MPa	low wear loss of 3.77 × 10^5 μm^3 under a load of 10 N	industrial	(Wang et al., 2021)
Fe-Cr-Mo-C-B-Y amorphous coatings on 35CrMo steel	high-density structure (porosity less than 1.5%), high hardness (vickers microhardness up to $1104HV_{0.3}$), and better thermal stability	good wear resistance with a wear rate of 5.59 × 10^{-6} $mm^3 \cdot m^{-1} \cdot N^{1}$(CS), 1.70 × 10^{-5} $mm^3 \cdot m^{-1} \cdot N^{-1}$ (HVOF).	industrial	(Xu et al., 2021)
$Pd_{43}Cu_{27}Ni_{10}P_{20}$	excellent mechanical properties	coefficient of friction (0.30 ± 0.02)	surface coatings	(Medina et al., 2020)
$Pt_{57.5}Cu_{14.7}Ni_{5.3}P_{22.5}$	excellent mechanical properties	coefficient of friction (0.58±0.08)	biomedical implant applications	(Medina et al., 2020)

Tribological Properties

A material's mechanical qualities can predict tribological properties, and one of the most common traits that make BMG material special is its low friction coefficient (Wang et al., 2004). In certain environments, low friction is desirable because this makes it easier to control mechanical forces and prevent the wear of other components. This is especially important when dealing with solids and liquids, as they have higher friction than solids. The strength of metallic glasses is not in their hardness but in their malleability. In other words, they are hard but malleable and strong yet flexible.

Further, one of the greatest benefits of metallic glasses is their excellent wear resistance. This makes them extremely useful for industrial applications where reliable lubrication and a long lifetime are crucial. Various metallic glasses have proven exceptionally wear-resistant (Hofmann et al., 2017). CuZr-based alloys are known for their excellent wear resistance, while the replacement of copper with the alloys copper-nickel, copper-aluminum, and so on has been repeatedly shown to have dramatically improved wear characteristics in several tribological applications (Fleury et al., 2004; Hofmann et al., 2017). Furthermore, Blau (Blau, 2001) found no evidence of amorphous-to-crystalline or crystalline-to-amorphous transitions on worn surfaces of ZrCuNiTiAl metallic glasses, Fu and coworkers (Fu et al., 2001) found evidence of both crystallization and re-amorphization in Zr-based BMGs. There have been comparisons between $Zr_{52.5}Ti_5Cu_{17.9}Ni_{14.6}Al_{10}$, aluminum 6061, and stainless steel AISI 304 in terms of friction coefficient. Using this block on a ring setup, friction and wear qualities can be studied (Parlar et al., 2008). Compared to aluminum and stainless steel, Zr-based BMG has a lower coefficient of friction. It is possible to reduce the wear rate during sliding by controlling the transfer of material between the counter surface material and the creation of shielding oxide layers on the Zirconium-based BMG surface. At lower weights, the coefficient of friction increases due to the destruction of significant surface asperities and then falls due to compacted and flattened wreckages. Although Zr-based BMG is the strongest and most wear-resistant, it has the highest wear rate. During abrasive wear operations, the Zr-based BMG is subjected to high tensile stress, increasing the wear rate (Prakash, 2005). Due to the rule of identical metal couples, Al (among Zr-based BMG, Al, and SS Steel) wears better than stainless steel.

The bonding that mechanical rubber materials exhibit with other materials is poor. Strong adhesion is essential when working with components that require high-impact loads. However, metal alloys offer several different properties that allow them to bond to various other substrates. Their high strength, low cost, corrosion resistance, and flexibility all help to boost their suitability for durable, low friction components. While rubber-metal alloys are often used for difficult mechanical parts, these materials tend to suffer from several compatibility issues and poor adhesion when they come into contact with many different substrates. This makes them less practical when working with less durable, more corrosive materials. Recently, Medina et al. reported a comparative investigation of the tribological properties of Pd-, Pt-, and Zr-based bulk metallic glasses ($Pt_{57.5}Cu_{14.7}Ni_{5.3}P_{22.5}$ (BMG-Pt), $Pd_{43}Cu_{27}Ni_{10}P_{20}$ (BMG-Pd), and $Zr_{35}Ti_{30}Cu_{8.25}Be_{26.75}$ (BMG-Zr)) under unlubricated conditions (Medina et al., 2020). In specifically, micro-tribometry is employed with a 52,100 steel ball, indicating that BMG-Pt exhibits a considerably greater coefficient of friction (COF) (0.58 ± 0.08) when linked with BMG-Pd (0.30 ± 0.02) and BMG-Zr (0.20 ± 0.03). The results of the atomic force microscope (AFM) on and off wear scars are inconsistent with the observed frictional behavior. For the other two samples, scanning electron microscopy (SEM) was used to show that the abrasive wear processes on BMG-Pd and BMG-Zr were different from those on BMG-Pt, which showed signs of adhesive wear. A possible explanation for high coefficients of friction observed on BMG-Pt is that it has more contact with steel than the other BMG samples, signifying more adhesive wear.

The most crucial ingredient in the formula for good tribological performance is lubricity. A poor lubrication level of lubricant's physical or chemical properties can contribute to poor mechanical performance or, in worst cases, even mechanical failure. However, a new and promising class of tribological lubricants has recently emerged: metallic glasses. These are semi-metallic in nature, which means that they contain various metallic atoms or ions, and typically appear as glasses. When used in the right way, they have proven to be an ideal lubricant for certain harsh environments and wear environments, where traditional greases are not suitable. Recently, Jiang et al. reported BMG/graphite composites with a 3D lubricating layer (Jiang et al., 2020). In this study, an in-situ integrated method is made $Zr_{48}Cu_{36}Al_8Ag_8$ BMG/graphite composites with a 3D lubricating layer. Micro-texture effect and graphite lubrication synergism could greatly minimize wear by smoothing out the friction coefficient curve. The corrosion of BMG/graphite composites, notwithstanding the lubricating effect of seawater, has a detrimental influence on their performance. It is possible to prevent galvanic events at the interface between the BMG and graphite and improve the composite material's wear resistance by introducing an external electric potential.

Surface Treatment and Coating

Surface modification is another aspect of metallic glasses that has gained importance lately (Fotovvati et al., 2019; Joshi et al., 2016). Metallics with a high crystallinity rate tend to exhibit a chalky appearance. However, it is possible to take care of this characteristic by various methods such as surface modification, surface coating, deposition, etc. (Li & Ma, 2021). The most common way is to sprinkle nickel into the molten metal that will form the bulk. The metallic glass will then transform into an oxide layer, which will make it very durable in the next step. Incidentally, it is also possible to modify the entire crystal structure without the need for a powder. Jinn P. Chu et al. show for the first time that BMGs can be made bendable by coating them with a thin, strong, flexible, and sticky coating (Chu et al., 2012). According to the experiment results, a thin bilayer coating placed on the tensile side of a 3 mm thick $Zr_{50}Cu_{30}Al_{10}Ni_{10}$ increased the bending ductility of the BMG from 0% to 13.7 percent. The bilayer is formed from two different materials to achieve the desired adhesion, strength, and elasticity: a $Zr_{53}Cu_{26}Al_{15}Ni_6$ metallic glass overlayer at 200 nm and a Ti adhesive layer at 25 nm. The application of bilayer coatings to the tensile side of BMGs reduced the number of starting points caused by surface imperfections, allowing for a higher density of more uniformly distributed shear bands to form during four-point bending deformation.

Furthermore, when selecting a coating process for a certain application, the capacity of the coating process to provide corrosion and wear protection is the most important factor to consider. This process is extremely versatile because it can use almost any conductive substance as a feedstock. It will be accomplished by using metal alloys such as Ni and Ti, as well as Al, Zn, Mo, and Mo, among other elements (Fotovvati et al., 2019). Coating quality, porosity, and adhesion effects on substrates such as stainless steel, titanium-based alloys, and ceramic materials are widely studied. Laser-based processing and thermal spray-based procedures are the most commonly used methods for producing amorphous coatings (Cheng et al., 2020; Joshi et al., 2016). Amorphous coatings are created by depositing the amorphous alloy or predepositing the alloy composition in powdered form and then changing that layer into an amorphous structure. Surface treatment of structural metallic materials can also enhance the properties of the materials they support. Energy is applied to the material's surface during the processing procedure to treat it. An advantage of laser-based techniques is that they can quickly and easily synthesize amorphous coatings with perfect control over process parameters and site-specific processing. Thermal spray techniques such as vaporization and plasma spraying are also regarded as appropriate for producing amorphous coatings and surface modification due to their extraordinarily rapid cooling rates. When vaporization-based procedures are used, the coating layer formed is a thin film with excellent corrosion and wear resistance, commonly used in tool coating and machine sliding component protection. The microstructure can be precisely controlled by altering this process.

The tribological properties of high-power laser surface-treated metallic glasses are of particular interest because of their non-abrasiveness and ability to be more resilient against abrasion. Engineers find them useful for applications such as bearing surfaces. These materials have a low friction coefficient and a low wear rate. This is because they have a microstructure consisting of many small crystals (coating) with an orientation that is very close to the surface. Many of today's BMGs are manufactured using the processing capability of a high-power Nd: YAG laser to achieve the intrinsically high cooling rates required. As a possible solution, the production of thick (($\geq$250 μm) amorphous surface layers has been investigated (Matthews et al., 2007). Because of an amorphous matrix interlayer approximately 50 μm thick, the coating's adhesion to the substrate is functionally graded. Matthews et al. discovered that using laser cladding to apply metallic glasses as surface layers to titanium substrates is a good technique for applying metallic glasses as surface layers (Matthews et al., 2007). A comparison of the hardness and nanoindentation profiles reveals coating hardnesses of up to 13 GPa throughout the coating's depth and an elastic modulus of around 150 GPa, which is equivalent to bulk metallic glass melt-spun ribbons. Tribological tests have also been performed, revealing that good wear qualities can be achieved, and that shear banding can be observed in the material's contact region. During scratch testing, shear band development is observed once more, indicating that the layers may have extremely low friction coefficients.

The qualities of BMGs based on Fe are superior to those of conventional metallic alloys in terms of hardness, elastic modulus/limit, and wear/corrosion resistance (Li et al., 2019). Because the microstructures generate compact, completely dense amorphous microstructures, Young's modulus, and yield stress are frequently lower than those of oxide glasses. In this view, A. Basu et al. used laser surface processing to develop an amorphous coating on an AISI 4140 substrate using $Fe_{48}Cr_{15}Mo_{14}Y_2C_{15}B_6$ bulk metallic glass then applied to an AISI 4140 substrate (Basu et al., 2008). It was discovered that the thickness of the coating was influenced by a direct link between laser power and time of contact. The microstructure and phase aggregation of the material and its microhardness and wear resistance were all evaluated using Vickers microhardness testers and ball-on-plate wear testing equipment. The microhardness of the coated layer rose from 240 VHN to 950 VHN, which is a significant improvement over the substrate's 240 VHN value. The best wear resistance is found on the surface of the material, and as it travels deeper into the material, the resistance declines.

Sohrabi et al. demonstrated that it is possible to achieve more than 99 percent relative densities by layer-wise laser additive manufacturing of a Zr-based metallic glass ($Zr_{59.3}Cu_{28.8}Al_{10.4}Nb_{1.5}$) on an aluminum substrate (Sohrabi et al., 2020). With less than 0.5 percent porosity content, $Zr_{59.3}Cu_{28.8}Al_{10.4}Nb_{1.5}$ powder was successfully cladded on an aluminum substrate to a thickness of 300 microns, and a laser power threshold for crack initiation at the interface was determined for the first time. According to these results, amorphous coatings have a wear resistance that is 20 times greater than that of their substrates, making them among the best potential coatings for aluminum alloys. To achieve both a highly amorphous layer and excellent bonding with the substrate, the laser parameters were adjusted from the interface to the bulk of the cladding. The employment of two separate sets of laser settings, one near the interface and one in bulk, was necessary to achieve a good crack-free coating. The first layer's energy density was higher to melt both the powder and the solid substrate, which resulted in excellent bonding at the interface between the two materials.

Ren L. W. et al. reported improvement of plasticity in Zirconium-based BMGs (Ren et al., 2015) by electrodeposition through the copper coating. It has been observed that copper coatings can improve the plasticity of $Zr_{52.5}Cu_{17.9}Ni_{14.6}Al_{10}Ti_5$ BMGs. On the BMGs, electrodeposition was utilized to coat copper films with film thicknesses of 71.5 μm and 161.1μm. It was found that copper plating increased the density of shear bands created during compression tests, which increased plasticity for BMGs coated with copper. Compared to the uncoated BMGs, coated BMGs with a coating thickness of 161.1 μm had a plastic strain of 6.1 percent. Copper film deformation reduced strain energy and slowed the spread of shear bands, forming many shear bands. This method suggests that nanostructured coatings and bulk materials can be created through electrodeposition, which confines shear bands and promotes overall plasticity at room temperature. Gi-Su Ham and colleagues employed various thermal spray methods to create Fe-Mo-Cr-C-B metallic glass coating layers and then evaluated the microstructure and wear parameters of the coating layers (Ham et al., 2020). To produce Fe-based metallic glass coating layers, they used metallic glass powder of newly designed $Fe_{46.8}$-$Mo_{30.6}$-$Cr_{16.6}$-$C_{4.3}$-$B_{1.7}$ composition and three representative thermal spray techniques, including atmospheric plasma (plasma), high-velocity oxygen plasma (HVOF), and vacuum plasma spray (VPS). According to the nano indentation experiment results, the average hardness of the plasma materials was 1242.8Hv. 746.0 Hv was found to exist between the maximum and minimum hardness of the coating layers generated by the atmospheric plasma spray technique, revealing a significant difference in mechanical properties between the two coating layers. With an average hardness of 1980.9 Hv and a difference of 516.1 Hv between the greatest and minimum hardnesses, the HVOF materials had a hardness of 1980.9 Hv. The VPS materials were found to have an extremely high hardness value, with an average hardness value of 2403.5 Hv. Their hardness deviation was only 13.9 Hv, indicating that their mechanical properties were generally constant throughout the study region. Microstructures were identified in the indentation microstructures, and the Cr depletion area that appeared in the plasma and HVOF materials had inferior mechanical properties to the non-depleted site. The findings of the wear tests revealed that the VPS coating layers exhibited the greatest wear resistance of all three coating layers, whereas the plasma material exhibited the least wear resistance of all three coating layers. Wear tests revealed that the Fe-based metallic glass coating layers thermally sprayed onto the substrate had a smaller wear volume than the substrate. The wear resistance of the Fe-based metallic glass coating layers under the condition of 49.00 N, in particular, was approximately 8.8 times more (plasma), 124.2 times greater (HVOF), and 271.0 times greater (VPS) than that of the substrate. When the wear resistance of Fe-based metallic glass coating layers was evaluated, it was found that the VPS materials had the best wear resistance under all wear load situations under consideration. The VPS materials had approximately 30.7 and 2.18 times more wear resistance than the plasma and high-temperature oxygen-free materials. Because of the decreased binding force between the particles in the plasma material discovered through worn surface and cross-section observations, the material had particle boundary delamination. Microcracks were produced along the particle boundaries directly beneath the worn-out surface of the HVOF coating layers, which was now beneath the worn-out surface of the HVOF coating layers. Directly beneath the worn-out surface of the VPS materials, on the other hand, were discovered to have no flaws at all. The microstructure of Fe-based metallic glass coating layers created using various thermal spray techniques was investigated to understand their wear behavior and durability.

In summary, the use of BMGs is ever-expanding. The tribological performance of BMGs is becoming more and more promising with the improvement of new technology such as 3D printing. BMGs are five or six major elements mixed in a glass-forming process. They have good metal properties but also have the advantages of plasticity, low weight, high strength, low friction, low wear rate, and excellent heat resistance. At present, the most popular BMGs are zinc-aluminum-nickel BMGs. Zinc-aluminum-cobalt BMGs are considered one of the next-generation materials that can replace high chromium steel in some cases. In addition, this technology is used in many different fields, including medical devices, aerospace engineering, and mechanical engineering. The properties of BMGs make them very attractive for these applications because they are strong yet lightweight; corrosion-resistant; can be engineered to be electrically conductive, and can withstand high temperatures. These properties make them excellent for use in situations where weight is a concern or where strength needs to be maintained even at extreme temperature

CONCLUSION

Tribology is the study of motion at the sub-atomic and micro-physical levels in which the influence of external forces and structures are observed. It covers the wide spectrum of materials in their various forms and their applications in a wide range of industries, ranging from the automotive sector to civil engineering to the oil and gas sector. Whatever the application, one thing is common to every tribological system - precise control and management of contact forces and strains by coupling strengths of deformations with friction. Therefore, it is important to know the causes of tribological stresses and their variations and the relationship between stresses and strains to design tribological systems that meet the application's requirements and perform optimally. One of the most promising materials for tribological applications in BMGs. The development of BMGs is motivated by the need for new, robust materials that can operate under extreme operating conditions. Tribology is a key technology for these applications because it significantly influences the overall performance of machine components and systems. BMGs are attractive materials due to their outstanding mechanical properties compared to traditional engineering materials. The high strength, elasticity, hardness, wear resistance, toughness, and chemical inertness confers many benefits over conventional materials – all of which make them an attractive option for applications in the automotive industry. BMGs also offer biomedical engineering and nanotechnology prospects, such as smart surfaces. There is still much to be explored about this exciting new material.

REFERENCES

Abad, M., & Browne, D. (2020). An Investigation of the Tribological Behavior of Hf-Based Bulk Metallic Glass and Crystalline Alloys. *Journal of Tribology*, *142*(10), 101703. doi:10.1115/1.4046950

Amiya, K., & Inoue, A. (2008). Fe-(Cr, Mo)-(C, B)-Tm bulk metallic glasses with high strength and high glass-forming ability. *Reviews on Advanced Materials Science*, *18*, 27–29.

Bajpai, S., Nisar, A., Sharma, R. K., Schwarz, U. D., Balani, K., & Datye, A. (2021). Effect of fictive temperature on tribological properties of $Zr_{44}Ti_{11}Cu_{10}Ni_{10}Be_{25}$ bulk metallic glasses. *Wear*, *486*, 204075. doi:10.1016/j.wear.2021.204075

Basu, A., Samant, A., Harimkar, S., Majumdar, J. D., Manna, I., & Dahotre, N. B. (2008). Laser surface coating of Fe–Cr–Mo–Y–B–C bulk metallic glass composition on AISI 4140 steel. *Surface and Coatings Technology, 202*(12), 2623–2631. doi:10.1016/j.surfcoat.2007.09.028

Basu, J., & Ranganathan, S. (2003). Bulk metallic glasses: A new class of engineering materials. *Sadhana, 28*(3-4), 783–798. doi:10.1007/BF02706459

Bhatt, J., & Murty, B. (2008). On the conditions for the synthesis of bulk metallic glasses by mechanical alloying. *Journal of Alloys and Compounds, 459*(1-2), 135–141. doi:10.1016/j.jallcom.2007.04.242

Blau, P. J. (2001). Friction and wear of a Zr-based amorphous metal alloy under dry and lubricated conditions. *Wear, 250*(1-12), 431-434.

Calka, A., & Radlinski, A. (1991). Universal high performance ball-milling device and its application for mechanical alloying. *Materials Science and Engineering A, 134*, 1350–1353. doi:10.1016/0921-5093(91)90989-Z

Chen, H. (1980). Glassy metals. *Reports on Progress in Physics, 43*(4), 353–432. doi:10.1088/0034-4885/43/4/001

Chen, M. (2011). A brief overview of bulk metallic glasses. *NPG Asia Materials, 3*(9), 82–90. doi:10.1038/asiamat.2011.30

Cheng, J., Feng, Y., Yan, C., Hu, X., Li, R., & Liang, X. (2020). Development and characterization of Al-based amorphous coating. *JOM, 72*(2), 745–753. doi:10.100711837-019-03966-y

Chu, J. P., Greene, J., Jang, J. S., Huang, J., Shen, Y.-L., Liaw, P. K., Yokoyama, Y., Inoue, A., & Nieh, T. (2012). Bendable bulk metallic glass: Effects of a thin, adhesive, strong, and ductile coating. *Acta Materialia, 60*(6-7), 3226–3238. doi:10.1016/j.actamat.2012.02.037

Chueva, T., Dyakonova, N., Molokanov, V., & Sviridova, T. (2007). Bulk amorphous alloy $Fe_{72}Al_5Ga_2C_6B_4P_{10}Si_1$ produced by mechanical alloying. *Journal of Alloys and Compounds, 434*, 327–332. doi:10.1016/j.jallcom.2006.08.308

Conner, R., Dandliker, R., & Johnson, W. L. (1998). Mechanical properties of tungsten and steel fiber reinforced $Zr_{41.25}Ti_{13.75}Cu_{12.5}Ni_{10}Be_{22.5}$ metallic glass matrix composites. *Acta Materialia, 46*(17), 6089–6102. doi:10.1016/S1359-6454(98)00275-4

Cornuault, P.-H., Colas, G., Lenain, A., Daudin, R., & Gravier, S. (2020). On the diversity of accommodation mechanisms in the tribology of Bulk Metallic Glasses. *Tribology International, 141*, 105957. doi:10.1016/j.triboint.2019.105957

Czichos, H. (2009). *Tribology: a systems approach to the science and technology of friction, lubrication, and wear* (Vol. 1). Elsevier.

Das, J., Tang, M. B., Kim, K. B., Theissmann, R., Baier, F., Wang, W. H., & Eckert, J. (2005). "Work-hardenable" ductile bulk metallic glass. *Physical Review Letters, 94*(20), 205501. doi:10.1103/PhysRevLett.94.205501 PMID:16090260

Deng, Y.-L., Lee, J.-W., Lou, B.-S., Duh, J.-G., Chu, J. P., & Jang, J. S.-C. (2014). The fabrication and property evaluation of Zr–Ti–B–Si thin film metallic glass materials. *Surface and Coatings Technology, 259*, 115–122. doi:10.1016/j.surfcoat.2014.03.026

Eckert, J., Das, J., Pauly, S., & Duhamel, C. (2007). Mechanical properties of bulk metallic glasses and composites. *Journal of Materials Research, 22*(2), 285–301. doi:10.1557/jmr.2007.0050

Fleury, E., Lee, S., Ahn, H., Kim, W., & Kim, D. (2004). Tribological properties of bulk metallic glasses. *Materials Science and Engineering A, 375*, 276–279. doi:10.1016/j.msea.2003.10.065

Fotovvati, B., Namdari, N., & Dehghanghadikolaei, A. (2019). On coating techniques for surface protection: A review. *Journal of Manufacturing and Materials Processing, 3*(1), 28. doi:10.3390/jmmp3010028

Fu, X.-Y., Kasai, T., Falk, M. L., & Rigney, D. (2001). Sliding behavior of metallic glass: Part I. Experimental investigations. *Wear, 250*(1-12), 409-419.

Gibson, M. A., Mykulowycz, N. M., Shim, J., Fontana, R., Schmitt, P., Roberts, A., ... Bordeenithikasem, P. (2018). 3D printing metals like thermoplastics: Fused filament fabrication of metallic glasses. *Materials Today, 21*(7), 697–702. doi:10.1016/j.mattod.2018.07.001

Greer, A. L., & Ma, E. (2007). Bulk metallic glasses: At the cutting edge of metals research. *MRS Bulletin, 32*(8), 611–619. doi:10.1557/mrs2007.121

Guddla, G., Katta, V., Gandi, S., Ambadipudi, S., & Ravuri, B. (2021). The role of titanium content in $(Ni_{75}Cr_{15}Si_{10})_{100-x}Ti_x$ bulk metallic glass systems to elevate mechanical and corrosion properties. *Phase Transitions, 94*(10), 679–690. doi:10.1080/01411594.2021.1957103

Guddla, G. T., Gandi, S., Ambadipudi, S., & Ravuri, B. R. (2021). Design of Ni-based Bulk Metallic Glasses with Improved Mechanical and Corrosion Properties. *Current Applied Science and Technology*, 115-131.

Hajlaoui, K., Yavari, A., LeMoulec, A., Botta, W., Vaughan, F., Das, J., Greer, A. L., & Kvick, Å. (2007). Plasticity induced by nanoparticle dispersions in bulk metallic glasses. *Journal of Non-Crystalline Solids, 353*(3), 327–331. doi:10.1016/j.jnoncrysol.2006.10.011

Ham, G.-S., Kim, K.-W., Cho, G.-S., Kim, C. P., & Lee, K.-A. (2020). Fabrication, microstructure and wear properties of novel Fe-Mo-Cr-CB metallic glass coating layers manufactured by various thermal spray processes. *Materials & Design, 195*, 109043. doi:10.1016/j.matdes.2020.109043

Heinrich, J., Busch, R., & Nonnenmacher, B. (2012). Processing of a bulk metallic glass forming alloy based on industrial grade Zr. *Intermetallics, 25*, 1–4. doi:10.1016/j.intermet.2012.02.011

Hofmann, D. C., Andersen, L. M., Kolodziejska, J., Roberts, S. N., Borgonia, J. P., Johnson, W. L., Vecchio, K. S., & Kennett, A. (2017). Optimizing bulk metallic glasses for robust, highly wear-resistant gears. *Advanced Engineering Materials, 19*(1), 1600541. doi:10.1002/adem.201600541

Hua, N., Chen, W., Wang, Q., Guo, Q., Huang, Y., & Zhang, T. (2018). Tribocorrosion behaviors of a biodegradable $Mg_{65}Zn_{30}Ca_5$ bulk metallic glass for potential biomedical implant applications. *Journal of Alloys and Compounds, 745*, 111–120. doi:10.1016/j.jallcom.2018.02.138

Hutchings, I., & Shipway, P. (2017). *Tribology: friction and wear of engineering materials*. Butterworth-Heinemann.

Inoue, A., Shen, B., Koshiba, H., Kato, H., & Yavari, A. R. (2003). Cobalt-based bulk glassy alloy with ultrahigh strength and soft magnetic properties. *Nature Materials*, *2*(10), 661–663. doi:10.1038/nmat982 PMID:14502274

Jia, Q., Zhou, Q., Ren, Y., Du, Y., Zhao, X., Wang, X.-Z., Wang, H., Beake, B. D., & Zhou, F. (2021). Tribological characteristics of Ti-based bulk metallic glass via deep cryogenic-cycling treatment. *Materials Characterization*, *179*, 111356. doi:10.1016/j.matchar.2021.111356

Jiang, J. Z., Hofmann, D., Jarvis, D. J., & Fecht, H. J. (2015). Low-density high-strength bulk metallic glasses and their composites: A Review. *Advanced Engineering Materials*, *17*(6), 761–780. doi:10.1002/adem.201400252

Jiang, X., Song, J., Chen, S., Su, Y., Fan, H., Zhang, Y., & Hu, L. (2020). In-situ fabricated bulk metallic glass/graphite composites with a 3D lubricating layer: Tribological properties under dry sliding and in seawater. *Tribology International*, *148*, 106301. doi:10.1016/j.triboint.2020.106301

Jiang, X., Song, J., Su, Y., Fan, H., Zhang, Y., & Hu, L. (2019). Tribological behaviors of self–mated $Cu_{36}Zr_{48}Ag_8Al_8$ bulk metallic glass under H_2SO_4 conditions with different concentrations. *Tribology International*, *136*, 395–403. doi:10.1016/j.triboint.2019.04.006

Johnson, W. L. (1996). Bulk metallic glasses—A new engineering material. *Current Opinion in Solid State and Materials Science*, *1*(3), 383–386. doi:10.1016/S1359-0286(96)80029-5

Joshi, S. S., Katakam, S., Singh Arora, H., Mukherjee, S., & Dahotre, N. B. (2016). Amorphous coatings and surfaces on structural materials. *Critical Reviews in Solid State and Material Sciences*, *41*(1), 1–46. doi:10.1080/10408436.2015.1053602

Khan, M. M., Nemati, A., Rahman, Z. U., Shah, U. H., Asgar, H., & Haider, W. (2018). Recent advancements in bulk metallic glasses and their applications: A review. *Critical Reviews in Solid State and Material Sciences*, *43*(3), 233–268. doi:10.1080/10408436.2017.1358149

Khun, N., Yu, H., Chong, Z., Tian, P., Tian, Y., Tor, S., & Liu, E. (2016). Mechanical and tribological properties of Zr-based bulk metallic glass for sports applications. *Materials & Design*, *92*, 667–673. doi:10.1016/j.matdes.2015.12.050

Kim, Y., Bae, D., Kim, W., & Kim, D. (2003). Glass forming ability and crystallization behavior of Ti-based amorphous alloys with high specific strength. *Journal of Non-Crystalline Solids*, *325*(1-3), 242–250. doi:10.1016/S0022-3093(03)00327-2

Kim, Y., Kim, W., & Kim, D. (2004). A development of Ti-based bulk metallic glass. *Materials Science and Engineering A*, *375*, 127–135. doi:10.1016/j.msea.2003.10.115

Klement, W., Willens, R., & Duwez, P. (1960). Non-crystalline structure in solidified gold–silicon alloys. *Nature*, *187*(4740), 869–870. doi:10.1038/187869b0

Kruzic, J. J. (2016). Bulk metallic glasses as structural materials: A review. *Advanced Engineering Materials*, *18*(8), 1308–1331. doi:10.1002/adem.201600066

Laws, K. J., Gun, B., & Ferry, M. (2009). Influence of casting parameters on the critical casting size of bulk metallic glass. *Metallurgical and Materials Transactions. A, Physical Metallurgy and Materials Science*, *40*(10), 2377–2387. doi:10.100711661-009-9929-7

Lee, J., Bae, D., Yi, S., Kim, W., & Kim, D. (2004). Effects of Sn addition on the glass forming ability and crystallization behavior in Ni–Zr–Ti–Si alloys. *Journal of Non-Crystalline Solids*, *333*(2), 212–220. doi:10.1016/j.jnoncrysol.2003.10.011

Li, H., Lu, Z., Wang, S., Wu, Y., & Lu, Z. (2019). Fe-based bulk metallic glasses: Glass formation, fabrication, properties and applications. *Progress in Materials Science*, *103*, 235–318. doi:10.1016/j.pmatsci.2019.01.003

Li, K., Cai, Z., Du, P., Xiang, T., Yang, X., & Xie, G. (2021). Core-shell $Mg_{66}Zn_{30}Ca_4$ bulk metallic glasses composites reinforced by Fe with high strength and controllable degradation. *Intermetallics*, *138*, 107334. doi:10.1016/j.intermet.2021.107334

Li, N., Zhang, J., Xing, W., Ouyang, D., & Liu, L. (2018). 3D printing of Fe-based bulk metallic glass composites with combined high strength and fracture toughness. *Materials & Design*, *143*, 285–296. doi:10.1016/j.matdes.2018.01.061

Li, X. (2018). Additive Manufacturing of Advanced Multi-Component Alloys: Bulk Metallic Glasses and High Entropy Alloys. *Advanced Engineering Materials*, *20*(5), 1700874. doi:10.1002/adem.201700874

Li, Y., Poon, S., Shiflet, G., Xu, J., Kim, D., & Löffler, J. F. (2007). Formation of bulk metallic glasses and their composites. *MRS Bulletin*, *32*(8), 624–628. doi:10.1557/mrs2007.123

Li, Z., & Ma, J. (2021). Water-repellent surfaces of metallic glasses: Fabrication and application. *Materials Today Advances*, *12*, 100164. doi:10.1016/j.mtadv.2021.100164

Liao, Z., Hua, N., Chen, W., Huang, Y., & Zhang, T. (2018). Correlations between the wear resistance and properties of bulk metallic glasses. *Intermetallics*, *93*, 290–298. doi:10.1016/j.intermet.2017.10.008

Löffler, J. F. (2003). Bulk metallic glasses. *Intermetallics*, *11*(6), 529–540. doi:10.1016/S0966-9795(03)00046-3

Lu, Z., Li, Y., & Ng, S. (2000). Reduced glass transition temperature and glass forming ability of bulk glass forming alloys. *Journal of Non-Crystalline Solids*, *270*(1-3), 103–114. doi:10.1016/S0022-3093(00)00064-8

Lu, Z., & Liu, C. (2002). A new glass-forming ability criterion for bulk metallic glasses. *Acta Materialia*, *50*(13), 3501–3512. doi:10.1016/S1359-6454(02)00166-0

Lu, Z., & Liu, C. (2004). Role of minor alloying additions in formation of bulk metallic glasses: A review. *Journal of Materials Science*, *39*(12), 3965–3974. doi:10.1023/B:JMSC.0000031478.73621.64

Lu, Z., Tan, H., Li, Y., & Ng, S. (2000). *Correlation between reduced glass transition temperature and glass forming ability of bulk metallic glasses*. Academic Press.

Ludema, K. C., & Ajayi, O. O. (2018). *Friction, wear, lubrication: A textbook in tribology*. CRC Press. doi:10.1201/9780429444715

Marimuthu, K. P., Han, J., Jeong, U., Lee, K., & Lee, H. (2021). Study on tribological characteristics of Zr-based BMG via nanoscratch techniques. *Wear, 486*, 204067. doi:10.1016/j.wear.2021.204067

Matthews, D., Ocelik, V., & De Hosson, J. T. M. (2007). Tribological and mechanical properties of high power laser surface-treated metallic glasses. *Materials Science and Engineering A, 471*(1-2), 155–164. doi:10.1016/j.msea.2007.02.119

Medina, M. A., Acikgoz, O., Rodriguez, A., Meduri, C. S., Kumar, G., & Baykara, M. Z. (2020). Comparative Tribological Properties of Pd-, Pt-, and Zr-Based Bulk Metallic Glasses. *Lubricants (Basel, Switzerland), 8*(9), 85. doi:10.3390/lubricants8090085

Nishiyama, N., Takenaka, K., Miura, H., Saidoh, N., Zeng, Y., & Inoue, A. (2012). The world's biggest glassy alloy ever made. *Intermetallics, 30*, 19–24. doi:10.1016/j.intermet.2012.03.020

Park, E., Kang, H., Kim, W., & Kim, D. (2001). The effect of Ag addition on the glass-forming ability of Mg–Cu–Y metallic glass alloys. *Journal of Non-Crystalline Solids, 279*(2-3), 154–160. doi:10.1016/S0022-3093(00)00412-9

Park, E., & Kim, D. (2005). Design of bulk metallic glasses with high glass forming ability and enhancement of plasticity in metallic glass matrix composites: A review. *Metals and Materials International, 11*(1), 19–27. doi:10.1007/BF03027480

Park, E., Lim, H., Kim, W., & Kim, D. (2002). The effect of Sn addition on the glass-forming ability of Cu–Ti–Zr–Ni–Si metallic glass alloys. *Journal of Non-Crystalline Solids, 298*(1), 15–22. doi:10.1016/S0022-3093(01)01047-X

Parlar, Z., Bakkal, M., & Shih, A. J. (2008). Sliding tribological characteristics of Zr-based bulk metallic glass. *Intermetallics, 16*(1), 34–41. doi:10.1016/j.intermet.2007.07.011

Pauly, S., Schricker, C., Scudino, S., Deng, L., & Kühn, U. (2017). Processing a glass-forming Zr-based alloy by selective laser melting. *Materials & Design, 135*, 133–141. doi:10.1016/j.matdes.2017.08.070

Pole, M., Sadeghilaridjani, M., Shittu, J., Mahajan, C., Ghodki, N., & Mukherjee, S. (2021). Electrodeposited metallic glasses with superlative wear resistance. *Materials Science and Engineering A, 816*, 141315. doi:10.1016/j.msea.2021.141315

Prakash, B. (2005). Abrasive wear behaviour of Fe, Co and Ni based metallic glasses. *Wear, 258*(1-4), 217–224. doi:10.1016/j.wear.2004.09.010

Qiu, S.-B., & Yao, K.-F. (2008). Novel application of the electrodeposition on bulk metallic glasses. *Applied Surface Science, 255*(5), 3454–3458. doi:10.1016/j.apsusc.2008.07.077

Rahaman, M. L., Zhang, L., Liu, M., & Liu, W. (2015). Surface roughness effect on the friction and wear of bulk metallic glasses. *Wear, 332*, 1231–1237. doi:10.1016/j.wear.2014.11.030

Ren, L., Meng, M., Wang, Z., Yang, F., Yang, H., Zhang, T., & Qiao, J. (2015). Enhancement of plasticity in Zr-based bulk metallic glasses electroplated with copper coatings. *Intermetallics, 57*, 121–126. doi:10.1016/j.intermet.2014.10.009

Saida, J., Kato, H., Setyawan, A. D. H., & Inoue, A. (2005). Characterization and properties of nano-crystal-forming Zr-based bulk metallic glasses. *Reviews on Advanced Materials Science*, *10*(1), 34–38.

Salehan, R., Shahverdi, H. R., & Miresmaeili, R. (2019). Effects of annealing on the tribological behavior of $Zr_{60}Cu_{10}Al_{15}Ni_{15}$ bulk metallic glass. *Journal of Non-Crystalline Solids*, *517*, 127–136. doi:10.1016/j.jnoncrysol.2019.05.013

Sawyer, V., Tao, X., Dong, H., Dashtbozorg, B., Li, X., Sammons, R., & Dong, H.-S. (2020). Improving the tribological properties and biocompatibility of Zr-based bulk metallic glass for potential biomedical applications. *Materials (Basel)*, *13*(8), 1960. doi:10.3390/ma13081960 PMID:32331294

Schneider, S. (2001). Bulk metallic glasses. *Journal of Physics Condensed Matter*, *13*(34), 7723–7736. doi:10.1088/0953-8984/13/34/316

Schroers, J. (2010). Processing of bulk metallic glass. *Advanced Materials*, *22*(14), 1566–1597. doi:10.1002/adma.200902776 PMID:20496386

Siegrist, M. E., Amstad, E. D., & Löffler, J. F. (2007). Tribological properties of graphite-and ZrC-reinforced bulk metallic glass composites. *Intermetallics*, *15*(9), 1228–1236. doi:10.1016/j.intermet.2007.03.001

Sohrabi, N., Jhabvala, J., & Logé, R. E. (2021). Additive manufacturing of bulk metallic glasses—process, challenges and properties: A review. *Metals*, *11*(8), 1279. doi:10.3390/met11081279

Sohrabi, N., Panikar, R. S., Jhabvala, J., Buch, A. R., Mischler, S., & Logé, R. E. (2020). Laser coating of a Zr-based metallic glass on an aluminum substrate. *Surface and Coatings Technology*, *400*, 126223. doi:10.1016/j.surfcoat.2020.126223

Su, J., Kang, J., Yue, W., Ma, G., Fu, Z., Zhu, L., She, D., Wang, H., & Wang, C. (2019). Comparison of tribological behavior of Fe-based metallic glass coatings fabricated by cold spraying and high velocity air fuel spraying. *Journal of Non-Crystalline Solids*, *522*, 119582. doi:10.1016/j.jnoncrysol.2019.119582

Sun, Q., Miskovic, D. M., Laws, K., Kong, H., Geng, X., & Ferry, M. (2020). Transition towards ultrastable metallic glasses in Zr-based thin films. *Applied Surface Science*, *533*, 147453. doi:10.1016/j.apsusc.2020.147453

Suryanarayana, C., & Inoue, A. (2013). Iron-based bulk metallic glasses. *International Materials Reviews*, *58*(3), 131–166. doi:10.1179/1743280412Y.0000000007

Suryanarayana, C., & Inoue, A. (2017). *Bulk metallic glasses*. CRC Press.

Tamura, T., Amiya, K., Rachmat, R. S., Mizutani, Y., & Miwa, K. (2005). Electromagnetic vibration process for producing bulk metallic glasses. *Nature Materials*, *4*(4), 289–292. doi:10.1038/nmat1341 PMID:15750599

Tariq, N., Hasan, B., Akhter, J., & Ali, F. (2009). Mechanical and tribological properties of Zr–Al–Ni–Cu bulk metallic glasses. *Journal of Alloys and Compounds*, *469*(1-2), 179–185. doi:10.1016/j.jallcom.2008.02.002

Telford, M. (2004). The case for bulk metallic glass. *Materials Today*, *7*(3), 36–43. doi:10.1016/S1369-7021(04)00124-5

Trexler, M. M., & Thadhani, N. N. (2010). Mechanical properties of bulk metallic glasses. *Progress in Materials Science*, *55*(8), 759–839. doi:10.1016/j.pmatsci.2010.04.002

Turnbull, D. (1969). Under what conditions can a glass be formed? *Contemporary Physics*, *10*(5), 473–488. doi:10.1080/00107516908204405

Wang, D.-H., Xie, S.-H., Yang, H.-P., Qian, H.-X., & Zeng, X.-R. (2018). Wear behaviors of three typical bulk metallic glasses in bearing applications. *Metals*, *8*(12), 1005. doi:10.3390/met8121005

Wang, Q., Bai, X., Sun, B., Liu, J., Cai, Z., Liang, X., & Shen, B. (2021). Influence of Si on tribological behavior of laser cladded Fe-based amorphous/crystalline composite coatings. *Surface and Coatings Technology*, *405*, 126570. doi:10.1016/j.surfcoat.2020.126570

Wang, W. (2009). Bulk metallic glasses with functional physical properties. *Advanced Materials*, *21*(45), 4524–4544. doi:10.1002/adma.200901053

Wang, W.-H., Dong, C., & Shek, C. (2004). Bulk metallic glasses. *Materials Science and Engineering R Reports*, *44*(2-3), 45–89. doi:10.1016/j.mser.2004.03.001

Wu, H., Baker, I., Liu, Y., Wu, X., Munroe, P. R., & Zhang, J. (2013). Tribological studies of a Zr-based bulk metallic glass. *Intermetallics*, *35*, 25–32. doi:10.1016/j.intermet.2012.11.010

Xu, J., Kang, J.-J., Yue, W., Fu, Z.-Q., Zhu, L.-N., & She, D.-S. (2021). High-temperature tribological property of Fe-based amorphous alloy coating. *Journal of Non-Crystalline Solids*, *573*, 121136. doi:10.1016/j.jnoncrysol.2021.121136

Yang, W., Liu, Y., Hua, N., Pang, S., Li, Y., Liaw, P. K., & Zhang, T. (2021). Formation and properties of biocompatible Ti-based bulk metallic glasses in the Ti–Cu–Zr–Fe–Sn–Si–Ag system. *Journal of Non-Crystalline Solids*, *571*, 121060. doi:10.1016/j.jnoncrysol.2021.121060

Yokoyama, Y., Mund, E., Inoue, A., & Schultz, L. (2007). Production of $Zr_{55}Cu_{30}Ni_{5}Al_{10}$ glassy alloy rod of 30 mm in diameter by a cap-cast technique. *Materials Transactions*, *48*(12), 3190–3192. doi:10.2320/matertrans.MRP2007164

Yoon, S., Kim, J., Bae, G., Kim, B., & Lee, C. (2011). Formation of coating and tribological behavior of kinetic sprayed Fe-based bulk metallic glass. *Journal of Alloys and Compounds*, *509*(2), 347–353. doi:10.1016/j.jallcom.2010.09.024

Zhang, L., & Huang, H. (2019). Micro machining of bulk metallic glasses: A review. *International Journal of Advanced Manufacturing Technology*, *100*(1-4), 637–661. doi:10.100700170-018-2726-y

Zhang, Y., Zhao, D., Wang, R., & Wang, W. (2003). Formation and properties of Zr48Nb8Cu14Ni12Be18 bulk metallic glass. *Acta Materialia*, *51*(7), 1971–1979. doi:10.1016/S1359-6454(02)00602-X

Zhou, K., Chen, C., Liu, Y., Pang, S., Hua, N., Yang, W., & Zhang, T. (2017). Effects of lutetium addition on formation, oxidation and tribological properties of a Zr-based bulk metallic glass. *Intermetallics*, *90*, 81–89. doi:10.1016/j.intermet.2017.07.007

Zhou, Q., Du, Y., Jia, Q., Han, W., Zhao, X., Deng, Y., & Wang, H. (2020). A nanoindentation study of Ti-based high entropy bulk metallic glasses at elevated temperatures. *Journal of Non-Crystalline Solids*, *532*, 119878. doi:10.1016/j.jnoncrysol.2019.119878

Zhou, Q., Du, Y., Ren, Y., Kuang, W., Han, W., Wang, H., Huang, P., Wang, F., & Wang, J. (2019). Investigation into nanoscratching mechanical performance of metallic glass multilayers with improved nano-tribological properties. *Journal of Alloys and Compounds*, *776*, 447–459. doi:10.1016/j.jallcom.2018.10.270

Zhou, Q., Han, W., Luo, D., Du, Y., Xie, J., Wang, X.-Z., Zou, Q., Zhao, X., Wang, H., & Beake, B. D. (2021). Mechanical and tribological properties of Zr–Cu–Ni–Al bulk metallic glasses with dual-phase structure. *Wear*, *474*, 203880. doi:10.1016/j.wear.2021.203880

ADDITIONAL READING

Chen, M. (2011). A brief overview of bulk metallic glasses. *NPG Asia Materials*, *3*(9), 82–90. doi:10.1038/asiamat.2011.30

Fleury, E., Lee, S. M., Ahn, H. S., Kim, W. T., & Kim, D. H. (2004). Tribological properties of bulk metallic glasses. *Materials Science and Engineering A*, *375*, 276–279. doi:10.1016/j.msea.2003.10.065

Greer, A. L. (2014). Metallic glasses. *Physical Metallurgy*, 305-385.

Khan, M. M., Nemati, A., Rahman, Z. U., Shah, U. H., Asgar, H., & Haider, W. (2018). Recent advancements in bulk metallic glasses and their applications: A review. *Critical Reviews in Solid State and Material Sciences*, *43*(3), 233–268. doi:10.1080/10408436.2017.1358149

Kruzic, J. J. (2016). Bulk metallic glasses as structural materials: A review. *Advanced Engineering Materials*, *18*(8), 1308–1331. doi:10.1002/adem.201600066

Saji, V. S. (2018). Electrodeposition in bulk metallic glasses. *Materialia*, *3*, 1–11. doi:10.1016/j.mtla.2018.09.021

Suryanarayana, C., & Inoue, A. (2010). *Bulk metallic glasses*. CRC Press. doi:10.1201/9781420085976

Trexler, M. M., & Thadhani, N. N. (2010). Mechanical properties of bulk metallic glasses. *Progress in Materials Science*, *55*(8), 759–839. doi:10.1016/j.pmatsci.2010.04.002

Wang, W. H., Dong, C., & Shek, C. H. (2004). Bulk metallic glasses. *Materials Science and Engineering R Reports*, *44*(2-3), 45–89. doi:10.1016/j.mser.2004.03.001

KEY TERMS AND DEFINITIONS

Amorphous Alloy: Alloys that do not have a crystalline structure.

Bulk Metallic Glass: New class amorphous metals, typically formed by the combination of three or more elements.

Crystalline Alloy: A metal which has been mixed with other metals in order to create crystalline structures through alloying.

Glass: An amorphous, non-crystalline solid that is typically transparent or translucent.

Metallic Glass: An amorphous metal.

Microstructure: The smallest level of a substance (metal or other material).

Tribology: Study of friction, lubrication, and wear.

Chapter 15
Morphology and Functionalization of Metal Foils and Other Surfaces for Electrochemical Applications

Martin Rozman
FunGlass – Centre for Functional and Surface Functionalized Glass, Alexander Dubček University of Trenčín, Slovakia

Miha Lukšič
https://orcid.org/0000-0001-7190-4013
Faculty of Chemistry and Chemical Technology, University of Ljubljana, Slovenia

ABSTRACT

Electrochemical applications had their first major impact in the late 20th century with the development of improved energy storage and conversion systems such as lithium-ion batteries, organic-inorganic dye-sensitized solar cells, and even e-ink displays. Depending on the requirements, the electrodes can be made of different materials, such as metal or alloy sheets, foils, bars or conductive ceramics, conductive polymers, etc. In this chapter, methods for surface functionalization and characterization of metallic and non-metallic surfaces used as electrode substrates are presented. The focus is on the use of metal foils in lithium-ion batteries and especially in the novel architecture of optoelectronic devices – from electrochromic and photovoltaic devices to biosensors.

INTRODUCTION AND BACKGROUND

Metal foil technology flourished in the 20th century, especially in the automotive and aeronautic industries (Barroqueiro, Andrade-Campos, Valente, & Neto, 2019) (Orsato & Wells, 2007). The use of metal foils had its particularity in the 1950s, when it was widely used in both industries, as well as in the packaging industry, where the use of aluminum was widespread (Lamberti & Escher, 2007). This packaging

DOI: 10.4018/978-1-7998-9683-8.ch015

technology was soon rendered obsolete by the introduction of polymer materials, which eventually found their way also into the automotive and aeronautic industries (Muhammad, Rahman, Baini, & Bin Bakri, 2021). Nevertheless, metal foils remained indispensable in certain sectors, and the rise of advanced electrochemical applications opened up new uses for the old technology. A great example of this represents widely used lithium-ion batteries design, which commonly use aluminum and copper foil as a current collector as well as to improve mechanical characteristics of device (Goodenough, 2018) along with photo/electrocatalysis devices that can be used for either chemical processing (Marinko et al., 2021) or energy storage (Dinh et al., 2019). There was particular interest in the development of semiconductors that could be used for advanced electronic systems (microprocessors) that could be further miniaturized and used in various devices (Covington, 1994). With the advent of miniaturization, the demand for energy storage devices increased as the older lead-lead oxide batteries simply could not keep up with the demand for energy. In addition, the space race between the United States and USSR began in the early 1950s, requiring the development of energy conversion systems that could operate reliably under extreme conditions, even in the space vacuum (Prince, 1955). These energy conversion systems focused on the development of photovoltaics and fuel cells, which initially had relatively limited applications, but soon attracted a great deal of interest in the 1970s and 1980s, when the world economy was beset by various energy crises (e.g., 1973 oil crisis, 1979 Three Mile Island incident, 1986 Chernobyl accident) and demanded new solutions that would make the world energy redundant (Chowdhury, Sumita, Islam, & Bedja, 2014). Electrochemical applications had their first major impact in the late 20th century with the development of improved energy storage and conversion systems such as lithium-ion batteries, organic-inorganic dye-sensitized solar cells, and even display screens (e.g., light-emitting diodes and e-ink displays) (Whittingham, 1974) (O'Regan & Grätzel, 1991) (Fischer, von Brüning, & Labhart, 1976) (Maruska & Stevenson, 1974). With the increasing development of computing and entertainment technologies and the widespread use of portable devices such as smartphones and tablets, energy storage systems and various optoelectronic devices are of great interest, not only for scientific research, but also for industry and consumers. Due to the increased interest of entertainment industry and information sector, any improvement in this field is greatly encouraged both from scientific and financial aspect.

Most electrochemical devices have a similar structure: they consist of (i) at least two highly conductive surfaces that act as electrodes, (ii) an electrolyte, which can be liquid, solid or quasi-solid, and (iii) an insulating container that prevents the device from short-circuiting and ensures that the contents of the device are not released into the environment. Depending on the requirements, the electrodes may be made of various materials, such as metal or alloy sheets, foils, bars or form conductive ceramics and conductive polymers, with the substrate material having significant influence on the further selection of the coating process. While metal foils are considered durable, flexible and temperature resistant (aluminum foil is an excellent example), some materials have their own advantages and disadvantages. One such class of materials are optically transparent materials, which are commonly used in optoelectronic applications such as solar cells and e-ink displays. These materials can be made from conductive polymers, ceramic-coated glass, or even thin film metal deposits on glass or plastic substrates. While conductive polymers and coated polymer substrates have interesting mechanical properties such as flexibility and stretchability, they are less suitable for high temperature refinement (above 100 °C) and can become brittle at lower temperatures. Ceramic substrates, on the other hand, are very temperature resistant but have a rigid structure. As there is an increased requirement in preparation of new ways to produce large quantities of low-cost materials, that can be used in high technology sectors such as optoelectronics, biomedical applications and even specialized construction, researchers and engineers tend to find new methods and procedures to invent reliable methods that will enable repeatable production of new materials.

In this chapter, examples of surface functionalization using different techniques on various metallic and non-metallic surfaces used as electrode substrates will be presented. Various methods for preparing electrode surfaces and methods for studying the morphology used in batteries and optoelectronic devices will be discussed and compared.

THIN-LAYER GENERAL PREPARATION METHODS

There are various coating methods, ranging from processes suitable for large-scale use, such as dip coating, level drop (inverse dipping), spray coating, or roll-to-roll techniques, to more precise coating systems such as spin coating or inkjet printing. Simple methods such as casting and brushing can also be used, although these are usually too coarse to produce thin films less than 10 μm thick. For certain coating applications, even more precise and specialized methods are used, such as various sputtering methods that can produce extremely thin films, and atomic layer deposition (ALD), which can achieve thicknesses of one or more atomic layers. From the variety of methods for producing thin films, only a few will be mentioned here.

Dip Coating

This is one of the best known and most commonly used coating methods. The basic principle is that you dip an object to which you want to apply a thin film into the coating solution and then lift it out of the solution (Figure 1a). The thickness of the applied film is controlled by the speed at which an object is lifted, with an increase in speed increasing the thickness of the film (Quéré, 1999). In addition, changing the properties of the coating solutions can have a direct effect on the coating properties and is usually achieved by changing the surface tension or viscosity properties of the solution by introducing additives. This method is ideal for sol-gel processes that can be used for both energy storage and energy conversion devices as well as for anti-corrosion applications, where increased thickness of the films produced (greater than 200 nm) has less impact on device performance. This method is most commonly used for materials where the primary requirement is a thick, uniform coating that prevents contact with the substrate surface. The disadvantage of this method is that it requires a large amount of liquid in relation to the coated surface.

Figure 1. Schematic representation of the basic principles of various coating methods: (a) dip coating, (b) level drop, (c) spin coating, (d) spray coating, (e) inkjet printing, (f) roll-to-roll, and (g) vacuum deposition. The dark blue or pink arrow indicates the location where the coating is applied

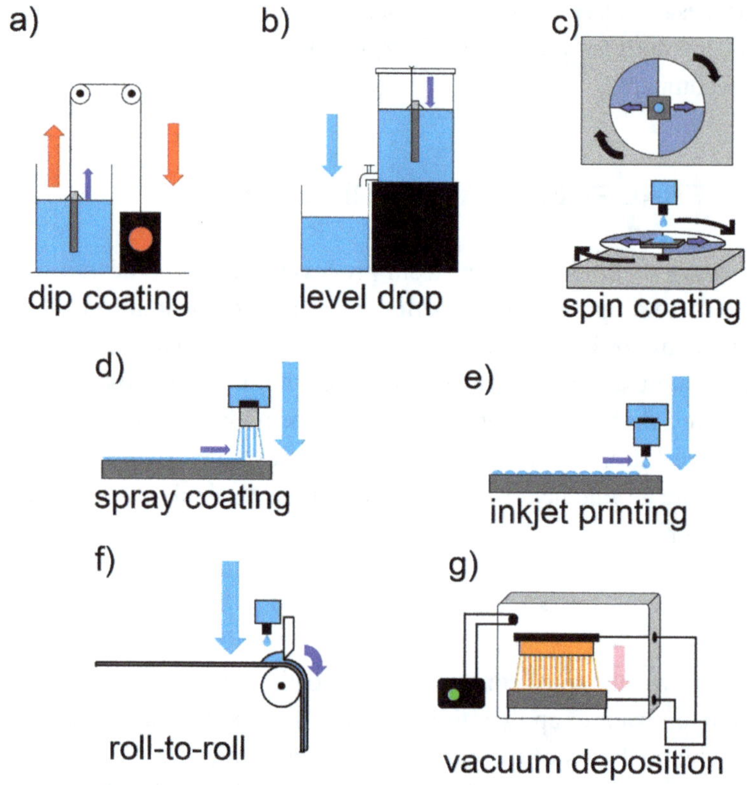

Level Drop

Even more simple than dip coating is the level drop method because of the simplicity of the coating equipment. Instead of lifting the material to be coated out of the liquid at a certain speed, the level of the liquid is steadily lowered by opening the valve at the bottom (Figure 1b) or by pumping the liquid out in case of more viscous liquids. The thickness of the film is controlled by the rate at which the liquid level is lowered. The level drop process has similar advantages and disadvantages to the dip coating process.

Spin Coating

The method uses centrifugal force to distribute the coating material evenly over the substrate surface (Sahu, Parija, & Panigrahi, 2009). Usually, the substrate is spun on a roller plate with the center of gravity of the substrate at the center of the spinner. The centrifugal force during spinning removes the excess material, leaving a thin layer of material on the substrate (Figure 1c). To achieve this, the substrate is placed on the center of the horizontal platen and held in place either by a vacuum or by other means such as clamps or tape. There are several ways to apply the coating material, as it can be applied before spinning or during spinning, depending on the properties of the coating solution, such as viscosity, density, surface tension etc. It is most commonly used for the preparation of smaller, flat surfaces, as the centrifugal force can have a large variation with larger radial differentials. This coating technique has been particularly popular for the preparation of inorganic-organic solar cells, since large photovoltaic panels can usually be built from smaller modules, along with excellent control of the film thickness, ranging from less than 10 nm to several hundred nanometers.

Spray Coating

This is one of the most widespread methods used in all kinds of applications, for example, in the construction and automotive industries. It is commonly used for the application of anti-corrosion coatings, as it can be quickly and easily applied to any size or shape of object where it is needed. Generally, spray coating involves forcing a substance through a nozzle, which produces a fine aerosol (Figure 1d).(Aziz & Ismail, 2015) In this process, the aerosol is cooled as it expands. This forms small droplets that are pressed onto the surface to which they adhere. The result is a film formed from many droplets that can reform into a solid surface depending on the thickness of the film applied. There are several ways to apply the coating to the surface, and certain parameters have a great influence on the quality of the film, such as: the distance between the substrate and the nozzle, the pressure with which the aerosol is forced out of the nozzle, the temperature of the coating material and the substrate, and the number of coatings applied. Since this coating method is best suited for substrates with complex shapes and the films produced are relatively thick (over 500 nm), this method is best suited for anti-corrosion and other protective films.

Inkjet Methods

Although developed in the mid-19th century, inkjet printing is considered a relatively new method, as its popularity increased in the 1950s with the gradual development of transistors and processors, which allowed much more precise control of the droplets applied to the surface, allowing the printing of complex patterns or even photographs (Cummins & Desmulliez, 2012). In the first practical iteration, the operation is quite similar to spray coating, where a coating material is forced through a nozzle, creating an aerosol, but on a much smaller scale (Figure 1e). This iteration is also known as thermal inkjet printing. In later iterations, droplet ejection and droplet size are controlled by a piezoelectric element. This type of printing is known as drop-on-demand (DOD), where droplets are ejected onto the substrate only when needed. Nowadays, this type of printing is mainly used in applications where high precision of coating or preparation of complex patterns is required. For example, it is used for printing electrical components such as printed circuit boards (PCBs) and even optoelectronic devices such as solar cells and smart windows.

Roll-to-Roll

This method is most commonly used for substrates of large-scale industrial production, which are in the form of foils. It provides the ability to coat large surfaces of material quickly, accurately, and with high repeatability. The main principle of the method is to apply a coating material which is evenly distributed on the substrate by rollers. Moreover, these rollers rotate to maintain the process of continuous deposition of the coating material on the substrate foil (Figure 1e). There are several variations of this method (e.g., knife edge, gravure coating, spray coating adapted for roll-to-roll etc.). Each of these techniques has different advantages and can be used selectively depending on the desired coating thickness, porosity and other properties of the precursor mainly viscosity and surface tension. Therefore, this technique is widely used for manufacturing batteries, solar cells, fuel cells, capacitors and other electrical products and their components.

Low Pressure and Vacuum Deposition Methods

These methods include physical or chemical vapor deposition, atomic layer deposition, and plasma spray deposition, where the main purpose is to apply coating with the help of electrostatic force. This is traditionally done for different metal and semiconductor deposition, where high precision and low consumption of coating material is preferred. The general schematics is given in Figure 1g. In most cases, these methods are considered extremely precise, with layer thicknesses of a few nanometers, sometimes as low as few atoms thick. Nevertheless, they usually require much more expensive instrumentation compared to more traditional methods such as dip, spray or spin coating. These methods are heavily used for the production of semiconductor materials and are therefore widely used in photosensing applications, as well as in other applications.

A comparison of the WO_3 thin film morphology on aluminum substrate obtained using various coating methods is shown in Figure 2. The deposited layer is the crudest in the case of dip coating, and the quality of deposition improves by using level-drop method. In case of spin coating the deposition is uniform with minimal cracks. The inkjet printing method also shows high uniformity of the deposited thin film.

Figure 2. Electron microscopy images (SEM) showing the quality of the WO_3 thin films on aluminum foil prepared by various coating methods: (a) dip coating, (b) level-drop, (c) spin coating, and (d) inkjet printing

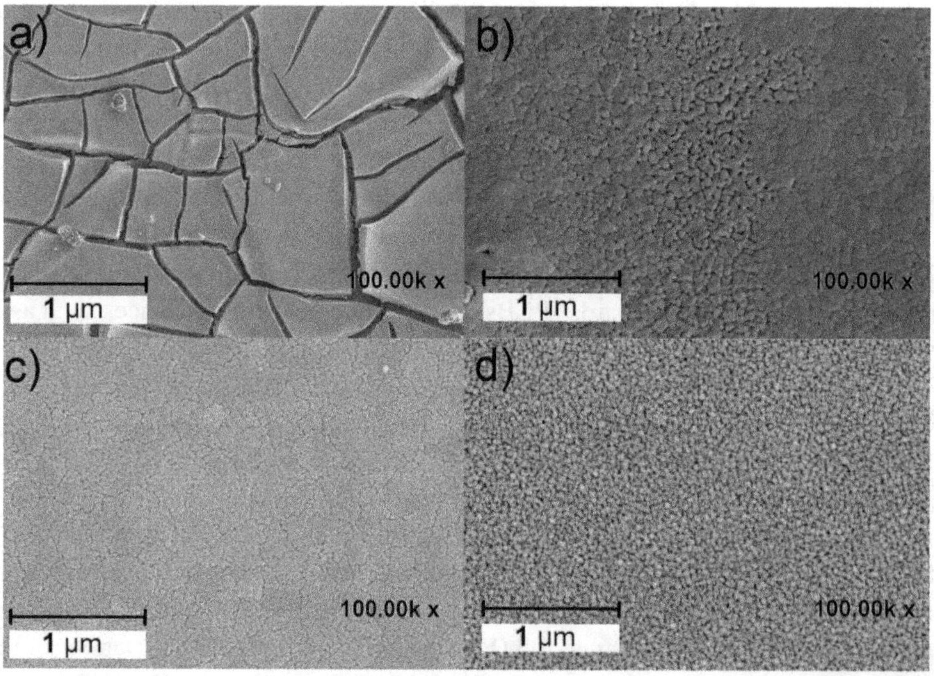

INVESTIGATION METHODS FOR THIN FILM SURFACE QUALITY CONTROL

To analyze the morphology of the functionalized surfaces, various microscopy and spectroscopy are used. A brief overview is given below:

Microscopy Methods

Optical Microscopy

This is perhaps the oldest method of studying microstructures, apart from using simple magnifying lenses. It was used as early as the 17th century and is directly responsible for advancements in anatomy and histology (Rochow & Tucker, 1994). Unlike the classical microscopes used to examine biological specimens, a metallurgical microscope or stereo microscope is designed to allow surfaces to be viewed without light (visible spectrum) passing through the sample (Nuttall, 1979). A metallurgical microscope allows magnification up to 500-1000x and usually provides higher resolution than a stereo microscope. Optical microscopy was one of the most important tools for examining materials for early signs of corrosion before it was superseded by more accurate electrochemical methods such as impedance spectroscopy (Usman, Scenini, & Curioni, 2020). However, it remains an important tool for process analysis in industry and with advances in digital imaging and other technologies such as machine learning and artificial intelligence, it is once again becoming an important tool for the efficient examination of complex structures for quality control of manufactured coatings (Booth, 2014).

Electron Microscopy

This technique, which includes scanning electron microscopy (SEM), transmission microscopy (TEM), and other derivatives, allows us to record the topography and surface structure of samples by electron irradiation. In the simplest case, this is done by a focused electron beam directed at the area of interest and moved across the surface (Figure 3a). This movement occurs in repeating lines called scans. The reflected or excited electrons (depending on the type of detector) are then scattered, with some hitting the detector. The electrons collected from multiple scan lines are formed into a map that can be further interpreted or displayed as an image. There is a difference between scanning electron microscopy (SEM) and transmission electron microscopy (TEM) (Inkson, 2016). In SEM the electron beam is reflected back from the surface to the detector, whereas in TEM the electrons pass through the material and then reach the detector, in principle similar to optical microscopy (Figure 3a). For this reason, TEM samples usually require much more preparation and can only be used with very thin samples (usually less than 100 nm thick), while SEM can be used with much thicker samples of a few millimeters in size or even larger. Therefore, SEM is considered one of the most important methods in materials science research, showing the microscopic behavior and even nano-properties of various materials such as metals, alloys, ceramics, semiconductors, and even organic materials. It offers a wide range of magnifications for imaging patterns (up to 1,000,000x), good resolution and depth of field, while TEM is reserved for smaller (thinner) samples where very high magnifications (above 500,000x) are required, such as when analyzing the structure of carbon nanotubes or studying the crystallinity of ceramic films. TEM is particularly useful for samples that transmit electrons in their natural state (e.g., nanoparticles, nanofibers, nanotubes). However, preparing thin samples from thick ones is a non-trivial task (Ayache, Beaunier, Boumendil,

Ehret, & Laub, 2010). There are a variety of thinning techniques, e.g. mechanical abrasion, controlled fracture, ion sputtering, chemical etching, electrochemical dissolution, etc. Different types of microscopes can be used depending on whether process analysis or detailed morphological examination is involved. In addition, TEM analysis enables much greater magnifications compared to SEM and also investigation of crystallinity of prepared thin films (Rozman, Mavrič, Kravanja, Valant, & Pakseresht, 2022).

Atomic Force Microscopy

This type of microscopy more known by its abbreviation AFM is another method of analyzing surfaces in which an image is produced by moving tipped cantilever across the surface of the sample (Figure 3b) (Khan, Wang, & Fitzpatrick, 2016). In general, the process of creating an image is similar to the way a gramophone works, where a needle moves across the surface of the disc, which is then converted to sound. The AFM can be used to take surface images at different magnification levels, depending on the size of the needle and the mode in which it is operating. There are different modes of operation: contact, tapping, and non-contact, with the different modes being more suitable for different magnifications and different analyses. This method of microscopy can even be used for samples that are immersed in liquid. It is a completely non-destructive method that allows measurements to be made even on surfaces that are electrochemically active or require careful handling in order to be used for further studies. This is particularly convenient for the study of batteries and other electrochemical devices, where measurements can even be made in-situ or on samples that generate or consume electricity in real time during observations (Maver, Žnidaršič, & Gaberšček, 2011). Thanks to its non-destructive nature, AFM can also perform sensitive measurements on biomedical samples and prototypes that are being prepared for or have just undergone medical testing (Maver, Velnar, Gaberšček, Planinšek, & Finšgar, 2016).

Figure 3. Basic principle of (a) electron microscopy (SEM, TEM), and (b) atomic force microscopy (AFM). In SEM electrons are scattered directly from the surface, while in TEM transmitted electrons are detected. In AFM the surface is scanned by the needle (either by direct contact or by electrostatic force). EDX denotes the energy dispersive X-ray spectroscopy

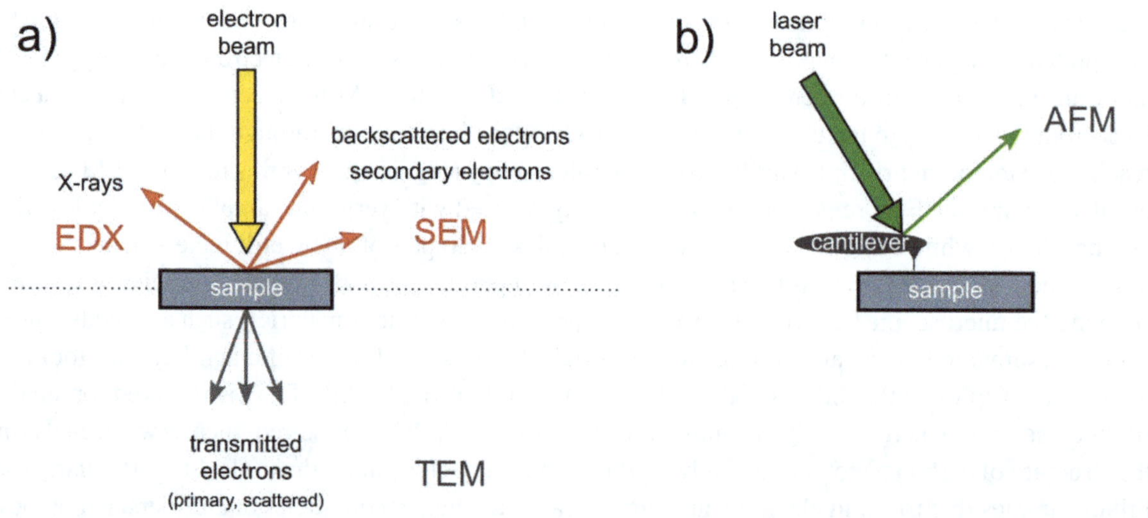

Spectroscopy Methods

Colorimetry and UV-Vis-NIR Reflectance Spectrophotometry

These methods are used to study the color response of functionalized surfaces and electrochemical devices. These methods are very useful in the study of electrochromic devices as they provide quantitative data on the coloration of the devices. In addition, these methods are important for the study of certain energy storage devices such as batteries and supercapacitors, as some of the charge/discharge mechanisms may involve intercalation or electrodeposition, which can result in color changes of the surface of the electrode under study.

Energy Dispersive X-ray Spectroscopy (EDS or EDX)

As a method, it is usually coupled with an electron microscope (either SEM or TEM) and allows the determination of certain elements in the sample under study and their relative proportions (Figure 3a) (Frankel & Aitken, 1970). It is a rapid, non-destructive point analysis method of surface analysis that can provide valuable information about the composition of the sample. This is particularly important in energy storage devices to investigate any impurities and can also help in the investigation of artefacts, such as dendrite growth in lithium ion batteries, which can pose a risk of battery short circuit and fire hazard.

Fourier IR Spectroscopy

This is a method most commonly used for the study of organic thin films (Putzig et al., 1994). This is particularly important in the analysis of surfaces that use combined organic-inorganic layers, such as in the preparation of photovoltaic devices, where precise control of layer deposition is critical. This is possible with this method as it provides data on various bonds present in the layer under investigation.

Electrochemical Methods

Electrochemical methods are considered rather unpleasant by most materials scientists because their results are difficult to interpret, especially to the untrained eye, even though they require relatively simple instrumentation. The development of miniaturized electronics in the late 1960s gave an enormous boost to electrochemical research, which included more complex types of measurement at high-frequency alternating current, such as impedance measurement methods. Today, electrochemical methods are becoming increasingly popular because potentiostats/galvanostats are relatively inexpensive compared to other equipment and the software for data processing is improving.

Potentiometric Methods

These methods are based on recording the potential, which shows a dependent response due to the current. Therefore, these methods are commonly used for the analysis of energy storage devices to obtain working parameters at which the device operates nominally. One such commonly used method is open circuit potentiometry (Sutija, Norby, & Björnbom, 1995) (Bisquert, Zaban, Greenshtein, & Mora-Seró, 2004). It is perhaps the simplest method, which is basically a potentiometry in which the current is set to zero, and the potential of an electrochemical system in its steady state is measured over a period of time. It used to be used as a method of investigating and predicting the long-term stability of the system, whereas today it is mostly used as a supporting measurement to ensure that the device is actually in its operating state or that there is no significant polarization of the material that could trigger corrosion. Open circuit potentiometry is a widely used method for checking battery capacity, as it is possible to predict the longevity of the battery by measuring the potential. It can provide information about corrosion potential and is a useful technique for checking the stability of electrochemical measurements. It is also widely used as a method of checking battery capacity, as it is possible to predict the longevity of the battery by measuring the potential.

Voltammetry Methods

They are in their basic principle electrochemical methods that record the change in current when a certain potential is applied, which can be either static, linear, cyclic, or any combination of these. An excellent example is linear sweep voltammetry (LSV), which has been used extensively for measuring corrosion properties and thickness of protective coatings (Yan, Bazant, Biesheuvel, Pugh, & Dawson, 2017). Another useful method is cyclic voltammetry (CV), but this is usually reserved for the analysis of electrolyte reactions rather than the study of electrolyte-electrode interaction (Margaretha & Tissot, 1979). Despite being gradually replaced by impedance methods voltametric methods still hold important place when it comes to study of assembled devices and the study of surface interaction.

Electrochemical Impedance Spectroscopy (EIS)

This is a type of impedance measurement that is specialized for investigation of electrolyte-electrode interaction (Chang & Park, 2010). As such, it is an invaluable method for corrosion studies as well as for studying the behavior of electroactive surfaces and deposited thin films on substrates that can serve as electrodes. In general, EIS provides information on electrode processes such as charge transfer and diffusion within an electrolyte, which can then be used to evaluate the stability or general operation of the electrochemical system under study (Macdonald, 1990). Although the method was first mentioned and discussed even in the first half of 20th century (Macdonald, 2006), it started to see wide spread use in the 1970s (Ferris & Rose, 1972), it has taken a long time due to difficulties in signal interpretation. Later, signal interpretation was greatly simplified by fitting the data to comparable electrical circuit equivalents (Alpen, Graf, & Hafendörfer, 1978). Today, there are modern electrical circuit simulation tools that can help researchers better understand the data obtained, which greatly facilitates the analysis. The method is considered non-destructive and is relatively easy to set up, requiring only a potentiostat/galvanostat with the option of frequency response analysis, which is relatively inexpensive compared to most microscopy methods such as SEM or AFM (Randviir & Banks, 2013).

Amperometry Methods

These are considered a special type of voltametric measurements where the potential is static and the current response is recorded as a function of time (Patel, 2020). Thus, these methods should not be confused with potentiometric methods in which the response to the potential is recorded. One of the best known methods for studying electrode material is chronoamperometry, which involves connecting an electrochemical cell (or device) to a specific potential and monitoring what current it draws (J. Wang & Zeng, 2021) (Tian et al., 2020). This method is particularly useful for analyzing electrochromic devices or energy storage devices because it mimics the practical behavior of the devices. The method itself is also very useful for studying electrolyte-electrode interactions such as intercalation, ion diffusion, electrodeposition, or other electrochemical surface interactions.

SURFACE FUNCTIONALIZATION IN ENERGY STORAGE DEVICES

Alloy foils are particularly common in battery production (Shobana, 2019) (Blomgren, 2016) because they allow an increase in electrode surface area while maintaining its low mass. Aluminum alloy foils coated with lithium cobalt oxide ($LiCoO_2$) are among the most commonly used cathodes in lithium-ion batteries (LIB). The thin film of $LiCoO_2$ acts as an energy storage crystal lattice (Li-ion intercalation) and has a high specific capacity (120-140 mAh/g or more when coated with metal oxides) and a high nominal voltage (3.7 V) (Shobana, 2019) (Blomgren, 2016). Due to the high order of lithium ions in the crystal structure, energy is released when the Li^+ ions leave the lattice. Besides $LiCoO_2$, a variety of other lithium transition metal oxides are commonly used in LIBs as active electrode materials (e.g., $LiMnO_2$, $LiMn_2O_4$, $LiNiO_2$, $LiFePO_4$, Li_2FeS_2, Li-Ni-Mn-Co oxides, Li-Ni-Co-Al oxides) (Shobana, 2019) (Blomgren, 2016) (Kendrick, 2019) (Zuo, Tian, Li, Chen, & Shu, 2017). All these thin film oxides are susceptible to water absorption (intercalation of water between the layers of transition metals), which leads to the formation of hydroxides and carbonates on the surface (Kendrick, 2019) and affects the durability of these components and the fabricated electrodes. Additionally, attention must be paid to electrode corrosion, which may require special preparation techniques, such as a controlled atmosphere environment. This may involve preparing the materials in a glovebox and then sintering in an oxygen-free or even reductive atmosphere. The peculiarity of battery electrodes is that they require a relatively large surface area to be efficient, and the coating must be done in such a way that there are no defects on the electrode surface. Defects could lead to a short-circuit and consequent failure of the battery. Recently, research in the field of cathodic materials for LIB has become more intense and the concomitant development of coating approaches to achieve the desired film thickness has become increasingly important. The advantage of battery electrodes is that they are less sensitive to the thickness of the coating layer compared to other devices. The coating can be made thicker than necessary to ensure that the electrode is adequately covered over the entire surface. This can obviously affect device performance, but device safety takes precedence over performance in the case of LIB. The energy and power density of batteries depends on the coating thickness: thicker electrode coatings increase the energy density. They provide better particle protection, but reduce ion/electron transport. On the other hand, thinner coatings are required for better battery power. Very thin coatings are difficult to fabricate and may not provide the desired protection (Zuo et al., 2017) (Guan et al., 2020). The active electrode coating can be further modified by a surface coating to stabilize the structure of the active material and improve the electrochemical performance of the electrode by preventing direct interaction between the electrolyte and the active electrode material and by improving the electrical conductivity of the coating. The result is a composite structure with separate layers of different coating materials. Commonly used materials for coating the active electrode material are metal oxides (e.g., Al_2O_3, MgO, CuO, TiO_2, ZnO, ZrO_2, SiO_2), phosphates (e.g., $AlPO_4$, $FePO_4$, $Co_3(PO_4)_2$, $Ni_3(PO_4)_2$, $LiPO_4$) and carbon (Zuo et al., 2017) (Guan et al., 2020). In order to achieve that the coating adheres to the metal foil (the current collector), binders must also be added (e.g. poly(vinylidene alcohol), poly(tetrafluoroethylene), poly(ethylene oxide) (Kendrick, 2019)). To increase the conductivity of the active layer, conductive additives are mixed into the coating mixture (e.g., graphite, carbon black, carbon nanotubes, acetylene black (Kendrick, 2019)). Most commonly used organic solvents are isopropyl alcohol, xylene, *N*-methyl-2-parrolidone and trimethyl benzene (Kendrick, 2019), however, water is replacing organic solvents due to low cost and nontoxicity. Use of water poses several problems, namely efficient wetting of the metallic surface on one hand, and slow drying on the other (potential cracking of thin layers). In practice, therefore, the electrode function depends on several interdependent factors involved in the electrode fabrication process (Zuo et al., 2017): the active material, the binder, the conductive additive, the solvent, and often surfactants, dispersants and acids (Kendrick, 2019). The exact composition of the coating formulation (slurry) is in most cases proprietary.

The basic schematic of a lithium-ion battery is shown in Figure 4a. During the charging cycle, Li^+ ions are released from the active material of the cathode and deposited into the anode (graphite-coated copper). The opposite occurs when the energy stored in the battery is released during the discharging cycle. The cathode and anode sections are separated by a membrane. The membrane prevents a short circuit between the electrodes, but allows Li^+ transport. Coating the active material improves electron conductivity and allows better Li^+ (de)intercalation while preventing the metal ions of the active material from dissolving or the electrolyte from corroding the cathode (Figure 4b).

Figure 4. Schematics of (a) lithium-ion battery structure and (b) of the active material with an additional coating in contact with electrolyte

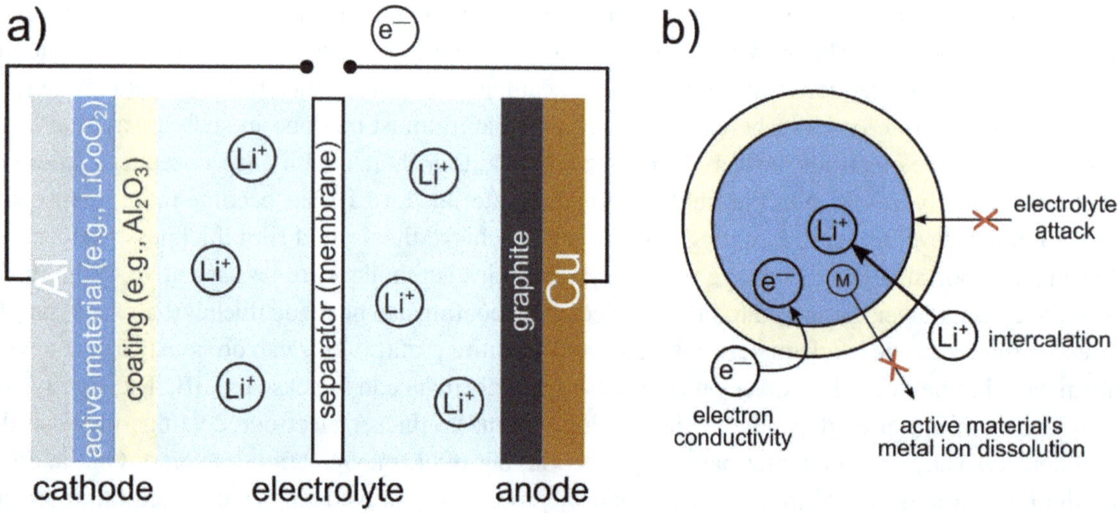

The choice of technique for coating can have a significant effect on the properties of the coating, since the microstructure of the coated layer depends on the technique used. Such differences lead to differences in Li^+ transport and electrochemical performance of the electrode. Tape casting techniques are most commonly used to coat the LIB cathodes (Kendrick, 2019) (Reynolds, Slater, Hare, Simmons, & Kendrick, 2021). Essentially, the electrode slurry is cast in a thin layer on a flat surface of the metal, dried, and sintered. For small scale coatings, a draw-down coater is used while for industrial production, a reel-to-reel coater is used. Different coating geometries exist, e.g. slot die, doctor blade, comma blade, tensioned web (Reynolds et al., 2021), which are self-metered or pre-metered. In a self-metered coater, the slurry is fed into a reservoir and the volume of slurry allowed onto the substrate is determined by the geometry of the coater (doctor blade, comma bar). The properties of the slurry, the web speed and the external properties (temperature, relative humidity) determine the coating. In pre-metered coaters, on the other hand, the slurry is fed by a pump that regulates the volume of the slurry by adjusting the flow rate (slot die). The flow rate, the coating speed and the width determine the coating thickness. The pre-metered techniques provide a much more controlled and definite coating thickness than self-metered techniques. However, they are more difficult to set up. Care must be taken to properly account for the edge gradient of the coatings (different film thickness at the edges of the coating) (Kendrick, 2019). After the slurry has been deposited, the electrode is dried. If drying is too rapid, coating adhesion is poor and surface defects can occur. The (hot or cold) coating is then densified by calendering (compression) and subjected to an additional vacuum drying.

Electrophoretic deposition can also be used to coat the LIB cathode. The concentration of particles here is much lower than in the coating techniques described above (Mazor, Golodnitsky, Burstein, Gladkich, & Peled, 2012) (Prasanna, Subburaj, Jo, & Lee, 2013). The fluid must have low viscosity, proper dielectric constant and proper conductivity, and the applied voltage must be optimized to obtain the desired film properties. As an alternative to solvent techniques, dry coating does not require a solvent (Ludwig, Zheng, Shou, Wang, & Pan, 2016). In a study of $LiCoO_2$ preparation by electrostatic spray deposition and slurry wet process, the better performance of the dry coated electrode was demonstrated. The method is more economical and environmentally friendly.

Different coating techniques have been applied to modify the active electrode materials (for a review, see references (Zuo et al., 2017) (Guan et al., 2020) (Kalluri et al., 2017)). For carbon coating, chemical vapor deposition or organic pyrolysis have been applied. For accurate control of layer thickness and uniformness atomic layer deposition or radio frequency magnetron sputtering techniques were documented. Co-precipitation coating or sol-gel methods can be used for coating metal oxides. The disadvantage of the wet processes (which are simpler and more economical) is rather difficult control of the uniformity of the coating.

As mentioned earlier, film thickness is a key factor in electrode production. Wet film thickness can be determined using a wet film comb, while dry film thickness can be determined using a dial gauge, micrometer, laser triangulation, and laser callipers (Reynolds et al., 2021). An accuracy of 1 μm and a repeatability of 0.2 μm have been demonstrated by using laser techniques (Schmitt et al., 2013), which allows the detection of defects (e.g., voids due to trapped air bubbles, ridges due to unstable flow, cracks formed during drying) and the study of edge effects. IR thermography is also useful for determining film thickness and detecting defects. For a more in-depth study of the microstructure, Raman spectroscopy and microscopy (optical, SEM and TEM) must be used, while EIS provides information on the processes occurring at the electrode.

SURFACE FUNCTIONALIZATION IN IN OPTOELECTRONIC DEVICES

Optoelectronics is a branch of science that deals with the conversion of a signal from electrical to light response and vice versa. Better known examples of optoelectronic modules are the photoelectric effect, which deals with the conversion of a light signal into an electrical signal, and electroluminescence, in which the material releases energy in the form of photons through radiative recombination under the influence of an electrical voltage (Alferov, 2013). This electrical voltage can then be used for different application such as energy production or photosensing applications. Although almost everyone uses the conversion of an electrical signal into a light response in the form of a screen, the conversion of a light signal into an electrical response is as important and useful in daily life as it may seem at first glance. Since the late 1980s, there has been a massive push to simplify optoelectronic devices, both in terms of the materials used and the means of production, to make them more affordable and useful for new applications (Pogue & Wilson, 2018) (Kholghi Eshkalak et al., 2019). This is important as novel applications and devices that are cheaper to produce could be accessible to a wider population. To achieve this, applied materials science and electrochemistry could be the answer to simplify optoelectronic device fabrication and shift production capacity from expensive vacuum deposition techniques (e.g., physical vapor deposition (PVD) and atomic layer deposition (ALD)) and UV-etching techniques to more traditional thin film techniques and simpler material preparation methods (such as base metallurgy). The combination with a more conventional electrochemical assembly approach involving roll-to-roll fabrication could enable dramatic cost reduction for optoelectronic devices (Khachatryan, Kim, Kim, & Kim, 2018) (Kim et al., 2020).

Electrochromic Devices

Of particular interest are electrochromic devices (ECDs). ECDs can change color under the influence of externally applied electrical voltage (or current) and retain that color for varying lengths of time after the power source is turned off (Mortimer, 2011). ECDs can be operated by various chemical reactions or physical phenomena: redox reactions (Kao et al., 2016), ion intercalation (Ohsuku & Hirai, 1982), proton transfer (Schmidt, Pridgen, Hammond, & Love, 2010), electrodeposition (Szmacinski & Lakowicz, 1995) and electrophoresis (Dalisa, 1977) (see Figure 5). Certain modes of operation can be used in different combinations and have different time responses and coloration characteristics (Kulesza, Malik, Zamponi, Berrettoni, & Marassi, 1995). The most commercially known is electronic ink, which was developed in the United States in the 1970s and is now found mainly in portable e-readers. It is a type of electrophoretic screen based on colored mineral oil with dispersed metal oxide microcapsules (C. Wang, Li, Zhang, & Fu, 2022). These come under the influence of high voltage near the optically transparent electrode, which causes a change in the color of the screen. Such a system has a very fast coloration response time of less than 50 ms. However, it requires relatively high voltages for the color change, typically between 30 V and 500 V (Somani & Radhakrishnan, 2003) (Amundson, 2016). The non-transparent back electrodes can sometimes be modified with various materials to achieve a paper-like appearance by applying various dyes, typically in the form of metal oxides, directly to the back electrode. Electrochromic materials that have gained popularity in the last two decades include transparent conductive polymers with added redox dyes (viologens) (X. W. Sun & Wang, 2008) for portable applications, although ceramic coated glass and thin film metal deposits coated on glass substrate are still the most widely used because glass is more stable and has better transparency.

Figure 5. Schematic representations of various electrochromic mechanisms: In panel (a), dark blue circles represent ions that intercalate into crystal lattice, in (b) the light brown circles indicate dissolved species, that are deposited onto electrode as dark brown circles via electrodeposition. Lastly, panel (c) shows redox reaction with blue circles representing dissolved redox active species, with green and white circles representing redox species that interact with electrodes and change coloration accordingly

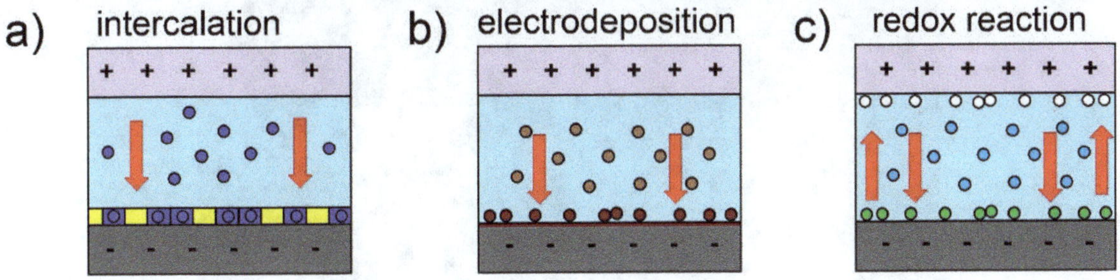

In general, the most popular type of smart glass that uses a thin film metal oxide uses intercalation as the main mechanism for color change. This can be done with the help of different types of metal oxides, such as WO_3, TiO_2, NbO_2 and CoO_2, to name a few (Rozman, Cetin, Bren, & Lukšič, 2021) (Weibin et al., 2013). The use of different metal oxides can lead to different coloration responses. The tinted state is usually a darker coloration than the original bleached state due to the change in the crystal lattice, which has a higher light absorption. Not all metal oxides change color only by intercalation. This can sometimes be achieved by a redox reaction, as in the case of NiO and V_2O_5 (Cerc Korošec et al., 2017) (Rozman, Cetin, Bren, & Lukšič, 2021). However, V_2O_5 is particularly interesting because its color change can occur by both types of reactions (ion intercalation and redox reaction). Finally, certain electrochromic metal oxide materials can change color by the principle of electrodeposition, PbO being an excellent example (Rozman et al., 2018).

In most cases, the metal oxides can be prepared by the sol-gel method, the differences being only in the starting material and the additives that allow adhesion to the surface as well as light transmission, porosity and conductivity properties. TiO_2 and NbO are most commonly prepared via organic precursors, usually in the form of alkoxides diluted with alcohol as solvent, while CoO_2, NiO, PbO and V_2O_5 are usually prepared from inorganic or organic salts such as nitrates or acetates (Polo da Fonseca, De Paoli, & Gorenstein, 1994) (Rozman et al., 2018). Microphotographs of various metal oxides deposited onto stainless steel are shown in Figure 6.

Figure 6. Photomicrographs of various thin films show (a) NiO and (b) PbO prepared using Ni or Pb acetate sol-gel, (c) V_2O_5 prepared using ammonium vanadate sol-gel, and (d) WO_3 prepared by oxidative dissolution of metal powder in hydrogen peroxide to prepare sol-gel

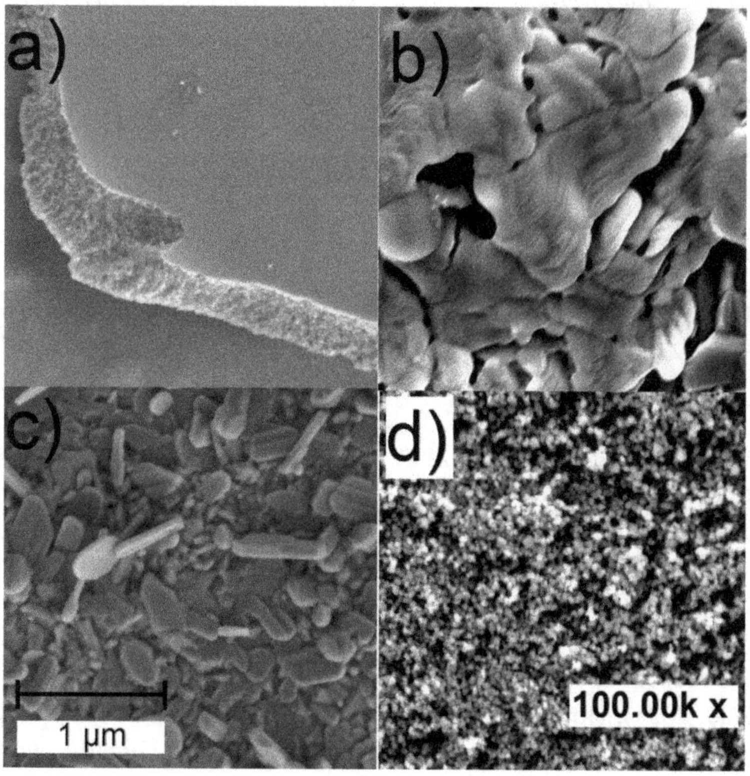

The salt precursor sol-gels are generally prepared as a mixture of water and alcohol, with a high water content to maintain solubility of the salts. Finally, some sol-gels can be prepared from metal powders, WO_3 being an excellent example where the W powder is dissolved in hydrogen peroxide (Rozman et al., 2020). In most cases, surfactants, small molecules or polymers are added to sol-gels to improve transparency or porosity, depending on the type of material required to change surface tension or viscosity of sol-gel. Some of the typical representatives are citric acid, commercial soaps and surfactants, such as Triton X-100, and even polymers, such as polyethylene glycol blocks, namely Pluronic P-123 and F-127 (Rozman et al., 2017) (Wu, Wang, Lin, Wang, & Huang, 2013).

The most common substrate for electrochromic devices is conductive glass-ceramics, i.e., an indium-doped (ITO) or fluorine-doped (FTO) thin film of SnO_2 or aluminum-doped ZnO (AZO) on plain glass. This glass can withstand high sintering temperatures (up to 600 °C) and can even be fabricated as larger windows (up to 50 cm x 50 cm) without the risk of breaking during the heating or cooling cycle (Bisht, Eun, Mehrtens, & Aegerter, 1999). An alternative to glass-ceramics are conductive polymers (Inzelt, 2012) (Das & Prusty, 2012) such as polyaniline, poly(3,4-ethylenedioxythiophene) (PEDOT), and PEDOT:PSS (PSS denoting polystyrene sulfonate), to which metal oxides can be deposited in pre-sintered powdered form, with the metal oxide powder sandwiched between solid or quasi-solid electrolyte and newly formed deposited polymer, usually in the form of a thin film on top of the metal powder (K. Sun et al., 2015). The thin film is finally protected by a stronger, transparent polymer, usually polyethylene, which serves as a backbone to prevent the newly formed device from breaking. Although conductive polymer devices typically have lower transparency and response time, their main advantage is that they are actually lightweight and flexible. For the development of flexible displays, the choice of transparent conductive polymers, especially PEDOT:PSS, proved to be particularly interesting as they allowed flexible devices that can be used in various applications, such as foldable displays (Miura et al., 2014).

The classical configuration of the ECD is the so-called "sandwich" configuration (Figure 7a). It requires at least one optically transparent electrode mentioned above, and the electrolyte is sandwiched between the electrodes. For some years relatively novel optoelectronic devices have been known in which the electrodes are arranged side by side, and only recently have appeared the first devices in which metal optically non-transparent electrodes are arranged one behind the other, the electrodes being separated by an insulator and connected laterally by an electrolyte. (Rozman, Gaberšček, Marolt, Bren, & Lukšič, 2019) This device uses a geometry known as an "inverted sandwich" (Figure 7c). Here, the electrodes are sandwiched between the electrolyte carrier. Such ECD uses solely optically non-transparent materials, but still allows one side of the device to be observed. Several different metal oxides have been tested on a stainless steel foil substrate, namely WO_3, TiO_2, and V_2O_5, as well as several conductive polymers deposited on the top of the metal substrate, with relatively good success (Rozman et al., 2021). The authors of one of the more extensive studies using WO_3 and its intercalation mechanism pointed out that the principle of this lateral electrode connection has so far only been successful with diffusion-based mechanisms, so it remains to be seen whether all materials can take advantage of most of the existing electrochromic color change principles (Rozman et al., 2020). While the "inverted sandwich" architecture has been tested mainly on electrochromic devices, it is possible that this concept could also work on certain types of batteries (e.g., LIB) since they have the same operating principle (intercalation of ions into the crystal lattice) (Meenakshi, Sivakumar, Sivanantharaja, & Sanjeeviraja, 2017).

Figure 7. Comparison of (a) a typical optoelectronic "sandwich" device with an optically transparent electrode, (b) a device with a metal mesh electrode, and (c) an "inverted sandwich" device. The use of a metal mesh makes it possible to build a device with all-metal electrodes, which simplifies fabrication. The dark grey area represents the metal electrode, with the dark blue area representing the electrochromic material. The yellow area represents the electrolyte placement, where the red arrows indicate the electrolyte connection between the electrodes

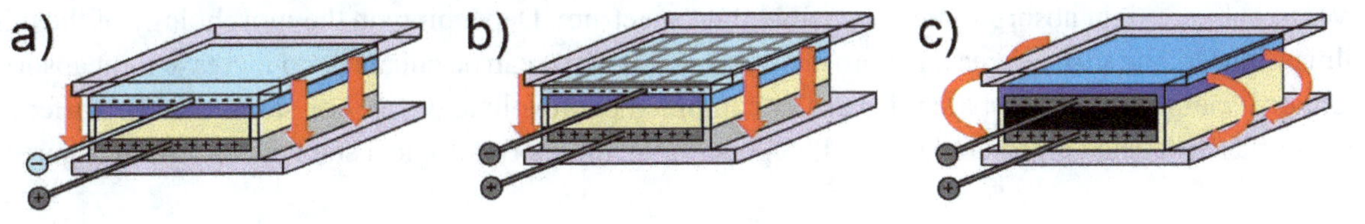

Photovoltaic Devices

Similar to electrochromic devices, photovoltaics uses a very comparable approach to device construction. This usually involves the use of at least one optically transparent electrode, which may be made of transparent conductive glass-ceramics such as ITO or FTO, or transparent conductive polymers, with the counter electrode usually made of solid metal to improve the overall conductivity. Although the vast majority of devices are based on semiconductor technologies such as polycrystalline silicon, CdTe, and GaAs solar cells (Bosio, Pasini, & Romeo, 2020), newer cell types use a hybrid organic-inorganic approach in which a semiconducting metal oxide such as titanium dioxide is deposited on a conductive substrate and coated with a dye that shifts electron excitation to a broader wavelength range, usually from the UV to the visible light range (Shalini et al., 2016) (Ng, Lim, Hayase, Zainal, & Huang, 2018). Typical representatives of such devices are dye-sensitized solar cells (DSSCs), which had their peak of popularity in the early 2000s. DSSCs are being replaced by much more (cost) effective perovskite solar cells (PSCs). In general, DSSCs and PSCs have a thin film of photoactive material deposited on a conductive substrate supported by a counter electrode and an electrolyte that blocks electron transfer from the anode to the cathode (Figure 8).

Figure 8. Schematic representations of (a) DSSC and (b) PSC photovoltaic devices. Both types use the same architecture and share some of the materials (transparent electrode material and photoactive electrode)

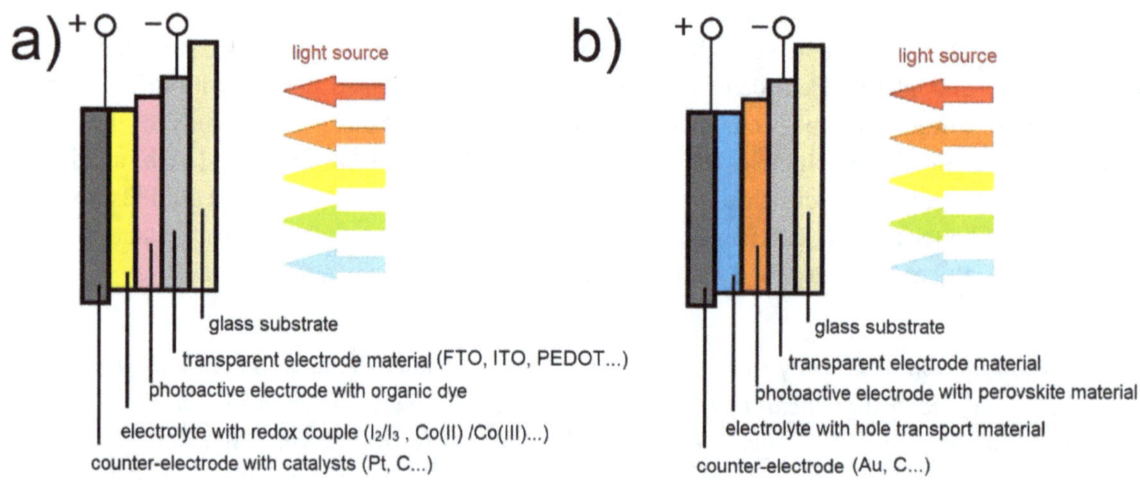

The most commonly used photoactive material is TiO_2, which is often used in combination with various dyes to enhance light absorption in the visible light spectrum. Depending on the morphology of the thin film produced, the surface area and porosity of the metal oxide can be enhanced to increase light absorption, or it can be made transparent for use in combined photovoltaic devices or stacked cells. Different types of TiO_2 (coated onto stainless steel) depending on the morphological state are shown in Figure 9.

Figure 9. Comparison between (a) mesoporous TiO_2 thin film and (b) transparent TiO_2 thin film on stainless steel substrate. Both materials were sintered at the same temperature (500 °C) and for the same period of time (15 min). The main difference is the addition of surfactants: Pluronic P-123 was used as a sol-gel precursor in case of mesoporous film and Triton X-100 was used in case of transparent film

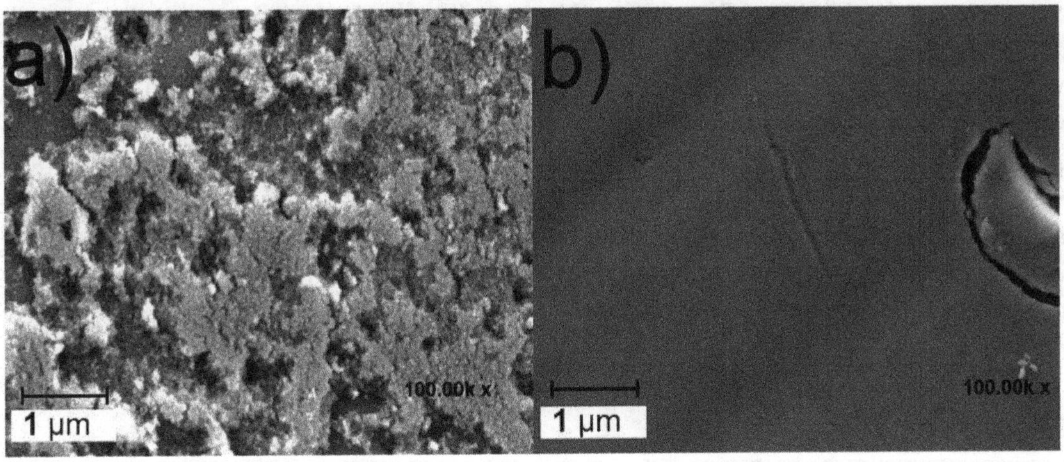

When photoactive material is struck by light, the electrons are released, creating a potential difference that can be used to generate electricity (Ng et al., 2018). Although the first generation of DSSCs often used TiO_2 thin films as photoactive bases, ruthenium complexes were also used as photoactive shifter dyes to achieve absorption of light in the visible region rather than the UV region, which was later replaced by porphyrin dyes (Higashino & Imahori, 2015), and then by even cheaper dyes with perovskite structure (Aftab & Ahmad, 2021). The dye component is not necessarily the most expensive, because in most cases the counter electrode is made of a chemically inert metal (typically gold, platinum, titanium or stainless steel) on which a thin film of a reduction catalyst (usually carbon nanoparticles or platinum) has been deposited. Due to the specifications for the counter electrode, the price remained quite high, especially for efficient catalysts such as thin platinum films, which are usually deposited by physical vapor deposition or in the form of planar Pt complexes or in the form of chloroplatinic acid. In most types of DSSCs, Pt thin film catalysts are crucial to act as an efficient reducing agent for the redox electrolyte (typically an I_2/I_3^- redox couple) to improve the voltage delivery (Stergiopoulos, Rozi, Karagianni, & Falaras, 2011). Different types of metal oxides, such as NiO and ZnO, have been tested as substitutes, but no great improvements were obtained for DSSCs (Vittal & Ho, 2017) (H. Wang, Wei, & Hu, 2014). However, for PSCs, some improvements were made to use a different type of counter electrode material such as Al/Mg or Mg coated with different thin film materials such as MgO, Au nanowires, and carbon nanoparticles, resulting in significantly lower cost compared to platinum catalysts (H. Wang et al., 2020). Despite the initial flop of inorganic-organic-based photovoltaics (mainly DSSCs), PSC devices are achieving excellent results in efficiency tests, where this type of solar cell has already achieved comparable efficiency results to silicon solar cells (Chen et al., 2019). The main reason for the delay in commercialization of PSCs is their sensitivity to moisture and UV light, so developments in recent years have focused precisely on protection against external influences.

In recent years, there have been several attempts to develop low-cost DSSC or PSC devices based on all-metal sheets or meshes, using the metal substrate as either the working electrode, the counter electrode, or even both. In the last decade, there has been a significant increase in interest in devices made of metal foils, such as stainless steel, titanium, and other chemically resistant metals (Li et al., 2014) (Jun, Kim, & Kang, 2007) (Rozman, Sygkridou, Fuchs Godec, Stathatos, & Bren, 2019). Initially, most of the research on the use of stainless steel electrodes was conducted only for DSSCs because they are relatively easy to assemble and stable under ambient conditions. The material preparation is similar to electrochromic devices where sol-gel is applied by spin coating, inkjet printing, or other coating method, except that the selected material can only be photoactive metal oxide such as ZnO or TiO_2. These metal oxides must be modified to have sufficient adhesion to the metal substrate and sufficient porosity to incorporate the photoactive dye into the structure. Some promising results were obtained with different metal foils, such as titanium foil with deposited TiO_2 (Xiao, Han, Zhou, & Wu, 2016). Despite the rather lower efficiency compared to typical PSCs or even DSSCs, the current development suggests the possibility of developing flexible photosensors that could be used in various applications, such as low-cost or disposable photosensors that could be used for military, medical, or analytical applications (Vuuren, Nunzi, & Givigi, 2021).

Biosensors

The last type of electrochemical devices that can be considered part of optoelectronics are biosensors. These are small and inexpensive analytical devices that convert a biological reaction into an electrical signal and provide us with information about the concentration of the target analyte (Monošík, Stred'anský, & Šturdík, 2012). Biosensors can be classified by the method of physicochemical conversion or by the type of biorecognition element. Based on the transducer, biosensors can be divided into electrochemical, optical and mechanical biosensors (Mehrotra, 2016). The electroactive substance located on the surface of the working electrode can be simple molecules (chemically active substance), proteins or enzymes (biochemically active substance), or living cells (biologically active substance). Although biosensors are generally used only as electrochemical or optically active devices, there are many systems in which the sensors are built to produce both electrical and optical signals, further increasing the accuracy and reliability of the system. To keep costs down and simplify production, the electrodes are made of metal foils. The metal chosen can vary, and either three electrodes of the same metal can be used or combined. In most cases, the electrodes for biosensors are manufactured so that they can be easily disposed of. This is usually done by using thin film electrodes made of gold nanoparticles sputtered onto a plastic substrate, or electrodes made of copper sheet. Recently, aluminum and stainless steel sheets and foils have also become desirable materials because they are easy to obtain and can be cut easily with inexpensive tools (Rezaei, Shams-Ghahfarokhi, Havakeshian, & Ensafi, 2016) (Rajapaksha, Uda, Hashim, Gopinath, & Fernando, 2018). The use of sheet metal electrodes also allows us to easily manipulate the surface of the working electrode by first making mechanical or chemical changes to the electrode (e.g., etching, thin film deposition etc.) before applying the final electroactive material. In general, most biosensors have various organic coatings on the surface. Mercaptans are quite common organic modifiers for gold surfaces because the bonding between sulfur and gold is relatively strong, which allows the deposition or application of additional bioactive layers by chemical reaction with the original organic film. One of the more well-known examples is the neurotoxin biosensor, which uses this principle to attach the enzyme acetylcholinesterase to the gold surface (Arduini, Guidone, Amine, Palleschi, & Moscone, 2013). The application of inorganic thin films is also not uncommon. In several cases, it has been shown that the modification of metal foils with metal oxide can improve the adhesion properties of the sensor substrate, which can be either an organic substance or even a living sample (Xia et al., 2017). In one of the papers, the authors showed that it is possible to improve the adhesion of living yeast cells by pre-coating the electrodes with WO_3 (Štukovnik, 2021). In general, metal oxides that are considered biocompatible, such as ZnO, TiO_2, WO_3, and other materials, are most commonly used because they both prevent electrode corrosion and prevent the bioactive material from losing its sensing properties (Jang et al., 2012) (Santos et al., 2016) (Arya et al., 2012). Therefore, in the case of biosensors, the porosity of the protective layer should be minimal, and the roughness of the layer depends on the adhesion properties of the bioactive material. In the case of simple organic molecules, roughness can increase adhesion, while in the case of living cells, sharp surfaces could damage or even rupture the cells (Groot & Rabone, 2001). Overall, biosensors are an interesting emerging technology that is becoming an important tool in medical diagnostics, environmental monitoring, and screening and identification of pathogens and toxins. A schematic representation of the two most common types of electrochemical biosensors is given in Figure 10.

Figure 10. Schematic representation of the biosensor with (a) living cells and (b) organic components (green) adhering to the electrode surface (gray/yellow). In the case of the living cell, the electrical connection (blue lines) is blocked until the cells adhere to the surface and the connection is made when a toxin or pathogen (red diamond) is added to the medium. In the case of an organic component, an antibody, protein, or other biochemically active element is applied as a film to the electrode surface, and electrical connection is disrupted when a pathogen or biochemical agent adheres to the surface of the electrode

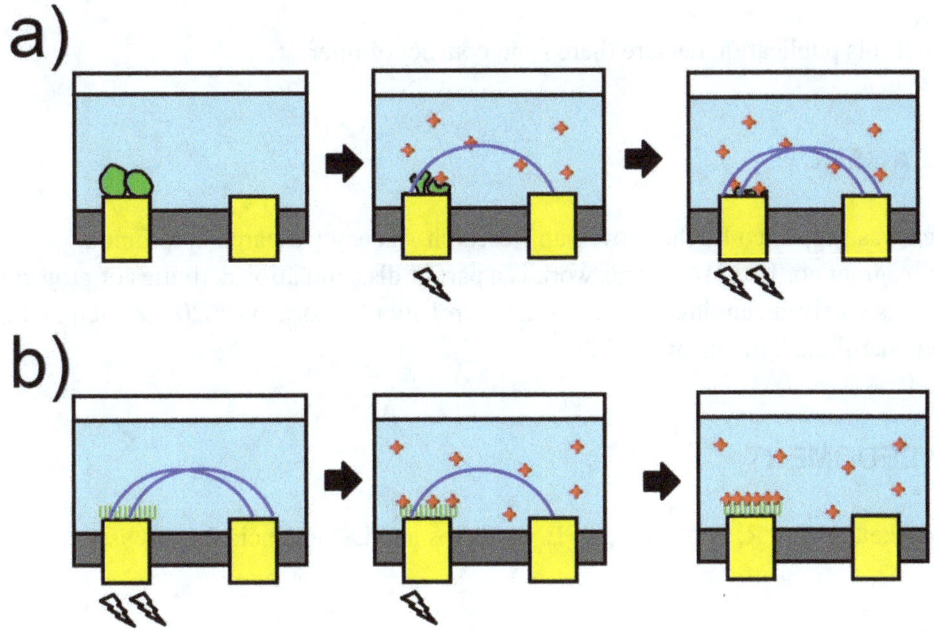

CONCLUSION

The functionalization of metal foils together with the study of their morphological properties has become an indispensable part of research and technology related to energy storage, optoelectronics, corrosion protection, and emerging applications of (bio)sensors. Over time, the development of new polymeric materials and other non-metallic materials (e.g., glass or carbon fibers) has displaced the use of metals to some extent. However, certain properties of metal foils such as high electrical conductivity, high flexibility, good tensile strength, to name a few, facilitate the use of metal foils over plastics, glass, etc. Metal foils thus represent a bridging material that can be used for a wide range of applications, from mechanical engineering and construction to sophisticated optoelectronic devices. In addition, metal foils can be one to two orders of magnitude cheaper compared to conventional materials, which makes them interesting from an economic point of view as well. In this chapter, the main methods for thin film preparation, surface and electrochemical characterization, and application of metal foils in electrochromic and photovoltaic applications, batteries, and biosensors were presented. Challenges for future development of metal foil-based devices certainly remain, especially in improving and optimizing device performance.

The main purpose of this chapter is to give a unifying perspective on the use of (surface functionalized) metal foils as electrodes in devices based on different principles: from batteries to optoelectronic devices. Recent architectural solution for electrochromic devices has enabled the use of metal foils in place of conventional optically transparent electrodes; this novel geometric solution has also been successful in other types of devices; dye-sensitized solar cells and also biosensors, with possible extension to other energy conversion or storage devices.

CONFLICT OF INTEREST

The authors of this publication declare there is no conflict of interest.

FUNDING AGENCY

This research was supported by the Slovenian Research Agency (research core funding No. P1-0201), and the VEGA grant no. 1/0171/21. This work is a part of dissemination activities of project FunGlass. This project has received funding from the European Union´s Horizon 2020 research and innovation programme under grant agreement No 739566.

ACKNOWLEDGMENT

M. R. would like to thank R. Fuchs-Godec, E. Stathatos and L. Stepančič for discussion and guidance.

REFERENCES

Aftab, A., & Ahmad, M. I. (2021). A review of stability and progress in tin halide perovskite solar cell. *Solar Energy*, *216*, 26–47. doi:10.1016/j.solener.2020.12.065

Alferov, Z. (2013). Heterostructures for optoelectronics: History and modern trends. *Proceedings of the IEEE*, *101*(10), 2176–2182. doi:10.1109/JPROC.2013.2274912

Alpen, U., Graf, K., & Hafendörfer, M. (1978). Data acquisition system for electrochemical applications. *Journal of Applied Electrochemistry*, *8*(6), 557–562. doi:10.1007/BF00610802

Amundson, K. (2016). Electrophoretic displays. In J. Chen, W. Cranton, & M. Fihn (Eds.), *Handbook of Visual Display Technology* (pp. 2405–2422). Springer International Publishing. doi:10.1007/978-3-319-14346-0_101

Arduini, F., Guidone, S., Amine, A., Palleschi, G., & Moscone, D. (2013). Acetylcholinesterase biosensor based on self-assembled monolayer-modified gold-screen printed electrodes for organophosphorus insecticide detection. *Sensors and Actuators. B, Chemical*, *179*, 201–208. doi:10.1016/j.snb.2012.10.016

Arya, S. K., Saha, S., Ramirez-Vick, J. E., Gupta, V., Bhansali, S., & Singh, S. P. (2012). Recent advances in ZnO nanostructures and thin films for biosensor applications [Review]. *Analytica Chimica Acta*, *737*, 1–21. doi:10.1016/j.aca.2012.05.048 PMID:22769031

Ayache, J., Beaunier, L., Boumendil, J., Ehret, G., & Laub, D. (2010). *Preliminary preparation techniques sample preparation handbook for transmission electron microscopy: Techniques*. Springer New York.

Aziz, F., & Ismail, A. F. (2015). Spray coating methods for polymer solar cells fabrication: A review. *Materials Science in Semiconductor Processing*, *39*, 416–425. doi:10.1016/j.mssp.2015.05.019

Barroqueiro, B., Andrade-Campos, A., Valente, R. A. F., & Neto, V. (2019). Metal additive manufacturing cycle in aerospace industry: A comprehensive review. *Journal of Manufacturing and Materials Processing*, *3*(3), 52. doi:10.3390/jmmp3030052

Bisht, H., Eun, H. T., Mehrtens, A., & Aegerter, M. A. (1999). Comparison of spray pyrolyzed FTO, ATO and ITO coatings for flat and bent glass substrates. *Thin Solid Films*, *351*(1), 109–114. doi:10.1016/S0040-6090(99)00254-0

Bisquert, J., Zaban, A., Greenshtein, M., & Mora-Seró, I. (2004). Determination of rate constants for charge transfer and the distribution of semiconductor and electrolyte electronic energy levels in dye-sensitized solar cells by open-circuit photovoltage decay method. *Journal of the American Chemical Society*, *126*(41), 13550–13559. doi:10.1021/ja047311k PMID:15479112

Blomgren, G. E. (2016). The development and future of lithium ion batteries. *Journal of the Electrochemical Society*, *164*(1), A5019–A5025. doi:10.1149/2.0251701jes

Booth, M. J. (2014). Adaptive optical microscopy: The ongoing quest for a perfect image. *Light, Science & Applications*, *3*(4), e165–e165. doi:10.1038/lsa.2014.46

Bosio, A., Pasini, S., & Romeo, N. (2020). The history of photovoltaics with emphasis on CdTe solar cells and modules. *Coatings*, *10*(4), 522. doi:10.3390/coatings10040344

Cerc Korošec, R., Felicijan, M., Žener, B., Pompe, M., Dražić, G., Padežnik Gomilšek, J., Pihlar, B., & Bukovec, P. (2017). The role of thermal analysis in optimization of electrochromic effect of nickel oxide thin films, prepared by the sol-gel method: Part III. *Thermochimica Acta*, *655*, 344–350. doi:10.1016/j.tca.2017.07.010

Chang, B.-Y., & Park, S.-M. (2010). Electrochemical impedance spectroscopy. *Annual Review of Analytical Chemistry (Palo Alto, Calif.), 3*(1), 207–229. doi:10.1146/annurev.anchem.012809.102211 PMID:20636040

Chen, Z., Turedi, B., Alsalloum, A. Y., Yang, C., Zheng, X., Gereige, I., AlSaggaf, A., Mohammed, O. F., & Bakr, O. M. (2019). Single-crystal $MAPbI_3$ perovskite solar cells exceeding 21% power conversion efficiency. *ACS Energy Letters, 4*(6), 1258–1259. doi:10.1021/acsenergylett.9b00847

Chowdhury, S., Sumita, U., Islam, A., & Bedja, I. (2014). Importance of policy for energy system transformation: Diffusion of PV technology in Japan and Germany. *Energy Policy, 68*, 285–293. doi:10.1016/j.enpol.2014.01.023

Covington, A. K. (1994). Terminology and conventions for microelectronic ion-selective field effect transistor devices in electrochemistry (IUPAC Recommendations 1994). *Pure and Applied Chemistry, 66*(3), 565–569. doi:10.1351/pac199466030565

Cummins, G., & Desmulliez, M. P. Y. (2012). Inkjet printing of conductive materials: A review. *Circuit World, 38*(4), 193–213. doi:10.1108/03056121211280413

Dalisa, A. L. (1977). Electrophoretic display technology. *IEEE Transactions on Electron Devices, 24*(7), 827–834. doi:10.1109/T-ED.1977.18838

Das, T. K., & Prusty, S. (2012). Review on conducting polymers and their applications. *Polymer-Plastics Technology and Engineering, 51*(14), 1487–1500. doi:10.1080/03602559.2012.710697

Dinh, K. N., Liang, Q., Du, C.-F., Zhao, J., Tok, A. I. Y., Mao, H., & Yan, Q. (2019). Nanostructured metallic transition metal carbides, nitrides, phosphides, and borides for energy storage and conversion. *Nano Today, 25*, 99–121. doi:10.1016/j.nantod.2019.02.008

Ferris, C. D., & Rose, D. R. (1972). An operational amplifier 4-electrode impedance bridge for electrolyte measurements. *Medical & Biological Engineering, 10*(5), 647–654. doi:10.1007/BF02476082 PMID:5076429

Fischer, J. K., von Brüning, D. M., & Labhart, H. (1976). Light modulation by electrochromism. *Applied Optics, 15*(11), 2812–2816. doi:10.1364/AO.15.002812 PMID:20165493

Frankel, R. S., & Aitken, D. W. (1970). Energy-Dispersive X-Ray Emission Spectroscopy. *Applied Spectroscopy, 24*(6), 557–566. doi:10.1366/000370270774372308

Goodenough, J. B. (2018). How we made the Li-ion rechargeable battery. *Nature Electronics, 1*(3), 204–204. doi:10.103841928-018-0048-6

Groot, R. D., & Rabone, K. L. (2001). Mesoscopic simulation of cell membrane damage, morphology change and rupture by nonionic surfactants. *Biophysical Journal, 81*(2), 725–736. doi:10.1016/S0006-3495(01)75737-2 PMID:11463621

Guan, P., Zhou, L., Yu, Z., Sun, Y., Liu, Y., Wu, F., Jiang, Y., & Chu, D. (2020). Recent progress of surface coating on cathode materials for high-performance lithium-ion batteries. *Journal of Energy Chemistry, 43*, 220–235. doi:10.1016/j.jechem.2019.08.022

Higashino, T., & Imahori, H. (2015). Porphyrins as excellent dyes for dye-sensitized solar cells: Recent developments and insights. *Dalton Transactions (Cambridge, England)*, *44*(2), 448–463. doi:10.1039/C4DT02756F PMID:25381701

Inkson, B. J. (2016). 2 - Scanning electron microscopy (SEM) and transmission electron microscopy (TEM) for materials characterization. In G. Hübschen, I. Altpeter, R. Tschuncky, & H.-G. Herrmann (Eds.), *Materials Characterization Using Nondestructive Evaluation (NDE) Methods* (pp. 17–43). Woodhead Publishing. doi:10.1016/B978-0-08-100040-3.00002-X

Inzelt, G. (2012). Applications of conducting polymers. In G. Inzelt (Ed.), *Conducting polymers: A new era in electrochemistry* (pp. 245–293). Springer Berlin Heidelberg. doi:10.1007/978-3-642-27621-7_7

Jang, H. D., Kim, S. K., Chang, H., Roh, K.-M., Choi, J.-W., & Huang, J. (2012). A glucose biosensor based on TiO2–Graphene composite. *Biosensors & Bioelectronics*, *38*(1), 184–188. doi:10.1016/j.bios.2012.05.033 PMID:22705409

Jun, Y., Kim, J., & Kang, M. G. (2007). A study of stainless steel-based dye-sensitized solar cells and modules. *Solar Energy Materials and Solar Cells*, *91*(9), 779–784. doi:10.1016/j.solmat.2007.01.007

Kalluri, S., Yoon, M., Jo, M., Liu, H. K., Dou, S. X., Cho, J., & Guo, Z. (2017). Feasibility of cathode surface coating technology for high-energy lithium-ion and beyond-lithium-ion batteries. *Advanced Materials*, *29*(48), 1605807. doi:10.1002/adma.201605807 PMID:28251710

Kao, S.-Y., Lu, H.-C., Kung, C.-W., Chen, H.-W., Chang, T.-H., & Ho, K.-C. (2016). Thermally cured dual functional viologen-based all-in-one electrochromic devices with panchromatic modulation. *ACS Applied Materials & Interfaces*, *8*(6), 4175–4184. doi:10.1021/acsami.5b11947 PMID:26807824

Kendrick, E. (2019). *CHAPTER 11 Advancements in manufacturing Future Lithium-ion Batteries*. The Royal Society of Chemistry.

Khachatryan, H., Kim, D.-J., Kim, M., & Kim, H.-K. (2018). Roll-to-Roll fabrication of ITO thin film for flexible optoelectronics applications: The role of post-annealing. *Materials Science in Semiconductor Processing*, *88*, 51–56. doi:10.1016/j.mssp.2018.07.033

Khan, M. K., Wang, Q. Y., & Fitzpatrick, M. E. (2016). 1 - Atomic force microscopy (AFM) for materials characterization. In G. Hübschen, I. Altpeter, R. Tschuncky, & H.-G. Herrmann (Eds.), *Materials Characterization Using Nondestructive Evaluation (NDE) Methods* (pp. 1–16). Woodhead Publishing. doi:10.1016/B978-0-08-100040-3.00001-8

Kholghi Eshkalak, S., Khatibzadeh, M., Kowsari, E., Chinnappan, A., Jayathilaka, W. A. D. M., & Ramakrishna, S. (2019). Overview of electronic ink and methods of production for use in electronic displays. *Optics & Laser Technology*, *117*, 38–51. doi:10.1016/j.optlastec.2019.04.003

Kim, Y. Y., Yang, T.-Y., Suhonen, R., Kemppainen, A., Hwang, K., Jeon, N. J., & Seo, J. (2020). Roll-to-roll gravure-printed flexible perovskite solar cells using eco-friendly antisolvent bathing with wide processing window. *Nature Communications*, *11*(1), 5146. doi:10.103841467-020-18940-5 PMID:33051454

Kulesza, P. J., Malik, M. A., Zamponi, S., Berrettoni, M., & Marassi, R. (1995). Electrolyte-cation-dependent coloring, electrochromism and thermochromism of cobalt(II) hexacyanoferrate(III, II) films. *Journal of Electroanalytical Chemistry (Lausanne, Switzerland)*, *397*(1), 287–292. doi:10.1016/0022-0728(95)04187-8

Lamberti, M., & Escher, F. (2007). Aluminium Foil as a food packaging material in comparison with other materials. *Food Reviews International*, *23*(4), 407–433. doi:10.1080/87559120701593830

Li, H., Zhao, Q., Dong, H., Ma, Q., Wang, W., Xu, D., & Yu, D. (2014). Highly-flexible, low-cost, all stainless steel mesh-based dye-sensitized solar cells. *Nanoscale*, *6*(21), 13203–13212. doi:10.1039/C4NR03999H PMID:25254313

Ludwig, B., Zheng, Z., Shou, W., Wang, Y., & Pan, H. (2016). Solvent-free manufacturing of electrodes for lithium-ion batteries. *Scientific Reports*, *6*(1), 23150. doi:10.1038rep23150 PMID:26984488

Macdonald, D. D. (1990). Review of mechanistic analysis by electrochemical impedance spectroscopy. *Electrochimica Acta*, *35*(10), 1509–1525. doi:10.1016/0013-4686(90)80005-9

Macdonald, D. D. (2006). Reflections on the history of electrochemical impedance spectroscopy. *Electrochimica Acta*, *51*(8), 1376–1388. doi:10.1016/j.electacta.2005.02.107

Margaretha, P., & Tissot, P. (1979). Observation of reduction product-electrolyte-electrode interaction by cyclic voltammetry. *Journal of Electroanalytical Chemistry and Interfacial Electrochemistry*, *99*(1), 127–131. doi:10.1016/S0022-0728(79)80419-2

Marinko, Ž., Suhadolnik, L., Šetina Batič, B., Šelih, V. S., Majaron, B., Kovač, J., & Čeh, M. (2021). Toward a flexible and efficient TiO_2 photocatalyst immobilized on a titanium foil. *ACS Omega*, *6*(36), 23233–23242. doi:10.1021/acsomega.1c02862 PMID:34549124

Maruska, H. P., & Stevenson, D. A. (1974). Mechanism of light production in metal-insulator-semiconductor diodes; GaN:Mg violet light-emitting diodes. *Solid-State Electronics*, *17*(11), 1171–1179. doi:10.1016/0038-1101(74)90161-0

Maver, U., Velnar, T., Gaberšček, M., Planinšek, O., & Finšgar, M. (2016). Recent progressive use of atomic force microscopy in biomedical applications. *Trends in Analytical Chemistry*, *80*, 96–111. doi:10.1016/j.trac.2016.03.014

Maver, U., Žnidaršič, A., & Gaberšček, M. (2011). An attempt to use atomic force microscopy for determination of bond type in lithium battery electrodes. *Journal of Materials Chemistry*, *21*(12), 4071–4075. doi:10.1039/c0jm04481d

Mazor, H., Golodnitsky, D., Burstein, L., Gladkich, A., & Peled, E. (2012). Electrophoretic deposition of lithium iron phosphate cathode for thin-film 3D-microbatteries. *Journal of Power Sources*, *198*, 264–272. doi:10.1016/j.jpowsour.2011.09.108

Meenakshi, M., Sivakumar, R., Sivanantharaja, A., & Sanjeeviraja, C. (2017). Electrochromic performance of RF sputtered WO3 thin films by Li ion intercalation and de-intercalation. *AIP Conference Proceedings*, *1832*(1), 080003. doi:10.1063/1.4980463

Mehrotra, P. (2016). Biosensors and their applications – A review. *Journal of Oral Biology and Craniofacial Research*, *6*(2), 153–159. doi:10.1016/j.jobcr.2015.12.002 PMID:27195214

Miura, H., Fukuyama, Y., Sunda, T., Lin, B., Zhou, J., Takizawa, J., Ohmori, A., & Kimura, M. (2014). Foldable textile electronic devices using all-organic conductive fibers. *Advanced Engineering Materials*, *16*(5), 550–555. doi:10.1002/adem.201300461

Monošík, R., Stred'anský, M., & Šturdík, E. (2012). Application of electrochemical biosensors in clinical diagnosis. *Journal of Clinical Laboratory Analysis*, *26*(1), 22–34. doi:10.1002/jcla.20500 PMID:24833531

Mortimer, R. J. (2011). Electrochromic materials. *Annual Review of Materials Research*, *41*(1), 241–268. doi:10.1146/annurev-matsci-062910-100344

Muhammad, A., Rahman, M. R., Baini, R., & Bin Bakri, M. K. (2021). 8 - Applications of sustainable polymer composites in automobile and aerospace industry. In M. R. Rahman (Ed.), *Advances in Sustainable Polymer Composites* (pp. 185–207). Woodhead Publishing. doi:10.1016/B978-0-12-820338-5.00008-4

Ng, C. H., Lim, H. N., Hayase, S., Zainal, Z., & Huang, N. M. (2018). Photovoltaic performances of mono- and mixed-halide structures for perovskite solar cell: A review. *Renewable & Sustainable Energy Reviews*, *90*, 248–274. doi:10.1016/j.rser.2018.03.030

Nuttall, R. H. (1979). Andrew Pritchard's contribution to metallurgical microscopy. *Technology and Culture*, *20*(3), 569–577. doi:10.2307/3103817

O'Regan, B., & Grätzel, M. (1991). A low-cost, high-efficiency solar cell based on dye-sensitized colloidal TiO2 films. *Nature*, *353*(6346), 737–740. doi:10.1038/353737a0

Ohsuku, T., & Hirai, T. (1982). An electrochromic display based on titanium dioxide. *Electrochimica Acta*, *27*(9), 1263–1266. doi:10.1016/0013-4686(82)80146-1

Orsato, R. J., & Wells, P. (2007). U-turn: The rise and demise of the automobile industry. *Journal of Cleaner Production*, *15*(11), 994–1006. doi:10.1016/j.jclepro.2006.05.019

Patel, B. A. (2020). Amperometry and potential step techniques. In B. Patel (Ed.), *Electrochemistry for Bioanalysis* (pp. 9–26). Elsevier. doi:10.1016/B978-0-12-821203-5.00009-9

Pogue, B. W., & Wilson, B. C. (2018). Optical and x-ray technology synergies enabling diagnostic and therapeutic applications in medicine. *Journal of Biomedical Optics*, *23*(12), 121610. doi:10.1117/1.JBO.23.12.121610 PMID:30350489

Polo da Fonseca, C. N., De Paoli, M.-A., & Gorenstein, A. (1994). Electrochromism in cobalt oxide thin films grown by anodic electroprecipitation. *Solar Energy Materials and Solar Cells*, *33*(1), 73–81. doi:10.1016/0927-0248(94)90291-7

Prasanna, K., Subburaj, T., Jo, Y. N., & Lee, C. W. (2013). Optimization of electrophoretic suspension to fabricate Li[Ni1/3Co1/3Mn1/3]O2 based positive electrode for Li-ion batteries. *Electrochimica Acta*, *95*, 295–300. doi:10.1016/j.electacta.2013.01.102

Prince, M. B. (1955). Silicon solar energy converters. *Journal of Applied Physics*, *26*(5), 534–540. doi:10.1063/1.1722034

Putzig, C. L., Leugers, M. A., McKelvy, M. L., Mitchell, G. E., Nyquist, R. A., Papenfuss, R. R., & Yurga, L. (1994). Infrared spectroscopy. *Analytical Chemistry*, *66*(12), 26–66. doi:10.1021/ac00084a003 PMID:8092469

Quéré, D. (1999). Fluid coating on a fiber. *Annual Review of Fluid Mechanics*, *31*(1), 347–384. doi:10.1146/annurev.fluid.31.1.347

Rajapaksha, R. D. A. A., Uda, M. N. A., Hashim, U., Gopinath, S. C. B., & Fernando, C. A. N. (2018). *Impedance based aluminium interdigitated electrode (Al-IDE) biosensor on silicon substrate for salmonella detection.* Paper presented at the 2018 IEEE International Conference on Semiconductor Electronics (ICSE). 10.1109/SMELEC.2018.8481324

Randviir, E. P., & Banks, C. E. (2013). Electrochemical impedance spectroscopy: An overview of bioanalytical applications. *Analytical Methods*, *5*(5), 1098–1115. doi:10.1039/c3ay26476a

Reynolds, C. D., Slater, P. R., Hare, S. D., Simmons, M. J. H., & Kendrick, E. (2021). A review of metrology in lithium-ion electrode coating processes. *Materials & Design*, *209*, 109971. doi:10.1016/j.matdes.2021.109971

Rezaei, B., Shams-Ghahfarokhi, L., Havakeshian, E., & Ensafi, A. A. (2016). An electrochemical biosensor based on nanoporous stainless steel modified by gold and palladium nanoparticles for simultaneous determination of levodopa and uric acid. *Talanta*, *158*, 42–50. doi:10.1016/j.talanta.2016.04.061 PMID:27343576

Rochow, T. G., & Tucker, P. A. (1994). *A brief history of microscopy introduction to microscopy by means of light, electrons, X rays, or acoustics*. Springer US.

Rozman, M., Bren, U., Lukšič, M., Godec, R. F., Bokias, G., Kalarakis, A. N., & Stathatos, E. (2018). Electrochromic cell with hydrogel-stabilized water-based electrolyte using electrodeposition as a fast color changing mechanism. *Electrochimica Acta*, *283*, 1105–1114. doi:10.1016/j.electacta.2018.07.052

Rozman, M., Cerar, J., Lukšič, M., Uršič, M., Mourtzikou, A., Spreizer, H., Kozjek Škofic, I., & Stathatos, E. (2017). Electrochromic properties of thin nanocrystalline TiO2 films coated electrodes with adsorbed Co(II) or Fe(II) 2,2′-bipyridine complexes. *Electrochimica Acta*, *238*, 278–287. doi:10.1016/j.electacta.2017.04.030

Rozman, M., Cetin, N., Bren, U., & Lukšič, M. (2021). Use of Different Metal Oxide Coatings in Stainless Steel Based ECDs for Smart Textiles. *Electronics (Basel)*, *10*(20), 2529. doi:10.3390/electronics10202529

Rozman, M., Gaberšček, M., Marolt, G., Bren, U., & Lukšič, M. (2019). An inverted sandwich electrochromic device architecture does not require optically transparent electrodes. *Advanced Materials Technologies*, *4*(9), 1900389. doi:10.1002/admt.201900389

Rozman, M., Mavrič, A., Kravanja, G., Valant, M., & Pakseresht, A. (2022). Ultra-low-cost, flexible and durable electrochromic tape device based on aluminum foil. *Electrochimica Acta*, *404*, 139760. doi:10.1016/j.electacta.2021.139760

Rozman, M., Sygkridou, D., Fuchs Godec, R., Stathatos, E., & Bren, U. (2019). Novel geometric approach for photosensor construction based on dye-sensitization of TiO_2 nanoparticles on stainless steel. *Sensors and Actuators. A, Physical*, *295*, 51–58. doi:10.1016/j.sna.2019.05.034

Rozman, M., Žener, B., Matoh, L., Godec, R. F., Mourtzikou, A., Stathatos, E., Bren, U., & Lukšič, M. (2020). Flexible electrochromic tape using steel foil with WO_3 thin film. *Electrochimica Acta*, *330*, 135329. doi:10.1016/j.electacta.2019.135329

Sahu, N., Parija, B., & Panigrahi, S. (2009). Fundamental understanding and modeling of spin coating process: A review. *Indian Journal of Physics*, *83*(4), 493–502. doi:10.100712648-009-0009-z

Santos, L., Silveira, C. M., Elangovan, E., Neto, J. P., Nunes, D., Pereira, L., Martins, R., Viegas, J., Moura, J. J. G., Todorovic, S., Almeida, M. G., & Fortunato, E. (2016). Synthesis of WO3 nanoparticles for biosensing applications. *Sensors and Actuators. B, Chemical*, *223*, 186–194. doi:10.1016/j.snb.2015.09.046

Schmidt, D. J., Pridgen, E. M., Hammond, P. T., & Love, J. C. (2010). Layer-by-Layer Assembly of a pH-Responsive and Electrochromic Thin Film. *Journal of Chemical Education*, *87*(2), 208–211. doi:10.1021/ed800045r

Schmitt, M., Baunach, M., Wengeler, L., Peters, K., Junges, P., Scharfer, P., & Schabel, W. (2013). Slot-die processing of lithium-ion battery electrodes—Coating window characterization. *Chemical Engineering and Processing*, *68*, 32–37. doi:10.1016/j.cep.2012.10.011

Shalini, S., Balasundaraprabhu, R., Kumar, T. S., Prabavathy, N., Senthilarasu, S., & Prasanna, S. (2016). Status and outlook of sensitizers/dyes used in dye sensitized solar cells (DSSC): A review. *International Journal of Energy Research*, *40*(10), 1303–1320. doi:10.1002/er.3538

Shobana, M. K. (2019). Metal oxide coated cathode materials for Li ion batteries – A review. *Journal of Alloys and Compounds*, *802*, 477–487. doi:10.1016/j.jallcom.2019.06.194

Somani, P. R., & Radhakrishnan, S. (2003). Electrochromic materials and devices: Present and future. *Materials Chemistry and Physics*, *77*(1), 117–133. doi:10.1016/S0254-0584(01)00575-2

Stergiopoulos, T., Rozi, E., Karagianni, C.-S., & Falaras, P. (2011). Influence of electrolyte co-additives on the performance of dye-sensitized solar cells. *Nanoscale Research Letters*, *6*(1), 307. doi:10.1186/1556-276X-6-307 PMID:21711833

Štukovnik, Z. B., & Rozman, M. (2021). Model electrochemical biosensor for the detection of methanol in aqueous solutions with yeast cells. *Acta Chimica Slovenica*, *68*, 8. doi:10.17344/acsi.2020.6545

Sun, K., Zhang, S., Li, P., Xia, Y., Zhang, X., Du, D., Isikgor, F. H., & Ouyang, J. (2015). Review on application of PEDOTs and PEDOT:PSS in energy conversion and storage devices. *Journal of Materials Science Materials in Electronics*, *26*(7), 4438–4462. doi:10.100710854-015-2895-5

Sun, X. W., & Wang, J. X. (2008). Fast switching electrochromic display using a viologen-modified ZnO nanowire array electrode. *Nano Letters*, *8*(7), 1884–1889. doi:10.1021/nl0804856 PMID:18564881

Sutija, D. P., Norby, T., & Björnbom, P. (1995). Transport number determination by the concentration-cell/open-circuit voltage method for oxides with mixed electronic, ionic and protonic conductivity. *Solid State Ionics*, *77*, 167–174. doi:10.1016/0167-2738(94)00268-W

Szmacinski, H., & Lakowicz, J. R. (1995). Fluorescence lifetime-based sensing and imaging. *Sensors and Actuators. B, Chemical*, *29*(1), 16–24. doi:10.1016/0925-4005(95)01658-9 PMID:33867678

Tian, R., King, P. J., Coelho, J., Park, S.-H., Horvath, D. V., Nicolosi, V., O'Dwyer, C., & Coleman, J. N. (2020). Using chronoamperometry to rapidly measure and quantitatively analyse rate-performance in battery electrodes. *Journal of Power Sources*, *468*, 228220. doi:10.1016/j.jpowsour.2020.228220

Usman, B. J., Scenini, F., & Curioni, M. (2020). Corrosion testing of anodized aerospace alloys: Comparison between immersion and salt spray testing using electrochemical impedance spectroscopy. *Journal of the Electrochemical Society*, *167*(4), 041505. doi:10.1149/1945-7111/ab74e3

Vittal, R., & Ho, K.-C. (2017). Zinc oxide based dye-sensitized solar cells: A review. *Renewable & Sustainable Energy Reviews*, *70*, 920–935. doi:10.1016/j.rser.2016.11.273

Vuuren, R. D. J., Nunzi, J. M., & Givigi, S. N. (2021). Frontiers in photosensor materials and designs for new image sensor applications. *IEEE Sensors Journal*, *21*(10), 11339–11348. doi:10.1109/JSEN.2020.3043288

Wang, C., Li, J., Zhang, L., & Fu, S. (2022). Preparation and property optimization of bistable electrochromic microcapsules. *Dyes and Pigments*, *197*, 109936. doi:10.1016/j.dyepig.2021.109936

Wang, H., Li, H., Cao, S., Wang, M., Chen, J., & Zang, Z. (2020). Interface modulator of ultrathin magnesium oxide for low-temperature-processed inorganic $CsPbIBr_2$ perovskite solar cells with efficiency over 11%. *Solar RRL*, *4*(9), 2000226. doi:10.1002olr.202000226

Wang, H., Wei, W., & Hu, Y. H. (2014). NiO as an efficient counter electrode catalyst for dye-sensitized solar cells. *Topics in Catalysis*, *57*(6), 607–611. doi:10.100711244-013-0218-8

Wang, J., & Zeng, H. (2021). Recent advances in electrochemical techniques for characterizing surface properties of minerals. *Advances in Colloid and Interface Science*, *288*, 102346. doi:10.1016/j.cis.2020.102346 PMID:33383471

Weibin, Z., Weidong, W., Xueming, W., Xinlu, C., Dawei, Y., Changle, S., Liping, P., Yuying, W., & Li, B. (2013). The investigation of NbO_2 and Nb_2O_5 electronic structure by XPS, UPS and first principles methods. *Surface and Interface Analysis*, *45*(8), 1206–1210. doi:10.1002ia.5253

Whittingham, M. S. (1974). *Fast ion transport materials and batteries*. Paper presented at the Conference on Electrical Insulation & Dielectric Phenomena - Annual Report 1974.

Wu, C.-L., Wang, C.-K., Lin, C.-K., Wang, S.-C., & Huang, J.-L. (2013). Electrochromic properties of nanostructured tungsten oxide films prepared by surfactant-assisted sol–gel process. *Surface and Coatings Technology*, *231*, 403–407. doi:10.1016/j.surfcoat.2012.01.061

Xia, S., Zhu, P., Pi, F., Zhang, Y., Li, Y., Wang, J., & Sun, X. (2017). Development of a simple and convenient cell-based electrochemical biosensor for evaluating the individual and combined toxicity of DON, ZEN, and AFB1. *Biosensors & Bioelectronics*, *97*, 345–351. doi:10.1016/j.bios.2017.06.002 PMID:28623817

Xiao, Y., Han, G., Zhou, H., & Wu, J. (2016). An efficient titanium foil based perovskite solar cell: Using a titanium dioxide nanowire array anode and transparent poly(3,4-ethylenedioxythiophene) electrode. *RSC Advances*, *6*(4), 2778–2784. doi:10.1039/C5RA23430A

Yan, D., Bazant, M. Z., Biesheuvel, P. M., Pugh, M. C., & Dawson, F. P. (2017). Theory of linear sweep voltammetry with diffuse charge: Unsupported electrolytes, thin films, and leaky membranes. *Physical Review. E*, *95*(3), 033303. doi:10.1103/PhysRevE.95.033303 PMID:28415284

Zuo, D., Tian, G., Li, X., Chen, D., & Shu, K. (2017). Recent progress in surface coating of cathode materials for lithium ion secondary batteries. *Journal of Alloys and Compounds*, *706*, 24–40. doi:10.1016/j.jallcom.2017.02.230

KEY TERMS AND DEFINITIONS

Biosensor: An optoelectronic device that uses biochemical or biological component to sense or detect physical, biological, or chemical change in the environment and convert that sensation into an optical or electrical response.

Coating Methods: Methods specialized in the application of thin layers on various substrates (metal, glass, plastic).

Electrochemistry: Branch of chemistry specializing in chemical reactions based on the conversion of energy from chemical energy to electricity and vice versa.

Electrochromism: A phenomenon of reversible change in the optical properties of a material (colour, opacity) under the influence of an external electric current or applied voltage.

Lithium-Ion Battery: A rechargeable battery device based on Li-ion transport through the electrolyte between the cathode and anode and Li-ion intercalation into the cathode material.

Metal Foils: Materials that are made from single metal or mixture of metal (alloys) where the material is shaped into a thin sheet, that enables material to become fully or partially flexible.

Morphology: The study of surfaces, their shapes, forms, and the presence of surface anomalies. In materials science, it is usually associated with the study of the physical properties of surfaces, roughness, and the presence of patterns.

Photovoltaics: A phenomenon of conversion of light into electricity. Photovoltaic devices are a group of optoelectronic devices specialized in light-to-electricity conversion for the purpose of light sensing or energy conversion.

Spectroscopic Methods: Methods based on the interaction of light with the material, useful for the study of material properties (consistency, structure) in different light spectra (UV-Vis, IR, X-rays, etc.).

Compilation of References

(1994). Thermal Spray Coatings. InCotell, C. M., Sprague, J. A., & Smidt, F. A. (Eds.), *Surface Engineering* (pp. 497–509). ASM International. doi:10.31399/asm.hb.v05.a0001282

Abad, M., & Browne, D. (2020). An Investigation of the Tribological Behavior of Hf-Based Bulk Metallic Glass and Crystalline Alloys. *Journal of Tribology*, *142*(10), 101703. doi:10.1115/1.4046950

Abbaszadeh, S., Pakseresht, A. H., Omidvar, H., & Shafiei, A. (2020). Investigation of the High-Temperature Oxidation Behavior of the Al0.5CoCrFeNi High Entropy Alloy. *Surfaces and Interfaces*, *21*, 100724. doi:10.1016/j.surfin.2020.100724

Abdullah, C. W. (2013). Hydroxyapatite-Coated Magnesium-Based Biodegradable Alloy: Cold Spray Deposition and Simulated Body Fluid Studies. *Journal of Materials Engineering and Performance*, *22*(10), 2997–3004. doi:10.100711665-013-0589-9

Aftab, A., & Ahmad, M. I. (2021). A review of stability and progress in tin halide perovskite solar cell. *Solar Energy*, *216*, 26–47. doi:10.1016/j.solener.2020.12.065

Ahmad Alidokht, S., Vo, P., Yue, S., & Chromik, R. R. (2017). Erosive wear behavior of Cold-Sprayed Ni-WC composite coating. *Wear*, *376–377*, 566–577. doi:10.1016/j.wear.2017.01.052

Ahmad, I., Fay, M., Kennedy, A., & Zhu, Y. (2009). Interfacial investigations and mechanical properties of carbon nanotube reinforcing Al2O3 nanocomposites. *ICCM International Conferences on Composite Materials.*

Aissani, L., Fellah, M., Radjehi, L., Nouveau, C., Montagne, A., & Alhussein, A. (2019). Effect of annealing treatment on the microstructure, mechanical and tribological properties of chromium carbonitride coatings. *Surface and Coatings Technology*, *359*, 403–413. doi:10.1016/j.surfcoat.2018.12.099

Akhtar, K., & Ragab, S. (2014). Numerical Simulations of a Particle-Laden Rectangular Supersonic Jet Impinging on a Solid Wall. *AIAA Science Tech. Forum, 13-17 January 2014, National Harbor, Maryland, 52nd Aerospace Sciences Meeting.*

Akhter, R., Zhou, Z., Xie, Z., & Munroe, P. (2021). Enhancing the Adhesion Strength and Wear Resistance of Nano-structured NiCrN Coatings. *Applied Surface Science*, *541*(March), 148533. doi:10.1016/j.apsusc.2020.148533

Akpan, U. G., & Hameed, B. H. (2010). The advancements in sol–gel method of doped-TiO_2 photocatalysts. *Applied Catalysis A, General*, *375*(1), 1–11. doi:10.1016/j.apcata.2009.12.023

Albrektsson, T., & Johansson, C. (2001). Osteoinduction, osteoconduction and osseointegration. *European Spine Journal*, *10*(0), S96–S101. doi:10.1007005860100282 PMID:11716023

Alehojat, M., Jafari, R., Karimi, P., & Sadeghi, E. (2020). Electron beam-powder bed fusion of Alloy 718: Effect of hot isostatic pressing and thermal spraying on microstructural characteristics and oxidation performance. *Surface and Coatings Technology*, *404*, 126626. doi:10.1016/j.surfcoat.2020.126626

Alferov, Z. (2013). Heterostructures for optoelectronics: History and modern trends. *Proceedings of the IEEE*, *101*(10), 2176–2182. doi:10.1109/JPROC.2013.2274912

Aliofkhazraei, M. (2014). *Anti-abrasive nanocoatings*. Elsevier.

Alpen, U., Graf, K., & Hafendörfer, M. (1978). Data acquisition system for electrochemical applications. *Journal of Applied Electrochemistry*, *8*(6), 557–562. doi:10.1007/BF00610802

Alsaran, A. (2003). Determination of Tribological Properties of Ion-nitrided AISI 5140 Steel. *Materials Characterization*, *49*(2), 171–176. doi:10.1016/S1044-5803(03)00008-1

Amateau, M. F., & Glaeser, W. A. (1964). Survey of Materials for High Temperature Bearing and Sliding Applications. *Wear*, *7*(5), 385–418. doi:10.1016/0043-1648(64)90134-6

Amin, S., & Panchal, H. (2016). A Review on Thermal Spray Coating Processes. *International Journal of Current Trends in Engineering & Research*, *2*, 556 – 563.

Amin, S., & Panchal, H. A. (2016). Review on thermal spray coating process. *International Journal of Current Trends in Engineering & Research*, *2*(4), 556–563.

Amiya, K., & Inoue, A. (2008). Fe-(Cr, Mo)-(C, B)-Tm bulk metallic glasses with high strength and high glass-forming ability. *Reviews on Advanced Materials Science*, *18*, 27–29.

Ammar, S., Ramesh, K., Vengadaesvaran, B., Ramesh, S., & Arof, A. K. (2016). Amelioration of anticorrosion and hydrophobic properties of epoxy/PDMS composite coatings containing nano ZnO particles. *Progress in Organic Coatings*, *92*, 54–65. doi:10.1016/j.porgcoat.2015.12.007

Amundson, K. (2016). Electrophoretic displays. In J. Chen, W. Cranton, & M. Fihn (Eds.), *Handbook of Visual Display Technology* (pp. 2405–2422). Springer International Publishing. doi:10.1007/978-3-319-14346-0_101

Anderson, J. D. (2003). *Modern compressible flow: with historical perspective*. McGraw-Hill.

Ang, A. S. M., & Sanpo, N. (2013). Thermal Spray Maps: Material Genomics of Processing Technologies. Journal of Thermal Spray. *Technology*, *22*(7), 1170–1183. doi:10.100711666-013-9970-3

Antonov, M., Afshari, H., Baronins, J., Adoberg, E., Raadik, T., & Hussainova, I. (2018). The Effect of Temperature and Sliding Speed on Friction and Wear of Si3N4, Al2O3, and ZrO2 Balls Tested against AlCrN PVD Coating. *Tribology International*, *118*(February), 500–514. doi:10.1016/j.triboint.2017.05.035

Antonov, M., Hussainova, I., Veintha, R., & Pirso, J. (2010). Effect of Temperature and Load on Three Body Abrasion of Cermets and Steel. *Tribology International*, *46*(1), 261–268. doi:10.1016/j.triboint.2011.06.029

Aouadi, K., Tlili, B., Nouveau, C., Besnard, A., Chafra, M., & Souli, R. (2019). Influence of substrate bias voltage on corrosion and wear behavior of physical vapor deposition CrN coatings. *Journal of Materials Engineering and Performance*, *28*(5), 2881–2891. doi:10.100711665-019-04033-y

Archard, J. F., & Hirst, W. (1956). The Wear of Metals Under Unlubricated Conditions. *Proc. R. Soc. Lond. Series A*, *236*, 397-410.

Archard, J. F. (1953). Contact and rubbing of flat surfaces. *Journal of Applied Physics*, *24*(8), 981–988. doi:10.1063/1.1721448

Arduini, F., Guidone, S., Amine, A., Palleschi, G., & Moscone, D. (2013). Acetylcholinesterase biosensor based on self-assembled monolayer-modified gold-screen printed electrodes for organophosphorus insecticide detection. *Sensors and Actuators. B, Chemical, 179*, 201–208. doi:10.1016/j.snb.2012.10.016

Ariharan, S., & Balani, K. (2021). Fretting wear behaviour and frictional force mapping of Al2O3 based thermal barrier coatings. *International Journal of Refractory Metals & Hard Materials, 98*, 105525. doi:10.1016/j.ijrmhm.2021.105525

Ariharan, S., Gupta, A., Keshri, A., Agarwal, A., & Balani, K. (2013). Size effect of yttria stabilized zirconia addition on fracture toughness and thermal conductivity of plasma sprayed aluminum oxide composite coatings. *Nanoscience and Nanotechnology Letters, 4*(3), 323–332. doi:10.1166/nnl.2012.1317

Ariharan, S., & Maurya, R. (2021). Assessment of plasma sprayed carbon nanotube reinforced Al2O3 -based nano-composite with micro-scratching. *Surface and Coatings Technology, 418*, 127216. doi:10.1016/j.surfcoat.2021.127216

Ariharan, S., Nisar, A., Balaji, N., Aruna, S. T., & Balani, K. (2017). Carbon nanotubes stabilize high temperature phase and toughen Al2O3 -based thermal barrier coatings. *Composites. Part B, Engineering, 124*, 76–87. doi:10.1016/j.compositesb.2017.05.032

Ariharan, S., Wangaskar, B., Xavier, V., Venkateswaran, T., & Balani, K. (2019). Process induced alignment of carbon nanotube decreases longitudinal thermal conductivity of Al2O3 based porous composites. *Ceramics International, 45*(15), 18951–18964. doi:10.1016/j.ceramint.2019.06.133

Arya, S. K., Saha, S., Ramirez-Vick, J. E., Gupta, V., Bhansali, S., & Singh, S. P. (2012). Recent advances in ZnO nanostructures and thin films for biosensor applications [Review]. *Analytica Chimica Acta, 737*, 1–21. doi:10.1016/j.aca.2012.05.048 PMID:22769031

Assadi, H., Gärtner, T., Stoltenhoff, H. K., & Kreye, H. (2003). Bonding mechanism in cold gas spraying. *Acta Materialia, 51*(15), 4379–4394. doi:10.1016/S1359-6454(03)00274-X

Assadi, H., Schmidt, T., Richter, H., Kliemann, J.-O., Binder, K., Gärtner, F., Klassen, T., & Kreye, H. (2011). On Parameter Selection in Cold Spraying. *Journal of Thermal Spray Technology., 20*(6), 1161–1176. doi:10.100711666-011-9662-9

Atrens, A., Shi, Z., Mehreen, S. U., Chen, X., Johnston, S., Song, G.-L., Chen, X., & Pan, F. (2022). Corrosion of Mg Alloys. Encyclopedia of Materials: Metals and Allloys, 1, 46-74.

Ayache, J., Beaunier, L., Boumendil, J., Ehret, G., & Laub, D. (2010). *Preliminary preparation techniques sample preparation handbook for transmission electron microscopy: Techniques*. Springer New York.

Aziz, F., & Ismail, A. F. (2015). Spray coating methods for polymer solar cells fabrication: A review. *Materials Science in Semiconductor Processing, 39*, 416–425. doi:10.1016/j.mssp.2015.05.019

Babu, P. S., Basu, B., & Sundararajan, G. (2011). The influence of erodent hardness on the erosion behavior of detonation sprayed WC-12Co coatings. *Wear, 270*(11), 903–913. doi:10.1016/j.wear.2011.02.019

Backman, D. G., & Williams, J. C. (1992). Advanced Materials for Aircraft Engine Applications. *Science, 255*(5048), 1082–1087. doi:10.1126cience.255.5048.1082 PMID:17817782

Badisch, E., Katsich, C., Winkelmann, H., Franek, F., & Roy, M. (2010). Wear Behaviour of Hardfaced FeCrC Alloy And Austenitic Steel Under 2-Body and 3-Body Conditions at Elevated Temperature. *Tribology International, 43*(7), 1234–1244. doi:10.1016/j.triboint.2010.01.008

Baharin, A. F. S., Ghazali, M. J., & Wahab, J. A. (2016). Laser surface texturing and its contribution to friction and wear reduction: A brief review. *Industrial Lubrication and Tribology, 68*(1), 57–66. doi:10.1108/ILT-05-2015-0067

Bahri, A., Guermazi, N., Elleuch, K., & Ürgen, M. (2015). Tribological performance of TiN coatings deposited on 304L stainless steel used for olive-oil extraction. *Wear, 342–343*, 77–84.

Bajpai, S., Nisar, A., Sharma, R. K., Schwarz, U. D., Balani, K., & Datye, A. (2021). Effect of fictive temperature on tribological properties of $Zr_{44}Ti_{11}Cu_{10}Ni_{10}Be_{25}$ bulk metallic glasses. *Wear, 486*, 204075. doi:10.1016/j.wear.2021.204075

Baker, H., & Avedesian, M. M. (1999). *Magnesium and Magnesium Alloys*. ASM International.

Bakhsheshi-Rad, H. R., Hamzah, E., Ismail, A. F., Kasiri-Asgarani, M., Daroonparvar, M., Parham, S., Iqbal, N., & Medraj, M. (2016). Novel bilayered nanostructured SiO2/Ag-FHAp coating on biodegradable magnesium alloy for biomedical applications. *Ceramics International, 42*(10), 11941–11950. doi:10.1016/j.ceramint.2016.04.119

Bakshi, S. R., Balani, K., & Agarwal, A. (2008a). Thermal Conductivity of Plasma-Sprayed Aluminum Oxide-Multi-walled Carbon Nanotube Composites. *Journal of the American Ceramic Society, 91*(3), 942–947. doi:10.1111/j.1551-2916.2007.02081.x

Balaceanu, M., Braic, M., Braic, V., & Pavelescu, G. (2005). Properties of arc plasma deposited TiCN/ZrCN superlattice coatings. *Surface and Coatings Technology, 200*(1–4), 1084–1087. doi:10.1016/j.surfcoat.2005.01.077

Balaceanu, M., Braic, V., Braic, M., Vladescu, A., Zoita, C. N., Grigorescu, C. E. A., Grigore, E., & Ripeanu, R. (2009). Characteristics of Ti-Nb, Ti-Zr and Ti-Al containing hydrogenated carbon nitride films. *Solid State Sciences, 11*(10), 1773–1777. doi:10.1016/j.solidstatesciences.2008.12.001

Balaceanu, M., Braic, V., Kiss, A., Zoita, C. N., Vladescu, A., Braic, M., Tudor, I., Popescu, A., Ripeanu, R., Logofatu, C., & Negrila, C. C. (2008). Characteristics of arc plasma deposited TiAlZrCN coatings. *Surface and Coatings Technology, 202*(16), 3981–3987. doi:10.1016/j.surfcoat.2008.02.005

Balaceanu, M., Parau, A. C., Braic, M., Vladescu, A., Luculescu, C. R., Logofatu, C., & Braic, V. (2015). Growth and characterization of arc evaporated TiSiC-Ni coatings. *Tribology Letters, 58*(43), 1–9. doi:10.100711249-015-0521-6

Balamurugan, G. M., Duraiselvam, M., & Anandakrishnan, V. (2012). Comparison of high temperature wear behaviour of plasma sprayed WC–Co coated and hard chromium plated AISI 304 austenitic stainless steel. *Materials & Design, 35*, 640–646. doi:10.1016/j.matdes.2011.10.012

Bala, N., Singh, H., Karthikeyan, J., & Prakash, S. (2013). Cold spray coating process for corrosion protection: A review. *Surface Engineering, 30*(6), 414–421. doi:10.1179/1743294413Y.0000000148

Balani, K., Bakshi, S. R., Chen, Y., Laha, T., & Agarwal, A. (2007). Role of powder treatment and carbon nanotube dispersion in the fracture toughening of plasma-sprayed aluminum oxide-carbon nanotube nanocomposite. *Journal of Nanoscience and Nanotechnology, 7*(10), 3553–3562. doi:10.1166/jnn.2007.851 PMID:18330173

Balani, K., Harimkar, S. P., Keshri, A., Chen, Y., Dahotre, N. B., & Agarwal, A. (2008). Multiscale wear of plasma-sprayed carbon-nanotube-reinforced aluminum oxide nanocomposite coating. *Acta Materialia, 56*(20), 5984–5994. doi:10.1016/j.actamat.2008.08.020

Baragetti, S., Lusvarghi, L., Bolelli, G., & Tordini, F. (2009). Fatigue behaviour of 2011-T6 aluminium alloy coated with PVD WC/C, PA-CVD DLC and PE-CVD SiOx coatings. *Surface and Coatings Technology, 203*(20), 3078–3087. doi:10.1016/j.surfcoat.2009.03.040

Barrau, O., Boher, C., Gras, R., & Rezai-Aria, F. (2003). Analysis of the Friction and Wear Behaviour of Hot Work Tool Steel for Forging. *Wear, 255*(7-12), 1444–1454. doi:10.1016/S0043-1648(03)00280-1

Barroqueiro, B., Andrade-Campos, A., Valente, R. A. F., & Neto, V. (2019). Metal additive manufacturing cycle in aerospace industry: A comprehensive review. *Journal of Manufacturing and Materials Processing*, *3*(3), 52. doi:10.3390/jmmp3030052

Barshilia, H. C. (2021). Surface modification technologies for aerospace and engineering applications: Current Trends. *Challenges and Future Prospects. Trans Indian Natl. Acad. Eng.*, *6*(2), 173–188. doi:10.100741403-021-00208-z

Barshilia, H. C., Ghosh, M., Shashidhara, Ramakrishna, R., & Rajam, K. S. (2010). Deposition and characterization of TiAlSiN nanocomposite coatings prepared by reactive pulsed direct current unbalanced magnetron sputtering. *Applied Surface Science*, *256*(21), 6420–6426. doi:10.1016/j.apsusc.2010.04.028

Bart, J. C. J., Gucciardi, E., & Cavallaro, S. (2013). Lubricants: Properties and Characteristics. In *Biolubricants: Science and Technology* (pp. 24–73). Woodhead Publishing. doi:10.1533/9780857096326.24

Basile, F., Benito, P., Fornasari, G., Monti, M., Scavetta, E., Tonelli, D., & Vaccari, A. (2010). A novel electrochemical route for the catalytic coating of metallic supports. *Studies in Surface Science and Catalysis*, *175*, 51–58. doi:10.1016/S0167-2991(10)75007-2

Bastwros, M., Kim, G.-Y., Zhu, C., Zhang, K., Wang, S., Tang, X., & Wang, X. (2014). Effect of ball milling on graphene reinforced Al6061 composite fabricated by semi-solid sintering. *Composites. Part B, Engineering*, *60*, 111–118. doi:10.1016/j.compositesb.2013.12.043

Basu, A., Samant, A., Harimkar, S., Majumdar, J. D., Manna, I., & Dahotre, N. B. (2008). Laser surface coating of Fe–Cr–Mo–Y–B–C bulk metallic glass composition on AISI 4140 steel. *Surface and Coatings Technology*, *202*(12), 2623–2631. doi:10.1016/j.surfcoat.2007.09.028

Basu, J., & Ranganathan, S. (2003). Bulk metallic glasses: A new class of engineering materials. *Sadhana*, *28*(3-4), 783–798. doi:10.1007/BF02706459

Baudet, E., Sergent, M., Němec, P., Cardinaud, C., Rinnert, E., Michel, K., Jouany, L., Bureau, B., & Nazabal, V. (2017). Experimental Design Approach for Deposition Optimization of RF Sputtered Chalcogenide Thin Films Devoted to Environmental Optical Sensors. *Scientific Reports*, *7*(1), 3500. doi:10.103841598-017-03678-w PMID:28615650

Baykara, M. Z., Vazirisereshk, M. R., & Martini, A. (2018). Emerging Superlubricity: A Review of the State of the Art and Perspectives on Future Research. *Applied Physics Reviews*, *5*(4), 041102. doi:10.1063/1.5051445

Bazyari, A., Khodadadi, A. A., Mamaghani, A. H., Beheshtian, J., Thompson, L. T., & Mortazavi, Y. (2016). Microporous Titania–Silica Nanocomposite Catalyst-Adsorbent for Ultra-Deep Oxidative Desulfurization. *Applied Catalysis B: Environmental*, *180*(January), 65–77. doi:10.1016/j.apcatb.2015.06.011

Bedi, T. S., Kumar, S., & Kumar, R. (2019). Corrosion performance of hydroxyapatite and hydroxyapatite/titanium bond coating for biomedical applications. *Materials Research Express*, *7*(1), 015402. doi:10.1088/2053-1591/ab5cc5

Beer, P., Djouadi, M. A., Marchal, R., Sokolowska, A., Lambertin, M., Czyzniewski, A., & Precht, W. (1999). Antiabrasive Coatings in a New Application-Wood Rotary Peeling Process. *Vacuum*, *53*(1–2), 363–366. doi:10.1016/S0042-207X(99)00110-4

Beltowska-Lehman, E., Indyka, P., Bigos, A., Szczerba, M. J., & Kot, M. (2015). Ni–W/ZrO_2 nanocomposites obtained by ultrasonic DC electrodeposition. *Materials & Design*, *80*, 1–11.

Ben Saada, F., & Elleuch, K. (2015). Damage of stainless steel components by olive paste. *Tribology Transactions*, *59*, 856–864.

Bergmann, C. P., & Vicenzi, J. (2011). *Protection against Erosive Wear Using Thermal Sprayed Cermet: A Review* (C. P. Bergmann & J. Vicenzi, Eds.). Springer Berlin Heidelberg. doi:10.1007/978-3-642-21987-0

Berman, D., Erdemir, A., & Sumant, A. V. (2014). Graphene: A New Emerging Lubricant. *Materials Today, 17*(1), 31–42. doi:10.1016/j.mattod.2013.12.003

Berman, D., Erdemir, A., & Sumant, A. V. (2018). Approaches for Achieving Superlubricity in Two-Dimensional Materials. *ACS Nano, 12*(3), 2122–2137. doi:10.1021/acsnano.7b09046 PMID:29522673

Berns, H., & Franc, S. (1997). Effect of Coarse Hard Particles on High Temperature Sliding Abrasion of New Metal Matrix Composites. *Wear, 203-204*, 608–614. doi:10.1016/S0043-1648(96)07427-3

Bhatt, J., & Murty, B. (2008). On the conditions for the synthesis of bulk metallic glasses by mechanical alloying. *Journal of Alloys and Compounds, 459*(1-2), 135–141. doi:10.1016/j.jallcom.2007.04.242

Bhosale, D. G., Prabhu, T. R., Rathod, W. S., Patil, M. A., & Rukhande, S. W. (2020). High temperature solid particle erosion behaviour of SS 316L and thermal sprayed WC-Cr_3C_2–Ni coatings. *Wear, 462–463*, 203520. doi:10.1016/j.wear.2020.203520

Bhosale, D. G., & Rathod, W. S. (2020a). Investigation on wear behaviour of SS 316L, atmospheric plasma and high velocity oxy-fuel sprayed WC-Cr_3C_2-Ni coatings for fracturing tools. *Surface and Coatings Technology, 390*, 125679. doi:10.1016/j.surfcoat.2020.125679

Bhosale, D. G., & Rathod, W. S. (2020b). Tribological behaviour of atmospheric plasma and high velocity oxy-fuel sprayed WC-Cr_3C_2-Ni coatings at elevated temperatures. *Ceramics International, 46*(8), 12373–12385. doi:10.1016/j.ceramint.2020.01.288

Bhosale, D. G., & Rathod, W. S. (2021). Tribo-behaviour of APS and HVOF sprayed WC–Cr_3C_2–Ni coatings for gears. *Surface Engineering, 37*(1), 80–90. doi:10.1080/02670844.2020.1742988

Bhosale, D. G., Rathod, W. S., Ghorpade, U. S., & Rukhande, S. W. (2020). Nickel alloy C-263 protection by WCCr3C2Ni coatings against high-temperature wear in nuclear applications. *Surfaces and Interfaces, 21*, 100689. doi:10.1016/j.surfin.2020.100689

Bhosale, D. G., Rathod, W. S., & Rukhande, S. W. (2019). Sliding wear behavior of high velocity oxy-fuel sprayed WC-Cr_3C_2-Ni coating for automotive applications. *Materials Today: Proceedings, 19*, 339–343. doi:10.1016/j.matpr.2019.07.609

Bhushan, B. (2002). Introduction to Tribology. John Wiley & Sons Inc.

Bhushan, B. (1996). *Tribology and Mechanics of Magnetic Storage Devices* (2nd ed.). Springer-Verlag. doi:10.1007/978-1-4612-2364-1

Bhushan, B. (2000). *Modern Tribology Handbook, Two Volume Set. Modern Tribology Handbook, Two Volume Set.* CRC Press. doi:10.1201/9780849377877

Bikramjit, B., & Kantesh, B. (2013). Advanced Structural Ceramics. *Journal of Chemical Information and Modeling, 53*(9).

Billard, A., Maury, F., & Aubry, P. (2018). Emerging processes for metallurgical coatings and thin films. Comptes Rendus Physique. *Elsevier Masson, 19*(8), 755–768. doi:10.1016/j.crhy.2018.10.005ff

Birol, Y. (2010). Response to Thermal Cycling of Plasma Nitrided Hot Work Tool Steel at Elevated Temperatures. *Surface and Coatings Technology, 205*(2), 597–602. doi:10.1016/j.surfcoat.2010.07.035

Bisht, H., Eun, H. T., Mehrtens, A., & Aegerter, M. A. (1999). Comparison of spray pyrolyzed FTO, ATO and ITO coatings for flat and bent glass substrates. *Thin Solid Films, 351*(1), 109–114. doi:10.1016/S0040-6090(99)00254-0

Bisquert, J., Zaban, A., Greenshtein, M., & Mora-Seró, I. (2004). Determination of rate constants for charge transfer and the distribution of semiconductor and electrolyte electronic energy levels in dye-sensitized solar cells by open-circuit photovoltage decay method. *Journal of the American Chemical Society*, *126*(41), 13550–13559. doi:10.1021/ja047311k PMID:15479112

Blau, P. J. (1992). Scale Effects in Sliding Friction: An Experimental Study, in Fundamentals of Friction: Macroscopic and Microscopic Processes. Kluwer Academic.

Blau, P. J. (2001). Friction and wear of a Zr-based amorphous metal alloy under dry and lubricated conditions. *Wear, 250*(1-12), 431-434.

Blau, P. J. (2010). Elevated temperature tribology of metallic materials. *Tribology International*, *43*(7), 1203–1208. doi:10.1016/j.triboint.2010.01.003

Blomgren, G. E. (2016). The development and future of lithium ion batteries. *Journal of the Electrochemical Society*, *164*(1), A5019–A5025. doi:10.1149/2.0251701jes

Bobzin, K., Bagcivan, N., Goebbels, N., Yilmaz, K., Hoehn, B.-R., Michaelis, K., & Hochmann, M. (2009). Lubricated PVD CrAlN and WC/C coatings for automotive applications. *Surface and Coatings Technology*, *204*(6), 1097–1101. doi:10.1016/j.surfcoat.2009.07.045

Boidi, G., Grützmacher, P., & Varga, M. (2021). *Tribological Performance of Random Sinter Pores vs. Deterministic Laser Surface Textures: An Experimental and Machine Learning Approach*. Friction. doi:10.5772/intechopen.100245

Bolelli, G., Berger, L.-M., Bonetti, M., & Lusvarghi, L. (2014). Comparative study of the dry sliding wear behaviour of HVOF-sprayed WC–(W, Cr) 2C–Ni and WC–CoCr hardmetal coatings. *Wear*, *309*(1–2), 96–111. doi:10.1016/j.wear.2013.11.001

Bolelli, G., Berger, L.-M., Börner, T., Koivuluoto, H., Lusvarghi, L., Lyphout, C., Markocsan, N., Matikainen, V., Nylén, P., Sassatelli, P., Trache, R., & Vuoristo, P. (2015). Tribology of HVOF-and HVAF-sprayed WC–10Co4Cr hardmetal coatings: A comparative assessment. *Surface and Coatings Technology*, *265*, 125–144. doi:10.1016/j.surfcoat.2015.01.048

Bolelli, G., Berger, L.-M., Börner, T., Koivuluoto, H., Matikainen, V., Lusvarghi, L., Lyphout, C., Markocsan, N., Nylén, P., Sassatelli, P., Trache, R., & Vuoristo, P. (2016). Sliding and abrasive wear behaviour of HVOF-and HVAF-sprayed Cr_3C_2–NiCr hardmetal coatings. *Wear*, *358*, 32–50. doi:10.1016/j.wear.2016.03.034

Bolelli, G., Hulka, I., Koivuluoto, H., Lusvarghi, L., Milanti, A., Niemi, K., & Vuoristo, P. (2014). Properties of WC–FeCrAl coatings manufactured by different high velocity thermal spray processes. *Surface and Coatings Technology*, *247*, 74–89. doi:10.1016/j.surfcoat.2014.03.021

Bolleddu, V., Racherla, V., & Bandyopadhyay, P. P. (2017). Characterization of air plasma-sprayed yttria-stabilized zirconia coatings deposited with nitrogen. *International Journal of Advanced Manufacturing Technology*, *90*(9–12), 3437–3449. doi:10.100700170-016-9613-1

Boon, D., & Frenken, J. (2019). Tunable Superlubricity of 2-Dimensional Materials. *Proceedings of the National Academy of Sciences of the United States of America*, *116*(49), 24386–24387. doi:10.1073/pnas.1918084116 PMID:31723046

Booth, M. J. (2014). Adaptive optical microscopy: The ongoing quest for a perfect image. *Light, Science & Applications*, *3*(4), e165–e165. doi:10.1038/lsa.2014.46

Bose, S. (2007). High temperature coatings. Elsevier. doi:10.1016/B978-075068252-7/50007-X

Bose, S. (2007). *High temperature coatings*. Elsevier Butterworth-Heinemann.

Bosio, A., Pasini, S., & Romeo, N. (2020). The history of photovoltaics with emphasis on CdTe solar cells and modules. *Coatings*, *10*(4), 522. doi:10.3390/coatings10040344

Bowden, F. P., Gregory, J. N., & Tabor, D. (1945). Lubrication of Metal Surfaces by Fatty Acids. *Nature*, *156*(3952), 97–101. doi:10.1038/156097a0

Braic, M., Balaceanu, M., Parau, A. C., Dinu, M., & Vladescu, A. (2015). Investigation of multilayered TiSiC/NiC protective coatings. *Vacuum*, *120*(A), 60–66. doi:10.1016/j.vacuum.2015.06.019

Braic, M., Balaceanu, M., Vladescu, A., Zoita, C. N., & Braic, V. (2011). Study of (Zr,Ti)CN, (Zr,Hf)CN and (Zr,Nb)CN films prepared by reactive magnetron sputtering. *Thin Solid Films*, *519*(12), 4092–4096. doi:10.1016/j.tsf.2011.01.375

Braic, M., Balaceanu, M., Vladescu, A., Zoita, C. N., & Braic, V. (2013). Deposition and characterization of multi-principal-element (CuSiTiYZr)C coatings. *Applied Surface Science*, *284*, 671–678. doi:10.1016/j.apsusc.2013.07.152

Braic, M., Braic, V., Balaceanu, M., Vladescu, A., Zoita, C. N., Lungu, C. P., Grigorescu, C. E. A., Grigore, E., & Logoftu, C. (2011). (Ti,Cr,Nb)CN coatings deposited on nitrided high-speed steel by cathodic arc method. *Surface and Coatings Technology*, *205*, S209–S213. doi:10.1016/j.surfcoat.2011.03.030

Braic, M., Braic, V., Balaceanu, M., Zoita, C. N., Kiss, A., Vladescu, A., Popescu, A., & Ripeanu, R. (2011). Structure and properties of Zr/ZrCN coatings deposited by cathodic arc method. *Materials Chemistry and Physics*, *126*(3), 818–825. doi:10.1016/j.matchemphys.2010.12.036

Braic, M., Braic, V., Balaceanu, M., Zoita, C. N., Vladescu, A., & Grigore, E. (2010). Characteristics of (TiAlCrNbY)C films deposited by reactive magnetron sputtering. *Surface and Coatings Technology*, *204*(12–13), 2010–2014. doi:10.1016/j.surfcoat.2009.10.049

Braic, M., Vladescu, A., Balaceanu, M., Luculescu, C., Padmanabhan, S. C., Constantin, L., Morris, M. A., Braic, V., Ana Grigorescu, C. E., Ionescu, P., Dracea, M. D., & Logofatu, C. (2017). A comparative study of the structural, mechanical and tribological characteristics of TiSiC-Cr coatings prepared in CH4 and C2H2 reactive atmosphere by cathodic vacuum arc. *Applied Surface Science*, *400*, 318–328. doi:10.1016/j.apsusc.2016.12.160

Braic, M., Vladescu, A., Parau, A. C., Pruncu, C. I., & Braic, V. (2020). Tribological properties of alloyed TiSi-stainless steel carbide coatings deposited by reactive cathodic arc method. *Wear*, *460–461*, 203456–203466. doi:10.1016/j.wear.2020.203456

Braic, V., Balaceanu, M., Braic, M., Vladescu, A., Panseri, S., & Russo, A. (2012). Characterization of multi-principal-element (TiZrNbHfTa)N and (TiZrNbHfTa)C coatings for biomedical applications. *Journal of the Mechanical Behavior of Biomedical Materials*, *10*, 197–205. doi:10.1016/j.jmbbm.2012.02.020

Braic, V., Parau, A. C., Pana, I., Braic, M., & Balaceanu, M. (2014). Effects of substrate temperature and carbon content on the structure and properties of (CrCuNbTiY)C multicomponent coatings. *Surface and Coatings Technology*, *258*, 996–1005. doi:10.1016/j.surfcoat.2014.07.044

Braic, V., Vladescu, A., Balaceanu, M., Luculescu, C. R., & Braic, M. (2012). Nanostructured multi-element (TiZrNbHfTa)N and (TiZrNbHfTa)C hard coatings. *Surface and Coatings Technology*, *211*, 117–121. doi:10.1016/j.surfcoat.2011.09.033

Bramowicz, M., Braic, L., Ak, F., Kulesza, S., Birlik, I., & Vladescu, A. (2016). Mechanical properties and fractal analysis of the surface texture of sputtered hydroxyapatite coatings. *Applied Surface Science*, *379*, 338–346. doi:10.1016/j.apsusc.2016.04.077

Braun, J. (2017). Additives. In T. Mang & W. Dresel (Eds.), *Lubricants and Lubrication* (3rd ed., pp. 119–144). Wiley-VCH. doi:10.1002/9783527645565.ch6

Brusciotti, F., Snihirova, D., Xue, H., Montemor, M. F., Lamaka, S., & Ferreira, M. G. S. (2013). Hybrid epoxy–silane coatings for improved corrosion protection of Mg alloy. *Corrosion Science*, *67*, 82–90. doi:10.1016/j.corsci.2012.10.013

Bull, S.J., & Berasetegui E. (2006). An overview of the potential of quantitative coating adhesion measurement by scratch testing. *Tribology Interface Engineering*, 99-114.

Bull, S. J. (2006). Physical Vapour Deposition Methods for Protection against Wear. *Surface Coatings for Protection Against Wear*, (January), 146–183. doi:10.1533/9781845691561.146

Bull, S. J., & Jones, A. M. (1996, January). Multilayer Coatings for Improved Performance. *Surface and Coatings Technology*, *78*(1-3), 173–184. doi:10.1016/0257-8972(94)02407-3

Bull, S. J., Rickerby, D. S., Matthews, A., Leyland, A., Pace, R., & Valli, J. (1988). The use of scratch adhesion testing for the determination of interfacial adhesion: The importance of frictional drag. *Surface and Coatings Technology*, *36*(1–2), 503–517. doi:10.1016/0257-8972(88)90178-8

Bunshah & Rointan. (1994). *Handbook of deposition technologies for films and coatings science, technology and applications* (2nd ed.). Noyes Publications.

Burnett, P. J., & Rickerby, D. S. (1987). The relationship between hardness and scratch adhession. *Thin Solid Films*, *154*(1–2), 403–416. doi:10.1016/0040-6090(87)90382-8

Calka, A., & Radlinski, A. (1991). Universal high performance ball-milling device and its application for mechanical alloying. *Materials Science and Engineering A*, *134*, 1350–1353. doi:10.1016/0921-5093(91)90989-Z

Cao, X. Q., Vassen, R., & Stoever, D. (2004). Ceramic Materials for Thermal Barrier Coatings. *Journal of the European Ceramic Society*, *24*(1), 1–10. doi:10.1016/S0955-2219(03)00129-8

Capa, M., Tamer, M., Gulmez, T., & Bodur, C. T. (2000). Life Enhancement of Hot-Forging Dies by Plasma-Nitriding. *Turk J Engin Environ Sci.*, *24*, 111–117.

Carreño, F., Gude, M. R., Calvo, S., Rodriguez de la Fuente, O., & Carmona, N. (2020). Design and development of icephobic coatings based on sol-gel/modified polyurethane paints. *Materials Today. Communications*, *25*, 101616. Advance online publication. doi:10.1016/j.mtcomm.2020.101616

Carruthers, W. D., van Roode, M., Becher, P. F., Ferber, M. K., & Pollinger, J. (2002). Advances in the Development of Silicon Nitride and Other Ceramics. *Turbo Expo 2002*, 67–76. doi:10.1115/GT2002-30504

Carvalho, O., Buciumeanu, M., Madeira, S., Soares, D., Silva, F. S., & Miranda, G. (2016). Mechanisms governing the mechanical behavior of an AlSi–CNTs–SiCp hybrid composite. *Composites. Part B, Engineering*, *90*, 443–449. doi:10.1016/j.compositesb.2016.01.032

Cashon, E. P. (1975). Wear resistant coatings applied by the detonation gun. *Tribology International*, *8*(3), 111–115. doi:10.1016/0301-679X(75)90028-6

Cerc Korošec, R., Felicijan, M., Žener, B., Pompe, M., Dražić, G., Padežnik Gomilšek, J., Pihlar, B., & Bukovec, P. (2017). The role of thermal analysis in optimization of electrochromic effect of nickel oxide thin films, prepared by the sol-gel method: Part III. *Thermochimica Acta*, *655*, 344–350. doi:10.1016/j.tca.2017.07.010

Cernuschi, F., Lorenzoni, L., Capelli, S., Guardamagna, C., Karger, M., Vaßen, R., von Niessen, K., Markocsan, N., Menuey, J., & Giolli, C. (2011). Solid particle erosion of thermal spray and physical vapour deposition thermal barrier coatings. *Wear*, *271*(11–12), 2909–2918. doi:10.1016/j.wear.2011.06.013

Chakradhar, B. T. (2012). Studies on the evaluation of ice adhesion strength of different hydrophobic and superhydrophobic coatings. *National Aerospace Laboratories, PD SE 1224*, 1–30.

Chandrasekar, M. S., & Pushpavanam, M. (2008). Pulse and pulse reverse plating – conceptual, advantages and applications. *Electrochimica Acta*, *53*(8), 3313–3322. doi:10.1016/j.electacta.2007.11.054

Chang, B.-Y., & Park, S.-M. (2010). Electrochemical impedance spectroscopy. *Annual Review of Analytical Chemistry (Palo Alto, Calif.)*, *3*(1), 207–229. doi:10.1146/annurev.anchem.012809.102211 PMID:20636040

Chang, Y. N., & Wei, F. I. (1989). High temperature Oxidation of Low Alloy Steels. *Journal of Materials Science*, *24*(1), 14–22. doi:10.1007/BF00660927

Charitidis, C. A. (2010). Nanomechanical and Nanotribological Properties of Carbon-Based Thin Films: A Review. *International Journal of Refractory Metals & Hard Materials*, *28*(1), 51–70. doi:10.1016/j.ijrmhm.2009.08.003

Chattopadhyay, R. (2004). Advanced thermally assisted surface engineering processes. Kluwer Academic Publishers.

Chavda, M. R., Dave, D. P., Chauhan, K. V., & Rawal, S. K. (2016). Tribological Characterization of TiN Coatings Prepared by Sputtering. *Procedia Technology*, *23*(January), 36–41. doi:10.1016/j.protcy.2016.03.070

Chen, Q., Yu, M., Cao, K., & Chen, H. (2022). Thermal conductivity and wear resistance of cold sprayed Cu-ceramic phase composite coating. *Surface and Coatings Technology*.

Chen,, W.W., & Gao, W. (2011). Microstructures and properties of sol-enhanced nanostructured metal-oxide composite coatings. *Progress in Material Science*, *21*, 355–362.

Chen, C., Xie, Y., Yan, X., Huang, R., Kuang, M., Ma, W., Zhao, R., Wang, J., Liu, M., Ren, R., & Liao, H. (2019). Cold Sprayed WC Reinforced Maraging Steel 300 Composites: Microstructure Characterization and Mechanical Properties. *Journal of Alloys and Compounds*, *785*, 499–511. doi:10.1016/j.jallcom.2019.01.135

Cheng, J., Feng, Y., Yan, C., Hu, X., Li, R., & Liang, X. (2020). Development and characterization of Al-based amorphous coating. *JOM*, *72*(2), 745–753. doi:10.100711837-019-03966-y

Cheng, Y. H., Browne, T., Heckerman, B., Bowman, C., Gorokhovsky, V., & Meletis, E. I. (2010). Mechanical and Tribological Properties of TiN/Ti Multilayer Coating. *Surface and Coatings Technology*, *205*(1), 146–151. doi:10.1016/j.surfcoat.2010.06.023

Chen, H. (1980). Glassy metals. *Reports on Progress in Physics*, *43*(4), 353–432. doi:10.1088/0034-4885/43/4/001

Chen, H., Ding, C., Zhang, P., La, P., & Lee, S. W. (2003). Wear of plasma-sprayed nanostructured zirconia coatings against stainless steel under distilled-water conditions. *Surface and Coatings Technology*, *173*(2–3), 144–149. doi:10.1016/S0257-8972(03)00311-6

Chen, H., Zhang, Y., & Ding, C. (2002). Tribological properties of nanostructured zirconia coatings deposited by plasma spraying. *Wear*, *253*(7–8), 885–893. doi:10.1016/S0043-1648(02)00221-1

Chen, L. B. (2006). Yttria-stabilized zirconia thermal barrier coatings— A review. *Surface Review and Letters*, *13*(05), 535–544. doi:10.1142/S0218625X06008670

Chen, L., Yang, M. C., Song, C. F., Yu, B. J., & Qian, L. M. (2013). Is 2 Nm DLC Coating Enough to Resist the Nanowear of Silicon. *Wear*, *302*(1–2), 909–917. doi:10.1016/j.wear.2013.01.088

Chen, M. (2011). A brief overview of bulk metallic glasses. *NPG Asia Materials*, *3*(9), 82–90. doi:10.1038/asiamat.2011.30

Chen, X., Gu, L., Zou, B., Wang, Y., & Cao, X. (2012). New functionally graded thermal barrier coating system based on LaMgAl11O19/YSZ prepared by air plasma spraying. *Surface and Coatings Technology*, *206*(8–9), 2265–2274. doi:10.1016/j.surfcoat.2011.09.076

Chen, X., He, M. Y., Spitsberg, I., Fleck, N. A., Hutchinson, J. W., & Evans, A. G. (2004). Mechanisms governing the high temperature erosion of thermal barrier coatings. *Wear*, *256*(7–8), 735–746. doi:10.1016/S0043-1648(03)00446-0

Chen, X., Wang, R., Yao, N., Evans, A. G., Hutchinson, J. W., & Bruce, R. W. (2003). Foreign object damage in a thermal barrier system: Mechanisms and simulations. *Materials Science and Engineering A*, *352*(1–2), 221–231. doi:10.1016/S0921-5093(02)00905-X

Chenxi, L., Narendra, S., Austin, A., Bernard, A., Thomas, E., & Christopher, J. (2019). Tribological properties of wear-resistant coatings obtained by cold gas dynamic spray. *International Journal of Heat and Mass Transfer*, *129*, 1161–1171.

Chenxi, L., Singh, N., Andrews, A., Olson, B., Schwartzentruber, T., & Hogan, C. Jr. (2019). Mass, momentum, and energy transfer in supersonic aerosol deposition processes. *International Journal of Heat and Mass Transfer*, *129*, 1161–1171. doi:10.1016/j.ijheatmasstransfer.2018.10.028

Chen, Y., Balani, K., & Agarwal, A. (2007). Modified Eshelby tensor modeling for elastic property prediction of carbon nanotube reinforced ceramic nanocomposites. *Applied Physics Letters*, *91*(3), 31903. doi:10.1063/1.2756360

Chen, Y., Samant, A., Balani, K., Dahotre, N. B., & Agarwal, A. (2009). Effect of laser melting on plasma-sprayed aluminum oxide coatings reinforced with carbon nanotubes. *Applied Physics. A, Materials Science & Processing*, *94*(4), 861–870. doi:10.100700339-008-4990-4

Chen, Z., Turedi, B., Alsalloum, A. Y., Yang, C., Zheng, X., Gereige, I., AlSaggaf, A., Mohammed, O. F., & Bakr, O. M. (2019). Single-crystal $MAPbI_3$ perovskite solar cells exceeding 21% power conversion efficiency. *ACS Energy Letters*, *4*(6), 1258–1259. doi:10.1021/acsenergylett.9b00847

Chi, L. (2010). Nanotechnology: Vol. 8. *Nanostructured Surfaces*. John Wiley & Sons.

Cho, D.-H., Kim, J.-S., Kwon, S.-H., Lee, C., & Lee, Y.-Z. (2013). Evaluation of Hexagonal Boron Nitride Nano-Sheets as a Lubricant Additive in Water. *Wear*, *302*(1-2), 981–986. doi:10.1016/j.wear.2012.12.059

Choi, W. S., Heo, J., Chung, I., & Hong, B. (2005). The Effect of RF Power on Tribological Properties of the Diamond-like Carbon Films. *Thin Solid Films*, *475*(1–2), 287–290. doi:10.1016/j.tsf.2004.07.033

Chowdhurya, M. A., & Nuruzzamanb, D. M. (2013). Experimental Investigation on Friction and Wear Properties of Different Steel Materials. *Tribology in Industry.*, *35*, 42–50.

Chowdhury, S., & Laugier, M. T. (2004). Non-Contact AFM with a Nanoindentation Technique for Measuring the Mechanical of Thin Films. *Nanotechnology*, *15*(8), 1017–1022. doi:10.1088/0957-4484/15/8/027

Chowdhury, S., Sumita, U., Islam, A., & Bedja, I. (2014). Importance of policy for energy system transformation: Diffusion of PV technology in Japan and Germany. *Energy Policy*, *68*, 285–293. doi:10.1016/j.enpol.2014.01.023

Choy, K. L. (2003). Chemical vapour deposition of coatings. *Progress in Materials Science*, *48*(2), 57–170. doi:10.1016/S0079-6425(01)00009-3

Chromik, R. R., Alidokht, S. A., Shockley, J. M., & Zhang, Y. (2018). *Tribological coatings prepared by cold spray*. Cold-Spray Coatings. doi:10.1007/978-3-319-67183-3_11

Chueva, T., Dyakonova, N., Molokanov, V., & Sviridova, T. (2007). Bulk amorphous alloy $Fe_{72}Al_5Ga_2C_6B_4P_{10}Si_1$ produced by mechanical alloying. *Journal of Alloys and Compounds*, *434*, 327–332. doi:10.1016/j.jallcom.2006.08.308

Chu, J. P., Greene, J., Jang, J. S., Huang, J., Shen, Y.-L., Liaw, P. K., Yokoyama, Y., Inoue, A., & Nieh, T. (2012). Bendable bulk metallic glass: Effects of a thin, adhesive, strong, and ductile coating. *Acta Materialia*, *60*(6-7), 3226–3238. doi:10.1016/j.actamat.2012.02.037

Chun-Hong, C., & Hideo, A. (2007). Temperature dependence of mechanical properties of aluminum titanate ceramics. *Journal of the European Ceramic Society*, *27*(1), 13–18. doi:10.1016/j.jeurceramsoc.2006.04.182

Clarke, D. R., & Phillpot, S. R. (2005). Thermal barrier coating materials. *Materials Today*, *8*(6), 22–29. doi:10.1016/S1369-7021(05)70934-2

Conner, R., Dandliker, R., & Johnson, W. L. (1998). Mechanical properties of tungsten and steel fiber reinforced $Zr_{41.25}Ti_{13.75}Cu_{12.5}Ni_{10}Be_{22.5}$ metallic glass matrix composites. *Acta Materialia*, *46*(17), 6089–6102. doi:10.1016/S1359-6454(98)00275-4

Constantin, L. R., Parau, A. C., Balaceanu, M., Dinu, M., & Vladescu, A. (2019). Corrosion and tribological behaviour in a 3.5% NaCl solution of vacuum arc deposited ZrCN and Zr–Cr–Si–C–N coatings. *Proceedings of the Institution of Mechanical Engineers. Part J, Journal of Engineering Tribology*, *233*(1), 158–169. doi:10.1177/1350650118774132

Cooke, T. F. (1991). Inorganic Fibers-A Literature Review. *Journal of the American Ceramic Society*, *74*(12), 2959–2978. doi:10.1111/j.1151-2916.1991.tb04289.x

Corengia, P., Ybarra, G., & Moina, C. (2005). Microstructural AISI 4140 Low-alloy Steel. *Surface and Coatings Technology*, *200*, 2391–2397. doi:10.1016/j.surfcoat.2005.01.060

Cornuault, P.-H., Colas, G., Lenain, A., Daudin, R., & Gravier, S. (2020). On the diversity of accommodation mechanisms in the tribology of Bulk Metallic Glasses. *Tribology International*, *141*, 105957. doi:10.1016/j.triboint.2019.105957

Cotrut, C. M., Balaceanu, M., Titorencu, I., Braic, V., & Braic, M. (2012). ZrNbCN thin films as protective layers in biomedical applications. *Surface and Coatings Technology*, *211*, 57–61. doi:10.1016/j.surfcoat.2011.08.016

Cotrut, C. M., Braic, V., Balaceanu, M., Titorencu, I., Braic, M., & Parau, A. C. (2013). Corrosion resistance, mechanical properties and biocompatibility of Hf-containing ZrCN coatings. *Thin Solid Films*, *538*, 48–55. doi:10.1016/j.tsf.2012.12.100

Covington, A. K. (1994). Terminology and conventions for microelectronic ion-selective field effect transistor devices in electrochemistry (IUPAC Recommendations 1994). *Pure and Applied Chemistry*, *66*(3), 565–569. doi:10.1351/pac199466030565

Crowe, C., Sommerfeld, M., & Tsuji, Y. (1998). *Multiphase Flows with Droplets and Particles*. CRC Press.

Cui, T., Wang, J., Guan, R., Chen, L., & Qiu, G. (2007). Microstructures and Properties of Thermal Barrier Coatings Plasma-Sprayed by Nanostructured Zirconia. *Journal of Iron and Steel Research International*, *14*(5), 116–120. doi:10.1016/S1006-706X(08)60063-1

Cummins, G., & Desmulliez, M. P. Y. (2012). Inkjet printing of conductive materials: A review. *Circuit World*, *38*(4), 193–213. doi:10.1108/03056121211280413

Czichos, H. (2009). *Tribology: a systems approach to the science and technology of friction, lubrication, and wear* (Vol. 1). Elsevier.

Dai, W., Kwon, S. H., Wang, Q., & Liu, J. (2018). Influence of Frequency and C2H2 Flow on Growth Properties of Diamond-like Carbon Coatings with AlCrSi Co-Doping Deposited Using a Reactive High Power Impulse Magnetron Sputtering. *Thin Solid Films*, *647*(February), 26–32. doi:10.1016/j.tsf.2017.12.016

Dalisa, A. L. (1977). Electrophoretic display technology. *IEEE Transactions on Electron Devices*, *24*(7), 827–834. doi:10.1109/T-ED.1977.18838

Das, J., Tang, M. B., Kim, K. B., Theissmann, R., Baier, F., Wang, W. H., & Eckert, J. (2005). "Work-hardenable" ductile bulk metallic glass. *Physical Review Letters*, *94*(20), 205501. doi:10.1103/PhysRevLett.94.205501 PMID:16090260

Das, K., Joseph, A., Ghosh, M., & Mukherjee, S. (2016). Microstructure and Wear Behaviour of Pulsed Plasma Nitrided AISI H13 Tool Steel. *Canad. Meta. Quar.*, *55*(4), 402–408.

Das, T. K., & Prusty, S. (2012). Review on conducting polymers and their applications. *Polymer-Plastics Technology and Engineering*, *51*(14), 1487–1500. doi:10.1080/03602559.2012.710697

De Pellegrin, D. V., Torrance, A. A., & Haran, E. (2009). Wear mechanisms and scale effects in two-body abrasion. *Wear*, *266*(1–2), 13–20. doi:10.1016/j.wear.2008.05.015

Dearnley, P. A., Neville, A., Turner, S., Scheibe, H. J., Tietema, R., Tap, R., Stüber, M., Hovsepian, P., Layyous, A., & Stenbom, B. (2010). Coatings Tribology Drivers for High Density Plasma Technologies. *Surface Engineering*, *26*(1–2), 80–96. doi:10.1179/174329409X451218

del Angel-López, D., Domínguez-Crespo, M. A., Torres-Huerta, A. M., Flores-Vela, A., Andraca-Adame, J., & Dorantes-Rosales, H. (2013). Analysis of degradation process during the incorporation of ZrO2:SiO2 ceramic nanostructures into polyurethane coatings for the corrosion protection of carbon steel. *Journal of Materials Science*, *48*(3), 1067–1084. doi:10.100710853-012-6839-7

Delprete, C., & Razavyka, A. (2018). Piston Dynamics, Lubrication and Tribological Performance Evaluation: A Review. *International Journal of Engine Research*, *21*(15), 725–741. doi:10.1177/1468087418787610

Deng, Y.-L., Lee, J.-W., Lou, B.-S., Duh, J.-G., Chu, J. P., & Jang, J. S.-C. (2014). The fabrication and property evaluation of Zr–Ti–B–Si thin film metallic glass materials. *Surface and Coatings Technology*, *259*, 115–122. doi:10.1016/j.surfcoat.2014.03.026

Derelizade, K., Rincon, A., Venturi, F., Wellman, R. G., Kholobystov, A., & Hussain, T. (2022). High temperature (900 °C) sliding wear of CrNiAlCY coatings deposited by high velocity oxy fuel thermal spray. *Surface and Coatings Technology*, *432*, 128063. doi:10.1016/j.surfcoat.2021.128063

Deshpande, M., & Altan, T. (2011). Selection of Die Materials and Surface Treatments for Increasing Die Life in Hot and Warm Forging. FIA tech Conference, 1-23.

Di Girolamo, G., & Serra, E. (2015). Thermally sprayed nanostructured coatings for anti-wear and TBC applications. In *Anti-Abrasive Nanocoatings* (pp. 513–541). Elsevier. doi:10.1016/B978-0-85709-211-3.00020-0

Di Maggio, R., Scardi, P., & Tomasi, A. (1990). Characterization of Ceria Stabilized Zirconia Coatings on Metal Substrates Materials Research Society symposia proceedings. *Materials Research Society*, *180*, 481. doi:10.1557/PROC-180-481

Dinelli, M., Fabbri, E., & Bondioli, F. (2011a). TiO2–SiO2 hard coating on polycarbonate substrate by microwave assisted sol–gel technique. *Journal of Sol-Gel Science and Technology*, *58*(2), 463–469. doi:10.100710971-011-2413-z

Ding, X. Z., & Zeng, X. T. (2005). Structural, Mechanical and Tribological Properties of CrAlN Coatings Deposited by Reactive Unbalanced Magnetron Sputtering. *Surface and Coatings Technology*, *200*(5–6), 1372–1376. doi:10.1016/j.surfcoat.2005.08.072

Dinh, K. N., Liang, Q., Du, C.-F., Zhao, J., Tok, A. I. Y., Mao, H., & Yan, Q. (2019). Nanostructured metallic transition metal carbides, nitrides, phosphides, and borides for energy storage and conversion. *Nano Today*, *25*, 99–121. doi:10.1016/j.nantod.2019.02.008

Dinu, M., Cojocaru, M., Braic, V., Tarcolea, M., Braic, M., Miculescu, F., Vladescu, A., & Cotrut, C. M. (2015). Improvement of the tribological performance in corrosive environment of CoCr alloy by TiSiON coatings. *Applied Surface Science*, *332*, 295–299. doi:10.1016/j.apsusc.2015.01.221

Dinu, M., Ivanova, A. A., Surmeneva, M. A., Braic, M., Tyurin, A. I., Braic, V., Surmenev, R. A., & Vladescu, A. (2017). Tribological behaviour of RF-magnetron sputter deposited hydroxyapatite coatings in physiological solution. *Ceramics International*, *43*(9), 6858–6867. doi:10.1016/j.ceramint.2017.02.106

Djafer, A. Z. A., Saoula, N., Madaoui, N., & Zerizer, A. (2014). Deposition and characterization of titanium carbide thin films by magnetron sputtering using Ti and TiC targets. *Applied Surface Science*, *312*, 57–62. doi:10.1016/j.apsusc.2014.05.084

Dobrzański, L. A., Pakuła, D., & Hajduczek, E. (2004). Structure and Properties of the Multi-Component TiAlSiN Coatings Obtained in the PVD Process in the Nitride Tool Ceramics. *Journal of Materials Processing Technology*, *157–158*, 331–340. doi:10.1016/j.jmatprotec.2004.09.052

Donaldson, C., & Snedeker, R. S. (1971). A Study of Free Jet Impingement, Part 1: Mean Properties of Free and Impinging Jets. *Journal of Fluid Mechanics*, *45*(2), 281–319. doi:10.1017/S0022112071000053

Dong, M., Zhu, Y., Duan, J., Wang, C., Guo, W., Li, J., & Wang, L. (2021). Understanding wear mechanisms of TiSiCN/Zr(C)N coatings at elevated temperatures. *Materials Characterization*, *180*, 111411. doi:10.1016/j.matchar.2021.111411

Donnet, C. (1998). *Recent Progress on the Tribology of Doped Diamond-like and Carbon Alloy Coatings: A Review. In Surface and Coatings Technology*. Elsevier. doi:10.1016/S0257-8972(97)00611-7

Dorfman, M. R. (2018). Thermal Spray Coatings. In *Handbook of Environmental Degradation of Materials* (pp. 469–488). Elsevier. doi:10.1016/B978-0-323-52472-8.00023-X

Dorfman, M. R., & Sharma, A. (2013). Challenges and strategies for growth of thermal spray markets: The six-pillar plan. *Journal of Thermal Spray Technology*, *22*(5), 559–563. doi:10.100711666-013-9906-y

Dorozhkin, S. V. (2002). A review on the dissolution of calcium apatites. *Progress in Crystal Growth and Characterization of Materials*, *44*(1), 45–61. doi:10.1016/S0960-8974(02)00004-9

Dorozhkin, S. V. (2009). Calcium Orthophosphates in Nature, Biology and Medicine. *Materials (Basel)*, *2*(2), 399–498. doi:10.3390/ma2020399

dos Santos, F. C., Harb, S., Menu, M.-J., Turq, V., Pulcinelli, S. H., Santilli, C., & Hammer, P. (2015). On the structure of high performance anticorrosive PMMA–siloxane–silica hybrid coatings. *RSC Advances*, *5*(129), 106754–106763. doi:10.1039/C5RA20885H

Dosta, S., Cinca, N., Vilardell, A., Cano, I., & Guilemany, J. M. (2017). Cold Spray Coatings for Biomedical Applications. In Cold Spray Coatings: Recent Trends and Future Perspectives. Springer International Publishing.

Ducheyne, P. (2017). *Comprehensive Biomaterials* (2nd ed.). Elsevier.

Du, H., Hau, W., & Liu, J. (2015). Influence of process variable on the quality of detonation gun sprayed WC-Co coatings. *Materials Science and Engineering*, *408*(1-2), 202–210. doi:10.1016/j.msea.2005.08.008

Du, L., Xu, B., Dong, S., Zhang, W., Zhang, J., Yang, H., & Wang, H. (2008). Sliding wear behavior of the supersonic plasma sprayed WC–Co coating in oil containing sand. *Surface and Coatings Technology*, *202*(15), 3709–3714. doi:10.1016/j.surfcoat.2008.01.009

Dwivedi, N., Kumar, S., & Malik, H. K. (2012). Superhard Behaviour, Low Residual Stress, and Unique Structure in Diamond-like Carbon Films by Simple Bilayer Approach. *Journal of Applied Physics*, *112*(2), 023518. Advance online publication. doi:10.1063/1.4739287

Dyett, B., & Lamb, R. (2016). Correlating Material Properties with the Wear Behavior of Sol–Gel Derived Superhydrophobic Films. *Advanced Materials Interfaces*, *3*(13), 1500680. doi:10.1002/admi.201500680

Dykhuizen, D. C., & Smith, M. F. (1998). Gas Dynamic Principles of Cold Spray. *Journal of Thermal Spray Technology*, *7*(2), 205–212. doi:10.1361/105996398770350945

Dykhuizen, R. C., Smith, M. F., Gilmore, D. L., Neiser, R. A., Jiang, X., & Sampath, S. (1999). Impact of High Velocity Cold Spray Particles. *Journal of Thermal Spray Technology*, *8*(4), 559–564. doi:10.1361/105996399770350250

Eakins, E., Jayaseelan, D. D., & Lee, W. E. (2011). Toward Oxidation Resistant Zrb_2-SiC Ultra High Temperature Ceramics. *Metallurgical and Materials Transactions. A, Physical Metallurgy and Materials Science*, *42A*(4), 878–887. doi:10.100711661-010-0540-8

Eaton, H. E., & Novak, R. C. (1987). Particulate erosion of plasma-sprayed porous ceramic. *Surface and Coatings Technology*, *30*(1), 41–50. doi:10.1016/0257-8972(87)90006-5

Ebrahimzadeh, I., & Ashrafizadeh, F. (2014). High Temperature Wear and Frictional Properties of Duplex-Treated Tool Steel Sliding Against a Two Phase Brass. *Ceramics International*, *40*(10), 16429–16439. doi:10.1016/j.ceramint.2014.07.151

Eckert, J., Das, J., Pauly, S., & Duhamel, C. (2007). Mechanical properties of bulk metallic glasses and composites. *Journal of Materials Research*, *22*(2), 285–301. doi:10.1557/jmr.2007.0050

Edenhofer, B. (1973). Part 2. Industrial Applications of the Process (conclusion). *Trait. Therm.*, *73*, 31–43.

Elashnikov, R., Ulbrich, P., Vokatá, B., Pavlíčková, V. S., Švorčík, V., Lyutakov, O., & Rimpelová, S. (2021). Physically Switchable Antimicrobial Surfaces and Coatings: General Concept and Recent Achievements. *Nanomaterials (Basel, Switzerland)*, *11*(11), 3083. doi:10.3390/nano11113083 PMID:34835852

Emley, E. F. (1966). *Principles of Magnesium Technology*. Pergamon Press.

Erdemir, A., & Voevodin, A. A. (2010). Nanocomposite coatings for severe applications. In Handbook of deposition technologies for films and coatings. William Andrew Publishing.

Erdemir, A. (2000). A crystal-chemical approach to lubrication by solid oxides. *Tribology Letters*, *8*(2), 97–102. doi:10.1023/A:1019183101329

Erdemir, A. (2005). A crystal chemical approach to the formulation of self-lubricating nanocomposite coatings. *Surface and Coatings Technology*, *200*(5–6), 1792–1796. doi:10.1016/j.surfcoat.2005.08.054

Erdemir, A., & Donnet, C. (2006). *Tribology of Diamond-like Carbon Films: Recent Progress and Future Prospects. Journal of Physics D: Applied Physics*. doi:10.1088/0022-3727/39/18/R01

Erfanmanesh, M., Shoja-Razavi, R., Abdollah-Pour, H., Mohammadian-Semnani, H., Barekat, M., & Hashemi, S. H. (2019). Friction and wear behavior of laser cladded WC-Co and Ni/WC-Co deposits at high temperature. *International Journal of Refractory Metals & Hard Materials*, *81*, 137–148. doi:10.1016/j.ijrmhm.2019.02.025

Fabbri, P., Messori, M., Toselli, M., Veronesi, P., Rocha, J., & Pilati, F. (2009). Enhancing the scratch resistance of polycarbonate with poly(ethylene oxide)-silica hybrid coatings. *Advances in Polymer Technology*, *27*(2), 117–126. doi:10.1002/adv.20122

Fang, W., Cho, T. Y., Yoon, J. H., Song, K. O., Hur, S. K., Youn, S. J., & Chun, H. G. (2009). Processing optimization, surface properties and wear behavior of HVOF spraying WC–CrC–Ni coating. *Journal of Materials Processing Technology, 209*(7), 3561–3567. doi:10.1016/j.jmatprotec.2008.08.024

Fan, L. S., & Zhu, C. (1998). *Principles of Gas-Solid Flows*. Cambridge University Press. doi:10.1017/CBO9780511530142

Fan, P., & Zhang, G. (2020). Study on process optimization of WC-Co50 cermet composite coating by laser cladding. *International Journal of Refractory Metals & Hard Materials, 87*, 105133. doi:10.1016/j.ijrmhm.2019.105133

Fattah-alhosseini, A., & Chaharmahali, R. (2021). Enhancing corrosion and wear performance of PEO coatings on Mg alloys using graphene and graphene oxide additions: A review. *FlatChem, 27*, 100241. doi:10.1016/j.flatc.2021.100241

Fauchais, P. (2004). Understanding plasma spraying. *Journal of Physics. D, Applied Physics, 37*(9), R86–R108. doi:10.1088/0022-3727/37/9/R02

Fauchais, P. L., Heberlein, J. V. R., & Boulos, M. I. (2014). Introduction. In *Thermal Spray Fundamentals*. Springer. doi:10.1007/978-0-387-68991-3_1

Fauchais, P., Coudert, J. F., Vardelle, M., Vardelle, A., & Denoirjean, A. (1992). Diagnostics of thermal spraying plasma jets. *J. Thermal Spray Technology, 1*(2), 117–128. doi:10.1007/BF02659011

Fauchais, P., & Vardelle, A. (2000). Heat, Mass and Momentum Transfer in Coating Formation by Plasma Spraying. *International Journal of Thermal Sciences, 39*(9–11), 852–870. doi:10.1016/S1290-0729(00)01195-9

Federici, M., Menapace, C., Moscatelli, A., Gialanella, S., & Straffelini, G. (2017). Pin-on-disc study of a friction material dry sliding against HVOF coated discs at room temperature and 300 C. *Tribology International, 115*, 89–99. doi:10.1016/j.triboint.2017.05.030

Felde, N., Coriand, L., Duparré, A., & Tünnermann, A. (2016). Wear- Resistant Nanostructured Sol- Gel Coatings for Functional Applications. *Journal of Coating Science and Technology, 3*(3), 100–108. doi:10.6000/2369-3355.2016.03.03.1

Feng, Y., Zhu, S., Wang, L., Chang, L., Yan, B., Song, X., & Guan, S. (2017). Characterization and corrosion property of nano-rod-like HA on fluoride coating supported on Mg-Zn-Ca alloy. *Bioactive Materials, 2*(2), 63–70. doi:10.1016/j.bioactmat.2017.05.001 PMID:29744413

Fernandes, F. A. P., Heck, S. C., Picone, C. A., & Casteletti, L. C. (2020). On the wear and corrosion of plasma nitrided AISI H13. *Surface and Coatings Technology, 381*.

Ferrari, A. C., & Basko, D. M. (2013). Raman spectroscopy as a versatile tool for studying the properties of graphene. *Nature Nanotechnology, 8*(4), 235–246. doi:10.1038/nnano.2013.46 PMID:23552117

Ferris, C. D., & Rose, D. R. (1972). An operational amplifier 4-electrode impedance bridge for electrolyte measurements. *Medical & Biological Engineering, 10*(5), 647–654. doi:10.1007/BF02476082 PMID:5076429

Firouzeh, A., Ranjbar, K., Lari Baghal, S. M., Heidari Kaidan, A., & Mohemi, E. (2018). Failure assessment of ASTM A213-T12 superheater boiler tubes in a natural gas liquid plant. *Engineering Failure Analysis, 89*, 15–27. doi:10.1016/j.engfailanal.2018.03.005

Fischer, J. K., von Brüning, D. M., & Labhart, H. (1976). Light modulation by electrochromism. *Applied Optics, 15*(11), 2812–2816. doi:10.1364/AO.15.002812 PMID:20165493

Fleury, E., Lee, S., Ahn, H., Kim, W., & Kim, D. (2004). Tribological properties of bulk metallic glasses. *Materials Science and Engineering A, 375*, 276–279. doi:10.1016/j.msea.2003.10.065

Fnides, B., Yallese, M. A., & Aouici, H. (2008). Hard Turning of Hot Work Steel AISI H11: Evaluation of Cutting Pressures, Resulting Force and Temperature. *Mechanic., 72*(4), 59–63.

Forati, H. R., Amadeh, A., & Moradi, H. (2011). Wear Assessment of Plasma Nitrided AISI H11 steel. *Materials & Design, 32*(5), 2635–2643. doi:10.1016/j.matdes.2011.01.027

Fotovvati, Namdari, & Dehghanghadikolaei. (2019). On Coating Techniques for Surface Protection: A Review. *Journal of Manufacturing and Materials Processing, 28*(3), 28. . doi:10.3390/jmmp3010028

Frankel, R. S., & Aitken, D. W. (1970). Energy-Dispersive X-Ray Emission Spectroscopy. *Applied Spectroscopy, 24*(6), 557–566. doi:10.1366/000370270774372308

Frank, I. W., Tanenbaum, P. M., & van der Zande, A. M. (2007). Mechanical properties of suspended graphene sheets. *Journal of Vacuum Science & Technology B. Microelectronics and Nanometer Structures Processing, 25*, 2558–25610.

Franz, S., Rammelt, S., Scharnweber, D., & Simon, J. C. (2011). Immune responses to implants – A review of the implications for the design of immunomodulatory biomaterials. *Biomaterials, 32*(28), 6692–6709. doi:10.1016/j.biomaterials.2011.05.078 PMID:21715002

Fu, X.-Y., Kasai, T., Falk, M. L., & Rigney, D. (2001). Sliding behavior of metallic glass: Part I. Experimental investigations. *Wear, 250*(1-12), 409-419.

Gaard, A., Hallback, N., Krakhmalev, P., & Bergström, J. (2010). Temperature Effects on Adhesive Wear in Dry Sliding Contacts. *Wear, 268*(7-8), 968–975. doi:10.1016/j.wear.2009.12.007

Gadow, R., Friedrich, C., Killinger, A., & Floristán, M. (2010). Development of Atmospheric Plasma Sprayed Dielectric Ceramic Coatings for High Efficiency Tubular Ozone Generators. *Journal of Water Resource and Protection, 2*(9), 799–808. doi:10.4236/jwarp.2010.29094

Gammer, K., Stoiber, M., Wagner, J., Hutter, H., Kullmer, R., & Mitterer, C. (2004). Investigations on the Effects of Plasma-Assisted Pre-Treatment for Plasma-Assisted Chemical Vapour Deposition Tin Coatings on Tool Steel. *Thin Solid Films, 461*(2), 277–281. doi:10.1016/j.tsf.2004.02.013

Ganbavle, V., Bangi, U. K. H., Latthe, S. S., Mahadik, S. A., & Rao, A. V. (2011). Self-cleaning silica coatings on glass by single step sol-gel route. *Surface and Coatings Technology, 205*(23–24), 5338–5344. doi:10.1016/j.surfcoat.2011.05.055

Ganvir, A., Calinas, R. F., Markocsan, N., Curry, N., & Joshi, S. (2019). Experimental visualization of microstructure evolution during suspension plasma spraying of thermal barrier coatings. *Journal of the European Ceramic Society, 39*(2), 470–481. doi:10.1016/j.jeurceramsoc.2018.09.023

Gao, Y. (2011). Surface modification of TA2 pure titanium by low energy high current pulsed electron beam treatments. *Applied Surface Science, 257*(17), 7455–7460. doi:10.1016/j.apsusc.2011.03.005

Gao, Y., Huang, L., & Zhang, H. (2016). Study on anti-freezing functional design of phase change and temperature control composite bridge decks. *Construction & Building Materials, 122*, 714–720. doi:10.1016/j.conbuildmat.2016.06.065

Garcia, E., Lee, H., & Sampath, S. (2018). Phase and microstructure evolution in plasma sprayed Yb2Si2O7 coatings. *Journal of the European Ceramic Society, 39*(4), 1477–1486. doi:10.1016/j.jeurceramsoc.2018.11.018

Gawali, S. A., Mahadik, S. A., Pedraza, F., Bhosale, C. H., Pathan, H. M., & Jadkar, S. R. (2017). Synthesis of zinc oxide nanorods from chemical bath deposition at different pH solutions and impact on their surface properties. *Journal of Alloys and Compounds, 704*, 788–794. Advance online publication. doi:10.1016/j.jallcom.2017.01.228

Geiger, M., Merklein, M., & Lechler, J. (2008). Determination of Tribological Conditions within Hot Stamping. *Production Engineering, 2*(3), 269–276. doi:10.100711740-008-0110-8

Geng, Z., Hou, S., Shi, G., Duan, D., & Li, S. (2016). Tribological behaviour at various temperatures of WC-Co coatings prepared using different thermal spraying techniques. *Tribology International, 104*, 36–44. doi:10.1016/j.triboint.2016.08.025

Geng, Z., Li, S., Duan, D. L., & Liu, Y. (2015). Wear behaviour of WC–Co HVOF coatings at different temperatures in air and argon. *Wear, 330*, 348–353. doi:10.1016/j.wear.2015.01.035

Ghasemi, R., Shoja-Razavi, R., Mozafarinia, R., & Jamali, H. (2013). Comparison of microstructure and mechanical properties of plasma-sprayed nanostructured and conventional yttria stabilized zirconia thermal barrier coatings. *Ceramics International, 39*(8), 8805–8813. doi:10.1016/j.ceramint.2013.04.068

Ghelichi, R., & Guagliano, M. (2009). Coating by the Cold Spray Process: A State of the Art. *Frattura Ed Integrità Strutturale, 3*(8), 30–44. doi:10.3221/IGF-ESIS.08.03

Ghosh, S. K., Limaye, P. K., Bhattacharya, S., Soni, N. L., & Grover, A. K. (2007). Effect of Ni sublayer thickness on sliding wear characteristics of electrodeposited Ni/Cu multilayer coatings. *Surface and Coatings Technology, 201*(16-17), 7441–7448. doi:10.1016/j.surfcoat.2007.02.014

Ghosh, S. K., Limaye, P. K., Srivastava, C., & Tewari, R. (2010). Comparison of sliding wear behaviour of pulse electrodeposited Ni–Cu nanocrystalline alloys and Ni–Cu/Cu multilayers. *Transactions of the IMF, 88*, 158-162. 10.1179/174591910X12692576434653

Gibson, M. A., Mykulowycz, N. M., Shim, J., Fontana, R., Schmitt, P., Roberts, A., ... Bordeenithikasem, P. (2018). 3D printing metals like thermoplastics: Fused filament fabrication of metallic glasses. *Materials Today, 21*(7), 697–702. doi:10.1016/j.mattod.2018.07.001

Gilewicz, A., & Warcholinski, B. (2014). Tribological Properties of CrCN/CrN Multilayer Coatings. *Tribology International, 80*, 34–40. doi:10.1016/j.triboint.2014.06.012

Gill, B. J., & Tucker, R. C. (1986). Plasma spray coating processes. *Materials Science and Technology, 2*(3), 207–213. doi:10.1179/mst.1986.2.3.207

Gilmore, D. L., Dykhuizen, R. C., Neiser, R. A., Roemer, T. J., & Smith, M. F. (1999). Particle velocity and deposition efficiency in the cold spray process. *Journal of Thermal Spray Technology, 8*(4), 576–582. doi:10.1361/105996399770350278

Gomez-Navarro, C., Burgard, M., & Ken, K. (2008). Elastic properties of chemically derived single graphene sheets. *Nano Letters, 8*(7), 2045–2049.

Gong, H., Yu, C., Zhang, L., Xie, G., Gui, D., & Luo, J. (2020). Intelligent Lubricating Materials: A Review. *Composites. Part B, Engineering, 202*, 108450. doi:10.1016/j.compositesb.2020.108450

Gong, T., Yao, P., Zuo, X., Zhang, Z., Xiao, Y., Zhao, L., Zhou, H., Deng, M., Wang, Q., & Zhong, A. (2016). Influence of WC carbide particle size on the microstructure and abrasive wear behavior of WC–10Co–4Cr coatings for aircraft landing gear. *Wear, 362*, 135–145. doi:10.1016/j.wear.2016.05.022

González, M. A., Rodriguez, E. T., Mojardín, E., Jimenez, O., Guillen, H., & Ibarra, J. (2017). Study of the erosive wear behaviour of cryogenically and tempered WC-CoCr coating deposited by HVOF. *Wear, 376-377*, 595–607. doi:10.1016/j.wear.2016.12.061

Goodenough, J. B. (2018). How we made the Li-ion rechargeable battery. *Nature Electronics, 1*(3), 204–204. doi:10.103841928-018-0048-6

Gray, J. E., & Luan, B. (2002). Protective coatings on magnesium and its alloys-a critical review. *Journal of Alloys and Compounds*, *336*(1-2), 1–2, 88–113. doi:10.1016/S0925-8388(01)01899-0

Greene, J. E. (2017). Review Article: Tracing the Recorded History of Thin-Film Sputter Deposition: From the 1800s to 2017. *Journal of Vacuum Science & Technology A: Vacuum, Surfaces, and Films*, *35*(5), 05C204. . doi:10.1116/1.4998940

Greer, A. L., & Ma, E. (2007). Bulk metallic glasses: At the cutting edge of metals research. *MRS Bulletin*, *32*(8), 611–619. doi:10.1557/mrs2007.121

Gröber, U., Schmidt, J., & Kisters, K. (2015). Magnesium in Prevention and Therapy. *Nutrients*, *7*(9), 8199–8226. doi:10.3390/nu7095388 PMID:26404370

Groot, R. D., & Rabone, K. L. (2001). Mesoscopic simulation of cell membrane damage, morphology change and rupture by nonionic surfactants. *Biophysical Journal*, *81*(2), 725–736. doi:10.1016/S0006-3495(01)75737-2 PMID:11463621

Grujicic, M., Cao, G., & Gersten, B. (2004). Atomic-scale computations of the lattice contribution to thermal conductivity of single-walled carbon nanotubes. *Materials Science and Engineering B*, *107*(2), 204–216. doi:10.1016/j.mseb.2003.11.012

Grujicic, M., Zhao, C. L., Tong, C., DeRosset, W. S., & Helfritch, D. (2004). Analysis of the impact velocity of powder particles in the cold-gas dynamic-spray process. *Materials Science and Engineering A*, *368*(1), 222–230. doi:10.1016/j.msea.2003.10.312

Grum, J., & Bergant, Z. (2008). The optimization of powder flame-spraying parameters using a Taguchi method. *International Journal of Microstructure & Materials Properties*, *3*(4/5), 4–5. doi:10.1504/IJMMP.2008.022043

Guan, P., Zhou, L., Yu, Z., Sun, Y., Liu, Y., Wu, F., Jiang, Y., & Chu, D. (2020). Recent progress of surface coating on cathode materials for high-performance lithium-ion batteries. *Journal of Energy Chemistry*, *43*, 220–235. doi:10.1016/j.jechem.2019.08.022

Guan, R.-G., Johnson, I., Cui, T., Zhao, T., Zhao, Z.-Y., Li, X., & Liu, H. (2012). Electrodeposition of hydroxyapatite coating on Mg-4.0Zn-1.0Ca-0.6Zr alloy and in vitro evaluation of degradation, hemolysis, and cytotoxicity. *Journal of Biomedical Research Part A*, *100*(4), 999–1015. doi:10.1002/jbm.a.34042 PMID:22307984

Guddla, G. T., Gandi, S., Ambadipudi, S., & Ravuri, B. R. (2021). Design of Ni-based Bulk Metallic Glasses with Improved Mechanical and Corrosion Properties. *Current Applied Science and Technology*, 115-131.

Guddla, G., Katta, V., Gandi, S., Ambadipudi, S., & Ravuri, B. (2021). The role of titanium content in $(Ni_{75}Cr_{15}Si_{10})_{100-x}Ti_x$ bulk metallic glass systems to elevate mechanical and corrosion properties. *Phase Transitions*, *94*(10), 679–690. doi:10.1080/01411594.2021.1957103

Guglielmi, M. (1997). Sol-Gel Coatings on Metals. *Journal of Sol-Gel Science and Technology*, *8*(1-3), 443–449. doi:10.1007/BF02436880

Guo, D., Wang, Y., Fernandez, R., Zhao, L., & Jodoin, B. (2021). Cold spray for production of in-situ nanocrystalline MCrAlY coatings – Part I: Process analysis and microstructure characterization. *Surface and Coatings Technology*, *409*, 126854. doi:10.1016/j.surfcoat.2021.126854

Guo, H., Xu, H., Bi, X., & Gong, S. (2002). Preparation of Al2O3–YSZ composite coating by EB-PVD. *Materials Science and Engineering A*, *325*(1–2), 389–393. doi:10.1016/S0921-5093(01)01476-9

Guo, S., & Liu, C. T. (2011). Phase stability in high entropy alloys: Formation of solid-solution phase or amorphous phase. *Progress in Natural Science*, *21*(6), 433–446. doi:10.1016/S1002-0071(12)60080-X

Guruvenket, S., Li, D., Klemberg-Sapieha, J. E., Martinu, L., & Szpunar, J. (2009). Mechanical and tribological properties of duplex treated TiN, nc-TiN/a-SiNx and nc-TiCN/a-SiCN coatings deposited on 410 low alloy stainless steel. *Surface and Coatings Technology*, *203*(19), 2905–2911. doi:10.1016/j.surfcoat.2009.03.009

Gyftou, P., Stroumbouli, M., Pavlatou, E. A., Asimidis, P., & Spyrellis, N. (2004). Tribological study of Ni matrix composite coatings containing nano and micro SiC particles. *Electrochimica Acta*, *50*(23), 4544–4550.

Hajlaoui, K., Yavari, A., LeMoulec, A., Botta, W., Vaughan, F., Das, J., Greer, A. L., & Kvick, Å. (2007). Plasticity induced by nanoparticle dispersions in bulk metallic glasses. *Journal of Non-Crystalline Solids*, *353*(3), 327–331. doi:10.1016/j.jnoncrysol.2006.10.011

Hallab & Jacobs. (2020). Orthopedic Applications. *Biomaterials Science,* 1079–1118.

Ham, G.-S., Kim, K.-W., Cho, G.-S., Kim, C. P., & Lee, K.-A. (2020). Fabrication, microstructure and wear properties of novel Fe-Mo-Cr-CB metallic glass coating layers manufactured by various thermal spray processes. *Materials & Design*, *195*, 109043. doi:10.1016/j.matdes.2020.109043

Hamid, A., Kreye, G. F., & Klassen, T. (2016). Cold spraying: A material perspective. *Acta Materialia, 116*, 382-407. doi:10.1016/j.actamat.2016.06.034

Hamid, Z. A., & Ghayad, I. M. (2002). Characteristics of electrodeposition of Ni-polyethylene composite coatings. *Materials Letters*, *53*(4-5), 238–243.

Hammer, P., dos Santos, F. C., Cerrutti, B. M., Pulcinelli, S. H., & Santilli, C. (2012). Highly corrosion resistant siloxane-polymethyl methacrylate hybrid coatings. *Journal of Sol-Gel Science and Technology*, *63*(2), 266–274. doi:10.100710971-011-2672-8

Hammer, P., Schiavetto, M. G., dos Santos, F. C., Benedetti, A. V., Pulcinelli, S. H., & Santilli, C. V. (2010). Improvement of the corrosion resistance of polysiloxane hybrid coatings by cerium doping. *Journal of Non-Crystalline Solids*, *356*(44–49), 2606–2612. doi:10.1016/j.jnoncrysol.2010.05.013

Harb, S. V., Cerrutti, B. M., Pulcinelli, S. H., Santilli, C. V., & Hammer, P. (2015). Siloxane–PMMA hybrid anti-corrosion coatings reinforced by lignin. *Surface and Coatings Technology*, *275*, 9–16. doi:10.1016/j.surfcoat.2015.05.002

Harb, S. v., Trentin, A., Torrico, R. F. O., Pulcinelli, S. H., Santilli, C. v., & Hammer, P. (2017). Organic-Inorganic Hybrid Coatings for Corrosion Protection of Metallic Surfaces. In *New Technologies in Protective Coatings*. InTech. doi:10.5772/67909

Harb, S., dos Santos, F. C., Caetano, B. L., Pulcinelli, S. H., Santilli, C., & Hammer, P. (2015). Structural properties of cerium doped siloxane–PMMA hybrid coatings with high anticorrosive performance. *RSC Advances*, *5*(20), 15414–15424. doi:10.1039/C4RA15974H

Harb, S., Pulcinelli, S. H., Santilli, C., Knowles, K. M., & Hammer, P. (2016). A Comparative Study on Graphene Oxide and Carbon Nanotube Reinforcement of PMMA-Siloxane-Silica Anticorrosive Coatings. *ACS Applied Materials & Interfaces*, *8*(25), 16339–16350. doi:10.1021/acsami.6b04780 PMID:27266403

Hardell, J., Kassfeldt, E., & Prakash, B. (2007). Friction and Wear Behaviour of High Strength Boron Steel at Elevated Temperatures of up to 800°C. *Wear*, *264*(9-10), 788–799. doi:10.1016/j.wear.2006.12.077

Hardell, J., & Prakash, B. (2008). High-Temperature Friction and Wear Behaviour of Different Tool Steels During Sliding Against Al-Si Coated High-Strength Steel. *Tribology International*, *41*(7), 663–671. doi:10.1016/j.triboint.2007.07.013

Hardell, J., & Prakash, B. (2010). Tribological Performance of Surface Engineered Tool Steel at Elevated Temperatures. *International Journal of Refractory Metals & Hard Materials*, *28*(1), 106–114. doi:10.1016/j.ijrmhm.2009.07.009

Hardell, J., Prakash, B., & Steinhoff, K. (2009). High Temperature Tribological Studies on Surface Engineered Tool Steel and High Strength Boron Steel. *Steel Research International*, *80*, 665–670.

Hardwicke, C. U., & Lau, Y.-C. (2013). Advances in thermal spray coatings for gas turbines and energy generation: A review. *Journal of Thermal Spray Technology*, *22*(5), 564–576. doi:10.100711666-013-9904-0

Hardy, W. B., & Doubleday, I. (1922). Boundary Lubrication – The Paraffin Series. *Proceedings of the Royal Society of London. Series A, Containing Papers of a Mathematical and Physical Character*, *100*(707), 550–574. doi:10.1098/rspa.1922.0017

Hartwig, A. (2001). Role of magnesium in genomic stability. *Mutation Research*, *475*(1-2), 113–121. doi:10.1016/S0027-5107(01)00074-4 PMID:11295157

Hassanzadeh, M., Saremi, M., Valefi, Z., & Pakseresht, A. H. (2018). Investigation on Improved-Durability Thermal Barrier Coatings: An Overview of Nanostructured, Multilayered, and Self-Healing TBCs. *Production, Properties, and Applications of High Temperature Coatings*, 60–78.

Hassanzadeh, M., Saremi, M., Valefi, Z., & Pakseresht, A. H. (2018). Investigation on improved-durability thermal barrier coatings- an overview of nanostructured, multilayered and self-healing TBCs. In A. H. Pakseresht (Ed.), *Production, Properties, and Applications of High Temperature Coatings* (pp. 60–78). IGI Global. doi:10.4018/978-1-5225-4194-3.ch003

Hazra, S., Sabiruddin, K., & Bandyopadhyay, P. P. (2012). Plasma and HVOF sprayed WC–Co coatings as hard chrome replacement solution. *Surface Engineering*, *28*(1), 37–43. doi:10.1179/1743294410Y.0000000009

He, G., Bian, Z., & Chen, G. (2001). Corrosion behavior of a Zr-base bulk glassy alloy and its crystallized counterparts. *Materials Transactions*, *42*(6), 1109–1111. doi:10.2320/matertrans.42.1109

Heinrich, J., Busch, R., & Nonnenmacher, B. (2012). Processing of a bulk metallic glass forming alloy based on industrial grade Zr. *Intermetallics*, *25*, 1–4. doi:10.1016/j.intermet.2012.02.011

He, L., & Hassani, M. A. (2020). Review of the Mechanical and Tribological Behavior of Cold Spray Metal Matrix Composites. *Journal of Thermal Spray Technology*, *29*(7), 1565–1608. doi:10.100711666-020-01091-w

Helmersson, U., Lattemann, M., Böhlmark, J., Ehiasarian, A. P., & Gudmundsson, J. T. (2006). Ionized physical vapor deposition (IPVD): A review of technology and applications. *Thin Solid Films*, *513*(1-2), 1–24. doi:10.1016/j.tsf.2006.03.033

Hepplestone, S. P., Ciavarella, A. M., Janke, C., & Srivastava, G. P. (2006). Size and temperature dependence of the specific heat capacity of carbon nanotubes. *Surface Science*, *600*(18), 3633–3636. doi:10.1016/j.susc.2005.12.070

Herman, H., Sampath, S., & McCune, R. (2000). Thermal Spray: Current Status and Future Trends. *MRS Bulletin*, *25*(7), 17–25. doi:10.1557/mrs2000.119

Hernandez, S., Hardell, J., Courbon, C., Winkelmann, H., & Prakash, B. (2014). High Temperature Friction and Wear Mechanism Map For Tool Steel and Boron Steel Tribopair. *Tribology-Materials. Surfaces and Interfaces*, *82*(2), 74–84. doi:10.1179/1751584X13Y.0000000049

Higashino, T., & Imahori, H. (2015). Porphyrins as excellent dyes for dye-sensitized solar cells: Recent developments and insights. *Dalton Transactions (Cambridge, England)*, *44*(2), 448–463. doi:10.1039/C4DT02756F PMID:25381701

Higuera, H. V., Belzunce, V. J., Menéndez, C. A., & Martinez, P. S. (2001). High temperature erosion wear of flame and plasma-sprayed nickel–chromium coatings under simulated coal-fired boiler atmospheres. *Wear*, *247*(2), 214–222. doi:10.1016/S0043-1648(00)00540-8

Hirano, M., & Shinjo, K. (1990). Atomistic Locking and Friction. *Physical Review B: Condensed Matter, 41*(17), 11837–11851. doi:10.1103/PhysRevB.41.11837 PMID:9993633

Hitoki, G., Takata, T., Kondo, J. N., Hara, M., Kobayashi, H., & Domen, K. (2002). An oxynitride, TaON, as an efficient water oxidation photocatalyst under visible light irradiation ($\lambda \leq 500$ nm). *Chemical Communications (Cambridge), 2*(16), 1698–1699. doi:10.1039/B202393H PMID:12196955

Hod, O., Meyer, E., Zheng, Q., & Urbakh, M. (2018). Structural Superlubricity and Ultralow Friction across the Length Scales. *Nature, 563*(7732), 485–492. doi:10.103841586-018-0704-z PMID:30464268

Hofmann, D. C., Andersen, L. M., Kolodziejska, J., Roberts, S. N., Borgonia, J. P., Johnson, W. L., Vecchio, K. S., & Kennett, A. (2017). Optimizing bulk metallic glasses for robust, highly wear-resistant gears. *Advanced Engineering Materials, 19*(1), 1600541. doi:10.1002/adem.201600541

Holleck, H., & Schier, V. (1995a). Multilayer PVD Coatings for Wear Protection. *Surface and Coatings Technology, 76–77*(1–3 pt 1), 328–336. doi:10.1016/0257-8972(95)02555-3

Holmberg, K., & Erdemir, A. (2019). The Impact of Tribology on Energy Use and CO_2 Emission Globally and in Combustion Engine and Electric Cars. *Tribology International, 135*, 389–396. doi:10.1016/j.triboint.2019.03.024

Holubar, P., Jilek, M., & Sima, M. (2000). Present and Possible Future Applications of Superhard Nanocomposite Coatings. *Surface and Coatings Technology, 133–134*(November), 145–151. doi:10.1016/S0257-8972(00)00956-7

Hong, S., Wu, Y., Wang, B., Zhang, J., Zheng, Y., & Qiao, L. (2017). The effect of temperature on the dry sliding wear behavior of HVOF sprayed nanostructured WC-CoCr coatings. *Ceramics International, 43*(1), 458–462. doi:10.1016/j.ceramint.2016.09.180

Hong, S., Wu, Y., Wang, B., Zheng, Y., Gao, W., & Li, G. (2014). High-velocity oxygen-fuel spray parameter optimization of nanostructured WC–10Co–4Cr coatings and sliding wear behavior of the optimized coating. *Materials & Design, 55*, 286–291. doi:10.1016/j.matdes.2013.10.002

Hoornaert, T., Hua, Z. K., & Zhang, J. H. (2009). Hard Wear-Resistant Coatings: A Review. *Advances in Tribology*, 774–779. doi:10.1007/978-3-642-03653-8_257

Hou, G.-L., Zhou, H.-D., An, Y.-L., Liu, G., Chen, J.-M., & Chen, J. (2011). Microstructure and high-temperature friction and wear behavior of WC-(W, Cr) 2C-Ni coating prepared by high velocity oxy-fuel spraying. *Surface and Coatings Technology, 206*(1), 82–94. doi:10.1016/j.surfcoat.2011.06.047

Howarth, G. A., & Manock, H. L. (1997). Water-borne polyurethane dispersions and their use in functional coatings. *Surface Coatings International, 80*(7), 324–328. doi:10.1007/BF02692680

Hua, N., Chen, W., Wang, Q., Guo, Q., Huang, Y., & Zhang, T. (2018). Tribocorrosion behaviors of a biodegradable $Mg_{65}Zn_{30}Ca_5$ bulk metallic glass for potential biomedical implant applications. *Journal of Alloys and Compounds, 745*, 111–120. doi:10.1016/j.jallcom.2018.02.138

Huang, R. Z., & Fukanuma, H. (2009) The Influence of Spray Conditions on Deposition Characteristics of Aluminum Coatings in Cold Spraying. *Thermal Spray 2009: Proceedings of the International Thermal Spray Conference.*

Huang, B., Zhang, C., Zhang, G., & Liao, H. (2019). Wear and corrosion resistant performance of thermal-sprayed Fe-based amorphous coatings: A review. *Surface and Coatings Technology, 377*, 124896. doi:10.1016/j.surfcoat.2019.124896

Huang, S., Mutyala, K. C., Sumant, A. V., & Mochalin, V. N. (2021). Achieving Superlubricity with 2D Transition Metal Carbides (MXenes) and MXene/graphene Coatings. *Materials Today Advances, 9*, 100133. doi:10.1016/j.mtadv.2021.100133

Hu, J. J., Muratore, C., & Voevodin, A. A. (2007). Silver Diffusion and High-Temperature Lubrication Mechanisms of YSZ–Ag–Mo Based Nanocomposite Coatings. *Composites Science and Technology*, *67*(3–4), 336–347. doi:10.1016/j.compscitech.2006.09.008

Hulka, I., Utu, D., Serban, V. A., Negrea, P., Lukáč, F., & Chráska, T. (2020). Effect of Ti addition on microstructure and corrosion properties of laser cladded WC-Co/NiCrBSi (Ti) coatings. *Applied Surface Science*, *504*, 144349. doi:10.1016/j.apsusc.2019.144349

Hutchings, I. M. (1992). *Tribology-Friction and Wear of Engineering Materials*. Butterworth-Heinemann. doi:10.1016/0261-3069(92)90241-9

Hutchings, I., & Shipway, P. (2017). *Tribology: friction and wear of engineering materials*. Butterworth-Heinemann.

Hu, Y., Cai, C., Wang, Y., Yu, H., Zhou, Y., & Zhou, G. (2018). YSZ/NiCrAlY interface oxidation of APS thermal barrier coatings. *Corrosion Science*, *142*, 22–30. doi:10.1016/j.corsci.2018.06.035

Hwang, D. K., Moon, J. H., Shul, Y. G., Jung, K. T., Kim, D. H., & Lee, D. W. (2003). Scratch resistant and transparent UV-protective coating on polycarbonate. *Journal of Sol-Gel Science and Technology*, *26*(1–3), 783–787. doi:10.1023/A:1020774927773

Ichikawa, Y., Horiuchi, S., Ogawa, K., Oikawa, M., Tatsuki, T., & Yamazaki, H. (2017). Microstructural Change and Delamination Resistance of the Thermal Barrier Coatings with Cold Sprayed Ce-Content Bond Coats. *Journal of the Society of Materials Science, Japan*, *66*(2), 142–149. doi:10.2472/jsms.66.142

Ilieva, G. I. (2016). Erosion failure mechanisms in turbine stage with twisted rotor blade. *Engineering Failure Analysis*, *70*, 90–104. doi:10.1016/j.engfailanal.2016.07.008

Indira, K., Mudali, U. K., Nishimura, T., & Rajendran, N. (2015). A Review on TiO2 Nanotubes: Influence of Anodization Parameters, Formation Mechanism, Properties, Corrosion Behavior, and Biomedical Applications. *Journal of Bio- and Tribo-Corrosion*, *1*(4), 28. doi:10.100740735-015-0024-x

Inkson, B. J. (2016). 2 - Scanning electron microscopy (SEM) and transmission electron microscopy (TEM) for materials characterization. In G. Hübschen, I. Altpeter, R. Tschuncky, & H.-G. Herrmann (Eds.), *Materials Characterization Using Nondestructive Evaluation (NDE) Methods* (pp. 17–43). Woodhead Publishing. doi:10.1016/B978-0-08-100040-3.00002-X

Inman, I. A., Datta, P. K. M., Du, H. L., Gray, J. S. B., Pierzgalski, S., & Luo, Q. (2005). Studies of High Temperature Sliding Wear of Metallic Dissimilar Interfaces. *Tribology International*, *38*(9), 812–823. doi:10.1016/j.triboint.2005.02.007

Innocenzi, P. C., Guglielmi, M., & Gobin, M. (1992). Coating of metals by the sol-gel dip-coating method. *J. Europ. Ceram. Soc.*, *10*, 431. (92) 90018-9 doi:10.1016/0955-2219

Inoue, A., Shen, B., Koshiba, H., Kato, H., & Yavari, A. R. (2003). Cobalt-based bulk glassy alloy with ultrahigh strength and soft magnetic properties. *Nature Materials*, *2*(10), 661–663. doi:10.1038/nmat982 PMID:14502274

International, A. S. M. (1990). Author. *ASM Metal Handbook*, *1*, 439–444.

Inzelt, G. (2012). Applications of conducting polymers. In G. Inzelt (Ed.), *Conducting polymers: A new era in electrochemistry* (pp. 245–293). Springer Berlin Heidelberg. doi:10.1007/978-3-642-27621-7_7

Ishikawa, Y., Kuroda, S., Kawakita, J., Sakamoto, Y., & Takaya, M. (2007). Sliding wear properties of HVOF sprayed WC–20% Cr_3C_2–7% Ni cermet coatings. *Surface and Coatings Technology*, *201*(8), 4718–4727. doi:10.1016/j.surfcoat.2006.10.007

Izumi, K., Murakami, M., & Deguchi, T. (1989). Zirconia Coating on Stainless Steel Sheets from Organozirconium Compounds. *Journal of American Ceramic Society*, *72*(8), 1465–1468. doi:10.1111/j.1151-2916.1989.tb07677

J.R. Davis & Associates & ASM International. (2004). *Handbook of thermal spray technology*. ASM International. https://app.knovel.com/hotlink/toc/id:kpHTST0006/handbook-of-thermal

Jacobs, J. J., Skipor, A. K., Patterson, L. M., Hallab, N. J., Paprosky, W. G., Black, J., & Galante, J. O. (1998). Metal release in patients who have had a primary total hip arthroplasty. *Journal of Bone & Joint Surgery*, *80*(10), 1447–1458. doi:10.2106/00004623-199810000-00006 PMID:9801213

Jacobson, S., & Hogmark, S., (2005). *Tribologi-friction notning och smorjning. andra upplagan*. Academic Press.

Jadidi, M., Moghtadernejad, S., & Dolatabadi, A. (2015). A comprehensive review on fluid dynamics and transport of suspension/liquid droplets and particles in high-velocity oxygen-fuel (HVOF) thermal spray. *Coatings*, *5*(4), 576–645. doi:10.3390/coatings5040576

Jafari, M., Han, J.-C., Seol, J.-B., & Park, C.-G. (2017). Tribological properties of HVOF-sprayed WC-Co coatings deposited from Ni-plated powders at elevated temperature. *Surface and Coatings Technology*, *327*, 48–58. doi:10.1016/j.surfcoat.2017.08.026

Jafari, R., & Sadeghi, E. (2019). High-temperature corrosion performance of HVAF-sprayed NiCr, NiAl, and NiCrAlY coatings with alkali sulfate/chloride exposed to ambient air. *Corrosion Science*, *160*, 108066. doi:10.1016/j.corsci.2019.06.021

Jahodova, V., Ding, X. Z., Seng, D. H. L., Gulbinski, W., & Louda, P. (2013). Mechanical, tribological and corrosion properties of CrBN films deposited by combined direct current and radio frequency magnetron sputtering. *Thin Solid Films*, *544*, 335–340. doi:10.1016/j.tsf.2013.02.103

Jamesh, M., Kumar, S., & Sankara Narayanan, T. S. N. (2012). Electrodeposition of hydroxyapatite coating on magnesium for biomedical applications. *Journal of Coatings Technology and Research*, *9*(4), 495–502. doi:10.100711998-011-9382-6

Jang, C. S., Jeon, J. H., Song, P. K., Kang, M. C., & Kim, K. H. (2005). Synthesis and mechanical properties of TiAlCxN1-x coatings deposited by arc ion plating. *Surface and Coatings Technology*, *200*(5–6), 1501–1506. doi:10.1016/j.surfcoat.2005.08.065

Jang, H. D., Kim, S. K., Chang, H., Roh, K.-M., Choi, J.-W., & Huang, J. (2012). A glucose biosensor based on TiO2–Graphene composite. *Biosensors & Bioelectronics*, *38*(1), 184–188. doi:10.1016/j.bios.2012.05.033 PMID:22705409

Janos, B. Z., Lugscheider, E., & Remer, P. (1999). Effect of thermal aging on the erosion resistance of air plasma sprayed zirconia thermal barrier coating. *Surface and Coatings Technology*, *113*(3), 278–285. doi:10.1016/S0257-8972(99)00002-X

Jansson, U., & Lewin, E. (2013). Sputter deposition of transition-metal carbide films - A critical review from a chemical perspective. *Thin Solid Films*, *536*, 1–24. doi:10.1016/j.tsf.2013.02.019

Jansson, U., Lewin, E., Råsander, M., Eriksson, O., André, B., & Wiklund, U. (2011). Design of carbide-based nanocomposite thin films by selective alloying. *Surface and Coatings Technology*, *206*(4), 583–590. doi:10.1016/j.surfcoat.2010.06.017

Jhi, S.-H., Ihm, J., Louie, S. G., & Cohen, M. L. (1999). Electronic mechanism of hardness enhancement in transition-metal carbonitrides. *Nature*, *399*(6732), 132–134. doi:10.1038/20148

Jiang, J. Z., Hofmann, D., Jarvis, D. J., & Fecht, H. J. (2015). Low-density high-strength bulk metallic glasses and their composites: A Review. *Advanced Engineering Materials*, *17*(6), 761–780. doi:10.1002/adem.201400252

Jiang, J., Hao, J., Pang, X., Wang, P., & Liu, W. (2010). Structure and characteristics of amorphous (Ti,Si)-C:H films deposited by reactive magnetron sputtering. *Diamond and Related Materials*, *19*(10), 1172–1177. doi:10.1016/j.diamond.2010.05.005

Jiang, X., Song, J., Chen, S., Su, Y., Fan, H., Zhang, Y., & Hu, L. (2020). In-situ fabricated bulk metallic glass/graphite composites with a 3D lubricating layer: Tribological properties under dry sliding and in seawater. *Tribology International*, *148*, 106301. doi:10.1016/j.triboint.2020.106301

Jiang, X., Song, J., Su, Y., Fan, H., Zhang, Y., & Hu, L. (2019). Tribological behaviors of self–mated $Cu_{36}Zr_{48}Ag_8Al_8$ bulk metallic glass under H_2SO_4 conditions with different concentrations. *Tribology International*, *136*, 395–403. doi:10.1016/j.triboint.2019.04.006

Jia, Q., Zhou, Q., Ren, Y., Du, Y., Zhao, X., Wang, X.-Z., Wang, H., Beake, B. D., & Zhou, F. (2021). Tribological characteristics of Ti-based bulk metallic glass via deep cryogenic-cycling treatment. *Materials Characterization*, *179*, 111356. doi:10.1016/j.matchar.2021.111356

Ji, Z., Zhang, L., Xie, G., Xu, W., Guo, D., Luo, J., & Prakash, B. (2020). Mechanical and Tribological Properties of Nanocomposites incorporated with Two-Dimensional Materials. *Friction*, *8*(5), 813–846. doi:10.100740544-020-0401-4

Johnson, K. L. (2003). *Contact Mechanics* (204th ed.). Cambridge University Press.

Johnson, L. J. S., Rogström, L., Johansson, M. P., Odén, M., & Hultman, L. (2010). Microstructure evolution and age hardening in (Ti,Si)(C,N) thin films deposited by cathodic arc evaporation. *Thin Solid Films*, *519*(4), 1397–1403. doi:10.1016/j.tsf.2010.08.150

Johnson, W. L. (1996). Bulk metallic glasses—A new engineering material. *Current Opinion in Solid State and Materials Science*, *1*(3), 383–386. doi:10.1016/S1359-0286(96)80029-5

Jones, M., Cockburn, A., Sparkes, M., O'Niell, W., & Lupoi, R. (2014). Supersonic Laser Deposition of Tungsten. *International Manufacturing Science and Engineering Conference, 45806*, V001T03A022.

Jones, L. C., & Llewellyn, R. J. (2009). Sliding Abrasion Resistance Assessment of Metallic Materials for Elevated Temperature Mineral Processing Conditions. *Wear*, *267*(11), 2010–2017. doi:10.1016/j.wear.2009.06.023

Joshi, S. S., Katakam, S., Singh Arora, H., Mukherjee, S., & Dahotre, N. B. (2016). Amorphous coatings and surfaces on structural materials. *Critical Reviews in Solid State and Material Sciences*, *41*(1), 1–46. doi:10.1080/10408436.2015.1053602

Joshi, S., & Nylen, P. (2019). Advanced Coatings by Thermal Spray Processes. *Technologies*, *7*(79), 1–14. doi:10.3390/technologies7040079

Jun, C., & Fengyuan, Y. (2013). Corrosive wear performance of Monel K500 alloy in artificial seawater. *Tribology Transactions*, *56*(5), 848–856. doi:10.1080/10402004.2013.804967

Jun, Y., Kim, J., & Kang, M. G. (2007). A study of stainless steel-based dye-sensitized solar cells and modules. *Solar Energy Materials and Solar Cells*, *91*(9), 779–784. doi:10.1016/j.solmat.2007.01.007

Kabaya, T., & Iwabuchi, A. (1981). The Fretting Wear of 0.45% C Steel and Austenitic Stainless Steel from 20 to 650 °C in Air. *Wear*, *74*(2), 229–245. doi:10.1016/0043-1648(81)90165-4

Kahar, S., Singh, A., Vala, U., Desai, A., & Desai, S. (2020). Thermal sprayed coating using zinc: A review. *International Research Journal of Engineering and Technology*, *6*, 6497–6505.

Kalita, G., & Tanemura, M. (2017). Fundamentals of Chemical Vapor Deposited Graphene and Emerging Applications. In *Graphene Materials - Advanced Applications*. IntechOpen. Available from: https://www.intechopen.com/chapters/54395 doi:10.5772/67514

Kallawar, G. A., & Bhanvase, B. A. (2021). Nanomaterial-based photocatalytic membrane for organic pollutants removal. Handbook of Nanomaterials for Wastewater Treatment, 699–737. doi:10.1016/B978-0-12-821496-1.00007-6

Kallel, M., Antar, Z., Masseoud, M., Vesco, S., Barletta, M., & Elleuch, K. (2021). Comparative investigation of scratch resistance and tribological performance of Ni–B–TiO2 composite coatings prepared by conventional and novel processing methods. *Ceramics International, 47*(10) Part A, 14438-14454. doi:10.1016/j.ceramint.2021.02.023

Kallel, M., Masseoud, M., Antar, Z., Fridrici, V., Barletta, M., & Elleuch, K. (2021). Tribological properties of Ni-B-TiO_2 sol composite coating elaborated by sol-enhanced process: abrasive wear and impact wear. *Journal of Materials Research and Technology, 1*(3), 857 – 871.

Kallel, M., Zouch, F., Antar, Z., Bahri, A., & Elleuch, K. (2017). Hammer premature wear in mineral crushing process. *Tribology International, 115*, 493–505.

Kalluri, S., Yoon, M., Jo, M., Liu, H. K., Dou, S. X., Cho, J., & Guo, Z. (2017). Feasibility of cathode surface coating technology for high-energy lithium-ion and beyond-lithium-ion batteries. *Advanced Materials, 29*(48), 1605807. doi:10.1002/adma.201605807 PMID:28251710

Kannan & Orr. (2011). In vitro mechanical integrity of hydroxyapatite coated magnesium alloy. *Biomedical Materials, 6*(4), 11.

Kannan, M. B., & Singh Raman, R. K. (2008). In vitro degradation and mechanical integrity of calcium-containing; magnesium alloys in modified-simulated body fluid. *Biomaterials, 29*(15), 2306–2314. doi:10.1016/j.biomaterials.2008.02.003 PMID:18313746

Kant, S., Kumar, M., Chawla, V., & Singh, S. (2020). Study of high temperature oxidation behavior of wire arc sprayed coatings. *Materials Today: Proceedings, 21*(4), 1741–1748. doi:10.1016/j.matpr.2020.01.226

Kao, S.-Y., Lu, H.-C., Kung, C.-W., Chen, H.-W., Chang, T.-H., & Ho, K.-C. (2016). Thermally cured dual functional viologen-based all-in-one electrochromic devices with panchromatic modulation. *ACS Applied Materials & Interfaces, 8*(6), 4175–4184. doi:10.1021/acsami.5b11947 PMID:26807824

Karami, M. (1991). An Investigation of the Properties and Wear Behaviour of Plasma Nitrided Hot-Working Steel (H13). *Wear, 150*(1-2), 331–342. doi:10.1016/0043-1648(91)90327-Q

Karami, M., & Staines, A. (1989). An Evaluation of the Response of 722M24 Material to High Temperature Plasma Nitriding Treatments. *Heat Treatment Metals., 3*, 79–82.

Karamis, M. B., Baki, M., & Aydın, G. C. (2012). Sliding/rolling Wear Performance of Plasma Nitrided H11 Hot Working Steel. *Tribology International, 51*, 18–24. doi:10.1016/j.triboint.2012.02.005

Karthikeyan, J. (2007). The advantages and disadvantages of the cold spray coating process. In V. K. Champagne (Ed.), *Woodhead Publishing Series in Metals and Surface Engineering, The Cold Spray Materials Deposition Process* (pp. 62–71). Woodhead Publishing.

Kattimani, Kondaka, & Lingamaneni. (2016). Hydroxyapatite—Past, Present, and Future in Bone Regeneration. *Bone and Tissue Regeneration Insights, 7*, 9-19.

Kaur, M., Singh, H., & Prakash, S. (2012). High-temperature behaviour of a high-velocity oxy-fuel sprayed Cr3C2-NiCr coating. *Metallurgical and Materials Transactions. A, Physical Metallurgy and Materials Science*, *43*(8), 2979–2993. doi:10.100711661-012-1118-4

Kavale, M. S. M. S., Mahadik, S. A. A., Mahadik, D. B. B., Parale, V. G. G., Rao, A. V., Vhatkar, R. S. S., Wagh, P. B. B., & Gupta, S. C. C. (2012). Enrichment in hydrophobicity and scratch resistant properties of silica films on glass by grafted microporosity of the network. *Journal of Sol-Gel Science and Technology*, *64*(1), 9–16. doi:10.100710971-012-2822-7

Kazempour-Liacy, H., Abouali, S., & Akbari-Garakani, M. (2011). Failure analysis of a first stage gas turbine blade. *Engineering Failure Analysis*, *18*(1), 517–522. doi:10.1016/j.engfailanal.2010.09.040

Kazmanli, M. K., Ürgen, M., & Cakir, A. F. (2003). Effect of nitrogen pressure, bias voltage and substrate temperature on the phase structure of Mo–N coatings produced by cathodic arc PVD. *Surface and Coatings Technology*, *167*(1), 77–82. doi:10.1016/S0257-8972(02)00866-6

Kchaou, M., Alimi, A., Elleuch, R., & Desplanques, Y. (2013). Characterisation of Oxidation and Wear Oxidized Surfaces of H13 Steel/ Brass in Dry Sliding Conditions. *International Journal of Microstructural and Material Properties.*, *8*(4/5), 373–384. doi:10.1504/IJMMP.2013.057073

Kendrick, E. (2019). *CHAPTER 11 Advancements in manufacturing Future Lithium-ion Batteries*. The Royal Society of Chemistry.

Keskin, İ. Ç., Türemiş, M., Katı, M. İ., Kibar, R., Şirin, K., Çipiloğlu, M. A., Kuş, M., Büyükçelebi, S., & Çetin, A. (2017). The Radioluminescence and Optical Behaviour of Nanocomposites with CdSeS Quantum Dot. *Journal of Luminescence*, *185*(May), 48–54. doi:10.1016/j.jlumin.2016.12.048

Khachatryan, H., Kim, D.-J., Kim, M., & Kim, H.-K. (2018). Roll-to-Roll fabrication of ITO thin film for flexible optoelectronics applications: The role of post-annealing. *Materials Science in Semiconductor Processing*, *88*, 51–56. doi:10.1016/j.mssp.2018.07.033

Khanchaiyaphum, S., Saikaew, C., Wisitsoraat, A., & Surinphong, S. (2017). Wear Behaviours of Filtered Cathodic Arc Deposited TiN, TiAlSiN and TiCrAlSiN Coatings on AISI 316 Stainless Steel Fishing Net-Weaving Machine Components under Dry Soft-Sliding against Nylon Fibres. *Wear*, *390–391*, 146–154. doi:10.1016/j.wear.2017.07.018

Khan, M. A., Sundarrajan, S., Duraiselvam, M., Natarajan, S., & Kumar, A. S. (2017). Sliding wear behaviour of plasma sprayed coatings on nickel based superalloy. *Surface Engineering*, *33*(1), 35–41. doi:10.1179/1743294415Y.0000000087

Khan, M. K., Wang, Q. Y., & Fitzpatrick, M. E. (2016). 1 - Atomic force microscopy (AFM) for materials characterization. In G. Hübschen, I. Altpeter, R. Tschuncky, & H.-G. Herrmann (Eds.), *Materials Characterization Using Nondestructive Evaluation (NDE) Methods* (pp. 1–16). Woodhead Publishing. doi:10.1016/B978-0-08-100040-3.00001-8

Khan, M. M., Nemati, A., Rahman, Z. U., Shah, U. H., Asgar, H., & Haider, W. (2018). Recent advancements in bulk metallic glasses and their applications: A review. *Critical Reviews in Solid State and Material Sciences*, *43*(3), 233–268. doi:10.1080/10408436.2017.1358149

Khelifa, F., Druart, M.-E., Habibi, Y., Bénard, F., Leclère, P., Olivier, M., & Dubois, P. (2013). Sol–gel incorporation of silica nanofillers for tuning the anti-corrosion protection of acrylate-based coatings. *Progress in Organic Coatings*, *76*(5), 900–911. doi:10.1016/j.porgcoat.2013.02.005

Kholghi Eshkalak, S., Khatibzadeh, M., Kowsari, E., Chinnappan, A., Jayathilaka, W. A. D. M., & Ramakrishna, S. (2019). Overview of electronic ink and methods of production for use in electronic displays. *Optics & Laser Technology*, *117*, 38–51. doi:10.1016/j.optlastec.2019.04.003

Khun, N., Yu, H., Chong, Z., Tian, P., Tian, Y., Tor, S., & Liu, E. (2016). Mechanical and tribological properties of Zr-based bulk metallic glass for sports applications. *Materials & Design, 92*, 667–673. doi:10.1016/j.matdes.2015.12.050

Kim, Y. Y., Yang, T.-Y., Suhonen, R., Kemppainen, A., Hwang, K., Jeon, N. J., & Seo, J. (2020). Roll-to-roll gravure-printed flexible perovskite solar cells using eco-friendly antisolvent bathing with wide processing window. *Nature Communications, 11*(1), 5146. doi:10.103841467-020-18940-5 PMID:33051454

Kim, Y., Bae, D., Kim, W., & Kim, D. (2003). Glass forming ability and crystallization behavior of Ti-based amorphous alloys with high specific strength. *Journal of Non-Crystalline Solids, 325*(1-3), 242–250. doi:10.1016/S0022-3093(03)00327-2

Kim, Y., Kim, W., & Kim, D. (2004). A development of Ti-based bulk metallic glass. *Materials Science and Engineering A, 375*, 127–135. doi:10.1016/j.msea.2003.10.115

Kirkland, N. T., Lespagnol, J., Birbilis, N., & Staiger, M. P. (2010). A survey of bio-corrosion rates of magnesium alloys. *Corrosion Science, 52*(2), 287–291. doi:10.1016/j.corsci.2009.09.033

Kirkland, N. T., Staiger, M. P., Nisbet, D., Davies, C. H. J., & Birbilis, N. (2011). Performance-driven design of Bio-compatible Mg alloys. *Journal of the Minerals Metals & Materials Society, 63*(6), 28–34. doi:10.100711837-011-0089-z

Klement, W., Willens, R., & Duwez, P. (1960). Non-crystalline structure in solidified gold–silicon alloys. *Nature, 187*(4740), 869–870. doi:10.1038/187869b0

Klimczak, T., & Jonasson, M. (1994). Analysis of real contact area and change of surface texture on deep drawn steel sheets. *Wear, 179*(1-2), 129–135. doi:10.1016/0043-1648(94)90230-5

Klinkov, S. V., Kosarev, V. F., & Shikalov, V. S. (2019). Control of cold spray process by changing of nozzle setting angle. *AIP Conference Proceedings,* 21-25. doi:10.1063/1.5117382

Klinkov, S., & Kosarev, V. (2016). Fundamentals of cold spraying. *AIP Conference Proceedings, 1770*. Advance online publication. doi:10.1063/1.4963926

Klobcar, D., Tusek, J., & Taljat, B. (2008). Thermal Fatigue of Materials for Die-Casting Tooling. Materials Science and Engineering, 472(A), 198-207.

Koivuluoto, H., & Vuoristo, P. (2017). Structure and corrosion properties of cold sprayed coatings: A review. *Surface Engineering, 30*(6), 404–413. doi:10.1179/1743294413Y.0000000201

Kokini, K., DeJonge, J., Rangaraj, S., & Beardsley, B. (2002). Thermal shock of functionally graded thermal barrier coatings with similar thermal resistance. *Surface and Coatings Technology, 154*(2–3), 223–231. doi:10.1016/S0257-8972(02)00031-2

Kokubo, T., & Takadama, H. (2006). How useful is SBF in predicting in vivo bone bioactivity? *Biomaterials, 27*(15), 2907–2915. doi:10.1016/j.biomaterials.2006.01.017

Kong, H., Yoon, E., & Kwon, O. (1995). Self-Formation of Protective Oxide Films at Dry Sliding Mild Steel Surfaces under a Medium Vacuum. *Wear, 181*, 183325–183333. doi:10.1016/0043-1648(94)07042-3

Kong, Y., Tian, X., Gong, C., & Chu, P. K. (2018). Enhancement of Toughness and Wear Resistance by CrN/CrCN Multilayered Coatings for Wood Processing. *Surface and Coatings Technology, 344*(June), 204–213. doi:10.1016/j.surfcoat.2018.03.027

Kosarev, V. F., Klinkov, S. V., Alkimov, A. P., & Papyrin, A. N. (2003). On Some Aspects of Gas Dynamics of the Cold Spray Process. *Journal of Thermal Spray Technology., 12*(2), 265–281.

Kossman, S., Chicot, D., Decoopman, X., Iost, A., van Gorp, A., Meillot, E., Puchi-Cabrera, E. S., Santana, Y. Y., & Staia, M. H. (2014). Sliding Wear Response of Nanostructured YSZ Suspension Plasma-Sprayed Coating. *Journal of Thermal Spray Technology*, *23*(8), 1350–1361. doi:10.100711666-014-0146-6

Kovac, H., & Secer, Y. (2020). Improved tribological performance of AISI 316L stainless steel by a combined surface treatment: Surface texturing by selective laser melting and plasma nitriding. *Surface and Coatings Technology, 400.*

Krella, A. (2020). Resistance of PVD Coatings to Erosive and Wear Processes: A Review. *Coatings,* 10(10), 921. . doi:10.3390/coatings10100921

Krenkel, W. (Ed.). (2008). *Ceramic Matrix Composites: Fiber Reinforced Ceramics and their Applications* (1st ed.). Wiley. doi:10.1002/9783527622412

Krishnamurthy, N., Prashanthareddy, M. S., Raju, H. P., & Manohar, H. S. (2012). A Study of Parameters Affecting Wear Resistance of Alumina and Yttria Stabilized Zirconia Composite Coatings on Al-6061 Substrate. *ISRN Ceramics*, *2012*, 1–13. doi:10.5402/2012/585892

Kruzic, J. J. (2016). Bulk metallic glasses as structural materials: A review. *Advanced Engineering Materials*, *18*(8), 1308–1331. doi:10.1002/adem.201600066

Kulesza, P. J., Malik, M. A., Zamponi, S., Berrettoni, M., & Marassi, R. (1995). Electrolyte-cation-dependent coloring, electrochromism and thermochromism of cobalt(II) hexacyanoferrate(III, II) films. *Journal of Electroanalytical Chemistry (Lausanne, Switzerland)*, *397*(1), 287–292. doi:10.1016/0022-0728(95)04187-8

Kumar, R., Kumar, R. and Kumar, S. (2018). Erosion corrosion study of HVOF sprayed thermal sprayed coating on boiler tubes: A Review. *IJSMS*, *1*(3), 1–6. doi:10.51386/25815946/ijsms-v1i3p101

Kumar, S., Handa, A., & Kumar, R. (2019). Overview of wire arc spray process: a review. *Journal of Composition Theory*, *12*(7), 900-907. http://jctjournal.com/gallery/106-july2019.pdf

Kumar, S., Maity, S. R., & Patnaik, L. (2020). Friction and tribological behavior of bare nitrided, TiAlN and AlCrN coated MDC-K hot work tool steel. Ceramics International, 46(11A), 17280-17294.

Kumar, A. (2020). Anodization of Titanium Alloy (Grade 5) to Obtain Nanoporous Surface Using Sulfuric Acid Electrolyte. *Journal of the Institution of Electronics and Telecommunication Engineers*, 1–7. Advance online publication. doi:10.1080/03772063.2020.1780958

Kumar, A., Kaur, M., Joseph, A., & Jhala, G. (2020). Surface engineering analysis of plasma-nitrided die steels. *Proceedings of the Institution of Mechanical Engineers. Part J, Journal of Engineering Tribology*, *234*(6), 917–931. doi:10.1177/1350650119873237

Kumar, A., Kaur, M., Singh, S., Joseph, A., Jhala, G., & Bhandari, S. (2018). High-temperature tribological studies of plasma-nitrided tool steels. *Surface Engineering*, *34*(8), 620–633. doi:10.1080/02670844.2017.1341223

Kumar, D. D., Kumar, N., Kalaiselvam, S., Radhika, R., Rabel, A. M., & Jayavel, R. (2017). Tribo-mechanical properties of reactive magnetron sputtered transition metal carbide coatings. *Tribology International*, *114*, 234–244. doi:10.1016/j.triboint.2017.04.031

Kumar, D., Wu, X., Fu, Q., Ho, J. W. C., Kanhere, P. D., Li, L., & Chen, Z. (2015). Development of durable self-cleaning coatings using organic–inorganic hybrid sol–gel method. *Applied Surface Science*, *344*, 205–212. doi:10.1016/j.apsusc.2015.03.105

Kumar, R., Bhardwaj, D., & Sharma, Y. C. (2018). A Review on Plasma Ion Nitriding (PIN) Process. *Journal of Materials Science*, *6*(1), 31–44. doi:10.26634/jms.6.1.14000

Kumar, R., & Kumar, S. (2018). Comparative Parabolic Rate Constant and Coating Properties of Nickel, Cobalt, Iron and Metal Oxide Based Coating: A Review. *I-Manager's Journal on Material Science*, *6*(1), 45–56. doi:10.26634/jms.6.1.14379

Kumar, R., & Kumar, S. (2018). Thermal Spray Coating Process: A Study. *International Journal of Engineering Sciences & Research Technology*, *7*(3), 610–617. doi:10.5281/zenodo.1207005

Kumar, R., Prakash, R., Jospeh, A. J., Jain, J., Pareek, A., Rayjada, P. A., Raole, P. M., & Mukherjee, S. (2012). Impact of Forging Conditions on Plasma Nitrided Hot-Forging Dies and Punches. *Journal of Material Science and Research.*, *1*(4), 11–18. doi:10.5539/jmsr.v1n4p11

Kumar, R., Singh, R., & Kumar, S. (2018). Erosion and hot corrosion phenomena in thermal power plant and their preventive methods: A study. *Asian Journal of Mechanical Engineering*, *7*(1), 38–45.

Kumar, S., Handa, A., Chawla, V., & Kumar, R. (2021). Performance of thermal-sprayed coatings to combat hot corrosion of coal-fired boiler tube and effects of process parameters and post coating heat treatment on coating performance: A review. *Surface Engineering*, *37*(5), 1–28. doi:10.1080/02670844.2021.1924506

Kumar, S., Kumar, M., & Handa, A. (2018). Combating hot corrosion of boiler tubes–A study. *Engineering Failure Analysis*, *94*, 379–395. doi:10.1016/j.engfailanal.2018.08.004

Kumar, S., Kumar, M., & Handa, A. (2019). Comparative Study of High Temperature Oxidation Behavior of Wire Arc Sprayed Ni-Cr and Ni-Al Coatings. *Engineering Failure Analysis*, *106*, 104173–104189. doi:10.1016/j.engfailanal.2019.104173

Kumar, S., Kumar, M., & Handa, A. (2020). Erosion corrosion behavior and mechanical property of wire arc sprayed Ni-Cr and Ni-Al coating on boiler steels in actual boiler environment. *Materials at High Temperatures*, *37*(6), 1–15. doi:10.1080/09603409.2020.1810922

Kumar, S., Kumar, M., & Jindal, N. (2020). Overview of Cold Spray Coatings Applications and Comparisons: A Critical Review. *World Journal of Engineering*, *17*(1), 27–51. doi:10.1108/WJE-01-2019-0021

Kumar, S., & Kumar, R. (2021). Gas Dynamic Cold Spraying: A Review on Materials, Parameters, Applications and Challenges. *I-manager's Journal on Future Engineering and Technology*, *16*(2), 43–56. doi:10.26634/jfet.16.2.17624

Kumar, S., & Kumar, R. (2021). Influence of processing conditions on the properties of thermal sprayed coating: A review. *Surface Engineering*, *37*(11), 1339–1372. doi:10.1080/02670844.2021.1967024

Kumar, S., Kumar, R., & Singh, S. (2020). The role of thermal spray coating to combat hot corrosion of boiler tubes: A study. *Journal of Xidian University*, *14*(5), 229–239. doi:10.37896/jxu14.5/024

Kumar, V., & Balasubramanian, K. (2016). Progress update on failure mechanisms of advanced thermal barrier coatings: A review. *Progress in Organic Coatings*, *90*, 54–82. doi:10.1016/j.porgcoat.2015.09.019

Kuptsov, K. A., Kiryukhantsev-Korneev, P. V., Sheveyko, A. N., & Shtansky, D. V. (2013). Comparative study of electrochemical and impact wear behavior of TiCN, TiSiCN, TiCrSiCN, and TiAlSiCN coatings. *Surface and Coatings Technology*, *216*, 273–281. doi:10.1016/j.surfcoat.2012.11.058

Kutz, M. (Ed.). (2018). Handbook of environmental degradation of materials (3rd ed.). William Andrew/Elsevier.

Lackner, J. M., Waldhauser, W., Ebner, R., Bakker, R. J., & Scho, T. (2004). Room temperature pulsed laser deposited (Ti,Al)Cx N1-x coatings - chemical, structural, mechanical and tribological properties. *Thin Solid Films*, *468*(1–2), 125–133. doi:10.1016/j.tsf.2004.05.089

Laguna-Camacho, J. R., Villagrán-Villegas, L. Y., Martínez-García, H., Juárez-Morales, G., Cruz-Orduña, M. I., Vite-Torres, M., Ríos-Velasco, L., & Hernández-Romero, I. (2016). A study of the wear damage on gas turbine blades. *Engineering Failure Analysis*, *61*, 88–99. doi:10.1016/j.engfailanal.2015.10.002

Lakshmi, R. V., Bharathidasan, T., & Basu, B. J. (2011). Superhydrophobic sol–gel nanocomposite coatings with enhanced hardness. *Applied Surface Science*, *257*(24), 10421–10426. doi:10.1016/j.apsusc.2011.06.122

Lamaka, S. V., Montemor, M. F., Galio, A. F., Zheludkevich, M. L., Trindade, C., Dick, L. F., & Ferreira, M. G. S. (2008). Novel hybrid sol–gel coatings for corrosion protection of AZ31B magnesium alloy. *Electrochimica Acta*, *53*(14), 4773–4783. doi:10.1016/j.electacta.2008.02.015

Lamberti, M., & Escher, F. (2007). Aluminium Foil as a food packaging material in comparison with other materials. *Food Reviews International*, *23*(4), 407–433. doi:10.1080/87559120701593830

Landolt, D., Mischler, S., Stemp, M., & Barril, S. (2004). Third body effects and material fluxes in tribocorrosion systems involving a sliding contact. *Wear*, *256*(5), 517–524. doi:10.1016/S0043-1648(03)00561-1

Laws, K. J., Gun, B., & Ferry, M. (2009). Influence of casting parameters on the critical casting size of bulk metallic glass. *Metallurgical and Materials Transactions. A, Physical Metallurgy and Materials Science*, *40*(10), 2377–2387. doi:10.100711661-009-9929-7

Leclercq, B., Mevrel, R., Liedtke, V., & Hohenauer, W. (2003). Thermal conductivity of zirconia-based ceramics for thermal barrier coating. *Materialwissenschaft und Werkstofftechnik*, *34*(4), 406–409. doi:10.1002/mawe.200390084

Lee, H. G., Yoon, H. M., & Lee, J. S. (2019). Anisotropic Nanoscale and Sub-Nanoscale Friction Behaviors between Phosphorene and Silicon Tip. *Applied Surface Science*, *481*, 1573–1584. doi:10.1016/j.apsusc.2019.01.204

Lee, J., Bae, D., Yi, S., Kim, W., & Kim, D. (2004). Effects of Sn addition on the glass forming ability and crystallization behavior in Ni–Zr–Ti–Si alloys. *Journal of Non-Crystalline Solids*, *333*(2), 212–220. doi:10.1016/j.jnoncrysol.2003.10.011

Lee, K., Mo, C., Park, S. B., & Hong, S. H. (2011). Mechanical and electrical properties of multiwalled CNT-alumina nanocomposites prepared by a sequential two-step processing of ultrasonic spray pyrolysis and spark plasma sintering. *Journal of the American Ceramic Society*, *94*(11), 3774–3779. doi:10.1111/j.1551-2916.2011.04689.x

Lee, W. Y., Stinton, D. P., Berndt, C. C., Erdogan, F., Lee, Y.-D., & Mutasim, Z. (1996). Concept of Functionally Graded Materials for Advanced Thermal Barrier Coating Applications. *Journal of the American Ceramic Society*, *79*(12), 3003–3012. doi:10.1111/j.1151-2916.1996.tb08070.x

LeGeros, R. Z. (2008). Calcium Phosphate-Based Osteoinductive Materials. *Chemical Reviews*, *108*(11), 4742–4753. doi:10.1021/cr800427g PMID:19006399

Leib, E. W., Vainio, U., Pasquarelli, R. M., Kus, J., Czaschke, C., Walter, N., Janssen, R., Müller, M., Schreyer, A., Weller, H., & Vossmeyer, T. (2015). Synthesis and thermal stability of zirconia and yttria-stabilized zirconia microspheres. *Journal of Colloid and Interface Science*, *448*, 582–592. doi:10.1016/j.jcis.2015.02.049 PMID:25796070

Lengauer, W. (2000). Transition metal carbides, nitrides and carbonitrides. In *Handbook of Ceramic Hard Materials Edited*. WILEY-VCH Verlag GmbH. doi:10.1002/9783527618217.ch7

Lenhard, A., Damasio, S., Milke, A., & Schaeffer, L. (2006). Method to Estimate Workpiece-Die Heat Transfer Coefficient on Precision Warm Forging Process. *4th JSTP International Seminar on Precision Forging*.

Lewei, H., & Hassani, M. (2020). A Review of the Mechanical and Tribological Behavior of Cold Spray Metal Matrix Composites. *Journal of Thermal Spray Technology*, *29*, 1565–1608. https://doi.org/10.1007/s11666-020-01091-w

Lhotka, C., Szekeres, T., Steffan, I., Zhuber, K., & Zweymüller, K. (2003). Four-year study of cobalt and chromium blood levels in patients managed with two different metal-on-metal total hip replacements. *Journal of Orthopaedic Research, 21*(2), 189–195. doi:10.1016/S0736-0266(02)00152-3 PMID:12568948

Li, G., Liang, Y., Sun, H., & Cao, Y. (2020). Nitriding behavior and mechanical properties of carburizing and nitriding duplex treated M50NiL steel. *Surface and Coatings Technology, 384.*

Liang, S., Chang, Z., Tsai, D., Lin, Y., Sung, H., Deng, M.-J., & Shieu, F.-S. (2011). Effects of substrate temperature on the structure and mechanical properties of (TiVCrZrHf)N coatings. *Applied Surface Science, 257*(17), 7709–7713. doi:10.1016/j.apsusc.2011.04.014

Liang, Y., Ju, J., Deng, N., Zhou, X., Yan, J., Kang, W., & Cheng, B. (2018). Super-hydrophobic self-cleaning bead-like SiO2@PTFE nanofiber membranes for waterproof-breathable applications. *Applied Surface Science, 442*, 54–64. doi:10.1016/j.apsusc.2018.02.126

Liao, H., Normand, B., & Coddet, C. (2000). Influence of coating microstructure on the abrasive wear resistance of WC/Co cermet coatings. *Surface and Coatings Technology, 124*(2), 235–242. doi:10.1016/S0257-8972(99)00653-2

Liao, Z., Hua, N., Chen, W., Huang, Y., & Zhang, T. (2018). Correlations between the wear resistance and properties of bulk metallic glasses. *Intermetallics, 93*, 290–298. doi:10.1016/j.intermet.2017.10.008

Li, C. J., Li, W. Y., & Liao, H. (2006). Examination of the Critical Velocity for deposition of Particles in Cold Spraying. *Journal of Thermal Spray Technology, 15*(2), 212–222.

Li, C.-J., Yang, G.-J., & Ohmori, A. (2006). Relationship between particle erosion and lamellar microstructure for plasma-sprayed alumina coatings. *Wear, 260*(11–12), 1166–1172. doi:10.1016/j.wear.2005.07.006

Li, H., Lu, Z., Wang, S., Wu, Y., & Lu, Z. (2019). Fe-based bulk metallic glasses: Glass formation, fabrication, properties and applications. *Progress in Materials Science, 103*, 235–318. doi:10.1016/j.pmatsci.2019.01.003

Li, H., Wang, J., Gao, S., Chen, Q., Peng, L., Liu, K., & Wei, X. (2017). Superlubricity between MoS_2 Monolayers. *Advanced Materials, 29*(27), 1701474. doi:10.1002/adma.201701474 PMID:28497859

Li, H., Zhao, Q., Dong, H., Ma, Q., Wang, W., Xu, D., & Yu, D. (2014). Highly-flexible, low-cost, all stainless steel mesh-based dye-sensitized solar cells. *Nanoscale, 6*(21), 13203–13212. doi:10.1039/C4NR03999H PMID:25254313

Li, J. F., Liao, H., Wang, X. Y., Normand, B., Ji, V., Ding, C. X., & Coddet, C. (2004). Improvement in wear resistance of plasma sprayed yttria stabilized zirconia coating using nanostructured powder, *Tribology International, 37*(1), 77–84. doi:10.1016/S0301-679X(03)00138-5

Li, J. J., & Luo, J. B. (2013). Advancements in Superlubricity. *Science China. Technological Sciences, 56*(12), 2877–2887. doi:10.100711431-013-5387-y

Li, J., Han, P., Ji, W., Song, Y., Zhang, S., Chen, Y., Zhao, C., Zhang, F., Zhang, X., & Jiang, Y. (2011). The in vitro indirect cytotoxicity test and in vivo interface bioactivity evaluation of biodegradable FHA coated Mg–Zn alloys. *Materials Science and Engineering B, 176*(20), 1785–1788. doi:10.1016/j.mseb.2011.05.029

Li, J., Li, J., & Luo, J. (2018). Superlubricity of Graphite Sliding against Graphene Nanoflake under Ultrahigh Contact Pressure. *Advancement of Science, 5*(11), 1800810. doi:10.1002/advs.201800810 PMID:30479926

Li, K., Cai, Z., Du, P., Xiang, T., Yang, X., & Xie, G. (2021). Core-shell $Mg_{66}Zn_{30}Ca_4$ bulk metallic glasses composites reinforced by Fe with high strength and controllable degradation. *Intermetallics, 138*, 107334. doi:10.1016/j.intermet.2021.107334

Li, M., Li, Y., Xue, F., & Jing, X. (2019). A robust and versatile superhydrophobic coating: Wear-resistance study upon sandpaper abrasion. *Applied Surface Science*, *480*, 738–748. doi:10.1016/j.apsusc.2019.03.001

Lima, R. S., Karthikeyan, J., Kay, C. M., Lindemann, J., & Berndt, C. C. (2002). Microstructural characteristics of cold-sprayed nanostructured WC–Co coatings. *Thin Solid Films*, *416*(1–2), 129–135. doi:10.1016/S0040-6090(02)00631-4

Li, N., Zhang, J., Xing, W., Ouyang, D., & Liu, L. (2018). 3D printing of Fe-based bulk metallic glass composites with combined high strength and fracture toughness. *Materials & Design*, *143*, 285–296. doi:10.1016/j.matdes.2018.01.061

Lindquist, M., Wilhelmsson, O., Jansson, U., & Wiklund, U. (2009). Tribofilm formation and tribological properties of TiC and nanocomposite TiAlC coatings. *Wear*, *266*(3–4), 379–387. doi:10.1016/j.wear.2008.04.046

Lin, S., LeGeros, R. Z., & LeGeros, J. P. (2003). Adherent octacalciumphosphate coating on titanium alloy using modulated electrochemical deposition method. *Journal of Biomedical Materials Research. Part A*, *66*(4), 819–828. doi:10.1002/jbm.a.10072 PMID:12926034

Lin, Y., Yuan, G., Sheehan, S., Zhou, S., & Wang, D. (2011). Hematite-based solar water splitting: Challenges and opportunities. *Energy & Environmental Science*, *4*(12), 4862. doi:10.1039/c1ee01850g

Li, S., Qin, H., Zuo, R., & Bai, Z. (2015). Friction Properties of La-doped Mg/Al Layered Double Hydroxide and Intercalated Product as Lubricant Additives. *Tribology International*, *91*, 60–66. doi:10.1016/j.triboint.2015.06.012

Liu, L., Zhang, Y., Qiao, Y., Tan, S., Feng, S., Ma, J., Liu, Y., & Luo, J. (2021). 2D Metal-Organic Frameworks with Square Grid Structure. A Promising New-Generation Superlubricating Material. *Nano Today*, *40*, 101262. doi:10.1016/j.nantod.2021.101262

Liu, L., Zhou, M., Jin, L., Li, L., Mo, Y., Su, G., Li, X., Zhu, H., & Tian, Y. (2019). Recent Advances in Friction and Lubrication of Graphene and other 2D Materials: Mechanisms and Applications. *Friction*, *7*(3), 199–216. doi:10.100740544-019-0268-4

Liu, R. L., & Yan, M. F. (2010). Improvement of Wear and Corrosion Resistances of 17-4PH Stainless Steel by Plasma Nitro-carburizing. *Materials & Design*, *31*(5), 2355–2359. doi:10.1016/j.matdes.2009.11.069

Liu, S., Liu, X., Latthe, S. S., Gao, L., An, S., Yoon, S. S., Liu, B., & Xing, R. (2015). Self-cleaning transparent superhydrophobic coatings through simple sol–gel processing of fluoroalkylsilane. *Applied Surface Science*, *351*, 897–903. doi:10.1016/j.apsusc.2015.06.016

Liu, S., Sun, D., Fan, Z., Yu, H., & Meng, H. (2008). The influence of HVAF powder feedstock characteristics on the sliding wear behaviour of WC–NiCr coatings. *Surface and Coatings Technology*, *202*(20), 4893–4900. doi:10.1016/j.surfcoat.2008.03.014

Liu, X., Chu, P. K., & Ding, C. (2004). Surface modification of titanium, titanium alloys, and related materials for biomedical applications. *Materials Science and Engineering R Reports*, *47*(3-4), 9–12. doi:10.1016/j.mser.2004.11.001

Liu, X., He, D., Zhou, Z., Wang, G., Wang, Z., & Guo, X. (2019). Effect of post-heat treatment on the microstructure of micro-plasma sprayed hydroxyapatite coatings. *Surface and Coatings Technology*, *367*, 225–230. doi:10.1016/j.surfcoat.2019.03.056

Li, X. (2018). Additive Manufacturing of Advanced Multi-Component Alloys: Bulk Metallic Glasses and High Entropy Alloys. *Advanced Engineering Materials*, *20*(5), 1700874. doi:10.1002/adem.201700874

Li, Y., Li, B., Zhao, X., Tian, N., & Zhang, J. (2018). Totally Waterborne, Nonfluorinated, Mechanically Robust, and Self-Healing Superhydrophobic Coatings for Actual Anti-Icing. *ACS Applied Materials & Interfaces*, *10*(45), 39391–39399. doi:10.1021/acsami.8b15061 PMID:30351901

Li, Y., Poon, S., Shiflet, G., Xu, J., Kim, D., & Löffler, J. F. (2007). Formation of bulk metallic glasses and their composites. *MRS Bulletin, 32*(8), 624–628. doi:10.1557/mrs2007.123

Li, Y., Zhang, Z., Zhu, X., Men, X., Ge, B., & Zhou, X. (2015). Fabrication of a superhydrophobic coating with high adhesive effect to substrates and tunable wettability. *Applied Surface Science, 328*, 475–481. doi:10.1016/j.apsusc.2014.12.086

Li, Z., & Ma, J. (2021). Water-repellent surfaces of metallic glasses: Fabrication and application. *Materials Today Advances, 12*, 100164. doi:10.1016/j.mtadv.2021.100164

Löffler, J. F. (2003). Bulk metallic glasses. *Intermetallics, 11*(6), 529–540. doi:10.1016/S0966-9795(03)00046-3

Loganathan, P. S., Narasimalu, S., Kannappan, L., & Sekar, M. (2017). Supersonic particle deposition coatings for improved Tribological performance of off-shore wind turbine gears. *Proceedings of Asian Conference on Energy, Power and Transportation Electrification (ACEPT).*

Lombardi, A. N., Casteletti, L. C., & Totten, G. E. (2013). Thermal spray technologies: an overview. In Q. J. Wang & Y. W. Chung (Eds.), *Encyclopedia of Tribology*. Springer. doi:10.1007/978-0-387-92897-5_684

López, M. F., Gutiérrez, A., Jiménez, J. A., Martinesi, M., Stio, M., & Treves, C. (2010). Thermal oxidation of vanadium-free Ti alloys: An X-ray photoelectron spectroscopy study. *Materials Science and Engineering C, 30*(3), 465–471. doi:10.1016/j.msec.2010.01.004

Lopez-Ortega, A., Arana, J. L., Rodriguez, E., & Bayon, R. (2018). Corrosion, wear and tribocorrosion performance of a thermally sprayed aluminum coating modified by plasma electrolytic oxidation technique for offshore submerged components protection. *Corrosion Science, 143*, 258–280. doi:10.1016/j.corsci.2018.08.001

Lu, Z., Tan, H., Li, Y., & Ng, S. (2000). *Correlation between reduced glass transition temperature and glass forming ability of bulk metallic glasses*. Academic Press.

Ludema, K. C., & Ajayi, O. O. (2018). *Friction, wear, lubrication: A textbook in tribology*. CRC Press. doi:10.1201/9780429444715

Ludwig, B., Zheng, Z., Shou, W., Wang, Y., & Pan, H. (2016). Solvent-free manufacturing of electrodes for lithium-ion batteries. *Scientific Reports, 6*(1), 23150. doi:10.1038rep23150 PMID:26984488

Lugscheider, E., & Weber, T. (1990). Plasma Spraying—An Innovative Coating Technique: Process Variants and Applications. *IEEE Transactions on Plasma Science, 18*(6), 968–973. doi:10.1109/27.61511

Lu, J., Yu, H., & Chen, C. (2018). Biological properties of calcium phosphate biomaterials for bone repair: A review. *RSC Advances, 8*(4), 2015–2033. doi:10.1039/C7RA11278E

Luo, F., Cockburn, A., Sparkes, M., Lupoi, R., Chen, Z., O'Neill, W., Yao, J., & Liu, R. (2015). Performance characterization of Ni60-WC coating on steel processed with supersonic laser deposition. *Defence Technology, 11*(1), 35–47. doi:10.1016/j.dt.2014.09.003

Luo, G., Jin, Z., Dong, Y., Huang, J., Zhang, R., Wang, J., Li, M., Shen, Q., & Zhang, L. (2018). Preparation and performance enhancements of wear-resistant, transparent PU/SiO 2 superhydrophobic coating. *Surface Engineering, 34*(2), 139–145. doi:10.1080/02670844.2016.1236068

Lu, P., & Wood, R. J. K. (2020). Tribological performance of surface texturing in mechanical applications—A review. *Surface Topography: Metrology and Properties, 8*(4), 043001. Advance online publication. doi:10.1088/2051-672X/abb6d0

Lu, Y. P., Chen, Y. M., Li, S. T., & Wang, J. H. (2008). Surface nanocrystallization of hydroxyapatite Coating. *Acta Biomaterialia, 4*(6), 1865–1872. doi:10.1016/j.actbio.2008.05.016 PMID:18567551

Lu, Z., & Liu, C. (2002). A new glass-forming ability criterion for bulk metallic glasses. *Acta Materialia*, *50*(13), 3501–3512. doi:10.1016/S1359-6454(02)00166-0

Lu, Z., & Liu, C. (2004). Role of minor alloying additions in formation of bulk metallic glasses: A review. *Journal of Materials Science*, *39*(12), 3965–3974. doi:10.1023/B:JMSC.0000031478.73621.64

Lu, Z., Li, Y., & Ng, S. (2000). Reduced glass transition temperature and glass forming ability of bulk glass forming alloys. *Journal of Non-Crystalline Solids*, *270*(1-3), 103–114. doi:10.1016/S0022-3093(00)00064-8

Macdonald, D. D. (1990). Review of mechanistic analysis by electrochemical impedance spectroscopy. *Electrochimica Acta*, *35*(10), 1509–1525. doi:10.1016/0013-4686(90)80005-9

Macdonald, D. D. (2006). Reflections on the history of electrochemical impedance spectroscopy. *Electrochimica Acta*, *51*(8), 1376–1388. doi:10.1016/j.electacta.2005.02.107

Maev, R. G., & Leshchynsky, V. (2006). Air gas dynamic spraying of powder mixtures: Theory and application. *Journal of Thermal Spray Technology*, *15*(2), 198–205. doi:10.1361/105996306X108048

Mahadik, S. A. S. A., Fernando, P. D., Hegade, N. D. N. D., Wagh, P. B. P. B., & Gupta, S. C. S. C. (2013). Durability and restoring of superhydrophobic properties in silica-based coatings. *Journal of Colloid and Interface Science*, *405*, 262–268. doi:10.1016/j.jcis.2013.04.042 PMID:23746435

Mahadik, S. A. S. A., Mahadik, D. B. B., Parale, V. G. G., Wagh, P. B. B., Gupta, S., & Venkateswara Rao, A. (2012). Recoverable and thermally stable superhydrophobic silica coating. *Journal of Sol-Gel Science and Technology*, *62*(3), 490–494. doi:10.100710971-012-2753-3

Mahadik, S. A., Kavale, M. S., Mukherjee, S. K., & Rao, A. V. (2010). Transparent Superhydrophobic silica coatings on glass by sol–gel method. *Applied Surface Science*, *257*(2), 333–339. doi:10.1016/j.apsusc.2010.06.062

Mahadik, S. A., Mahadik, D. B., Kavale, M. S., Parale, V. G., Wagh, P. B., Barshilia, H. C., Gupta, S. C., Hegde, N. D., & Rao, A. V. (2012b). Thermally stable and transparent superhydrophobic sol-gel coatings by spray method. *Journal of Sol-Gel Science and Technology*, *63*(3), 580–586. Advance online publication. doi:10.100710971-012-2798-3

Mahadik, S. A., & Mahadik, S. S. (2021). Surface morphological and topographical analysis of multifunctional superhydrophobic sol-gel coatings. *Ceramics International*, *47*(20), 29475–29482. Advance online publication. doi:10.1016/j.ceramint.2021.07.115

Mahadik, S. A., Parale, V., Vhatkara, R. S., Mahadik, D. B., Kavale, M. S., Wagh, P. B., Gupta, S., & Gurav, J. (2013). Superhydrophobic silica coating by dip coating method. *Applied Surface Science*, *277*, 67–72. doi:10.1016/j.apsusc.2013.04.001

Mahadik, S. A., Pedraza, F. D., Relekar, B. P., Parale, V. G., Lohar, G. M., & Thorat, S. S. (2016). Synthesis and characterization of superhydrophobic–superoleophilic surface. *Journal of Sol-Gel Science and Technology*, *78*(3), 475–481. doi:10.100710971-016-3974-7

Mahadik, S. A., Pedraza, F., & Mahadik, S. S. (2017). Comparative studies on water repellent coatings prepared by spin coating and spray coating methods. *Progress in Organic Coatings*, *104*, 217–222. doi:10.1016/j.porgcoat.2016.11.006

Mahadik, S. A., Pedraza, F., Mahadik, S. S., Relekar, B. P., & Thorat, S. S. (2017). Biocompatible superhydrophobic coating material for biomedical applications. *Journal of Sol-Gel Science and Technology*, *81*(3), 791–796. doi:10.100710971-016-4244-4

Mahadik, S. A., Pedraza, F., & Vhatkar, R. S. (2016). Silica based superhydrophobic coating for long-term industrial and domestic applications. *Journal of Alloys and Compounds*, *663*, 487–493. Advance online publication. doi:10.1016/j.jallcom.2015.12.016

Mahadik, S., Pedraza, F., Parale, V. G., & Park, H.-H. (2016). Organically modified silica aerogel with different functional silylating agents and effect on their physico-chemical properties. *Journal of Non-Crystalline Solids*, *453*, 164–171. doi:10.1016/j.jnoncrysol.2016.08.035

Mahato, N., Sharma, S., Keshri, A. K., Simpson, A., Agarwal, A., & Balani, K. (2013). Nanomechanical properties and thermal conductivity estimation of plasma-sprayed, solid-oxide fuel cell components: Ceria-doped, yttria-stabilized zirconia electrolyte. *JOM*, *65*(6), 749–762. doi:10.100711837-013-0601-8

Makwana & Bhingole. (2018). Electrochemical and Plasma Surface Modification of Magnesium and its Alloy: Review. *Materials Today: Proceedings*, *5*(9), 18260-18267.

Malek, M. H. A., Saad, N. H., Abas, S. K., & Shah, N. B. M. (2014). Critical process and performance parameters of thermal arc spray coating. *International Journal of Materials Engineering Innovation*, *5*(1), 12–27. doi:10.1504/IJMATEI.2014.059489

Malmberg, S., & Heberlein, J. (1993). Effect of plasma spray operating conditions on plasma jet characteristics and coating properties. *Journal of Thermal Spray Technology*, *2*(4), 339–344. doi:10.1007/BF02645862

Mandes, A., Vladoiu, R., Prodan, G., Dinca, V., Porosnicu, C., & Dinca, P. (2018). The properties of binary and ternary Ti based coatings produced by Thermionic Vacuum Arc (TVA) technology. *Coatings*, *8*(3), 114–125. doi:10.3390/coatings8030114

Manjaiah, M., & Laubscher, R. F. (2017). Effect of anodizing on surface integrity of Grade 4 titanium for biomedical applications. *Surface and Coatings Technology*, *310*, 263–272. doi:10.1016/j.surfcoat.2016.12.038

Manova, D., Gerlach, J. W., & Mändl, S. (2010). Thin film deposition using energetic ions. *Materials (Basel)*, *3*(8), 4109–4141. doi:10.3390/ma3084109

Ma, P.-C., Siddiqui, N. A., Marom, G., & Kim, J.-K. (2010). Dispersion and functionalization of carbon nanotubes for polymer-based nanocomposites: A review. *Composites. Part A, Applied Science and Manufacturing*, *41*(10), 1345–1367. doi:10.1016/j.compositesa.2010.07.003

Margaretha, P., & Tissot, P. (1979). Observation of reduction product-electrolyte-electrode interaction by cyclic voltammetry. *Journal of Electroanalytical Chemistry and Interfacial Electrochemistry*, *99*(1), 127–131. doi:10.1016/S0022-0728(79)80419-2

Marimuthu, K. P., Han, J., Jeong, U., Lee, K., & Lee, H. (2021). Study on tribological characteristics of Zr-based BMG via nanoscratch techniques. *Wear*, *486*, 204067. doi:10.1016/j.wear.2021.204067

Marinko, Ž., Suhadolnik, L., Šetina Batič, B., Šelih, V. S., Majaron, B., Kovač, J., & Čeh, M. (2021). Toward a flexible and efficient TiO_2 photocatalyst immobilized on a titanium foil. *ACS Omega*, *6*(36), 23233–23242. doi:10.1021/acsomega.1c02862 PMID:34549124

Martin, J. M., & Erdemir, A. (2018). Superlubricity: Friction's Vanishing Act. *Physics Today*, *71*(4), 40–46. doi:10.1063/PT.3.3897

Maruska, H. P., & Stevenson, D. A. (1974). Mechanism of light production in metal-insulator-semiconductor diodes; GaN:Mg violet light-emitting diodes. *Solid-State Electronics*, *17*(11), 1171–1179. doi:10.1016/0038-1101(74)90161-0

Mathapati, M., Ramesh, M. R., & Doddamani, M. (2017). High temperature erosion behavior of plasma sprayed NiCrAlY/WC-Co/cenosphere coating. *Surface and Coatings Technology*, *325*, 98–106. doi:10.1016/j.surfcoat.2017.06.033

Matthews, A., Franklin, S., & Holmberg, K. (2007). Tribological coatings: Contact mechanisms and selection. *Journal of Physics. D, Applied Physics*, *40*(18), 5463–5475. doi:10.1088/0022-3727/40/18/S07

Matthews, D., Ocelik, V., & De Hosson, J. T. M. (2007). Tribological and mechanical properties of high power laser surface-treated metallic glasses. *Materials Science and Engineering A*, *471*(1-2), 155–164. doi:10.1016/j.msea.2007.02.119

Maver, U., Velnar, T., Gaberšček, M., Planinšek, O., & Finšgar, M. (2016). Recent progressive use of atomic force microscopy in biomedical applications. *Trends in Analytical Chemistry*, *80*, 96–111. doi:10.1016/j.trac.2016.03.014

Maver, U., Žnidaršič, A., & Gaberšček, M. (2011). An attempt to use atomic force microscopy for determination of bond type in lithium battery electrodes. *Journal of Materials Chemistry*, *21*(12), 4071–4075. doi:10.1039/c0jm04481d

Ma, X. G., Komvopoulos, K., Wan, D., Bogy, D. B., & Kim, Y. S. (2003). Effects of film thickness and contact load on nanotribological properties of sputtered amorphous carbon thin films. *Wear*, *254*(10), 1010–1018. doi:10.1016/S0043-1648(03)00307-7

Mayrhofer, E., Janka, L., Mayr, W. P., Norpoth, J., Ripoll, M. R., & Gröschl, M. (2015). Cracking resistance of Cr_3C_2–NiCr and WC–Cr_3C_2–Ni thermally sprayed coatings under tensile bending stress. *Surface and Coatings Technology*, *281*, 169–175. doi:10.1016/j.surfcoat.2015.09.002

Mazor, H., Golodnitsky, D., Burstein, L., Gladkich, A., & Peled, E. (2012). Electrophoretic deposition of lithium iron phosphate cathode for thin-film 3D-microbatteries. *Journal of Power Sources*, *198*, 264–272. doi:10.1016/j.jpowsour.2011.09.108

Mebarki, N., Delagnes, D., Lamesle, P., Delmas, F., & Levaillant, C. (2002). Relationship between Microstructure and Mechanical Properties of a 5% Cr hot work tool steel. *Proceedings of the Sixth International Conference on Tooling.*

Medina, M. A., Acikgoz, O., Rodriguez, A., Meduri, C. S., Kumar, G., & Baykara, M. Z. (2020). Comparative Tribological Properties of Pd-, Pt-, and Zr-Based Bulk Metallic Glasses. *Lubricants (Basel, Switzerland)*, *8*(9), 85. doi:10.3390/lubricants8090085

Meekhanthong, K., & Wirojanupatump, S. (2014). Characterization and comparison of thermally sprayed hard coatings as alternative to hard chrome plating. *Advanced Materials Research*, *974*, 183–187. doi:10.4028/www.scientific.net/AMR.974.183

Meenakshi, M., Sivakumar, R., Sivanantharaja, A., & Sanjeeviraja, C. (2017). Electrochromic performance of RF sputtered WO3 thin films by Li ion intercalation and de-intercalation. *AIP Conference Proceedings*, *1832*(1), 080003. doi:10.1063/1.4980463

Mehrotra, P. (2016). Biosensors and their applications – A review. *Journal of Oral Biology and Craniofacial Research*, *6*(2), 153–159. doi:10.1016/j.jobcr.2015.12.002 PMID:27195214

Mejía, V., Hernán, D., Perea, D., & Gilberto Bejarano, G. (2020). Development and Characterization of TiAlN (Ag, Cu) Nanocomposite Coatings Deposited by DC Magnetron Sputtering for Tribological Applications. *Surface and Coatings Technology*, *381*(January), 125095. Advance online publication. doi:10.1016/j.surfcoat.2019.125095

Meng, Y., & Liu, C. (2021). Spatiotemporal Manipulation of Boundary Lubrication by Electro-Charging and Electrochemical Methods. In Superlubricity (2nd ed., pp. 499-516). doi:10.1016/B978-0-444-64313-1.00025-9

Meng, E. C., Guan, S. K., Wang, H. X., Wang, L. G., Zhu, S. J., Hu, J. H., Ren, C. X., Gao, J. H., & Feng, Y. S. (2011). Effect of electrodeposition modes on surface characteristics and corrosion properties of fluorine-doped hydroxyapatite coatings on Mg–Zn–Ca alloy. *Applied Surface Science*, *257*(11), 4811–4816. doi:10.1016/j.apsusc.2010.12.073

Meng, R., Deng, J., Liu, Y., Duan, R., & Zhang, G. (2018). Improving tribological performance of cemented carbides by combining laser surface texturing and W-S-C solid lubricant coating. *International Journal of Refractory Metals & Hard Materials*, *72*, 163–171. doi:10.1016/j.ijrmhm.2017.12.024

Messaoud, M., Houmard, M., Briche, S., Roussel, F., & Langlet, M. (2010). Hydrophobic functionalization of cotton-based textile fabrics through a non-fluorinated sol–gel route. *Journal of Sol-Gel Science and Technology*, *55*, 243–254.

Michelmore, A. (2016). Thin Film Growth on Biomaterial Surfaces. *Thin Film Coatings for Biomaterials and Biomedical Applications*, (January), 29–47. doi:10.1016/B978-1-78242-453-6.00002-X

Mikell, P. G. (2010). Fundamentals of modern manufacturing materials, processes, and systems. John Wiley and Sons, Inc.

Miller, R. A. (1987). Current status of thermal barrier coatings—An overview. *Surface and Coatings Technology*, *30*(1), 1–11. doi:10.1016/0257-8972(87)90003-X

Mi, P., Zhao, H., Wang, T., & Ye, F. (2018). Sliding wear behavior of HVOF sprayed WC-(nano-WC-Co) coating at elevated temperatures. *Materials Chemistry and Physics*, *206*, 1–6. doi:10.1016/j.matchemphys.2017.09.066

Mishra, S. B., & Prakash, S. (2015). Erosion –corrosion behaviour of Ni-20Cr plasma coationg in actual boiler environment. *Surface Engineering*, *31*(1), 29–38. doi:10.1179/1743294414Y.0000000338

Mittal, M., Sardar, S., & Janta, A. (2021). Nanofabrication techniques for semiconductor chemical sensors. Handbook of Nanomaterials for Sensing Applications, Micro and Nano Technology, 119-137. doi:10.1016/B978-0-12-820783-3.00023-3

Mitterer, C., Fateh, N., & Munnik, F. (2011). Microstructure-property relations of reactively magnetron sputtered VCxNy films. *Surface and Coatings Technology*, *205*(13–14), 3805–3809. doi:10.1016/j.surfcoat.2011.01.037

Miura, H., Fukuyama, Y., Sunda, T., Lin, B., Zhou, J., Takizawa, J., Ohmori, A., & Kimura, M. (2014). Foldable textile electronic devices using all-organic conductive fibers. *Advanced Engineering Materials*, *16*(5), 550–555. doi:10.1002/adem.201300461

Mohajeri, S., Dolati, A., & Ghorbani, M. (2016). An investigation on the electrodeposition mechanism of Ni-TiO_2 nano-composite coatings. *Journal of Ultrafine Grained Nanostructure Materials*, *49*, 51–63.

Mo, J. L., & Zhu, M. H. (2009). Tribological Oxidation Behaviour of PVD Hard Coatings. *Tribology International*, *42*(11–12), 1758–1764. doi:10.1016/j.triboint.2009.04.026

Molaei, M., Babaei, K., & Fattah-alhosseini, A. (2021). Improving the wear resistance of plasma electrolytic oxidation (PEO) coatings applied on Mg and its alloys under the addition of nano- and micro-sized additives into the electrolytes: A review. *Journal of Magnesium and Alloys*, *9*(4), 1164–1186. doi:10.1016/j.jma.2020.11.016

Monošík, R., Stred'anský, M., & Šturdík, E. (2012). Application of electrochemical biosensors in clinical diagnosis. *Journal of Clinical Laboratory Analysis*, *26*(1), 22–34. doi:10.1002/jcla.20500 PMID:24833531

Monteiro, M. M., Campos da Rocha, N. C., Rossi, A. M., & Gloria de Almeida, S. (2003). Dissolution properties of calcium phosphate granules with different compositions in simulated body fluid. *Journal of Biomedical Materials Research. Part A*, *65A*(2), 299–305. doi:10.1002/jbm.a.10479 PMID:12734825

Morsi, S., & Alexander, A. (1972). An Investigation of Particle Trajectories in Two-Phase Flow Systems. *Journal of Fluid Mechanics*, *55*(2), 193-208.

Mortimer, R. J. (2011). Electrochromic materials. *Annual Review of Materials Research, 41*(1), 241–268. doi:10.1146/annurev-matsci-062910-100344

Mosa, J., Rosero-Navarro, N. C., & Aparicio, M. (2016). Active corrosion inhibition of mild steel by environmentally-friendly Ce-doped organic–inorganic sol–gel coatings. *RSC Advances, 6*(46), 39577–39586. doi:10.1039/C5RA26094A

Moseler, M., Cumbsch, P., Casiraghi, C., Ferrari, A. C., & Robertson, J. (2005). The Ultrasmoothness of Diamond-like Carbon Surfaces. *Science, 309*(5740), 1545–1548. doi:10.1126cience.1114577 PMID:16141070

Mostajeran, A., Shoja-Razavi, R., Hadi, M., Erfanmanesh, M., Barekat, M., & Firouzabadi, M. S. (2020). Evaluation of the mechanical properties of WC-FeAl composite coating fabricated by laser cladding method. *International Journal of Refractory Metals & Hard Materials, 88*, 105199. doi:10.1016/j.ijrmhm.2020.105199

Mouritz. (Ed.). (2012). Introduction to Aerospace Materials. Academic Press.

Mozetič, M. (2019). Surface Modification to Improve Properties of Materials. *Materials (Basel, Switzerland), 12*(3), 441. doi:10.3390/ma12030441 PMID:30709009

Mozgovoy, S., Alik, L., Hardell, J., & Prakash, B. (2019). Material Transfer during High Temperature Sliding of Al-Si Coated 22MnB5 Steel against PVD Coatings with and without Aluminium. *Wear, 426–427*(April), 401–411. doi:10.1016/j.wear.2018.12.042

Mozgovoy, S., Hardell, J., Deng, L., Oldenburg, M., & Prakash, B. (2014). Effect of Temperature on Friction and Wear of Pre Hardened Tool Steel during Sliding Against 22MnB5 Steel. *Tribology-Materials. Surfaces and Interfaces, 82*, 65–73. doi:10.1179/1751584X13Y.0000000056

MRS Bulletin. (2009). *Technology Advances, 34*(792).

Muhammad, A., Rahman, M. R., Baini, R., & Bin Bakri, M. K. (2021). 8 - Applications of sustainable polymer composites in automobile and aerospace industry. In M. R. Rahman (Ed.), *Advances in Sustainable Polymer Composites* (pp. 185–207). Woodhead Publishing. doi:10.1016/B978-0-12-820338-5.00008-4

Mülazim, Y., Kahraman, M. V., Apohan, N. K., Kızıltaş, S., & Güngör, A. (2011). Preparation and characterization of UV-curable, boron-containing, transparent hybrid coatings. *Journal of Applied Polymer Science, 120*(4), 2112–2121. doi:10.1002/app.33358

Müller, W. D., Nascimento, M. L., Zeddies, M., Córsico, M., Gassa, L. M., Fernández, M. A., & de Mele, L. (2007). Magnesium and its alloys as degradable biomaterials. Corrosion studies using potentiodynamics and EIS electrochemical techniques. *Journal of Materials Research, 10*(1), 5–10. doi:10.1590/S1516-14392007000100003

Murthy, J. K. N., & Venkataraman, B. (2006). Abrasive wear behaviour of WC–CoCr and Cr_3C_2–20 (NiCr) deposited by HVOF and detonation spray processes. *Surface and Coatings Technology, 200*(8), 2642–2652. doi:10.1016/j.surfcoat.2004.10.136

Murugesan, P., Kumar, A. B., Kambhampati, A. T., Pillai, S., Chandrasekar, G. C., Srikrishnan, A. R., & Velamati, R. K. (2020). Numerical Study of Characteristics of Underexpanded Supersonic Jet. *Journal of Aerospace Technology and Management*, 12.

Musalek, R., Tesar, T., Medricky, J., Lukac, F., Chraska, T., & Gupta, M. (2020). Microstructures and Thermal Cycling Properties of Thermal Barrier Coatings Deposited by Hybrid Water-Stabilized Plasma Torch. *Journal of Thermal Spray Technology, 29*(3), 444–461. doi:10.100711666-020-00990-2

Musil, J. (2000). Hard and Superhard Nanocomposite Coatings. *Surface and Coatings Technology, 125*(1–3), 322–330. doi:10.1016/S0257-8972(99)00586-1

Musil, J., Karvánková, P., & Kasl, J. (2001). Hard and Superhard Zr–Ni–N Nanocomposite Films. *Surface and Coatings Technology, 139*(1), 101–109. doi:10.1016/S0257-8972(01)00989-6

Musztyfaga-Staszuk, M. Czupryński, A., & Kciuk, M. (2018). Investigation of mechanical and anti-corrosion properties of flame sprayed coatings. *Advances in Materials Science, 18*(4), 42-53. . doi:10.1515/adms-2017-0049

Myalska, H., Lusvarghi, L., Bolelli, G., Sassatelli, P., & Moskal, G. (2019). Tribological behavior of WC-Co HVAF-sprayed composite coatings modified by nano-sized TiC addition. *Surface and Coatings Technology, 371*, 401–416. doi:10.1016/j.surfcoat.2018.09.017

Nagano, M., Nagashima, S., Maeda, H., & Kato, A. (1999). Sintering behaviour of Al_2TiO_5 base ceramics and their thermal properties. *Ceramics International, 25*(8), 681–687. doi:10.1016/S0272-8842(98)00083-2

Nagels, J., Stokdijk, M., & Rozing, P. M. (2003). Stress shielding and bone resorption in shoulder arthroplasty. *Journal of Shoulder and Elbow Surgery, 12*(1), 35–39. doi:10.1067/mse.2003.22 PMID:12610484

Nakamuraa, M., Hirao, K., Yamauchi, Y., & Kanzaki, S. (2002). Tribological Behaviour of Unidirectional Aligned Silicon Nitride against Steel. *Wear, 252*(5-6), 484–490. doi:10.1016/S0043-1648(02)00005-4

Nan, C.-W., Liu, G., Lin, Y., & Li, M. (2004). Interface effect on thermal conductivity of carbon nanotube composites. *Applied Physics Letters, 85*(16), 3549–3551. doi:10.1063/1.1808874

Nan, C.-W., Shi, Z., & Lin, Y. (2003). A simple model for thermal conductivity of carbon nanotube-based composites. *Chemical Physics Letters, 375*(5–6), 666–669. doi:10.1016/S0009-2614(03)00956-4

Narayanan, R., Dutta, S., & Seshadri, S. K. (2006). Hydroxy apatite coatings on Ti-6Al-4V from seashell. *Surface and Coatings Technology, 200*(16–17), 4720–4730. doi:10.1016/j.surfcoat.2005.04.040

Narayanan, R., Seshadri, S. K., Kwon, T. Y., & Kim, K. H. (2008). Calcium phosphate-based coatings on titanium and its alloys. *Journal of Biomedical Materials Research. Part B, Applied Biomaterials, 85*(1), 297–299. doi:10.1002/jbm.b.30932 PMID:17853421

Narayan, R., & Colombo, P. (2009). Advances in Bioceramics and Porous Ceramics. *Ceramic Engineering and Science Proceedings, 52*. doi:10.1002/9780470584354

Nath, S., Manna, I., & Majumdar, J. D. (2014). Kinetics and mechanism of isothermal oxidation of compositionally graded yttria stabilized zirconia (YSZ) based thermal barrier coating. *Corrosion Science, 88*, 10–22. doi:10.1016/j.corsci.2014.06.050

Ng, C. H., Lim, H. N., Hayase, S., Zainal, Z., & Huang, N. M. (2018). Photovoltaic performances of mono- and mixed-halide structures for perovskite solar cell: A review. *Renewable & Sustainable Energy Reviews, 90*, 248–274. doi:10.1016/j.rser.2018.03.030

Nguyen, X. H., Kim, I. K., Jang, B. K., & Oh, Y. S. (2013). Effect of carbon content on the tribological behavior of TiCxN1-x films prepared by arc-vapor deposition. *Journal of the Ceramic Society of Japan, 121*(1420), 961–967. doi:10.2109/jcersj2.121.961

Nieto, A., Kim, J., Penkov, O. V., Kim, D.-E., & Schoenung, J. M. (2017). Elevated temperature wear behavior of thermally sprayed WC-Co/nanodiamond composite coatings. *Surface and Coatings Technology, 315*, 283–293. doi:10.1016/j.surfcoat.2017.02.048

Niinomi, M. (2002). Recent metallic materials for biomedical applications. *Metallurgical and Materials Transactions. A, Physical Metallurgy and Materials Science, 33*(3), 477–486. doi:10.100711661-002-0109-2

Nishiyama, N., Takenaka, K., Miura, H., Saidoh, N., Zeng, Y., & Inoue, A. (2012). The world's biggest glassy alloy ever made. *Intermetallics*, *30*, 19–24. doi:10.1016/j.intermet.2012.03.020

Nolan, D., Leskovsek, V., & Jenko, M. (2006). Estimation of Fracture Toughness of Nitride Compound Layers on Tool Steel by Application of Vickers Indentation Method. *Surface and Coatings Technology*, *201*(1-2), 182–188. doi:10.1016/j.surfcoat.2005.11.077

Nuttall, R. H. (1979). Andrew Pritchard's contribution to metallurgical microscopy. *Technology and Culture*, *20*(3), 569–577. doi:10.2307/3103817

O'Regan, B., & Grätzel, M. (1991). A low-cost, high-efficiency solar cell based on dye-sensitized colloidal TiO2 films. *Nature*, *353*(6346), 737–740. doi:10.1038/353737a0

Ochoa, E. A., Wisnivesky, D., Minea, I., Ganciu, M., Tauziede, C., Chapon, P., & Alvarez, F. (2009). Microstructure and Properties of the Compound Layer Obtained by Pulsed Plasma Nitriding in Steel Gears. *Surface and Coatings Technology*, *203*(10-11), 1457–1461. doi:10.1016/j.surfcoat.2008.11.025

Ogihara, H., Wang, H., & Saji, T. (2014). Electrodeposition of Ni–B/SiC composite films with high hardness and wear resistance. *Applied Surface Science*, *296*, 108–113.

Ohsuku, T., & Hirai, T. (1982). An electrochromic display based on titanium dioxide. *Electrochimica Acta*, *27*(9), 1263–1266. doi:10.1016/0013-4686(82)80146-1

Oksa, M., Turunen, E., Suhonen, T., Varis, T., & Hannula, S. P. (2011). Optimization and Characterization of High Velocity Oxy-fuel Sprayed Coatings: Techniques, Materials, and Applications. *Coatings*, *1*(1), 17–52. doi:10.3390/coatings1010017

Oliver, W. C., & Pharr, G. M. (1992). An improved technique for determining hardness and elastic modulus using load and displacement sensing indentation experiments. *Materials Research Society. Journal of Materials Research*, *7*(6). Advance online publication. doi:10.1557/JMR.1992.1564

Orsato, R. J., & Wells, P. (2007). U-turn: The rise and demise of the automobile industry. *Journal of Cleaner Production*, *15*(11), 994–1006. doi:10.1016/j.jclepro.2006.05.019

Ozgurluk, Y., Doleker, K. M., & Karaoglanli, A. C. (2018). Hot corrosion behavior of YSZ, $Gd_2Zr_2O_7$ and YSZ/$Gd_2Zr_2O_7$ thermal barrier coatings exposed to molten sulfate and vanadate salt. *Applied Surface Science*, *438*, 96–113. doi:10.1016/j.apsusc.2017.09.047

Pakseresht, A. H., Kimiayi, A., Alizadeh, M., Nuranian, H., & Faeghinia, A. (2020). Microstructural study and hot corrosion behavior of bimodal thermal barrier coatings under laser heat treatment. *Ceramics International, 46*(11), 19217-19227.

Pakseresht, A. H., Rahimipour, M. R., Alizadeh, M., Hadavi, S. M. M., & Shahbazkhan, A. (2016). Concept of advanced thermal barrier functional coatings in high temperature engineering components. In Research perspectives on functional micro-and nanoscale coatings (pp. 396–419). IGI Global.

Pakseresht, Rahimipour, Alizadeh, Hadavi, & Shahbazkhan. (2016). Concept of Advanced Thermal Barrier Functional Coatings in High Temperature Engineering Components. In Research Perspectives on Functional Micro- and Nanoscale Coatings (pp. 396–419). IGI Global. doi:10.4018/978-1-5225-0066-7.ch015

Pakseresht, A. H. (Ed.). (2018). *Production, Properties, and Applications of High Temperature Coatings*. IGI Global. doi:10.4018/978-1-5225-4194-3

Pakseresht, A. H., Kimiayi, A., Alizadeh, M., Nuranian, H., & Faeghinia, A. (2020). Microstructural study and hot corrosion behavior of bimodal thermal barrier coatings under laser heat treatment. *Ceramics International*, *46*(11), 19217–19227. doi:10.1016/j.ceramint.2020.04.259

Pakseresht, A. H., Mousavi, S. M., Saremi, M., Ghasali, E., & Rajaei, H. (2021). Microstructure and mechanical properties of YSZ-alumina composites designed for thermal barrier coatings. *Materials at High Temperatures*, *38*(1), 23–30. doi:10.1080/09603409.2020.1837414

Pakseresht, A. H., Rahimipour, M. R., Vaezi, M. R., & Salehi, M. (2015). Effect of splat morphology on the microstructure and dielectric properties of plasma sprayed barium titanate films. *Applied Surface Science*, *324*, 797–806. doi:10.1016/j.apsusc.2014.11.041

Pakseresht, A. H., Rahimipour, M. R., Vaezi, M. R., & Salehi, M. (2016). Thermal plasma spheroidization and spray deposition of barium titanate powder and characterization of the plasma sprayable powder. *Materials Chemistry and Physics*, *173*, 395–403. doi:10.1016/j.matchemphys.2016.02.028

Pakseresht, A. H., Saremi, M., Omidvar, H., & Alizadeh, M. (2019). Micro-structural study and wear resistance of thermal barrier coating reinforced by alumina whisker. *Surface and Coatings Technology*, *366*, 338–348. doi:10.1016/j.surfcoat.2019.03.059

Pana, I., Vladescu, A., Constantin, L. R., Sandu, I. G., Dinu, M., & Cotrut, C. M. (2020). In vitro corrosion and tribocorrosion performance of biocompatible carbide coatings. *Coatings*, *10*(7), 654–670. doi:10.3390/coatings10070654

Pandey, A. K., & Biswas, K. (2016). Effect of hydrothermal treatment on tribological properties of alumina and zirconia based bioceramics. *Ceramics International*, *42*(2), 2306–2316. doi:10.1016/j.ceramint.2015.10.026

Papyrin, A. (2001). Cold spray technology. *Advanced Materials & Processes*, *159*(9), 49.

Papyrin, A. N., Kosarev, V. F., Klinkov, S. V., Alkhimov, A. P., & Fomin, V. (2007). *Cold Spray Technology*. Elsevier.

Parale, V. G. G., Mahadik, D. B. B., Kavale, M. S. S., Mahadik, S. A. A., Rao, A. V. V., & Mullens, S. (2013). Sol-gel preparation of PTMS modified hydrophobic and transparent silica coatings. *Journal of Porous Materials*, *20*(4), 733–739. doi:10.100710934-012-9648-0

Parale, V. G., Mahadik, D. B., Mahadik, S. A., Kavale, M. S., Wagh, P. B., Gupta, S. C., & Rao, A. V. (2013). OTES modified transparent dip coated silica coatings. *Ceramics International*, *39*(1), 835–840. Advance online publication. doi:10.1016/j.ceramint.2012.05.079

Parau, A. C., Vitelaru, C., Balaceanu, M., Braic, V., Constantin, L. R., Braic, M., & Vladescu, A. (2016). TiSiC, TiSiC-Zr, and TiSiC-Cr coatings—Corrosion resistance and tribological performance in saline solution. *Tribology Transactions*, *59*(1), 72–79. doi:10.1080/10402004.2015.1077406

Park, E., Kang, H., Kim, W., & Kim, D. (2001). The effect of Ag addition on the glass-forming ability of Mg–Cu–Y metallic glass alloys. *Journal of Non-Crystalline Solids*, *279*(2-3), 154–160. doi:10.1016/S0022-3093(00)00412-9

Park, E., & Kim, D. (2005). Design of bulk metallic glasses with high glass forming ability and enhancement of plasticity in metallic glass matrix composites: A review. *Metals and Materials International*, *11*(1), 19–27. doi:10.1007/BF03027480

Park, E., Lim, H., Kim, W., & Kim, D. (2002). The effect of Sn addition on the glass-forming ability of Cu–Ti–Zr–Ni–Si metallic glass alloys. *Journal of Non-Crystalline Solids*, *298*(1), 15–22. doi:10.1016/S0022-3093(01)01047-X

Park, J. B., & Kim, Y. K. (2003). *Metallic biomaterials. Biomaterials principles and application.* CRC Press.

Parlar, Z., Bakkal, M., & Shih, A. J. (2008). Sliding tribological characteristics of Zr-based bulk metallic glass. *Intermetallics*, *16*(1), 34–41. doi:10.1016/j.intermet.2007.07.011

Patel, B. A. (2020). Amperometry and potential step techniques. In B. Patel (Ed.), *Electrochemistry for Bioanalysis* (pp. 9–26). Elsevier. doi:10.1016/B978-0-12-821203-5.00009-9

Patscheider, J. (2003). Nanocomposite Hard Coatings for Wear Protection. *MRS Bulletin*, *28*(03), 180–183. doi:10.1557/mrs2003.59

Pattison, J., Celotto, S., Khan, A., & O'Neill, W. (2008). Standoff distance and bow shock phenomena in the cold spray process. *Surface and Coatings Technology*, *202*(8), 1443–1454.

Pauly, S., Schricker, C., Scudino, S., Deng, L., & Kühn, U. (2017). Processing a glass-forming Zr-based alloy by selective laser melting. *Materials & Design*, *135*, 133–141. doi:10.1016/j.matdes.2017.08.070

Pawlak, W., & Wendler, B. (2009). Multilayer, hybrid PVD coatings on Ti6Al4V titanium alloy. *The Journal of Achievements in Materials and Manufacturing Engineering*, *37*(2), 660–667.

Pawlowski, L. (2009). Suspension and solution thermal spray coatings. *Surface and Coatings Technology*, *203*(19), 2807–2829. doi:10.1016/j.surfcoat.2009.03.005

Pedraza, F., Mahadik, S. A., & Bouchaud, B. (2015). Synthesis of ceria based superhydrophobic coating on Ni20Cr substrate via cathodic electrodeposition. *Physical Chemistry Chemical Physics*, *17*(47), 31750–31757. doi:10.1039/C5CP04723D PMID:26562006

Pei, Y. T., Chen, C. Q., Shaha, K. P., De Hosson, J. T. M., Bradley, J. W., Voronin, S. A., & Čada, M. (2008). Microstructural control of TiC/a-C nanocomposite coatings with pulsed magnetron sputtering. *Acta Materialia*, *56*(4), 696–709. doi:10.1016/j.actamat.2007.10.025

Pelcastre, L., Hardell, J., Herrera, N., & Prakash, B. (2012). Investigations into the Damage Mechanisms of Form Fixture Hardening Tools. *Engineering Failure Analysis*, *25*, 219–226. doi:10.1016/j.engfailanal.2012.05.014

Pellizzari, M., Zadra, M., & Molinari, A. (2007). Tribological Properties of Surface Engineered Hot Work Tool Steel for Aluminum Extrusion Dies. *Surface Engineering*, *23*(3), 165–168. doi:10.1179/174329406X150477

Peng, P., Kumar, S., Voelcker, N. H., Szili, E., Smart, R. S. C., & Griesser, H. J. (2006). Thin calcium phosphate coatings on titanium by electrochemical deposition in modified simulated body fluid. *Journal of Biomedical Materials Research*, *76*(2), 347–355. doi:10.1002/jbm.a.30514 PMID:16270340

Peters, A. M., & Nastasi, M. (2002). Effect of carrier gas on the deposition of titanium carbo-nitride coatings by a novel organo-metallic plasma immersion ion processing technique. *Vacuum*, *67*(2), 169–175. doi:10.1016/S0042-207X(02)00108-2

Pfender, E. (1994). Plasma jet behavior and modeling associated with the plasma spray process. *Thin Solid Films*, *238*(2), 228–241. doi:10.1016/0040-6090(94)90060-4

Phulera, N., Kaushik, S., Kshetri, R., Kanojia, N., Rawat, K., & Uniyal, V. (2021). Effects of plasma nitriding on mechanical, tribological and corrosion properties of friction stir welded joints of Al 2024. Materials Today: Proceedings, 4(15), 6726-6732.

Pintilei, L. (2008). *The science and engineering of thermal spray coatings* (2nd ed.). Wiley. doi:10.1002/9780470754085

Podgornik, B., Vižintin, J., Jacobson, S., & Hogmark, S. (2004). Tribological behaviour of WC/C coatings operating under different lubrication regimes. *Surface and Coatings Technology*, *177–178*, 558–565. doi:10.1016/S0257-8972(03)00927-7

Podgornik, B., Zuzek, B., Kafexhiu, F., & Leskovšek, V. (2016). Effect of Si Content on Wear Performance of Hot Work Tool Steel. *Tribology Letters*, *63*(5), 5. Advance online publication. doi:10.100711249-016-0695-6

Pogrebnjak, A. D., & Beresnev, V. M. (2012). *Hard Nanocomposite Coatings, Their Structure and Properties. Nanocomposites - New Trends and Developments* (Vol. 1). InTech Publishers.

Pogue, B. W., & Wilson, B. C. (2018). Optical and x-ray technology synergies enabling diagnostic and therapeutic applications in medicine. *Journal of Biomedical Optics*, *23*(12), 121610. doi:10.1117/1.JBO.23.12.121610 PMID:30350489

Polcar, T., Novák, R., & Široký, P. (2006). The tribological characteristics of TiCN coating at elevated temperatures. *Wear*, *260*(1–2), 40–49. doi:10.1016/j.wear.2004.12.031

Polcar, T., Vitu, T., Cvrcek, L., Vyskocil, J., & Cavaleiro, A. (2010). Effects of carbon content on the high temperature friction and wear of chromium carbonitride coatings. *Tribology International*, *43*(7), 1228–1233. doi:10.1016/j.triboint.2009.12.010

Pole, M., Sadeghilaridjani, M., Shittu, J., Mahajan, C., Ghodki, N., & Mukherjee, S. (2021). Electrodeposited metallic glasses with superlative wear resistance. *Materials Science and Engineering A*, *816*, 141315. doi:10.1016/j.msea.2021.141315

Polo da Fonseca, C. N., De Paoli, M.-A., & Gorenstein, A. (1994). Electrochromism in cobalt oxide thin films grown by anodic electroprecipitation. *Solar Energy Materials and Solar Cells*, *33*(1), 73–81. doi:10.1016/0927-0248(94)90291-7

Poot, M., & van der Zant, H. S. J. (2008). Nanomechanical properties of few-layer graphene membranes. *Applied Physics Letters*, *92*(6), 063111.

Posner, A. S., & Betts, F. (1975). *Synthetic Amorphous Calcium Phosphate and Its Relation to Bone Mineral Structure* (Vol. 8). Bone Mineral Structure.

Posti, E., & Nieminen, I. (1989). Influence of Coating Thickness on the Life of TiN-Coated High Speed Steel Cutting Tools. *Wear*, *129*(2), 273–283. doi:10.1016/0043-1648(89)90264-0

Pougoum, F., Qian, J., Laberge, M., Martinu, L., Klemberg-Sapieha, J., Zhou, Z., Li, K. Y., Savoie, S., & Schulz, R. (2018). Investigation of Fe3Al-Based PVD/HVOF Duplex Coatings to Protect Stainless Steel from Sliding Wear against Alumina. *Surface and Coatings Technology*, *350*(September), 699–711. doi:10.1016/j.surfcoat.2018.07.070

Prakash, B. (2005). Abrasive wear behaviour of Fe, Co and Ni based metallic glasses. *Wear*, *258*(1-4), 217–224. doi:10.1016/j.wear.2004.09.010

Prasanna, K., Subburaj, T., Jo, Y. N., & Lee, C. W. (2013). Optimization of electrophoretic suspension to fabricate Li[Ni1/3Co1/3Mn1/3]O2 based positive electrode for Li-ion batteries. *Electrochimica Acta*, *95*, 295–300. doi:10.1016/j.electacta.2013.01.102

Pratap, B., Bhatt, V., & Chaudhary, V. (2015). A review on thermal spray coating. *International Journal of Scientific and Engineering Research*, *6*(5), 53–59.

Prchlik, L., & Sampath, S. (2007). Effect of the Microstructure of Thermally Sprayed Coatings on Friction and Wear Response under Lubricated and Dry Sliding Conditions. *Wear*, *262*(1-2), 11–23. doi:10.1016/j.wear.2006.03.042

Premphet, P., Prasoetsri, M., Boonyawan, D., Supruangnet, R., Udomsom, S., & Leksakul, K. (2017). Optimization of DC Magnetron Sputtering Deposition Process and Surface Properties of HA-TiO2 Film. *Materials Today: Proceedings*, *4*(5), 6372–6380. doi:10.1016/j.matpr.2017.06.141

Prince, M. B. (1955). Silicon solar energy converters. *Journal of Applied Physics*, *26*(5), 534–540. doi:10.1063/1.1722034

Pruncu, C. I., Braic, M., Dearn, K. D., Farcau, C., Watson, R., Constantin, L. R., Balaceanu, M., Braic, V., & Vladescu, A. (2017). Corrosion and tribological performance of quasi-stoichiometric titanium containing carbo-nitride coatings. *Arabian Journal of Chemistry*, *10*(7), 1015–1028. doi:10.1016/j.arabjc.2016.09.009

Pruncu, C. I., Vladescu, A., Parau, A. C., Braic, M., Dearn, K. D., Constantin, L. R., & Braic, V. (2019). Multifunctional Ti based carbonitride coatings for applications in severe environments. *Thin Solid Films*, *682*, 63–75. doi:10.1016/j.tsf.2019.04.052

Purohit, P., & Vagge, S. T. (2016). Evaluation of alumina incorporated combined ceramic layer thermal barrier coating. *Surface and Coatings Technology*, *307*, 871–878. doi:10.1016/j.surfcoat.2016.10.022

Putzig, C. L., Leugers, M. A., McKelvy, M. L., Mitchell, G. E., Nyquist, R. A., Papenfuss, R. R., & Yurga, L. (1994). Infrared spectroscopy. *Analytical Chemistry*, *66*(12), 26–66. doi:10.1021/ac00084a003 PMID:8092469

Pye, D. (2003). *Practical Nitriding and Ferritic Nitrocarburizing*. ASM International. doi:10.31399/asm.tb.pnfn.9781627083508

Qiao, Y., Li, W., Wang, G., Zhang, X., & Cao, N. (2015). Application of ordered mesoporous silica nanocontainers in an anticorrosive epoxy coating on a magnesium alloy surface. *RSC Advances*, *5*(59), 47778–47787. doi:10.1039/C5RA05266A

Qi, C., Zheng, Y., Cao, L., Gao, J., & Wan, Y. (2016). Preparation and performance of sol–gel-derived alumina film modified by stearic acid. *Journal of Sol-Gel Science and Technology*, *78*(3), 641–646. doi:10.100710971-016-3989-0

Qiu, S.-B., & Yao, K.-F. (2008). Novel application of the electrodeposition on bulk metallic glasses. *Applied Surface Science*, *255*(5), 3454–3458. doi:10.1016/j.apsusc.2008.07.077

Quéré, D. (1999). Fluid coating on a fiber. *Annual Review of Fluid Mechanics*, *31*(1), 347–384. doi:10.1146/annurev.fluid.31.1.347

Quinn, T. F. J. (1967). The Effect of Hot Spot Temperatures on the Un-lubricated Wear of Steels. *Transasle.*, *10*, 158–168.

Quinn, T. F. J. (1971). Oxidational Wear. *Wear*, *18*(5), 413–419. doi:10.1016/0043-1648(71)90005-6

Quinn, T. F. J. (1992). Oxidational Wear Modelling: I. *Wear*, *153*(1), 179–200. doi:10.1016/0043-1648(92)90269-E

Quinn, T. F. J. (1994). Oxidational Wear Modelling: Part II. The General Theory of Oxidational Wear. *Wear*, *175*(1-2), 199–208. doi:10.1016/0043-1648(94)90183-X

Quint, M., & Kopech, H. (1999). High Energy Plasma Ceramic Coating Optimization. In *23rd Annual Conference on Composites, Advanced Ceramics, Materials, and Structures: B: Ceramic Engineering and Science Proceedings* (pp. 335–345). Academic Press.

Radha, R., & Sreekanth, D. (2017). Insight of magnesium alloys and composites for orthopedic implant applications – a review. *Journal of Magnesium and Alloys*, *5*(3), 286–312. doi:10.1016/j.jma.2017.08.003

Rahaman, M. L., Zhang, L., Liu, M., & Liu, W. (2015). Surface roughness effect on the friction and wear of bulk metallic glasses. *Wear*, *332*, 1231–1237. doi:10.1016/j.wear.2014.11.030

Rahmana, M. S., Ding, J., Beheshti, A., Zhang, X., & Polycarpou, A. (2018). Elevated temperature tribology of Ni alloys under helium environment for nuclear reactor applications. *Tribology International*, *123*, 372–384. doi:10.1016/j.triboint.2018.03.021

Rahman, M., Li, Y., & Wen, C. (2020). HA coating on Mg alloys for biomedical applications: A review. *Journal of Magnesium and Alloys*, *8*(3), 929–943. doi:10.1016/j.jma.2020.05.003

Rajapaksha, R. D. A. A., Uda, M. N. A., Hashim, U., Gopinath, S. C. B., & Fernando, C. A. N. (2018). *Impedance based aluminium interdigitated electrode (Al-IDE) biosensor on silicon substrate for salmonella detection.* Paper presented at the 2018 IEEE International Conference on Semiconductor Electronics (ICSE). 10.1109/SMELEC.2018.8481324

Rajendran, R. (2012). Gas turbine coatings-An overview. *Engineering Failure Analysis*, *26*, 355–369. doi:10.1016/j.engfailanal.2012.07.007

Ramakrishna, S., Ramalingam, M., Kumar, T. S. S., & Soboyejo, W. O. (2010). *Biomaterials: A Nano Approach. CRC Press*. Taylor & Francis Group.

Randviir, E. P., & Banks, C. E. (2013). Electrochemical impedance spectroscopy: An overview of bioanalytical applications. *Analytical Methods*, *5*(5), 1098–1115. doi:10.1039/c3ay26476a

Rangaraj, S., & Kokini, K. (2003). Interface thermal fracture in functionally graded zirconia–mullite–bond coat alloy thermal barrier coatings. *Acta Materialia*, *51*(1), 251–267. doi:10.1016/S1359-6454(02)00396-8

Rao, A. V. V., Latthe, S. S. S., Mahadik, S. A. S. A., & Kappenstein, C. (2011). Mechanically stable and corrosion resistant superhydrophobic sol-gel coatings on copper substrate. *Applied Surface Science*, *257*(13), 5772–5776. doi:10.1016/j.apsusc.2011.01.099

Rathod, W. S., Khanna, A. S., Rathod, R. C., & Sapate, S. G. (2014). Wear and Corrosion Behavior of CoNiCrAlY Bond Coats. *Journal of The Institution of Engineers (India): Series C*, *95*(3), 261–271.

Rathod, S. W., Khanna, A. S., Karthikeyan, J., & Rathod, C. R. (2014). Effect of N_2 and He Carrier Gases on Oxidation Behavior of Cold Sprayed CoNiCrAlY Powder to Deposit Bond Coats. *Transactions of the Indian Institute of Metals*, *67*(2), 247–262. doi:10.100712666-013-0344-9

Ratia, V. L., Zhang, D., Daure, J. L., Shipway, P. H., McCartney, D. G., & Stewart, D. A. (2019). Sliding wear of a self-mated thermally sprayed chromium oxide coating in a simulated PWR water environment. *Wear*, *426-427*, 1466-1473.

Ren, X., Yang, X., Xie, G., He, F., Wang, R., Zhang, C., Guo, D. & Luo, J. (2021). Superlubricity under Ultrahigh Contact Pressure Enabled by Partially Oxidized Black Phosphorus Nanosheets. *NPJ 2D Materials and Applications*, *5*, 44. doi:10.1038/s41699-021-00225-0

Ren, L., Meng, M., Wang, Z., Yang, F., Yang, H., Zhang, T., & Qiao, J. (2015). Enhancement of plasticity in Zr-based bulk metallic glasses electroplated with copper coatings. *Intermetallics*, *57*, 121–126. doi:10.1016/j.intermet.2014.10.009

Rester, M., Neidhardt, J., Eklund, P., Emmerlich, J., Ljungcrantz, H., Hultman, L., & Mitterer, C. (2006). Annealing studies of nanocomposite Ti-Si-C thin films with respect to phase stability and tribological performance. *Materials Science and Engineering A*, *429*(1–2), 90–95. doi:10.1016/j.msea.2006.05.053

Rettig, R., & Virtanen, S. (2008). Time-dependent electrochemical characterization of the corrosion of a magnesium rare-earth alloy in simulated body fluids. *Journal of Biomedical Materials Research. Part A*, *85*(1), 167–175. doi:10.1002/jbm.a.31550 PMID:17688266

Reynolds, C. D., Slater, P. R., Hare, S. D., Simmons, M. J. H., & Kendrick, E. (2021). A review of metrology in lithium-ion electrode coating processes. *Materials & Design*, *209*, 109971. doi:10.1016/j.matdes.2021.109971

Rezaei, B., Shams-Ghahfarokhi, L., Havakeshian, E., & Ensafi, A. A. (2016). An electrochemical biosensor based on nanoporous stainless steel modified by gold and palladium nanoparticles for simultaneous determination of levodopa and uric acid. *Talanta*, *158*, 42–50. doi:10.1016/j.talanta.2016.04.061 PMID:27343576

Richer, P., Yandouzi, M., Beauvais, L., & Jodoin, B. (2010). Oxidation behaviour of CoNiCrAlY bond coats produced by plasma, HVOF and cold gas dynamic spraying. *Surface and Coatings Technology*, *204*(24), 3962–3974. doi:10.1016/j.surfcoat.2010.03.043

Righeira Carnegie, M., Sherine, A., Sivagami, D., & Sakthivel, S. (2016). Anti-reflection coatings with enhanced abrasion and scratch resistance properties. *Journal of Sol-Gel Science and Technology*, *78*(1), 176–186. doi:10.100710971-015-3924-9

Riva, R., Eugenio, I. P.-S., Lak, A., Pellegrino, T., Pérez-Juste, J., & Mattoli, V. (2017). Plasmonic/Magnetic Nanocomposites: Gold Nanorods-Functionalized Silica Coated Magnetic Nanoparticles. *Journal of Colloid and Interface Science*, *502*(September), 201–209. doi:10.1016/j.jcis.2017.04.089 PMID:28486141

Robertson, J. (2002). Diamond-like Amorphous Carbon. *Materials Science and Engineering R Reports*, *37*(4–6), 129–281. doi:10.1016/S0927-796X(02)00005-0

Rochow, T. G., & Tucker, P. A. (1994). *A brief history of microscopy introduction to microscopy by means of light, electrons, X rays, or acoustics*. Springer US.

Rodriguez, A., Jaman, M. S., Acikgoz, O., Wang, B., Yu, J., Grützmacher, P. G., Rosenkranz, A., & Baykara, M. Z. (2021). The Potential of $Ti_3C_2T_x$ Nano-sheets (MXenes) for Nanoscale Solid Lubrication Revealed by Friction Force Microscopy. *Applied Surface Science*, *535*, 147664. doi:10.1016/j.apsusc.2020.147664

Rodríguez-González. (2009). *Biomaterials in Orthopaedic Surgery*. ASM International.

Rokhlin, L. L. (2003). *Magnesium Alloys Containing Rare Earth Metals: Structure and Properties*. Taylor & Francis. doi:10.1201/9781482265163

Ropp, R. C., & Libowitz, G. G. (1978). The nature of the alumina-rich phase in the System La_2O_3-$A1_2O_3$. *Journal of the American Ceramic Society*, *61*(11-12), 473–475. doi:10.1111/j.1151-2916.1978.tb16119.x

Rosenkranz, A., Costa, H. L., Baykara, M. Z., & Martini, A. (2020). Synergetic Effects of Surface Texturing and Solid Lubricants to Tailor Friction and Wear - A Review. *Tribology International*, *155*, 106792. doi:10.1016/j.triboint.2020.106792

Rosenkranz, A., Grützmacher, P. G., Espinoza, R., Fuenzalida, V. M., Blanco, E., Escalona, N., Gracia, F. J., Villarroel, R., Guo, L., Kang, R., Mücklich, F., Suarez, S., & Zhang, Z. (2019). Multi-Layer $Ti_3C_2T_x$-Nanoparticles (MXenes) as Solid Lubricants – Role of Surface Terminations and Intercalated Water. *Applied Surface Science*, *494*, 13–21. doi:10.1016/j.apsusc.2019.07.171

Rosenkranz, A., Liu, Y., Yang, L., & Chen, L. (2020). 2D Nano-Materials beyond Graphene: From Synthesis to Tribological Studies. *Applied Nanoscience*, *10*(9), 3353–3388. doi:10.100713204-020-01466-z

Rovatti, L., Lecis, N., Dellasega, D., Russo, V., & Gariboldi, E. (2018). Influence of aging in the temperature range 250–350° C on the tribological performance of a WC-CoCr coating produced by HVOF. *International Journal of Refractory Metals & Hard Materials*, *75*, 218–224. doi:10.1016/j.ijrmhm.2018.04.017

Rozman, M., Bren, U., Lukšič, M., Godec, R. F., Bokias, G., Kalarakis, A. N., & Stathatos, E. (2018). Electrochromic cell with hydrogel-stabilized water-based electrolyte using electrodeposition as a fast color changing mechanism. *Electrochimica Acta*, *283*, 1105–1114. doi:10.1016/j.electacta.2018.07.052

Rozman, M., Cerar, J., Lukšič, M., Uršič, M., Mourtzikou, A., Spreizer, H., Kozjek Škofic, I., & Stathatos, E. (2017). Electrochromic properties of thin nanocrystalline TiO2 films coated electrodes with adsorbed Co(II) or Fe(II) 2,2′-bipyridine complexes. *Electrochimica Acta*, *238*, 278–287. doi:10.1016/j.electacta.2017.04.030

Rozman, M., Cetin, N., Bren, U., & Lukšič, M. (2021). Use of Different Metal Oxide Coatings in Stainless Steel Based ECDs for Smart Textiles. *Electronics (Basel), 10*(20), 2529. doi:10.3390/electronics10202529

Rozman, M., Gaberšček, M., Marolt, G., Bren, U., & Lukšič, M. (2019). An inverted sandwich electrochromic device architecture does not require optically transparent electrodes. *Advanced Materials Technologies*, *4*(9), 1900389. doi:10.1002/admt.201900389

Rozman, M., Mavrič, A., Kravanja, G., Valant, M., & Pakseresht, A. (2022). Ultra-low-cost, flexible and durable electrochromic tape device based on aluminum foil. *Electrochimica Acta*, *404*, 139760. doi:10.1016/j.electacta.2021.139760

Rozman, M., Sygkridou, D., Fuchs Godec, R., Stathatos, E., & Bren, U. (2019). Novel geometric approach for photosensor construction based on dye-sensitization of TiO_2 nanoparticles on stainless steel. *Sensors and Actuators. A, Physical*, *295*, 51–58. doi:10.1016/j.sna.2019.05.034

Rozman, M., Žener, B., Matoh, L., Godec, R. F., Mourtzikou, A., Stathatos, E., Bren, U., & Lukšič, M. (2020). Flexible electrochromic tape using steel foil with WO_3 thin film. *Electrochimica Acta*, *330*, 135329. doi:10.1016/j.electacta.2019.135329

Saeidi, F., Parlinska-Wojtan, M., & Hoffmann, P. (2021). *Effects of laser surface texturing on the wear and failure mechanism of grey cast iron reciprocating against steel under starved lubrication conditions.* . doi:10.1016/j.wear.2017.05.015

Saffar, M. A., Eshaghi, A., & Dehnavi, M. R. (2021). Fabrication of superhydrophobic, self-cleaning and anti-icing ZnO/PTFE-SiO2 nano-composite thin film. *Materials Chemistry and Physics*, *259*, 124085. doi:10.1016/j.matchemphys.2020.124085

Sahith, M. S., Giridhara, G., & Kumar, R. S. (2018). Development and analysis of thermal barrier coatings on gas turbine blades – A Review. *Materials Today: Proceedings*, *5*(1), 2746–2751. doi:10.1016/j.matpr.2018.01.060

Sahoo, S. (2021). Self-Lubricating Composites with 2D Materials as Reinforcement: A New Perspective. *Reinforced Plastics*, *65*(2), 101–103. doi:10.1016/j.repl.2020.06.007

Sahu, N., Parija, B., & Panigrahi, S. (2009). Fundamental understanding and modeling of spin coating process: A review. *Indian Journal of Physics*, *83*(4), 493–502. doi:10.100712648-009-0009-z

Saida, J., Kato, H., Setyawan, A. D. H., & Inoue, A. (2005). Characterization and properties of nanocrystal-forming Zr-based bulk metallic glasses. *Reviews on Advanced Materials Science*, *10*(1), 34–38.

Saikia, U., Sahariah, M. B., & Pandey, R. (2016). Stability of Cu–Nb Layered Nanocomposite from Chemical Bonding. *Chemical Physics Letters*, *655–656*(July), 59–65. doi:10.1016/j.cplett.2016.05.022

Salehan, R., Shahverdi, H. R., & Miresmaeili, R. (2019). Effects of annealing on the tribological behavior of $Zr_{60}Cu_{10}Al_{15}Ni_{15}$ bulk metallic glass. *Journal of Non-Crystalline Solids*, *517*, 127–136. doi:10.1016/j.jnoncrysol.2019.05.013

Salman, S. A., Kuroda, K., & Okido, M. (2013). Preparation and Characterization of Hydroxyapatite Coating on AZ31 Mg Alloy for Implant Applications. *Hindawi Publishing Corporation, Bioinorganic Chemistry and Applications. Article ID*, *175756*, 6.

Sanjay, L. H., Bhamre, V. G., & Londhe, B. C. (2017). A Review on plasma spray coatings and its characterization. *IRJET*, *4*(7), 1627–1631.

Santecchia, E., Hamouda, A. M. S., Musharavati, F., Zalnezhad, E., Cabibbo, M., & Spigarelli, S. (2015). Wear resistance investigation of titanium nitride-based coatings. *Ceramics International*, *41*(9), 10349–10379. doi:10.1016/j.ceramint.2015.04.152

Santos, L., Silveira, C. M., Elangovan, E., Neto, J. P., Nunes, D., Pereira, L., Martins, R., Viegas, J., Moura, J. J. G., Todorovic, S., Almeida, M. G., & Fortunato, E. (2016). Synthesis of WO3 nanoparticles for biosensing applications. *Sensors and Actuators. B, Chemical*, *223*, 186–194. doi:10.1016/j.snb.2015.09.046

Sapate, S. G., & Manish, R. (2015). Solid Particle Erosion of Thermal Sprayed Coatings. In M. Roy & J. P. Davim (Eds.), *Thermal Sprayed Coatings and their Tribological Performances* (pp. 193–226). IGI Global. doi:10.4018/978-1-4666-7489-9.ch007

Saris, N. E., Mervaala, E., Karppanen, H., Khawaja, J. A., & Lewenstam, A. (2000). Magnesium: An update on physiological, clinical and analytical aspects. *Clinica Chimica Acta*, *394*(1-2), 1–26. doi:10.1016/S0009-8981(99)00258-2 PMID:10727669

Sawyer, V., Tao, X., Dong, H., Dashtbozorg, B., Li, X., Sammons, R., & Dong, H.-S. (2020). Improving the tribological properties and biocompatibility of Zr-based bulk metallic glass for potential biomedical applications. *Materials (Basel)*, *13*(8), 1960. doi:10.3390/ma13081960 PMID:32331294

Schmidt, D. J., Pridgen, E. M., Hammond, P. T., & Love, J. C. (2010). Layer-by-Layer Assembly of a pH-Responsive and Electrochromic Thin Film. *Journal of Chemical Education*, *87*(2), 208–211. doi:10.1021/ed800045r

Schmitt, M., Baunach, M., Wengeler, L., Peters, K., Junges, P., Scharfer, P., & Schabel, W. (2013). Slot-die processing of lithium-ion battery electrodes—Coating window characterization. *Chemical Engineering and Processing*, *68*, 32–37. doi:10.1016/j.cep.2012.10.011

Schneider, H., Schreuer, J., & Hildmann, B. (2008). Structure and properties of mullite—A review. *Journal of the European Ceramic Society*, *28*(2), 329–344. doi:10.1016/j.jeurceramsoc.2007.03.017

Schneider, S. (2001). Bulk metallic glasses. *Journal of Physics Condensed Matter*, *13*(34), 7723–7736. doi:10.1088/0953-8984/13/34/316

Schroers, J. (2010). Processing of bulk metallic glass. *Advanced Materials*, *22*(14), 1566–1597. doi:10.1002/adma.200902776 PMID:20496386

Schwarzer, N., & Richter, F. (2006). On the determination of film stress from substrate bending: STONEY's formula and its limits. *Materials Science*, *1*, 1–17.

Scully, J., Gebert, A., & Payer, J. (2007). Corrosion and related mechanical properties of bulk metallic glasses. *Journal of Materials Research*, *22*(2), 302–313. doi:10.1557/jmr.2007.0051

Sebastian, D., Yao, C.-W., Nipa, L., Lian, I., & Twu, G. (2021). Corrosion Behavior and Mechanical Properties of a Nanocomposite Superhydrophobic Coating. *Coatings*, *11*(6), 652. doi:10.3390/coatings11060652

Seiler, H. G., Sigel, H., & Sigel, A. (1988). *Handbook on Toxicity of Inorganic Compounds*. Marcel Dekker.

Selva, K. N., Krupanidhi, S. B., & Barshilia, H. C. (2014). Carbon nano tubes based tandem absorber with tunable spectral selectivity: Transition from near-perfect blackbody absorber to solar selective absorber. *Advanced Materials*, *26*(16), 2552–2557. doi:10.1002/adma.201305070 PMID:24474148

Seo, D., Ogawa, K., Nakao, Y., Miura, H., & Shoji, T. (2009). Influence of high-temperature creep stress on growth of thermally grown oxide in thermal barrier coatings. *Surface and Coatings Technology*, *203*(14), 1979–1983. doi:10.1016/j.surfcoat.2009.01.029

Sergevnin, V. S., Blinkov, I. V., Belov, D. S., Smirnov, N. I., Volkhonskii, A. O., & Kuptsov, K. A. (2018). Wear and Erosion of Arc-PVD Multilayer Ti-Al-Mo-N Coatings under Various Conditions of Friction and Loading. *International Journal of Advanced Manufacturing Technology*, *98*(1–4), 593–601. doi:10.100700170-018-2235-z

Shahmohammadi, P., & Khazaei, B. A. (2021). Characterization of Zn/Mg-enriched calcium phosphate coating produced by the two-step pulsed electrodeposition method on titanium substrate. *Surfaces and Interfaces*, *22*, 100819. doi:10.1016/j.surfin.2020.100819

Shalini, S., Balasundaraprabhu, R., Kumar, T. S., Prabavathy, N., Senthilarasu, S., & Prasanna, S. (2016). Status and outlook of sensitizers/dyes used in dye sensitized solar cells (DSSC): A review. *International Journal of Energy Research*, *40*(10), 1303–1320. doi:10.1002/er.3538

Sharifahmadian, O., & Mahboubi, F. (2019). A comparative study of microstructural and tribological properties of N-DLC/DLC double layer and single layer coatings deposited by DC-pulsed PACVD process. *Ceramics International*, *45*(6), 7736–7742. doi:10.1016/j.ceramint.2019.01.076

Sharma, S. (2012). High Temperature Erosive Wear Study of NiCrFeSiB Flame Sprayed Coatings. *Journal of The Institution of Engineers (India): Series D, 93*(1), 7–12. . doi:10.1007/s40033-012-0006-9

Sharma, A. K. (1992). Anodizing titanium for space applications. *Thin Solid Films*, *208*(1), 48–54. doi:10.1016/0040-6090(92)90946-9

Shaw. (2003). Corrosion resistance of magnesium alloys. *ASM handbook volume 13a: Corrosion: Fundamentals, testing and protection.*

Shekhar, H., & Dumpala, R. (2021). Overcoming Friction and Steps towards Superlubricity: A Review of Underlying Mechanisms. *Applied Surface Science Advances*, *6*, 100175. doi:10.1016/j.apsadv.2021.100175

Shen, J., Yin, W., Wei, Q., Li, Y., Liu, J., & An, L. (2013). Effect of ceramic nanoparticle reinforcements on the quasi-static and dynamic mechanical properties of magnesium-based metal matrix composites. *Journal of Materials Research*, *28*, 1835–1852.

Shenogin, S., Xue, L., Ozisik, R., Keblinski, P., & Cahill, D. G. (2004). Role of thermal boundary resistance on the heat flow in carbon-nanotube composites. *Journal of Applied Physics*, *95*(12), 8136–8144. doi:10.1063/1.1736328

Shobana, M. K. (2019). Metal oxide coated cathode materials for Li ion batteries – A review. *Journal of Alloys and Compounds*, *802*, 477–487. doi:10.1016/j.jallcom.2019.06.194

Shtansky, D. V., Kuptsov, K. A., Kiryukhantsev-Korneev, P. V., Sheveiko, A. N., Fernandez, A., & Petrzhik, M. I. (2011). Comparative investigation of Al- and Cr-doped TiSiCN coatings. *Surface and Coatings Technology*, *205*(19), 4640–4648. doi:10.1016/j.surfcoat.2011.04.012

Shum, P. W., Zhou, Z. F., & Li, K. Y. (2013). To Increase the Hydrophobicity and Wear Resistance of Diamond-like Carbon Coatings by Surface Texturing Using Laser Ablation Process. In *Thin Solid Films* (Vol. 544, pp. 472–476). Elsevier. doi:10.1016/j.tsf.2013.02.075

Sidhu, B.S., & Prakash, S. (2006). Performance of Ni-CrAlY, Ni-Cr, stellite –6, and Ni3 Al coating in Na2SO4-60%V_2O_5 environment at 900°C under cyclic conditions. *Surface and Coating Technology, 201*(3), 1643–1654. doi:10.1016/j.surfcoat.2006.02.035

Sidhu, T. S., Prakash, S., & Agrawal, R. D. (2006a). Hot corrosion resistance of high-velocity oxy-fuel sprayed coatings on a nickel-base super alloy in molten salt environment. *Thermal Spray Technology.*, *15*(3), 387–399. doi:10.1361/105996306X124392

Sidhu, T. S., Prakash, S., & Agrawal, R. D. (2006b). Hot corrosion behaviour of HVOF sprayed NiCrBSi coatings on Ni- and Fe- base super alloys in Na2SO4-60%V2O5 environment at 9000C. *Acta Materialia (Elsevier)*, *54*(3), 773–784. doi:10.1016/j.actamat.2005.10.009

Siegrist, M. E., Amstad, E. D., & Löffler, J. F. (2007). Tribological properties of graphite-and ZrC-reinforced bulk metallic glass composites. *Intermetallics*, *15*(9), 1228–1236. doi:10.1016/j.intermet.2007.03.001

Singer, I. L., Fayeulle, S., & Ehni, P. D. (1991). Friction and Wear Behavior of TiN in Air: The Chemistry of Transfer Films and Debris Formation. *Wear*, *149*(1–2), 375–394. doi:10.1016/0043-1648(91)90386-9

Singh, H., Kumar, S., & Kumar, R. (2021). Overview of Corrosion and its Control: A Critical Review. *Proceedings on Engineering Sciences.*, *3*(1), 13–24. Advance online publication. doi:10.24874/PES03.01.002

Singh, H., Prakash, S., Puri, D., & Phase, D. M. (2006). Cyclic oxidation behavior of some plasma-sprayed coatings in Na2SO4-60%V2O5 environment. *Materials Engineering and Performance.*, *15*(6), 729–741. doi:10.1361/105994906X150858

Singh, J., & Wolfe, D. E. (2005). Review Nano and Macro-Structured Component Fabrication by Electron Beam-Physical Vapor Deposition (EB-PVD). *Journal of Materials Science*, *40*(1), 1–26. doi:10.100710853-005-5682-5 PMID:15754137

Singh, L., Chawla, V., & Grewal, J. S. (2012). A review on detonation gun spray coating. *Journal of Minerals & Materials Characterization & Engineering*, *11*(03), 243–265. doi:10.4236/jmmce.2012.113019

Singh, S. K., Chattopadhyaya, S., Pramanik, A., Kumar, S., & Gupta, N. (2018). Influence of Nano-particle on the Wear behaviour of Thin Film Coatings (A Review). *International Journal of Applied Engineering Research.*, *13*(6), 4053–4058.

Siwal, S. S., Saini, A. K., Rarotra, S., Zhang, Q., & Thakur, V. K. (2021). Recent advancements in transparent carbon nanotube films: Chemistry and imminent challenges. *Journal of Nanostructure in Chemistry*, *11*(1), 93–130. doi:10.100740097-020-00378-2

Skapin, S., Kolar, D., & Suvorov, D. (1993). X-ray diffraction and microstructural investigation of the Al_2O_3-La_2O_3-TiO_2 system. *Journal of the American Ceramic Society*, *76*(9), 2359–2362. doi:10.1111/j.1151-2916.1993.tb07777.x

Smith, M. F. (2007). Comparing cold spray with thermal spray coating technologies. *The Cold Spray Materials Deposition Process*, 43–61. doi:10.1533/9781845693787.1.4

Smith, D. S., Fayette, S., Grandjean, S., Martin, C., Telle, R., & Tonnessen, T. (2003). Thermal resistance of grain boundaries in alumina ceramics and refractories. *Journal of the American Ceramic Society*, *86*(1), 105–111. doi:10.1111/j.1151-2916.2003.tb03285.x

Sohrabi, N., Jhabvala, J., & Logé, R. E. (2021). Additive manufacturing of bulk metallic glasses—process, challenges and properties: A review. *Metals*, *11*(8), 1279. doi:10.3390/met11081279

Sohrabi, N., Panikar, R. S., Jhabvala, J., Buch, A. R., Mischler, S., & Logé, R. E. (2020). Laser coating of a Zr-based metallic glass on an aluminum substrate. *Surface and Coatings Technology*, *400*, 126223. doi:10.1016/j.surfcoat.2020.126223

Somani, P. R., & Radhakrishnan, S. (2003). Electrochromic materials and devices: Present and future. *Materials Chemistry and Physics*, *77*(1), 117–133. doi:10.1016/S0254-0584(01)00575-2

Song, G. L., & Atrens, A. (2000). Corrosion mechanisms of magnesium alloys. *Advanced Engineering Materials*, •••, 11–33.

Song, Y. W., Shan, D. Y., & Han, E. H. (2008). Electrodeposition of hydroxyapatite coating on AZ91D magnesium. *Materials Letters*, *62*(17–18), 3276–3279. doi:10.1016/j.matlet.2008.02.048

Song, Y., Mandelli, D., Hod, O., Urbakh, M., Ma, M., & Zheng, Q. (2018). Robust Microscale Superlubricity in Graphite/Hexagonal Boron Nitride Layered Heterojunctions. *Nature Materials*, *17*(10), 894–899. doi:10.103841563-018-0144-z PMID:30061730

Song, Y., Zhang, S., Li, J., Zhao, C., & Zhang, X. (2010). Electrodeposition of Ca–P coatings on biodegradable Mg alloy: In vitro biomineralization behavior. *Acta Biomaterialia*, *6*(5), 1736–1742. doi:10.1016/j.actbio.2009.12.020 PMID:20018262

Sreenivasulu, V., & Manikandan, M. (2019). Hot corrosion studies of HVOF sprayed carbide and metallic powder coatings on alloy 80A at 900°C. *Materials Research Express*, *5*, 42–53. doi:10.1088/2053-1591/aaf65d

Sridharan, S., Xie, L., Jordan, E. H., Gell, M., & Murphy, K. S. (2005). Damage evolution in an electron beam physical vapor deposited thermal barrier coating as a function of cycle temperature and time. *Materials Science and Engineering A*, *393*(1-2), 51–62. doi:10.1016/j.msea.2004.09.037

Srinivasan, Kulkarni, & Anand. (2007). Thermal Stability and High-Temperature Wear of Ti-TiN and TiN-CrN Nanomultilayer Coatings under Self-Mated Conditions. *Tribology International*, *40*(2), 266–77. . doi:10.1016/j.triboint.2005.09.035

Staia, M. H., Mejias, A., La Barbera, G., Puchi-Cabrera, E. S., Villalobos-Gutierrez, C., Santana, Y. Y., Montagne, A., Iost, A., & Rodriguez, M. A. (2020). Mechanical properties of WC/Co-CNT HVOF sprayed coatings. *Surface Engineering*, *36*(11), 1156–1164. doi:10.1080/02670844.2018.1529285

Staiger, M. P., Pietak, A. M., Huadmai, J., & Dias, G. (2006). Magnesium and its alloys as orthopedic biomaterials: A review. *Biomaterials*, *27*(9), 1728–1734. doi:10.1016/j.biomaterials.2005.10.003 PMID:16246414

Staines, A., & Bell, T. (1979). Plasma Nitriding of High Alloy Steels. In *Proceedings of the Conference on Heat Treatment-Methods and Media*. Institute of Met.

Stergiopoulos, T., Rozi, E., Karagianni, C.-S., & Falaras, P. (2011). Influence of electrolyte co-additives on the performance of dye-sensitized solar cells. *Nanoscale Research Letters*, *6*(1), 307. doi:10.1186/1556-276X-6-307 PMID:21711833

Steve, O., & Bhadeshia, H. K. D. H. (2012). Duplex hardening of steels for aero engine Bearings. *ISIJ International*, *52*(11), 1927–1934. doi:10.2355/isijinternational.52.1927

Stott, F. H., Glascott, J., & Wood, G. C. (1985). Models for the Generation of Oxides during Sliding Wear. *Proc. R. Soc. Lond. A*, 167-186.

Stott, F. H. (2002). High Temperature Sliding Wear of Metals. *Tribology International*, *35*(8), 489–495. doi:10.1016/S0301-679X(02)00041-5

Stott, F. H., & Jordan, M. P. (2001). The Effects of Load and Substrate Hardness on the Development and Maintenance of Wear Protective Layers During Sliding at Elevated Temperatures. *Wear*, *250*(1-12), 391–400. doi:10.1016/S0043-1648(01)00601-9

Straffelini, G., Trabucco, D., & Molina, A. (2001). Oxidative Wear of Heat Treated Steels. *Wear*, *250*(1-12), 485–491. doi:10.1016/S0043-1648(01)00661-5

Strauss, H. W., Chromik, R. R., Hassani, S., & Klemberg-Sapieha, J. E. (2011). In situ tribology of nanocomposite Ti-Si-C-H coatings prepared by PE-CVD. *Wear*, *272*(1), 133–148. doi:10.1016/j.wear.2011.08.001

Štukovnik, Z. B., & Rozman, M. (2021). Model electrochemical biosensor for the detection of methanol in aqueous solutions with yeast cells. *Acta Chimica Slovenica*, *68*, 8. doi:10.17344/acsi.2020.6545

Sudharshan Phani, P., Srinivasa Rao, D., Joshi, S., & Sundararajan, G. (2007). Effect of process parameters and heat treatments on properties of cold sprayed copper coatings. *Journal of Thermal Spray Technology*, *16*(3), 425–434. doi:10.100711666-007-9048-1

Su, J., Kang, J., Yue, W., Ma, G., Fu, Z., Zhu, L., She, D., Wang, H., & Wang, C. (2019). Comparison of tribological behavior of Fe-based metallic glass coatings fabricated by cold spraying and high velocity air fuel spraying. *Journal of Non-Crystalline Solids*, *522*, 119582. doi:10.1016/j.jnoncrysol.2019.119582

Sundaresan, C., Rajasekaran, B., & Rao, D. S. (2020). Hot corrosion behavior of plasma and D-Gun sprayed coatings on T91 steel used in boiler applications. *IOP Conference Series. Materials Science and Engineering*, *872*(1), 45–56. doi:10.1088/1757-899X/872/1/012092

Sun, K., Zhang, S., Li, P., Xia, Y., Zhang, X., Du, D., Isikgor, F. H., & Ouyang, J. (2015). Review on application of PEDOTs and PEDOT:PSS in energy conversion and storage devices. *Journal of Materials Science Materials in Electronics*, *26*(7), 4438–4462. doi:10.100710854-015-2895-5

Sun, L., Yuan, G., Gao, L., Yang, J., Chhowalla, M., Gharahcheshmeh, M. H., & Liu, Z. (2021). Chemical vapour deposition. *Nature Reviews Methods Primers*, *1*(5), 5. Advance online publication. doi:10.103843586-020-00005-y

Sun, Q., Miskovic, D. M., Laws, K., Kong, H., Geng, X., & Ferry, M. (2020). Transition towards ultrastable metallic glasses in Zr-based thin films. *Applied Surface Science*, *533*, 147453. doi:10.1016/j.apsusc.2020.147453

Sun, X. W., & Wang, J. X. (2008). Fast switching electrochromic display using a viologen-modified ZnO nanowire array electrode. *Nano Letters*, *8*(7), 1884–1889. doi:10.1021/nl0804856 PMID:18564881

Sun, Y., Bell, T., & Wood, G. (1994). Wear Behaviour of Plasma Nitrided Martensitic Stainless Steel. *Wear*, *178*(1-2), 131–138. doi:10.1016/0043-1648(94)90138-4

Suresh Babu, P., Chanikya Rao, P., Jyothirmayi, A., Sudharshan Phani, P., Rama Krishna, L., & Srinivasa Rao, D. (2018). Evaluation of microstructure, property and performance of detonation sprayed WC-(W,Cr)2C-Ni coatings. *Surface and Coatings Technology*, *335*, 345–354. doi:10.1016/j.surfcoat.2017.12.055

Suresh Babu, P., Srinivasa Rao, D., Rama Krishna, L., & Sundararajan, G. (2017). Weibull analysis of hardness distribution in detonation sprayed nano-structured WC-12Co coatings. *Surface and Coatings Technology*, *319*, 394–402. doi:10.1016/j.surfcoat.2017.04.028

Surmeneva, M. A., Chaikina, M. V., Zaikovskiy, V. I., Pichugin, V. F., Buck, V., Prymak, O., Epple, M., & Surmenev, R. A. (2013). The structure of an RF-magnetron sputter-deposited silicate-containing hydroxyapatite-based coating investigated by high-resolution techniques. *Surface and Coatings Technology*, *218*, 39–46. doi:10.1016/j.surfcoat.2012.12.023

Surmeneva, M. A., Tyurin, A. I., Mukhametkaliyev, T. M., Pirozhkova, T. S., Shuvarin, I. A., Syrtanov, M. S., & Surmenev, R. A. (2015). Enhancement of the mechanical properties of AZ31 magnesium alloy via nanostructured hydroxyapatite thin fi lms fabricated via radio-frequency magnetron sputtering. *Journal of the Mechanical Behavior of Biomedical Materials*, *46*, 127–136. doi:10.1016/j.jmbbm.2015.02.025

Surmenev, R. A. (2012). A review of plasma-assisted methods for calcium phosphate-based coatings fabrication. *Surface and Coatings Technology*, *206*(8–9), 2035–2056. doi:10.1016/j.surfcoat.2011.11.002

Surmenev, R. A., Surmeneva, M. A., Evdokimov, K. E., Pichugin, V. F., Peitsch, T., & Epple, M. (2011). The influence of the deposition parameters on the properties of an rf-magnetron-deposited nanostructured calcium phosphate coating and a possible growth mechanism. *Surface and Coatings Technology*, *205*(12), 3600–3606. doi:10.1016/j.surfcoat.2010.12.039

Suryanarayana, C., & Inoue, A. (2013). Iron-based bulk metallic glasses. *International Materials Reviews*, *58*(3), 131–166. doi:10.1179/1743280412Y.0000000007

Suryanarayana, C., & Inoue, A. (2017). *Bulk metallic glasses*. CRC Press.

Sutija, D. P., Norby, T., & Björnbom, P. (1995). Transport number determination by the concentration-cell/open-circuit voltage method for oxides with mixed electronic, ionic and protonic conductivity. *Solid State Ionics*, *77*, 167–174. doi:10.1016/0167-2738(94)00268-W

Szmacinski, H., & Lakowicz, J. R. (1995). Fluorescence lifetime-based sensing and imaging. *Sensors and Actuators. B, Chemical*, *29*(1), 16–24. doi:10.1016/0925-4005(95)01658-9 PMID:33867678

Szymański, K., Góral, M., Kubaszek, T., & Monteiro, P. C. (2015). Microstructure of TBC Coatings Deposited by HVAF and PS-PVD Methods. *Diffusion and Defect Data, Solid State Data. Part B, Solid State Phenomena*, *227*, 373–376. doi:10.4028/www.scientific.net/SSP.227.373

Takeda, M., Onishi, T., Nakakubo, S., & Fujimoto, S. (2009). Physical Properties of Iron Oxide Scales on Si Containing Steels at High Temperature. *Materials Transactions*, *50*(9), 2242–2246. doi:10.2320/matertrans.M2009097

Talib, R. J., Saad, S., & Toff, M. R. M. (2003). Thermal spray coating technology a review. *Solid State Science and Technology*, *11*, 109–117.

Tamura, T., Amiya, K., Rachmat, R. S., Mizutani, Y., & Miwa, K. (2005). Electromagnetic vibration process for producing bulk metallic glasses. *Nature Materials*, *4*(4), 289–292. doi:10.1038/nmat1341 PMID:15750599

Tang, G., Su, F., Xu, X., & Chu, P. K. (2020). 2D Black Phosphorus Dotted with Silver Nanoparticles: An Excellent Lubricant Additive for Tribological Applications. *Chemical Engineering Journal*, *523*, 146503. doi:10.1016/j.cej.2019.123631

Tang, W., Zhou, Y., Zhu, H., & Yang, H. (2013). The effect of surface texturing on reducing the friction and wear of steel under lubricated sliding contact. *Applied Surface Science*, *273*, 199–204. doi:10.1016/j.apsusc.2013.02.013

Tang, Y., Zhang, Q., Zhan, X., & Chen, F. (2015). Superhydrophobic and anti-icing properties at overcooled temperature of a fluorinated hybrid surface prepared via a sol–gel process. *Soft Matter*, *11*(22), 4540–4550. doi:10.1039/C5SM00674K PMID:25966370

Tan, H. (2011). Preparation of long alumina fibers by sol-gel method using tartaric acid. *International Journal of Minerals Metallurgy and Materials*, *18*(6), 691–694. doi:10.100712613-011-0498-6

Tan, P., Fu, L., Teng, J., Zhu, J., Yang, W., Li, D., & Zhou, L. (2019). Effect of Texture on Wear Resistance of Tantalum Nitride Film. *Tribology International*, *133*(May), 126–135. doi:10.1016/j.triboint.2019.01.001

Tarasi, F., Medraj, M., Dolatabadi, A., Oberste-Berghaus, J., & Moreau, C. (2011). High-Temperature Performance of Alumina-Zirconia Composite Coatings containing Amorphous Phases. *Advanced Functional Materials*, *21*(21), 4143–4151. doi:10.1002/adfm.201100471

Tariq, N., Hasan, B., Akhter, J., & Ali, F. (2009). Mechanical and tribological properties of Zr–Al–Ni–Cu bulk metallic glasses. *Journal of Alloys and Compounds*, *469*(1-2), 179–185. doi:10.1016/j.jallcom.2008.02.002

Taurino, R., Fabbri, E., Pospiech, D., Synytska, A., & Messori, M. (2014). Preparation of scratch resistant superhydrophobic hybrid coatings by sol–gel process. *Progress in Organic Coatings*, *77*(11), 1635–1641. doi:10.1016/j.porgcoat.2014.05.009

Telford, M. (2004). The case for bulk metallic glass. *Materials Today*, *7*(3), 36–43. doi:10.1016/S1369-7021(04)00124-5

Thakur, L., & Vasudev, L. (2021). *Thermal Spray Coatings* (1st ed.). CRC Press. doi:10.1201/9781003213185

Thorpe & Merle. (1993). Thermal spray: Industry in transition. *Advanced Materials & Processes*, *143*(5), 50–56.

Thorpe, M. L. (1993). Thermal Spray: Industry in Transition. *Advanced Materials & Processes*, *143*(5), 50–56.

Tian, R., King, P. J., Coelho, J., Park, S.-H., Horvath, D. V., Nicolosi, V., O'Dwyer, C., & Coleman, J. N. (2020). Using chronoamperometry to rapidly measure and quantitatively analyse rate-performance in battery electrodes. *Journal of Power Sources*, *468*, 228220. doi:10.1016/j.jpowsour.2020.228220

Tillmann, W., & Dildrop, M. (2017). Influence of Si content on mechanical and tribological properties of TiAlSiN PVD coatings at elevated temperatures. *Surface and Coatings Technology*, *321*, 448–454. doi:10.1016/j.surfcoat.2017.05.014

Tisza, M. (2002). *Physical Metallurgy for Engineers*. ASM International Materials Park and Freund Publishing House Ltd.

Tojo, T., Atake, T., Mori, T., & Yamamura, H. (1999). Excess heat capacity in yttria stabilized zirconia. *Journal of Thermal Analysis and Calorimetry*, *57*(2), 447–458. doi:10.1023/A:1010159807127

Tojo, T., Atake, T., Mori, T., & Yamamura, H. (1999). Heat capacity and thermodynamic functions of zirconia and yttria-stabilized zirconia. *The Journal of Chemical Thermodynamics*, *31*(7), 831–845. doi:10.1006/jcht.1998.0481

Tokarz, A., Wieczorek, P., Fraczek, T., & Nitkiewicz, Z. (2008). Preparation, structural and mechanical properties of electrodeposited Co/Cu multilayers. *Physica Status Solidi*, *5*(c), 3526–3529. doi:10.1002/pssc.200779431

Tomala, A., Pauschitz, A., & Roy, M. (2013). Nanotribology of Pulsed Direct Current Magnetron Sputtered Diamond like Carbon Films. *Surface Science*, *616*(October), 60–70. doi:10.1016/j.susc.2013.06.008

Topcu, A. S. K., Erdogan, E., & Cengiz, U. (2018). Preparation of stable, transparent superhydrophobic film via one step one pot sol-gel method. *Colloid & Polymer Science*, *296*(9), 1523–1532. doi:10.100700396-018-4377-9

Trabelsi, O., Tighzert, L., Jbara, O., & Hadjadj, A. (2011). Synthesis via sol–gel process and characterization of novel organic–inorganic coatings. *Journal of Non-Crystalline Solids*, *357*(24), 3910–3916. doi:10.1016/j.jnoncrysol.2011.08.005

Trexler, M. M., & Thadhani, N. N. (2010). Mechanical properties of bulk metallic glasses. *Progress in Materials Science*, *55*(8), 759–839. doi:10.1016/j.pmatsci.2010.04.002

Triantou, K. I., Mergia, K., Perez, B., Florez, S., Stefan, A., Ban, C., Pelin, G., Ionescu, G., Zuber, C., Fischer, W. P. P., & Barcena, J. (2017). Thermal shock performance of carbon-bonded carbon fiber composite and ceramic matrix composite joints for thermal protection re-entry applications. *Composites. Part B, Engineering*, *111*, 270–278. doi:10.1016/j.compositesb.2016.12.020

Tsukakoshi, S., Itatani, K., & Koda, S. (2011). Properties of silicious film on polycarbonate substrate prepared by vacuum ultraviolet irradiation: Effect of intermediate silane layer. *Journal of the European Ceramic Society*, *31*(14), 2489–2496. doi:10.1016/j.jeurceramsoc.2011.06.011

Tsyntsaru, N. I., Bobanova, Z. I., Kroitoru, D. M., Cheban, V. F., Poshtaru, G. I., & Dikusar, A. I. (2010). Effect of a multilayer structure and lubrication on the tribological properties of coatings of Fe-W alloys. *Surface Engineering and Applied Electrochemistry*, *46*(6), 538–546. doi:10.3103/S1068375510060025

Tucker. (1994). Thermal Spray Coatings. ASM Handbook, Journal of Surface Engineering, 5, 497-509.

Tulsyan, R., Shivpuri, R., & Altan, T. (1993). *Investigation of Die Wear in Extrusion and Forging of Exhaust Valves*. ERC for Net Shape Manufacturing, Report No. B-93-28. The Ohio State University.

Turnbull, D. (1969). Under what conditions can a glass be formed? *Contemporary Physics*, *10*(5), 473–488. doi:10.1080/00107516908204405

Tyagi, A., Walia, R. S., Murtaza, Q., Pandey, S. M., Tyagi, P. K., & Bajaj, B. (2019). A critical review of diamond like carbon coating for wear resistance applications. *International Journal of Refractory Metals & Hard Materials*, *78*, 107–122. doi:10.1016/j.ijrmhm.2018.09.006

Ulianitsky, V., Batraev, I., Dudina, D., & Smurov, I. (2017). Enhancing the properties of WC/Co detonation coatings using two-component fuels. *Surface and Coatings Technology*, *318*, 244–249. doi:10.1016/j.surfcoat.2016.08.008

Usman, B. J., Scenini, F., & Curioni, M. (2020). Corrosion testing of anodized aerospace alloys: Comparison between immersion and salt spray testing using electrochemical impedance spectroscopy. *Journal of the Electrochemical Society*, *167*(4), 041505. doi:10.1149/1945-7111/ab74e3

Usmani, S., Sampath, S., Houck, D. L., & Lee, D. (1997). Effect of carbide grain size on the sliding and abrasive wear behavior of thermally sprayed WC-Co coatings. *Tribology Transactions*, *40*(3), 470–478. doi:10.1080/10402009708983682

Uyulgan, B., Dokumaci, E., Celik, E., Kayatekin, I., Ak Azem, N. F., Ozdemir, I., & Toparli, M. (2007). Wear behavior of thermal flame sprayed FeCr coatings on plain carbon steel substrate. *Journal of Materials Processing Technology*, *190*(1-3), 204–210. doi:10.1016/j.jmatprotec.2007.02.044

Van Steenkiste, T. H., Smith, J. R., Teets, R. E., Moleski, J. J., Gorkiewicz, D. W., Tison, R. P., Marantz, D. R., Kowalsky, K. A., Riggs, W. L. II, Zajchowski, P. H., Pilsner, B., McCune, R. C., & Barnett, K. J. (1999). Kinetic spray coatings. *Surface and Coatings Technology*, *111*(1), 62–71. doi:10.1016/S0257-8972(98)00709-9

Vanossi, A., Bechinger, C., & Urbakh, M. (2020). Structural Lubricity in Soft and Hard Matter Systems. *Nature Communications*, *11*(1), 4657. doi:10.103841467-020-18429-1 PMID:32938930

Vardavoulias, M., Jeandin, M., Velasco, F., & Torralba, J. M. (1996). Dry Sliding Wear Mechanism for P/M Austenitic Stainless Steels and Their Composites Containing Al_2O_3 and Y_2O_3 Particles. *Tribology International*, *29*(6), 499–506. doi:10.1016/0301-679X(95)00110-P

Vardelle, A., Moreau, C., Akedo, J., Ashrafizadeh, H., Berndt, C. C., Berghaus, J. O., Boulos, M., Brogan, J., Bourtsalas, A. C., Dolatabadi, A., Dorfman, M., Eden, T. J., Fauchais, P., Fisher, G., Gaertner, F., Gindrat, M., Henne, R., Hyland, M., Irissou, E., ... Vuoristo, P. (2016). The 2016 Thermal Spray Roadmap. *Journal of Thermal Spray Technology*, *25*(8), 1376–1440. doi:10.100711666-016-0473-x

Varga, M., Rojacz, H., Winkelman, H., Mayer, H., & Badisch, E. (2013). Wear Reducing Effects and Temperature Dependence of Tribo Layers Formation in Harsh Environment. *Tribology International*, *65*, 190–199. doi:10.1016/j.triboint.2013.03.003

Vashishtha, N., Khatirkar, R. K., & Sapate, S. G. (2017). Tribological behaviour of HVOF sprayed WC-12Co, WC-10Co-4Cr and Cr3C2−25NiCr coatings. *Tribology International*, *105*, 55–68. doi:10.1016/j.triboint.2016.09.025

Vassen, R., Cao, X., Tietz, F., Basu, D., & Stöver, D. (2000). Zirconates as New Materials for Thermal Barrier Coatings. *Journal of the American Ceramic Society*, *83*(8), 2023–2028. doi:10.1111/j.1151-2916.2000.tb01506.x

Vasudev, H., Singh, G., Bansal, A., Vardhan, S., & Thakur, L. (2019). Microwave heating and its applications in surface engineering: A review. *Materials Research Express*, *6*(10), 52–59. doi:10.1088/2053-1591/ab3674

Venkatesan, K., Subramanian, C., & Surnmerville, E. (1997). Three Body Abrasion of Surface Engineered Die Steel at Elevated Temperatures. *Wear*, *203-204*, 129–138. doi:10.1016/S0043-1648(96)07442-X

Veprek, S., & Argon, A. S. (2001). Mechanical Properties of Superhard Nanocomposites. *Surface and Coatings Technology*, *146–147*(September), 175–182. doi:10.1016/S0257-8972(01)01467-0

Veprek, S., Niederhofer, A., Moto, K., Bolom, T., Männling, H.-D., Nesladek, P., Dollinger, G., & Bergmaier, A. (2000). Composition, Nanostructure and Origin of the Ultrahardness in Nc-TiN/a-Si3N4/a- and Nc-TiSi2 Nanocomposites with HV=80 to ≥105 GPa. *Surface and Coatings Technology*, *133–134*(November), 152–159. doi:10.1016/S0257-8972(00)00957-9

Vepřek, S., & Reiprich, S. (1995). A Concept for the Design of Novel Superhard Coatings. *Thin Solid Films, 268*(1–2), 64–71. doi:10.1016/0040-6090(95)06695-0

Vergne, C., Boher, C., Gras, R., & Levaillant, C. (2006). Influence of Oxides on Friction in Hot Rolling: Experimental Investigations and Tribological Modelling. *Wear, 260*(9-10), 957–975. doi:10.1016/j.wear.2005.06.005

Verlinden, B. (2005). Severe Plastic Deformation, Metalurgija-. *Journal of Metallurgy, 11*(3), 1–17. doi:10.30544/380

Verstak, A., Andrew, A., & Baranovski. (2004). Deposition of carbides by Activated Combustion HVAF spraying. *Proceedings of the International Thermal Spray Conference*, 551–555.

Verstak, A., & Baranovski, V. (2003). Activated Combustion HVAF Coatings for Protection against Wear and High Temperature Corrosion. In B. R. Marple & C. Moreau (Eds.), *ASM International, Volunme 1* (pp. 535–541).

Vicenzi, J., Villanova, D. L., Lima, M. D., Takimi, A. S., Marques, C. M., & Bergmann, C. P. (2006). HVOF-coatings against high temperature erosion (~300 °C) by coal fly ash in thermoelectric power plant. *Materials & Design, 27*(3), 236–242. doi:10.1016/j.matdes.2004.10.008

Vinayo, M. E., Kassabji, F., Guyonnet, J., & Fauchais, P. (1985). Plasma sprayed WC–Co coatings: Influence of spray conditions (atmospheric and low pressure plasma spraying) on the crystal structure, porosity, and hardness. *Journal of Vacuum Science & Technology. A, Vacuum, Surfaces, and Films, 3*(6), 2483–2489. doi:10.1116/1.572863

Vipin, S., Ujjwal, P., & Manoj, K. B. V. (2015). Surface composites by friction stir processing: A review. *Journal of Materials Processing Technology, 224*, 117–134. doi:10.1016/j.jmatprotec.2015.04.019

Visuttipitukul, P. C., & Kuwahara, H. (2006). Characterization of Plasma Nitrided AISI H13 Tool Steel. *Acta Metall Slov., 12*, 264–274.

Viswanathan, V., Dwivedi, G., & Sampath, S. (2014). Engineered Multilayer Thermal Barrier Coatings for Enhanced Durability and Functional Performance. *Journal of the American Ceramic Society, 97*(9), 2770–2778. doi:10.1111/jace.13033

Vitelaru, C., Balaceanu, M., Parau, A., Luculescu, C. R., & Vladescu, A. (2014). Investigation of nanostructured TiSiC-Zr and TiSiC-Cr hard coatings for industrial applications. *Surface and Coatings Technology, 251*, 21–28. doi:10.1016/j.surfcoat.2014.04.001

Vittal, R., & Ho, K.-C. (2017). Zinc oxide based dye-sensitized solar cells: A review. *Renewable & Sustainable Energy Reviews, 70*, 920–935. doi:10.1016/j.rser.2016.11.273

Vitu, T., Polcar, T., Cvrcek, L., Novak, R., Macak, J., Vyskocil, J., & Cavaleiro, A. (2008). Structure and tribology of biocompatible Ti-C:H coatings. *Surface and Coatings Technology, 202*(22–23), 5790–5793. doi:10.1016/j.surfcoat.2008.06.040

Vladescu, A., Braic, M., Azem, F. A., Titorencu, I., Braic, V., Pruna, V., Kiss, A., Parau, A. C., & Birlik, I. (2015). Effect of the deposition temperature on corrosion resistance and biocompatibility of the hydroxyapatite coatings. *Applied Surface Science, 354*(B), 373–379. doi:10.1016/j.apsusc.2015.05.059

Vladescu, A., Braic, V., Braic, M., & Balaceanu, M. (2013). Arc plasma deposition of TiSiN/Ni nanoscale multilayered coatings. *Materials Chemistry and Physics, 138*(2–3), 500–506. doi:10.1016/j.matchemphys.2012.12.010

Voevodin, A. A., O'Neill, J. P., & Zabinski, J. S. (1999). Nanocomposite Tribological Coatings for Aerospace Applications. *Surface and Coatings Technology, 116–119*(September), 36–45. doi:10.1016/S0257-8972(99)00228-5

Vranceanu, D. M., Cotrut, C. M., Bramowicz, M., Titorencu, I., Kulesza, S., Kiss, A., Berbecaru, A., Pruna, V., Branzei, M., & Vladescu, A. (2016). Osseointegration of sputtered SiC-added hydroxyapatite for orthopaedic applications. *Ceramics International, 42*(8), 10085–10093. doi:10.1016/j.ceramint.2016.03.114

Vuoristo, P. (2014). Comprehensive materials processing, 1st edition Volume 4: Coatings and films. In D. Cameron (Ed.), Comprehensive materials processing, Volume 4: Coatings and films (4th ed., pp. 229–276). Academic Press.

Vuuren, R. D. J., Nunzi, J. M., & Givigi, S. N. (2021). Frontiers in photosensor materials and designs for new image sensor applications. *IEEE Sensors Journal*, *21*(10), 11339–11348. doi:10.1109/JSEN.2020.3043288

Wang, C. B., Wang, D. L., Chen, W. X., & Wang, Y. Y. (2002). Tribological properties of nanostructured WC/CoNi and WC/CoNiP coatings produced by electro-deposition. *Wear*, *253*(5–6), 563–571. doi:10.1016/S0043-1648(02)00173-4

Wang, C., Li, J., Zhang, L., & Fu, S. (2022). Preparation and property optimization of bistable electrochromic microcapsules. *Dyes and Pigments*, *197*, 109936. doi:10.1016/j.dyepig.2021.109936

Wang, D. S., Tian, Z. J., Wang, S. L., & Shen, L. D. (2012). The Friction and Wear Properties of Conventional and Nanostructured ZrO_2-7wt.%Y_2O_3 Thermal Barrier Coatings Deposited on TiAl Intermetallic Alloy by Plasma Spraying. *Advanced Materials Research*, *538–541*, 336–339. doi:10.4028/www.scientific.net/AMR.538-541.336

Wang, D. Y., Shu, D. L., & Guo, X. C. (1987). Effect of Microstructure and Properties on High Temperature Wear Characteristics of $3Cr_2W_8V$ (H21) steel. *Wear*, *119*(1), 101–117. doi:10.1016/0043-1648(87)90101-3

Wang, D., & Bierwagen, G. P. (2009). Sol–gel coatings on metals for corrosion protection. *Progress in Organic Coatings*, *64*(4), 327–338. doi:10.1016/j.porgcoat.2008.08.010

Wang, D.-H., Xie, S.-H., Yang, H.-P., Qian, H.-X., & Zeng, X.-R. (2018). Wear behaviors of three typical bulk metallic glasses in bearing applications. *Metals*, *8*(12), 1005. doi:10.3390/met8121005

Wang, F.-F., Liu, Z., & Cheng, Z.-L. (2020). Ultrasonic-assisted Exfoliation for 2D Zn(Bim)(OAc) Nanosheets used as an Oil-Soluble Additive in Lubricants. *Applied Organometallic Chemistry*, *34*(11), e5950. doi:10.1002/aoc.5950

Wang, H. X., Guan, S. K., Wang, X., Ren, C. X., & Wang, L. G. (2010). In vitro degradation and mechanical integrity of Mg–Zn–Ca alloy coated with Ca-deficient hydroxyapatite by the pulse electrodeposition process. *Acta Biomaterialia*, *6*(5), 1743–1748. doi:10.1016/j.actbio.2009.12.009 PMID:20004746

Wang, H., Guan, S., Wang, Y., Liu, H., Wang, H., Wang, L., Ren, C., Zhua, S., & Chend, K. (2011). In vivo degradation behavior of Ca-deficient hydroxyapatite coated Mg–Zn–Ca alloy for bone implant application. *Colloids and Surfaces. B, Biointerfaces*, *88*(1), 254–259. doi:10.1016/j.colsurfb.2011.06.040 PMID:21783346

Wang, H., Li, H., Cao, S., Wang, M., Chen, J., & Zang, Z. (2020). Interface modulator of ultrathin magnesium oxide for low-temperature-processed inorganic $CsPbIBr_2$ perovskite solar cells with efficiency over 11%. *Solar RRL*, *4*(9), 2000226. doi:10.1002olr.202000226

Wang, H., & Liu, Y. (2020). Superlubricity achieved with Two-Dimensional Nano-Additives to Liquid Lubricants. *Friction*, *8*(6), 1007–1024. doi:10.100740544-020-0410-3

Wang, H., Liu, Y., Liu, W., Wang, R., Wen, J., Sheng, H., Peng, J., Erdemir, A., & Luo, J. (2017). Tribological Behavior of NiAl-Layered Double Hydroxide Nanoplatelets as Oil-based Lubricant Additives. *ACS Applied Materials & Interfaces*, *9*(36), 30891–30899. doi:10.1021/acsami.7b10515 PMID:28841291

Wang, H., Wei, W., & Hu, Y. H. (2014). NiO as an efficient counter electrode catalyst for dye-sensitized solar cells. *Topics in Catalysis*, *57*(6), 607–611. doi:10.100711244-013-0218-8

Wang, H., Zhu, S., Wang, L., Feng, Y., Ma, X., & Guan, S. (2014). Formation mechanism of Ca-deficient hydroxyapatite coating on Mg-Zn-Ca alloy for orthopaedic implant. *Applied Surface Science*, *307*, 92–100. doi:10.1016/j.apsusc.2014.03.172

Wang, J., & Zeng, H. (2021). Recent advances in electrochemical techniques for characterizing surface properties of minerals. *Advances in Colloid and Interface Science*, *288*, 102346. doi:10.1016/j.cis.2020.102346 PMID:33383471

Wang, L., Huang, Y., Yuan, Y., Jia, C., & Yang, L. (2022). Microstructure, wear and oxidation resistance of Al-doped Ti–Si3N4 coatings by laser cladding. *Surface and Coatings Technology*, *429*, 127942. doi:10.1016/j.surfcoat.2021.127942

Wang, Q., Bai, X., Sun, B., Liu, J., Cai, Z., Liang, X., & Shen, B. (2021). Influence of Si on tribological behavior of laser cladded Fe-based amorphous/crystalline composite coatings. *Surface and Coatings Technology*, *405*, 126570. doi:10.1016/j.surfcoat.2020.126570

Wang, Q., Hou, T., Wang, W., Zhang, G., Gao, Y., & Wang, K. (2021). *Tribological Behavior of Black Phosphorus Nanosheets as Water-based Lubrication Additives*. Friction. doi:10.100740544-020-0465-1

Wang, S. Q., Wei, M. X., & Zhao, Y. T. (2010). Effects of the Tribo Oxide and Matrix on Dry Sliding Wear Characteristics and Mechanisms of A Cast Steel. *Wear*, *269*(5-6), 424–434. doi:10.1016/j.wear.2010.04.028

Wang, T., & Ye, F. (2018). The elevated-temperature wear behavior evolution of HVOF sprayed tungsten carbide coatings: Respond to heat treatment. *International Journal of Refractory Metals & Hard Materials*, *71*, 92–100. doi:10.1016/j.ijrmhm.2017.11.007

Wang, W. (2009). Bulk metallic glasses with functional physical properties. *Advanced Materials*, *21*(45), 4524–4544. doi:10.1002/adma.200901053

Wang, W.-H., Dong, C., & Shek, C. (2004). Bulk metallic glasses. *Materials Science and Engineering R Reports*, *44*(2-3), 45–89. doi:10.1016/j.mser.2004.03.001

Wang, W., Xie, G., & Luo, J. (2018). Black Phosphorous as a New Lubricant. *Friction*, *6*(1), 116–142. doi:10.100740544-018-0204-z

Wang, X., Guo, L., Peng, H., Zheng, L., Guo, H., & Gong, S. (2015). Hot-corrosion behavior of a La2Ce2O7/YSZ thermal salt at 900 C barrier coating exposed to Na2SO4+V2O5 or V2O5. *Ceramics International*. Advance online publication. doi:10.1016/j.ceramint.2015.01.107

Wang, Y., Guo, H., & Gong, S. (2009). Thermal shock resistance and mechanical properties of La2Ce2O7 thermal barrier coatings with segmented structure. *Ceramics International*, *35*(7), 2639–2644. doi:10.1016/j.ceramint.2009.02.025

Wang, Y., Wang, S. J., Shu, X., Gao, W., & Lu, W. (2014). Preparation and property of sol-enhanced Ni–B–TiO_2 nano-composite coatings. *Journal of Alloys and Compounds*, *617*, 472–478.

Wänstrand, O., Larsson, M., & Hedenqvist, P. (1999). Mechanical and tribological evaluation of PVD WC/C coatings. *Surface and Coatings Technology*, *111*(2–3), 247–254. doi:10.1016/S0257-8972(98)00821-4

Weibin, Z., Weidong, W., Xueming, W., Xinlu, C., Dawei, Y., Changle, S., Liping, P., Yuying, W., & Li, B. (2013). The investigation of NbO_2 and Nb_2O_5 electronic structure by XPS, UPS and first principles methods. *Surface and Interface Analysis*, *45*(8), 1206–1210. doi:10.1002ia.5253

Wei, R. (2008). Plasma enhanced magnetron sputter deposition of Ti-Si-C-N based nanocomposite coatings. *Surface and Coatings Technology*, *203*(5–7), 538–544. doi:10.1016/j.surfcoat.2008.05.019

Wellman, R. G., Deakin, M. J., & Nicholls, J. R. (2005). The effect of TBC morphology on the erosion rate of EB PVD TBCs. *Wear*, *258*(1–4), 349–356. doi:10.1016/j.wear.2004.04.011

Wellman, R. G., & Nicholls, J. R. (2007). A review of the erosion of thermal barrier coatings. *Journal of Physics. D, Applied Physics*, *40*(16), R293–R305. doi:10.1088/0022-3727/40/16/R01

Wen, C., Guan, S., Peng, L., Ren, C., Wang, X., & Hu, Z. (2009). Characterization and degradation behavior of AZ31 alloy surface modified by bone-like hydroxyapatite for implant applications. *Applied Surface Science*, *255*(13-14), 6433–6438. doi:10.1016/j.apsusc.2008.09.078

Whittingham, M. S. (1974). *Fast ion transport materials and batteries.* Paper presented at the Conference on Electrical Insulation & Dielectric Phenomena - Annual Report 1974.

Williamson, R. S., Disegi, J., Griggs, J. A., & Roach, M. D. (2013). Nanopore formation on the surface oxide of commercially pure titanium grade 4 using a pulsed anodization method in sulfuric acid. *Journal of Materials Science. Materials in Medicine*, *24*(10), 2327–2335. doi:10.100710856-013-4985-3 PMID:23807314

Winkelmann, H., Badisch, E., Varga, M., & Danninger, H. (2010). Wear Mechanisms at High Temperatures. Part 3: Changes of the Wear Mechanism in the Continuous Impact Abrasion Test with Increasing Testing Temperature. *Tribology Letters*, *37*(2), 37419–37429. doi:10.100711249-009-9534-3

Witte, F. (2015). Reprint of: The history of biodegradable magnesium implants: A review. *Acta Biomaterialia, 23*(S), S28–S40. doi:10.1016/j.actbio.2015.07.017

Witte, F., Kaese, V., Haferkamp, H., Switzer, E., Meyer-Lindenberg, A., Wirth, C. J., & Windhagen, H. (2005). In vivo corrosion of four magnesium alloys and the associated bone response. *Biomaterials*, *26*(17), 3557–3563. doi:10.1016/j.biomaterials.2004.09.049 PMID:15621246

Wolfe, D., & Singh, J. (1998).. . *Journal of Materials Science*, *33*(14), 3677–3692. doi:10.1023/A:1004675900887

Woods, C. R., Britnell, L., Eckmann, A., Ma, R. S., Lu, J. C., Guo, H. M., Lin, X., Yu, G. L., Cao, Y., Gorbachev, R. V., Kretinin, A. V., Park, J., Ponomarenko, L. A., Katsnelson, M. I., Gornostyrev, N., Watanabe, K., Tanigucho, T., Casiraghi, C., Gao, H.-J., ... Novoselov, K. S. (2014). Commensurate-Incommensurate Transition in Graphene on Hexagonal Boron Nitride. *Nature Physics*, *10*(6), 451–456. doi:10.1038/nphys2954

Wouters, M. E. L., Wolfs, D. P., van der Linde, M. C., Hovens, J. H. P., & Tinnemans, A. H. A. (2004). Transparent UV curable antistatic hybrid coatings on polycarbonate prepared by the sol–gel method. *Progress in Organic Coatings*, *51*(4), 312–319. doi:10.1016/j.porgcoat.2004.07.020

Wu, C.-L., Wang, C.-K., Lin, C.-K., Wang, S.-C., & Huang, J.-L. (2013). Electrochromic properties of nanostructured tungsten oxide films prepared by surfactant-assisted sol–gel process. *Surface and Coatings Technology*, *231*, 403–407. doi:10.1016/j.surfcoat.2012.01.061

Wu, D., Liu, Y., Li, D., Zhao, X., & Liu, Y. (2016). Tribo-corrosion properties of WC-10Co-4Cr coating in natural silt-laden waters when sliding against Si3N4. *International Journal of Refractory Metals & Hard Materials*, *58*, 143–151. doi:10.1016/j.ijrmhm.2016.04.019

Wu, H., Baker, I., Liu, Y., Wu, X., Munroe, P. R., & Zhang, J. (2013). Tribological studies of a Zr-based bulk metallic glass. *Intermetallics*, *35*, 25–32. doi:10.1016/j.intermet.2012.11.010

Wu, K. R., Bayer, R. G., Engel, P. A., & Sun, D. C. (1998). Wear of Physical Vapor Deposition Tin Coatings Sliding against Cr-Steel and WC Counterbodies. *Journal of Tribology*, *120*(3), 482–488. doi:10.1115/1.2834576

Wu, L. Y. L., Chwa, E., Chen, Z., & Zeng, X. T. (2008). A study towards improving mechanical properties of sol–gel coatings for polycarbonate. *Thin Solid Films*, *516*(6), 1056–1062. doi:10.1016/j.tsf.2007.06.149

Wu, W., Liu, J., Li, Z., Zhao, X., Liu, G., Liu, S., Ma, S., Li, W., & Liu, W. (2021). Surface-functionalized nanoMOFs in Oil for Friction and Wear Reduction and Antioxidation. *Chemical Engineering Journal*, *410*, 128306. doi:10.1016/j.cej.2020.128306

Wu, X., & Chen, Z. (2018a). A mechanically robust transparent coating for anti-icing and self-cleaning applications. *Journal of Materials Chemistry. A, Materials for Energy and Sustainability*, *6*(33), 16043–16052. doi:10.1039/C8TA05692G

Wu, X., Fu, Q., Kumar, D., Ho, J. W. C., Kanhere, P., Zhou, H., & Chen, Z. (2016). Mechanically robust superhydrophobic and superoleophobic coatings derived by sol–gel method. *Materials & Design*, *89*, 1302–1309. doi:10.1016/j.matdes.2015.10.053

Wu, X., Tang, Y., Silberschmidt, V. V., Wilson, P., & Chen, Z. (2018). Mechanically Robust Transparent Anti-Icing Coatings: Roles of Dispersion Status of Titanate Nanotubes. *Advanced Materials Interfaces*, *5*(18), 1800773. doi:10.1002/admi.201800773

Wyatt, B. C., Rosenkranz, A., & Anasori, B. (2021). 2D MXenes: Tunable Mechanical and Tribological Properties. *Advanced Materials*, *33*(17), 2007973. doi:10.1002/adma.202007973 PMID:33738850

Xiao, Y., Han, G., Zhou, H., & Wu, J. (2016). An efficient titanium foil based perovskite solar cell: Using a titanium dioxide nanowire array anode and transparent poly(3,4-ethylenedioxythiophene) electrode. *RSC Advances*, *6*(4), 2778–2784. doi:10.1039/C5RA23430A

Xia, S., Zhu, P., Pi, F., Zhang, Y., Li, Y., Wang, J., & Sun, X. (2017). Development of a simple and convenient cell-based electrochemical biosensor for evaluating the individual and combined toxicity of DON, ZEN, and AFB1. *Biosensors & Bioelectronics*, *97*, 345–351. doi:10.1016/j.bios.2017.06.002 PMID:28623817

Xie, M., Zhang, S., & Li, M. (2013). Comparative investigation on HVOF sprayed carbide-based coatings. *Applied Surface Science*, *273*, 799–805. doi:10.1016/j.apsusc.2013.03.010

Xie, X., Guo, H., & Gong, S. (2010). Mechanical properties of $LaTi_2Al_9O_{19}$ and thermal cycling behaviours of plasma-sprayed $LaTi_2Al_9O_{19}$/YSZ thermal barrier coatings. *Journal of Thermal Spray Technology*, *19*(6), 1179–1185. doi:10.100711666-010-9529-5

Xie, X., Guo, H., Gong, S., & Xu, H. (2011). Thermal cycling behavior and failure mechanism of $LaTi_2Al_9O_{19}$/YSZ thermal barrier coatings exposed to gas flame. *Surface and Coatings Technology*, *205*(17-18), 4291–4298. doi:10.1016/j.surfcoat.2011.03.047

Xing, Y. Z., Wang, G., Zhang, Y., Chen, Y.-N., & Dargusch, M. (2017). Development in plasma surface diffusion techniques of Ti-6Al-4V alloy: A review. *International Journal of Advanced Manufacturing Technology*, *92*(5-8), 1901–1912. doi:10.100700170-017-0302-5

Xiong, H.-P., Kawasaki, A., Kang, Y.-S., & Watanabe, R. (2005). Experimental study on heat insulation performance of functionally graded metal/ceramic coatings and their fracture behavior at high surface temperatures. *Surface and Coatings Technology*, *194*(2–3), 203–214. doi:10.1016/j.surfcoat.2004.07.069

Xiu, Y., Hess, D. W., & Wong, C. P. (2007). A Novel Method to Prepare Superhydrophobic, Self-Cleaning and Transparent Coatings for Biomedical Applications. *2007 Proceedings 57th Electronic Components and Technology Conference*, 1218–1223. 10.1109/ECTC.2007.373949

Xiu-Tian, Y., & Yongdong, X. (2010) *Chemical vapour deposition an integrated engineering design for advanced materials*. Springer-Verlag London Limited.

Xue, C., Zhang, P., Wei, D., Hu, H., Li, F., & Yang, K. (2020). Corrosion and tribocorrosion behaviors for TA3 in Ringer's solution after implantation of Nb ions. *Applied Sciences (Basel, Switzerland)*, *10*(23), 8329–8339. doi:10.3390/app10238329

Xue, L., Wang, W., Guo, Y., Liu, G., & Wan, P. (2017). Flexible Polyaniline/Carbon Nanotube Nanocomposite Film-Based Electronic Gas Sensors. *Sensors and Actuators. B, Chemical, 244*(June), 47–53. doi:10.1016/j.snb.2016.12.064

Xu, H., & Guo, H. (Eds.). (2011). *Thermal barrier coatings*. Woodhead Pub Ltd. doi:10.1533/9780857090829

Xu, H., Guo, H., & Gong, S. (2008). Thermal barrier coatings. In *Developments in High Temperature Corrosion and Protection of Materials* (pp. 476–491). Elsevier. doi:10.1533/9781845694258.2.476

Xu, J., Kang, J.-J., Yue, W., Fu, Z.-Q., Zhu, L.-N., & She, D.-S. (2021). High-temperature tribological property of Fe-based amorphous alloy coating. *Journal of Non-Crystalline Solids, 573*, 121136. doi:10.1016/j.jnoncrysol.2021.121136

Xu, S., Chen, L., Gong, M., Hu, X., Zhang, X., & Zhou, Z. (2017). Characterization and engineering application of a novel ceramic composite insulation material. *Composites. Part B, Engineering, 111*, 143–147. doi:10.1016/j.compositesb.2016.12.010

Yalamaç, E., Sutcu, M., & Basturk, S. B. (2017). Ceramic fibers. In *Fiber Technology for Fiber-Reinforced Composites* (pp. 187–207). Elsevier. doi:10.1016/B978-0-08-101871-2.00009-6

Yan, D., Bazant, M. Z., Biesheuvel, P. M., Pugh, M. C., & Dawson, F. P. (2017). Theory of linear sweep voltammetry with diffuse charge: Unsupported electrolytes, thin films, and leaky membranes. *Physical Review. E, 95*(3), 033303. doi:10.1103/PhysRevE.95.033303 PMID:28415284

Yang, F., Zhao, X., & Xiao, P. (2010). Thermal conductivities of YSZ/Al2O3 composites. *Journal of the European Ceramic Society, 30*(15), 3111–3116. doi:10.1016/j.jeurceramsoc.2010.07.007

Yang, H.-S., Bai, G.-R., Thompson, L. J., & Eastman, J. A. (2002). Interfacial thermal resistance in nanocrystalline yttria-stabilized zirconia. *Acta Materialia, 50*(9), 2309–2317. doi:10.1016/S1359-6454(02)00057-5

Yang, J., Chen, B., Song, H., Tang, H., & Li, C. (2014). Synthesis, Characterization, and Tribological Properties of Two-Dimensional Ti_3C_2. *Crystal Research and Technology, 49*(11), 926–932. doi:10.1002/crat.201400268

Yang, J., Liu, Y., Ye, Z., Yang, D., & He, S. (2011). Microstructural and Tribological Characterization of Plasma and Gas-Nitrided 2Cr13 Steel in Vacuum. *Materials & Design, 32*(2), 808–814. doi:10.1016/j.matdes.2010.07.022

Yang, L. W., Zhang, X. S., Liu, H. T., & Zu, M. (2017). Thermal resistant, mechanical and electrical properties of a novel ultrahigh-content randomly-oriented CNTs reinforced SiC matrix composite-sheet. *Composites. Part B, Engineering, 119*, 10–17. doi:10.1016/j.compositesb.2017.03.039

Yang, L., & Zhang, E. (2009). Biocorrosion behavior of magnesium alloy in different simulated fluids for biomedical application. *Materials Science and Engineering C, 29*(5), 1691–1696. doi:10.1016/j.msec.2009.01.014

Yang, Q., Senda, T., & Hirose, A. (2006). Sliding wear behavior of WC–12% Co coatings at elevated temperatures. *Surface and Coatings Technology, 200*(14–15), 4208–4212. doi:10.1016/j.surfcoat.2004.12.032

Yang, S. H., Kong, H., Lee, K. R., Park, S., & Kim, D. E. (2002). Effect of Environment on the Tribological Behavior of Si-Incorporated Diamond-like Carbon Films. *Wear, 252*(1–2), 70–79. doi:10.1016/S0043-1648(01)00856-0

Yang, W., Liu, Y., Hua, N., Pang, S., Li, Y., Liaw, P. K., & Zhang, T. (2021). Formation and properties of biocompatible Ti-based bulk metallic glasses in the Ti–Cu–Zr–Fe–Sn–Si–Ag system. *Journal of Non-Crystalline Solids, 571*, 121060. doi:10.1016/j.jnoncrysol.2021.121060

Yang, X. K., Li, Q., Zhang, S. Y., Zhong, X. K., Dai, Y., & Luo, F. (2010). Electrochemical corrosion behaviors and protective properties of Ni–Co–TiO_2 composite coating prepared on sintered NdFeB magnet. *Materials and Corrosion, 61*(7), 618–625.

Yang, X., & Zhang, Y. (2012). Prediction of high-entropy stabilized solid-solution in multi-component alloys. *Materials Chemistry and Physics*, *132*(2–3), 233–238. doi:10.1016/j.matchemphys.2011.11.021

Yan, P., Deng, J., Wu, Z., Li, S., Xing, Y., & Zhao, J. (2012). Friction and wear behavior of the PVD (Zr,Ti)N coated cemented carbide against 40Cr hardened steel. *International Journal of Refractory Metals & Hard Materials*, *35*, 213–220. doi:10.1016/j.ijrmhm.2012.06.003

Yan, Z., Jiang, D., Gao, X., Hu, M., Wang, D., Fu, Y., Sun, J., Feng, D., & Weng, L. (2018). Friction and Wear Behavior of TiN Films against Ceramic and Steel Balls. *Tribology International*, *124*(August), 61–69. doi:10.1016/j.triboint.2018.03.031

Yao, J., Zhang, J., Wu, G., Wang, L., Zhang, Q., & Liu, R. (2018). Microstructure and wear resistance of laser cladded composite coatings prepared from pre-alloyed WC-NiCrMo powder with different laser spots. *Optics & Laser Technology*, *101*, 520–530. doi:10.1016/j.optlastec.2017.12.007

Yao, S. H., Su, Y. L., & Lai, Y. C. (2017). Antibacterial and tribological performance of carbonitride coatings doped with W, Ti, Zr, or Cr deposited on AISI 316L stainless steel. *Materials (Basel)*, *10*(10), 1189–1205. doi:10.3390/ma10101189

Yashar, P. C., & Sproul, W. D. (1999). Nanometer scale multilayered hard coatings. *Vacuum*, *55*(3), 179–190. doi:10.1016/S0042-207X(99)00148-7

Yeh, J. W., Chen, S. K., Lin, S. J., Gan, J. Y., Chin, T. S., Shun, T. T., Tsau, C. H., & Chang, S. Y. (2004). Nanostructured high-entropy alloys with multiple principal elements: Novel alloy design concepts and outcomes. *Advanced Engineering Materials*, *6*(5), 299–303. doi:10.1002/adem.200300567

Yin, Z.-Z., Qi, W.-C., Rong-Chang, Z. X.-B. C., Gu, C.-D., Guan, S.-K., & Zheng, Y.-F. (2020). Advances in coatings on biodegradable magnesium alloys. *Journal of Magnesium and Alloys*, *8*(1), 42–65. doi:10.1016/j.jma.2019.09.008

Yokoyama, Y., Mund, E., Inoue, A., & Schultz, L. (2007). Production of $Zr_{55}Cu_{30}Ni_5Al_{10}$ glassy alloy rod of 30 mm in diameter by a cap-cast technique. *Materials Transactions*, *48*(12), 3190–3192. doi:10.2320/matertrans.MRP2007164

Yoon, S., Kim, J., Bae, G., Kim, B., & Lee, C. (2011). Formation of coating and tribological behavior of kinetic sprayed Fe-based bulk metallic glass. *Journal of Alloys and Compounds*, *509*(2), 347–353. doi:10.1016/j.jallcom.2010.09.024

Young, E. J., Mateeva, E., Moore, J. J., Mishra, B., & Loch, M. (2000). Low pressure plasma spray coatings. *Thin Solid Films*, *377–378*, 788–792. doi:10.1016/S0040-6090(00)01452-8

Yuan, F. H., Chen, Z. X., Huang, Z. W., Wang, Z. G., & Zhu, S. J. (2008). Oxidation behavior of thermal barrier coatings with HVOF and detonation-sprayed NiCrAlY bond coats. *Corrosion Science*, *50*(6), 1608–1617. doi:10.1016/j.corsci.2008.02.002

Yuan, K., Peng, R. L., Li, X.-H., Talus, A., Johansson, S., & Wang, Y.-D. (2015). Hot corrosion of MCrAlY coatings in sulphate and SO2 environment at 900 C: Is SO2 necessarily bad? *Surface and Coatings Technology*, *261*, 41–53. doi:10.1016/j.surfcoat.2014.11.065

Yuan, S., Lin, N., Zou, J., Lin, X., Liu, Z., Yu, Y., Wang, Z., Zeng, Q., Chen, W., Tian, L., Qin, L., Xie, R., Li, B., Zhang, H., Wang, Z., Tang, B., & Wu, Y. (2020). In-Situ Fabrication of Gradient Titanium Oxide Ceramic Coating on Laser Surface Textured Ti6Al4V Alloy with Improved Mechanical Property and Wear Performance. *Vacuum*, *176*(June), 109327. doi:10.1016/j.vacuum.2020.109327

Zaman, A., & Meletis, E. I. (2017). Microstructure and Mechanical Properties of TaN Thin Films Prepared by Reactive Magnetron Sputtering. *Coatings*, *7*(12), 209. doi:10.3390/coatings7120209

Zavareh, M. A., Diaa, A. A. Sarhan, M., Abd Razak, B. B., & Basirun, W. J. (2014). Plasma thermal spray of ceramic oxide coating on carbon steel with enhanced wear and corrosion resistance for oil and gas applications. *Ceramics International, 40*(9), 14267-14277.

Zhai, W., Bai, L., Zhou, R., Fan, X., Kang, G., Liu, Y., & Zhou, K. (2021). *Recent Progress on Wear-Resistant Materials: Designs, Properties, and Applications. In Advanced Science*. John Wiley and Sons Inc. doi:10.1002/advs.202003739

Zhang B. P., Wang Y. & Geng L. (2011). Research on Mg-Zn-Ca Alloy as Degradable Biomaterial. *Biomaterials-Physics and Chemistry*, 183-204.

Zhang, B. Y., Zhou, Y. J., Lin, J. P., Chen, G. L., & Liaw, P. K. (2008). Solid-solution phase formation rules for multi-component alloys. *Advanced Engineering Materials*, *6*(6), 534–538. doi:10.1002/adem.200700240

Zhang, E., Yang, L., Xu, J., & Chen, H. (2010). Microstructure, mechanical properties and bio-corrosion properties of Mg-Si(-Ca, Zn) alloy for biomedical application. *Acta Biomaterialia*, *6*(5), 1756–1762. doi:10.1016/j.actbio.2009.11.024 PMID:19941979

Zhang, G., Ke, Y., He, J., Qin, M., Shen, H., Lu, S., & Xu, J. (2015). Effects of Organo-Modified Montmorillonite on the Tribology Performance of Bismaleimide-Based Nanocomposites. *Materials & Design*, *86*(December), 138–145. doi:10.1016/j.matdes.2015.07.090

Zhang, G., Xie, G., Si, L., Wen, S., & Guo, D. (2017). Ultralow Friction Self-Lubricating Nanocomposites with Mesoporous Metal-Organic Frameworks as Smart Nanocontainers for Lubricants. *ACS Applied Materials & Interfaces*, *9*(43), 38146–38152. doi:10.1021/acsami.7b12591 PMID:28994574

Zhang, J., & Meng, Y. (2015). Boundary Lubrication by Adsorption Film. *Friction*, *3*(2), 115–147. doi:10.100740544-015-0084-4

Zhang, L., & Huang, H. (2019). Micro machining of bulk metallic glasses: A review. *International Journal of Advanced Manufacturing Technology*, *100*(1-4), 637–661. doi:10.100700170-018-2726-y

Zhang, S., Hou, Y., Li, S., Liu, L., Zhang, Z., Feng, X.-Q., & Li, Q. (2019b). Tuning Friction to a Superlubric State via in-Plane Straining. *Proceedings of the National Academy of Sciences of the United States of America*, *116*(49), 24452–24456. doi:10.1073/pnas.1907947116 PMID:31659028

Zhang, S., Ma, T., Erdemir, A., & Li, Q. (2019a). Tribology of Two-Dimensional Materials: From Mechanisms to Modulating Strategies. *Materials Today*, *26*, 67–86. doi:10.1016/j.mattod.2018.12.002

Zhang, S., Sun, D., Fu, Y., & Du, H. (2003). Recent Advances of Superhard Nanocomposite Coatings: A Review. *Surface and Coatings Technology*, *167*(2–3), 113–119. doi:10.1016/S0257-8972(02)00903-9

Zhang, W., Liu, L., Zhang, M., Huang, G., Liang, J., Xian, L. I., & Zhang, L. (2015). Comparison between WC–10Co–4Cr and Cr_3C_2–25NiCr coatings sprayed on H13 steel by HVOF. *Transactions of Nonferrous Metals Society of China*, *25*(11), 3700–3707. doi:10.1016/S1003-6326(15)64011-0

Zhang, X.-F., Zhao, J.-P., & Hu, J.-M. (2017). Abrasion-Resistant, Hot Water-Repellent and Self-Cleaning Superhydrophobic Surfaces Fabricated by Electrophoresis of Nanoparticles in Electrodeposited Sol-Gel Films. *Advanced Materials Interfaces*, *4*(13), 1700177. doi:10.1002/admi.201700177

Zhang, X., Li, Q., Li, L., Zhang, P., Wang, Z., & Chen, F. (2012). Fabrication of hydroxyapatite/stearic acid composite coating and corrosion behavior of coated magnesium alloy. *Materials Letters*, *88*, 76–78. doi:10.1016/j.matlet.2012.08.011

Zhang, Y., Dong, B., Wang, S., Zhao, L., Wan, L., & Wang, E. (2017). Mechanically robust, thermally stable, highly transparent superhydrophobic coating with low-temperature sol–gel process. *RSC Advances*, *7*(75), 47357–47365. doi:10.1039/C7RA08578H

Zhang, Y., Li, P., Ji, L., Wan, H., Chen, L., Li, H., & Jin, Z. (2021). Tribological Properties of MoS_2 Coating for Ultra-Long Wear-Life and Low Coefficient of Friction combined with Additive g-C_3N_4 in Air. *Friction*, *9*(4), 789–801. doi:10.100740544-020-0374-3

Zhang, Y., Zhao, D., Wang, R., & Wang, W. (2003). Formation and properties of Zr48Nb8Cu14Ni12Be18 bulk metallic glass. *Acta Materialia*, *51*(7), 1971–1979. doi:10.1016/S1359-6454(02)00602-X

Zhao, J., Huang, Y., Li, Y., Gao, T., Dou, Z., Mao, J., Wang, H., He, Y., Li, S., & Luo, J. (2020). Superhigh-Exfoliation Graphene with a Unique Two-Dimensional (2D) Microstructure for Lubrication Application. *Applied Surface Science*, *513*, 145608. doi:10.1016/j.apsusc.2020.145608

Zhao, X.-Q., Zhou, H.-D., & Chen, J.-M. (2006). Comparative study of the friction and wear behavior of plasma sprayed conventional and nanostructured WC–12% Co coatings on stainless steel. *Materials Science and Engineering A*, *431*(1–2), 290–297. doi:10.1016/j.msea.2006.06.009

Zheng, G., Sano, H., Suzuki, K., Kobayashi, K., Uchiyama, Y., & Cheng, H.-M. (1999). A TEM study of microstructure of carbon fiber/polycarbosilane-derived SiC composites. *Carbon*, *37*(12), 2057–2062. doi:10.1016/S0008-6223(99)00098-6

Zheng, J., Wang, Q. D., Jin, Z. L., & Peng, T. (2010). Effect of Sm on the microstructure, mechanical properties and creep behavior of Mg-0.5Zn-0.4Zr based alloys. *Materials Science and Engineering A*, *527*(7–8), 1677–1685. doi:10.1016/j.msea.2009.10.067

Zheng, Q., & Liu, Z. (2014). Experimental Advances in Superlubricity. *Friction*, *2*(2), 182–192. doi:10.100740544-014-0056-0

Zheng, Y. (2013). Plating hard chrome plating alternative technologies-HVOF tungsten carbide coating. *Advanced Materials Research*, *712*, 395–398. doi:10.4028/www.scientific.net/AMR.712-715.395

Zhong, H., Dai, L., Yue, Y., Wang, B., Zhang, X., Tan, C., Ma, M., & Liu, R. (2016). Tribological Properties of Plasma-Nitrided AISI 4340 Steel In Vacuum. *Materials Science and Technology*, *32*(4), 275–281. doi:10.1080/02670836.2015.1121341

Zhong, X., Li, Q., Hu, J., Zhang, S., Chen, B., Xu, S., & Luo, F. (2010). A novel approach to heal the sol–gel coating system on magnesium alloy for corrosion protection. *Electrochimica Acta*, *55*(7), 2424–2429. doi:10.1016/j.electacta.2009.11.063

Zhou, H., & Lee, J. (2011). Nanoscale hydroxyapatite particles for bone tissue engineering. *Acta Biomaterialia*, *7*(7), 2769–2781. doi:10.1016/j.actbio.2011.03.019 PMID:21440094

Zhou, J., & Kong, D. (2021). Friction–wear performances and oxidation behaviors of Ti_3AlC_2 reinforced Co–based alloy coatings by laser cladding. *Surface and Coatings Technology*, *408*, 126816. doi:10.1016/j.surfcoat.2020.126816

Zhou, K., Chen, C., Liu, Y., Pang, S., Hua, N., Yang, W., & Zhang, T. (2017). Effects of lutetium addition on formation, oxidation and tribological properties of a Zr-based bulk metallic glass. *Intermetallics*, *90*, 81–89. doi:10.1016/j.intermet.2017.07.007

Zhou, K., Xie, F., Wu, X., & Wang, S. (2021). Fretting wear behavior of nano ZrO_2 doped plasma electrolytic oxidation composite coatings on TC21 titanium alloy. *Surface and Coatings Technology*, *429*, 127429. doi:10.1016/j.surfcoat.2021.127429

Zhou, Q., Du, Y., Jia, Q., Han, W., Zhao, X., Deng, Y., & Wang, H. (2020). A nanoindentation study of Ti-based high entropy bulk metallic glasses at elevated temperatures. *Journal of Non-Crystalline Solids*, *532*, 119878. doi:10.1016/j.jnoncrysol.2019.119878

Zhou, Q., Du, Y., Ren, Y., Kuang, W., Han, W., Wang, H., Huang, P., Wang, F., & Wang, J. (2019). Investigation into nanoscratching mechanical performance of metallic glass multilayers with improved nano-tribological properties. *Journal of Alloys and Compounds*, *776*, 447–459. doi:10.1016/j.jallcom.2018.10.270

Zhou, Q., Han, W., Luo, D., Du, Y., Xie, J., Wang, X.-Z., Zou, Q., Zhao, X., Wang, H., & Beake, B. D. (2021). Mechanical and tribological properties of Zr–Cu–Ni–Al bulk metallic glasses with dual-phase structure. *Wear*, *474*, 203880. doi:10.1016/j.wear.2021.203880

Zhu, D., & Miller, R. A. (2000). Thermal conductivity and elastic modulus evolution of thermal barrier coatings under high heat flux conditions. *Journal of Thermal Spray Technology*, *9*(2), 175–180. doi:10.1361/105996300770349890

Zin, V., Miorin, E., Deambrosis, S. M., Montagner, F., & Fabrizio, M. (2018). Mechanical properties and tribological behaviour of Mo-N coatings deposited via high power impulse magnetron sputtering on temperature sensitive substrates. *Tribology International*, *119*, 372–380. doi:10.1016/j.triboint.2017.11.007

Zou, Y. S., Wang, W., Song, G. H., Du, H., Gong, J., Huang, R. F., & Wen, L. S. (2004). Influence of the Gas Atmosphere on the Microstructure and Mechanical Properties of Diamond-like Carbon Films by Arc Ion Plating. *Materials Letters*, *58*(26), 3271–3275. doi:10.1016/j.matlet.2004.06.017

Zuo, D., Tian, G., Li, X., Chen, D., & Shu, K. (2017). Recent progress in surface coating of cathode materials for lithium ion secondary batteries. *Journal of Alloys and Compounds*, *706*, 24–40. doi:10.1016/j.jallcom.2017.02.230

About the Contributors

Amirhossein Pakseresht is an associate professor and head of coating department at FunGlass Centre for Functional and Surface Functionalized Glass, Alexander Dubček University of Trencin, Slovakia. He received his undergraduate study in materials science and engineering from Tehran University which is the biggest and oldest university in Iran. Pakseresht completed his Ph.D. in materials science and engineering at Materials and Energy Research Center (MERC). During his PhD, he also worked as a researcher in MERC and involved in so many industrial projects. After completion of his Ph.D. He joined the Tehran University as a postdoctoral research associate in the Department of material engineering. In 2019, he has joined Amirkabir University (Tehran Polytechnic) as an adjunct research assistant professor and involved in so many projects about coating and composite. Generally speaking, his research interests lie in the area of plasma spraying, with a focus on splat morphology and new thermal barrier coatings. He also has collaborated actively in multidisciplinary materials science, particularly surface science, anti corrosion coating and composite materials. Dr. Pakseresht authored a book titled production, properties, and application of high-temperature coatings in IGI Global, USA. He also authored a book titled Physical chemistry and thermodynamic of materials (in Persian). He has published more than 50 journal papers, 5 patents and 5 international book chapters which were cited 1500 times (h-index 21) until 2021. He is a member of the editorial board and reviewer of more than 10 international scientific journals.

Omid Sharifahmadian received his Ph.D. in Materials and Metallurgical Engineering (coating and surface engineering) from Amirkabir University of Technology (Tehran Polytechnic), Iran in 2019. During his Ph. D, he could join as a scholar in the Applied Plasma and Surface Engineering Research Group at University of Sydney, Australia. He has a good background in the field of surface engineering, tribology, plasma processing, and coating. Dr. Omid Sharifahmadian has experienced on the working with a variety of coating processes such as DC- pulsed PECVD, RF-PECVD, PVD, and thermal spray methods. He joined department of coating in FunGlass – Centre for Functional and Surface Functionalized Glass at Alexander Dubček University of Trenčín, Slovakia from April 2021. He focused on a variety of issues such as tribological behavior, structural evolution, mechanical properties and hydrophobicity in the fields of thin film and coatings. Moreover, during his research in FunGlass center, he could join as a scholar in Nano Research Infrastructure group at Central European Institute of Technology (CEITEC) in Brno, Czech Republic. He has published many papers in high impact journals which were cited by many authors.

* * *

Srikrishnan A. R. did his Ph.D. in the field of experimental studies on the mixing of supersonic flows. He has both industrial and academic experience in the domains related to high-speed flows, heat transfer, particle-laden jets and reacting flows. He worked for 16 years with a global leader in engineering simulation holding several leadership positions in the organization. For the last 10 years, he has been working as a faculty member in the Department of Aerospace Engineering, Amrita University, India where he has been focusing on teaching as well as funded research in several areas including supersonic flows, heat transfer, fluid dynamics and combustion. His publications in international journals include studies on supersonic cold spray deposition of particles and various other applications of jet dynamics.

Mahdi Alebrahim is an ardent researcher in material science. Currently, he is working on the thermal shock behavior of thermal barrier coatings reinforced by YSZ fibers at Alexander Dubček University of Trenčín (TnUAD). Ceramic materials, coating development, corrosion testing, and composites are some of his research interests.

Kamalan Kirubaharan Amirtharaj Mosas has 10 years of research experience in the field of geopolymer based high temperature coatings, coatings for corrosion protection, thin films, and phase transformations of materials. Expertise in handling PVD techniques such as, thermal, EBPVD, RF/DC sputtering and various characterization techniques. The area of expertise is as follows: 1. Silicate based heat resistant coatings 2. Thermal barrier coatings prepared through atmospheric plasma spray and electron beam physical vapor deposition techniques. 3. Optical and hard coatings prepared through PVD techniques.

Amir Bahri is a doctor of materials engineering and a materials engineer who graduated from the National Engineering School of Sfax-Tunisia (ENIS-Tunisia). I was a substitute teacher in the same university for 6 years. I taught undergraduate students the numerical simulation of the mechanical behavior of materials. I supervised many students in their final project (Tribology, mechanical behavior, numerical simulation, design…) and I have more than 22 communication and publications. I made a national patent in the field of tribology in particular erosion wear issues in olive oil extraction. I'm a reviewer in the American Journal of Mechanical and Materials Engineering. I was a part of several international research projects like the European project Oil and Sugar EU FP7-PEOPLE- 2011-IRSES. I was a Post-doctor at the Laboratory for materials testing at FH Bielefeld, University of Applied Sciences: DAAD fellowship holder within the cooperation project "DAAD 2018. I obtained the Georg Forster Research Fellowship for Researchers in 2020.

Amiya Bhaumik is the President & Vice-Chancellor of Lincoln University College, Malaysia. He is an Executive Vice President of the International Education Consulting Group, St. Louis, the USA since 1999. Dr. Amiya Bhaumik was a Research Fellow of UNESCO, he has travelled to more than 100 countries. He has served as a professor of Business Administration at the University of Lucknow, India and in the University of Malaya and many other universities overseas. Dr. Bhaumik has been awarded the prestigious BCM (Bintang Cemerlang Melaka) from the government of Melaka, Malaysia. He has authored numerous book chapters and has a huge number of publications in many national and international journals and guided PhD students in Engineering and Management domain. Dr. Amiya is an Editor In Chief for 'Asia-Pacific Journal of Management and Technology (AJMT)'. Dr. Bhaumik is a

Visionary Leader to develop society through education, training and skill development, he believes that Education and Training will help us to manifest our own divine strength and power.

Digvijay G. Bhosale is a surface engineer with research experience of more than four years in high temperature tribology, additive manufacturing and thermal spray coatings. He graduated with a Master in mechanical engineering and a specialization in automotive engineering at Veermata Jijabai Technological Institute (VJTI, Govt Autonomous Institute), Mumbai in 2011, and hold a Ph.D. in Mechanical Engineering (Tribology and Surface Engineering) from VJTI Mumbai. He focused on the development of wear and corrosion resistant thermal spray coatings. Skills Tribology Wear and corrosion Material Characterization and evaluation Process implementation and improvement Thermal spray materials and processes Thermal and environmental barrier coatings Additive Manufacturing Powertrain design and analysis Finite Element Analysis and optimization.

Amol Bajarang Chavan is a research scholar perusing PhD at Shivaji University, Kolhapur, M. S., India. His research work is related with hydroxyapatite coated magnesium based biodegradable medical implants. He graduated with a Master in mechanical-production engineering at Government College of engineering, Karad, M. S., India in 2014.

Pankaj Chhabra has submitted his PhD in the year 2021 in Mechanical Engineering. He has published two papers in international journals from his work. He has a keen interest in surface coatings and surface treatments. He is working on the wear and friction behaviour of uncoated and coated/treated tool die steels.

Dinesh Kumar Devarajan is currently working as Scientist at Centre for Nanoscience and Nanotechnology, Sathyabama Institute of Science and Technology, Chennai, India. His research interest is in the field of coatings tribology.

Mihaela Dinu is a Scientific Researcher at National Institute of Research and Development for Optoelectronics INOE 2000, Research Centre for Advanced Surface Processing and Analysis by Vacuum Technologies. Professional experience includes surface characterization of thin films in mono and multilayer structures (nitrides, carbides, oxides and oxynitrides) deposited by systems based on plasma and ultra-high vacuum technologies for various applications (biomaterials, tribology, optoelectronics, etc.). The main research activities concern structural characterisation of thin films by XRD, surface morphology characterisation by profilometry and SEM, adhesion strength assessment of coating–substrate systems by scratch test, potentiodynamic techniques and electrochemical characterisation by EIS and data modelling, wear and friction coefficient in dry/wet environment. She obtained her PhD degree in 2015 and her experience includes several achievements with increased international visibility: 4 awards at international conferences, research experience abroad, 27 scientific articles and 7 book chapters.

Khaled Elleuch has been full professor in mechanical engineering since 2014. In 2002 he integrated Sfax National engineering school as associate professor in mechanical engineering since 2011 and till 2017 he was the dean of material engineering department and since 2017 till now he is the director of internships and international relations. Pr. Elleuch was the founder of Material Engineering Master and the international conference of material environment and durability (MED). he is the head of Mechanics of material and Durability groups in the research laboratory of Engineering Material and Environment. He has supervised more than 20 PHD thesis, 30 Masters and 200 final engineering study thesis. He has more than 100 papers in recognized scientific journals. Pr. Elleuch has participated in many national and international research project. he was invited as visitor professor at many European universities in France, Greece, Belgium, Germany, Italy, Spain, Turkey, and Portugal.

Dušan Galusek is the professor of materials science at the Alexander Dubček University of Trenčín (TnUAD), Slovakia, the director of the Centre for functional and surface-functionalized glass (FunGlass), and the head of the Join glass centre of the Institute of Inorganic Chemistry of the Slovak Academy of Sciences (IIC SAS), and the TnUAD. Prof. Galusek obtained his master degree from the Slovak University of Technology in 1991. There he also earned his PhD in inorganic technology in 1995. He habilitated in 2006 and became full professor and doctor of sciences in 2013. The start of his career was associated with the IIC SAS in Bratislava, where he worked as a researcher in 1996-1999, and then as the head of the Joint glass centre until 2005. He also worked at various institutions abroad, including the Brunel University (1993), University of Leeds (1998, 2000, 2001), Darmstadt University of Technology (1994, 1996, 1997, 2003-2004), and Karlsruhe University of Technology (2007), under the DAAD, NATO-Royal Society Fellowship, and Alexander von Humboldt Fellowship schemes. Since 1997 he connected his professional life with the TnUAD, where he later founded the Centre for ceramics, glass and silicate materials. After receiving funding of 25 M€ from the Horizon 2020 programme in 2017, he further developed the centre to the now existing FunGlass, an international establishment with 90 employees, focused on research of glass, ceramics, and coatings. His R&D interests cover advanced ceramic materials and composites, polymer derived ceramics, aluminate glasses, bioactive glass and ceramics, and corrosion of glass and ceramics. He co-authored 190 research papers in peer-reviewed journals, and more than 300 conference papers (30+ invited). He earned numerous prizes and awards, including the Scientist of the year 2019, Head of the scientific team of the year of the Slovak republic in 2008 and 2018, and the nomination for the Crystal Wing Award in the category Medicine and Science in 2017. He is an active member of many professional societies and serves as the chairman of the Slovak Glass society, and vice-chairman of the Slovak Silicate Society. Since 2019 serves as an editor of the Journal of the European Ceramic Society.

Sanjaykumar Gawade is Professor and PG and PhD Conveyor at R. I. T. Rajaramnagar. He has 30 years of teaching and research experience. His major fields of research are vibration engineering, smart materials and experimental mechanics. He has received research grant of Rupees 9 lac from AICTE, New Delhi. He has published more than 50 research papers in reputed journals and conferences. He has guided more than 75 M. Tech students, 2 PhD students under his guidance. He is member of various committees in Shivaji University, Kolhapur. His research and administrative contribution is remarkable.

Subhenjit Hazra works as Scientist at Centre for Nanoscience and Nanotechnology, Sathyabama Institute of Science and Technology, Chennai. He completed his Ph.D. from Birla Institute of Technology and Science, Pilani. He worked as UGC- Dr. D.S.Kothari Postdoctoral fellow at University of Calcutta, India. His research interests are Nanoscience and Nanotechnology, Naocomposites, heterogeneous catalysis, mesoporous materials, sensor and organic synthesis. He authored 30 research article in peer reviewed international journals.

Reza Jafari is a doctoral researcher in Materials Science and Environmental Engineering in the Faculty of Engineering and Natural Sciences. He is passionate about materials development and application for demanding environments. Currently, he is working on durable icephobic cold-sprayed coatings in both the coatings development stage and comprehensive microstructural characterization at the Tampere Microscopy Center, as well as the Icing Research group. His fields of interest are materials characterization, coating development, surface treatment, and corrosion evaluation.

Ghanshyam Jhala has been associated with the plasma nitriding process both for research and industrial applications for more than two decades and is currently working in Institute for Plasma Research, Gandhinagar, Gujarat. He has obtained his M.Sc degree from Madurai Kamraj University. He has more than 21 publications in peer-reviewed journals. He is currently working with different variants of plasma nitriding processes.

Alphonsa Joseph has both academic and industrial experience in the field of plasma nitriding and its variants processes. She has obtained her PhD degree from IIT Bombay, India and is currently working in Institute for Plasma Research, Gandhinagar, Gujarat India for more than two decades. She has about more than 50 publications in peer-reviewed journals. Her research interests also include magnetron sputtering coatings.

Gobi Saravanan Kaliaraj is Scientist - C at the Centre for Nanoscience and Nanotechnology, Sathyabama Institute of Science and Technology, Chennai. Currently, he is working on biomaterials, surface modification, antimicrobial coatings, corrosion protection of implant materials, and bioactive and biocompatibility materials. He has published 30 papers in reputed international journals and many papers in national/international conferences and has two patents. In addition, he has attracted funding from various government funding agencies like Department of Biotechnology, Indian Space Research Organization, etc.

Mouna Kallel is a doctor of materials engineering and a materials engineer who graduated from the National Engineering School of Sfax-Tunisia (ENIS-Tunisia). I am also a researcher at Laboratory of Material Engineering and Environment at the same university. My research interests include surface treatment, composite coating, tribology and mechanical behavior of Materials. I was a substitute teacher in National Engineering School of Sfax-Tunisia (ENIS-Tunisia) for 2 years. I am now contractual assistant Professor Higher Institute of Applied Sciences and Technology of Kairouan-Tunisia. I taught undergraduate students the mechanical behavior of materials, Materials sciences and design. I supervised many students in their final project (Tribology, mechanical behavior, electrodeposition, ...) and I have 4 publications and 7 communications. I was a part of an international research projects the European project Oil and Sugar EU FP7-PEOPLE- 2011-IRSES.

Manpreet Kaur is presently working as Associate Professor in the Mechanical Engineering Department at Baba Banda Singh Bahadur Engineering College, Fatehgarh Sahib, Punjab India. She has both academic and research experience. She has been working in the field of Surface Engineering for the last 15 years. She has obtained PhD degree in Mechanical Engineering in the year 2011 from IKGPTU, Jalandhar. She has published more than 40 papers in international journals and presented more than 40 papers in international/national conferences. She is actively working in the field of surface coatings/

surface treatments. Her research interests are in analyzing High-temperature erosion-corrosion and wear problems in the industries. Under her supervision, 06 students have completed M. Tech Thesis and 5 students have completed PhD.

Ashish Kumar has completed his PhD in the year 2020 Mechanical Engineering. His PhD title was, 'High-Temperature Tribological Behaviour of Plasma Nitrided Die Steels'. He has published two papers in international journals from his work. He has worked as JRF in one of the DST projects.

Rakesh Kumar is working as Research Engineer, in the Department of Regulatory affair, Auxein Medical Private Limited, Sonipat, Haryana, India. In addition, He is also pursuing his Ph.D. in Mechanical Engineering from Chandigarh University, Punjab, India. His research interest includes Surface Engineering, Additive Manufacturing (orthopedic implant) and Biodiesel Production. He has more than 12 years' experience in teaching and industry. In addition, he has published many technical research papers, books and patents.

Santosh Kumar is working as Associate Professor in the Department of Mechanical Engineering at Chandigarh Group of Colleges, Landran, Mohali, Punjab (India). He has done his Ph.D from IK Gujral Punjab Technical University Kapurthala in the field of Surface Engineering (Thermal Spray Coatings). He has done his Ph.D., M.Tech., B.Tech and Diploma in Mechanical Engineering with distinction. He has more than 10 years of teaching and research experience. His research interests includes Surface Engineering, Rapid Prototyping, Nanotechnology and Bio fuels on which he has published more than 50 technical papers, 2 books and 4 patents.

Motilal Lakavat is currently working as a Senior Lecturer with the Department of Mechanical & Industrial Engineering, College of Engineering, National University of Science & Technology(Prior name of the Institute is Caledonian College of Engineering, affiliated with Glasgow Caledonian University, UK), Sultanate of Oman, Oman. Mr Motilal Lakavat has a total experience of 19 plus years in teaching, research and industry. Mr Motilal Lakavat is involved in Teaching and Research and is an International Internship Coordinator for the College of Engineering. Mr Motilal Lakavat's area of research interest is tribological studies of materials and their applications, manufacturing, materials properties, and its application effect studies. Mr Motilal Lakavat has expertise in Mechanical Engineering, Manufacturing Engineering, Material Engineering and Nano Technology, Industrial Engineering programs and tailored made Industrial based training programs related to Mechanical and Industrial Engineering. Mr Motilal Lakavat pursuing his PhD from Lincoln University College, Malesia in the field of Materials, Nano Coatings and Corrosion Engineering in Automotive Applications. Mr Motilal Lakavat has completed his M. Tech in Industrial Engineering & Management from Jawaharlal Nehru Technological University, Hyderabad, Telangana, India in 2007, and B. Tech in Mechanical Engineering from Acharya Nagarjuna University, Guntur, Andhra Pradesh, India in 2002. Mr Motilal Lakavat has more than 25 publications to his credit in journals and conferences of international repute and supervised many graduates and postgraduate students. Mr Motilal Lakavat is associated as the Principal Investigator/CO-Investigator on many research projects and is also involved in research areas. Also, offered significant contributions as a peer reviewer for reputed international journals and conferences. He was awarded "Best Teacher" by the Nishitha College of Engineering & Technology, Hyderabad, Telangana, India in the year 2009.

Miha Lukšič studied chemistry at the Faculty of Chemistry and Chemical Technology (FCCT), University of Ljubljana, Slovenia. In 2010, he received his PhD in physical chemistry from the same university. He was a visiting scientist at the University of Regensburg (Germany), the National Autonomous University of Mexico (Mexico) and the University of California, San Francisco. From 2012 to 2014, he worked as a postdoctoral fellow at the Laufer Center for Physical and Quantitative Biology at Stony Brook University. He is currently an associate professor of physical chemistry at FCCT, University of Ljubljana. His research interests are in statistical mechanical theory of water and hydration, partly-quenched systems, solutions of (poly)electrolytes, protein aggregation and protein-protein interactions, and electrochromic cell devices.

Satish A Mahadik received his PhD degree of Material Science in 2012 from Shivaji University, Kolhapur (India). After completing PhD, he moved to La Rochelle University, La Rochelle (France) to take up postdoctoral experience with the prestigious European Commision Erasmus Mundus Fellowship. Also, he has research experience from University of Seoul, Seoul (South korea). Currently he is working as assistant Professor at Department of Physics, Sanjay Ghodawat University Kolhapur. He is interested in research for multifunctional superhydrophobic coating with consideration of integrated sol-gel chemistry. So far he has published over 23 refereed journal papers and received over 1225 citations.

Sarika S. Mahadik is master student at school nanoscience and nanotechnology at Shivaji University Kolhapur.

Rita Maurya is currently serving as an Assistant Professor at the department of Materials Science & Engineering at NIT-Hamirpur, HP, India. She earned her doctorate in Materials Science & Engineering from IIT-Kanpur, India, in 2018. She earned her Masters of Technology in Metallurgical & Materials Engineering from NIT-Durgapur in 2012 and Bachelor of Technology in Material Science & Metallurgical Engineering from UIET, Kanpur, India in 2010. Her area of research concentrated on physical metallurgy, surface engineering of Mg, Li and Al-based light-weight metal alloys and advanced materials characterization. She is expertise in mechanical, tribological and corrosion studies on the engineered materials. She is also worked on the development of advanced materials with super-hydrophobic surfaces and filed an Indian Patent as well. She has published 15 articles (h-index: 9, i-10 index: 8) with 300+ citations in peer-reviewed journals. She is reviewer of journals from Elsevier, Wiley, Springer and Taylor & Francis publishers. Also, she is associated with various society, such as Electron Microscopy Society of India, Indian Institute of Metals and Powder Metallurgical Society of India.

Iulian Pana is a scientific researcher at National Institute of Research and Development in Optoelectronics – INOE 2000 (Research Centre for Advanced Surface Processing and Analysis by Vacuum Technologies - ReCAST) and has worked in the research area in the last 10 years. His research interests are focused on modelling and simulation of optical and physical systems as well as on the deposition and investigation of the thin films properties produced by Physical Vapor Deposition (PVD) techniques. Dr. Iulian Pana has received his Bachelor (2011) and Master Degrees (2013) in Physics-Informatics at Faculty of Physics, University of Bucharest (Romania). In 2018, Dr. Iulian Pana obtained the Ph.D. in Physics and up to now has published 24 scientific articles as first and co-author in ISI journals.

Saran Srihari Sripada Panda received the B.Sc. degree in Physics and the M.Sc. degree in Electronics science from Berhampur University, Berhampur, Odisha, in 2015 and 2017, respectively. He is currently pursuing the Ph.D. degree with the Department of Applied Science, National Institute of Technology Goa, India. His current research includes microwave photonic radar and Material sciences.

Anca Constantina Parau obtained her PhD degree in Materials Science and Engineering in 2013 and works at National Institute of Research and Development for Optoelectronics, Department for Advanced Surface Processing and Analysis by Vacuum Technologies since 2011. Professional activities include surface characterization of thin films deposition used in different domains. The main research activities included scanning electron microscopy (SEM) analyses, structural characterisation of thin films by X-ray diffraction (XRD), surface morphology and roughness characterisation by atomic force microscopy (AFM) and surface profilometry, hardness, wear and friction properties by nanoindentation and tribological analyses in different environments, corrosion evaluation by potentiodynamic methods using different solutions, etc. Her activities includes also research experience abroad and various conference participations with experimental results highlighted by 37 papers published in ISI journals and 2 patents. Personal page: orcid.org/0000-0002-4661-8362.

Saidi Reddy Parne is an Associate Professor in the Department of Applied Sciences at National Institute of Technology Goa. He earned his B.Sc. from Osmania University, his M.Sc. (Engineering Physics & Instrumentation) from Kakatiya University, and his Ph.D. (Fiber optic sensors) from the Department of Physics at the National Institute of Technology Warangal. His research areas include Fiber-optic sensors, photonics, and energy storage systems, smart and advanced-functional materials for sustainability. Dr. Reddy worked as a sensor specialist at Pricol Technologies Limited in Coimbatore before joining NIT Goa. He has over 60 research articles published in prestigious international journals and standard conferences. He is currently engaged in two national research projects and one international research project (Indo-Norwegian). He has more than five international patents and has written one book and two book chapters. He is a senior member of the international societies SPIE and IEEE. Dr. Reddy is currently Editorial Board Member for various international journals. He serves as peer reviewer for various leading journals, and is a Top Peer Reviewer in Scopus, Web of Science.

Fernando Pedraza is Full Professor in Chemistry of Materials at the University of La Rochelle (France). After a European PhD in Chemistry from University Complutense of Madrid (Spain), Prof. Pedraza spent 1 year as post-doc fellow at La Rochelle University and then gained the prestigious Marie Curie Fellowship for 18 months at SIFCO Turbine (Ireland). He became Full Professor at La Rochelle University in 2011. His main research topics involve high temperature (diffusion, slurries and electroceramics) and low temperature (superhydrophobic and corrosion resistant sol-gel coatings and electrodeposits) coatings and repair technologies (electrochemical and chemical stripping). Prof. Pedraza is currently leading a research group "durability, microstructure, protection and coatings" with 16 permanent staff and 20 PhD and Post-docs. He is PI and participates in various academic and industrial R&D programmes (including EU-funded). Presently, Prof. Pedraza has published more than 140 scientific and technical papers and is inventor of various patents.

Ram Prabhu received B.Tech degree in Metallurgy from NIT, Trichy in 2005, M. E in Materials from IISc Bangalore in 2008 and Ph.D. from IIT, Madras in 2014 . Presently he is working as a joint director in DRDO, Bengaluru. His research interest includes Tribology, High Temperature Materials, Additive Manufacturing, Corrosion, Powder Metallurgy.

Pritee Purohit is working as Assistant Professor at Army Institute of Technology Pune India and Research scholar at College of Engineering, Pune, India.

Walmik Rathod received B.E. degree in Metallurgy and material Science from College of Engineering, Pune in 1998, M. E in Production Engineering from VJTI in 2004 and Ph.D. from IIT, Bombay, Mumbai in 2014 . Presently he is working as an Assistant Professor in Department of Mechanical Engineering, VJTI, Mumbai. His research interest includes Physical Metallurgy, High Temperature Coating, Corrosion, Powder Metallurgy, Casting, Biomaterials.

Martin Rozman finished his Master thesis at Faculty of Chemistry and Chemical Technology (FCCT), University of Ljubljana, Slovenia in 2017. He continued with his PhD at Faculty of Chemistry and Chemical Technology (FCCT), University of Maribor, Slovenia from the subject of electrochemistry which he finished in 2020. From 2021, he works as a postdoctoral fellow at Centre for Functional and Surface Functionalized Glass, Alexander Dubček University of Trenčín, Slovakia. His research interests are in material science, electrochemistry cell architecture, electrochromic cell devices, photovoltaics, and biosensors.

Ariharan S. earned his doctorate in Materials Science and Engineering from IIT-Kanpur, India, in 2018. He earned his postgraduate in Masters of Technology (Materials and Metallurgical Engineering) from IIT-Kanpur in 2011 and Masters of Science (Materials Science) from College of Engineering Guindy (Anna University, Chennai) in 2009. His undergraduate degree is in Bachelors of Science (Physics) from Pioneer Kumaraswamy College (Manonmanium Sundaranar University, Tirunelveli), Nagercoil in 2006. He is the all India rank holder of Graduate Aptitude Test in Engineering (GATE-2009) and Joint Admission test for M.Sc. (JAM-2007), conducted by IITs and Indian Institute of Science-Bangalore, India. He is recipient of Nanomaterials and Energy Prize-2019 by the journal-Nanomaterials and Energy and National Post-Doctoral Fellowship-2018 sponsored by Science and Engineering Research Board (SERB), Government of India. He pursued his post-doctoral research (February 2019-August 2021) in the development of carbonaceous materials reinforced ZrO2 based advanced materials from the department of Metallurgical and Materials Engineering, IIT-Madras, India. His area of research concentrated on the role of carbon nanotube dispersion in enhancing the mechanical, tribological and thermal properties of ceramics based thermal spray coatings and oxidative protective coatings. He also worked on ultra-high temperature ceramics based nanocomposites for thermal protection system and hydroxyapatite based biomaterials for biomedical applications. He has published more than 25 research articles (h-index of 12, i-10 index of 13) with 450+ citations in peer-reviewed journals and conference proceedings. Also, has presented over 10 lectures at international conferences. His research interests include electron microscopy, materials for extreme environments and tribology of nanocomposites. He is reviewer of journals from Elsevier, Wiley, Springer, ICE (Institution of Civil Engineers) and Taylor & Francis publishers. Also, he is the life time member of Electron Mocroscopy Society of India (EMSI) and Powder Metallurgical Society of India (PMSI).

Fariborz Sharifian Jazi received his Ph.D. degree in material engineering under the supervision of Prof. Nader Parvin at Amirkabir University, Iran. In 2021, he joined the University of Georgia, Tbilisi as a faculty member. His current research interests are dedicated to the development of novel synthetic methods for the synthesis of nanomterials and biomaterials. To date, he has published around 44 scientific publications with an h-index of 23.

Suman Gandi is an Assistant Professor in the Department of Applied Sciences at National Institute of Technology Goa, India. His research and development is focused on new battery technologies that could significantly impact the future of energy. He received his M.Sc (Tech) Engineering degree from the National Institute of Technology, Warangal in 2012 and his Ph.D. in Physics from the GITAM Deemed to be University, Hyderabad in 2019, where his doctoral research focused on glass and glass-ceramic electrode and electrolyte materials for sodium and lithium-ion batteries. Dr. Suman Gandi has authored and co-authored over 23 research publications published in prestigious international journals and has presented 35 papers at international/national conferences. He is the author of two international book chapters, one book, and four patents. Dr. Suman Gandi research interests included: glass and glass-ceramic materials for energy storage system applications; glasses with special functional properties for reflection, and anti-reflection coatings for solar energy production and optoelectronic applications; mass transport in ceramics (particularly at grain boundaries); high-temperature corrosion; sol-gel processing of ceramics; self-cleaning/antibacterial coatings of glasses; bulk metallic glasses; high entropy alloys; catalysts and adsorbents for environmental pollution abatement; and the mechanical properties of glasses.

Venkata Satya Chidambara Swamy Vaddadi graduated from Andhra University, Visakhapatnam, Andhra Pradesh, India in 2011. He received his master's degree (M. Sc.) in Physics from Andhra University, Visakhapatnam, in 2013. Later he joined as physics lecturer in private organisation from 2013 to 2016. Since January 2017 presently he is pursuing PhD in National Institute of Technology Goa, India. His research mainly focused on fiber optic sensors for temperature and pressure measurement in ocean applications. Also, he is working on room temperature gas sensors for ammonia gas detection.

Shashikant Tukaram Vagge is a professor in Metallurgy and Materials Science at Government college of Engineering Pune.

Rafael Vargas-Bernal received a Bachelor's degree in Communications and Electronics Engineering from the University of Guanajuato in 1995 from Mexico, and the degrees of Master of Science and Doctorate in Science with Specialty in Electronics from the National Institute of Astrophysics, Optics and Electronics (INAOE) in 1997 and 2000 from Mexico, respectively. Since January 2002, he has been a professor-researcher at the Higher Technological Institute of Irapuato (ITESI), Mexico and, particularly since 2006, he has worked in the Department of Materials Engineering where he has established himself as a researcher. He has authored 15 articles in journals, 26 chapters in books and about 100 conference articles. He is a member of the National System of Researchers (SNI-Mexico). He regularly serves as a reviewer of scientific articles in RSC Advances and Royal Society Open Science as well as Standards in Semiconductor Equipment and Materials International (SEMI). His research interests include two-dimensional materials, nanomaterials, space materials, composite materials, gas sensors and biosensors.

Alina Vladescu has completed her PhD in Materials Science and Engineering from the University Politehnica of Bucharest, Romania. She works at National Institute for Optoelectronics, Department for Advanced Surface Processing and Analysis by Vacuum Technologies since 2002. She is also associate professor at University Politehnica of Bucharest since 2015 and research scientist at National Research Tomsk Polytechnic University, Russia, since 2016. She has published more than 123 papers in reputed journals, 14 patents, 4 books. She was Guest Editor in Frontiers in Materials (2016-2017), Composite Interfaces (2016), Coatings (2019-2021).

Index

D

E

F

G

H

I

J

L

M

T

W

X

Y

www.ingramcontent.com/pod-product-compliance
Lightning Source LLC
LaVergne TN
LVHW080532300926
849162LV00007B/1588
* 9 7 8 1 7 9 9 8 9 6 8 3 8 *